"An impressively researched, authoritative, and absolutely mind-boggling survey. Even a casual dipping into his text, which will no doubt become a primary source for future mind-body investigation, will reveal a world of inspiring wonders."

—*Kirkus Reviews*

"*The Future of the Body* is a work that should be shared with our children, to ensure that they grow up with a vision of life that is large enough to accommodate even the most questioning mind and the most restless heart."

—*Yoga Journal*

"Not since William James' *The Varieties of Religious Experience* has there appeared such a galvanizing probe into uncommon human capacities."

—Stephen Phillips, professor of philosophy, University of Texas at Austin

"Murphy has accomplished an extraordinary feat in categorizing and synthesizing findings from more than 10,000 studies and reports which collectively provide evidence for the possibility of human transformation, and which may presage the next phase of human evolution."

—*Noetic Sciences Review*

"*The Future of the Body* will broaden the intellectual horizon of anyone interested in what an authentic incarnational spirituality ought to mean in our time."

—Catholic Book Club

"Cases of extraordinary human ability tend to be dismissed by conventional intellectuals looking for a way to keep their philosophical apple carts upright. But in *The Future of the Body* Murphy threatens to send apples tumbling."

—*New Age Journal*

"*The Future of the Body* is a magnificent tour de force in documenting and interpreting practically the entire range of extraordinary physical and psychic capacities with which our species seems to be so richly endowed."

—Georg Feuerstein for *The Quest*

THE
FUTURE
OF THE
BODY

*Explorations
Into the Further Evolution
of Human Nature*

MICHAEL MURPHY

A Jeremy P. Tarcher/Putnam Book
published by
G. P. PUTNAM'S SONS
New York

OTHER BOOKS BY MICHAEL MURPHY

An End to Ordinary History
Golf in the Kingdom
Jacob Atabet

A Jeremy P. Tarcher/Putnam Book
Published by G. P. Putnam's Sons
Publishers Since 1838
200 Madison Avenue
New York, NY 10016

Jeremy P. Tarcher, Inc.
5858 Wilshire Blvd., Suite 200
Los Angeles, CA 90036

Published simultaneously in Canada.

Library of Congress Cataloging-in-Publication Data

Murphy, Michael, date.
   The future of the body : explorations into the further evolution of human
nature / Michael Murphy.
      p.   cm.
   Includes bibliographical references and index.
   ISBN 0-87477-730-5
   1. Parapsychology and science.   2. Human evolution—Miscellanea.
3. Mind and body.   4. Supernatural.   I. Title.
BF1045.S33M87   1992   92-5950 CIP
133.8—dc20

Cover design by Susan Shankin
Text design by Tanya Maiboroda
Production editor: Paul Murphy

Manufactured in the United States of America
        5   6   7   8   9   10

This book is printed on acid-free paper.

For my son,
MacKenzie Wilmott Murphy

CONTENTS

PART 3

Transformative Practices

APPENDICES

ACKNOWLEDGMENTS

OF THE MANY people who contributed to this book, three deserve special thanks. George Leonard read at least two versions of every chapter, providing evidence to support my speculations, editorial advice, and a war cry against doubts and detractors. Margaret Livingston searched libraries, interviewed scientists and scholars, traveled to Europe for material that my inquiry demanded, and helped me separate facts from fictions. Steven Donovan for many years housed and paid for the archives from which the book grew and encouraged me from the day I started writing. All three have deepened my understanding of friendship.

Several other friends read portions of the book and provided helpful criticism. I want especially to acknowledge Don Johnson, Don Michael, Stephen Phillips, Sam Keen, Jay Ogilvy, David Griffin, Rhea White, James Hickman, William Braud, George Solomon, Herbert Benson, Charles Tart, Philip Novak, Richard Baker-Roshi, Frank Barron, Huston Smith, Robert McDermott, Jacob Needleman, John B. Cobb, Jr., Francisco Ayala, David Deamer, Jerry Solfvin, Etzel Cardena, Roger Walsh, Glen Albaugh, Jeanne Achterberg, Michael Grosso, Stuart Miller, Lacey Fosburgh, James Cutsinger, Brendan O'Reagan, Tom Hurley, Mike Maliszewski, Karen Cameron, Steven Matias, and Aidan Kelly.

Laurance Rockefeller, my mother Marie Murphy, the trustees and staff of Esalen Institute, Winston Franklin, and Jeremy Tarcher provided constant financial and moral support.

And for general advice, encouragement, and *joie de vivre,* I am grateful to Dennis Murphy, Fred Hill, Keith Thompson, Nancy Lunney, Ron Brown, David Harris, Lew Reid, Lawrence Chickering, Bruce Nelson, Mike Brown, Dulce Murphy, and my principal mentor, Frederic Spiegelberg.

A Note
to the Reader

CHAPTER SECTIONS ARE numbered so that references to them can be made without citing their titles: 6.3, for example, refers to chapter 6, section 3. In referring the reader to a particular chapter or section, I will not ordinarily say "see chapter 18.2," but will simply place the chapter and section numbers in parentheses.

Whenever an author's name is cited in the text and followed by a date in parentheses, an article or book by that author, published on the date given, can be found in the bibliography. I sometimes use this method of reference instead of a numbered footnote, usually because the work referred to has an important place in the history of a given field.

A glossary of frequently used terms may be found on page 587.

Possibilities for Extraordinary Life

My intention is to tell
of bodies changed
to different forms.

The heavens and all below them,
Earth and her creatures,
All change,
And we, part of creation,
Also must suffer change.
OVID—*Metamorphoses*

Man has no body distinct from his Soul: for that called Body is a portion of Soul discern'd by the five senses, the chief inlets of Soul in this age.

WILLIAM BLAKE

We must assume our existence as broadly as we in any way can; everything, even the unheard-of, must be possible in it. This is at bottom the only courage that is demanded of us: to have courage for the most strange, the most inexplicable.

RAINER MARIA RILKE

1

Introduction

WE LIVE ONLY part of the life we are given. Growing acquaintance with once-foreign cultures, new discoveries about our subliminal depths, and the dawning recognition that each social group reinforces just some human attributes while neglecting or suppressing others have stimulated a worldwide understanding that all of us have great potentials for growth. Perhaps no culture has ever possessed as much publicly available knowledge as we do today regarding the transformative capacities of human nature.

Unfortunately, however, professional specialization and divergent belief systems, along with the information explosion, make it difficult to bring such knowledge into a single purview. Like the unassembled pieces of a great jigsaw puzzle, discoveries about our developmental possibilities are scattered across the intellectual landscape, isolated from one another in separate fields of inquiry. In this book, I bring some of these pieces together to see what picture of human possibility they present.

More specifically, I argue that by gathering data from many fields—including medical science, anthropology, sports, the arts, psychical research, and comparative religious studies—we can identify extraordinary versions of most, if not all, of our basic attributes, among them sensorimotor, kinesthetic, communication, and cognitive abilities; sensations of pain and pleasure; love; vitality; volition; sense of self; and various bodily processes. These grace-laden analogues of our normal attributes, which arise spontaneously or as products of particular practices, can be cultivated. Indeed, the evidence assembled here suggests that we harbor a range of capacities that no single philosophy or psychology has fully embraced, and that these can be developed by practicing certain virtues and disciplines and by building institutions to support them. Though every enduring religion has affirmed something analogous to Judeo-Christian doctrines of grace, none has acknowledged the larger spectrum of grace that a collection of this kind begins to reveal.

3

Furthermore, these many extraordinary attributes exhibit an apparent continuity with features of animal nature. In subsequent chapters, I suggest that they are part of a richly complex development that began with the earliest forms of life, and that they point toward further human advance. Their cultivation, in other words, would carry forward the earth's evolutionary adventure.

❧

My general proposals about human possibility are presented in Part One. In Part Two, evidence supporting those proposals is arranged according to a spectrum of activities that runs, roughly, from sickness to healing to religious and other life-encompassing practices. I employ this scheme of presentation to show: first, that dramatic transformations of body and mind occur in all sorts of conditions, from disease to saintly living; and second, that significant changes all along this spectrum of experience are mediated in part by specifiable activities, or transformative modalities, that are commonly utilized in programs for human growth. In Part Three, I discuss these constituent elements of transformative discipline and suggest ways in which they can be joined to promote balanced human development. Briefly stated, my central observations and proposals include the following.

□ The observation that cultural conditioning powerfully shapes (or extinguishes) metanormal capacities is noted in most chapters and emphasized in 6.1, 7.1, 15.1, 19.5, and 25.

□ The proposition that we cannot comprehend our potentials for extraordinary life without a synoptic, or integral, empiricism that involves many fields of inquiry and different kinds of knowing (including sense perception, inference, and mystical insight) is developed in chapters 2.1 and 25.3.

□ The idea that there exist extraordinary versions of most, if not all, human attributes is developed in chapters 3, 4, and 5. I elaborate this idea by focusing upon perception of external events, somatic awareness, communication abilities, vitality, movement abilities, capacities to manipulate the environment directly, feelings of pain and pleasure, cognition, volition, sense of self, love, and bodily structures.

□ The suggestion that metanormal capacities, when viewed in their entirety, point toward extraordinary types of embodiment is developed in chapters 3–5 and 7–9.

◻ The notion that a widespread realization of extraordinary capacities would constitute an evolutionary transcendence analogous to the rise of living species from inorganic matter, or of humankind from its hominid ancestors, is presented in chapter 3.

◻ The proposal that all or most instances of significant human development are produced by a limited number of identifiable activities such as disciplined self-observation, visualization of desired capacities, and caring for others is noted in chapters 11–23, and summarized in 24.1. The proposal that these activities can be incorporated into transformative practices is developed in Part Three.

◻ The proposition that a wide range of extraordinary capacities, though produced by specifiable activities, arise without benefit of formal discipline is discussed in chapters 4.3 and 14.

◻ The idea that metanormal capacities and experiences are mediated by superordinary agencies or processes (which in Christianity are called the "graces of God," in Buddhism "the workings of Buddha Nature," in Taoism the "way of the Tao") is discussed in chapters 3, 5, 7.1, 24.3, 25.2, and 26.2.

◻ The observation that transformative practices can produce imbalanced development, inhibiting certain capacities while promoting others, is discussed in chapters 7.1, 25.1, and 25.2; and the allied observation that all human capacities and virtues depend on one another, either directly or indirectly, is discussed in chapters 4.1 and 25.2.

◻ The idea that a balanced development of our various attributes is possible through integral practices is discussed in chapters 24–26.

◻ The idea that the evidence for extraordinary human attributes supports some sort of panentheism is developed in chapter 7.2. *Panentheism* (in distinction to *pantheism*) is the doctrine that Divinity is both immanent in and transcendent to the universe.

◻ The suggestion that all human attributes might develop after death, in some sort of extrasomatic body, is presented in chapter 10.

Some of these observations and proposals are unarguable. It is obvious, for example, that cultural conditioning has powerful effects upon individual functioning, whether that functioning is ordinary or extraordinary. On the other hand, some of the suggestions noted above, and others in this book, are highly speculative. For example, the nature of life after death—if there is indeed any afterlife at all—must remain uncertain for thinking persons. Still, all my speculations are based upon evidence of the kind presented in this book. In subsequent chapters, I frequently indicate where I believe such evidence is uncertain and where it is strong.

I firmly believe, for example, that all of us can realize at least some of the extraordinary possibilities described here. I am convinced that men and women young and old, in widely disparate situations, at times experience the unitive awareness, selfless love, and redeeming joy that crown human life. I have come to believe that virtually everyone of us has experienced, and that everyone of us can cultivate, moments when the ordinary becomes extraordinary, when mind and body are graced by something beyond themselves.

2

The Varieties of Evidence for Human Transformative Capacity

WE FACE SOME daunting obstacles in trying to make sense of the evidence for human transformative capacity. Among these are the fragmentation of knowledge caused by professional specialization, the uneven quality of evidence for extraordinary functioning, and a widespread resistance among scientists and academic philosophers to mystical truth claims and data regarding paranormal phenomena. In this chapter I describe these and other hindrances to our understanding of human nature and propose a general method of inquiry that can help overcome them.

[2.1]
Synoptic, or Integral, Empiricism

Each field of inquiry that informs this book has its own ways of providing evidence for extraordinary human functioning. Some, such as medical science, depend heavily upon controlled experiments and elaborate instruments. Others, such as anthropology, rely primarily upon field observations and subjective reports (as well as subsequent checking of those among different investigators). Still others, such as comparative religious studies and psychical research, depend upon reliable testimonies about experiences or events that are not always repeatable upon command, and the systematic comparison of such testimonies with one another. Each of these fields has developed its own methods of inquiry in response to its distinctive subject matter. Each is *empirical* in the sense that it depends upon disciplined acquaintance with its data, whether that acquaintance is established by controlled experiment, by observation of naturally occurring events, or by comparisons of subjective reports and reliable testimonies to unusual phenomena.

Psychology and medical science give us many ways to explore both normal and extraordinary human functioning. Their experimental devices, observational methods, and introspective techniques provide us with increasing knowledge about human nature, including more information about bodily processes than any culture had before. Physicians of former times, for example, could not measure alterations in white-cell count produced by placebos. They did not know about neuropeptides that mediate mood changes induced by meditation. They could not specify the beneficial effects of exercise with the precision of today's medical scientists.

Furthermore, contemporary scholarship has given us unprecedented access to the esoteric lore of many religious traditions. The *Tibetan Book of the Dead*, a document once reserved for a small group of Buddhist monks, is available now in paperback editions. The Upanishads, once transmitted orally in Hindu culture, have been published in many translations. Secrets about dervish practices previously reserved for Sufi initiates are discussed in popular books. This publicizing of the esoteric has helped to broaden many people's perspectives about religious experience. While we have lost certain insights and disciplines once nurtured by the sacred traditions, it is arguable that no single culture has possessed as much publicly accessible lore about shamanic and contemplative capacities as we do today.

Contemporary science and scholarship, then, give us many ways to explore human nature, often providing information about our functioning that was unavailable in previous times. But as I have said, professional specialization, divergent (or conflicting) conceptual systems, and the information explosion make it difficult to gather the evidence for human transformative capacity so that it can be seen as a whole. Experts in a given field typically have less than expert acquaintance with knowledge produced outside their own scholarly or scientific domain. Relatively few physicists and biologists, for example, appreciate how much evidence there is for paranormal events or the great care with which such evidence has been gathered by psychical researchers. Few scholars of contemplative activity know much about discoveries of medical science that bear upon mystical realization. Few scholars in any field recognize the wide range of metanormal experiences that are triggered by athletic performance. In spite of such fragmentation, however, the different kinds of evidence for human transformative capacity can be viewed together so that we can find patterns that connect them. A synoptic acquisition of soundly verified data that draws at once upon the natural and human sciences, psychical research, religious studies, and other fields is the general method I use in this book.

But let me emphasize the term *soundly verified*. Because accounts of extraordinary human experience have various degrees of plausibility, we must approach them with care. Whether any Tibetan lama has in fact levitated is far less certain, for example, than the evidence that brain waves can be slowed by meditation. Whether any saint has lit up a room through prayer is far less certain than the evidence that many people today enjoy quasi-mystical illuminations (for that evidence see chapter 4.3 and appendix A.4). To explore the further reaches of human nature, in short, we must be open-minded but keep an ear for tall tales. To evaluate reports of extraordinary human feats, we need both prudence and imagination, both discrimination and a willingness to suspend judgment. For an inquiry as wide-ranging as this, we must be bold and yet employ the critical distance that characterizes good science. In this empirical spirit, we can divide supposed metanormalities into three categories: first, those, such as spiritual healing, that have been reported again and again in many cultures and that have been verified by good tests of their authenticity; second, those, such as telekinesis, for which we do not have certain evidence but which nevertheless are supported by testimonies that are hard to dismiss; and third, those for which there is little or no support. In this book, I reject members of the third class but do consider members of the second.

But despite the great access we have to evidence for human transformative capacity, we are deprived of certain understandings available to previous cultures. With the rise of secular materialism in the modern world, many religious lineages have been broken so that we cannot now observe or participate in their ecstatic practices. Rene Guenon, Frithjof Schuon, S. H. Nasr, Gershom Scholem, and other religious scholars have described (and lamented) the loss of contemplative knowledge in Judaism, Christianity, and Islam; and several authorities on shamanism and Eastern religions have noted the disappearance of esoteric lore in China, India, and Tibet.[1] However, this loss of sacred tradition is compensated for to some extent by the publication of once-esoteric texts and by modern studies of extraordinary human functioning. Frederic Myers, William James, Herbert

Thurston, and other scholars whom I will discuss in subsequent chapters have compared observations and introspective reports of metanormal capacities. Their work has stimulated subsequent research and has suggested lines of further inquiry. They have helped to create a new kind of natural history, as it were, showing that specimens of extraordinary functioning can be collected for comparative analysis. Like naturalists who by gathering biological specimens helped reveal the fact of evolution, these researchers have prepared the way for new understandings of our human potential. But here, too, we find broken lineages. Few people today appreciate Myers's work, though he was a principal founder of modern psychical research. Few students of religious experience read Thurston, even among Roman Catholics, though he was perhaps the leading Catholic expert on paranormal phenomena in the first half of the twentieth century (22.1). Few contemporary hypnosis researchers are acquainted with the experiments and theories of nineteenth-century mesmerists who pioneered the systematic study of altered states (15.1). In subsequent chapters, I often cite the work of such largely forgotten people.

Today we possess a great variety of evidence regarding metanormal capacities because people in different disciplines have winnowed accurate from distorted perceptions of extraordinary events and have worked to discriminate good data from bad. As I have said, in gathering this evidence, people in different fields have needed different methods of data acquisition and verification. Roman Catholic scholars, in their assessment of healing and saintly powers, have relied upon testimony that they have checked through cross-examination of witnesses and comparisons with third-party reports of the same phenomena (13.2, 22.1). Psychical researchers have supplemented reliable accounts of paranormal events with many kinds of experiments (2.2). Contemplatives have tested their mystical insights against the experience of their peers and teachers. Philosopher Ken Wilber has described the process of data acquisition and verification in the religious life.

> At the very heart and foundation of Zen is not a theory, a dogma, a belief, or a proposition, but, as in any true knowledge quest, an *injunction*. This injunctive strand—zazen or causal-insight meditation—requires years of specialized training and critical discipline. . . . Zazen is simply the injunctive tool for possible cognitive disclosure, and a person must be developmentally adequate to that disclosure or there is, in fact, no disclosure. . . . Not surprisingly, then, the zazen injunction is

always of the form, "If you want to know whether there is Buddha Nature, you must first do this." That is *experimental* and *experiential* injunction.

After that strand is mastered, the investigator is opened to the . . . intuitive apprehension of the object domain addressed by the injunction, in this case, the data of the transcendental sphere. This intuitive apprehension—an immediate experiential impact or perception—is known in Zen as *satori* or *kensho,* both meaning, in essence, a "direct seeing into one's spiritual nature"—as perfectly direct as looking into the microscope to see the cell nucleus, with the all-important proviso in each case: only a trained eye need look.

But, of course, any particular individual's apprehensions of the transcendental object domain may be less than sound or initially mistaken, and therefore Zen, at every stage, has recourse to . . . careful confirmation by both the Zen Master and the community of participant meditators. This is no merely automatic pat on the back and mutual agreement society; it is a vigorous *test,* and it constitutes a potentially powerful rebuff and *non-verification.* . . . Both in private, intense interaction with the Zen Master (*dokusan*) and in exacting public participation in rigorous tests of authenticity (*shosan*), all apprehensions are struck against the community of those whose cognitive eyes are adequate to the transcendent, and such apprehensions are soundly nonverified if they do not match the facts of transcendence as disclosed by the community of like-spirited (and this includes past apprehensions once judged, by the standards of the times, to be true but now . . . found to be partial by more sophisticated experience).[2†]

Virtually all contemplative traditions have claimed that objects of mystical insight such as Buddha Nature, God, or Brahman are realities that exist independently of any human experience; they have also held that these objective realities can be apprehended through particular practices that produce experiences (or data) that can be confirmed by the contemplative's mentors or fellow seekers. In this they are, broadly speaking, empirical. Philosopher Stephen Phillips has called this position "mystic empiricism," and has argued like Wilber

---

† Taking contemplative experience to be a kind of data in the sense that Wilber suggested, one can argue that it has been subjected to growing numbers of communal tests as global communications have improved and once-esoteric religious teachings have become publicly accessible. Countless experiences of shamans, saints, and mystics have been publicized in modern times; metaphysical doctrines associated with them have been translated into contemporary languages; similarities and differences between them have been widely discussed. Such comparisons, I suggest, comprise a kind of data verification in the sense that they implicitly involve tests of each realization's richness, comprehensiveness, and power to account for world and self.

that there is a parallel between the evidentiality of sensory experience and that of mystical realization.[†]

Granting this general parallelism, we must acknowledge two more similarities between scientific and contemplative empiricisms. First, both can be distorted by bad practice. Second, in both, discovery can be inhibited by strict adherence to particular (scientific or religious) beliefs. Just as some laboratories perform bad science, certain religious communities have breakdowns in their disciplines.[3] And, just as potentially significant data are sometimes rejected because they cannot be explained by the conceptual system currently dominant in a particular science, extraordinary experiences that deviate from a particular religious model are sometimes neglected or suppressed. In chapters 7.1, 9, and 25.1, I argue that each religious tradition is limited by its dominant models of the good so that it nurtures only part of our metanormal potential.

Because we suffer perceptual distortions caused by fatigue, sensory malfunctions, or unexamined needs and motives; because such distortions can produce unwarranted, incoherent, and sometimes destructive beliefs or behavior; because our grasp of any data may be hampered by limiting expectations; and for other reasons, we need help in clarifying our data apprehension in any domain of experience. Our ways of knowing, whether sensory, rational, or contemplative, have to be freed from their distorting tendencies. The natural and human sciences, psychical research, and religious practice all include procedures for doing this. These procedures include peer review and replication of results in science, experimental procedures and cross-checking of anecdotal accounts in psychical research, and scrutiny by fellow seekers and religious masters in the contemplative life.

But despite the disciplined tests of experience provided by psychical research and the contemplative traditions, many scientists and philosophers today reject mystical truth claims and the evidence for paranormal phenomena. Such rejection is frequently prompted by lack of acquaintance with the data of psychical research, by ignorance about the enduring testimony to mystical experience, or by automatic associations of such data and testimony with outmoded or

---

[†] Phillips, S. H., 1986, pp. 5–53. In a preface to Phillips's book, philosopher Robert Nozick wrote:

> [The Indian philosopher] Aurobindo is a mystic empiricist in that he builds upon his mystic experiences, offering us descriptions of them, hypotheses that stick rather closely to them, and also bold speculations which reach far beyond the experiences themselves in order to place them in a coherent world picture.

superstitious beliefs. It is prompted, too, by the violation of common-sense or scientific assumptions that paranormal and mystical experience often exhibit. For example, many laypeople and scientists assume that minds cannot produce effects upon inanimate objects at a distance (psychokinetically) without the mediations of specifiable physical agencies; or that organisms cannot know the feelings or thoughts of another organism (telepathically) without sensory cues; or that humans cannot have direct contact with dimensions of existence beyond the material world. (For a detailed description of such assumptions by the British philosopher C. D. Broad, see appendix B.)

However, given the ancient witness to contemplative knowing and the great abundance of evidence for psi phenomena, many sources of which I cite in subsequent chapters, it is a great mistake to exclude such things from our accounts of human nature. That is why many, if not most, great thinkers since antiquity have given paranormal events and mystical truth claims a central place in their philosophies. In *The Republic,* for example, Plato explored the relations between different kinds of knowing, including *dianoia,* discursive reason, and *noesis,* contemplative apprehension of the Good. Plotinus related sense data, discursive intellect, and mystical illumination in his encompassing view of the world, and like Plato discussed occult connections between humans and supraphysical entities. In similar fashion, Western thinkers of late antiquity and the Middle Ages—among them Neo-Platonists such as Porphyry, Iamblichus, and Proclus; Christians such as Origen, Augustine, John Scotus, Bonaventura, Meister Eckhart, and Thomas Aquinas; Jews such as Philo, Moses Maimonides, Abraham Abulafia, and the anonymous author of the *Zohar*; Moslems such as Avicenna, Suhrawardi, and Ibn Arabi—all held that contemplative illumination delivered testable truths about God and human nature. In modern times, mystical knowledge has been viewed sympathetically by Henri Bergson (who regarded mystics to be forerunners of human evolution), Vladimir Solovyov, Nicholas Berdyaev, Schopenhauer, William James, Alfred North Whitehead, and Sri Aurobindo; and many thinkers—among them Bergson, James, C. D. Broad, and H. H. Price—have believed that paranormal events do occur. In Eastern philosophy, mystical truth claims as well as psi functioning have been accepted for three millennia. In the Upanishads, Brahma sutras, and Bhagavad Gita; in the works of Sankara, Ramanuja, and other Vedantic philosophers; in the writings of Nagarjuna, Dogen, and other Buddhists; in the texts of Lao-tze, Chuang-Tze, and other Taoists there are various formulations by which sensory, rational, and contemplative knowledge are related to one another. According to certain Indian philosophers, for example, sense perception, rational inference, and yogic illumination are all deemed to be *pramanas,* "sources or means of acquiring new knowledge," and all are deemed to be instrinsically valid.[4] In the words of British philosopher C. D. Broad:

> If we can judge what philosophy *is* by what great philosophers have *done* in the past, its business is by no means confined to accepting without question, and trying to analyze, the beliefs held in common by contemporary European and North American plain men. Judged by that criterion, philosophy involves at least two other closely connected activities, which I call *Synopsis* and *Synthesis.* Synopsis is the deliberate viewing together of aspects of human experience which, for one reason or another, are generally kept apart by the plain man and even by the professional scientist or scholar. The object of synopsis is to try to find out how these various aspects are inter-related. Synthesis is the attempt to supply a coherent set of concepts and principles which cover satisfactorily all the regions of fact which have been viewed synoptically.[5]

This book is written in the spirit of Broad's synopsis and synthesis, as it joins different fields and kinds of human experience, and attempts to specify certain relations among them. Without data from many domains of inquiry, without various *kinds* of knowing, our understanding of human development will be incomplete. Without a deliberate viewing together of our physiological, emotional, and cognitive processes, without a coherent set of concepts to relate our normal and metanormal functioning, we cannot comprehend many possibilities for further human advance. To serve this multidisciplinary approach, theories of course must stand or fall with the data they are meant to interpret. General systems of belief, however honored in particular traditions, must be attuned to the facts produced by science, psychical research, comparative studies of religious experience, and other fields.

In this book, then, I draw upon many fields of experience and various kinds of data acquisition. For the best thinking about evolution I turn to widely recognized evolutionary theorists; for the best medical science, to its leading authorities; for the best paranormal data, to competent psychical researchers; for the best insights of religious schools, to their acknowledged masters. Each of these activities—and others I draw upon—have solid traditions, time-tested methods, exemplars, and considerable peer review (whether scientific, parapsychological, or religious), which have produced much knowledge to support exploration of high-level change. If we need a term for this approach, we might call it a synoptic, multidisciplinary, or integral empiricism (remembering, of course, that *empiricism* usually refers to data acquisition and verification limited to sensory experience).

[2.2]
THE DATA OF PSYCHICAL RESEARCH
AND PARAPSYCHOLOGY

In the 1880s, Frederic Myers and other founders of the (British) Society for Psychical Research publicly separated themselves from unwarranted claims and deceptive methods prevalent among mediums and occultists. With other scientists and academics in Europe and the United States, they established a disciplined study of paranormal phenomena clearly separate from Spiritualism and other movements notable for incoherent speculation, fuzzy thinking, and outright deception. For more than a century, the best psychical researchers and parapsychologists have continued to distance themselves from the fraudulence, superstition, and self-deception that characterize much activity related to the occult. Though many scientists and academic philosophers still argue that paranormal events do not occur, we have considerable evidence for such events provided by competent researchers. The *Journals* and *Proceedings* of the British and American Societies for Psychical Research contain several thousand accounts of psi phenomena gathered with great care to eliminate trickery and observational error. Furthermore, since J. B. Rhine opened his Parapsychology Laboratory at Duke University in the 1930s, experimental studies of psi have been conducted with statistical methods, control groups, machines to register PK effects, and other devices to improve the quality of evidence. Experimental data produced by parapsychologists have been presented in several credible journals for more than 50 years. Modern evidence for paranormal events is also supported by its obvious correspondence with reports about telepathic, clairvoyant, and psychokinetic powers in the shamanic and religious traditions, and by surveys during recent decades in which large percentages of various subject groups have reported spontaneous experiences of ESP (see chapter 4.3). There is, in short, considerable evidence for paranormal phenomena. To assist those readers who want to examine such evidence for themselves, I have assembled the reading lists in appendices A.1–A.6.

In an attempt to overcome scientific objections to the findings of psychical research, J. B. Rhine developed an experimental approach, which he called parapsychology, to the study of extrasensory perception and psychokinesis. By using controlled procedures modeled largely upon those of experimental psychology, he hoped to convince academic and other skeptics that paranormal phenomena could be regularly produced in the laboratory. His studies, which were widely publicized, have been thought by many experimenters to comprise a methodological advance in the field of psychical research.[†] Recently, however, some students of the field have come to believe that the experimental approach has not greatly strengthened the evidence for paranormal events, and have criticized parapsychology for its procedural deficiencies. Philosopher Stephen Braude, for example, wrote:

> My own view is that Rhine's so-called revolution has failed, and in just those respects in which it was expected to succeed. Skeptics have rarely (if ever) been converted, or even deeply impressed by laboratory experiments. Nor have they been more effectively swayed than they were in the past by seances or anecdotal reports. The reasons all concern the fact that the laboratory evidence is quantitative. Most successful experiments boast odds against chance, at best, of 100 to 1,000 to 1. And although some reach staggering levels of significance, they are rare enough for the skeptic or parapsychological fence-sitter to wonder whether they represent nothing more than a statistical anomaly, rather than occasional spasms of outstanding—or conspicuous—psi. Certainly, laboratory results in PK are never as viscerally compelling or mind-boggling as experiences—or even mere reports—of large-scale physical phenomena (e.g., object levitations or materializations).[6]

With other students of paranormal phenomena, Braude has suggested that experimental parapsychology is severely limited because psi functioning depends upon psychological processes that are hard to influence upon demand in the contrived circumstances of a labora-

---

[†] Nils Wiklund of Stockholm's Karolinska Institute asked several past presidents of the Parapsychological Association (which is based in the United States with members from various nations) to list the best parapsychology experiments. Eight groups of experiments were nominated in response to Wiklund's request, of which seven were also chosen by members (not presidents) of the same association. In appendix A.2, I reference articles describing the seven sets of experiments that were chosen in both surveys. They exemplify the sophistication of method developed in the field since Rhine's first experiments, and/or provide exceptional evidence for psi (Wiklund 1984).

tory. The experiments of Rhine and his followers were prompted, after all, by the occurrence of psi in real life where it appears to serve basic human needs. Half a century of laboratory tests have produced relatively few robust phenomena equivalent to those reported by early psychical researchers who observed them in natural circumstances. "Parapsychology would be in a sorry state indeed," Braude wrote, "if the experimental evidence were the only evidence [for psi] available."[7]

Though controlled experiments have sometimes produced good evidence that psi occurs, they rarely produce evidence as strong as the spontaneous cases studied by the founders of the American and British Societies for Psychical Research. Robust phenomena typically happen, as Braude and others have noted, in response to powerful needs and desires rather than to (frequently boring) experimental procedures. Studies such as Rhine's, perhaps, are not altogether adequate to their data domain, not entirely appropriate to the capacities they were designed to study. With some exceptions, such as those produced by the studies listed in appendix A.2, robust paranormal events do not happen in laboratories. They are often, however, observed during shamanic rituals, ecstatic prayer, intense athletic events, and other circumstances conducive to them. We take it for granted that many phenomena that science explores cannot be produced in controlled experiments. Anthropology, ethnology, and other fields draw upon evidence that cannot be produced at will. Rare animals and rock formations are discovered in the field, not upon experimental demand. To observe paranormal events in their more vivid forms, we must do so when and where they happen.

The constraints upon extraordinary functioning that we find in the parapsychology laboratory occur as well in other fields. In studies of hypnosis, biofeedback, meditation, and mental training in sport, experimental procedures can weaken results by their preoccupation with devices meant to enhance scientific precision. Discussing this problem in hypnosis studies, psychologist Ronald Shor wrote:

> The attitude of disciplined skepticism so essential for building a realistic science must not become such a blinding preoccupation that the investigator thereby becomes an inept hypnotist. But equally important, the hypnotist's exuding of confident persuasiveness, so essential for properly catalyzing the hypnotic processes, must not become such a blinding preoccupation that the investigator thereby loses his scientific objectivity. Thus, taking the "magic" out of hypnosis debilitates the phenomena but taking the "magic" too seriously deludes the investigator.
>
> Investigators in the academic experimentalist's tradition have

generally been most vulnerable to the danger of insufficient catalyst; investigators in the clinical practitioner's tradition have generally been most vulnerable to the danger of insufficient skepticism. The experimentalists have been mainly concerned with rigorous method and the practitioners mainly with improving their clinical effectiveness. Attempts to understand and share each other's objectives and points of view unfortunately have frequently been hampered by clannish loyalties and polemics.[8]

Researchers Elmer and Alyce Green, Steve Fahrion, Solomon Steiner, and W. Dince have noted similar problems in the field of biofeedback research (16.6). To be effective, biofeedback training must last long enough and offer sufficient preliminary reward to promote lasting self-regulation skills. Poorly motivated undergraduates with just a few sessions of practice, for example, cannot reasonably be expected to achieve significant control of their brain waves or other autonomic processes. To produce experimental evidence for extraordinary human functioning, researchers must complement disciplined controls with persuasive confidence, with sufficient practice time, or with other procedural catalysts.

But the study of extraordinary functioning suffers from more than methodological shortcomings. Resistance to psychical research, for example, arises not only from valid criticisms of its procedures, and not only from the apparent violation of basic scientific assumptions its findings exhibit, but also from irrational fears and attitudes—among its opponents and proponents alike. Psychiatrist Jule Eisenbud, who has explored the operations of psi in his therapeutic practice (17.3), has discussed this problem from a psychoanalytic perspective.

> Most of the interferences with and obstacles in the way of actual verification in parapsychology seem to arise directly from the oversights and the omissions, the goofs and the gaffes that, however much they may in the aggregate be accountable for in terms of a Gaussian statistic of error, we really have no one to blame for but ourselves. I am thinking not only of the witnesses not interrogated, of the notes *not* made at the time they should have been, of the precautions *not* taken against simple errors of observation or recording or even counting, or against alternative hypotheses which after the fact may appear perfectly obvious (and I can speak from personal experience with every one of these), but also of the avenues of investigation only glimpsed and then never followed up, of the preliminary work left hanging, of all the experiments not pushed to their logically demanded next stages (classical example: the work of the 1880s on hypnosis at a distance).

We can hardly any longer claim for ourselves extraterritoriality from the findings of sixty years of investigation into the unconscious purposefulness behind events of this character, as if these findings didn't apply to our error and our failures.[9]

If parapsychologists themselves are impelled to bungle their own experiments by largely unconscious resistance to the very evidence they seek, we can suppose that some people who openly dismiss the field's work are similarly motivated. Scientists, like everyone else, have blind spots and prejudices, and their work is sometimes influenced by unexamined needs and motives. Some physicists, biologists, and psychologists who attack psychical research do so with a vehemence and a denial of the field's good work that suggests something more than scientific virtue.

[2.3]
EVIDENCE THAT THE NERVOUS, ENDOCRINE, AND IMMUNE SYSTEMS AFFECT ONE ANOTHER

A number of discoveries in recent decades have revealed linkages among the nervous, endocrine, and immune systems that have an important bearing upon my proposals in this book.

Experimental conditioning of the immune system, for example, was discovered serendipitously by Robert Ader and Nicholas Cohen, who were studying taste aversion in animals by pairing a saccharin-flavored solution (as a conditioned stimulus) with a nausea-producing drug that incidentally suppressed the immune system. After conditioning, when the drug was *no longer given,* the animals became sick and exhibited a higher mortality rate after they were given saccharin.[10] Conditioning of the immune system was subsequently demonstrated with other animals and other conditioned stimuli.[11] By showing that the central nervous system interacts with and can directly affect the immune system, such demonstrations have confirmed the long-standing belief of many doctors, philosophers, and spiritual teachers that mental images, attitudes, and emotions help determine sickness and well-being. Life-giving exercises of the imagination, for example, were described in Egyptian books of the dead, the Vedas, and other religious scriptures. In the *Phaedrus, Phaedo,* and *Symposium,* Plato celebrated the power of images to stimulate health. Aristotle recognized the intimate connections between mental pictures, volition, sensation, and other bodily functions (see, for example, *de Anima*, Book III.3). Renaissance physicians believed that images were connected to impulses that travel through the nerves from the brain, stirring up humors that lead to changes in cognitive, emotional, and somatic functioning.[12]

During the early 1970s, endorphins and other opioid peptides (complex molecules secreted by the brain, spinal cord, glands, visceral tissue, and other organs) were discovered. These substances circulate in the blood and lymph systems and attach to specific molecules, or receptor sites, located in the brain and other body parts, causing alterations of mood, pain, and pleasure. Several studies have shown that opioid peptides help mediate analgesia, hormonal changes, and other responses to stress and illness. After it was discovered (in 1979) that T-cell lymphocytes, which are fundamental components of the immune system, have receptors for methionine-enkaphalin,[13] there was an enormous growth of understanding about the role of opioids as immunomodulators.[14]

In 1983, researchers found that the thymus gland, an organ of the immune system, secretes a substance, thymosin fraction 5, that powerfully stimulates certain adrenal hormones that in turn affect the central nervous system.[15] Further experiments have shown that other components of the immune system modulate brain and other nervous activity (which in turn affects the endocrine and immune systems). Many processes by which the immune system modulates the central nervous system, either directly or through the endocrine system, have now been discovered.

Indeed, our body produces as yet undetermined numbers of messenger substances, some of them secreted by the central nervous system, some by the endocrine system, and some by the immune system, including neurotransmitters such as serotonin, neuropeptides such as the endorphins, hormones such as adrenaline, and lymphokines or cytokines such as interferon and interleukin-1 and -2. These various substances promote resistance to viruses, bacteria, and other potentially noxious agents as well as changes in psychological state and behavior.[16]

Studies of peptide molecules have shown that particular types of molecules fit particular receptor molecules attached to particular organs, in a lock-and-key fashion. This particularity enables peptides to work with precision in various body parts, and is one reason why images and emotions can influence the body with great specificity.

Burgeoning research on relations between the nervous, endocrine, and immune systems has revealed the body's great redundancy of healing (or homeostasis-maintaining) process. Because of this redundancy, particular kinds of physiological repair or mood alteration can be mediated in more than one way. (The tendency among some medical and psychological researchers to think that there is just one way to achieve a particular physical change might be called "the fallacy of single mediation.")

➔

In short, many *bidirectional* pathways between the nervous, endocrine, and immune systems are being specified now so that disciplines such as immunology and endocrinology that were traditionally separated must work in concert for optimal results. Thus, the term *psychoimmunology*, introduced by research psychiatrist George Solomon in the early 1960s[17] and expanded to *psychoneuroimmunology* by Robert Ader in 1975, is now commonly accepted, as are the terms *psychoendocrinology* and *psychobiology*.[18] The understanding of human nature inspired by these interdisciplinary fields resembles the view of ancient Greek and Renaissance doctors that mental images and the flesh are intimately connected. For example, Hippocrates and Galen, the most influential physicians of Greek antiquity, believed that the human organism was a single hierarchical complex in which imagination, vital spirits, nerves, humors, blood, and muscle were closely joined. Our thoughts and feelings, they assumed, moved within the body's physical elements, affecting all our functioning by immediate contact. Renaissance doctors generally held the same belief, subscribing to the Hippocratic maxim that imagery and physical processes "tread in a ring," constantly influencing one another. Particular images and emotions, according to Renaissance medical opinion, stimulated particular body states. Indeed, certain images and their resulting emotions could kill. Gloucester died in *King Lear* because

> . . . his flaw'd heart,
> Alack, too weak the conflict to support!
> 'Twixt two extremes of passion, joy and grief,
> Burst smilingly (Act 5, Scene 3).

Powerful pictures in the mind, it was widely believed, could produce lasting changes in bodily structure. Parents could even influence their offspring at conception by the thoughts they held. As one Renaissance writer put it, imagination "marks and deformes, nay, sometimes kills Embryos in the womb, hastens Births, or causes Abortions."[19] Because images were considered to be the primary causes of many afflictions, Renaissance physicians often addressed their patients' imagination. In *Approved Directions for Health* (1612), William Vaughan outlined this therapeutic procedure: "The physician must invent and devise some spiritual pageant to fortify and help the imaginative faculty, which is corrupted and depraved; yea, he must endeavour to deceive and imprint another conceit, whether it be wise or foolish, in the patient's braine, thereby to put out all former phantasies."

This holistic understanding was almost entirely displaced, however, by the dualistic view of human functioning developed by post-Renaissance thinkers. According to Descartes, the mind "consists entirely in thinking, and for its existence, has no need of place, and is not dependent on any material thing."[20] Descartes's dualistic conception of human nature supported the mechanistic physiology that emerged in the late 1600s. In this new understanding, mind lost its intimate connections with the body and was thought to operate in parallel with physiological processes rather than through constant interplay with them. The "vital spirits," by which imagination stirred the flesh, began to disappear from medical thinking. The body was assigned to anatomists and physiologists, the mind to philosophers and psychologists. The interactive psychophysiology of Aristotle, Hippocrates, and the Renaissance was replaced by a mind-body parallelism that separated organic and functional afflictions. Psychosomatic medicine suffered from the reductionism implicit in the idea that functional afflictions were only imaginary; and by the middle of the eighteenth century, mechanistic notions had almost entirely supplanted holistic ideas about human functioning. During the nineteenth century, most researchers studied body and mind as if they were separate domains.[21]

In the twentieth century, however, interactionist views have reappeared in psychobiology, psychoneuroimmunology, somatic education, and the wellness or holistic health movements; and imagery has become a legitimate object of study by experimental psychologists. Since 1960 especially, scientific opinion has been generally favorable to both clinical and experimental studies that take an interactionist approach to images and their physiological correlates. This holistic view, like the one held by most Greek and Renaissance doctors, is more consonant with the picture of human possibility presented in this book than are mechanistic models of the mind-body complex generally accepted by medical scientists from the eighteenth century until recent decades.

The findings of psychoneuroimmunology and related fields reveal: (1) the highly interactive, feedback-laced nature of psychophysical functioning; (2) multiple ways in which particular alterations of consciousness, behavior, bodily structure and process are mediated; and (3) the immense specificity with which significant changes are happening, moment by moment, throughout the nervous, endocrine, and immune systems. A body so mutable and interactive, operating with such redundancy and specificity, seems more capable of the transformations described in this book than a tightly compartmentalized, poorly articulated organism would be.

In summary, then, we have unprecedented access now to evidence for human transformative capacity. With critical discernment we can gather such evidence to better appreciate our possibilities for further development, but to do this we must reject scientific, religious, and other prejudices against certain time-tested data from the contemplative traditions, psychical research, anthropological studies of shamanism, and other fields. In this multidisciplinary organization of knowledge, we can draw upon the physical, biological, and human sciences, including new fields such as psychoneuroimmunology, that are revealing human nature's capacity for creative change. In the following chapters, I will argue that a wide range of extraordinary human attributes revealed by this synoptic approach are rooted in functioning developed during the long course of animal and human evolution.

3

EVOLUTION AND EXTRAORDINARY FUNCTIONING

IN ASTRONOMY AND ASTROPHYSICS the term *evolution* refers to the production of elements, stars, galaxies, and other configurations of inorganic matter. In biology, where its use is most prominent, it implies adaptation of organisms to their environment through natural selection and refers to the long-term process of change in plant and animal species. And the term is employed as well, in a very general sense, to denote different kinds of human growth. Though the kinds of development that occur in the physical, biological, and psychosocial domains are shaped by different processes and have different patterns, they proceed in sequences that are called evolutionary.[†]

[†] Lamarck did not use the term *evolution* at all, and Darwin spoke of "descent with modification," using the word *evolved* just once in his first (1859) edition of *The Origin of Species,* as the very last word in the book. Biologists in the eighteenth and nineteenth centuries who held a pre-formationist view of life's development had used the term to denote the "unrolling of parts" contained in the homunculus, or miniature adult form, they believed to be present in the embryos of animal and plant organisms. In the 1860s, the term *evolution* still was not commonly used as it is today.

Herbert Spencer was primarily responsible for the word's transformation. In 1862, in his *First Principles,* he defined evolution to be "a change from an indefinite, incoherent homogeneity, to a definite, coherent heterogeneity, through continuous differentiations and integrations." Spencer's break with previous usage and the aspect of his definition that permitted a later extension to all organic change came with his contention that evolution was not controlled by a preset, internal program, but depended upon interaction with external forces. Subsequently, biologists appropriated his word as a succinct substitute for Darwin's "descent with modification." In its more general usage today, however, the term *evolution* refers to inorganic and psychosocial as well as biological development. For a brief history of the word's use, see: Gould 1977, pp. 28–32.

[3.1]
EVOLUTIONARY TRANSCENDENCE

Life on earth arose from the physical elements and depends upon them for its sustenance; and we humans are rooted in biological processes inherited from our animal ancestors. But the three kinds of evolution—inorganic, biological, and psychosocial—though having many features in common, operate according to separate principles. While there are many kinds of interaction between them, physical, organic, and human development each has unique patterns that are best understood on their own terms. For example, inorganic chemistry cannot specify a cell's life processes, though each cell depends upon chemical interactions; and zoology cannot adequately account for human culture, though humans share many features with their primate forebears. Some scientists have wanted to explain all living processes in terms of physics or chemistry, or to explain human consciousness in terms of biological process, in hopes that such explanations might help to integrate separate domains of scientific inquiry as classical mechanics did during the eighteenth and nineteenth centuries, but no one has put forth theories adequate to do that.[1] From time to time, a relatively autonomous theory in some field of biological or human science is absorbed by—or reduced to—a more inclusive theory; and certain features of mind or life are sometimes accounted for in terms of physics, chemistry, or statistical concepts (as for example when it was found that random changes in DNA of the genome produce mutations). But in spite of these simplifying moves, we need to study and explain organisms or human consciousness on their own terms. Helpful scientific reductions do not erase the fact that inorganic, biological, and psychosocial evolutions proceed according to their own distinctive patterns.[†]

[†] In his even-tempered and highly influential analysis of scientific explanation *The Structure of Science*, philosopher Ernest Nagel wrote:

> There is no more reason for rejecting a biological theory (e.g., the gene theory of heredity) because it is not a mechanistic one . . . than there is for discarding some physical theory (e.g., modern quantum theory) on the ground that it is not reducible to a theory in another branch of physical science (e.g., to classical mechanics). A wise strategy of research may indeed require that a given discipline be cultivated as a relatively independent branch of science, at least during a certain period of its development, rather than as an appendage to some other discipline, even if the theories of the latter are more inclusive and better established than are the explanatory principles of the former. The protest of organismic biology against the dogmatism often associated with the mechanistic standpoint in biology is salutary (Nagel 1979, pp. 444–45).

Inorganic matter, animal and plant species, and human nature, then, can be said to comprise three levels or kinds of existence, each of which is organized according to separate principles. These three levels comprise an evolutionary triad in which the first two have transcended themselves, inorganic elements producing living species, animals giving rise to humanity. Evolutionary theorists Theodosius Dobzhansky and Francisco Ayala have called these two stupendous events instances of "evolutionary transcendence" because in each of them there arose a new order of existence. "Inorganic evolution went beyond the bounds of [its] previous physical and chemical patternings when it gave rise to life," Ayala wrote. "In the same sense, biological evolution transcended itself when it gave rise to man."[2]

These two epochal transitions, the appearance of life and the rise of humankind, marked the beginnings of new evolutionary eras. They were made possible, though, by dramatic changes that preceded them. The creation of new elements in exploding stars, for example, and the formation of complex molecules on earth some four billion years ago set the stage for living cells; and the evolution of terrestrial vertebrates from fishlike ancestors opened new realms on land for the mammalian evolutions that resulted in homo sapiens. Paleontologist George Gaylord Simpson called certain changes of this kind among living species instances of "quantum evolution" because they involved relatively abrupt alterations of adaptive capacity or bodily structure and left little or no evidence in the fossil record of the transitions between them.[†] And these big jumps in the development of life and inorganic matter were made possible in turn by a great many smaller steps that prepared the way for them. G. Ledyard Stebbins, a principal architect of modern evolutionary theory, described certain differences between large and small steps in organic evolution, distinguishing minor from major advances in grade of plants and animals. There have been about 640,000 of the former, he estimated, and from 20 to 100 of the latter during the several hundred million years of eukaryote evolution.[‡] Though Stebbins's estimates only approxi-

---

[†] In the words of Francisco Ayala: "Since [species] appear abruptly and last millions of years without change, it follows that most of the morphological change in [them] was associated with their origin. In them, evolution proceeded by bursts separated by periods of stasis when little change occurred, rather than by a gradual shift in characters" (Ayala & Valentine 1979, p. 261). This feature of evolution has prompted the punctuational model of species origination developed by Niles Eldridge and Stephen Jay Gould (Eldridge & Gould 1972).

[‡] Eukaryote cells have a nucleus, organelles, and convoluted intracellular membrane called the endoplasmic reticulum, which distinguish them from their primitive ancestors, the prokaryote bacteria and cyanobacteria, or blue-green algae.

mate the actual number of graduations, or advances in grade, which have occurred among living species, they reflect the great complexity of evolutionary progress, the many steps by which life gave rise to homo sapiens.

I cite Stebbins to suggest that humanity has also evolved by means of minor and major steps toward yet another epochal transition. For certain types of extraordinary human development, I believe, herald a third evolutionary transcendence. With them, a new level of existence has begun to appear on earth, one whose patterns cannot be adequately specified by physics, biology, or mainstream social science. As life developed from inorganic elements and humankind from its primate ancestors, a new evolutionary domain is tentatively rising in the human race, both spontaneously and by transformative practice, and it was made possible by quantum jumps in development such as the discovery of fire, the emergence of language, and the birth of religious awareness. In chapters 4 and 5, I describe 12 sets of human attributes that characterize this emergent level of development:

1. Extraordinary perceptions of things outside the organism, including apprehensions of a numinous beauty in familiar objects, voluntary clairvoyance, and contact with entities or events that are inaccessible to the ordinary senses.

2. Extraordinary somatic awarenesses and self-regulation.

3. Extraordinary communication abilities, including the transmission of thoughts, volitions, and ecstatic states through extrasomatic modalities.

4. Superabundant vitality that is difficult to account for in terms of ordinary bodily processes.

5. Extraordinary movement abilities.

6. Extraordinary capacities to alter the environment, including exceptional hand-eye skills and abilities to influence things at a dis-

---

The term *grade* is used among biologists to denote a set of characteristics or abilities that clearly give descendent species certain advantages over their ancestors. Evolution from one grade to another requires net evolutionary progress; i.e., improvement in some feature (or features) *on the average* among later members of an evolutionary sequence, as for example when mammals achieved warm-bloodedness as they developed from reptiles. Birds also have advanced to the warm-blooded grade, but independently of mammals since they arose from a different reptile ancestor. A grade, then, may be *polyphyletic*, i.e., occupied by animals with diverse ancestral backgrounds. According to Stebbins, the development of pollination mechanisms in milkweeds and orchids is an example of minor advances in grade, while the appearance of the digestive tube, central nervous system, elaborate sense organs, vertebrate limbs, and elaborate social behavior represent major advances (Stebbins 1969).

tance without reliance upon direct physical action, as for example in spiritual healing.

7. Self-existent delight, which does not depend upon the satisfaction of needs or desires in the manner of ordinary pleasure, and which persists through sickness and other adversity.

8. The supreme intellectual capacities evident in some works of genius, by which great artistic or other productions are apprehended *tout ensemble*, all at once; and the unitive knowledge inherent in mystical experience, which differs radically from ordinary thinking, described, for example, by Plato, Plotinus, and other Neo-Platonist philosophers, by the authors of the Upanishads and other Indian seers, by Christian mystics, and by countless sages of the Kabbalistic, Hasidic, Sufi, Buddhist, and Taoist traditions.

9. Volition exceeding ordinary will, which unifies separate impulses to produce extraordinary actions.

10. Personhood that simultaneously transcends and fulfills one's ordinary sense of self while revealing one's fundamental unity with others; and individuality based upon the extraordinary capacities listed above.

11. Love that transcends ordinary needs and reveals a fundamental unity with others.

12. Alterations in bodily structures, states, and processes that support the experiences and capacities just noted.

Most of these attributes are evident, however briefly, in the course of everyday life (4.3), but their sustained realization comprises a break with ordinary human activity, and their lasting integration by many people would constitute a new kind of life on this planet. They are also frequently marked off from ordinary functioning by a sense they convey of something beyond the familiar patterns of our existence. Jewish, Christian, and Moslem mystics, for example, typically attribute them to God's blessedness, mercy, or grace; Buddhists to the omnipresent Buddha-Mind; Hindus to Divinity's *shakti*, or world-power; and Taoists to the Way, or Tao. When some athletes experience them, they say they are "zoned." Martial artists have called them the action of *sunyam*, or Emptiness, "through human hands and feet" (20.1). And more commonly, when we have a spiritual insight that lifts us to certitudes we have not experienced before, or when we surprise ourselves by accomplishing some extraordinary deed, we might say that "something came over us," that we were "carried away." The recognition of ego-transcendent powers is reflected in religious terms and in our common language. That recognition of a Something beyond, I propose, coupled with our inability to specify its operations in us, points toward a new kind of human development. We don't know where our new vision, love, or joy came from, or how we effected our marvelous deed, precisely because such things are unfamiliar and because their mediations are related to something emergent in us.

Their radical novelty and nonordinary causes, then, suggest that the extraordinary capacities listed above are instances of a new type of evolution that has patterns which distinguish it from ordinary psychosocial development. Conversely, however, it might be concluded that because they resist verification by standard scientific procedures and often appear to violate certain assumptions of physics and biology, some of these extraordinary human attributes do not—or cannot—exist. And indeed, that is what many scientists and philosophers argue: the fact that telepathic empathy cannot be demonstrated in controlled experiments with the reliability of known physical processes, or that spiritual healing seems to occur independently of observable agencies proves ipso facto that such things are figments of the imagination. Even mystical experience is denied its claim to knowledge, given its "subjectivity." But such arguments can be turned on their head. The same resistance to laboratory experiments or verification through the physical senses, and the same apparent violation of natural laws that scientists invoke against various kinds of extraordinary functioning, can be taken as signs that such functioning is part of a new domain, one that transcends ordinary human activity and methods developed to study it. The same conceptual and methodological difficulties that prompt many scientists to deny metanormal capacities can be seen as marks of a fundamental transition to a new kind of human existence.

This reversal of scientists' frequent complaint about mystical truth claims and reports of psi phenomena seems entirely legitimate to me. Should an ancient body of testimony supported by religious figures and philosophers with radically different backgrounds since the beginnings of recorded history be denied because it causes difficulty for certain scientists? Every domain that science has illumined has required unique approaches and appropriate instruments. Astronomy needed the optical telescope and Newton's calculus; zoology required the collection of animal specimens and their classification according to form and function; depth psychology has depended upon subjective report. None of these fields could have developed without concepts and methods adequate to them. And the same holds for extraordinary human attributes. They, too, must be studied and developed with appropriate methods and theories. As I said in chapter 2.1, they can be explored through anthropological field studies, psychical research, contemplative discipline, and other empirical approaches that have time-tested procedures for data acquisition and verification. The fact that certain human events apparently violate assumptions of contemporary science does not require us to deny the evidence for them. Such violation can be taken instead to suggest that such events are part of a new evolutionary domain that has its own distinctive features and patterns.

But need we imagine that a widespread realization of extraordinary capacities would comprise an evolutionary transcendence analogous to the origination of living species or homo sapiens? Does the evidence for metanormal activity require such a grand projection of the world's advance? Perhaps any widely shared development of metanormality would be more aptly and simply described as a product of ordinary psychosocial development. Though we can conceive such progress in this more economical manner, I am led by data such as those presented in this book to believe that the self-evident break with normal consciousness and behavior, the transcendence of certain needs, and the self-mastery of mind and flesh characteristic of metanormal functioning would, if realized by enough people, create a new kind of life on this planet. This new life, it seems to me, would involve new types of social interaction, new styles of energy consumption, greater care for the physical environment, more wisdom in dealing with human aggressiveness, new rituals of work and play. As it began to appear among large groups, such functioning might not appear at first so dramatic that it comprised a new kind of evolution, but it would, I believe, eventually exhibit features and regularities we cannot predict from the patterns of ordinary human existence.†

But all this is only speculation. Even if we harbor potentials for

---

† It is likely that metanormal capacities appeared among our paleolithic ancestors. Certain figures in cave drawings at Lascaux and other locations, for example, suggest that Cro-Magnon people supported some kind of shamanism; and there is evidence that certain caves of the Pyrenees housed initiatory ceremonies like those practiced among stone-age people today (Campbell 1959, pp. 342–45). That metanormal development is taking thousands of years to establish itself should not surprise us. Inorganic molecules that gave rise to the first living cells tooks millions of years to develop, while the emergence of humankind took place by degrees during the long course of hominid evolution. The two instances of evolutionary transcendence on earth seem abrupt only in relation to the multibillion-year history of our universe.

extraordinary living, their lasting realization by large numbers of people on earth is by no means assured. In the next section, I emphasize this fact while reminding the reader that evolution is not always progressive.

[3.2]
EVOLUTION AND PROGRESS

If a greater life is latent in the human race, its permanent establishment is far from being guaranteed. Nuclear war, overpopulation, ecological disaster, social upheaval, or other catastrophes could so diminish life on earth that few people would have the will or resources to cultivate metanormal capacities. Such events could destroy the conditions for any kind of widespread human progress, let alone a third evolutionary transcendence. Furthermore, even favored groups do not always produce lasting moral or spiritual growth. Human cultures, like animal species, often remain static or regress. Neither human nor animal evolution is automatically progressive.

Indeed, evolution as it is commonly understood in biology is not the same thing as progress. Progress occurs when there is directional change toward a *better* condition, however that improvement may be defined, while biological or human evolution is sometimes regressive and can lead to the extinction of a species or culture.† In the words of George Gaylord Simpson:

> Evolution is not invariably accompanied by progress, nor does it really seem to be characterized by [it] as an essential feature. Progress has occurred within it but is not of its essence. Aside from the broad tendency for the expansion of life, which is also inconstant, there is no sense in which it can be said that evolution *is* progress.[3]

---

† Francisco Ayala defined biological progress as "systematic change in at least one feature belonging to all the members of a historical sequence, which is such that later members of the sequence exhibit an improvement of that feature. More simply, it could be defined as directional change towards the better." Ayala has refined this definition by making two further distinctions: first, between *net* progress or improvement on the average of some feature among later members of an evolutionary sequence, and *uniform* progress, or improvement in every descendent of a species; and second, between general progress, or improvement "in all historical sequences of a given domain . . . from the beginning of the sequences until their end," and *particular* progress, which occurs "in one or several, but not in all, historical sequences of a given domain . . . during part but not all the duration of the sequence." Using these definitions, there is no standard by which uniform or general progress has taken place through evolution. There has only been net progress; that is, improvement among some members of certain species, and only particular progress along certain lines during some periods in the past (Ayala 1974, pp. 341–43, 346).

Simpson, Francisco Ayala, and other evolutionary theorists have suggested criteria by which organisms can be judged to have progressed, among them an increase of genetic information stored in the members of a species; broadening of adaptive behavioral repertoires; development of efficient sensory organs; increase of general energy level (as in the warm-bloodedness of birds and mammals); growth of information-processing skills; improvements in care for the young; expansion into new environments; increased specialization to make a species more adaptive and efficient in a given environment; and progress in individualization.[4] Similarly, there are many criteria by which to judge individual human development, whether emotional, moral, cognitive, or spiritual, as well as standards for the progress of human cultures, such as their stimulation of individual liberties, social justice, prosperity, artistic achievement, or religious expression. There may be disagreement about the relative importance or worth of such criteria, but no matter which we choose, many individuals and cultures will be judged not to have developed significantly beyond their predecessors, and by any standard, some have clearly regressed.[†] Nor would progress be assured in the domain of metanormal events and processes. Long experience in the religious traditions has shown that ecstasies, illuminations, and extraordinary powers are no guarantee of lasting goodness or growth. Whenever I argue that progress *might* or *could* occur along certain lines, I do not mean that it necessarily will. Nor do I mean that progress is bound to occur along a single line of metanormal development. It is conceivable that very different transformative disciplines could arise, with different kinds of supporting institutions, which in time would produce different configurations of physical, emotional, and cognitive experience among their participants. Along with stasis or regression, then, a fourth evolutionary domain might give rise to several kinds of progress.[‡]

[†] However, the two instances of evolutionary transcendence can be regarded as a supreme form of progress toward the extraordinary activity I am exploring. More than any developments within domains, the appearance of living cells and of human beings exhibited a clear advance in information-processing skills, motility, and other activities that point toward a further development of life upon earth. Indeed, these two epochal transitions might be called instances of *macro-progress*. In surpassing other evolutionary advances, they dramatize our world's journey toward a greater existence.

[‡] It is possible, too, that metanormal capacities have been established already in certain populations, then lost when their host culture changed or vanished. In remote parts of India or Tibet, or among wandering groups that flourished in ancient stone-age cultures, mystical cognitions, readily accessible clairvoyance, or other kinds of extraordinary functioning might have been

[3.3]
EVOLUTION AND ITS MEDIATING PROCESSES

"Today the proposition that [biological] evolution has occurred is no longer in doubt," wrote Francisco Ayala and James Valentine, reflecting the opinion of virtually all life scientists. "The evidence in favor of [it] is overwhelming."[5] Many discoveries cohere in an overall theory that illuminates the development of living forms—among them fossil finds in many parts of the earth; precise reckonings of fossil ages by carbon dating and other methods; new insights concerning the nature of genetic material, its recombination, mutation, and "drift"; and specifications of climatic and geological changes that affect the adaptive advantages of various organisms. As this well-confirmed information has accumulated, a marvelously detailed picture of life's advance has appeared, highly articulated in many details, beautifully connected in its general patterns. As Ayala and other biologists tell us, there is overwhelmingly persuasive evidence that organic evolution has occurred on this planet. In spite of this evidence, however, creationists still argue that evolution is unproven.[6] In answer to them, paleontologist Stephen Jay Gould wrote:

> In the American vernacular, "theory" often means "imperfect fact"— part of a hierarchy of confidence running downhill from fact to theory to hypothesis to guess. Thus, creationists can (and do) argue: evolution is "only" a theory, and intense debate now rages about many aspects of the theory. If evolution is less than a fact, and scientists can't even make up their minds about the theory, then what confidence can we have in it?
>
> Well, evolution is a theory. It is also a fact. And facts and theories are different things, not rungs in a hierarchy of increasing certainty. Facts are the world's data. Theories are structures of ideas that explain and interpret facts. Facts do not go away while scientists debate rival theories for explaining them. Einstein's theory of gravitation replaced Newton's, but apples did not suspend themselves in mid-air pending the outcome. And human beings evolved from apelike ancestors whether they did so by Darwin's proposed mechanism or by some other, yet to be discovered.

widely realized. Though we have no proof, it is conceivable that such capacities were prevalent in a few societies lost to modern memory. Capacities of this kind are metanormal now because they surpass types of functioning that are well established among contemporary peoples.

Evolutionists have been clear about this distinction between fact and theory from the very beginning, if only because we have always acknowledged how far we are from completely understanding the mechanisms (theory) by which evolution (fact) occurred. Darwin continually emphasized the difference between his two great and separate accomplishments; establishing the fact of evolution, and proposing a theory—natural selection—to explain the mechanism of evolution. He wrote in *The Descent of Man*: "I have two distinct objects in view; firstly, to show that species had not been separately created, and secondly, that natural selection has been the chief agent of change. . . . Hence if I have erred in having exaggerated its [natural selection's] power I have at least, as I hope, done good service in aiding to overthrow the dogma of separate creations."

Thus Darwin acknowledged the provisional nature of natural selection while affirming the fact of evolution. The fruitful theoretical debate that Darwin initiated has never ceased.[7]

Building upon Darwin's discovery of natural selection, biology now provides a richly detailed account of various mechanisms that help determine the origination and survival of species. This explanatory theory is one of the great achievements of modern science. But it is not complete. Contrasting the understanding of evolution at each of the three Darwinian centennials (1909, the hundredth anniversary of Darwin's birth; 1959, the hundredth year since publication of *The Origin of Species;* and 1982, the hundredth anniversary of Darwin's death), Gould wrote:

Although no thinking person doubted the fact of evolution by 1909, Darwin's own theory about its mechanism—natural selection—was not then at the height of its popularity. Indeed, 1909 marked the acme of confusion about how evolution happened in the midst of complete confidence that it had occurred.

By 1959, confusion had ceded to the opposite undesired state of complacency. Strict Darwinism had triumphed. . . . Nearly all evolutionary biologists had concluded that natural selection, after all, provided the creative mechanism of evolutionary change. At age 150, Darwin had triumphed. Yet, in the flush of victory, his latter-day disciples devised a version of his theory far narrower than anything Darwin himself would have allowed.

At the second centennial [in 1959], some experts even declared that the immense complexity of evolution had yielded to final resolution. One leader remarked in a famous essay: "Differences of opinion on relatively minor points naturally persist and many details remain to be filled in, but the essentials of the explanation of the history of life have probably now been achieved."

Now, at the third centennial [in 1982], Darwinian theory is in a vibrantly healthy state. Confidence in the basic mechanism of natural selection provides a theoretical underpinning and point of basic agreement that carries us beyond the pessimistic anarchy of 1909. But the constraints of an over-zealous strict version, so popular in 1959, are loosening. Exciting discoveries in molecular biology and in the study of embryological development have reemphasized the integrity of organic form and hinted at modes of change different from the cumulative, gradual alteration emphasized by strict Darwinians. Direct study of fossil sequences has also challenged gradualistic biases (the "punctuated equilibrium" pattern of long-term stasis within species and geologically rapid origin of new species) and asserted the idea of explanatory hierarchy in identifying species as discrete and active evolutionary agents.[8]

In short, though the fact of evolution is established, evolutionary theory is still developing. I emphasize this point because my proposals here presume that organic evolution has occurred, but they do not stand or fall with the changes in evolutionary theory that are bound to come with new discoveries in the biological sciences. And the same holds for the physical and social sciences. There, too, we must distinguish fact from theory. That evolutions take place in the psychosocial, organic, and inorganic domains is certain, but explanations of them are still incomplete.

4

ANALOGIES BETWEEN THE PRODUCTS OF SEPARATE EVOLUTIONARY DOMAINS

IN CHAPTER 3, I emphasized the fact that each evolutionary domain operates according to its own regularities and patterns, its own distinctive nature. In this chapter, however, I will look at some analogous kinds of animal, ordinary human, and extraordinary development, emphasizing their similarities instead of their differences. Though they are produced in different ways, such analogues comprise a continuum of general evolutionary progress.

[4.1]
CONTINUITIES BETWEEN ANIMAL, NORMAL HUMAN, AND EXTRAORDINARY HUMAN DEVELOPMENT

In the listing that follows, 12 sets of animal attributes are aligned with corresponding features of ordinary and extraordinary human life. Lined up in this way, each set of attributes exhibits an apparent continuity, a vector as it were by which we can discern a possible outcome of further human advance. Even if we do not suppose that the extraordinary phenomena listed here comprise a new kind of evolution, but simply an extension of ordinary psychosocial development, they can be seen as part of a developmental continuum stretching back into animal life. For example, a frog's dim perception of light, the enhanced human vision produced by sensory training, and the apprehensions of extraordinary color and vibrancy reported by contemplatives seem to form a clear line of progress, produced in turn by natural selection, sensory education, and the ego-transcending gifts, or graces, of contemplative practice. Even though these three kinds of experience (and the psychophysical modifications upon which they depend) come into existence by different processes, they appear to be stages of a single development from rudimentary sentience to metanormal perception.[†]

The ability to perceive environmental stimuli, in short, continues to develop even though it is shaped in different ways in different domains of evolution. And the same principle is evident among other evolutionary products. Somatic awareness, movement abilities, information-processing, and other capacities that are shaped by natural selection during animal evolution can be amplified by human discipline and at times, perhaps, by superordinary agencies, developing all the while even though they are produced by different processes at each succeeding level. The mediation-transcending—indeed, domain-transcending—growth of these various abilities constitutes a dramatic progress. Such advance suggests that evolution is influenced by purposes or agencies that to some extent subsume the mechanisms presently described by mainstream science. It invites us to wonder whether nature has a *telos,* or creative tendency, to manifest the activities I've labeled metanormal, a drive or attraction toward greater ends that appropriates the processes of any domain to produce more developed capacities. Paraphrasing a statement made by physicist Joseph Ford, if the randomness evident in all evolution to date is characterized as a kind of dice-rolling, the extraordinary advances described here suggest that the dice are loaded.[1]

But whether or not we believe that this world has some sort of governing purpose or final cause, the capacities listed below constitute a sustained, richly complex development that began with single cells, continues in humans, and points toward higher levels still. However, such development does not necessarily entail that most or even many humans will graduate to higher kinds of functioning in the foreseeable future, for we can just as easily find correspondences

---

[†] As noted in chapter 3.1, though development within each evolutionary domain has its own distinctive patterns, it depends as well upon various types of activity native to the domain that preceded it. Animal functioning, for example, depends upon inorganic elements that interact as they did before life appeared upon earth. Ordinary human functioning depends upon biological processes that appeared among our animal ancestors. And metanormal activity is based upon capacities such as self-reflection that have been developed and passed on to succeeding generations in the course of psychosocial development.

between the stasis or regression of species and the innumerable failures of human individuals and cultures. By any criterion, life meanders more than it progresses. Relatively few animal organisms evolve to higher grades, and few individuals as numbered against the entire human population have exhibited fully developed metanormal capacities. Nevertheless, life upon earth has advanced through countless graduations of living species from single-celled organisms to humankind, and human cultures have given rise to many kinds of extraordinary functioning. Against the backdrop of life's vast meandering, many animal species, human individuals, and cultures have embodied a new complexity of cooperation, structure, and awareness, a new vitality, and a freedom from many of their ancestors' constraints that set the stage for further progress. The complex ascent of living forms, lasting for several hundred million years, is a stupendous event. Though it does not prove life will progress any further, it certainly suggests that it can. Do we really believe that humans have lost life's ancient capacity for transcendence?

The following inventory, then, compares 12 sets of evolutionary products according to their appearance in the domains of (1) animal development, where they are produced by natural selection, differing rates of reproduction, and so on; (2) psychosocial development, where they are shaped by social reinforcement, education, and transformative practice; and (3) metanormal development, where they are evoked by self-discipline, supporting institutions, and conceivably by processes beyond those involved in ordinary human functioning. This catalogue, which is elaborated in chapter 5, could be organized in other ways but is adequate to picture a many-sided continuity of living forms through successive levels of animal, normal human, and metanormal development.

A cross-matrix such as this serves a number of purposes. First, by reminding us that humans exhibit a wide range of extraordinary abilities, it warns us against limited conceptions of our possible development and highest good, including those embedded in philosophies we honor. In chapter 7.1, for example, I use it to argue that various religious traditions fail to appreciate certain capacities their disciplines inadvertently promote. With such an inventory, we can begin to develop an understanding of human nature that more adequately reflects our possibilities for further advance than do highly ascetic (as well as strictly materialist) ideas about human development.

Second, by orienting us toward a many-sided realization, it can help guide our imagination, research, aspirations, and practices toward integral development. It places us on a path toward the fullest actualization of our greater potentials. By helping us assess the comprehensiveness, as well as the comparative strengths and limitations, of various institutions, teachers, and practices, it can help us find practical methods to achieve a balanced progress toward extraordinary life.[†]

Third, it can be employed as a map to help us identify capacities we do not presently recognize. I have used it, for example, to identify extraordinary abilities that arise spontaneously, outside the context of formal discipline, among people who do not seek or expect them. In chapter 4.3, I list some of these, which I call metanormalities of everyday life. Conversely, we can identify extraordinary capacities by looking for analogues of animal and normal human functioning. In this manner, I have come to appreciate the astonishing physical feats of certain Catholic contemplatives. Our search for extraordinary attributes, in other words, can be pattern-driven as well as data-driven. We can discover possibilities for further growth deductively as well as inductively.

Fourth, it can help us identify family resemblances among phenomena that occur in different cultures, or that have been given different names, or that for some other reason are not thought to be the same sort of experience. Establishing the category *metanormal vitality,* for example, has helped me recognize the close similarity, or identity, of the Roman Catholic *incendium amoris,* the Tibetan *tumo,* the "boiling *n/um*" of the Kalahari bushmen, certain inexplicable energies of shamans and athletes, and many kinds of Hindu-Buddhist kundalini episodes (5.4). By showing that particular metanormal attributes occur in different cultures or in different social contexts within the same culture, this classificatory scheme strongly suggests that those attributes are universally latent in the human race.

Fifth, such an inventory suggests an explanation for the fact that many extraordinary capacities tend to emerge *at once* in the course of transformative practice. Quasi-mystical illuminations sometimes occur during athletic endeavors, for example, even though their

---

[†] We might, for example, compare contemplative, martial-arts, and other adepts to see how many metanormal capacities they realized. According to his biography *Abundant Peace* (see Stevens, J., 1987) and other accounts, aikido master Morehei Uyeshiba exhibited or reported all 12 major types of extraordinary attributes outlined in my inventory, and more than 20 subtypes; while according to various accounts of their lives, Saint John of the Cross and the great Indian sage Ramana Maharshi realized fewer. There is a basis here, it seems, for a (somewhat blasphemous) comparative study of metanormality.

recipients have no theory of self with which to interpret them; while conversely, many contemplatives exhibit astonishing physical capacities in conjunction with their mystical illuminations (21.2, 21.3, 21.5, 21.8, and 22). Such capacities and experiences emerge in clusters, perhaps, because we are attracted, consciously or not, toward a many-sided realization of extraordinary attributes. We want to experience different kinds of metanormal capacity simply because they are latent in us. This attraction or drive is apparent, it seems to me, in the impulse toward wholeness evident in many programs for human growth. I return to these observations in subsequent chapters and especially in Part Three.

But another word of introduction to this inventory. Because it is framed in general terms, it does not reflect many differences between capacities included within the same categories. My intention here is simply to indicate broad classes of animal, normal human, and extraordinary development. Furthermore, certain types of experience listed here could be assigned to more than one category. For example, those epiphanies I call "perceptions of the numinous in the physical world" (5.1) can also be called types of cognition. "Kundalini-type" excitement includes both metanormal vitality and metanormal delight. Spiritual healing might be classified as a kind of extraordinary communication or as an extraordinary influence upon the environment. Because human attributes are too fluid to map with the same specificity and exactitude we can bring to a table of the physical elements or a classification of animal species, this inventory is only suggestive. It does not include all animal or human capacities, or exhaust the evidence for extraordinary functioning. Beyond the 12 continuities of development it contains, others could be added. Still, it is comprehensive enough, I believe, to indicate the broad spectrum of extraordinary capacities that are available to us.

And a further note. If some of the capacities discussed here seem improbable, I ask you to suspend judgment until you read discussions of them in later chapters. I am skeptical about some myself but include them because there has been a lasting witness to them in more than one culture or sacred tradition. As I said in chapter 2.1, we can divide supposed metanormalities into three classes: first, those such as mystical illumination that have been attested to by people in many times and places and that have been confirmed again and again by well-developed tests of their authenticity; second, those such as telekinesis for which we do not have indubitable evidence but which nevertheless are described in reports that are difficult to completely dismiss; and third, those for which we have little or no good testimony. In the listing below (and its elaboration in chapter 5), I exclude members of the third class but include members of the second. Though the inclusion of some is more "pattern-driven" than "data-driven," we do have some good evidence that every kind of experience listed below might indeed occur.

I. *Perception of External Events*

PRODUCTS OF ANIMAL EVOLUTION

Sensory abilities mediated by organs such as the canine ear, the feline olfactory bulb, and the human eye.

PRODUCTS OF ORDINARY PSYCHOSOCIAL DEVELOPMENT

Improved sensory awareness produced by somatic education, sports, martial-arts training, contemplative practice, and other disciplines. For example, after a long search to find art objects he could "enjoy with complete abandon," art historian Bernard Berenson had the following experience:

> One morning as I was gazing at the leafy scrolls carved on the door jams of S. Pietro outside Spoleto, suddenly stem, tendril and foliage became alive, and in becoming alive, made me feel as if I had emerged into the light after long groping in the darkness of initiation. I felt as one illumined, and beheld a world where every outline, every edge, and every surface was in a living relation to me and not as hitherto, in a merely cognitive one. Since that morning, nothing visible has been indifferent (Berenson 1950, pp. 68–70).

PRODUCTS OF METANORMAL DEVELOPMENT

a. Metanormal sensory perception.
   i. Extraordinary vision, hearing, touch, smell, and taste.
   ii. Extraordinary synesthesias.

iii. Perception of the numinous in the physical world. Williams James cited a woman's account of such experience:

When rising from my knees it was like entering another world, a new state of existence. Natural objects were glorified, my vision was so clarified that I saw beauty in every material object . . . the woods were vocal with heavenly music (*The Varieties of Religious Experience*, Lecture IX).

b. Deliberate clairvoyant perception of physical events.

c. Perception of events or entities beyond the normally perceivable world, including:

i. Auras or scintillae around people and things. Religious scholar Henri Corbin, for example, described the "divine sparks" that sometimes appear in the course of contemplative practice. "In the beginning these lights are manifested as ephemeral flashes," he wrote. "The more perfect the transparency of the [aspirant] the more they grow, the longer they last, the more diverse they become."[2]

ii. Phantom figures. In chapter 5.1, I cite firsthand accounts of encounters with disembodied entities by Charles Lindbergh; Joshua Slocum, the first person to sail alone around the world; two members of a British team that climbed Mount Everest; and Ernest Shackleton, the famous explorer of Antarctica.

iii. Auditions of supremely beautiful music for which no physical sources are apparent. Parapsychologist Raymond Bayless reported his own experience of this kind:

. . . the music was totally unearthly and inconceivably beautiful and majestic. At the time I believed it to be associated with religious matters, and I still believe this to be so. It seemed to be produced by vast numbers of players and singers: I do not know, but the sense of a great number of units was felt. I cannot say that the music was vocal or that it was instrumental; it was on an inconceivably higher level than such distinction, and all that can be said is that it was incredibly beautiful, clearly superhuman, and could not possibly originate from earthly instruments or voices. In spite of the time that has passed, the memory of the experience is powerful and unforgettable.[3]

iv. Metanormal touch, including tactile impressions that appear to accompany telepathic exchanges with living persons or disembodied entities (see 5.1).

II. *Somatic Awareness and Self-Regulation*

PRODUCTS OF ANIMAL EVOLUTION

Capacities for kinesthetic awareness and self-regulation.

PRODUCTS OF ORDINARY PSYCHOSOCIAL DEVELOPMENT

Enhanced kinesthetic awareness produced, for example, by somatic disciplines, sports training, meditation, biofeedback, martial arts, or hatha yoga. Such awareness is mediated by visual imagery, auditions, or an immediate feeling of the body's processes.

PRODUCTS OF METANORMAL DEVELOPMENT

a. Extraordinary somatic awareness mediated by the central nervous system.
b. Perception of somatic events that does not depend upon bodily organs. This extrasomatic awareness, or internal clairvoyance, is said to accompany certain types of yogic experience. It is characterized as the *anudrishti siddhi,* or *animan siddhi,* for example, and is said to produce awareness of cells, molecules, and atomic patterns within the body.

III. *Communication Abilities*

PRODUCTS OF ANIMAL EVOLUTION

Complex signaling abilities by which information is transmitted between organisms.

PRODUCTS OF ORDINARY PSYCHOSOCIAL DEVELOPMENT

Exceptional verbal and nonverbal communication abilities produced by psychotherapy, religious practice, and other transformation practices, including the capacity to sense another person's psychological state and basic interests.

PRODUCTS OF METANORMAL DEVELOPMENT

a. Extraordinary communication abilities mediated by sensory cues.
b. Extraordinary communication abilities mediated by telepathic interaction.
  i. Sustained telepathic rapport.
  ii. Direct transmission of spiritual illumination.
  iii. Shared ecstasy or illumination that arises spontaneously in a couple or group.

IV. *Vitality*

PRODUCTS OF ANIMAL EVOLUTION

Sustained energy levels, exemplified by warm-bloodedness among birds and mammals.

PRODUCTS OF ORDINARY PSYCHOSOCIAL DEVELOPMENT

Enhanced vital capacity such as the exceptional fitness produced by endurance sports and the ability to survive extreme deprivation produced by religious asceticism.

PRODUCTS OF METANORMAL DEVELOPMENT

Extraordinary vitality evident, for example, in the rising of *N/um* among the Kung bushmen of the Kalahari desert, the *tumo* produced by Tibetan yoga, the rising kundalini of Indian yoga traditions, and the *incendium amoris* of Catholic saints. A Kung bushman described the rising of *N/um* in this manner:

> You dance, dance, dance. Then *N/um* lifts you in your belly and lifts you in your back, and then you start to shiver. *N/um* makes you tremble; it's hot . . . then *N/um* enters every part of your body, right to the tip of your feet and your hair. In your backbone you feel a pointed something and it works its way up. Then the base of your spine is tingling, tingling, and then it makes your thoughts nothing in your head (Katz, 1982b, p. 42).

Mutations of vitality similar, or identical, to the Kung bushmens' have frequently been described by Roman Catholic contemplatives, by Hindu and Buddhist yogis, and by modern Westerners.

V. Movement Abilities

PRODUCTS OF ANIMAL EVOLUTION

Specialized organs to facilitate movement of the body, such as wings, legs, or flippers.

PRODUCTS OF ORDINARY PSYCHOSOCIAL DEVELOPMENT

Enhancements of agility, coordination, grace, and stamina produced by somatic disciplines, sports, and the martial arts. Japanese Ninja, for example, could cover about 300 miles in three days on foot.[4]

PRODUCTS OF METANORMAL DEVELOPMENT

a. Extraordinary physical movement exhibited, for example, by certain shamans, martial artists, and Tibetan lamas. Lama Govinda, a scholar of Tibetan contemplative lore, described his own experience of this kind:

It was no longer possible to pick my way between the boulders that covered the ground; night had completely overtaken me; and yet to my amazement I jumped from boulder to boulder without slipping or missing a foothold, in spite of wearing only a pair of flimsy sandals on my bare feet. And then I realized that a strange force had taken over, a consciousness that was no more guided by my eyes or my brain. My limbs moved as in a trance, with an uncanny knowledge of their own, though their movement seemed almost mechanical . . . my own body had become distant, quasi-detached from my will-power. I was like an arrow that pursued its course by the force of its initial impetus, and the only thing I knew was that on no condition must I break the spell that had seized me.

It was only later that I realized what had happened: that unwittingly and under the stress of circumstances and acute danger I had become a *lung-gom-pa,* a trance walker, who, oblivious of all obstacles and fatigue, moves on towards his contemplated aim, hardly touching the ground, which might give a distant observer the impression that the *lung-gom-pa* was borne by the air (*lung*), merely skimming the surface of the earth.[5]

b. Though the evidence for it is inconclusive, the levitation attributed to Catholic mystics, Indian yogis, Taoist sages, and adepts of other religious traditions.

c. Out-of-body experience involving perceptions of the physical world.

d. Movement into extraphysical worlds during lucid dreams, deep meditation, hypnotic trance, and other altered states of mind.

VI. *Abilities to Alter the Environment Directly*

PRODUCTS OF ANIMAL EVOLUTION

Specialized organs to manipulate the environment, such as beak, talons, and hands.

PRODUCTS OF ORDINARY PSYCHOSOCIAL DEVELOPMENT

Enhancements of dexterity, hand-eye coordination, and specific hand skills produced by somatic education, athletic training, and the martial arts.

PRODUCTS OF METANORMAL DEVELOPMENT

a. Extraordinary hand-eye coordination and dexterity evident, for example, in great athletic feats.

b. Extrasomatic influence upon the environment.

　i. Deliberate influence of a mind at a distance upon living tissue (telergy), as in spiritual healing.

　ii. Deliberate mental influence upon inanimate objects at a distance (telekinesis), without apparent physical connections between agent and target, as in apparent alterations of a ball trajectory that appear to be caused by mental intention alone.

iii. The direct modification of some portion of space by mental influence, as in the apparent creation by mystics and saints of a
special joy or presence in their place of contemplation and
worship.

VII. Pain and Pleasure

PRODUCTS OF ANIMAL EVOLUTION

Sensations of pain and pleasure mediated by the vertebrate nervous
system.

PRODUCTS OF ORDINARY PSYCHOSOCIAL DEVELOPMENT

Increased capacity for pain and pleasure produced by psychotherapy
and somatic disciplines. Enhancement of pleasure and ability to control pain produced by hypnotic suggestion, athletic training, martial-
arts training, or religious practice.

PRODUCTS OF METANORMAL DEVELOPMENT

Self-existent delight, which subsumes ordinary pain and pleasure, described in statements such as these:

> For who could live or breathe if there were not this delight or existence
> as the ether in which we dwell? (*Taittiriya Upanishad,* II.9)

> Before losing myself entirely in contemplation of an unbounded, glow-
> ing, conscious void, I felt an incomparably blissful sensation in all my
> nerves moving from the tips of my fingers and toes and other parts of
> the trunk and limbs towards the spine, where, concentrated and inten-
> sified, it mounted upwards [into] the brain in a rapturous and exhilarat-
> ing stream (Gopi Krishna 1971).

> In the mind, it translates into a calm intense delight of perception and
> vision and knowledge, in the heart into a passionate delight of union,
> love and sympathy. In the will and vital parts it is felt as delight in action
> or a beatitude of the senses perceiving and meeting the One every-
> where (Sri Aurobindo 1970, *The Life Divine,* pt. 2, chap. 27. Vol. 19 in
> *The Collected Works*).

VIII. Cognition

PRODUCTS OF ANIMAL EVOLUTION

Specialized organs and internal networks to transmit information within the organism, culminating in human symbol-making and self-reflection mediated by the central nervous system.

PRODUCTS OF ORDINARY PSYCHOSOCIAL DEVELOPMENT

Cognitive skills developed by intellectual training, logic, and stimulation of the imagination through art and philosophy.

PRODUCTS OF METANORMAL DEVELOPMENT

a. Mystical illumination described in statements such as these:

The knower and the known are one. God and I, we are one in knowledge.[6] (Meister Eckhart)

In that which is the subtle essence, all that exists has its self. That is the True, that is the Self, and Thou, Svetaketu, art That (*Chandogya Upanishad*, VI.8.7).

Until his death, the French philosopher Blaise Pascal carried with him these words, which he composed after his religious conversion:

From about half past ten in the evening to
about half an hour after midnight.
> *Fire.*
God of Abraham, God of Isaac, God of Jacob,
Not the God of philosophers and scholars.
Absolute Certainty: Beyond reason.
> *Joy. Peace.*
Forgetfulness of the world and everything
> *but God.*
The world has not known thee,
>> *but I have known thee.*
>> *Joy! joy! joy! tears of joy!*

b. Creative works marked by extraordinary immediacy, ease, and completeness, which come ready-made as if from powers beyond ordinary consciousness. Mozart, for example, said that he saw many of his compositions tout ensemble, all at once, and Blake claimed he received poems by "dictation." In Platonic, Sufi, Kabbalistic, and Vedantic traditions, inspired works of this kind are said to come from God, the gods, the One, or Brahman.

IX. *Volition*

PRODUCTS OF ANIMAL EVOLUTION

Complex chains of purposeful action culminating in human will.

PRODUCTS OF ORDINARY PSYCHOSOCIAL DEVELOPMENT

Refinement, clarification, strengthening, and integration of volition produced by psychotherapy, hypnotic suggestion, sports, martial-arts training, religious practice, and other transformative disciplines.

PRODUCTS OF METANORMAL DEVELOPMENT

Ego-transcending volition reflected, for example, by the Taoist doctrine of *wu wei*, noninterference with the Tao; the Judeo-Christian-Islamic statement "Not my will, but Thine be done"; descriptions of the *ishatva* and *vashitva* siddhis, by which a yogi exercises his will with constant success because he is in harmony with the Divine intention; and testimonies to athletic performance "in the zone." Such volition involves extraordinary self-mastery, including metanormal abilities to regulate autonomic processes.

X. *Individuation and Sense of Self*

PRODUCTS OF ANIMAL EVOLUTION

Marked individualization, especially among higher vertebrates.

PRODUCTS OF ORDINARY PSYCHOSOCIAL DEVELOPMENT

Individuation and a healthy sense of identity produced by parental influence on early development, psychotherapy, religious practice, and other transformative practices.

PRODUCTS OF METANORMAL DEVELOPMENT

a. The realization of an ego-transcendent identity that perceives its oneness with all things while remaining a unique center of awareness and action.

b. The cognitive-emotional-behavioral complexity and uniqueness evident, for example, in religious figures such as Francis of Assisi and Sri Ramakrishna.

XI. Love

PRODUCTS OF ANIMAL EVOLUTION

Loving devotion to others exemplified, for example, in cetacean care for the young and Neanderthal protection of wounded comrades.

PRODUCTS OF ORDINARY PSYCHOSOCIAL DEVELOPMENT

The clarification of projections, identifications, and dependencies by psychotherapy and other forms of sustained introspection, which helps to liberate and mobilize concern for others. Empathy and interpersonal creativity produced by affective education. The loving service evoked by religious practice.

PRODUCTS OF METANORMAL DEVELOPMENT

Love that transcends normal needs and motives, revealing a unity among people and things more fundamental than any differences between them. It is its own reward. "Love seeks no cause beyond itself," wrote St. Bernard of Clairveaux. "It is its own fruit, its own enjoyment. I love because I love; I love in order that I may love." Or it may focus upon a transcendent principle. "God cannot be seen with these physical eyes," said Sri Ramakrishna. "In the course of spiritual discipline one gets a 'love body' endowed with 'love eyes,' and 'love ears.' One sees God with those 'love eyes.' One hears the voice of God with those 'love ears'... and with this 'love body' the soul communes with God" (Nikhilananda 1942).

XII. Bodily Structures, States, and Processes

PRODUCTS OF ANIMAL EVOLUTION

Structures such as the spine to support the organism and separate its components so that they do not crowd each other. Limbs to grasp or execute skillful movement. Other organs to facilitate life in the physical world.

PRODUCTS OF ORDINARY PSYCHOSOCIAL DEVELOPMENT

Improved articulation of muscles and myofascia produced by Rolfing. Improved posture and carriage produced by the Alexander Technique, hatha yoga, and the martial arts. Expanded chest and abdominal cavities produced by Reichian Therapy. Strengthened muscles, tendons, ligaments, and skeleton produced by fitness training.

PRODUCTS OF METANORMAL DEVELOPMENT

Extraordinary alterations of bodily process and structure to facilitate the metanormal capacities noted above. Such alterations are represented, for example, by the *nadi drishti siddhi* through which a yogi transforms the nervous centers and other internal structures of his body; by the Taoists' "opening of the Golden Flower"; or by the activation of *chakras* and kundalini described in Tantric lore.

[4.2]
GENERAL ANALOGIES BETWEEN ACTIVITIES IN
DIFFERENT EVOLUTIONARY ARENAS

The developmental analogues listed above do not comprise all the similarities among evolutionary domains or stages. For example, James Grier Miller has argued that living systems, at seven levels from cells to human organizations, require 19 critical processes to function successfully (see appendix J.1). He wrote:

> As more complex cells evolved, they had more complex subsystems, but still the same 19 basic processes. Similarly, as cells evolved into more complex systems—organs, organisms, and so on—their subsystems shredded out into increasingly complicated units carrying out more complicated and often more effective processes. If at any single point in the entire evolutionary sequence any one of the 19 subsystem processes had ceased, the system would not have endured. That explains why the same 19 subsystems are found at each level from cell to supranational system. And it explains why it is possible to discover, observe, and measure cross-level formal identities.[7]

The triple-leveled scheme proposed here, though it includes aspects of existence outside Miller's conceptual framework, is compatible, I believe, with general systems theory. If Dr. Miller will forgive me, his seven-tiered hierarchy is compatible with the analogies presented in this chapter.

Like Miller, several philosophers and general system theorists have identified patterns common to different evolutionary levels (7.2). Though these different levels or arenas have their own distinctive regularities, they also have many analogous features. In appendix J.2, I briefly note some of these analogies. Viewed together, they suggest a grand continuity in the world's development through its inorganic, biological, psychosocial, and metanormal domains.

[4.3]
METANORMALITIES OF EVERYDAY LIFE

In chapter 4.1, I listed twelve sets of attributes produced by animal, ordinary human, and metanormal development. In this section, I propose that there exist kinds of experience, which I will call "metanormalities of everyday life," that are nascent expressions of fully developed extraordinary attributes. These, I suggest, are first versions of metanormal capacities available to all of us.

➤

Responding to various surveys conducted during the 1970s and 1980s, many people reported their own mystical or paranormal experiences (see appendix A.4). For example, in polls conducted at the University of Chicago's National Opinions Research Council by sociologist Andrew Greeley and his colleagues, 58 percent of American survey subjects in 1973, and 67 percent in 1984, said they had experienced ESP. In 1973, 8 percent, and in 1984, 29 percent, reported they had had some sort of "vision." In 1973, 27 percent, and in 1984, 42 percent, claimed to have had contact with the dead.[8] Greeley attributed the apparent increase in such experience between 1973 and 1984 to a growing willingness among Americans to talk about paranormal events. His data are supported by the Gallup Organization, which found that 43 percent of the subjects in a 1985 survey reported "an unusual or inexplicable spiritual experience."[9] In *The New York Times Magazine*, Greeley and his colleague William McCready wrote:

> Such extraordinary experiences—intense, overwhelming, indescribable— are recorded at every time in history and in every place on the globe and are widespread in American society today. No one familiar with history or anthropology or psychology can deny [they] occur. They are a form of "altered state of consciousness"—to use the current approved phrase—something like intoxication or delirium or hypnotic trance but different in their intensity, their joyfulness and their "lifting out" dimension. In some cases, such experiences were triggered by drugs, in others by ritual dances, in still others by disciplined meditation; but most such "ecstatic interludes" about which we have accounts seem to be purely spontaneous—a man casually walking by a tennis court suddenly is caught up in a wave of peace and joy "as time stands still."[10]

My own inquiries among friends, acquaintances, and seminar audiences during the past thirty years have convinced me that many, or perhaps most, people have extraordinary experiences that are not triggered by formal discipline. Some of these are occasioned by personal crisis, lovemaking, childbirth, intense parent-child relations, near-death episodes, the concentrated rituals of work, or psychoactive drugs, etc., whereas others—as Greeley and McCready noted—

appear to be entirely spontaneous. (Some experiences that seem to be spontaneous, however, might be triggered by activities such as deep concentration that are central to formal disciplines.) Furthermore, such experiences exhibit more variety than I once guessed, corresponding in many ways with the metanormal phenomena (siddhis or charisms) recognized by the religious traditions. This correspondence becomes evident when such experiences are grouped as follows according to the twelve classes of developing attributes outlined in chapters 4.1 and 5.

Perception of External Events

- Synesthesias, or crossing of the senses ("seeing" sounds, for example, or "hearing" flowers), in which a numinous beauty is revealed or an extrasensory perception is dramatized.
- "Feeling" people in a house though you cannot see or hear them.
- Feeling that someone is watching you, after which you turn to meet his or her gaze.
- Feeling a basic compatibility (or incompatibility) with people before you meet them.
- Correctly guessing the location of water or other materials by extrasensory processes.
- Spontaneous, unexpected perceptions of distant events.
- Hearing music or other sounds, for which there is no discoverable physical source, that enliven particular thoughts or emotions.
- Correctly sensing the location of lost objects without the help of sensory cues.
- Opening books to the exact passage you are searching for.
- Watching someone's face reveal—as if in slow motion—unsuspected feeling, traits, or possibilities for development.
- Sensing a numinous presence during meditation, intimate conversation, or other circumstances.
- Seeing lights around people or inanimate objects for which there are no apparent sources.
- Looking at something familiar and seeing it as if for the first time.

- Spontaneously apprehending the presence of someone physically distant or dead, by direct and vivid contact.

Somatic Awareness and Self-Regulation

- Experiencing vivid images of arteries, capillaries, or other bodily structures that immediately seem to be your own, and sensing they may be damaged or in a process of recovery from injury.
- Determining by spontaneous tastes or smells your level of stress during extreme exertion.
- Picturing what appear to be *chakras* or other entities depicted in esoteric teachings.
- Hearing melodies that seem to reflect your physical condition.

Communication Abilities

- Correctly intuiting someone's (negative or positive) feelings or thoughts about you or a third party.
- Telepathically stimulating loving or hateful, serene or agitated moods in another person.
- Saying something unexpected in unison with someone else.
- Writing to a friend you haven't communicated with for several years at approximately the same time he or she writes to you.
- Sensing correctly who is calling on the telephone, even though the caller hasn't communicated for a long time, or thinking about someone who then calls you.
- Feeling no separation at all from your lover during sexual intercourse.
- Feeling the pains of a distant friend, then discovering he or she is ill or injured.
- Sharing sensations of well-being with friends though there is no physical communication between you.
- "Feeling" what someone else is thinking.
- Sensing the mood and intention of a pet or other animal.
- Having the same dream a friend does.
- Accurately sensing someone's prayers in your behalf.

Vitality

- Feeling great warmth on cold days, without benefit of extra clothing.
- Experiencing immense energy, sometimes frightening in its intensity, for which there is no apparent cause.
- Sensing a rush of electricity up the spine, or radiating out from the abdomen, accompanied by mental illuminations or great strength.
- Remaining free of infection in spite of contagious diseases among those around you.
- Going without normal amounts of sleep for extended periods without loss of clarity, vitality, or physical strength.

Movement Abilities

- Executing moves in sports beyond your normal ability while sensing a new power or "self."
- Sensing physical levitation during strenuous physical exercise, prayer, or lovemaking.
- Experiencing flight as if in a subtle body, during an especially vivid dream or state of creative absorption.
- Out-of-body experience (during which you may see your own body) after which you report events that could not be known to you in ordinary circumstances.
- Experiencing an extraordinary pleasure in movement accompanied by an apparent release of new energies in particular body parts.

Abilities to Alter the Environment Directly

- Appearing to alter another person's mood at a distance, as if by extrasensory influence.
- Appearing to correct a machine's malfunction by mental intention alone.
- Appearing to alter the flight of a ball by mental intention.

▫ Altering ambient room temperature as if by psychokinesis.

▫ Leaving a strong mood—whether loving or hateful, serene or agitated—in an empty room.

▫ Promoting or inhibiting plant growth in extraordinary fashion, as if by some sort of "green thumb."

▫ Experiencing a powerful mood or image while taking a photograph, then finding an unexplainable object or light on the picture you have taken.

▫ Feeling that you have invisible hands that touch another person, after which that person responds as if he or she had been touched.

Pain and Pleasure

▫ Eliminating pain by simply willing it away.

▫ Feeling inexplicable pleasure, or a stream of vitality, that seems localized in the spine, the solar plexus, or some other body part.

▫ Experiencing sudden shivers of ecstasy that arise without apparent stimulation.

▫ Experiencing profound joy during a routine task or in the midst of pain or discomfort, joy that does not seem to depend upon the satisfaction of a particular need or desire. This state may be attributed simply to the "joy of living," and may have a contagious effect upon others.

Cognition

▫ Correctly sensing unexpected danger.

▫ Correctly anticipating a melody before it comes on the radio, or a dramatic event before it happens, or a sentence before a companion says it.

▫ Apprehending a situation or place as if you had experienced it before.

▫ Apprehending an exceptionally complex and original set of ideas all at once, in conjunction with great excitement and joy.

▫ Remembering extremely complex material with perfect detail.

▫ Correctly determining historical events connected with a particular location or object, as if by some sort of clairvoyance.

□ Experiencing a sense of mystical contact with God, perhaps at work, or during a festive event, or while engaged in intimate conversation, or in the midst of suffering.

Volition

□ Waking from sleep at a designated moment without assistance from an alarm.

□ Shedding pain by willing it away.

□ Overcoming the effects of spoiled food or other toxins by an immediate act of will.

□ Spontaneously throwing off the effects of injury or disease, or (on the darker side) psychokinetically triggering some affliction in others.

□ Accomplishing some deed requiring strength or endurance beyond your usual capacity, in a crisis perhaps, during a sports competition, or to inflict punishment upon others.

□ Spontaneously adapting to great heat, cold, or other adverse conditions.

□ Deliberately rising above thirst or hunger, with no apparent loss of strength.

□ Spontaneously exerting subliminal influence upon others in such ways as harmonizing conflicting parties, bringing peace to potentially violent situations, or—conversely—causing discord and suffering.

Individuation and Sense of Self

□ Awakening to a witness self that is fundamentally distinct from particular thoughts, impulses, feelings, or sensations.

□ Feeling for a moment as if your body is only a small part of yourself, or that it is located at a specific point in the field of awareness.

□ Feeling a new substantiality, as if you are somehow larger, stronger, and more solid.

□ Spontaneously realizing a new and profound self-confidence that adverse criticism does not affect.

- Feeling as if you are suddenly more real, more authentic, more truly yourself.
- Momentarily apprehending all objects of perception as if they are contained within you.
- Experiencing an identity that self-evidently existed before your birth and that will outlast your body's death.

Love

- Seeing new beauty and possibilities for growth in someone of long acquaintance.
- Experiencing a love that allows you to feel a friend's suffering, deep intentions, or personal conflicts.
- Experiencing love that removes all sense of boundaries between you and a loved one, as if you and the other were a single person or body.
- Experiencing a love for someone who is physically distant that appears to elevate that person's self-esteem and sense of well-being.

Bodily Structures, States, and Processes

- Pleasurable streaming sensations that envelop the body and seem to involve a significant enhancement of health.
- Sudden and unaccountable loss of weight that is not accompanied by disability.
- A subtle effervescence from head to toes during sleep, through which you sense that your bodily processes are being altered.
- Sensing an opening in the body—located perhaps between the eyes, around the heart, near the navel, or at the base of the spine—through which energy is flowing.
- Spontaneous rushes of energy up and down the spine, spiraling around the torso, or rising from the soles of the feet.
- Sensing an extraordinary lightness while moving or at rest, or a sense of elevation from the ground.
- Agility beyond your ordinary range of movement that seems related to a new and extraordinary elasticity of tendons, ligaments, and muscles.

▫ Radical alterations of body image, as if you are much taller or shorter, for example, or shaped like a sphere, column, diamond, or point.

▫ Extraordinary ability to withstand blows, in sports or other circumstances, as if you are suddenly heavier and harder.

▫ Sensing that new structures, which seem to be made of subtle matter, are forming inside your body or on your skin.

In *The Psychopathology of Everyday Life,* Freud taught us that many common behaviors, among them jokes and slips of the tongue, expressed half-conscious or unconscious thoughts and motives. The experiences listed above, I believe, also express a largely unrecognized aspect of human life, namely its latent (positive or negative) metanormality. The fact that such experiences occur in all sorts of people, even though they are neither sought nor expected, indicates that they are deeply rooted in human nature. They do not seem to be entirely created by social processes, nor can they be completely suppressed, even in cultures that do not celebrate them.

As I have collected reports about such events, I have discovered that they share certain features with activities that are clearly metanormal. Comparing them with the 12 sets of extraordinary attributes outlined in chapters 4.1 and 5, we find correspondences that lead us straightforwardly to suppose that they are first expressions of fully developed metanormal capacities. That so many of these experiences appear among us without benefit of formal discipline—in people with different backgrounds, temperaments, and cultural conditionings—strongly suggests that there is a continuum between ordinary and extraordinary human functioning.

It is difficult to estimate how many extraordinary attributes are available to us, given our general ignorance today about the further reaches of human nature. We do not know how many human potentials are neglected or suppressed in this (and every) culture. But it would not be difficult, I believe, to enlarge the inventories in this and the following chapter through interviews, questionnaire surveys, and informal conversations in which unusual capacities were deemed worthwhile. Greeley found, for example, that word of his research gave many people permission to reveal their own quasi-mystical or extrasensory experiences.[11] Furthermore, a great variety of spontaneous metanormalities are recorded in the *Proceedings, Journals,* and archives of the British and American Societies for Psychical Research; in publications of the Alister Hardy Research Centre at Oxford University (originally called the Religious Experience Research Unit); in Frederic Myers's *Human Personality*, Edmund Gurney's *Phantasms of the Living*, William James's *Varieties of Religious Experience*, Marghanita Laski's *Ecstasy*, Raynor Johnson's *Watcher on the Hill*, and other studies of exceptional experience. A systematic listing derived from such literature would reveal a great many extraordinary attributes beyond those mentioned here, providing a basis for further research and guidance for the integral practices discussed in Part Three.

[4.4]
THE UNCERTAINTIES OF HUMAN DEVELOPMENT

Though they bring new goodness and joy to life, metanormal capacities can serve many kinds of destructive behavior. No animal slaughters its own kind with the extraordinary calculation exhibited by certain humans, for example, and no animal is inspired to hatred by religious fervor. Our intelligence and manipulative powers have made us the world's most dangerous creature, to each other and to all living things, and transformative practices sometimes support our most lethal impulses. In Part Three, I discuss some ways in which therapeutic and religious disciplines reinforce or conceal common human weaknesses.

Evolution's uncertainty, of course, is not peculiar to homo sapiens. For every species that exists today, hundreds have passed away. For each culture still operating, dozens have perished. Because evolution meanders more than it progresses, there is no guarantee that modern societies—or the human species itself—will last long enough to establish metanormal development as a widespread feature of life upon earth. Indeed, our chronic perversity and suffering have led some thinkers to suppose that humans have already evolved in a fatal direction. Arthur Koestler wrote:

> The evidence from man's past record and from contemporary brain research strongly suggests that at some point during the last explosive stages of the biological evolution of homo sapiens something went wrong; that there is a flaw, some potentially fatal engineering error built into our native equipment—more specifically, into the circuits of our nervous system—which would account for the streak of paranoia running through our history. This is the hideous but plausible hypothesis which any serious inquiry into man's condition has to face.[12]

In diagnosing the causes of human destructiveness, Koestler was influenced by Paul MacLean, a neurophysiologist who suggested that evolution produced the human neocortex without ensuring adequate coordination between it and older brain structures we inherit from our animal ancestors.[13] MacLean called our nervous disability *schizophysiology*, which he defined as

> a dichotomy in the function of the phylogenetically old and new cortex that might account for differences between emotional and intellectual behavior. While our intellectual operations are carried on in the newest and most highly developed part of the brain, our affective behavior continues to be dominated by a relatively crude and primitive system, by archaic structures in the brain whose fundamental pattern has undergone but little change in the whole course of evolution from mouse to man.[14]

Koestler compared our faulty construction to other evolutionary mistakes, such as the brain's placement around the esophagus among the arthropods and the marsupials' failure to develop a corpus callosum. Arthropod brains could not develop without choking off food intake, and marsupials could not give rise to the human neocortex, though they closely resembled their mammalian relatives, because their brain halves were too poorly coordinated. Koestler wrote:

> Homo sapiens, too, might be a victim of faulty brain design. We, thank God, have a solid corpus callosum which integrates the right and left halves, horizontally; but in the vertical direction, from the seat of conceptual thought to the spongy depths of instinct and passion, all is not so well.[15]

Such speculation, however, has been dismissed by many students of human physiology. For example, Karl Pribram, a leader in the field of brain research, has attributed human dysfunction to social causes rather than faulty brain design; and with other brain scientists, he has criticized MacLean for overemphasizing the old brain's role in destructive emotion. Nevertheless, most psychologists and medical researchers—like most philosophers and religious seers—agree that with our capacities for love, knowledge, and further development, we have profound disabilities, whether these are caused by our genes or cultural conditioning. Indeed, most programs for human betterment address our shortcomings, either directly or indirectly, by therapeutic intervention, psychosomatic education, ethical training, religious discipline, or the reconstruction of social institutions. In subsequent chapters, I discuss some of our common deficiencies and their potential reduction by transformative practice.

5

DEVELOPING HUMAN
ATTRIBUTES

IN CHAPTER 4.1, I listed 12 classes of extraordinary human attributes that are analogous to kinds of animal and ordinary human functioning. In this chapter, I will elaborate upon that listing, giving examples in each of the 12 classes and breaking most down into subcategories. But, as I have said, this classificatory scheme is only suggestive. It neither includes all the evidence for metanormal functioning nor identifies every type of extraordinary human activity. Beyond these 12 kinds of developing attributes, others could be added. Furthermore, some of the experiences listed here can be assigned to more than one category. Because human nature is fluid and extremely complex, its various features can be classified in more than one way. This inventory is inclusive enough, however, to suggest the great range of attributes our further growth might encompass.

[5.1]
PERCEPTION OF EXTERNAL EVENTS

Evolution has endowed us with many abilities to sense chemical, electromagnetic, and mechanical stimuli from the external world. Although some animals have sensory capacities superior to ours (such as the canine sense of smell, or an eagle's distance vision), we have a great variety of visual, auditory, tactile, gustatory, and olfactory abilities, all of which can be strengthened through practices such as hypnotic suggestion (15.11), somatic education (18), martial-arts training (20.1), and meditation (23). Furthermore, there is considerable evidence that extrasensory perceptions can be cultivated too (see appendix A.1 and chapters 20.1, 21, and 22). Both sensory and extrasensory capacities have many kinds of metanormal expression, which I will group here in three broad classes: first, extraordinary perceptions mediated by the physical senses; second, readily accessible clairvoyant apprehensions of physical things; and third, perceptions of qualities, events, or entities beyond the normally perceivable world.

Extraordinary Sensory Perception

Extraordinary vision. Human vision, an immensely complex product of animal evolution, can be developed in various ways. For example, a Boer girl trained under hypnosis by the South African naturalist Eugene Marais learned to distinguish (with the help of binoculars) acquaintances who were three miles away. Visual (and other sensory) capacity, Marais suggested, is normally inhibited by mental activity but can be enhanced by hypnotic inductions that quiet the mind.[1] Indeed, improvements of visual acuity and field awareness have been noted in many realms of activity, including the martial arts (20.1) and modern sport (19). Some athletes, for instance, learn to recognize moving targets in a swarm of once-confusing events. John Brodie, one of the National Football League's best quarterbacks for several years, described occasions during a game when

> time seems to slow way down, in an uncanny way, as if everyone were moving in slow motion. It seems as if I have all the time in the world to watch the receivers run their patterns and yet I know the defensive line is coming at me as fast as ever.[2]

In his playoff victory over Ben Hogan during the 1955 U.S. Open Golf Championship, Jack Fleck had a startling perception. "I can't exactly describe it," he wrote, "but as I looked at my putt, the hole looked as big as a wash tub. I suddenly became convinced I couldn't miss. All I tried to do was keep the sensation by not questioning it."[3] Pittsburgh Steelers safety Paul Martha also described an extraordinary improvement of visual perception. "All of a sudden, midway through the 1967 season," he wrote, "I realized I was following the quarterback all the way—and the receiver too. It just happened. It was like I had stepped into an entirely new dimension."[4] In more dramatic terms, running back MacArthur Lane described his own improvement of field awareness, claiming to see with an "extra eye," at times from a point above his head.[5]

Like contemplative illuminations, experiences such as these: (1) seem to be freely given, though they are triggered by intense discipline; (2) involve a new kind of functioning (a "new dimension" as Martha wrote, or a "point above the head" as Lane described it); and (3) require a focused surrender (Fleck's "not questioning").

Extraordinary hearing. A hypnotized subject described by Eugene Marais could hear a hiss of constant volume at a distance of 230 yards, although nonhypnotized people typically could not detect the same sound until they were within 30 yards of its source.[6] Bushmen of the Kalahari desert learn to hear approaching animals, even in their sleep, by lying with an ear to the ground.[7] Some orchestra conductors can discern a single violin's wrong note during a symphony in which more than 100 instruments are playing. And meditation enables some people to hear many sounds at once or to detect auditory stimuli that normally escape their attention. In the great contemplative treatise of Theravada Buddhism, *Visuddhimagga* (chap. 13.1–7), it is said that one can learn to apprehend an immense range of sounds inaccessible to ordinary hearing. After training in concentration, a meditation student can turn

> first to the loud sounds normally within range of hearing; the sound in the forest of lions, etc., or in the monastery the sound of a gong, the sound of a drum, the sound of a conch, the sound of recitation by novices and young *bhikkus* reciting with full vigour, the sound of their ordinary talk such as "What, venerable sir?", "What, friend?", etc., the sound of birds, the sound of wind, the sound of footsteps, the fizzing sound of boiling water, the sound of palm leaves drying in the sun, the sound of ants, and so on. Beginning in this way with quite gross sounds, he should successively [turn] to more and more subtle sounds. He should give attention to . . . sounds in the eastern direction, in the western direction, in the northern direction, in the southern direction, in the upper direction, in the lower direction, in the eastern intermediate direction, in the western intermediate direction, in the northern intermediate direction, and in the southern intermediate direction.[8]

During a further stage of meditation, the student can delimit an area of a single finger-breadth, saying to himself,

> I will hear sounds within this area, then two finger-breadths, four finger-breadths, eight finger-breadths, a span, a *ratana* (24 finger-breadths), the interior of the room, the verandah, the building, the sur-

rounding walk, the park belonging to the community, the alms-resort village, the district, and so on. This is how he should extend it by delimited stages.[9]

Extraordinary tactile experience. Touch, too, can be refined. Eugene Marais said that one of his hypnotized subjects could tell by touch alone whether a metal bar was magnetized.[10] In their comprehensive study *The Senses of Animals and Men,* Lorus and Margery Milne described "cloth feelers" who can identify fabrics by tapping them just once with a stick or can judge a cloth's quality even if their fingers are coated with collodion.[11] Helen Keller could distinguish the sounds of different musical instruments by stroking a turning phonograph record (14), and some blind people, it seems, can recognize the colors of fabrics they hold in their hands.[12] Tactile capacities have also been trained in the laboratory. At the University of Virginia, researchers devised an alphabetic code and transmitted it through a set of three vibrators applied to the chest. The subjects of this research learned to distinguish three intensities of stimulation and three lengths of signal, of which the briefest lasted just a tenth of a second. After 65 hours of training, one subject could understand with 90 percent accuracy entire sentences received as tactile vibrations transmitted at 38 five-letter words per minute. In chapter 18, I describe somatic disciplines that increase receptivity to tactile sensations. Through the Sensory Awareness training developed by Charlotte Selver, for example, one can find new information and pleasure through touch.

Extraordinary sense of smell. The English poet and essayist Peter Redgrove has described several kinds of olfactory prodigy. A researcher working with Saturnid moths, for example, learned to distinguish previously imperceptible moth odors and locate streams of them winding through a house. A lay analyst said that he detected complex emotions in his patients through scent: there were, he said, I-am-about-to-recall-a-dream smells, panic smells, insight smells, and disagreement smells. And navigating through icy waters, Eskimos are guided by odors of wind and snow.[13] In chapter 14, I describe the extraordinary olfactory abilities of Helen Keller, whose blindness and deafness helped catalyze her great sensitivity. Her sense of smell, Keller wrote, indicated

a coming storm hours before [it was] visible. I notice first a throb of expectancy, a slight quiver, a concentration in my nostrils. As the storm draws near my nostrils dilate, the better to receive the flood of earth odors which seem to multiply and extend, until I feel the splash of rain against my cheek.[14]

Cosmetic companies depend upon people who have been gifted with or who have developed great sensitivity to smells. According to the calculations of one manufacturer, an expert can distinguish more than 30,000 nuances of scent.[15] One such person, Russian-born Sophia Grojsman, told Diane Ackerman that when she was designing a new perfume

> you will have simple fragrances, simple accords made from two or three items, and it will be like a two- or three-piece band. And then you have a multiple accord put together, and it becomes a big modern orchestra. In a strange way, creating a fragrance is similar to composing music.
>
> You can't plagiarize. You have to start from scratch. But there are accords you might return to as themes, as a kind of shortcut. I make approximately five hundred to seven hundred formulas a year.[16]

Taste sensitivity. Human adults typically have about 10,000 taste buds, grouped on the tongue according to their types of sensitivity, while in each bud some 50 taste cells relay stimulation to the brain. We taste sweet things at the tip of the tongue, bitter things at the back, sour things at the sides, and salty things over the whole surface of the tongue. All the tastes we enjoy derive from combinations of just these four gustatory sensations plus a few minor ones, but such combinations are too numerous to be counted. Indeed, we override our instinctive aversions to bitter and foul-tasting things for the sake of sensory adventure. All of us cultivate our gustatory capacities to some extent, while tea tasters, wine tasters, and other professionals do so through lifelong practice. For a lyrical description of highly developed taste, read Diane Ackerman's "In Praise of Vanilla."[17]

Extraordinary synesthesias. Most of us occasionally experience synesthesia, the stimulation of one sense by another. For some people, however, this crossing of the senses is highly developed and the source of great enjoyment and creativity. When it is a constant enlivening feature of perception, we might deem it to be metanormal. For example, both Rimski-Korsakov and Scriabin vividly associated E major with blue, A flat with purple, and D major with yellow.[18] Baudelaire wrote a sonnet about correspondences between fragrances, colors, and sounds, and helped inspire the Symbolist movement's celebration of synesthesia. Rimbaud pictured particular colors in conjunction with each of the vowel sounds and said that his quest for artistic truth involved "an immense planned disordering of all the senses." And Vladimir Nabokov experienced "colored hearing," which he described in the following passage.

> Perhaps "hearing" is not quite accurate since the color sensation seems to be produced by the very act of my orally forming a given letter while I imagine its outline. The long *a* of the English alphabet . . . has for me the tint of weathered wood, but a French *a* evokes polished ebony. This black group also includes hard *g* (vulcanized rubber) and *r* (a sooty rag being ripped). Oatmeal *n*, noodle-limp *l,* and the ivory-backed hand mirror of *o* take care of the whites. . . . Passing on to the blue group, there is steely *x*, thundercloud *z*, and huckleberry *k*. Since a subtle interaction exists between sound and shape, I see *q* as browner than *k*, while *s* is not the light blue of *c*, but a curious mixture of azure and mother-of-pearl. Adjacent tints do not merge, and diphthongs do not have special colors of their own, unless represented by a single character in some other language (thus the fluffy-gray, three-stemmed Russian letter that stands for *sh*, a letter as old as the rushes of the Nile, influences its English representation). . . . The word for rainbow, a primary, but decidedly muddy, rainbow, is in my private language the hardly pronounceable: *kzspygu*. The first author to discuss *audition coloree* was, as far as I know, an albino physician in 1812, in Erlangen.
>
> The confessions of a synesthete must sound tedious and pretentious to those who are protected from such leaking and drafts by more solid walls than mine are. To my mother, though, this all seemed quite normal. The matter came up, one day in my seventh year, as I was using a heap of old alphabet blocks to build a tower. I casually remarked to her that their colors were all wrong. We discovered then that some of her letters had the same tint as mine and that, besides, she was optically affected by musical notes. These evoked no chromatisms in me whatsoever.[19]

Perceptions of the numinous in the physical world. The vivid synesthetic experience reported by many artists is frequently produced by religious practice. Saint Francis of Assisi saw bird sounds and heard the sun singing. A young American contemplative told me about shimmering forms she saw when she heard distant voices from her meditation retreat. But the extraordinary sensory experiences reported by religious ecstatics typically involve more than synesthesia. There is a type of perception that in its richness, power, and depth exceeds the sensory enhancements mentioned thus far. For example, during a Zen Buddhist *sesshin*, a retired Japanese government worker noticed a loquat tree.

> Its branches seemed possessed of truth! I told myself. My mind, I knew, had progressed in awareness, and I returned to my sitting with renewed vigor. In the evening at *dokusan*, I told Roshi what I had felt about the tree and asked him what it meant. "You have reached a decisive point. . . . This is the last evening of the sesshin. Do zazen all night."
>
> Customarily all lights are put out at 9 p.m., but this night, with the roshi's permission, I kept one small light burning. Mr. M——, the senior monitor, joined my sitting, and with his spiritual strength added to mine I felt stronger. Centering my energy, I began to feel exhilarated. Intently I watched the still shadow of my chin and head until I lost awareness of them in a deepening concentration. As the evening wore on, the pain in my legs became so grueling that even changing from full to half-lotus didn't lessen it. My only way of overcoming it was to pour all my energy into a single-minded concentration of Mu. Even with the fiercest concentration to the point of panting "Mu! Mu! Mu!" there was nothing I could do to free myself of the excruciating pain except to shift my posture a little. Abruptly the pains disappear, there's only Mu! . . . All is freshness and purity. Every single object is dancing vividly, inviting me to look. Every single thing occupies its natural place and breathes quietly. I notice zinnias in a vase on the altar, an offering to Monju, the Bodhisattva of Infinite Wisdom. They are indescribably beautiful![20]

William Wordsworth's *Lines Composed a Few Miles Above Tintern Abbey* was inspired by a quasi-mystical exaltation. "I began [the poem] upon leaving Tintern, after crossing the Wye," Wordsworth wrote, "and concluded it just as I was entering Bristol in the evening after a ramble of four or five days. Not a line of it was altered, and not any part of it written down till I reached Bristol."[21]

> Once again
> Do I behold these steep and lofty cliffs,
> That on a wild secluded scene impress
> Thoughts of more deep seclusion; and connect
> The Landscape with the quiet of the sky. . . .
> And I have felt
> A presence that disturbs me with the joy
> Of elevated thoughts; a sense sublime
> Of something far more deeply interfused,

Whose dwelling is the light of setting suns,
And the round ocean and the living air,
And the blue sky, and in the mind of man;
A motion and a spirit, that impels
All thinking things, all objects of all thought,
And rolls through all things.

These lines refer to a perception that appropriates sensory stimuli in a deeper, richer embrace of external events than ordinary vision or hearing. Graced by it, everyday sights and sounds reveal a world that stands, as the following passage has it, "from everlasting to everlasting." In his *Centuries of Meditations*, the English poet and priest Thomas Traherne wrote:

The corn was orient and immortal wheat, which never should be reaped, nor was ever sown. I thought it had stood from everlasting to everlasting. The dust and stones of the street were as precious as gold. The gates at first were the end of the world. The green trees, when I saw them first through one of the gates, transported and ravished me; their sweetness and unusual beauty made my heart to leap, and almost mad with ecstasy, they were such strange and wonderful things. The Men! Oh what venerable and reverent creatures did the aged seem! Immortal Cherubin! And young men glittering and sparkling angels, and maids strange seraphic pieces of life and beauty! Boys and girls tumbling in the street, and playing, were moving jewels. I knew not that they were born or should die.

Clairvoyance or Remote Viewing

In chapter 2.1, I cite evidence that extrasensory perception of physical events (clairvoyance or remote viewing) can occur without benefit of transformative practice. Its spontaneous occurrence among people of various temperaments, cultures, and religious beliefs has prompted some theorists to propose that it is a universal human (or even animal) capacity that is unconsciously utilized—either creatively or destructively—in the service of ordinary needs and desires. Inspired partly by Freud's interest in the unconscious shaping of telepathic interactions, Jan Ehrenwald, Jule Eisenbud, and other psychiatrists have studied various types of psi, including clairvoyance, as they are distorted by the largely unconscious processes that shape normal sensory impressions. There is now a large body of clinical evidence that ESP operates in our everyday lives in ways we do not typically recognize. In chapter 17.3, I present some of this case material.

But clairvoyance can be freed from unconscious distortions and made accessible to conscious control. There is evidence, for example, that shamans of stone-age cultures have used it for hunting and other purposes.[22] It is deemed a yogic power in many Hindu and Buddhist contemplative schools.[23] It has been attributed to Christian religious figures since antiquity: see, for example, the *Historia Monachorum*, a celebrated history of the desert fathers; and *Butler's Lives of the Saints*. [24] And it has been demonstrated with consistency by subjects in contemporary experiments: see reports by Russell Targ and Harold Puthoff of their remote viewing research, which for several years was supported by United States government grants at SRI International.[†] Testimony to clairvoyance in the religious traditions, and the dramatic results of modern experiments, have persuaded me that consciously accessible remote viewing is a latent human capacity.

Perceptions of Entities or Events Outside the Normally Perceivable World

In this class of experience I include perceptions of extraordinary luminosities, visions of disembodied entities, auditions for which no physical sources are apparent, and tactile sensations that seem to be caused by extrasomatic agencies. Other types of experience could be added to these, for there is a long-standing witness to many kinds of metanormal sensing. In Hindu-Buddhist yoga, such perception is deemed to be a product of the *indriyas*, which are attributes of the subtle body (*suksma sarira*); and in the writings of several Neo-Platonist, Christian, and Moslem thinkers it is said to be mediated by the spiritual senses. Origen, the great third-century theologian, as

---

[†] Targ & Puthoff 1977; Targ & Harary 1984; Tart, Puthoff, & Targ 1979. The success of their experiments led Targ and Puthoff to write: "We have not found a single person who could not do remote viewing to satisfaction. Of course, there are differences in each person's ability—as there are in the ability to sing or play the piano: some subjects are more consistently reliable, others get better faster. The indications are that this is a widespread human talent" (Targ & Puthoff 1977, p. 104).

See also: the *New York Times* (editorial) Nov. 6, 1974, "Paranormal Science"; and Targ, R., & H. Puthoff, 1974, Information transmission under conditions of sensory shielding, *Nature* (Oct. 18), pp. 602–7.

well as many Christian mystics and philosophers since, have gathered considerable testimony and scriptural reference to support the idea that there are metanormal analogues of ordinary seeing, hearing, touching, taste, and smell. In his *First Principles*, Origen wrote:

> This sense unfolds in various individual faculties: sight for the contemplation of immaterial forms, hearing for the discrimination of voices which do not echo in the empty air, taste in order to savour the living bread which came down from heaven to bring life to the world (John 6:33), and even a sense of smell, with which Paul perceived those realities which caused him to describe himself as a sweet odor of Christ (2 Cor. 2:15), and finally touch, which possessed John when he states that he has touched with his own hands the Word of life (John 1:1). This sense for the divine was discovered by the prophets. . . . Solomon already realized that there are two modes of sense perception, one mortal, transient, and human, and the other immortal, spiritual, and divine.[25]

This type of experience has been recognized by Christian mystics and philosophers from antiquity until modern times. In his monumental *Theological Investigations,* the Catholic theologian Karl Rahner outlined the understanding of spiritual senses by Origen and medieval thinkers;[26] and in *The Graces of Interior Prayer,* widely regarded to be the most authoritative account of Christian contemplative activity produced in the twentieth century, Father Augustin Poulain described such faculty, carefully distinguishing it from both ordinary and imagined sensing, and cited great mystics of the church who wrote about it.[27]

Perceptions of extraordinary luminosities. Perceptions of auras or halos around animals, plants, and humans; of luminous forms associated with particular thoughts or emotions; of sparks that seem to arise from empty space; and of unexplained lights that pervade a room or other location have long been described by shamans, mediums, and contemplatives. In modern times, such perceptions have been deemed important by Carl Jung, Wilhelm Reich, and their followers. Jung, for example, found references to *opintheres*, or scintillae, in alchemical texts; he linked them to Cabalistic ideas about world-pervading soul-sparks, to Gnostic doctrines regarding "atoms of Light," and to notions of Heraclitus and Democritus concerning "sparks of stellar essence."[28] Jung and other students of paranormal experience have argued that such ideas arise from perceptions of luminous phenomena that have no apparent physical cause. Physicist Edward Russell, president of a New York computer consulting firm, described his own experience of this kind in the journal *Quadrant*.

> Several years ago I realized that I rarely looked directly at anyone, but rather after a fleeting glance turned my eyes away from my subject, or blinked constantly. Since this was an unconscious mechanism, it disturbed me, and I began to struggle with it. With my first efforts to control the focal point of my vision, I began to see that which I had been doing my best not to see: little points-of-light forming around the observed subject. At that moment the idea occurred that my avoidance might be a repressive mechanism developed in early childhood, perhaps due to the strange emotion that accompanies their perception, or to the impossibility of integrating these visions into the world of my parents.
>
> In time, this avoidance has broken down to a degree. Although I usually have to make a conscious effort to perceive aura phenomena and a strange feeling still accompanies their perception, they have become, for me, a daily reality.

In his *Quadrant* article, Russell described several episodes during which he perceived these luminous phenomena. For example:

> Discussing a difficult design task with an employee, I must have asked the impossible, because a look of desperate, forlorn hopelessness came over him. Then his entire aura envelope lit up with hundreds of little luminosities when I asked him what was the matter, he responded, "All the lights on the switchboard came on at once!"
>
> An acquaintance was stifling a sneeze while he fetched a tissue from the desk drawer. A large luminous form of gas-like light, yellow in color, gathered before his forehead. He released the sneeze; simultaneously the cloud dissipated. Note the ancient custom associating a sneeze with an exchange of souls. Gesundheit!
>
> I had just completed a discussion about business contingencies and possibilities with a customer. He sat back and looked to one side. Then a dozen or so small bright points of light appeared and danced over his head. Amazed, I wondered what was happening. Soon the scintillations stopped, and he delivered an oral summary of the relations and priorities to be established for the many proposals we were discussing. The lucidity of his synthesis awed me.
>
> My companion was fishing for a thought. He said something to the effect that he just couldn't quite get it over the threshold. Then I noticed that he was motioning a small golden point toward himself

with his hand. He gave up, however, before getting his thought. The luminosity drifted away from him and faded from view.

Sometimes a bright spark exactly between and slightly above two persons precedes intimate rapport based on a shared idea. I have observed lovers who appear to be contained in a common "cloud of affection," which I see as a mist or haze enveloping and flowing around and between them. Or during an animated discussion (when the sparks fly!) I have seen luminosities literally flowing between the participants.

What I hope my examples show, above all, is the relative autonomy of the luminosities. Typically, there is no clear definition of to whom they belong. The probability of a luminosity "acting on" or "belonging to" an observed subject seems generally proportional to its distance from him; but when they are especially intense . . . even though the luminosity radiates directly from the subject's skin, I too feel the impact. It might be more accurate to think of them as autonomous entities, like "vibrations" present in the atmosphere. A luminosity may seem to belong to one person, yet the spark may jump at any moment to anyone who is susceptible.[29]

Reichian therapists have reported phenomena such as Russell described. The auras, or biofields, of living things have been studied by Soviet researchers. And several studies of the human aura by Western medical people have been published since the early twentieth century (see appendix F). Psychiatrist John Pierrakos described his own aura-reading in psychotherapy.

I use the information it provides to diagnose illness, locate skeletomuscular and characterological blocks that need to be dissolved, and guide the person toward internal integration. More than this: the aura communicates, sometimes in considerable detail, the nature of the person's unique gifts. . . . The person appears to swim in it, as in a fluid tinged with brilliant colors that constantly change hues, shimmer, and vibrate.[30]

In chapter 22.5, I discuss luminous phenomena associated with Catholic saints and mystics, and cite a report that Theresa Neumann, the famous German stigmatic, exhibited an aura around her left hand that was observed simultaneously by several bystanders and photographed. According to her biographer, Albert Schimberg, "when the photographic plate was developed, there appeared on it a bright, strong light, an aura as it were, about the left hand stigma."[31]

Paraphrasing a celebrated treatise of Iranian mysticism, the religious scholar Henri Corbin suggested that as a general rule the capacity to perceive suprasensory lights is proportionate to the degree of spiritual proficiency, or "polishing," that a contemplative has achieved. "In the beginning, such lights are manifested as ephemeral flashes," Corbin wrote. "The more perfect the transparency of the [aspirant], the more they grow, the longer they last, the more diverse they become, until they manifest the form of heavenly entities. As a general rule, the source where these lights take shape is the spiritual entity of the mystic."[32] Corbin cited the Islamic mystic Najm Razi:

> It may be asked whether all these theophanies take place in the inner, esoteric world or in the outer, exoteric world? [Najm Razi's] answer is that anyone who asks this kind of question remains far from the real situation where the two worlds meet and coincide. In one case it may be that the suprasensory perception is awakened and stimulated by a sensory perception. . . . In another instance, a direct perception of the suprasensory may come about without a sensory organ or physical support.

The luminous phenomena of mysticism, in this formulation, depend upon both the percept and the percipient. They are, in other words, objectively real but are perceptible only to those who are trained to see them.[33] However, some extrasensory lights have allegedly been seen by persons without significant perceptual gifts. The luminosities of some mystics, it is said, overcome our normal blindness to them (22.5).[34]

In the lore of Indian contemplative discipline, such luminosities as well as letters, words, and sentences (*lipi*), and faces, landscapes, or other forms (*rupa*) can be apprehended through *drishti*, or subtle sight. The Indian philosopher Sri Aurobindo kept records of his own subtle sight, noting his progress and regress in making it a reliable faculty; and he carefully described various qualities of the *lipi* and *rupa* he saw, including their vividness, duration, color, density, background, and mode of appearance. Some of these images are perceived with eyes open, some with eyes closed, either floating in air (*akasha*), or two-dimensionally against a discernible background (*citra*), or as if sculpted into a glass pane, wall, or other surface (*sthapatya*). Revelatory forms of this kind informed Aurobindo about distant events, philosophical doctrines, his own yoga practice, and other matters. For excerpts from records he kept of such experience, see the journal *Archives and Research*, April 1986, and subsequent issues published by the Sri Aurobindo Ashram, Pondicherry, India.

Perceptions of disembodied entities. Visions of angels, jinn, devas, and other extraphysical beings have been reported in nearly every culture since the beginnings of recorded history. Most of these can be dismissed, of course, as the fancies of superstitious folk, but some defy easy explanation. For example, apprehensions of phantom figures have been reported by ultradistance runners, sailors, explorers, and adventurers not given to occult experience. This was the case during the successful ascent of Mount Everest by a British team in 1975, whose members Doug Scott and Nick Estcourt sensed disembodied companions. Scott felt a presence that guided their party by some sort of mental speech and warned them about dangers ahead. "I had a mental chat with it," he wrote; ". . . it seemed like an extension of my mind outside my head."[35] Estcourt, however, had a different kind of experience. He reported:

> I was about 200 feet above the camp when I turned around. I can't remember why, but perhaps I had a feeling that someone was following me. Anyway, I turned around and saw this figure behind me. He looked like an ordinary climber, far enough behind so that I could not feel him moving up the fixed rope, but not all that far below. I could see his arms and legs and assumed that it was someone trying to catch me up.
>
> I stopped and waited for him. He then seemed to stop or to be moving very, very slowly; he made no effort to signal or wave; I shouted down, but got no reply, and so in the end I thought, "Sod it, I might as well press on." I wondered if perhaps it was Ang Phurba coming through from Camp 2, hoping to surprise us all by being at Camp 5 when we arrived that morning.
>
> I carried on and turned around three or four times between there and the old site of Camp 4 . . . and this figure was still behind me. It was definitely a figure with arms and legs, and at one stage I can remember seeing him behind a slight undulation in the slope, from the waist upwards as you would expect with the lower part of his body hidden in the slight dip.
>
> I turned round again as I reached the old site of Camp 4, and there was no one there at all. It seemed very eerie; I wasn't sure if anyone had fallen off or what; he couldn't possibly have had time to have turned back and drop back down the ropes out of sight, since I could see almost all the way back to Camp 4.[36]

Estcourt's vivid, sustained perception of a phantom companion, and Scott's persistent experience of a disembodied presence, while remarkable in themselves, were subsequently given a stranger twist when it was revealed that a well-known psychical researcher, C. J.

Williamson, apparently had received predictions of them. In 1974, during a session of automatic writing, Williamson had asked for information about the 1924 Everest expedition in which the British climbers Mallory and Irvine had died. The disembodied agent evoked by Williamson's question indicated that the two famous climbers had reached Everest's summit before they disappeared, then it made a prediction. "There were hints coming through," Williamson wrote, "that something psychic was being planned to happen on the mountain during the British 1975 expedition."[37] Subsequently, Williamson got another message through automatic writing,

> a message of such importance that on January 17, 1975, I lodged it in a sealed envelope with the Bank of Scotland Ltd., Lerwick, for safe keeping. This letter lay there untouched until, after the expedition had returned home, I heard through the Press and TV that something strange had indeed been experienced by some of the climbers.[38]

Upon hearing about the climbers' visions, Williamson contacted John Beloff of Edinburgh University, then president of the Society for Psychical Research, and asked him to open his sealed message. Beloff did so in the presence of a witness and read the following words.

> Bonington will not be disappointed even should they not reach the summit. They will come back with better news than that. On the mountain they will see others, not of their own party, others who simply could not be there in the physical body. He will tell you all when they arrive back. Will they reach the summit you ask? Weather will be difficult and ice crevasses. Disaster to some but Bonington will come back safely. Hastie or is it Haston? will be in very great danger and I doubt for his safety.[39]

The message was accurate in that expedition leader Chris Bonington did return safely and one member of the party died. Another member, Dougal Haston, was among the first to reach the summit but was forced to spend the night without oxygen in a bivouac which at that time was the highest ever attempted: though he survived, his life had indeed been in danger. The most striking feature of Williamson's message, however, was its prediction that the climbers would "see others, not of their own party . . . who simply could not be there in the physical body." In his own account, Bonington rejected the possibility that Estcourt hallucinated his phantom companion because he was not acclimatized, suggesting instead that Estcourt had a psychic experience related either to the death of his fellow climber (which would happen later that day) or to a past tragedy. Estcourt was climbing, Bonington wrote, near the place where a Sherpa "who had worked very closely with Nick . . . had perished in an avalanche."[40] In any case, Scott's vivid sense of an invisible companion, Estcourt's sustained perceptions of a phantom figure, and Williamson's sealed predictions coincided to suggest that something more than hallucination was involved in this remarkable episode.

American adventurer Joshua Slocum was the first person to sail alone around the world, taking his 37-foot yawl the *Spray* from Boston on April 24, 1895, to Gibraltar, then back to South America and around Cape Horn—once for 72 days without a port—and subsequently to South Africa, across the Atlantic and on to the United States, completing his voyage in Newport, Rhode Island, on June 27, 1898. Slocum's account *Sailing Alone Around the World,* first published in *The Century Illustrated Monthly* (1899–1900), is a classic tale of heroic adventure. In it Slocum told about an experience now famous in the annals of phantom sailors.

> I am a careful man at sea, but this night, in the coming storm, I swayed up my sails, which, reefed though they were, were still too much in such heavy weather; and I saw to it that the sheets were securely belayed. In a word, I should have laid to, but did not. I gave her the double-reefed mainsail and whole jib instead, and set her on her course. Then I went below, and threw myself upon the cabin floor in great pain. How long I lay there I could not tell, for I became delirious. When I came to, as I thought, from my swoon, I realized that the sloop was plunging into a heavy sea, and looking out of the companionway, to my amazement I saw a tall man at the helm. His rigid hand, grasping the spokes of the wheel, held them as in a vise. One may imagine my astonishment. His rig was that of a foreign sailor, and the large red cap he wore was cockbilled over his left ear, and all was set off with shaggy black whiskers. He would have been taken for a pirate in any part of the world. While I gazed upon his threatening aspect I forgot the storm, and wondered if he had come to cut my throat. This he seemed to divine. "Senor," said he, doffing his cap, "I have come to do you no harm." And a smile, the faintest in the world, but still a smile, played on his face, which seemed not unkind when he spoke. "I have come to do you no harm. I have sailed free," he said, "but was never worse than a *contrabandista*. I am one of Columbus's crew," he continued. "I am the pilot of the *Pinta* come to aid you. Lie quiet, senor captain," he added, "and I will guide your ship to-night. You have a *calentura* [fever], but you will be all right

to-morrow." I thought what a very devil he was to carry sail. Again, as if he read my mind, he exclaimed: "Yonder is the *Pinta* ahead; we must overtake her. Give her sail; give her sail! *Vale, vale, muy vale!*" Biting off a large quid of black twist, he said: "You did wrong, captain, to mix cheese with plums. White cheese is never safe unless you know whence it comes. *Quien sabe,* it may have been from *leche de Capra* and becoming capricious—"

"Avast, there!" I cried. "I have no mind for moralizing."

I made shift to spread a mattress and lie on that instead of the hard floor, my eyes all the while fastened on my strange guest, who, remarking again that I would have "only pains and *calentura,*" chuckled as he chanted a wild song:

> High are the waves, fierce, gleaming,
> High is the tempest roar!
> High the sea-bird screaming!
> High the Azore!

I suppose I was not on the mend, for I was peevish, and complained: "I detest your jingle. Your Azore should be at roost, and would have been were it a respectable bird!" I begged he would tie a rope-yarn on the rest of the song, if there was any more of it. I was still in agony. Great seas were boarding the *Spray,* but in my fevered brain I thought they were boats falling on deck, that careless draymen were throwing from wagons on the pier to which I imagined the *Spray* was now moored, and without fenders to breast her off. "You'll smash your boats!" I called out again and again, as the seas crashed on the cabin over my head. "You'll smash your boats, but you can't hurt the *Spray.* She is strong!" I cried.

I found, when my pains and *calentura* had gone, that the deck, now as white as a shark's tooth from seas washing over it, had been swept of everything movable. To my astonishment, I saw now at broad day that the *Spray* was still heading as I had left her, and was going like a race-horse. Columbus himself could not have held her more exactly on her course. The sloop had made ninety miles in the night through a rough sea. A meridian altitude and the distance on the patent log, which I always kept towing, told me that she had made a true course throughout the twenty-four hours. . . . I put my wet clothes out in the sun when it was shining, and lying down there myself, fell asleep. Then who should visit me again but my old friend of the night before, this time, of course, in a dream. "You did well last night to take my advice," said he, "and if you would, I should like to be with you often on the voyage, for the love of adventure alone." Finishing what he had to say, he again doffed his cap and disappeared as mysteriously as he came, returning, I suppose, to the phantom *Pinta.* I awoke much refreshed, and with the feeling that I had been in the presence of a friend and a seaman of vast experience. I gathered up my clothes, which by this time were dry, then, by inspiration, I threw overboard all the plums in the vessel.[41]

Slocum's experience, of course, can be interpreted in various ways. Perhaps he was simply lucky and hallucinated his tall sailor. But his vision corresponds with experiences reported by other adventurers. Charles Lindbergh, for example, encountered phantom figures during his pioneer flight across the Atlantic. Though he did not describe the encounter in his first book, *We, Pilot and Plane*, published in 1927, he did so in considerable detail in *The Spirit of St. Louis*, which appeared 26 years later.

> While I'm staring at the instruments, during an unearthly age of time, both conscious and asleep, the fuselage behind me becomes filled with ghostly presences—vaguely outlined forms, transparent, moving, riding weightless with me in the plane. I feel no surprise at their coming. . . . Without turning my head, I see them as clearly as though in my normal field of vision. There's no limit to my sight—my skull is one great eye, seeing everywhere at once.
>
> These phantoms speak with human voices—friendly, vapor-like shapes, without substance, able to vanish or appear at will, to pass in and out through the walls of the fuselage as though no walls were there. Now, many are crowded behind me. Now, only a few remain. First one and then another presses forward to my shoulder to speak above the engine's noise, and then draws back among the group behind. At times, voices come out of the air itself, clear yet far away, traveling through distances that can't be measured by the scale of human miles; familiar voices, conversing and advising on my flight, discussing problems of my navigation, reassuring me, giving me messages of importance unattainable in ordinary life.
>
> I'm on the border line of life and a greater realm beyond, as though caught in the field of gravitation between two planets, acted on by forces I can't control, forces too weak to be measured by any means at my command, yet representing powers incomparably stronger than I've ever known.[42]

Lindbergh, Slocum, Estcourt, and Scott had been taxed to the edge of endurance before their strange encounters. All had been fatigued yet struggled to maintain their alertness. This combination of stress and vigilance, conceivably, predisposed them to psychic visitations. Their ordeals appeared to alter their senses, opening them to unusual perceptions. During his famous exploration of the South Pole in 1916, Ernest Shackleton, too, sensed a phantom companion:

When I look back at those days I have no doubt that Providence guided us, not only across those snowfields, but across the stormwhite sea that separated Elephant Island from our landing-place on South Georgia. I know that during that long and racking march of thirty-six hours . . . it seemed to me often that we were four, not three. I said nothing to my companions on the point, but afterwards Worsley said to me, "Boss, I had a curious feeling on the march that there was another person with us." Crean confessed to the same idea. One feels "the dearth of human words, the roughness of mortal speech" in trying to describe things intangible, but a record of our journeys would be incomplete without a reference to a subject very near to our hearts.[43]

Shackleton's impression was shared by F. A. Worsley, who wrote that

each step of that journey comes back clearly, and even now I again find myself counting our party—Shackleton, Crean, and I and—who was the other? Of course, there were only three, but it is strange that in mentally reviewing the crossing we should always think of a fourth, and then correct ourselves.[44]

These participants in four celebrated adventures—Slocum, Lindbergh, Estcourt and Scott, Shackleton and other members of his party—all perceived phantom figures or felt invisible presences that haunted them afterward. With similar experiences reported by religious figures and other adventurers, their visions and the strangely coincidental events that sometimes accompanied them suggest that humans can indeed perceive disembodied entities. Such entities at times appear malevolent, at other times benign. They can, like Slocum's tall sailor, provide guidance in extreme circumstances. They sometimes lend comfort, challenge comfortable assumptions, convey information, suggest that life has dimensions beyond the range of the ordinary senses, or bring ecstasy to their percipient. Some people, it seems, have developed their capacity to perceive them. In chapter 5.5, I describe Emmanuel Swedenborg's and Stainton Moses's cultivation of such capacity.[†]

[†] There is long-standing witness to phantom figures in every religious tradition, a great many descriptions of them in the literature of psychical research, and a considerable body of speculation about their nature and causes. It has been proposed, for example, that because human perception is shaped by habitual mind-sets, disembodied entities are typically perceived as the angels, jinn, devas, elves, or (in our time) UFOs of the subject's culture. It is also proposed that such entities—if they do indeed exist—can shape themselves to fit their percipients' expectations and mental habits. See chapter 9.

Metanormal auditions. Hearing sounds with no apparent physical cause constitutes another class of nascent metanormal perception. In his *Phantasms of the Living,* the pioneering psychical researcher Edmund Gurney cited several cases in which one or more people heard music as a friend or relative was dying. A Mrs. Lucia Stone reported, for example, that

> on the 13th of January 1882, my eldest son, who had been paying us a visit, left by a morning train for his home, but I did not know the exact time at which he would reach his destination. In the afternoon of that day, my daughter having gone to the neighboring town . . . I was sitting at work by a window of which the upper ventilator was open. Suddenly I heard my son's voice distinctly; I could not mistake it; he was speaking eagerly, and as if bothered; the voice seemed wafted to me by an air current, but I could not distinguish words. I was startled, but not very much frightened; the voice did not seem to indicate accident or calamity. I looked at my watch, which pointed to three minutes past 3. In perhaps a few seconds, his voice began again, but soon became faint, and died away in the distance. . . . The next morning I was very thankful to get a post-card from my eldest son: "Arrived all right, train very punctual, just three minutes past 3; but to my annoyance, I found no carriage waiting for me, or my luggage, only Frank on his bicycle." He explained that they had made a mistake by looking at the station clock (which was an hour too slow), and had driven away again.[45]

A Mr. Fryer of Bath recounted the following experience for Gurney's collection:

> A brother of mine had been away from home for 3 or 4 days, when, one afternoon, at half-past 5 (as nearly as possible), I was astonished to hear my name called out very distinctly. I so clearly recognised my brother's voice that I looked all over the house for him; but not finding him, and indeed knowing that he must be distant some 40 miles, I ended by attributing the incident to a fancied delusion, and thought no more about the matter. On my brother's arrival home, however, on the sixth day, he remarked amongst other things that he had narrowly escaped an ugly accident. It appeared that, whilst getting out from a railway carriage, he missed his footing, and fell along the platform; by putting out his hands quickly he broke the fall, and only suffered a severe shaking. "Curiously enough," he said, "when I found myself falling I called out your name." This did not strike me for a moment, but on my asking him during what part of the day this happened, he gave me the time, which I found corresponded exactly with the moment I heard myself called.

Fryer explained to Gurney that he had frequently scolded his brother for alighting from trains in motion; and the automatic utterance of his name, on this occasion, might thus be accounted for by association. Fryer's brother confirmed the incident and agreed with Fryer's explanation of its possible cause.[46]

A different type of extraordinary audition was reported by friends of the dying Goethe. The following account is taken from Professor Ernesto Bozzano's *Phenomenes Psychiques au Moment de la Mort.*

On the 22nd day of March, 1832, about 10:00 in the evening, two hours before Goethe's death, a carriage stopped outside the great poet's house. A lady got out and hastened to enter, asking the servant in a trembling voice, "Is he still alive?" It was Countess V., an enthusiastic admirer of the poet, who always received her with pleasure because of the comforting vivacity of her conversation. While she was going up the stairs she suddenly stopped, listening to something, then she questioned the servant, "What! Music in this house? Good heavens, how can anyone play music here on such a day as this?" The man listened in turn, but he had become pale and trembling, and made no reply. Meanwhile, the Countess had crossed the drawing room and gone into the study, where only she had the privilege of entry. Frau von Goethe, the poet's sister-in-law, went to meet her: The two women fell into each other's arms, bursting into tears. Presently the Countess asked, "Tell me, Ottilie, while I was coming upstairs I heard music in the house. Why? Why? Or was I perhaps mistaken?"

"So you have heard it too?" replied Frau von Goethe. "It's inexplicable! Since dawn yesterday a mysterious music has resounded from time to time, getting into our ears, our bones." At this very instant there resounded from above, as if they came from a higher world, sweet and prolonged chords of music which weakened little by little until they faded away.

At this same moment, Jean, the valet, came out of the dying man's room, much moved by the sounds and asked, "Did you hear it, Madame? This time the music came from the garden, and sounded just at the level of the window."

"No," said the Countess. "It came from the room beside us."

They drew back the curtains and looked out at the garden. A light and almost silent wind blew across the bare branches of the trees; you could hear, far away, the sound of a cart going along the road; but there was nothing to be seen which could explain the origin of the mysterious music.

Then the two friends went into the drawing room, whence they thought the music must have come, but without observing anything unusual. While they were still busy searching, another series of marvelous harmonics was heard. This time they seemed to come from the study.

The Countess, going back into the drawing room, said, "I don't think I can be mistaken; it must be a quartet playing fragments of music some way off which reaches us from time to time."

But Frau von Goethe for her part remarked, "On the contrary, it seemed to me that I was hearing the sound of a piano, clear and close by. This morning I was so sure of it that I sent the servant to implore my neighbors to stop playing the piano, out of consideration for the dying man. But they all said the same thing; that they knew very well what condition the poet was in, and were too much distressed to dream of disturbing his last hours by playing the piano."

Suddenly the music burst out again, delicate and sweet; this time it seemed to arise in the room where they were; only, for one person it seemed to be the sound of an organ, for the other a choral chant, and for the third . . . the notes of a piano. A Mr. S., who was at that moment signing the medical report with Dr. B. in the hall, looked at his friend with surprise, asking him, "Is that a concertina playing?"

"It seems to be," the doctor replied. "Perhaps someone in the neighborhood is amusing himself."

"No," said Mr. S., "whoever is playing is definitely in this house."

It was thus that the mysterious music went on making itself heard up until the moment when Wolfgang Goethe breathed out his last sigh; sometimes recurring after long intervals, sometimes after the briefest of remissions, sometimes in one direction, sometimes in another, but always seeming to come from the house itself or from quite close by. All the searches and inquiries undertaken to solve the mystery were in vain.[47]

Beautiful and informative auditions are often reported among religious figures. Saint Guthlac was said to have heard angelic songs as he died; Saint Therese de Lisieux heard celestial music on her deathbed; Joseph of Copertino heard the sound of a bell summoning him to God on the day before his passing.[48] In his essay *The Fire of Love*, the English contemplative Richard Rolle (1290–1349) described music revealed to him by prayer.

For when I was sitting in the same chapel and was reciting psalms as well as I might before supper, I heard above me the noise of harpers, or rather of singers. And when with all my heart I attended to heavenly things in prayer, I perceived within me, I know not how, a melody and a most delightful harmony from heaven, which abode in my mind. For my thought was straightway changed into a song, and even when praying and singing psalms I gave forth the selfsame sound.[49]

Supraphysical or "celestial" music like Rolle's, called *nad* in Sanskrit, has been celebrated in Hindu-Buddhist contemplative lore. Its rhythms and harmonics, which religious devotion can reveal, rise from sublimity to sublimity, pervading the universe. All of us are informed by *nad*, according to certain yogic traditions, as if we were secretly in resonance with a constant background music. Sorrow, joy, anger, victory, indeed all our moods, are secretly amplified by it. Such doctrine is summarized in the Nada-Bindu Upanishad of the Rig Veda, according to which concentration upon supraphysical sound lifts us beyond ordinary consciousness, produces metanormal hearing, and leads us toward extraordinary life.[50]

Metanormal touch. Extraordinary tactile impressions have been described by modern psychical researchers and by religious ecstatics. Frederic Myers found, for example, that experiences of touch that had no apparent physical cause sometimes signaled telepathic messages. Edmund Gurney described a case of this kind reported by a Reverend and Mrs. Newnham.

> In March 1854, I was up at Oxford [the Reverend Newnham wrote] keeping my last term, in lodgings. I was subject to violent neuralgic headaches, which always culminated in sleep. One evening, about 8 p.m., I had an unusually violent one; when it became unendurable, about 9 p.m., I went into my bedroom, and flung myself, without undressing, on the bed, and soon fell asleep.
>
> I then had a singularly clear and vivid dream, all the incidents of which are still as clear to my memory as ever. I dreamed that I was stopping with the family of the lady who subsequently became my wife. All the younger ones had gone to bed, and I stood chatting to the father and mother, standing up by the fireplace. Presently I bade them goodnight, took my candle, and went off to bed. On arriving in the hall, I perceived that my fiancee had been detained downstairs, and was only then near the top of the staircase. I rushed upstairs, overtook her on the top step, and passed my two arms round her waist, under her arms, from behind. Although I was carrying my candle in my left hand when I ran upstairs, this did not, in my dream, interfere with this gesture.
>
> On this I woke, and a clock in the house struck 10 almost immediately afterwards.
>
> So strong was the impression of the dream that I wrote a detailed

account of it next morning to my fiancee. Crossing my letter, not in answer to it, I received a letter from the lady in question [in which she asked] "Were you thinking about me, very specially, last night just about 10 o'clock? For, as I was going upstairs to bed, I distinctly heard your footsteps on the stairs, and felt you put your arms round my waist."

The letters in question are now destroyed, but we verified the statement made therein some years later, when we read over our old letters, previous to their destruction, and we found that our personal recollections had not varied in the least degree therefrom. The above narratives may therefore be accepted as absolutely accurate.

Confirming her husband's account, Mrs. Newnham wrote:

I remember distinctly the circumstance which my husband has described as corresponding with his dream. I was on my way up to bed, as usual, about 10 o'clock, and on reaching the first landing I heard distinctly the footsteps of the gentleman to whom I was engaged, quickly mounting the stairs after me, and then I as plainly felt him put his arms round my waist. So strong an impression did this make upon me that I wrote the very next morning to the gentleman, asking if he had been particularly thinking of me at 10 o'clock the night before, and to my astonishment I received (at the same time that my letter would reach him) a letter from him describing his dream, in almost the same words that I had used in describing my impression of his presence.[51]

Supraphysical kinds of touch have been described by many Hindu, Moslem, and Catholic saints.[52] Describing her encounter with a fiery angel, Saint Teresa of Avila wrote that

in his hand I saw a long golden spear and at the end of the iron tip I seemed to see a point of fire. With this he seemed to pierce my heart several times so that it penetrated into my entrails. When he drew it out, I thought he was drawing them out with it and he left me completely afire with a great love for God. The pain was so sharp that it made me utter several moans; and so excessive was the sweetness caused me by this intense pain that one can never wish to lose it, nor will one's soul be content with anything less than God. It is not bodily pain, but spiritual, though the body has a share in it—indeed, a great share (Saint Teresa, *Autobiography*, chap. 29, Sec. 13. E. Allison Peers, trans.).

Though experiences such as the Newnhams' and Saint Teresa's might be purely imaginary or somatizations of telepathic interactions, it is conceivable that some are stimulated by phantoms projected in dream or by permanently disembodied entities, as most religious traditions insist.

Other kinds of experience could be added here, among them the perception of odors of sanctity (see, for example, 22.7). As I have said, this inventory does not include every type of extraordinary capacity.

[5.2]

SOMATIC AWARENESS AND SELF-REGULATION

All living organisms depend for their survival upon sensitive communication among their constituent parts. In humans, this sensitivity can be increased through biofeedback training, somatic education, martial arts, sports, meditation, and other disciplines. Furthermore, both long experience with such practices and experimental research have shown that humans can, by developing somatic awareness, increase their capacities for voluntary self-control. Today we have strong evidence that any aspect of bodily functioning, once brought to awareness, can be deliberately altered to some extent, for healing or the development of new abilities. Documentation that this is the case is presented in Part Two; see especially chapters 16, 18, and 23.1.

Such awareness and self-regulation, I propose, are based not only upon kinesthetic (proprioceptive) sensitivity made possible by the central nervous system but also upon extrasomatic contact with one's bodily parts. In this section, I will group somatic awareness and self-regulation skills into two classes: first, those mediated primarily by central nervous activity; and second, those that appear to be largely due to receptive and expressive psi—that is, clairvoyance and psychokinesis (see chapter 1 for definitions of these terms). As with other human attributes discussed here, such capacities can be gradually freed from unconscious distortions and made accessible to voluntary control.

Extraordinary Somatic Awareness Mediated by the Central Nervous System

Biofeedback training can be used to improve kinesthetic sensitivity and control of autonomic functioning. In chapter 16, I describe vol-

88

untary modulations of brain, heart, skin, gastrointestinal, and other bodily processes facilitated by instrumented feedback. Even a single nerve cell's activity can be modified through such training. And other types of practice, too, promote sensitivity to nervous and muscular functioning. The Alexander Technique, Feldenkrais Method, Progressive Relaxation, Autogenic Training, and Sensory Awareness, for example, enhance somatic awareness and self-regulation: several first-hand accounts of such enhancement are presented in chapter 18. Frank Jones, for example, described his experience with the Alexander Technique under the guidance of Alexander's brother, A. R.

> I had been aware of neck-muscle tension before but had not been aware that the tension increased in response to stimuli. Now the response pattern—the increment in tension—began to stand out against the newly induced background of postural tone so that there was a clear-cut figure-ground relation between them. What the procedures I learned from A. R. had done was to remove a great deal of the "noise" from the tonic "ground" so that the tensional "figure" was easier to perceive.[†] Once the figure was perceived for what it was—an increment of tension in response to a particular stimulus—it could be controlled; "inhibited" was the word A. R. used.
>
> After I had clearly perceived the pattern of neck-muscle tension and understood the part it played in one everyday movement, I began to notice it in other movements. It appeared when I started to sit down as well as when I started to stand up. I noticed it sharply in climbing stairs, in picking up a suitcase, in taking a deep breath, in writing a letter. Less sharply but unmistakably, the pattern appeared in everything I did. [But] the pattern was not confined to the neck. The neck was merely the distribution point at which the increase in tension began and from which it spread like a net to other parts of the body.
>
> Inhibition is a negative term, but it describes a positive process. By refusing to respond to a stimulus in a habitual way you release a set of reflexes that lengthen the body and facilitate movement. The immediate result of Alexandrian inhibition is a sense of freedom, as if a heavy garment that had been hampering all of your movements has been removed.[53]

Sports, dance, and the martial arts, too, promote kinesthesis and self-regulation. Accomplished distance runners creatively change

---

[†] Jones's use of the noise metaphor here resembles the noise-reduction model for self-regulation training proposed by some biofeedback researchers. In the Shurley-Lloyd experiments described in chapter 17.3, for example, external stimuli (conceived as noise) were reduced by placing subjects in a darkened, quiet chamber to enhance their attention to proprioceptive signals.

their racing stride by responding to fine shifts of muscle tone. Yachts-men can track a boat's course through subtle alterations of their bal-ance. Great rodeo riders stay on bucking bulls, gymnasts do hand-stands, and divers enter the water cleanly through extraordinary muscular feel. Consider these prodigies of somatic awareness and control:

- Indian Kathakali dancers who can simultaneously move their cheeks, forehead, ears, and eye muscles, make intricate gestures with their fingers, flex the muscles of their abdomen, and per-form complex steps in concert with fellow dancers.
- Amerindian drummers who simultaneously beat different rhythms on separate drums.
- A mountain climber who, while hanging under a protruding rock ledge, reels in rope with his left hand, holds on to a clamp with his right, pushes himself along with one foot, and balances with the other.
- A running back who when tackled at the knees does a midair somersault, dodges another tackler, and keeps on running.

But extraordinary kinesthesis is not limited to muscular feel or a sense of balance. It is also mediated by auditions, tastes, and smells. For example, some runners have told me that they monitor their level of stress by the taste of their saliva or the smell of their sweat. Many people claim that they hear, as if with an inner ear, sounds that reflect their somatic condition, such as rushing water, rolling surf, sighing, laughing, bells, horns, or drums. Sports people have described disem-bodied voices telling them what moves to make in a game or what paths to follow on a dangerous mountainside,[54] while others have re-ported music that suggests a style of performance. Bobby Jones, for example, said that when he grooved his swing to a melody rising spon-taneously in him, he played his finest golf, as if the melody captured his best rhythms for that occasion. These spontaneous sounds suggest vitality or fatigue, smooth or rough performance, a slow or rapid beat. Hindu and Buddhist yogas have a considerable lore about such audi-tions, or *sravanas*. The Nadu Bindu Upanishad, for example, suggests that we can detect sounds inaccessible to ordinary hearing.[55] With practice we can follow their rise and fall, their subtle transformations, and thus discover new aspects of mind and flesh.

Somatic awareness is also mediated by visual imagery. Some run-ners have told me, for example, that they had brief, vivid images during races or hard workouts that they took to be pictures of their own veins or capillaries.[56] During intense psychotherapeutic sessions, people have reported glimpses of their insides that they felt represented their actual physical state.[57] Physician Carl Simonton described a patient who drew accurate pictures of his throat cancer as it regressed in response to visualization and radiation therapy.[58] And students of aikido teacher George Leonard have reported a wide range of imagery, some of it resembling photographs in medical books, that seemed to represent their organs, tissues, and cells.[59] Such reports, provided by people with different temperaments and beliefs over a period of many years, have led me to think that we harbor capacities for inner seeing such as those attributed to adepts of Hindu, Buddhist, and Taoist yoga. In the next section, I discuss yogic powers of this kind.

Extraordinary Somatic Awareness Mediated by Internal Clairvoyance

There is reason to believe that body awareness has extrasomatic as well as purely physical components. This extrasomatic sensitivity, like the clairvoyance discussed in chapter 5.1, is metanormal or extraordinary, by the definition used here, when it is largely freed from distortions and is made accessible to conscious control. That it does indeed occur is indicated by much lore of Hindu, Buddhist, and Taoist yoga. For example, among the siddhis produced by yogic practice, it is said, there exist some by which we can perceive the smallest particles of matter, either inside or outside the body. The *anudrishti* siddhi—a term derived from the Sanskrit *anu,* atom, and *drishti,* insight—refers to a yogic apprehension of small, hidden, or remote things, including somatic structures. The *antara drishti* siddhi is said to produce an X-ray look into our bodily parts. And the *animan* siddhi, one of eight famous powers referred to in several yoga sutras, also involves such capacity.[†] Patanjali's Yoga Sutras refer to this power in the twenty-sixth sutra of Book Three, *Vibhuti Pada.*

[†] The *anudrishti* and *antara drishti* siddhis are listed in a catalogue of siddhis assembled by the philosopher Haridas Chaudhuri, founder of the California Institute of Integral Studies in San Francisco. Chaudhuri believed that the *animan* siddhi has various aspects, among them internal clairvoyance by which we can perceive our body's cells, molecules, and atomic patterns. During conversations with me, he described this capacity in some detail, drawing upon his long exposure to yogic lore. My novel *Jacob Atabet* includes fictionalized accounts of such clairvoyance based upon firsthand reports I collected from athletes, meditators, and yoga students.

Internal clairvoyance, if it does indeed exist, would be shaped in different ways by different people, like the extrasensory perception of external events. The experiments by Targ and Puthoff noted in chapter 5.1, for example, have shown that habitual psychological processes, whether conscious or unconscious, influence the imagery associated with clairvoyance, sometimes distorting what appears to be accurate information. One subject in the Targ-Puthoff experiments said that a designated target, actually a swimming pool, might be a purification tank, and mistook a drive-in movie with speaker posts for a parking lot full of meters. Another subject labeled a pedestrian overpass "a trough in the air," and described a video terminal as a "black box in the middle of a room, complete with glass porthole and light coming out," but incorrectly deemed it to be an oven or radiation machine.[60]

Experimental results such as these, as well as testimony from the Hindu, Buddhist, and Taoist yogic traditions, indicate that internal clairvoyance—like every other type of extraordinary functioning—is shaped by the mind-body complex in which it occurs, with all its genetic and cultural history. This is a theme I will return to again and again: *All of our capacities, whether normal or metanormal, somatic or extrasomatic, are subject to the limitations and distortions produced by our inherited and socially conditioned nature.* And it follows, then, that some internally clairvoyant perceptions would be more accurate than others, less filtered by half-conscious or unconscious psychological processes, and could be improved by transformative discipline, as Sutra 26 of Book Three in Patanjali's Yoga Sutras suggests. Some, or perhaps all, of us might perceive our own body parts through occasional psi, but through a glass darkly, while with practice we might see them more clearly.

Furthermore, there is good reason to think that in addition to occasional clairvoyant cognitions of our own insides, we enjoy an ongoing, though primarily unconscious, extrasomatic interaction with them. Several psychiatrists, among them Jule Eisenbud and Jan Ehrenwald, have presented evidence from their therapeutic practices that we, without knowing it, interact with others telepathically, and that such interaction frequently affects bodily processes (see chapter 17.3). Theologian David Griffin has gathered considerable data to support the idea that all living (as well as inorganic) things interact by expressive psi (PK) and receptive psi (telepathy and clairvoyance) in a mutually formative manner that operates mainly outside of consciousness (see chapter 7.2). I am persuaded by such findings that we do indeed relate to other persons and to our own body parts in this way, and that we can by transformative practice lift such interaction to ongoing awareness and a measure of deliberate control.

➤

Though they do not provide compelling evidence for clairvoyant perception of microscopic events, a series of experiments by the Theosophists Annie Besant and Charles Leadbeater suggest that the *animan* siddhi, or micro-psi, might be studied experimentally. In his book *Extra-Sensory Perception of Quarks,* physicist Stephen Phillips reviewed the Besant-Leadbeater experiments:

> From 1895 until 1933 Besant and Leadbeater examined all the elements from hydrogen to uranium, as well as an assortment of organic and inorganic compounds. In this vast task, they were assisted by their colleague C. Jinarajadasa, who acted as recorder during their experimental sessions. The well-known chemist Sir William Crookes, a friend of both the investigators, provided specimens of some elements. Various minerals were studied in a museum in Dresden, Germany.[61]

Phillips described some apparent agreements between the Besant-Leadbeater micro-psi observations and subsequent scientific discoveries. The two Theosophists claimed in 1908, for example, that a few elements, including neon, had more than one form—five years before the concept of isotopes was put forth by the physicist Soddy. They also appeared to anticipate the discovery of Larmor precession of spinning in atoms and the behavior of charged particles in magnetic fields. After reviewing the voluminous Besant-Leadbeater reports, published in numerous articles and books over the course of 33 years, Phillips concluded that the two Theosophists might well have demonstrated accurate micro-psi observations upon occasion.[62] Though these experiments were not directed toward atomic events located within living tissue, they lend support to the notion that clairvoyant perception of cells and their molecules might be extended to the submolecular, or even subatomic, levels of the body.

In summary, then, it is conceivable that there exists an internal clairvoyance through which we can develop a somatic scanning device, a microscope with zoom lens as it were, by which we can focus upon our organs or cells in the service of bodily transformation. Though there is less evidence for such clairvoyance than there is for some other types of extraordinary functioning, there is evidence nevertheless.

[5.3]
COMMUNICATION ABILITIES

In the long course of hominid and human development, primate signaling systems evolved into the complex repertoires of language, gesture, and facial expression by which we communicate with our fellows. Our abilities to convey information and mood develop from infancy and are shaped—for better or worse—by parents, teachers, and friends until the day we die. Through imitation and tutoring, we learn how to form new words and phrases, move our hands, adopt certain postures, and alter facial muscles to express our feelings and thoughts. To a greater or lesser degree, we learn to sense other people's mental-emotional states so that we can respond creatively to them. Through relationships with different kinds of people, we learn new styles of personal interaction. In chapter 6.1, I review several studies that demonstrate the great range and malleability of, as well as the pervasive cultural influence upon, our communication abilities. The elaboration of ways to communicate with others is a stupendous achievement of the human race.

The physical, emotional, and cognitive capacities upon which human communication depends can be improved by several kinds of practice. Psychotherapy, by promoting insight into our attitudes, motives, and impulses, can make it easier to achieve clear, sensitive, and effective expression (17.1, 17.2). Somatic education can release feelings that enrich both language and gesture, thus bringing new liveliness, subtlety, and range to our interactions with others (18). Sustained introspection, whether encouraged by a therapist, moral educator, or meditation teacher, can uncover strictures upon expressive abilities, help broaden behavioral repertoires, and deepen empathy, helping us reach out to our fellows with greater feeling and skill. These various outcomes of transformative discipline are described in many articles and books cited in subsequent chapters.

Here I will divide metanormal communication abilities into two classes: first, those that depend primarily on sensory cues; and second, those that appear to be mediated largely by telepathic interaction.

Extraordinary Communication Abilities
Mediated by Sensory Cues

Certain men and women convey humor by a magic that defies analysis. They induce laughter in us by a combination of postures, gestures, facial expressions, voice inflections, turns of phrase, mental images, and other means, the overall effect of which is difficult—or impossible—to explain. The abilities of Harpo Marx, W. C. Fields, Jonathan Winters, or Whoopi Goldberg to reveal the absurd, the ironic, or the ludicrous aspects of life have elements of genius. Great comedians and humorists, as well as less comically gifted people, sometimes come up with images that by any criteria are extraordinary. Such inspirations might heal a conflict, bring new understanding to a particular set of events, or through laughter illuminate an entire dimension of life. Like other kinds of extraordinary human activity, some inspired humor and wit seem to be given by powers beyond the ordinary self.

And the same indefinable ability to communicate is evident in other modalities of human exchange. Think of Mozart, Bach, or Beethoven. The mere recitation of their names calls up unique worlds that they—and only they—have imparted to us. Indeed, their musical communications are so important to the human race that they are repeated at every moment somewhere in the world today. The same observation holds for Sophocles and Dante, for authors of the Old and New Testaments, for the seers who composed the Upanishads. According to Indian yoga lore, the *Vak* siddhi is an extraordinary power to create and transmit human language. Those who possess it give birth to mantrams, poems, and scriptures that transform life on this planet.

It is indisputable that in every domain of human experience some people have extraordinary powers to communicate. Their messages may be musical, poetic, or humorous, religious or philosophic, either for the enrichment or dimunition of life. No matter what type of communication we think of, there are men and women with abilities that transcend the ordinary patterns of human connection.

Extraordinary Communication Abilities Mediated by Telepathic Interaction

Sustained telepathic rapport. There is good evidence that some intimate friends and family members develop remarkable extrasensory rapport. In a book entitled *Parent–Child Telepathy,* psychiatrist Berthold Schwarz described 505 episodes that appeared to involve telepathic exchanges between himself, his wife Ardis, their son Eric, and their daughter Lisa. Like Freud and other psychoanalysts who have studied telepathy (see 17.3), Schwarz noted several ways in which extrasensory communication is affected by unconscious psychological processes. His book also lists reviews of experimental research with parent–child telepathy. Like other neglected studies noted here, it deserves a wider audience among students of human development. Here are some excerpts from it:

(On November 24, 1958, Lisa celebrated her second birthday.)

December 29, 1958. Having used all our available liquor entertaining, Ardis was, perhaps half ruefully, thinking, "Well, we will not have any liquor with dinner." Simultaneously with this thought, Lisa started shouting, "Give me a drink of ginger ale!" Ardis almost invariably drinks ginger ale and bourbon, and at such times Lisa is given a small glass of ginger ale.

March 4, 1959. Ardis was thinking about whether she "should wear black suede or black patent leather shoes because of the snow." Lisa then said "Lisa going to find Indian shoes." Ardis's next thought was "I wonder where Eric's table is?" Lisa immediately responded telepathically by telling her where it was. At 6:00 P.M., at the conclusion of a TV show, two-year-old Lisa said, quite out of context, "Kugel [the babysitter] getting better—Papa get Kugel now." This was the time for Mrs. Kugel to come. Lisa did not know of our plans to go out and we didn't eat out frequently. Lisa seemed to be aware of the situation and of the specific person who was to take care of her.

July 23, 1959. Ardis's eye fleetingly caught a piece of chocolate frosting adhering to a bakery cake box. It was her intention to clean the box to make it more presentable. The box was behind pots and pans and completely hidden from the view of Lisa, who had just come into the room. Before Ardis could clean the box, Lisa, while looking out the window (on the opposite side of the room) at Dr. R's house, suddenly said, "Annie has chocolate outfit!" Annie, Dr. R's daughter, who is one year older than Lisa, was nowhere in sight. Lisa is fond of chocolate, but Ardis is reluctant to give it to her and ruin her appetite. By her condensation Lisa gave the valued chocolate (outfit) to her rival, Annie.

November 1, 1959. Ardis was looking at my appointment book for the coming month and her eye fell on November 24. As she thought of writing "Lisa's birthday" on the appropriate line, Lisa said, "Draw a birthday cake!" Lisa was in a friendly mood. Although she might have associated the appointment book with her embryonic concept of time, and that in turn with her birthday, it should be pointed out that on previous occasions when either her mother or father examined the appointment book and did not specifically think of Lisa's birthday, Lisa never commented. This might be another illustration of how telepathic communication is an extension of the usual means of thinking and perception. It would appear that telepathic reactions can work as harmoniously with the sense organs and reflexes as they work together.

January 1, 1963. In my room after a New Year's Eve party, I mused, "How tired one can get—does telepathy occur as frequently at such times? Met some good people last night; they'll have children; maybe, someday, Lisa will marry one of them." Lisa immediately entered the room and said, "How much does a house cost? Someday I will be a mother and will want to buy one." At first glance it might appear that the associations are telepathically transmitted by the father; however, who can say that the reverse wasn't the case and the father's thoughts amounted to his unmasking the telepathic exchange.[63]

Rapport with apparent telepathic elements is not limited to family members. It is evident, too, among people at work. Baseball pitcher Sandy Koufax, for example, recalled his most vivid memory of the 1963 World Series:

As I got the ball back and began to look in for the sign, I thought to myself: I'd like to take something off my curve ball. . . . Now why does a thought like that come to you? A change-up curve is exactly what you don't throw Mantle, particularly in a spot where it can cost you a ball game. Change-up curves are what Mantle hits out of ball parks. I hadn't thrown a change-up in the entire game, as far as I could remember. And at the same moment that the thought came into my mind there flickered the answering thought: But how will I explain why I threw it if he hits it out of here?

And while the thought was still half formed in my mind, I was looking down toward the plate, and John Roseboro was putting down two fingers, the sign for the curve. He was putting them down hesitantly, though, so hesitantly that I had the feeling there was something more he wanted to tell me, something that couldn't be communicated by means of a sign. Normally he'd pull the fingers right back. This time he left the fingers there for a couple of seconds and then, slowly, still hesitantly, he began to wiggle them, the sign to take something off it.

As soon as I saw the fingers wiggle, I began to nod my head emphatically. I could see John begin to smile behind the mask, and then the fingers began to wiggle faster, as if he were saying to me, "Sandy baby, you don't know how glad I am that you see it this way too."

As it was, I copped out just a little. I did take something off my normal curve, but I didn't throw it real slow. It was a good pitch, though. It broke right down in there for a called third strike.

As soon as we hit the clubhouse, I grabbed Roseboro. "What was the matter, John?" I said. "You seemed a little hesitant about wiggling the fingers on Mantle." And he grinned back and said, "I wanted to call it, but I was thinking: How are we ever going to explain a change-curve if he hits it out?"

That's how close the rapport between us can get. Not only did we have the same idea at the same moment, we even had the same thoughts about what could happen back in the clubhouse.[64]

In every religious tradition there is lore about saints and sages who possess mysterious powers of communication. According to his friends and fellow mendicants, Saint Francis of Assisi spoke with creatures of the forest. Yogis, it is said, can tame animals by their contagious serenity, know their intentions without sensory cues (by the *pasutwa* siddhi), and communicate with them with animal signs (by the *jantujna* siddhi). And if we can believe accounts such as those recorded in the *Historia Monachorum* (see chapter 21.2), contemplatives can reach out telepathically to nurture troubled souls.[65]

Telepathic transmission of illumined states. Some yogis and saints, it is said, can transmit states of mystical ecstasy by extrasensory influence. In the lore of Hinduism, such transmission, or *diksha,* can be effected by a glance, a gaze, a touch, a word, an embrace, or other gesture, as well as by communication that does not depend upon sensory cues. It can even occur when the recipient is physically distant or unaware it is happening. Sri Aurobindo, a yogi as well as a philosopher, wrote to a disciple: "My touch is always there; but you must learn to feel it not only with the outward contact as a medium . . . but in its direct action on the mind and heart and body."[66] The Bengali mystic Sri Ramakrishna initiated his famous disciple Narendra (later Swami Vivekananda) by a direct transmission of this kind. According to an account by Swami Nikhilananda:

Narendra, because of his Brahmin upbringing, considered it wholly blasphemous to look on man as one with his Creator. One day at the temple garden he said to a friend: "How silly. This jug is God. This cup is God. Whatever we see is God. And we too are God. Nothing could be more absurd." Sri Ramakrishna came out of his room and gently touched him. Spellbound [Narendra] immediately perceived that everything in the world was indeed God. A new universe opened around him. Returning home in a dazed state, he found there too that the food, the plate, the eater himself, the people around him, were all God. When he walked in the street, he saw that the cabs, the horses, the streams of people, the buildings, were Brahman. He could hardly go about his day's business. . . . And when the intensity of the experience abated a little, he saw the world as a dream. Walking in the public square, he would strike his head against the iron railings to know whether they were real. It took him a number of days to recover his normal self. He had a foretaste of the great experiences yet to come and realized that the words of the Vedanta were true.[67]

Shared ecstasy and illumination. Spiritual transmission is not always a one-way affair. Contemplatives, lovers, and friends have shared extraordinary states that arose spontaneously among them. The poet W. H. Auden described his own experience of this kind:

One summer night in June 1933 I was sitting on a lawn after dinner with three colleagues, two women and one man. We liked each other well enough but we were certainly not intimate friends, nor had any one of us a sexual interest in another. (Incidentally, we had not drunk any alcohol.) We were talking casually about everyday matters when, quite suddenly and unexpectedly, something happened. I felt myself invaded by a power which, though I consented to it, was irresistible and certainly not mine. For the first time in my life I knew exactly— because, thanks to the power, I was doing it—what it means to love one's neighbor as oneself. I was also certain, though the conversation continued to be perfectly ordinary, that my three colleagues were having the same experience. (In the case of one of them, I was able later to confirm this.) My personal feelings towards them were unchanged— they were still colleagues, not intimate friends—but I felt their existence as themselves to be of infinite value and rejoiced in it.[68]

In his *Watcher on the Hill: A Study of Some Mystical Experiences of Ordinary People,* physicist-philosopher Raynor Johnson presented the following account of simultaneous illumination.

There is a girl who has done me the greatest injury that one woman can do another, yet some higher power seemed to demand that I help this girl, even against my own will and my own material interests.

We were standing one day talking earnestly, when I felt this sense of power and began to speak. . . . I was in the doorway between two rooms; she was looking out at me, when suddenly she said, "When you had your hair shampooed, did you have a blue rinse?" I thought this was rather a prosaic interruption to a serious conversation, but I replied: "No, why do you ask?" She answered in a puzzled kind of way: "Because there is a blue light round your hair, and it is beautiful." . . . We both became silent, the vibration grew more powerful, circulating between us, and in that moment I felt compassion welling up in my heart, the forgiveness of Christ, and I knew that she felt it too, though she did not know what she felt. She flung her arms wide, her face upturned in a kind of ecstasy, . . . and she stammered: "It's wonderful . . . it's marvelous." What was working through me, I don't know, I can only guess, but I went and put my arms round her and said: "It's all right," and she replied: "I have never felt so wonderful, so happy." It affected us both profoundly.[69]

To this list of communication abilities, others could be added. It is said, for example, that some yogis, shamans, and saints can telepathically convey volitions and impulses. People in many cultures have received new vitality at a distance from healers or friends. Indeed, it is held in most sacred traditions that virtually any capacity can be communicated without sensory cues.[†] Such capacities, however, like other exceptional attributes discussed in this book, can be used destructively. The same religious traditions that celebrate metanormal transmission of illumined states also bear witness to communication abilities employed for egocentric, bullying, even monstrous purposes. There is a lore in virtually every religious culture about adepts who use their special powers of communication for selfish ends. This lore, it seems to me, is supported by modern research on suggestion at a distance. Frederic Myers, for example, described experiments observed by Pierre Janet and other trustworthy medical people during which an unsuspecting subject behaved in apparent compliance with hypnotic commands given to her from another location (see Appendix A.7). Such studies, and others like them conducted during the last hundred years, suggest that it is possible to manipulate others telepathically.[70] Furthermore, there is good reason to think that extrasensory communication operates destructively in everyday life. In chapter 17.3, I cite Sigmund Freud, Wilhelm Stekel, Jule Eisenbud, Jan Ehrenwald, and other psychotherapists who have argued from their own therapeutic experience that telepathic interaction frequently serves malicious impulses.

[†] Here again, extraordinary capacities can be assigned to more than one set of attributes. The communication of volitions or vitality, for example, can also be considered types of influence upon the environment (see 5.6).

[5.4]
VITALITY

Among evolutionary developments that can be called progressive, paleontologist George Gaylord Simpson included general increase of available energy, or warm-bloodedness. The typical mammalian system, he wrote, "has a higher vital minimum which is maintained at almost constant levels" that gives mammals independence from environmental changes not enjoyed by reptiles or amphibians.[71] This capacity , which we inherit from our primate ancestors, can be strengthened by various practices. Somatic disciplines and psychotherapy, for example, by resolving internal conflicts, unblocking defenses against strong feeling, and reducing chronic muscular tensions, can markedly increase the energy available for mental and physical activity (17.1 & 18). By improving blood circulation and cardiopulmonary capacity and by making metabolism more efficient, fitness training can increase vitality (19.2). By promoting mental alertness and emotional balance in stressful situations, the martial arts bring new liveliness to their practitioners (20). And religious practices, by liberating their subjects from draining worries and hostilities while helping to integrate their conflicting volitions, often produce an exceptional life-force. The ability to endure environmental extremes evident in certain shamans (21.3); the great fortitude of Christian desert fathers (21.2); the *incendium amoris* of Catholic mystics (22.6); the "boiling *n/um*" of the Kalahari bushmen (13.2); the resistance to cold generated by Tibetan yogis; and the kundalini effects triggered by some forms of yoga, all involve extraordinary gains in energy level.[72]

New energies triggered by the practices just noted confer marked independence from environmental vicissitudes and support great spiritual, mental, or physical activity. These increases of vital capacity, of course, result in part from processes understood by mainstream science. Athletic fitness involves cardiopulmonary changes described by contemporary medicine. *Tumo* is caused partly by vasodilation. "Boiling *n/um*" is produced to some extent by the violence of the bushmen's dancing. Every instance of vital increase among sportspeople and religious adepts, it seems to me, is caused to a large extent by processes that operate in ordinary human functioning. However, factors unspecified by contemporary science may also contribute to some of these energetic mutations. *Incendium amoris,* kundalini, *tumo,* boiling *n/um,* shamanic fortitude, and even the release of new energies in sports, are sometimes associated with religious ecstasies, apparent psychokinesis, and mystical insight. Given the fact that they trigger superordinary powers and states of mind, it is conceivable that *tumo, n/um,* and the like involve processes beyond those evident in normal functioning. And indeed, that is what shamans, yogis, bushmen, Catholic saints, and even some sportspeople say. "A man's spiritual consciousness is not awakened," said Sri Ramakrishna, "unless his kundalini is aroused."[73] Kung bushmen told anthropologist Richard Katz that *n/um* was "given by the gods."[74] For numerous Catholic saints, the *incendium amoris* is *gratie gratis datae,* a gift of the Holy Spirit.

In Dobu, wrote Mircea Eliade, magical heat accompanied the practice of sorcery, as it did on Rossel Island and the Solomons. The same ideas, Eliade continued,

> have survived in more complex religions. The Hindus give an especially powerful divinity the epithets *prakhar,* "very hot," *jajval,* "burning," or *jvalit,* "possessing fire." The Mohammedans of India believe that a man in communication with God becomes "burning." A man who performs miracles is called *sahib-josh, josh* meaning "boiling."
>
> All these myths and beliefs have their counterparts, we should note, in initiatory rituals that involve a real mastery over fire. The future Manchurian or Eskimo Shaman, like the Himalayan or tantric yogin, must prove his magical power by resisting the most severe cold or drying wet sheets on his bare body.[75]

In certain Hindu-Buddhist yoga traditions, the ability to withstand cold can be tested by sitting naked in snow.[76] In Tibet this ability, called *tumo,* is practiced by monks who wrap themselves in wet sheets at high altitudes. To explore such capacity, Herbert Benson of the Harvard Medical School and a group of colleagues tested three Tibetan lamas in Upper Dharmsala, India, in 1981, and subsequently described their experiments in the British journal *Nature.* During a 55-minute period of *tumo* meditation to raise his internal heat, their first subject, aged 59, exhibited a 5.9°C increase in finger temperature and a 7.0°C increase in toe temperature. His navel and lumbar temperatures increased by 1°C during the meditation period and his rectal temperature remained unchanged, while the air temperature in his meditation room increased from 22.0° to 23.5°C. During meditation, this subject's heart rate increased by two beats per minute, then returned to baseline levels. Finger temperature in the second subject, aged 46, increased 7.2°C, reaching its highest levels during his recovery period, at a time when he said that his experience of *tumo* continued though he had stopped meditating. His toe temperature increased by 4°C and stayed at that level after his 85-minute meditation was done, while his navel temperature increased 1.9°C and his nipple temperature 1.5°C, though his rectal temperature remained unchanged. Air temperature rose in his room as well, from 16 to 19.2°C. This lama's heart rate was essentially unchanged during his meditation and recovery periods. The third subject, aged 50, was studied in a cool hotel room in Upper Dharmsala, and exhibited a rise in finger temperature of 3.15°C and in toe temperature of 8.3°C, while his rectal temperature was unchanged. He demonstrated no other large changes in skin temperature, however, while air temperature in his room dropped from 20 to 18.5°C during his 40-minute meditation, then returned to 19.5°C during his recovery period. His heart rate dropped after he assumed his seated meditation position and continued to fall slowly during the entire experiment. Given the lamas' leanness and their normal resting heart rates during the experiment, Benson and his colleagues concluded that "the most likely mechanism to account for [their] increase in finger and toe temperature [was] vasodilation."[77]

The changes in temperature demonstrated by Benson's three lamas were not as dramatic as the mutations of bodily heat attributed to some saints and mystics. According to numerous witnesses, for example, the Catholic saint Philip Neri often felt a burning love of God, the *incendium amoris*, in the region of his heart. Herbert Thurston wrote:

[I]t sometimes extended over his whole body, and for all his age, thinness and spare diet, in the coldest days of winter it was necessary, even in the midst of the night, to open the windows, to cool the bed, to fan him while in bed, and in various ways to moderate the great heat. Sometimes it burned his throat, and in all his medicines something cooling was generally mixed to relieve him. Cardinal Crescenzi said that sometimes when he touched his hand, it burned as if the saint was suffering from a raging fever. . . . Even in winter he almost always had his clothes open from the girdle upwards, and sometimes when they told him to fasten them lest he should do himself some injury, he used to say he really could not because of the excessive heat he felt. One day, at Rome, when a great quantity of snow had fallen, he was walking in the streets with his cassock unbuttoned; and when some of his penitents who were with him were hardly able to endure the cold, he laughed at them and said it was a shame for young men to feel cold when old men did not.

There can be little doubt that the discovery which was made in the autopsy performed after St. Philip's death must be closely connected with the same intense fervour of divine love. During more than fifty years of his long life, he had suffered from a strange and inexplicable palpitation of the heart, which was noticed, not only by himself, but by many of his companions and friends whom in the tenderness of his affection for their souls he often pressed to his bosom. The surgeons, when they opened his body, found a swelling under his left breast, which proved to be due to the fact that two of his ribs were broken and thrust outwards. In view of the positive testimony of the surgeons, there can be no dispute that the injury was there and had been there for many years. His biographers seem therefore fully justified in tracing it to that strange incident of the coming to him of the Holy Ghost in 1544 under the guise of a glove of fire. "Thereupon," we are told, "he was suddenly surprised by such an ardour of love that, unable to bear it he threw himself down upon the ground, and, like one trying to cool himself, bared his breast, to temper in some measure the flame which he felt." Certain it is in any case that from that time forth his body was liable in moments of deep emotional feeling to tremble convulsively with intense palpitations, while he became conscious of the presence of a swelling on the left breast, the size of a man's fist. This he retained for all the rest of his life. [The fullest account of the autopsy will be found in the *Life* by Father Gallonio, St. Philip's intimate friend and disciple. See *Acta Sanctorum*, May, Vol. vi., p. 510.] It is curious that a displacement of the ribs, similar in cause and character, but apparently less in degree, is recorded in the case of St. Paul of the Cross, the founder of the Passionists, who lived two centuries later. An even more striking modern example is that of Gemma Galgani who died at Lucca in 1903.[78]

The Renaissance saint Catherine of Genoa was also consumed by the *incendium amoris,* and suffered bodily dislocations like St. Philip Neri's. Drawing upon various accounts of her life, Thurston wrote that

on August 28th, when the tragedy of suffering began to near its end, she was again all on fire. She cried aloud that "all the water which the world contains could not give me the least refreshment." Later her tongue and lips became so parched with the burning heat within that she could not move them or speak. At such times, if anyone touched a hair of her head, or even the edge of the bed, or the bedclothes, she would scream as if she had been stabbed. Her confessor sometimes hesitated to bring her Communion in this state, for she could swallow nothing, neither food or drink, "but, with a joyous face, she would make him a sign that she was not afraid, and then, when she had received, she remained with her countenance glowing and rosy, like that of a Seraph."[79]

Similar mutations of vitality occur among the Kung bushmen of the Kalahari desert (see chapter 13.3). "You dance, dance, dance," a Kung healer told anthropologist Richard Katz. "Then *n/um* lifts you up in your belly and lifts you in your back, and then you start to shiver. [It] makes you tremble, it's hot. *N/um* enters every part of your body right to the tip of your feet and even your hair."[80]

And the same type of experience is reported by contemporary Westerners. Psychiatrist Lee Sannella described this episode reported to him by a woman acquaintance:

> In 1973 [a] woman then in her 41st year [who had been engaged in various intensive group and meditative disciplines] noted the onset of heat in her head and chest while meditating. . . . There was a prickly itching heat all over her body, but she was not troubled because she believed that these sensations indicated successful and centered meditations and a flow between herself and others. She assumed that she was having the rise of the kundalini which she believed to be dangerous unless the "higher mind" was in control.
>
> A few months after this onset, during meditations, she felt as if she were two feet taller than her normal self and as if her eyes were looking out from above her head. At this time she was sure that she knew what people were thinking and many of her impressions were confirmed.
>
> She felt heat on one side of her back, and was convinced that unless it spread to both sides she would be in danger. Once she succeeded in spreading it, this crisis passed.
>
> Then a tingling started to move from her pelvis up her back to her neck. She began to see light inside her head. She was amazed to find that she could see this light all the way down her spine as well. The energy and tingling moved over her forehead and became focused under her chin. She felt as if there were a hole in the top of her head. Sleep was very difficult, and for the next six weeks meditation was the only thing that helped her. She felt that if she did not meditate, the heat flowing in her body would grow so intense as to damage her. Other people could feel excessive heat when they touched her lower back.[81]

Experiences of this kind are too charged with a special energy, excitement, and vision, I believe, to be dismissed as products of psychoneurosis, exercise, vasodilation, or social suggestion. They have an autonomous power, a relentless and overwhelming intensity that alters the body, brings their recipient ecstasy, and confers special powers; and their host traditions consistently attribute them to superordinary agencies. *N/um* is "given by the gods." The *incendium amoris* appears in response to Christ's "living presence." Kundalini is a secret power of *shakti,* the immanent divinity. All of these phenomena are sharply differentiated by those who experience them from ordinary mental or somatic events, and in their respective traditions are ascribed to forces beyond ordinary functioning.

But inexplicable energies also arise without benefit of religious discipline. In crisis, in sport, and in ordinary circumstances, unaccountable forces animate certain people, enabling them to meet unusual challenges or exceed themselves in astonishing ways. Psychiatrist Lee Sannella has described many experiences of this kind;[82] and accounts of similar episodes are frequently reported by martial artists and athletes.[83] Quarterback John Brodie, for example, wrote that

> there are times when an entire team will leap up a few levels. Then you feel that tremendous rush of energy across the field. When you have eleven men who know each other well and have every ounce of [intention] focused on a common goal, and all their energy flowing in the same direction, this creates a very special concentration of power. Everyone feels it. The people in the stands feel and respond to it, whether they have a name for it or not.[84]

Like other kinds of extraordinary functioning, such mutations of vital capacity might be more common than most of us think.

[5.5]
MOVEMENT ABILITIES

The mobility we inherit from our animal ancestors has several types of metanormal analogue, which I will group here into four classes:

- extraordinary physical movement;
- levitation, for which the evidence is uncertain;
- out-of-body experience, or traveling clairvoyance, which has been testified to in every culture since antiquity; and
- movement into "other worlds" during exceptionally vivid dreaming, near-death episodes, mediumistic trance, or religious vision.

Extraordinary Physical Movement

Since antiquity, people have found many ways to improve their movement abilities. Indeed, the human drive for physical self-exceeding is so great that it sometimes becomes a religious passion. Amerindian

runners, Tibetan yogis, Taoist monks (21.4 & 21.5), and martial artists (20) have drawn upon spiritual practices to move in new ways, and
in so doing have often surpassed their apparent limits of physical capacity. In this worldwide effort of self-exceeding, shamans, yogis,
martial artists, and modern athletes have discovered new agility,
strength, and coordination that they attribute to ego-transcendent
forces. Swordsmanship, for the Japanese swordmaster Yagyu Tajima,
became an activity in which Original Mind, or Emptiness, *k'ung,* or
sunyata, could "grasp with human hands, walk with human feet, see
with human eyes" (20.1). Native American runners have said that
gods or animal powers helped them run beyond their normal abilities.
The *lung-gom* practices of Tibetan yoga, it is said, can transform ordinary movement into something superhuman.[†] Lama Govinda, a European scholar of Tibetan contemplative practice, wrote that *lung-gom*
can be interpreted to mean

> "concentration on the dynamic vital principle." It reveals the dynamic
> nature of our physical organism and of all material states of aggrega-
> tion—not in the sense of a self-sufficient dynamism, but as something
> that depends on the co-operation and interaction of various forces and
> ultimately on the fundamental (and universal) faculties of conscious-
> ness. Thus, a direct influence is possible upon the bodily functions and
> their respective organs, so that a psycho-physical co-operation is estab-
> lished: a parallelism of thought and movement, and a rhythm that
> gathers all available forces into its service. If one has reached the point
> where the transformation of one force or state of materialization into
> another one is possible, one may produce various effects of an appar-
> ently miraculous nature, as, for instance, the transformation of psychic
> energy into bodily movement (a miracle that we perform on a smaller
> scale every moment, without being conscious of it), or the transforma-
> tion of matter into an active state of energy, resulting at the same time
> in a reduction of weight or the apparent elimination or reduction of
> the power of gravitation.[85]

Lama Govinda observed cells where such training was conducted. At the monastery of Nyang-to Kyi-phug, for example, seven
Buddhist monks were sealed inside small enclosures so that they could
not leave without their superior's permission. At the time of his visit,
Lama Govinda wrote,

[†] *Lung-gom* refers to yogic practices involving bodily movement. It is a term derived from the
Tibetan words *lung,* which signifies air, vital heat, or energy, and *gom,* meditation upon a particular thing until identity with it is realized.

one had been shut in for three years already. He was expected to leave his meditation cubicle in six years. Nobody is allowed to speak to the *lung-gom-pa* or to see any part of his person. The latter rule is to ensure his complete anonymity. When receiving alms through an opening near the bottom of the wall, next to the sealed entrance, even the hand of the hermit is covered with a sock or a cloth bag, so that he may not be recognized even by a scar or any other particular sign or shape of his hand. The same small opening, which I measured as being 9 in. × 10 in., is said to be used as an exit by the *lung-gom-pa* after completion of his nine years' practice in uninterrupted seclusion and perfect silence. It is said that his body by that time has become so light and subtle that he can . . . perform [a] pilgrimage to all the main shrines and sanctuaries of Central Tibet within an incredibly short time.[86]

Another student of Tibetan mysticism, Alexandra David-Neel, claimed to have seen a *lung-gom* adept on the northern plain of Tibet who appeared to be making the kind of pilgrimage Lama Govinda described. The man proceeded with an unusual gait, his gaze fixed on the distant horizon. "[He] did not run," David-Neel wrote. "[But] seemed to lift himself from the ground as if he had been endowed with the elasticity of a ball. His steps had the regularity of a pendulum." The Tibetans in David-Neel's party immediately identified the man as a *lama long-gom-pa* and bowed in respect as he passed them. With binoculars, David-Neel watched the man disappear into the mountains that bordered the steppe. Four days later, herdsmen told David-Neel's party that they had seen the *lung-gom-pa,* and David-Neel was able to estimate the speed at which he was traveling. To have reached the herdsmen when he did, the monk had moved rapidly through the mountains for two days without stopping.[87]

In recent years, many athletes have described "the zone," a condition in which their physical skills reach superordinary levels. Again, people with different temperaments and backgrounds claim that a particular capacity—in this case for bodily movement—is at times inspired by something beyond their ordinary functioning.

Levitation

Levitation—if it occurs—would transcend ordinary human movement and operate in ways unspecified by standard science, but no instance of it has been recorded with movie cameras. There are photographs of people said to be hovering off the ground, some of them produced by the Transcendental Meditation Society to publicize its siddhis courses, but none of these proves that the subjects were doing anything more than jumping. Nevertheless, belief in the phenomenon has persisted for at least two millennia in the shamanic and religious traditions. *Vayu stambham* is an ancient Sanskrit term for levitation; and shamans in many parts of the world have been said to rise from the ground in defiance of gravity.[88] But in spite of the worldwide lore about levitation, no one has levitated upon command for modern investigators. For example, when Herbert Benson and a group of colleagues (with the Dalai Lama's help) observed lamas who reputedly could rise from the ground supernormally, their attempt to confirm levitation failed completely. Benson wrote:

Two monks, the Venerable K.G., aged seventy, and the Venerable T.O., aged thirty-four, demonstrated their practice. They dressed only in loincloths and sat in a cross-legged position upon a small pile of carpets. They carried out a number of physical exercises in unison, including deep breathing; slapping their hands against their own chests, arms, and legs; and swaying. As these were performed, they chanted. They stood, rapidly crossed their legs into a cross-legged position, and fell to the ground with their legs remaining crossed. As they landed on the carpets, they slapped the outer portions of their legs outward and downward, thus creating a very loud noise. Then, as they sat crosslegged, they sucked their abdomens into their rib cages and produced a large cavity below their ribs. At this point, the older monk said that the younger would proceed alone because what was to follow was too strenuous for his age.

The young monk stood, bent his knees slightly, and jumped three to four feet into the air, with his legs straight. While in the air, he rapidly assumed the cross-legged position and fell to the ground while maintaining this position. He landed with a resounding crash as he slapped his crossed legs outward and downward. The monks completed their ritual by sitting together and apparently repeating many of their previous actions.

We asked whether what we had seen was the so-called "levitation meditation" and were told that indeed it was. We had witnessed a remarkable athletic performance, but there was no floating or hovering. The fact that only the younger monk could perform the exercise also suggested it was athletic, not spiritual. If spiritual, one would have suspected the older monk would be more proficient.

I later asked them whether it was possible to stay up in the air. The younger monk said his great-grandfather had been able to do so, but he knew of no one doing it today. I then asked the older man if he knew of anyone who could carry out such a feat. He said it was an ability that was present many hundreds of years ago, but not today. I asked the older monk if he would like to levitate, and with a twinkle in his eyes, he responded, "There is no need. We now have airplanes."[89]

No modern investigator has produced conclusive evidence that levitation occurs. Still, gravity defiance has been celebrated in virtually every religious tradition. It has been described, for example, by Roman Catholic authorities of the contemplative life and is discussed at length in some of the church's canonization records. Reputable witnesses swore under oath that they had seen Saint Joseph of Copertino rise from the ground without visible support, often before several onlookers (see chapter 22.10). Such testimony convinced Pope Benedict XIV, who established most of the criteria by which the church came to judge saintly marvels, that Joseph had actually levitated. "Eyewitnesses of unchallengeable integrity," Pope Benedict wrote, "gave evidence of the famous upliftings, and prolonged flights of [Saint Joseph] when rapt in ecstasy."[90]

Saint Teresa of Avila, renowned for her integrity and strong common sense, wrote that her raptures were sometimes so great that "my whole body has been . . . raised up from the ground."[91] Saint Teresa's words were definitely authored by the saint, as they appear in a facsimile edition of her autobiography written in her own hand as it was submitted to the Inquisition in her lifetime. Furthermore, depositions of various observers agree with her account. Testifying before the Congregation of Rites, 10 witnesses according to the *Acta Sanctorum* described their firsthand observations of Saint Teresa's levitations, sometimes in vivid detail. "As I was looking on," testified Sister Anne of the Incarnation, "[Saint Teresa] was raised . . . from the ground without her feet touching it. At this I was terrified and she, for her part, was trembling all over. [Afterward] she asked me . . . whether I had been there all the while. I said yes, and then she ordered me under obedience to say nothing of what I had seen."[92] In *The Physical Phenomena of Mysticism*, Herbert Thurston presented a list of levitated ecstatics and references to more detailed descriptions of them (see appendix K.2).

Taken together, the testimonies of Catholic saints and many who observed them provide a considerable witness to levitation in fairly recent times. Given such testimony and the great body of lore about the phenomenon in other sacred traditions, we have reason for caution in deciding whether humans may upon occasion rise supernormally from the ground. Still, we have no movies, television tapes, or other records that skeptics can trust. Until we do, many of us will consider levitation to be possible but unproven. Someday, perhaps, a high jumper, wide receiver, or basketball star might exceed gravity in front of television cameras. Maybe a highly trained dancer will leap so high—or stay suspended in midair so long—that videotapes of their feat will provide indisputable evidence for levitation. Author James Michener, for example, described a play he saw Hank Luisetti make in 1941.

> Somehow, Luisetti stayed up in the air, faked a shot at the basket, made the Denver center commit himself, and with a movement I had never seen before, simply extended his right arm an extra foot and banked a one-handed shot gently against the backboard and into the basket. It seemed as if he had been in the air a full minute, deceiving three different players, and ending with a delayed shot that was staggering in its beauty.[93]

With similar wonderment, a reporter for *Time* magazine described the dancing of Mikhail Baryshnikov.

> When he launches his perfectly arched body into the arc of one of his improbably sustained leaps—high, light, the leg beats blurring precision—he transcends the limits of physique and, it sometimes seems, those of gravity itself. If one goes by the gasps in the theater or the ecstasies of the critics, such moments turn Mikhail Baryshnikov, if not into a minor god, then into a major sorcerer.
>
> He is an unbelievable technician with invisible technique. Most dancers, even the great ones, make obvious preliminaries to leaps. He simply floats into confounding feats of acrobatics and then comes to still, collected repose. He forces the eye into a double take: did that man actually do that just now?[94]

Does a slight levitation occur in feats such as Luisetti's and Baryshnikov's? Might the long training, special gifts, and passionate commitment of certain athletes and dancers trigger extraordinary energies, or otherwise reconstitute relations between the elements to produce moments of gravity defiance? Friends of Nijinsky believed that the great dancer was the vehicle of occult powers; and basketball star Julius Erving entertained the idea that play "above the basket" occasionally transcends the constraints of gravity.[95]

Out-of-Body Experience

Out-of-body experiences during which the subject seems to perceive the physical environment have been described in many cultures and were studied by the early psychical researchers (see appendix E). In his *Human Personality,* for example, Frederic Myers presented the following account by one S. R. Wilmot, which was originally reported by Mrs. Henry Sidgwick in the *Proceedings of the Society for Psychical Research,* Volume 7:

> On October 3rd, 1863, I sailed from Liverpool for New York, on the steamer *City of Limerick,* of the Inman line, Captain Jones commanding. On the evening of the second day out, soon after leaving Kinsale Head, a severe storm began, which lasted for nine days. During this time we saw neither sun nor stars nor any vessel; but bulwarks on the weather bow were carried away, one of the anchors broke loose from its lashing, and did considerable damage before it could be secured, and several stout storm sails, though closely reefed, were carried away, and the booms broken.
>
> Upon the night following the eighth day of the storm the tempest moderated a little, and for the first time since leaving port I enjoyed refreshing sleep. Toward morning I dreamed that I saw my wife, whom I had left in the United States, come to the door of my state-room, clad in her night-dress. At the door she seemed to discover that I was not the only occupant of the room, hesitated a little, then advanced to my side, stooped down and kissed me, and after gently caressing me for a few moments, quietly withdrew.
>
> Upon waking I was surprised to see my fellow-passenger, whose berth was above mine, but not directly over it—owing to the fact that our room was at the stern of the vessel—leaning upon his elbow, and looking fixedly at me. "You're a pretty fellow," said he at length, "to have a lady come and visit you in this way." I pressed him for an explanation, which he at first declined to give, but at length related what he had seen while wide awake, lying in his berth. It exactly corresponded with my dream.
>
> This gentleman's name was William J. Tait, and he had been my roommate in the passage out, in the preceding July, on the Cunard steamer *Olympus;* a native of England, and son of a clergyman of the Established Church. He had for a number of years lived in Cleveland, in the State of Ohio, where he held the position of librarian of the Associated Library. He was at this time perhaps fifty years of age, by no means in the habit of practical joking, but a sedate and very religious man, whose testimony upon any subject could be taken unhesitatingly.
>
> The incident seemed so strange to me that I questioned him about

it, and upon three separate occasions, the last one shortly before reaching port, Mr. Tait repeated to me the same account of what he had witnessed. On reaching New York we parted, and I never saw him afterward.

The day after landing I went by rail to Watertown, Connecticut, where my children and my wife had been for some time visiting her parents. Almost her first question when we were alone together was, "Did you receive a visit from me a week ago Tuesday?" "A visit from you?" said I, "we were more than a thousand miles at sea." "I know it," she replied, "but it seemed to me that I visited you."

My wife then told me that on account of the severity of the weather and the reported loss of the *Africa,* which sailed for Boston on the same day that we left Liverpool for New York, and had gone ashore at Cape Race, she had been extremely anxious about me. On the night previous, the same night when, as mentioned above, the storm had just begun to abate, she had lain awake for a long time thinking of me, and about four o'clock in the morning it seemed to her that she went out to seek me. Crossing the wide and stormy sea, she came at length to a low, black steamship, whose side she went up, and then descending into the cabin, passed through it to the stern until she came to my state-room. "Tell me," said she, "do they ever have state-rooms like the one I saw, where the upper berth extends further back than the under one? A man was in the upper berth, looking right at me, and for a moment I was afraid to go in, but soon I went up to the side of your berth, bent down and kissed you, and embraced you, and then went away."

The description given by my wife of the steamship was correct in all particulars, though she had never seen it. I find by my sister's diary that we sailed October 4th; the day we reached New York, 22nd; home, 23rd.[96]

Wilmot's sister, who was also aboard the ship, recounted her own conversation with Tait in which he described the incident as her brother reported it; and Wilmot's wife provided written confirmations of her out-of-body experience. In considering such incidents, we need not believe that the people involved actually traveled in some sort of spirit-body. This case is compelling, however, because four people were involved, three of them directly, and their experiences corresponded in striking ways. Reports such as the Wilmots', moreover, have prompted experiments in which an agent deliberately tries to appear as a phantom figure to someone ignorant that they are the designated target. Edmund Gurney reported experiments of this kind in *Phantasms of the Living.* The following narrative was copied by Gurney from an account by a Mr. S. H. B.

On a certain Sunday evening in November 1881 having been reading of the great power which the human will is capable of exercising, I determined with the whole force of my being that I would be present in spirit in the front bedroom on the second floor of a house situated at 22 Hogarth Road, Kensington, in which room slept two ladies of my acquaintance, viz., Miss L. S. V. and Miss E. C. V., aged respectively 25 and 11 years. I was living at this time at 23 Kildare Gardens, a distance of about three miles from Hogarth Road, and I had not mentioned in any way my intention of trying this experiment to either of the above ladies, for the simple reason that it was only on retiring to rest upon this Sunday night that I made up my mind to do so. The time at which I determined I would be there was 1 o'clock in the morning, and I also had a strong intention of making my presence perceptible.

On the following Thursday I went to see the ladies in question, and in the course of conversation (without any allusion to the subject on my part) the elder one told me that on the previous Sunday night she had been much terrified by perceiving me standing by her bedside, and that she screamed when the apparition advanced towards her, and awoke her little sister, who saw me also.

I asked her if she was awake at the time and she replied most decidedly in the affirmative, and upon my inquiring the time of the occurrence, she replied, about 1 o'clock in the morning. This lady, at my request, wrote down a statement of the event and signed it.

This was the first occasion upon which I tried an experiment of this kind, and its complete success startled me very much. Besides exercising my power of volition very strongly, I put forth an effort which I cannot find words to describe. I was conscious of a mysterious influence of some sort permeating my body, and had a distinct impression that I was exercising some force with which I had been hitherto unacquainted, but which I can now at certain times set in motion at will.

The targets of this experiment, Misses L. S. and E. C. Verity, in writing confirmed the experiences Mr. S. H. B. described, and Gurney carefully cross-examined them. Gurney wrote:

There is not the slightest doubt that their mention of the occurrence to S. H. B. was spontaneous. They had not at first intended to mention it; but when they saw him their sense of its oddness overcame their resolution. Miss Verity is a perfectly soberminded and sensible witness, with no love of marvels, and with a considerable dread and dislike of [them].

At Gurney's prompting, Mr. S. H. B. tried again to project himself into the Verity sisters' presence. Miss L. S. Verity subsequently made this statement:

44 Norland Square, W

On Saturday night, March 22nd, 1884, at about midnight, I had a distinct impression that Mr. S. H. B. was present in my room, and I distinctly saw him whilst I was quite widely awake. He came towards me, and stroked my hair. I voluntarily gave him this information, when he called to see me on Wednesday, April 2nd, telling him the time and the circumstances of the apparition, without any suggestion on his part. The appearance in my room was most vivid, and quite unmistakable.

Mr. S. H. B's own account was as follows:

On Saturday, March 22nd, I determined to make my presence perceptible to Miss V., at 44 Norland Square, Notting Hill, at 12 midnight, and as I had previously arranged with Mr. Gurney that I should post him a letter on the evening on which I tried my next experiment (stating the time and other particulars), I sent a note to acquaint him with the above facts. About ten days afterwards I called upon Miss V., and she voluntarily told me, that on March 22nd, at 12 o'clock midnight, she had seen me so vividly in her room (whilst widely awake) that her nerves had been much shaken, and she had been obliged to send for a doctor in the morning.[97]

Experiences such as the Wilmots' and the Veritys' resemble a phenomenon long reported in the contemplative traditions, namely bilocation, the simultaneous appearance by a yogi or saint in more than one location. In *The New Catholic Encyclopedia,* bilocation is listed among the principal charisms of Christian devotion. It is also a result of the *akasha* and *moksha* siddhis of Indian yoga and is described in the lore of Taoism, Tibetan Buddhism, and Sufism. Rejecting the superstitious belief that such experience involves the physical body's existence in two places at once, we find witness to traveling clairvoyance among both modern and ancient peoples. But as we have seen, such events are not limited to religious adepts. They frequently occur spontaneously, during illness, near-death experience, vivid dreams, or intense adventure. A rock climber, for example, claimed that

about 15 or 20 feet above the ground, I slipped and fell. As I fell, I seemed to be about 5 or 10 feet out from the face, looking at my body falling. I vaguely recollect moving around to the other side of my body to look at it. Once I hit the ground, I was immediately preoccupied with my pain.[98]

A man who had been caught in a strong undertow described a similar experience. After a desperate struggle, he felt too tired to care.

> I felt peace settle over me. Well, I thought, I had tried, and I was so tired. It seemed that a wonderful transition occurred. I was high above the water looking down upon it. The sky, that had been so grey and lowering, was iridescent with indescribable beauty. There was music that I seemed to feel rather than hear. Waves of ecstatic and delicate color vibrated around me and lulled me to a sense of peace beyond comprehension.
>
> In the water beneath me, a boat came into view, with two men and a girl in it. Then I saw a blob of something in the water. A wave tossed it and rolled it over. I found myself looking into my own distorted face. What a relief, I thought, that that ungainly thing was no longer needed by me. Then men lifted the form into the boat, and my vision faded. The next thing I knew, it was dark and I was lying on the beach, cold and sick and sore. Men were working over me. I was told later that they worked over me for more than two hours.[99]

Movement into Other Worlds

Journeys into extraphysical worlds comprise yet another class of extraordinary locomotion. Like traveling clairvoyance, such experience has been studied by modern psychical researchers. The famous British medium Stainton Moses claimed, for example, that

> once or twice—once very lately in the Isle of Wight—my interior dormant faculties awoke and I lost the external altogether. For a day and a night I lived in another world, while dimly conscious of material surroundings. I saw my friends, the house, the room, the landscape, but dimly. I talked, and walked, and went about as usual, but through all, and far more clearly, I saw my spiritual surroundings, the friends I know so well, and many I had never seen before. The scene was clearer than the material landscape, yet blended with it in a certain way. I did not wish to talk. I was content to look and live among such surroundings. It was as I have heard Swedenborg's visions described.[100]

Moses's experience resembles Charles Lindbergh's visions during his epic crossing of the Atlantic (see chapter 5.1). Lindbergh, however, did not describe his visions in print until 26 years after the flight. A similar reticence prompted Emmanuel Swedenborg, the Swedish scientist and philosopher (1688–1772), to retire from public life to explore the extramundane worlds revealed by his intense introspection. Swedenborg described his spiritual excursions in some 282 Latin works, and like Stainton Moses a hundred years later, appeared to have clairvoyant powers associated with such experience. His visionary witness has influenced many artists and thinkers, among them Henry James, Sr., William Blake, and Baudelaire.[101]

There is a huge modern literature about extraphysical worlds, which each of us can assess for ourselves. It is highly uneven, of course, ranging from pure fabrication to sincere accounts such as Lindbergh's, but in it we can find many correspondences with shamanic-contemplative lore. In the yoga tradition, for example, the *manomaya* and *akashaloka* siddhis provide access to other dimensions of the universe. Siberian shamans go down to the underworld, ascend the sky, and enter worlds of ancestral spirits. In Iranian mysticism, Hurqalya, the celestial earth, is accessible to spiritual travel.[102] For a representative list of accounts regarding travel in other worlds, see appendix E.

In summary, then, we can identify four kinds of metanormal movement: the extraordinary agility attributed to certain athletes, dancers, shamans, and saints; partial or complete levitation; traveling clairvoyance (sometimes called bilocation); and entrance into other worlds. In the lore of yoga, there are separate siddhis for each of these, but some religious doctrines join them in a comprehensive form of extraordinary locomotion. Catholic doctrine, for example, attributes "agility" to risen bodies of the just. According to the *Catechism of the Council of Trent* 1.12.13, at the Resurrection the glorified body "will be freed from the heaviness that now presses it down and will [move] with the utmost ease and swiftness wherever the soul pleases."[103] I discuss this doctrine in chapter 8.1, arguing that it might express intuitive anticipations of metanormal movement.

[5.6]
ABILITIES TO ALTER THE ENVIRONMENT

Extraordinary Hand-Eye Coordination and Dexterity

Evolution has given rise to specialized organs such as the talon, the beak, and the primate hand with which animals manipulate their immediate environments. Among humans, hand skills are developed through occupations ranging from surgery to needlework, and by disciplines such as somatic education and the martial arts. Structural Integration, for example, can improve dexterity by freeing muscles and ligaments (18.5); Sensory Awareness training promotes sensitive, tender, loving touch (18.7); various martial arts produce abilities to strike with great accuracy, speed, and power (20.2). Anatomy, kinesiology, and other fields have produced much knowledge about such capacities, but there is an element of extraordinary hand-eye functioning that mainstream science has not described. In an essay entitled "Psi-Sensorimotor Interaction," Joseph Rush, a physicist interested in parapsychology, introduced the term *psi-enhancement.* Such enhancement, Rush proposed, is caused by the direct action of mind upon living matter, especially as "the deep self, or superconscious" affects muscular activity.[104] Heavyweight boxing champion Ingemar Johansson, for example, told an interviewer that

> there is something strange about my right hand, something hard to explain. It is almost as if it was not a part of me at all. I never know when it is coming. The arm works by itself. It is faster than the eye and I cannot even see it. Without my telling it to, the right goes and when it hits, there is this good feeling all down my arm and down through my body.[105]

Describing the final holes of his first tour championship, golfer Frank Beard wrote:

> I was still about two hundred yards from the middle of the green, and something popped into my mind, that good rule I still follow: when you're pumped up, always take less iron than you think you need because you'll hit it farther than you normally would. For two hundred yards, I'd normally take a three-iron. I took a five-iron. On a normal lie, under normal conditions, I couldn't hit a five-iron two hundred yards if my life depended on it. But I hit the ball right in the middle of the green, maybe twenty feet past the pin. If I'd hit a three-iron, I probably would've gone over the clubhouse.[106]

Modern sport has not produced a psychology to account for metanormal activity. In certain contemplative and martial-arts traditions, however, such activity is cultivated. For example, the German philosopher Eugen Herrigel studied archery with a Zen master in Japan, and was taught to wait while drawing the bow until something beyond his normal reflexes released the arrow. One day, immediately after he had made a shot, the Master made a deep bow and broke off the lesson. "Just then, 'It' shot?", he cried, as I stared at him bewildered. And when I at last understood what he meant I couldn't suppress a sudden whoop of delight.

"What I have said," the Master told me severely, "was not praise, only a statement that ought not to touch you. Nor was my bow meant for you, for you are entirely innocent of this shot. You remained this time absolutely self-oblivious and without purpose in the highest tension, so that the shot fell from you like a ripe fruit. Now go on practicing as if nothing had happened."

Only after a considerable time did more right shots occasionally come off, which the Master signalized by a deep bow. How it happened that they loosed themselves without my doing anything, how it came about that my tightly closed right hand suddenly flew back wide open, I could not explain then and I cannot explain today. The fact remains that it did happen, and that alone is important. But, at least I got to the point of being able to distinguish, on my own, the right shots from the failures. The qualitative difference is so great that it cannot be overlooked once it has been experienced.

But inwardly, for the archer himself, right shots have the effect of making him feel that the day has just begun. He feels in the mood for all right doing, and, what is perhaps even more important, for all right not-doing. Delectable indeed is this state. But he who has it, said the Master with a subtle smile, would do well to have it as though he did not have it. Only unbroken equanimity can accept it in such a way that it is not afraid to come back.[107]

Joseph Rush attributed the inspiration that Herrigel's teacher called *It* to the deeper self or superconscious.[†] Others, however, simply ascribe it to the efficient operation of nerves and muscles. Writing in *The New York Times Magazine,* author Lawrence Shainberg proposed that sports performances that athletes attribute to "the zone" might be mediated by highly trained motor memory rooted in the basal ganglia or other subcortical elements of the central nervous system. Drawing upon research by psychiatrist Monte Buchsbaum, which showed that there is a marked decrease in the brain's metabolic rate during periods of intense concentration, Shainberg suggested that meditation helps sensorimotor command centers operate more efficiently, unimpeded by unnecessary mental activity. When the Zen teacher told Herrigel to practice "as if nothing had happened," relinquishing both bad and happy memories of his past performance, he was echoing admonitions by traditional and modern teachers alike

[†] Herrigel's experience also involved extraordinary volition (see chapter 5.9).

that mental quietude permits the appropriate body parts involved in a particular task to do their job more effectively. Shainberg quoted baseball player Tim McCarver in this regard: "The mind's a great thing," said the retired catcher, "as long as you don't have to use it." Archery coach Tim Strickland noted: "The better your technique, the more you can anchor your mind in it. And the more you anchor your mind, the better your technique will be."

Denise Parker, a teenage archery prodigy, described her state when she became the first American female to score 1,300 points in a particular competition. "I don't know what happened," she said. "I wasn't concentrating on anything. It didn't feel like I was shooting my shots, but like they were shooting themselves. I try to remember what happened so I can get back to that place, but when I try to understand it, I only get confused. It's like thinking how the world began."[108] Psi-enhancement? Efficient basal ganglia? There are different theories to account for "the zone" and extraordinary dexterity. But whatever we call such phenomena, and however we account for them in scientific, philosophic, or religious terms, they are evident in many kinds of sensorimotor activity. Hand-eye skills, like other evolutionary products, are transformed at times in ways that astonish their authors and observers alike.

Extrasomatic Influence upon the Environment

But metanormal influence upon our surroundings is not limited to inspired dexterity and hand-eye coordination. It also includes these three types of activity:

1. Telergy, the direct influence of a mind upon brain or other living tissue without any mediations observable by the ordinary senses (or extensions of them such as microscopes), as for example in spiritual healing.
2. Telekinesis, influence upon inanimate objects at a distance from, and without material connection with, the motive cause or agent.
3. The direct modification of some portion of space by mental influence.

In the current terminology of parapsychologists, these three forms of mind-matter interaction are called types of psychokinesis, the direct influence of mind upon either living or inorganic matter.

Telergy. In chapter 13, I present a brief history of spiritual cures and a review of experimental evidence that mental healing can happen at a distance without sensory cues. Numerous testimonies by reliable witnesses to cures that are scientifically inexplicable, and many laboratory studies, indicate that humans can indeed help others through activities such as prayer that do not depend upon physical stimuli (see chapter 13 and appendix D.1). And telergy is evident, too, in the harmonizing, energizing influence, often exercised at a distance from its recipients, that is attributed to saints and sages. Such influence, which can also be classified as a type of metanormal communication (5.3), has been described again and again since humans began to record their religious experience. Christian desert fathers; shamans of Siberia, the Americas, and the South Seas; Indian and Tibetan yogis; Jewish mystics; Sufis and other religious adepts, it is said, can help others by their extrasomatic influence, often when the recipients of their healing power do not know how or why they are uplifted. And extrasomatic influence upon the flesh has been attributed as well to inspirational athletes,[109] physicians, musicians, lovers, parents, and friends. In chapter 17.3, I cite Wilhelm Stekel, Jule Eisenbud, and other psychiatrists who have studied the psychosomatic effects of telepathic interaction. The case histories provided by these psychiatrists indicate that telergy operates—for better or for worse—in all sorts of human circumstance.

Telekinesis. Telekinesis is recognized by the Catholic church as one of its principal charisms (see chapter 22.11). Herbert Thurston, for example, quoted this incredible testimony by the *Cure d'Ars,* a French priest of the nineteenth century who was famous for his saintly integrity:

> Two Protestant ministers came here the other day who disbelieved our Lord's real presence in the Blessed Sacrament. I said to them, "Do you believe that a piece of bread could detach itself and, of its own accord, place itself upon the tongue of a person who was approaching to receive it?"
>
> "No," [they said].
>
> [One of them] desired to believe and . . . prayed to the Blessed Virgin to obtain the gift of faith for him. Now listen well to what I am going to tell you. I do not say that it happened *somewhere or other,* but I say that it happened to me. When that man presented himself to receive Holy Communion, the Sacred Host detached Itself from my fingers while I was yet a good distance from him, and placed Itself upon the tongue of the man.[110]

According to Hindu lore, telekinesis is a notable siddhi. It has been attributed to shamans of stone-age cultures.[111] And it is reported today by martial artists (20.1) and athletes. Quarterback John Brodie, for example, told me that the power of mind over matter was evident in a touchdown pass he threw to wide receiver Gene Washington.

> MURPHY: When the play began it looked for a moment like the safety would make an interception. But then it seemed as if the ball went through or over his hands as he came in front of Washington.
>
> BRODIE: Pat Fischer, the cornerback, told reporters after the game that the ball seemed to jump over his hands as he went for it. When we studied the game films that week, it did look as if the ball kind of jumped over his hands into Gene's. Some coaches said it was the wind—and maybe it was.
>
> MURPHY: What do you mean by *maybe*?
>
> BRODIE: What I mean is that our sense of that pass was so clear and our intention so strong that the ball was bound to get there, come wind, cornerbacks, hell or high water.[112]

Many sports fans consciously or half-consciously feel that rooting has an effect that goes beyond mere encouragement to a contest's participants; witness the many hexes aimed at games via radios and television sets. Such behavior seems instinctive among fans in every nation. If rooting channels or triggers powers of mind over matter, it is no wonder that during certain contests balls take funny bounces and athletes jump higher than ever or stumble inexplicably. Many of us have sensed the uncanny at sports events, when an innocent game suddenly becomes a theater of the occult. Hexing has even been employed in the rarefied atmosphere of chess competition; witness the Korchnoi-Karpov world championship in Manila during which Karpov was assisted by a parapsychologist while Korchnoi hired meditation students to influence Karpov from a distance. Comic as such efforts seem, they demonstrate sportspeople's persistent belief in powers of mind over matter.

Psychokinetic alteration of physical spaces or objects. Extrasomatic alteration of the physical environment has a third modality, namely modification of some portion of space or the structure of physical objects. The healing power of many shrines has been attributed to their founders' influence (see 13.1). Many visitors to the rooms inhabited by the twentieth-century Indian mystics Ramana Maharshi and Sri

Aurobindo have experienced profound serenity there. Certain objects are said to carry the influence of remarkable people. When Herrigel's archery training was done, his Zen master gave him his best bow. "When you shoot with this," he said, "you will feel the spirit of the Master near you."[113]

There is also good evidence that some people can deliberately imprint images on photographic film by mental influence. Unlikely as this phenomenon seems, it has since the 1860s been demonstrated again and again in Europe, Japan, and the United States. For many years psychiatrist Jule Eisenbud studied an American, Ted Serios, who produced several thousand so-called thoughtographs in the presence of physicians, physicists, and other reliable observers under conditions that precluded trickery. Eisenbud has described his research with Serios in a book and several articles (see appendix A.8).

Frederic Myers defined a *phantasmogenetic center* as "a point in space so modified by the presence of a spirit that it becomes perceptible to persons materially present near it."[114] Amplifying the idea that such centers exist, Myers suggested that there is a spectrum of apparitional phenomena.

> At the spiritual end there may be what we have called "clairvoyant visions,"—pictures manifestly symbolical, and not located by the observer in ordinary three-dimensional space. These seem analogous to the views of the spiritual world which the sensitive enjoys during entrancement. Then comes that larger class of veridical apparitions where the figure seems to be externalized from the percipient's mind, some stimulus having actually been applied—whether by the agent's or percipient's spirit—to the appropriate brain-centre. These cases of "sensory automatism" resemble those experimental transferences of pictures of cards, etc. And beyond these again, on the physical or rather the ultra-physical side, come those collective apparitions which in my view involve some unknown kind of modification of a certain portion of space not occupied by any organism—as opposed to a modification of centers in one special brain. Here comes in, as I hold, the gradual transition from subjective to objective, as the portion of space in question is modified in a manner to affect a larger and larger number of percipient minds.[115]

In summary, then, extrasomatic influence upon the environment (including living tissue) operates in various ways, including: by spiritual healing and other forms of telergy; by telekinesis; and by the psychokinetic alteration of an object's structure or some portion of space. There is enduring testimony in most sacred traditions that such functioning can be cultivated and made accessible to conscious control. In noting such testimony, I refer you to the animistic interactionism proposed by David Griffin (see 7.2), in which perception and causal influence are, as Griffin put it, "simply two perspectives on the same event." Telergy, telekinesis, and the psychokinetic alteration of particular locations occur, perhaps, because the modifying agent and modified objects, however dissimilar, prehend one another.

[5.7]
PAIN AND PLEASURE

Among vertebrates, sensations of pain and pleasure are carried by the central nervous system and modulated by opioid peptides that circulate in the blood. In humans, they are also affected by culturally conditioned psychophysical processes that can be deliberately modified. During hypnosis, for example, cool stimuli can be made to feel hot (or vice versa), and surgical incisions can be made without anesthesia. By means of athletic or martial-arts training, many people learn to enjoy strenuous exercise or environmental extremes. Through ascetic practices, shamans and yogis acquire the ability to endure wounds with complete equanimity. There is now considerable scientific evidence that humans can alter their responses to potentially painful stimuli (see 15.8, 15.9, 17.2, 20.2, 22.3, 22.4, 23). Many clinical and experimental studies have shown that pain can be diminished, and in some cases eliminated, by hypnotic, psychotherapeutic, meditative, or other exercises.

Modern studies of pain control support the contention to be found in all sacred traditions that suffering can be overcome through certain virtues and disciplines. Sri Aurobindo reflected the teaching of countless philosophers and religious teachers when he wrote:

> We feel pleasure or pain in a particular contact because that is the habit our nature has formed, because that is the constant relation the recipient has established with the contact. [But] it is within our competence to return quite the opposite response, pleasure where we used to have pain, pain where we used to have pleasure.
>
> The nervous being in us, indeed, is accustomed to a certain fixedness, a false impression of absoluteness in these things. [However] there is something in us which takes delight impartially in all external being and enables us to persevere through all labors, suffering and ordeals.

In our ordinary life this truth is hidden from us or only glimpsed at times or imperfectly held or conceived. But if we learn to live within, we infallibly awaken to this presence which is our more real self, a presence profound, calm, joyous, of which the world is not the master.[116]

The self-existent delight that Aurobindo referred to sometimes emerges in athletic ordeals. Championship miler Herb Elliott said that his coach Percy Cerutty helped him to world records "not so much by improving my technique, but by releasing in my mind and soul a power that I only vaguely thought existed. Thrust against pain, Percy told me. Walk towards suffering. Love suffering. Embrace it."[117] Cerutty and Elliott claimed that mastering the mile made one a "bigger person." Both men believed there is a joy in sport greater than ordinary pleasure. The athletic embrace of suffering, which is difficult for some of us to appreciate, is more understandable if we remember the admonition that there is something in us profounder than ordinary consciousness, something that takes "delight impartially in all labors and suffering."

While studying effects of the Lamaze method, a system of natural childbearing, research psychologist Deborah Tanzer found that some mothers had rapturous experiences while giving birth,[118] and some doctors have reported that childbirth brings certain women a happiness that turns pain into pleasure. Physician Grantly Dick-Read wrote:

> Many women have written to me of this happiness: "Something no words can describe," and so on. What does it mean? I do not know what inference should be drawn from such . . . expressions. The thoughts and words of women who watch their babies being born are so constant that in spite of the different terms in which they are couched they must be included in the purposeful phenomena of labor. Young mothers with no pretensions to piety have unhesitatingly told me that they felt the nearness of God, or the presence of a superhuman being at the birth of their child, or "a heavenly feeling that they have never known before."[119]

Behavior modification therapies, wilderness training for the young, martial arts, and many sports embrace painful stimuli so that they can be mastered. Implicit in such activities is the belief that we can break our habitual responses to things that usually make us suffer. And this belief is central to much religious practice. For mystics of the Christian tradition, ordinary pains and pleasures are transcended through love for our fellows and contact with Christ's living presence.

For Buddhists, suffering is rooted in desire, which can be dissolved in nirvana. For Vedantists, unhappiness is overcome by the experience of *ananda,* self-existent delight. In spite of their differing metaphysics, religious teachers East and West have asserted that we can realize a joy that subsumes ordinary pains and pleasures.[†] "From delight all these creatures are born," says the Taittiriya Upanishad. "By delight they exist and grow. To delight they return. For who could live or breathe if there were not this delight of existence as the ether in which we dwell?"

Every sacred tradition celebrates a blessedness that does not depend upon the satisfaction of everyday needs and desires. But the acknowledgment of such blessedness does not release us from our biologically inherited and culturally conditioned responses to potentially painful stimuli. Only through transformative practice and self-sacrifice can we transform our habitual pains and pleasures into the lasting enjoyment described in the Taittiriya Upanishad.

However, such fulfillment does not always require formal discipline. On a sunny morning as we walk down a long familiar street thinking about nothing in particular, or through the glance of someone we love, or in a field at sunset, a sudden unaccountable gladness, something free and untouchable, is given to us. Or a redeeming buoyancy arises in the midst of our suffering, pervading our pain like a subtle presence, lifting us into its merciful sustenance, spreading to others with a life of its own. At such times we know, deeply and beyond all doubt, that there is a truth and a goodness, a redeeming joy in which this world rests. In these unexpected, unguarded moments, we experience that gift mystics describe, a delight that passes understanding.

[5.8]
COGNITION

The word *cognition* refers to various ways of knowing, including analysis and reasoning; pattern recognition through the use of meta-

---

[†] Religious life produces countless types of joy. A great many of these have been named in Sanskrit, among them *kamananda,* a sensuous feeling arising from the transmutation of sexual desire; *raudrananda,* which is produced by the conversion of pain to pleasure; *vaidyutananda,* an electric pleasure that comes as blissful shock; and *visayananda,* a yogic delight of the senses. Many more joys are specified in Sanskrit, and to a lesser extent in other languages. That there is an immense variety of extraordinary pleasures is an open secret of the sacred traditions.

phor; intuitive comprehension of another person's subjective state; problem solving that involves visual, auditory, or other imagery; and mystical illumination.[120] Here I do not attempt a comprehensive inventory of extraordinary cognitions, but focus upon the mystical knowledge celebrated by the contemplative traditions and the supreme intellection evident in certain inspired works, which is typically accompanied by a sense that one has received knowledge or creative power beyond one's normal capacities. These two kinds of cognition, like other human attributes, exhibit continuities with particular animal capacities, in this case primate information processing; they are both deemed by their subjects to surpass ordinary types of knowing; and both can be triggered by practices described in this book. Hypnotic suggestion and imagery practice, for example, can improve performance on visual memory tasks, evoke comprehensive images for problem solving, reveal memories and perceptions that words do not adequately represent, promote the ability to restructure conceptual systems, and facilitate nonanalytic, holistic strategies involved in creative inspiration (see 15.11 & 17.2). Psychotherapy can release intellectual inhibitions, stimulate new problem-solving styles, and relieve inner conflicts that impede intellectual work (17.1). And religious practice, by promoting concentration and access to the mind's subliminal depths, can stimulate mystical knowing. These disciplines, and others, can directly or indirectly facilitate extraordinary cognitions, either by strengthening particular mental capacities or by improving the perceptual, emotional, and volitional processes they involve.

Mystical Knowledge

Various kinds of mystical knowledge have been described in the sacred traditions and by modern students of religious experience. The Buddhist experience of an impersonal *sunyam,* or fertile void, for example, is different from apprehensions of a personal Beloved described by Hindu, Christian, and Islamic mystics. Cognitions of a world-transcendent Brahman (and consequent sense of the world as illusion) reported by Indian yogis are different from visions of Divinity in *this* world celebrated by the sages of many cultures. If mystical cognition is deemed to be a family of experience, it has many branches. But it is not my purpose to list the varieties of mystical knowing. I only want to emphasize the fact that it is a type of extraordinary cognition. As I said in chapter 2.1, virtually all contemplatives assert that objects of mystical insight, whether Buddha Nature, God, or Brahman, are realities that exist independently of any human experience.

In its classic form, mystical knowing reveals a fundamental reality, self, or divinity by direct experience, without analysis and reasoning. "Although [they are] similar to states of feeling," wrote William James, "mystical states seem to those who experience them to be also states of knowledge. They are states of insight into depths of truth unplumbed by the discursive intellect. They are illuminations, revelations, full of significance and importance, all inarticulate though they remain; and as a rule they carry with them a curious sense of authority for aftertime."[121] In *The Varieties of Religious Experience,* James presented this firsthand account of such illumination by the Canadian psychiatrist Richard Bucke.

> I had spent the evening in a great city with two friends reading and discussing poetry and philosophy. We parted at midnight. I had a long drive in a hansom to my lodging. My mind, deeply under the influence of the ideas, images and emotions called up by the reading and talk, was calm and peaceful. I was in a state of quiet, almost passive enjoyment, not actually thinking, but letting ideas, images, and emotions flow of themselves, as it were, through my mind. All at once, without warning of any kind, I found myself wrapped in a flame-colored cloud. For an instant I thought of fire, an immense conflagration somewhere close by in that great city; the next, I knew that the fire was within myself. Directly afterward there came upon me a sense of exultation, of immense joyousness accompanied or immediately followed by an intellectual illumination impossible to describe. Among other things, I did not merely come to believe, but I saw that the universe is not composed of dead matter, but is, on the contrary, a living Presence; I became conscious in myself of eternal life. It was not a conviction that I would have eternal life, but a consciousness that I possessed eternal life then; I saw that all men are immortal; that the cosmic order is such that without any peradventure all things work together for the good of each and all; that the foundation principle of the world, of all the worlds, is what we call love, and that the happiness of each and all is in the long run absolutely certain. The vision lasted a few seconds and was gone but the memory of it and the sense of the reality of what it taught has remained during the quarter of a century which has since elapsed. I knew that what the vision showed was true. I had attained to a point of view from

which I saw that it must be true. That view, that conviction, I may say that consciousness, has never, even during periods of the deepest depression, been lost.[†]

The certainty that extraordinary knowledge has been acquired is also evident in the following account provided by the Religious Experience Research Unit at Oxford University.

I don't know exactly how old I was. Certainly it was before the age of eight; I think perhaps I was six. I was taken to a park in the evening to enjoy a firework display. It was summer. There was a crowd of people round the lake. It was twilight, and against the darkening sky, before the fireworks were set alight, I remember seeing these trees, poplar trees they were, three of them. It's very difficult to say exactly what happened because the order of this experience is of its own kind. There was a breeze, and the leaves of the poplars vibrated, rustled. I believe I said to myself, "How beautiful, how wonderful those three trees are." I think there was awe and wonder, and I remember comparing the luminousness—that's a grown-up word of course—the marvelous beauty, the haunting, oppressive power of those trees with the artificiality of the surroundings, the people, the fireworks and so on. Oddly, I kind of knew that this was something extraordinary at the moment that it occurred. It was as simple as that, just seeing these trees; but it was *the* event of my childhood. All children have many strong impressions of all sorts stamped on their minds, and I have had lots of other experiences as a child which I remember. I have known often enough since then the elation derived from the reading of poetry, or listening to music, or seeing the beauty of nature, or from human love relationships; but love and artistic or literary delights in all their intensity are firmly distinguishable from that experience that I had as a child. They are fine ecstasies, but they are of a different order.

I see it now as an illumination of a religious kind, an argument for theism. To my young mind this experience had nothing to do with such rudimentary teaching in moral rectitude, such as about saying one's prayers and so forth. It didn't seem to have any relation to a child's

---

[†] Cited in James, W., 1902, lectures 16 & 17. This passage was taken from a pamphlet printed before the publication of Bucke's *Cosmic Consciousness*, in which the same account appeared in slightly different form. Bucke attributed experiences such as his to a new faculty emerging in human nature, which he called cosmic consciousness. This, he wrote, "in its more striking instances is not simply an expansion or extension of the self-conscious mind with which we are all familiar, but the superaddition of a function as distinct from any possessed by the average man as self-consciousness is distinct from any function possessed by the higher animals" (See: Bucke 1961, pp. 1–3).

Christianity. And even now I do not find it easy to relate this childhood event to the theology and the church which I now acknowledge and obey. That doesn't mean that there is no connection, but simply that I find it rather difficult to link them.

I might say, looking back, that it was some kind of assurance of the divine that I experienced; nothing like a sense of the Trinity, which nevertheless I accept. But they don't offend each other, that experience and my present belief. Whether that experience had any influence in my conversion to Catholicism at the age of twenty-seven I can't say: whether it suggested or precipitated that move—I think it's possible.[122]

Other subjects of the Oxford study distinguished illuminations such as those described above from their formal religious beliefs. Like most authorities on the contemplative life, these people—who were for the most part untrained in religious psychology—made a distinction between their spiritual cognitions and the particular conceptual systems, or overbeliefs, with which they are typically associated. Though they "in general assert a pretty distinct theoretic drift," wrote William James, mystical insights are "capable of forming matrimonial alliances with material furnished by the most diverse philosophies and theologies. . . . Here the prophets of all the different religions come, with their visions, voices, raptures, and other openings, supposed by each to authenticate his own peculiar faith."[123]

Mystical knowing, in short, like all metanormal experience, is subject to cultural shaping. And it can also be clouded by various pathologies. Because it can be distorted by unexamined motives, desires, or beliefs, it has to be disciplined. It "must be sifted and tested, and run the gauntlet of confrontation with the total context of experience," James wrote.[124] As I said in chapter 2.1, mystical cognitions can be confirmed by our own tests and the witness of others.

Scientific, Artistic, and Philosophic Inspiration

Creative people have frequently claimed that in their greatest works they radically transcended themselves, receiving inspiration from the gods, a higher self, daimon, or Divinity itself. In his *Ecce Homo*, Nietzsche described such inspiration, which led to his writing *Thus Spoke Zarathustra*.

Has anyone at the end of the nineteenth century a clear idea of what poets of strong ages have called inspiration? If not, I will describe it—If one had the slightest residue of superstition left in one's system, one could hardly reject altogether the idea that one is merely incarnation, merely mouthpiece, merely a medium of overpowering forces. The concept of revelation—in the sense that suddenly, with indescribable certainty and subtlety, something becomes visible, audible, something that shakes one to the last depths and throws one down—that merely describes the facts. One hears, one does not seek; one accepts, one does not ask who gives; like lightning, a thought flashes up, with necessity, without hesitation regarding its form—I never had any choice.

A rapture whose tremendous tension occasionally discharges itself in a flood of tears—now the pace quickens involuntarily, now it becomes slow; one is altogether beside oneself, with the distinct consciousness of subtle shudders and of one's skin creeping down to one's toes; a depth of happiness in which even what is most painful and gloomy does not seem something opposite but rather conditioned, provoked, a necessary color in such a superabundance of light; an instinct for rhythmic relationships that arches over wide spaces of forms—length, the need for a rhythm with wide arches, is almost the measure of the force of inspiration, a kind of compensation for its pressure and tension.

Everything happens involuntarily in the highest degree but as in a gale of a feeling of freedom, of absoluteness, of power, of divinity. The involuntariness of image and metaphor is strangest of all; one no longer has any notion of what is an image or a metaphor; everything offers itself as the nearest, most obvious, simplest expression. It actually seems, to allude to something Zarathustra says, as if the things themselves approached and offered themselves as metaphors.[125]

Frederic Myers proposed that works of genius involve a "subliminal uprush," an emergence into consciousness of ideas that the subject "has not consciously originated, but which have shaped themselves beyond his will, in profounder regions of his being."[126] To support his thesis, Myers cited Wordsworth's *The Prelude; or Growth of a Poet's Mind.* "We find Wordsworth insisting," Myers wrote, "upon the distinctive character of this subliminal uprush. . . . Of imagination he says (Book VI):"

> *That awful Power rose from the mind's abyss*
> *Like an unfathomed vapour that enwraps,*
> *At once, some lonely traveller. I was lost;*
> *Halted without an effort to break through;*
> *But to my conscious soul I now can say—*
> *"I recognise thy glory"; in such strength*
> *Of usurpation, when the light of sense*
> *Goes out, but with a flash that has revealed*
> *The invisible world, doth greatness make abode.*

This passage [Myers continued] expresses in the language of poetry the very relations between the supraliminal and the subliminal on which I have dwelt. The influence rises from no discoverable source; for a moment it may startle or bewilder the conscious mind; then it is recognised as a source of knowledge, arriving through inner vision; while the action of the senses is suspended in a kind of momentary trance. The knowledge gained, however, is simply a perception of "the invisible world"; there is no claim to any more definite revelation.

And since it is mainly by inward vision that these rememberable things are in truth discerned, there is a growing fusion between subjective and objective; between that which is generated in the seer himself and that of which the visible universe conveys the half-caught intimation (Book II):

> An auxiliar light
> Came from my mind, which on the setting sun
> Bestowed new splendour.

Subliminal uprushes, in other words, so far as they are intellectual . . . bring with them indefinite intimations of what I hold to be the great truth that the human spirit is essentially capable of a deeper than sensorial perception, of a direct knowledge of facts of the universe outside the range of any specialised organ or of any planetary view.[127]

For Nietzsche, Wordsworth, and Myers, there is a kind of cognition by which "everything happens involuntarily," with a completeness and power beyond ordinary thought. Such knowing is like vision: "everything offers itself as the nearest, most obvious, simplest expression." According to Myers, that is why great poets "are usually Platonists," acknowledging forms of beauty beyond and independent of, yet sometimes accessible to, human knowing. The following letter attributed to Mozart describes cognition of this kind.

When I am, as it were, completely myself, entirely alone, and of good cheer—say, travelling in a carriage, or walking after a good meal, or during the night when I cannot sleep; it is on such occasions that my ideas flow best and most abundantly. Whence and how they come, I know not; nor can I force them. Those ideas that please me I retain in memory, and am accustomed, as I have been told, to hum them to myself. If I continue in this way, it soon occurs to me how I may turn this or that morsel to account, so as to make a good dish of it, that is to say, agreeably to the rules of counterpoint, to the peculiarities of the various instruments, etc.

All this fires my soul, and, provided I am not disturbed, my subject enlarges itself, becomes methodised and defined, and the whole, though it be long, stands almost complete and finished in my mind, so that I can survey it, like a fine picture or a beautiful statue, at a glance. Nor do I hear in my imagination the parts *successively*, but I hear them, as it were, all at once (*gleich alles zusammen*). What a delight this is I cannot tell! All this inventing, this producing, takes place in a pleasing lively dream. Still the actual hearing of the *tout ensemble* is after all the best. What has been thus produced I do not easily forget, and this is perhaps the best gift I have my Divine Maker to thank for.

When I proceed to write down my ideas, I take out of the bag of my memory, if I may use that phrase, what has been previously collected into it in the way I have mentioned. For this reason the committing to paper is done quickly enough, for everything is, as I said before, already finished; and it rarely differs on paper from what it was in my imagination. At this occupation I can therefore suffer myself to be disturbed; for whatever may be going on around me, I write, and even talk, but only of fowls and geese, or of Gretel or Barbel, or some such matters. But why my productions take from my hand that particular form and style that makes them *Mozartish*, and different from the works of other composers, is probably owing to the same cause which renders my nose so large or so aquiline, or, in short, makes it Mozart's, and different from those of other people. For I really do not study or aim at any originality.[128]

The "all-at-onceness" of Mozart's inspirations distinguishes them from ordinary mental productions. The great mathematician Henri Poincare also received insights *tout ensemble*. In the following passage, he described his discovery of the Fuchsian functions. These are immensely complex, but appeared to him—as music did to Mozart—in a series of densely packed revelations.

For fifteen days I strove to prove that there could not be any functions like those I have since called Fuchsian functions. I was then very ignorant; every day I seated myself at my work table, stayed an hour or two, tried a great number of combinations and reached no results. One evening, contrary to my custom, I drank black coffee and could not sleep. Ideas rose in crowds; I felt them collide until pairs interlocked, so to speak, making a stable combination. By the next morning I had established the existence of a class of Fuchsian functions, those which come from the hypergeometric series; I had only to write out the results, which took but a few hours.

Just at this time I left Caen, where I was then living, to go on a geologic excursion under the auspices of the school of mines. The changes of travel made me forget my mathematical work. Having reached Coutances, we entered an omnibus to go some place or other. At the moment when I put my foot on the step the idea came to me, without anything in my former thoughts seeming to have paved the way for it, that the transformations I had used to define the Fuchsian functions were identical with those of non-Euclidean geometry. I did not verify the idea; I should not have had time, as, upon taking my seat in the omnibus, I went on with a conversation already commenced, but I felt a perfect certainty. On my return to Caen, for conscience's sake I verified the result at my leisure.

Then I turned my attention to the study of some arithmetical questions apparently without much success and without a suspicion of any connection with my preceding researches. Disgusted with my failure, I went to spend a few days at the seaside, and thought of something else. One morning, walking on the bluff, the idea came to me, with just the same characteristics of brevity, suddenness and immediate certainty, that the arithmetic transformations of indeterminate ternary quadratic forms were identical with those of non-Euclidean geometry.

Returned to Caen, I meditated on this result and deduced the consequences. The example of quadratic forms showed me that there were Fuchsian groups other than those corresponding to the hypergeometric series. . . . But all my efforts only served at first the better to show me the difficulty, which indeed was something. All this work was perfectly conscious.

Thereupon I left for Mont-Valerien, where I was to go through my military services; so I was very differently occupied. One day, going along the street, the solution of the difficulty which had stopped me suddenly appeared to me. I did not try to go deep into it immediately, and only after my service did I again take up the question. I had all the elements and had only to arrange them and put them together. So I wrote out my final memoir at a single stroke and without difficulty.[129]

Frederic Myers described yet another feature of artistic, philosophic, and mathematical inspiration. From our subliminal depths, he observed, come revelations that point us toward a greater existence. Criticizing the materialistic view of evolution gaining ascendancy in the late nineteenth century, he noted:

For the greater part of the time during which life has existed on earth it would have been thought chimerical to suggest that we could live in anything [besides water]. It was a great day for us when an ancestor crawled up out of the slowly-cooling sea;—or say rather when a previously unsuspected capacity for directly breathing air gradually revealed the fact that we had for long been breathing air in the water;—

and that we were living in the midst of a vastly extended environment,—the atmosphere of the earth. It was a great day again when another ancestor felt on his pigment-spot the solar ray;—or say rather when a previously unsuspected capacity for perceiving light revealed the fact that we had for long been acted upon by light as well as by heat; and that we were living in the midst of a vastly extended environment,—namely the illumined Universe that stretches to the Milky Way. It was a great day when the first skate (if skate he were) felt an unknown virtue go out from him towards some worm or mudfish;—or say rather when a previously unsuspected capacity for electrical excitation demonstrated the fact that we had long been acted upon by electricity as well as by heat and by light. . . . All this,—phrased perhaps in some other fashion—all men admit as true. May we not then suppose that there are yet other environments, other interpretations, which a further awakening of faculty still subliminal is yet fated by its own nascent response to discover? Will it be alien to the past history of evolution if I add: It was a great day when the first thought or feeling flashed into some mind of beast or man from a mind distant from his own?—when a previously unsuspected capacity of telepathic percipience revealed the fact that we had long been acted upon by telepathic as well as by sensory stimuli; and that we were living in an inconceivable and limitless environment,—a thought-world or spiritual universe charged with infinite life . . . up to what some have called World-Soul, and some God?

The higher gifts of genius—poetry, the plastic arts, music, philosophy, pure mathematics—all of these are precisely as much in the central stream of evolution—are perceptions of new truth and powers of new actions. There is, then, about those loftier interests nothing exotic, nothing accidental; they are an intrinsic part of that ever-evolving response to our surroundings which forms not only the planetary but the cosmic history of all our race.[130]

In Myers's view, apprehensions of supraphysical worlds (or new dimensions of the physical world) contribute to our evolutionary advance by revealing new territories for the human race to explore and inhabit. This revelatory and pioneering power, I propose, is another characteristic of metanormal cognition.

The inspirations described by Wordsworth, Nietzsche, Mozart, and Poincaré share a number of features, such as their speed and spontaneity, their joy and excitement, their exceeding of ordinary mental processes, and the beauty of their products. They also resemble each other in their dependence upon several extraordinary cognitive capacities, among them exceptional memory (Wordsworth, for example, did not write down his lengthy *Lines Written Above Tintern*

Abbey for several days), talent for bringing complex material into magnificent but elegant form, good integration of analysis and synthesis, and the ability to join abstract thought with great emotion. Though creative inspiration is not fully understood, modern psychology has specified several types of intellience upon which it depends.[†]

In most cases, however, metanormal cognitions do not last for long. Following the experience that prompted the passage cited above, Nietzsche reported his sickness in Genoa, "then came a melancholy spring in Rome."[131] In some instances, Myers wrote,

> we seem to see our subliminal perceptions and faculties acting truly in unity, truly as a Self;—co-ordinated into some harmonious "inspiration of genius," or some profound and reasonable hypnotic self-reformation, or some far-reaching supernormal achievement of clairvoyant vision or of self-projection into a spiritual world.
>
> But it seems that this degree of clarity, of integration, cannot be long preserved. Much oftener we find the subliminal perceptions and faculties acting in less co-ordinated, less coherent ways. We have products which, while containing traces of some faculty beyond our common scope, involve, nevertheless, something as random and meaningless as the discharge of the uncontrolled middle-level centres of arms and legs in the epileptic fit. We get, in short, a series of phenomena which the term dream-like seems best to describe.
>
> In the realm of genius,—or uprushes of thought and feeling fused beneath the conscious threshold into artistic shape,—we get no longer masterpieces but half-insanities,—not the Sistine Madonna, but Wiertz's Vision of the Guillotined Head; not *Kubla Khan*, but the disordered opium dream.[132]

Commenting upon these flawed inspirations, Myers wrote: "Hidden in the deep of our being is a rubbish-heap as well as a treasure-house;—degenerations and insanities as well as beginnings of higher development." The intermingling of metanormal with abnormal elements, of psychic rubbish with psychic treasure, occurs in religious and philosophic as well as artistic works. Indeed, inspiration in any field can be clouded by various pathologies.[133] To experience it over time without distortion, to develop a lasting capacity for it, most of us need disciplined commitment to creative work as well as developing self-knowledge. In Part Three, I discuss some practices that such work

---

[†] Psychologist Howard Gardner, for example, has proposed that there are at least six basic types of intelligence: linguistic, musical, logical-mathematical, spatial, bodily-kinesthetic, and personal (or social), all of which can be educated. See Gardner, H., 1985.

can draw upon, either to strengthen particular cognitive abilities or to enhance the perceptual, emotional, and bodily processes they involve.

Still, inspired cognitions occur at times without formal discipline. They can come when we least expect them, bringing new insight in ordinary circumstances. Though we do not ask for them, such illuminations can show us why we must suffer a particular situation or persevere in a difficult task. Freely given, they can eliminate misunderstanding or enable us finally to act. Frederic Myers's "subliminal up-rushes" are not limited to famous artists or philosophers. They happen for all of us. This is the case, I propose, because we harbor a greater life that informs our everyday functioning. Edward Robinson, who for many years directed Oxford University's Religious Experience Research Unit, called epiphanies of this kind "small happenings." Accounts such as the following, he wrote, reflect their "everydayness."

I was between eleven and twelve years old. It was a summer day and I was playing out back of the house, in an alley in the city where we lived. . . . A sudden storm came up and interrupted our play. I sat alone between garages behind the house, waiting for it to end. It was near noon. The rain ended almost as soon as it came, and the sun shone hot and bright once more. All at once I felt as if I were seeing everything for the first time. The light seemed like gold, the smell of the wet ground and foliage was like perfume, with the rain water shining and running about in little rivulets, the humming and buzzing of insects and bees was pleasant to my ears. Everywhere I looked there was beauty. In that dirty alley, wherever there was a leaf or blade of grass, it sparkled. . . . Now I watched a beetle going about its business, and then a small garden spider, and I was glowing with warmth. It was as if all that was ouside of me, I felt myself to be a part of it. Then a thought came. It said "See! Everything is alive, everything lives. That insect, it has a life, the grass, the air even!" And then I felt joy, and with the joy, love, and then a feeling of reverence. . . . I was part of all this, but I experienced myself also as an entity, distinct, conscious, of a higher consciousness than the grass and the creatures I was watching. I felt a loving obligation to be respectful and kind, to relate to all of life with a feeling of reverence and love. To be gentle, to never hurt anything because it all had life. My awareness of myself was acute. I felt that the life I was observing around me was as real as my own but distinctly different. That life around me was not as conscious of me as I was of it. This made me feel very big and filled with a sense, a loving sense of obligation. The whole experience lasted only a few minutes, and when I heard my mother call me to come to lunch I said nothing of this, but the glow of it remained for some time.[134]

[5.9]
VOLITION

Through self-observation we can clarify the nature of our shifting or enduring motives, some of them rooted in basic needs, some of them acquired by social conditioning, some healthy, some destructive, many of them unnoticed or repressed. The ongoing self-reflection enjoined by religious leaders and moral authorities of every culture, and by psychotherapists of modern times, can help to illumine secret or half-concealed attitudes that rule our thought and behavior. Recognizing our competing impulses more clearly, we can more freely choose among them, suppressing or sublimating some while acting upon others. To the extent that religious, moral, or therapeutic disciplines succeed, a person's many wills tend to become one will, single but articulated. As they are integrated, once-divergent intentions produce stronger results. Like a body in which each muscle functions in coordination with other body parts while retaining its own integrity, a well-articulated self can harmonize its separate volitions to achieve its deepest ends.

If successful, transformative practices extend the capacity for purposeful action produced by animal evolution. They can strengthen the capacity for one-pointed behaviors evident in earlier forms of life while providing more options for creative behavior. They can make us better animals and better humans at once, as it were, more single-minded when we choose to be, but more various in our realized intentions. In the words of psychiatrist Roberto Assagioli, it is possible to have a strong, a good, and a skillful will.[135]

Religious traditions East and West bear witness to the fact that volition can be cultivated, and they provide many accounts of its metanormal expression. According to Catholic teaching, for example, saintly power in charitable work arises from a surrender of egocentricity that makes possible activity given by God. Even if we do not accept their theology, we can find frequent acknowledgment of extraordinary volition in the works of Saint Teresa of Avila, Saint John of the Cross, and other Catholic authorities on the contemplative life. The same witness can be found in the Bhagavad Gita. By sacrificing his slavery to unexamined desire and social convention, Arjuna becomes a powerful instrument of God in the battle of Kurukshetra. By transcending his immediate inclinations, he expresses the will of Divinity. Embedded in this central parable of Hindu religious philosophy is recognition of a will beyond ordinary willing.

Taoism offers a similar doctrine. By concentrating our vital energies, quieting our thoughts, and disregarding external rewards, we can realize mastery in everyday work and express our deep nature, or way. The Taoist doctrine of *wu wei,* unblocked or unimpeded doing, refers to such activity. Thus according to Chuang-Tzu:

> Ch'ing, the chief carpenter, was carving wood into a stand for musical instruments. When finished, the work appeared to those who saw it as though of supernatural execution; and the Prince of Lu asked him, saying, "What mystery is there in your art?"
>
> "No mystery, Your Highness," replied Ch'ing. "And yet there is something. When I am about to make such a stand, I guard against any diminution of my vital power. I first reduce my mind to absolute quiescence. Three days in this condition, and I become oblivious of any reward to be gained. Five days, and I become oblivious of any fame to be acquired. Seven days, and I become unconscious of my four limbs and my physical frame. Then, with no thought of the Court present in my mind, my skill becomes concentrated, and all disturbing elements from without are gone. I enter some mountain forest, I search for a suitable tree. It contains the form required, which is afterwards elaborated. I see the stand in my mind's eye, and then set to work. Beyond that there is nothing. I bring my own native capacity into relation with that of the wood. What was suspected to be of supernatural execution in my work was due solely to this.[136]

All of the doctrines just noted are based upon activity common to peoples in many cultures, but they are embedded in quite different philosophies. According to Saint John of the Cross, the volition of saints is given by God; while on the other hand, Chuang-Tzu's carpenter says that his supernatural-seeming mastery is produced by natural causes. Both Catholic and Taoist contemplative traditions bear witness to extraordinary volition, but they account for it in different ways. Such volition, in short, like all metanormal capacities, is associated with different beliefs. We can, however, try to see what features it commonly exhibits. If we examine firsthand accounts of such willing from sports, the martial arts, and other disciplines—or indeed from any walk of life—we find that it is marked by one-pointed involvement; by disregard for immediate results; by spontaneity, freedom, and effortless mastery; and by a sense that the self is somehow larger and more complex, or conversely, that it disappears into something beyond itself. Distance runner Ian Jackson, for example, described a type of workout in which he exceeded his normal abilities.

We would be easing along at 7:00 [minutes] per mile and one of us (I never knew which one) would surge very slightly. Then we'd both be onto the pace of the surge. Later, there'd be another surge—just a little, almost unnoticeable increase in pace. But we kept pushing it up. Once we were moving, we didn't back off. Back and forth we'd play with the pace. He'd throw in a little more tempo. I'd match it and throw in a little of my own. Within a mile we'd go from 7:00 down to 6:30 pace. Two miles later, we'd be under 6:00. Another mile and we'd be down to about 5:30. It was so smooth you hardly knew it was happening. Finally, we'd be flying at 5:15 or 5:10, and the miles would reel by effortlessly.[137]

The extraordinary quality of Jackson's running was contagious, it seems, since Jackson's running partner, too, exhibited a spontaneity, ease, and mastery beyond his normal volition. However named, such command has been noted by many students of sports. The philosopher Michael Novak wrote:

This is one of the great secrets of sport. There is a certain point of unity within the self, and between the self and its world, a certain complicity and magnetic mating, a certain harmony, that conscious mind and will cannot direct. Perhaps analysis and the separate mastery of each element are required before the instincts are ready to assume command, but only at first. Command by instinct is swifter, subtler, deeper, more accurate, more in touch with reality than command by conscious mind. The discovery takes one's breath away.[138]

The American psychologist Mihalyi Csikszentmihalyi has studied experience of this kind for many years as part of an effort to understand a type of functioning he has named flow. Though flow, as Csikszentmihalyi conceives it, involves many human attributes, it clearly includes an exceptional type of volition. It is characterized, for example, by a marked concentration upon the activity at hand; by intrinsic motivation rather than a search for secondary benefits; by a reduction or absence of limiting self-consciousness; by a pronounced sense of mastery; by highly efficient psychophysical functioning; by positive moods; and by growth in complexity of the self. It breaks through to new levels of thought and behavior without social reinforcement, and perseveres without immediate rewards. Research on flow has been conducted with elderly Korean women, Indian and Thai adults, Tokyo teenagers, Navaho shepherds, farmers in the Italian Alps, and workers on assembly lines in Chicago.[139] The main dimensions of flow, Csikszentmihalyi wrote, including "deep concentration, clarity of goals, loss of a sense of time, lack of self-consciousness, and transcendence of a sense of self . . . are recognized in more or less the same form by people the world over."[140]

Extraordinary volition, then, has several features that have been studied by contemporary psychologists. But in some of its more dynamic expressions, it has attributes that mainstream science disregards. At times, for example, profound intention and highly focused activity seem to trigger those startling events Carl Jung called synchronicities, or meaningful coincidences. These, Jung proposed, involve a dramatic confluence of psychic and physical happenings that reveals a new path to be taken, a long-sought discovery, or something else of significance to those who experience them.[141] In his *Memories, Dreams, and Reflections,* Jung described events in his own life that exhibited this mysterious quality.

> Some years later (in 1927) I obtained confirmation of my ideas about the center and the self by way of a dream. I represented its essence in a mandala which I called "Window on Eternity." The picture is reproduced in *The Secret of the Golden Flower.* A year later I painted a second picture, likewise a mandala, with a golden castle in the center. When it was finished, I asked myself, "Why is this so Chinese?" I was impressed by the form and choice of colors, which seemed to me Chinese, although there was nothing outwardly Chinese about it. Yet that was how it affected me. It was a strange coincidence that shortly afterward I received a letter from Richard Wilhelm enclosing the manuscript of a Taoist-alchemical treatise entitled *The Secret of the Golden Flower,* with a request that I write a commentary on it. I devoured the manuscript at once, for the text gave me undreamed-of confirmation of my ideas about the mandala and the circumambulation of the center. That was the first event which broke through my isolation. I became aware of an affinity; I could establish ties with something and someone.
>
> On remembrance of this coincidence, this "synchronicity," I wrote underneath the picture which had made so Chinese an impression upon me: "In 1928, when I was painting this picture, showing the golden, well-fortified castle, Richard Wilhelm in Frankfurt sent me the thousand-year-old Chinese text on the yellow castle, the germ of the immortal body."[142]

Jung recorded other coincidences associated with important events in his life, and many others are described in the literature of Jungian psychology.[143] These synchronicities, it seems, confirm the joining of conscious and unconscious intentions and call attention to a purposiveness beyond ordinary volition.

[5.10]
INDIVIDUATION AND SENSE OF SELF

Paleontologist George Gaylord Simpson described the individualization produced by animal evolution, notably among higher vertebrates. Reactibility to different kinds of environmental stimuli, he wrote,

> is linked with adaptability, and often carries with it progress in individual versatility. This increase in the possible range and variety of reactions by the individual organism makes each more independent as a unit and more distinctive in its particular reactions and interrelationships. There may be, in short, an accompanying progress in individualization. A mammal or a bird is much more individualized than say, an oyster, not in structure or appearance, which may be more varied and more distinctive among oysters, but in reaction patterns, behavior, or general ability. Progress in versatilization of the individual and individualization within the species has been carried by human evolution to altogether new heights.[144]

Indeed, it is arguable that each of us has a unique identity that is apparent in all our activities. Essayist George Leonard wrote:

> The chance of two fingerprints being exactly the same is said to be less than one in 64 billion; matching a whole set of fingerprints with another whole set is impossible. Faces are equally distinctive and, except in the case of identical twins, easily recognizable. A good voiceprint—a printed replica of electronically recorded voice frequencies—effectively identifies the speaker. Handwriting can be forged, but only through the most consummate skill, and then not with unshakable fidelity. A bloodhound can pick up the scent of one person out of a million. Brain-wave patterns, as our means for analyzing them becomes more precise, are seen to be entirely distinctive. Human infants are born with unique and identifiable rhythms of sleeping and waking; recent research suggests that a newborn's breathing pattern is as distinctive as a thumbprint.
>
> The ability to recognize other individuals as members of the same species goes far back in evolution, certainly as far as the development of separate sexes. Social insects such as bees, ants, and termites have the additional ability to distinguish between different castes (queen bees from workers, for example), and also can recognize one life stage from another (eggs from larvae, and so on). Going up the evolutionary ladder, all vertebrates can distinguish among infants, juveniles, and adults of their own species. . . . The brain/sense mechanism, characteristi-

cally, comes up with varied and ingenious means for the recognition process. Some primates can recognize faces, just as people do. Many mammals, as every dog owner knows, use secretions as personal signatures; the ability of these animals to distinguish one individual from another by smell is phenomenal. The howl of a wolf communicates its emotional state, its location, and its spectographs show that wolves can make out extremely subtle differences in sound. Birds can distinguish individuals through the absolute frequency or pattern or "dialect" of their songs or calls, as well as through visual appearance.

All human societies, of course, are based around personal identity, and it is instructive to note the extreme concern with which this matter is treated from birth to death. There is something nightmarish and almost unthinkable about a human being without an identity. Since every cell in the body (except the blood cells) contains molecules of DNA on which is written the blueprint for the entire body, it is theoretically possible to create an exact copy of any human being—or a million copies—from one cell. Cloning, as such a process is known, is not only a staple of science fiction, but is actually under consideration by some scientists. Still, our common sense rebels at the notion, and we may find ourselves wishing it will turn out to be more difficult than some futurists believe, or perhaps impossible. For we all realize at the deepest level of our intuition that a million John Smiths, even a million Albert Einsteins, would be something other than human. Recent research, in fact, suggests that the evolutionary process itself involves mechanisms for producing and maintaining individual variations within each species.

To be human, it seems clear, is to have a personal identity. This identity is unique and irreversible. It provides our particular viewpoint of the universe. It expresses itself in numerous ways, subsuming what we call body, mind and spirit, memory and works.[145]

The uniqueness that is evident in all human beings becomes more pronounced as our capacities develop. The broadening of cognitive, emotional, and behavioral repertoires, the deepening of self-awareness, the creativity, and the articulation of human will noted in these pages contribute to the individuality to which animal and human development gives rise. This individuality, I propose, has at least two kinds of metanormal development: first, the realization of an ego-transcending identity reported by people in many walks of life but most notably by religious contemplatives; and second, the complex cognitive-emotional-behavioral repertoires, or subpersonalities, evident in highly creative people. Here I will briefly examine these two sides of our developing personhood.

Ego-Transcendent Identity

The realization of an identity beyond normal self-consciousness has been reported in cultures the world over. Though such experience is attributed to various causes, it is described in much the same manner by people with different backgrounds, philosophies, and temperaments. For example, it is said to reduce one's sense of separation from others while at the same time confering a greater personhood. One is simultaneously less and more than his or her former self, newly connected with the world at large but also more powerful, independent, and self-sufficient. Such experience typically confers a confluence of freedom and security that does not depend upon this or that set of ideas or behaviors. Indeed, this combination of liberation and confidence can become so intense that it is said to involve something self-existent or eternal. And it is often marked by a coincidence of newness with something remembered. While it lifts one beyond reminders of the ordinary self, it is immediately recognizable, so much so that for some people it appears to reveal their true identity or original nature. This type of experience, in short, typically joins detachment and solidarity with others, freedom and security that does not depend upon particular personality structures, and feelings of rebirth with a subjectivity that seems everlasting. Such ego-transcendence, however, is interpreted in different ways. According to William James:

> The mind-curers . . . have demonstrated that a form of regeneration by relaxing, by letting go, psychologically indistinguishable from the Lutheran justification by faith and the Wesleyan acceptance of free grace, is within the reach of persons who have no conviction of sin and care nothing for the Lutheran theology. It is but giving your little private convulsive self a rest, and finding that a greater Self is there. The results, slow or sudden, or great or small, of the combined optimism and expectancy, and the regenerative phenomena which ensue on the abandonment of effort, remain firm facts of human nature, no matter whether we adopt a theistic, a pantheistic-idealistic, or a medical-materialistic view of their ultimate causal explanation.[146]

The theistic explanation of such experience, James wrote, posits a

> divine grace, which creates a new nature within one the moment the old nature is sincerely given up. The pantheistic explanation [posits] the merging of the narrower private self into the wider or greater self, the spirit of the universe (which is your own "subconscious" self), the moment the isolating barriers of mistrust and anxiety are removed.[147]

In the view of Sri Aurobindo, our everyday self is only "a practical selection and limited conscious synthesis for the temporary utility of life in a particular body. Behind it there is a consciousness, a *purusha,* who is not determined or limited by this individualisation or by this synthesis but on the contrary determines, supports and yet exceeds it."[148] This *purusha,* or deeper self, can be known by witness meditation and other forms of introspection.† In its greater self-realization it is atman, at one with brahman, perceiving its oneness with everything but transcending all creations too. In its simultaneous immanence and transcendence, this higher or deeper self is analogous to the God of panentheism (see chapter 7.2). According to Aurobindo and other commentators upon the Vedas and Upanishads, it is symbolized in these lines of the Rig Veda (1.164.20):

> Two birds, beautiful of wing, friends and comrades, cling to a common tree, and one eats the sweet fruit, while the other regards him and eats not.

This verse, it is said, refers to the unity or fundamental connection between a world-participating self (eating the sweet fruit) and a world-transcending identity (who "regards him and eats not"). A similar image exists in this passage from the Swetaswatara Upanishad (2.7):

> The soul seated on the same tree of Nature is absorbed and deluded and has sorrow because it is not the Lord, but when it sees and is in union with that other self which is the Lord, then sorrow passes away from it.

In the Bhagavad Gita (15.7), Krishna, the Lord of Existence, says: "It is an eternal portion of Me that has become the living being in a world of living things." Commentators upon the Gita have interpreted this line to mean that a supreme personhood (*purushottoma*) supports the individual's action in the everyday world. Krishna, by this interpretation, symbolizes ultimate selfhood, immanent in the

---

† The *purusha,* or basic consciousness referred to here is a complex entity (or set of entities) in Aurobindo's psychology, embracing both atman, the eternal self, and a psychic being or soul that evolves from life to life. While atman is eternal in its oneness with brahman, the psychic being is its representative in time, developing gradually though it is essentially divine, with a unique identity that gives special flavor to each person's mind, life, and body. By transformative practice, this hidden core of the self "comes to the front," impressing its essential uniqueness more fully upon one's thoughts, emotions, and physical expression. See: Sri Aurobindo, *The Collected Works,* index in vol. 30, for entries under *psychic being, psychic entity,* and *jivatman.*

world while transcending all created things. And even in Buddhism with its doctrine of *anatta,* or no-self, there is a similar teaching. According to the Japanese Buddhist scholar Gadjin Nagao:

> The self is revived in Mahayana literature through the expression "great self" (*mahatmya*), a term which undoubtedly had affinity to the Universal Soul of *atman*-theory. The real awakening or the attainment of Buddhahood is explained as the annihilation of the "mean self" and the realization of "great self."[149]

The Mahayana designation of a great self is reflected, too, in the language of Shunryu Suzuki-roshi, the founder of San Francisco's Zen Center. Suzuki-roshi, for example, frequently used the term *big mind* in distinction to *small mind.* Thus:

> Water and waves are one. Big mind and small mind are one. When you understand your mind in this way, you have some security in your feeling. As your mind does not expect anything from outside, it is always filled. A mind with waves in it is not a disturbed mind, but actually an amplified one. Whatever you experience is an expression of big mind.
>
> The activity of big mind is to amplify itself through various experiences. In one sense our experiences coming one by one are always fresh and new, but in another sense they are nothing but a continuous or repeated unfolding of the one big mind.
>
> Dogen-zenji said, "Do not expect that all who practice zazen will attain enlightenment about this mind which is always with us." He meant if you think that big mind is somewhere outside yourself, outside of your practice, then that is a mistake. Big mind is always with us.[150]

Suzuki-roshi's imagery reflects Mahayana Buddhism's general emphasis upon the connection, rather than the distinction, between mean self and great self. Big mind and small mind are one. In contrast to yogas that emphasize detachment from things of this world, Zen practice orients us toward the unfolding of big mind in everyday experience.

Neo-Platonist seers, too, have described a secret connection between ordinary selfhood and our greater identity, though from a different perspective than that enjoyed by most Buddhist contemplatives. Plotinus, for example, held that each worldly self was based upon an eternal archetype. Before this birth, he asserted, we existed There as men different from those we are now, some of us even as gods, pure souls, intellect united with the whole of reality, parts of the world of *Nous*, not separated or cut off, belonging to the whole; and indeed we are not cut off even now.[151]

Describing his own experience of a higher self rooted in *Nous*, the Divine World of Forms, Plotinus wrote:

Often I have woken up out of the body to myself and have entered into myself, going out from all other things. I have seen a beauty wonderfully great and felt assurance that then most of all I belonged to the better part. I have lived to the full the best life and come to identify with the Divine. Set firm in It I have come to That Supreme Actuality, setting myself above all else in the realm of *Nous*. Then after that rest in the Divine, when I have come down from *Nous* to discursive reasoning, I am puzzled how I ever came down, and how my soul has come to be in the body when it is what it has shown itself to be by itself, even when it is in the body.[152]

The realization of an identity that transcends the constraints of ordinary selfhood has been celebrated as well by Christian and Islamic mystics. "My Me is God, nor do I recognize any other Me except my God himself," wrote Saint Catherine of Genoa.

To gauge the soul we must gauge it with God, for the Ground of God and the Ground of the Soul are one and the same. . . . The knower and the known are one. Simple people imagine that they should see God, as if He stood there and they here. This is not so. God and I, we are one in knowledge.

MEISTER ECKHART[153]

The supreme identity represented in these statements includes but does not depend upon the formations of normal self-consciousness, and is recognized by those who experience it to be the secret basis and fulfillment of ordinary personality. Evidently, that is why mystics in disparate cultures have described "a most real I" or "truest self." But this realization is only one side of our developing individuation. It is just part of our greater personhood.

Cognitive-Emotional-Behavioral Uniqueness

By discovering an identity beyond ordinary consciousness, we can exercise our different capacities more freely. We can enjoy countless states of mind and exercise different subpersonalities in the observing self, open space, emptiness, or big mind suggested by terms such as *purusha, sunyam,* and *mahatmya.* As we give up our compulsive cling-ing to particular types of activity, and as we realize that our essential being does not depend upon any one set of beliefs or behaviors, we have new freedom to broaden our experience. Relinquishing our need for an identity fundamentally separate from others, we find a security to act in new ways. That is why in Hindu lore saints rise above caste in their inspired but unconventional behaviors. The unpredictable ges-tures of love, the unique language, and the novel attitudes of mystics East and West are legendary. In their realization of God, brahman, *sunyam,* or the Tao, many saints become unforgettable personalities.

Broadly speaking, then, we can designate two aspects of meta-normal individuation: first, the realization represented by terms such as *atman* or *big mind*; and second, the unique combinations of ex-traordinary attributes that can develop in a particular person. Such personhood would constitute the metanormal analogue of individu-alization produced by earlier kinds of evolution.

[5.11]
LOVE

Loving behaviors are evident in many animal species, as for example in the self-sacrificing protection provided by cetacean mothers for their young, the mutual grooming of primates, and affectionate feline play. Both the animal and human worlds exhibit countless acts of car-ing and erotic delight. Indeed, human growth depends upon love from its very inception. No infant can survive without physical and emo-tional nurturance, nor enjoy anything approaching normal develop-ment without some sort of loving touch. No child can learn to talk, think, or be sociable without continuing affection from people around him. Love, like all human attributes, grows through its ex-ercise. Love grows through acts of love, both given and received.

And like our other capacities, love can flower in extraordinary ways, transforming this world to some extent while revealing new worlds to us. As we are conceived in love and brought forth from our mothers in love, we are opened by love to our greater possibilities. This fact of life has been celebrated since antiquity. In Plato's *Sym-posium* (210–212), Socrates describes love's progress from devotion to a single body to beauty in all bodies, and from devotion to bodies to beauty in laws, institutions, sciences, and wisdom, until on that shore he grows strong, and at last the vision is revealed to him of a single science, which is the science of beauty everywhere. . . . He who has been instructed thus far in the things of love and who has learned to see the beautiful in due order and succession, when he comes toward the end will suddenly perceive a nature of wondrous beauty, a nature which in the first place is everlasting, not growing and decaying, or waxing and waning; secondly, not fair in one point of view and foul in another, or at one time or in one relation or at one place fair, at another time or in another relation or at another place foul, as if fair to some and foul to others.

Remember how in that communion only, beholding beauty with the eye of the mind, he will be enabled to bring forth, not images of beauty, but realities (for he has hold not of an image but of a reality), and bringing forth and nourishing true virtue to become the friend of God and be immortal, if mortal man may.

Socrates' famous speech reflects certain features of love that are crucial to the human fulfillments described in this book. Central among these are love's entailment of other attributes, its transformation of those who give and receive it, and its power to bring forth "not images of beauty, but realities." Though the many varieties of love have unique properties, they all involve a complex set of capacities upon which our further advance depends. They all transform, even if momentarily, every person they touch, bringing new substance, beauty, and joy to lover, friend, child, or stranger.

Psychiatrist Rudolph von Urban, a student of Freud, described several kinds of erotic experience. One married couple told him, for example, that during moments of physical intimacy the wife was suddenly outlined with "a nimbus of greenish-blue light that radiated from her whole body." Another man and woman experienced an electrical flow through their skin: "a million sources of delight merged into one," they told von Urban.[154] Commenting upon such experience, the British poet Peter Redgrove quoted Saint Gregory of Nyssa's saying that "He, who has made his soul dry like the spider, has put on his aerial tunic. It extends from the head to the extremity of the feet."[155] Relating the saint's metaphor to an episode reported by a friend, Redgrove wrote:

He slept, and then woke a short time afterwards with a beautiful feeling from love-making, as though [his skin] were open and enlarged and no longer a barrier, and through it he could feel his wife sleeping by him

149

and interpenetrating his skin, as though their bodies had intermingled. . . . After lying there and enjoying this afterglow, he opened his eyes and found that the room was full of a golden-coloured gossamer arranged in a webwork that emanated from . . . centres of gold, and this webwork extended as if in care to the small bed of their daughter. He thought this was a dream reflecting his relaxed state until he saw the passage I have just quoted from St. Gregory.[156]

Many people say that during deep moments of love new energies or entities materialize in the space around them. Like Redgrove's friend, they sense new substance and feel their boundaries shifting. Essayist George Leonard described such experience. At dawn, he wrote, after a night of love,

consciousness itself began to change. Awareness of the different parts of our bodies which earlier in the evening had brought such delight had faded away, leaving only a generalized awareness of luminescent sinuosity. Separate acts had blended into a prolonged single movement. Even the divisions between waking and sleeping had become unclear. Please do not misunderstand me. I offer no expertise in these matters; such nights may be rarer for me than for you. But I must tell you that the moment did come when our own once separate and private emotions began to appear on each other's faces. Just that. Every flicker of feeling I might expect to originate inside me appeared instead on her face. . . . There was nothing metaphorical about this merging. In the faint light from another room, each of us could see our actual selves embodied in another—and we were terrified.

But that wasn't the end of it. The following afternoon a little miracle occurred. We were having a late lunch in a modest restaurant by the Pacific, seated at a table by a window overlooking a narrow strip of beach. It was a soft blue day with thin sheets of high white clouds. We merely toyed with our food. Gradually we became aware that the experience of the previous evening was coming back, only now it was happening gently and unthreateningly. We held hands across the table and surrendered. Little by little our separateness faded.

On the beach just outside our window, about seven or eight sandpipers were running back and forth looking for food. Sometimes they would run out on their incredibly fast legs after the receding water, then rush back just a fraction of an inch ahead of the incoming wave. But we didn't perceive them as separate from the wave. They and we and wave and sky and beach and sea and all in it and all existence was part of a single flow.

We stayed there all afternoon, not wishing to be anywhere else or to do anything else, seeing straight through the cardboard illusion of separateness. When the sun touched the Pacific the birds flew away. The tables were set for dinner. The cocktail crowd came in. And we were awakened from reality to the demi-existence of wanting and waiting that we all know so well.[157]

Erotic intimacy can be enjoyed with deliberate attention to states such as Redgrove and Leonard described. A woman who had practiced Tantric intercourse wrote that as she and her husband lay side by side,

> the energy between us was communicated [through our] eyes, and made me begin to cry. We were both feeling the same thing, without any words being spoken. It was as though we recognized one another, not merely from the objective point of view, but as though we were one entity, or one field. There was no obstruction or delineation between us. My heart felt like it had burst open.
>
> I was very vulnerable, very airy, very light. [My] whole body was breathing, in a certain sense. I was perceiving through the heart, not just through the eyes. It was like the . . . etheric being was in resonance with my husband's etheric being. Somehow this alignment occurred, and we were not limited to the flesh. . . .[158]

In erotic union, joy overflows, erasing boundaries between lover and beloved, deepening the care of each for the other. But this marriage of caring and delight does not depend upon sexual intimacy. It is a characteristic of love in all its expressions. In *The Varieties of Religious Experience,* James cited this account by Mrs. Jonathan Edwards of the love which arose from her spiritual illumination:

> Last night was the sweetest night I ever had in my life. I never before, for so long a time together, enjoyed so much of the light and rest and sweetness of heaven in my soul, but without the least agitation of body during the whole time. Part of the night I lay awake, sometimes asleep, and sometimes between sleeping and waking. But all night I continued in a constant, clear, and lively sense of the heavenly sweetness of Christ's excellent love, of his nearness to me, and of my dearness to him.
>
> When . . . I arose on the morning of the Sabbath, I felt a love to all mankind, wholly peculiar in its strength and sweetness, far beyond all that I had ever felt before. The power of that love seemed inexpressible. I thought, if I were surrounded by enemies, who were venting their malice and cruelty upon me, in tormenting me, it would still be impossible that I should cherish any feelings towards them but those of love, and pity, and ardent desires for their happiness. I never before felt so far from a disposition to judge and censure others, as I did that morning.[159]

"There is an organic affinity between joyousness and tenderness," James wrote. "Religious rapture, moral enthusiasm, ontological wonder, cosmic emotion, are all unifying states of mind, in which the sand and grit of selfhood incline to disappear, and tenderness to rule."[160] In a state such as Mrs. Edwards described, we give birth to energies, joy, and beauty beyond our normal experience. This regenerative, incarnational power of love is especially apparent in religious passion. "In the course of spiritual discipline," said Sri Ramakrishna, "one gets a 'love body' endowed with 'love eyes' and 'love ears.' One sees God with those love eyes. One hears the voice of God with those love ears. One even gets a sexual organ made of love . . . and with this love body the soul communes with God."[161]

In his own ecstatic devotion, Ramakrishna exhibited a striking physical radiance and highly contagious energy. During the initiation of his disciple Narendra (see chapter 5.3), and in the course of other meetings, people he touched felt a presence or force that had immediate physical effects. In his remarkable diary, Ramakrishna's follower "M." recorded several instances of such transmission, which appeared to involve something materialized by Ramakrishna's love of God.[162] People saw lights around the Indian saint, or felt a liberating power, or were enveloped by a palpable spiritual atmosphere, just as the persons noted above felt a new energy and presence around their lovers. Returning to the imagery of Socrates's speech, Ramakrishna brought forth not images of goodness and beauty but physical expressions of them. Indeed, religious ecstasies such as Ramakrishna's have been associated in most religions with the same bodily phenomena, including great beauty of voice and countenance, luminosities (5.1 & 22.5), superabundant vitality (5.4 & 22.6), and a sense of new boundaries or holy flesh (for example, Saint Gregory's "aerial tunic"). In every sacred tradition, certain men and women have through their love of God transformed their bodies and the atmosphere around them.

But the incarnational power of love is not limited to sexual or religious passion. It is evident as well in other kinds of relationship. Think of a struggling student you have known who exhibited new virtue or talent through a particular teacher's interest, or a normally glum acquaintance who was filled with new creativity by performing some generous deed, or a dispirited friend who was given new purpose and meaning by someone who appreciated his special gifts. There are many expressions of love, indeed many types of love, each with its own transformative power. Poets and philosophers have celebrated agape, goodwill, philanthropy, friendship, fellow feeling, empathy, congeniality, romance, married devotion, parental self-giving, and love for a work or set of ideals, as well as eros and religious devotion. Love takes many forms and has many kinds of effect but always gives birth to new life. In Romain Rolland's novel *Jean Christophe*, the protagonist loves another young man.

> During this honeymoon of their friendship, the first day of deep and silent rejoicing . . . they hardly spoke to each other, they dared hardly breathe a word; it was enough for them to feel each other's nearness, to exchange a look, a word in token that their thoughts, after long periods of silence, still ran in the same channel. Without probing or inquiring, without even looking at each other, yet unceasingly they watched each other. Unconsciously the lover takes for model the soul of the beloved: so great is his desire to give no hurt, to be in all things as the beloved, that with mysterious and sudden intuition he marks the imperceptible movements in the depths of his soul. One friend to another is crystal-clear: they exchange entities. Their features are assimilated. Soul imitates soul. . . .
>
> Christophe spoke in low tones, walked softly, tried hard to make no noise in his room, which was next to that of the silent Olivier: he was transfigured by his friendship: he had an expression of happiness, confidence, youth, such as he had never worn before. He adored Olivier. It would have been easy for the boy to abuse his power if he had not been so timorous in feeling that it was a happiness undeserved: for he thought himself much inferior to Christophe, who in his turn was no less humble. This mutual humility, the product of their great love for each other, was an added joy. It was a pure delight—even with the consciousness of unworthiness—for each to feel that he filled so great a room in the heart of his friend. Each to other they were tender and filled with gratitude.[163]

Such friendship, however, whether enjoyed by men or women, rarely continues in this state of grace. All honeymoons come to an end; every friendship needs to be cultivated. Love in all its greater expressions requires dedication as well as natural attraction, and steadfastness through many kinds of difficulty. But it is always renewable, and it can take root anywhere. Indeed, it is part of the genius of love that it can be summoned in situations where its existence at first seems impossible. In the novel *Incognito* by the Romanian writer Petru Dumitriu, this fact is revealed to the protagonist as he is brutally tortured. Why, he asks, had I needed to search so long? Why had I expected a teaching that would come from outside myself? Why had I expected the world to justify itself to me, and prove its meaning and purity? It was for me to justify the world by loving and forgiving it, to discover its meaning through love, to purify it through forgiveness.

They went on beating me, but I learned to pray while the screams issued mechanically from my ill-used body—wordless prayers to a universe that could be a person, a being, a multitude or something utterly strange, who could say?

This realization leads Dumitriu's protagonist to find possibilities for love in all circumstances, and thus to see everywhere the presence of Divinity. All events, all persons, are the incognitos of God.

If I love the world as it is, I am already changing it: a first fragment of the world has been changed, and this is my own heart. Through this first fragment the light of God, His goodness and His love penetrate into the midst of His anger and sorrow and darkness, dispelling them as the smile on a human face dispels the lowered brows and the frowning gaze.

Nothing is outside God. I have sought to love in as far as may be. I have tried to keep within the radiance of God, as far away as possible from His face of terror. We were not created to live in evil, any more than we can live in the incandescence that is at the heart of every star. Every contact with evil is indissolubly linked with its own chastisement, and God suffers. It is for us to ease His sufferings, to increase His joy and enhance His ecstasy. I made friends in the crowd, at meetings, in the sports stadium, and coming out of the cinema, as a rule only for a moment, linked with them by a friendly exchange of words, a smile, a look, even a moment of silence. And in closer contacts my discovery spread slowly but constantly from one person to another, in that dense and secret undergrowth which is wholly composed of personal events.[164]

But some of us will question Dumitriu's hero. Isn't it the case, we ask, that those who return only love for injustice surrender the world to thieves and murderers? Doesn't this imperfect life require that we oppose tyrants with force, cruelties with punishment? "We must frankly confess," wrote William James, "that in the world that actually is, the virtues of sympathy, charity, and non-resistance may be, and often have been, manifested in excess. The powers of darkness have systematically taken advantage of them."[165] To this concern, of course, we can respond as James did. If the world depended solely upon "hard-headed, hard-hearted, hard-fisted" methods, he wrote, it would be an unimaginable horror. Our everyday life depends upon countless acts of kindness. The overcoming of injustice requires love as well as strength. Without many kinds of charity, this world would not last for long. While love requires other virtues, among them courage in dealing with cruelty and aggression, it can bring forth goodness in all sorts of circumstance. It is the culmination of caring behavior evident in animal life, and our profoundest transformative act.

[5.12]
BODILY STRUCTURES, STATES, and PROCESSES

No matter which part of the psyche-soma spectrum it is focused upon, each transformative practice has physical effects,[†] many of which have been mapped through medical research. In subsequent chapters, I discuss beneficial somatic changes produced by hypnotic suggestion (15), biofeedback training (16), Autogenic Training (18.3), Structural Integration (18.5), Progressive Relaxation (18.6), physical fitness training (19.2), martial arts (20), yoga and Zen Buddhist practice (23.1), and meditation in a variety of secular settings (23.2). Numerous clinical and experimental studies have shown that each of these disciplines affects the flesh as well as the psyche. Medical science has demonstrated beyond reasonable doubt that each of the practices just noted can benefit our physical functioning as well as our mental-emotional life.

That we can creatively alter our muscles, organs, cells, and molecular processes is clearly established. Given that fact, we can wonder about the limits of self-induced bodily change. Might our bodies accommodate alterations beyond those presently mapped by medical science? Since new abilities among our animal ancestors were in many instances made possible by alterations of their bodies, we can suppose that analogous changes—developed through practice rather than natural selection—might accompany and support a lasting realization of metanormal capacities. This supposition is supported by a considerable religious and shamanic lore about extraordinary somatic processes. For example, the esoteric anatomies of some Indian yogas

---

[†] Depending upon its primary focus, a transformative practice might be deemed to be either psychosomatic or somatopsychic. Meditation and sustained imagery practice, for instance, are examples of the former, sports and fitness training examples of the latter; while martial arts that give equal emphasis to physical and mental training lie somewhere in between.

involve *nadis,* or energy channels; chakras, organizing centers; and kundalini power. By stimulating these, it is said, yogis produce radiance of eyes and skin, supernormal suppleness of joint and muscle, and other dramatic improvements of the flesh.

But in supposing that metanormal development will require new somatic structures and processes, I do not propose that we uncritically accept esoteric anatomies. Accounts of chakras, *nadis,* or kundalini may be imperfect representations of the psychophysical processes that facilitate extraordinary functioning. None of them has been confirmed by modern science, and different schools give different versions of them. In his scholarly *Layayoga,* for example, Shyam Sundara Goswami (1891–1978) catalogued accounts of *nadis* and chakras in the Vedas, Upanishads, and Tantra Shastras that differ from descriptions of such entities by other scholars. Satyananda Saraswati, for example, has described seven chakras below the seven traditionally assigned to points running from the base of the spine to the top of the head.[166] According to Saraswati, there are 14 major chakras and several minor ones. And how do we reconcile descriptions of different *ochemata,* or subtle bodies, such as the *soma pneumatikon, eidolon, imago, simulacrum, skia,* and *umbra* in mystical literature of Greco-Roman antiquity?[167] Or the different occult anatomies of Islamic mystics described, for example, in Henri Corbin's *Spiritual Body and Celestial Earth*?[168] Or the different accounts of Buddha Bodies in Chinese and Tibetan yoga lore?[169] Given the many differences among them, we cannot automatically accept esoteric teachings about metanormal embodiment.

Still, we can begin to picture somatic alterations that might accompany the further reaches of extraordinary functioning, supposing: first, that accounts of chakras, kundalini, and the like, though fanciful in some respects, might reflect actual developments of physical structures unrecognized by medical science; second, that metanormal capacities require distinctive types of supporting process; and third, that we can extrapolate from physiological changes revealed by modern research in imagining somatic developments required for high-level change. For example, we might guess that molecular and atomic interactions within cells are altered to facilitate extraordinary vitality and movement. The marvelous flexibility exhibited by sportspeople in certain inspired performances, the astonishing distortions of limbs and torso experienced by some Catholic saints (22.9), and the breathtaking agility of certain dancers and shamans indicate that muscles, tendons, and ligaments are, under some conditions, capable of extraordinary elasticity. Might such elasticity be taken to new levels still? As noted in chapter 19.2, a great many cellular changes now specified by medical research are produced by fitness training, while new levels of fitness continually facilitate new sports records. No one knows how far such bodily changes and record-breaking might go. Could a culture devoted to integral transformations of mind and flesh push our limits in analogous fashion? Kundalini-type experience, for example, may well involve the release of extraordinary forces from the body's most basic constituents. Do yogic luminosities and odors of sanctity involve energies that arise from some sort of elemental repatterning? Metanormal restructuring of the body, in short, might involve atomic or molecular reformations that would eventually change the look, feel, and capacities of tissues and cells.

Furthermore, most sacred traditions offer a long-standing witness to subtle bodies that resonates with such conjectures. There exists, for example, a lore in Tibetan Buddhism about the diamond body, and in Taoism about the spirit child (8.2), that can be created within the flesh to support and express metanormal capacities. Similarly, the *sariras, koshas,* and *dehas* of Indian yoga, the *jism* of Persian Neo-Platonism and Sufism, and analogous entities in other religious lineages can be developed so that they eventually alter the physical body. It is also said that our present flesh can open to materializations or infusions of subtle matter from the greater worlds in which we are embedded (see chapter 9). In *Spiritual Body and Celestial Earth,* Henri Corbin summarizes the teaching of Iranian seers about the influx of such matter from "the heavens of Hurqalya."[170] Here, once again, a long-established teaching corresponds with modern experience, for some people today report new substances, effervescences, and energies that seem to rise in their bodies as a result of transformative practice.[171] A systematic comparison of such reports with esoteric doctrines of subtle matter would, I believe, yield new insights about the body's capacity for dramatic restructuring (see appendix F).

[5.13]
DIFFERENCES IN AVAILABILITY AMONG METANORMAL CAPACITIES

No human attribute stands alone, even those that appear to be given by ego-transcending agencies. Though there is no simple English equivalent to the Stoics' term *antakolouthia,* mutual entailment of the virtues, we need the time-tested insight it stands for, which is embedded in many transformative practices. Christian contemplative discipline tries to correct egoistic volition by an emphasis upon surrender to God, Buddhism by doctrines of nonattainment, Taoism by encouraging reliance on the Tao. Sophisticated psychotherapists try to balance upsetting self-awareness with self-acceptance, openness with empathy, honesty with kindness. The best somatic educators resist fixed images of physical functioning, having learned that each body's unique complexity must be respected. Aikido, tai chi, and other martial arts join force and surrender, initiative and harmony, hardness and softness in their complex repertoires. Having observed the destructive results of one-sided aspiration, many contemplative teachers, psychotherapists, somatic educators, and martial artists have emphasized wholeness in their programs for growth. However, in practice wholeness is not always respected. Virtually every transformative program has been abused. To develop the attributes described here, we must learn from both the positive and negative results of religious, therapeutic, and other disciplines.

But destructive practices are balanced to some extent, I propose, by a general feature of human nature, namely that some human potentials are more available than others. Studies of metanormal experience such as those listed in appendices A.3 and A.4 suggest that those capacities most crucial to human goodness and growth—namely love, need-transcending joy, and perceptions of oneness with others—are more commonly experienced than potentially dangerous or distracting phenomena such as kundalini or internal clairvoyance. Furthermore, it is usually the case that these three attributes can be exercised with more immediate and rewarding results than the others described here. All of us can reach out with love in most circumstances: there are countless ways to perform acts of kindness. All of us can cultivate equanimity in response to events that usually make us suffer, and in so doing experience a regenerative delight. All of us can conceive our solidarity with others, imagine a oneness with the world at large, and thus awaken to the unitive knowing that crowns human cognition. In these, our most available potentials for growth, we find the surest way to our greater fulfillment. It is more difficult, more dangerous, and less rewarding to emphasize phenomena such as kundalini or metanormal kinesthesis at the beginning of transformative practice.

Differences in the availability of metanormal attributes is a blessing, it seems to me, another sign that a comprehensive wholeness informs living systems. We are tilted in the right way, as it were, toward balanced growth. In this respect, our species as a whole may have a fortunate orientation toward the future, though any one of us can lose our way through misguided practice, destructive social reinforcements, our darker motives, or sheer perversity.

6

CULTURE, GENES, AND EXTRAORDINARY FUNCTIONING

[6.1]

CULTURE'S INFLUENCE UPON INDIVIDUAL FUNCTIONING

The psychophysical alterations described in this book dramatize human nature's enormous capacity for regressive, progressive, or merely curious change. But accounts of these alterations, by their sheer oddity, often arouse our skepticism. Given the strangeness of such phenomena, it is useful to remember that most cultures produce forms of behavior and consciousness that seem strange to other peoples. Westerners observing Hindu festivals today are sometimes as amazed as their counterparts three hundred years ago at what they take to be excesses of Indian religious fervor. The ecstasies of certain shamans have alarmed some anthropologists who observed them. By their differences in physical deportment, valued emotions, and metaphysical assumptions, different peoples have long dumbfounded one another. Contemplating some of the extraordinary experiences described in this book, I have moderated my own skepticism by remembering that these experiences happen in cultures that are different from mine. To appreciate certain potentials for growth, we need to suspend disbelief and some of our aesthetic predispositions.

But there is another, more fundamental reason for emphasizing culture's formative role in human development, namely that *exceptional abilities develop most fully in cultures that prize them*. With some notable exceptions, they are valued by the groups in which they flourish. And conversely, such abilities are often distorted or inhibited by social conditioning. Some athletes, for example, exhibit a sensitive self-control that suggests they might be accomplished yogis if they lived in Hindu culture; and certain fantasy-prone personalities today exhibit a responsiveness to imagery that might have made them gifted contemplatives in other times and places (15.6). Indeed, no aspect of human nature is immune to social influence, not even those mystical illuminations that appear to transcend all conditionings. Like all our other capacities, metanormal cognition is subject to culture's formative agencies. Because culture constantly shapes us, reinforcing or extinguishing our greater possibilities, we need to know what science and scholarship tell us about its formative power.

From Aristotle to Hegel, Durkheim, and Marx, scholars and philosophers have observed that we are social animals, and that different milieus produce different kinds of human functioning. This enduring insight has been enlarged upon by twentieth-century anthropologists who have found that parts of our makeup once thought to be universal are acquired by socialization. Even gestures and facial expressions once thought by most biologists and social scientists to be inherited from our animal ancestors are now seen to be shaped to a significant extent by social practice.† Our very physiques are affected by the expectations and habits of our nation, family, and class.

In 1935, Marcel Mauss, a student of Emile Durkheim, published an influential paper *Les Techniques du Corps*, in which he proposed a three-leveled study of human nature: "We are everywhere faced with physio-psycho-sociological assemblages of series of actions," he wrote. "It is the triple viewpoint, that of the 'total man', that is needed."[1] Since Mauss published his essay, social scientists in Europe, England, and the United States have explored culture's formative influence upon *l'homme total*. The anthropologists Jane Below, Flora Bailey, and George Devereaux have shown that certain physical behaviors among the Balinese, the Navaho, and the Mohave are unique.[2] Margot Astrov has correlated Navaho motor habits with the unique language and mythology of Navaho culture.[3] Weston Labarre has described gestures and their related meanings among peoples throughout the

---

† In *The Expression of the Emotions in Man and Animals* (1872), Charles Darwin developed the idea that most psychophysical expression is invariant across cultures, being transmitted from generation to generation through biological inheritance. "As far as we can judge, only a few expressive movements . . . are learnt by each individual," he wrote, "that is, were consciously and voluntarily performed during the early years of life for some definite object, or in imitation of others, and then became habitual. The far greater number of the movements of expression, and all the more important ones, are, as we have seen, innate or inherited; and such cannot be said to depend upon the will of the individual" (p. 352). Contemporary anthropological fieldwork, however, has shown that human gestures and facial expression are more culturally determined than Darwin realized; see, for example, Hinde 1972.

world.[4] And David Efron has used large sample populations to show that assimilated Jews and Italians in New York City differ greatly in their gestures from unassimilated members of the same populations.[5]

Posture, too, has been shown to be culturally influenced. In his wide-ranging survey "World Distribution of Certain Postural Habits" (1955), Gordon Hewes compared styles of sitting, squatting, kneeling, and standing among peoples in Europe, Asia, Africa, and the Americas. Hewes said that his work explored the borderline between culture and biology. Although some human postures may be archetypical or precultural, he wrote,

> others have probably been diffused like other items of culture. The extent and vigour of postural etiquette evidently varies from one culture or area to another, with some societies going to great lengths to ensure postural propriety on all public occasions. Many cultures maintain careful distinctions in posture on the basis of sex, and there are others which emphasize age and status considerations in the manner of sitting or standing. Postural conformity is enforced as a rule by the same methods as conformity to other rules of etiquette—by ridicule, verbal scolding or by physical punishment where deviation from the postural norms verges on lese-majesty or deliberate indignity to a superior. While our culture has perhaps relaxed its postural codes since the nineteenth century, certain areas of it preserve archaic postural etiquettes backed up by formidable sanction as in the military drill regulations.[6]

The British ethnographer Raymond Firth studied culture's effect upon physical attitudes; and Edward Sapir proposed that gesture is a code that must be learned to be successful.[7] Theory and fieldwork developed by these and other social scientists contributed to the study of proxemic behavior (the use of spatial arrangements by humans in their relationships), established by American anthropologist Edward Hall; and to kinesics, the systematic study of body language developed by American communications theorist Ray Birdwhistell.[8] But social conditioning affects more than our postures, gestures, and spoken language. Every attribute of *l'homme total* is shaped to some extent by culture. Appreciating this fact of human functioning, English sociologist Mary Douglas promoted the interdisciplinary research proposed by Marcel Mauss. Joining biological, psychological, and sociological perspectives in her own work, Douglas developed the idea that the human body is an image of society.

The body communicates information for and from the social system in which it is a part. It should be seen as mediating the social situation in at least three ways. It is itself the field in which a feedback interaction takes place. It is itself available to be given as the proper tender for some of the exchanges which constitute the social situation. And further, it mediates the social structure by itself becoming its image.[9]

The social body constrains the way the physical body is perceived. The physical experience of the body, always modified by the social categories through which it is known, sustains a particular view of society. . . . As a result of this interaction the body itself is a highly restricted medium of expression.[10]

In the spirit of Marcel Mauss and Mary Douglas, French sociologist Pierre Bourdieu described some effects of social norms upon the bodies and manners of French working people. Bourdieu has shown that differences in gesture, clothing, eating habits, and physique are central in establishing one's place in a given culture.

While the working classes are more attentive to the strength of the (male) body than its shape, and tend to go for products that are both cheap and nutritious, the professions prefer products that are tasty, health-giving, light and not fattening. Taste, a class culture turned into nature, that is, embodied, helps to shape the class body. It is an incorporated principle of classification which governs all forms of incorporation, modifying everything that the body ingests and digests and assimilates, physiologically and psychologically. It follows that the body is the most indisputable materialization of class taste, which it manifests in several ways. It does this first in the seemingly most natural features of the body, the dimensions (volume, height, weight) and shapes (round or square, stiff or supple, straight or curved) of its visible forms, which express in countless ways a whole relation of the body, i.e., a way of treating it, caring for it, feeding it, maintaining it. . . .

The quasi-conscious representation of the approved form of the perceived body, and in particular its thinness or fatness, is not the only mediation through which the social definition of appropriate foods is established. At a deeper level, the whole body schema, in particular the physical approach to the act of eating, governs the selection of certain foods. For example, in the working classes, fish tends to be regarded as an unsuitable food for men, not only because it is a light food, insufficiently "filling," which would only be cooked for health reasons, i.e., for invalids and children, but also because, like fruit (except bananas) it is one of the "fiddly" things which a man's hands cannot cope with and which make him childlike (the woman, adopting a maternal role, as in all similar cases, will prepare the fish on the plate or peel the pear); but above all, it is because fish has to be eaten in a way which totally contradicts the masculine way of eating, that is, with restraint, in small mouthfuls, chewed gently, with the front of the mouth, on the tips of the teeth (because of the bones). The whole masculine identity—what is called virility—is involved in these two ways of eating, nibbling and picking, as befits a woman, or with whole-hearted male gulps and mouthfuls, just as it is involved in the two (perfectly homologous) ways of talking, with the front of the mouth or the whole mouth, especially the back of the mouth, the throat. . . .

This opposition can be found in each of the uses of the body, especially in the most insignificant-looking ones, which, as such, are predisposed to serve as "memory joggers" charged with the group's deepest values, its most fundamental "beliefs." It would be easy to show, for example, that Kleenex tissues, which have to be used delicately, with a little sniff from the tip of the nose, are to the big cotton handkerchief, which is blown into sharply and loudly, with the eyes closed and the nose held tightly, as repressed laughter is to a belly laugh, with wrinkled nose, wide-open mouth and deep breathing ("doubled up with laughter"), as if to amplify to the utmost an experience which will not suffer containment, not least because it has to be shared and therefore clearly manifested for the benefit of others.

And the practical philosophy of the male body as a sort of power, big and strong, with enormous, imperative, brutal needs, which is asserted in every male posture, especially when eating, is also the principle of the division of foods between the sexes, a division which both sexes recognize in their practices and their language. It behooves a man to drink and eat more, and to eat and drink stronger things. Thus, men will have two rounds of aperitifs (more on special occasions), big ones in big glasses (the success of Ricard or Pernod is no doubt partly due to its being a drink both strong and copious—not a dainty "thimbleful"), and they leave the tid-bits (savoury biscuits, peanuts) to the children and the women, who have a small measure (not enough to "get tipsy") of home-made aperitif (for which they swap recipes).

The body, a social product which is the only tangible manifestation of the "person," is commonly perceived as the most natural expression of innermost nature. There are no merely "physical" facial signs; the colour and thickness of lipstick, or expressions, as well as the shape of the face or the mouth, are immediately read as indices of a "moral" physiognomy, socially characterized, i.e., of a "vulgar" or "distinguished" mind, naturally "natural" or naturally "cultivated."

Thus one can begin to map out a universe of class bodies, which (biological accidents apart) tends to reproduce . . . the social structure.[11]

I have quoted Bourdieu at length to suggest the many ways in which cultural influences work upon us. Consider, for example, the sources of your own food selection and favored images of physique. What expectations of family or friends have shaped your exercise, manners, and eating habits? And might you be influenced by activities in other cultures? A neighbor of mine back from climbing in the Himalayas said that his Sherpa guides carried tape recorders, listened to Michael Jackson, hiked in Nike running shoes, and wore shirts from California. Cultural conditioning is mediated now by worldwide communications.

Research psychiatrists and psychologists, too, have explored culture's effect upon human development. Seymour Fisher and Sidney Cleveland, for example, described variations of body image in several cultures, and differences in body-boundary maintenance among different groups.[12] Wilhelm Reich proposed certain relations between character formation, "body armor," and social conditioning, placing work with the body at the center of his therapy (18.8). Since Freud, the links between personal psychodynamics and enculturation have been explored from many angles, so that today social scientists generally agree that our body, our feelings, and our thoughts are shaped by the norms, expectations, and rewards of our social milieus.[13]

And culture shapes metanormal capacities. Mystical illuminations, for example, are subject to social influence, even if they are rooted in a primordial Ground that transcends all conditionings. Such cognitions take different forms in different religious traditions, while exhibiting certain features that appear to be universal. Like types of ordinary consciousness, they seem to have both socially determined and invariant attributes. Some writers, among them Aldous Huxley, Walter Stace, and Frithjof Schuon, have emphasized the universal features of mystical experience, while others such as William James have examined its varieties as well as its universality,[14] and still others have emphasized its radical differences from culture to culture. The religious scholar Steven Katz, for example, has argued that there is no "perennial philosophy."

> Neither mystical experience nor more ordinary forms of experience give any indication, or any grounds for believing, that they are unmediated. That is to say, all experience is processed through, organized by, and makes itself available to us in extremely complex ways. The notion of unmediated experience seems, if not self-contradictory, at best empty. This epistemological fact seems to me to be true because of the

sorts of beings we are, even with regard to the experiences of those ultimate objects of concern with which mystics have intercourse, e.g. God, Being, nirvana, etc. This "mediated" aspect of all our experience seems an inescapable feature of any epistemological inquiry, including the inquiry into mysticism, which has to be properly acknowledged if our investigation of experience, including mystical experience, is to get very far.

The Hindu mystic does not have an experience of x which he then describes in the, to him, familiar language and symbols of Hinduism, but rather he has a Hindu experience, i.e., his experience is not an unmediated experience of x but is itself the, at least partially, pre-formed anticipated Hindu experience of Brahman. Again, the Christian mystic does not experience some unidentified reality, which he then conveniently labels God, but rather has the at least partially prefigured Christian experiences of God, or Jesus, or the like.

The significance of these considerations is that the forms of consciousness which the mystic brings to experience set structured and limiting parameters on what the experience will be, i.e. on what will be experienced, and rule out in advance what is "inexperienceable" in the particular given, concrete context. Thus, for example, the nature of the Christian mystic's pre-mystical consciousness informs the mystical consciousness such that he experiences the mystic reality in terms of Jesus, the Trinity, or a personal God, etc., rather than in terms of the non-personal . . . Buddhist doctrine of nirvana.[15]

Katz edited a collection of essays, *Mysticism and Philosophical Analysis*, which emphasized the fact that mystics in different traditions have different types of spiritual realization. That this is the case is obvious to most students of religious life, including philosophers who believe that there is a universal core in all mystical realization.[†] That mystical experience is culturally conditioned does not, however, mean that it gives birth only to illusions. Establishing culture's role in such experience, Katz wrote,

is not to evaluate either the truth-claims of the sought and reported experiences in Judaism and Buddhism, etc., or to presume to rank them in terms of better or worse. What I wish to show is only that there is a

---

† Philosopher Huston Smith, who has argued in favor of the perennial philosophy or universal core position, wrote, for example, that "mystics in different traditions, and to some extent in different pockets of the same traditions, 'see' different things. This is overwhelmingly the case, of course. The question is whether, amidst these manifold differences, which no one disputes, there is one form of mystical experience that *is* cross-culturally identical" (Smith, H., 1987).

clear causal connection between the religious and social structure one brings to experience and the nature of one's actual religious experience.[16]

But I do not wish to enter the debate between scholars who, like Katz, emphasize the culturally determined differences between religious realizations, and philosophers, such as Schuon and Stace, who emphasize the universal core in all mysticism. While it is obvious that there are many differences between the cultural determinations, realizations, and reports of mystics in different traditions, it is obvious as well that adepts in different religions report and exhibit many similar or identical metanormal capacities. That this is the case becomes especially evident when we consider the wide range of phenomena described in this book. Similar kinds of extraordinary perception, kinesthesis, communication, vitality, movement, environmental influence, delight, cognition, volition, love, and sense of self are reported in different sacred traditions. Furthermore, many such experiences are reported by athletes and martial artists (19.6, 20.1) and by people who do not engage in any transformative practice (4.3, 14). Testimonies by people in different cultures to the 12 classes of metanormal functioning described in chapter 5 have many striking similarities. These undeniably similar, and sometimes identical, reports strongly suggest that all humans share the same potentials for extraordinary life.

At the same time, however, the metanormalities described in this book also demonstrate cultural shaping. In virtually all sacred traditions, for example, mystical cognition is honored more than are extraordinary vitality, kinesthesis, or motor abilities. Indeed, as noted in chapter 7.1, certain capacities evoked through religious practice (siddhis, charisms, etc.) are neglected or actively suppressed. This is a point I return to in subsequent chapters. Every religious school—like every society—reinforces just some of our potentials while extinguishing others, and thus develops its own favored profile of exemplary functioning. Like the working-class people described by Bourdieu, saints and mystics are powerfully shaped by their respective cultures. Indian yogis, for example, do not exhibit stigmata resembling Christ's crucifixion wounds, nor for that matter do Eastern Orthodox monks (22.3). Only in Western Christendom, and most particularly in Roman Catholicism, have there been numerous ecstatics whose hands and feet bleed on Good Friday. Nor do Roman Catholic saints typically exhibit physical agility like the *lung-gom* walkers of Tibet (5.6) or Zen-inspired martial artists (20.1). I emphasize this point because a many-sided growth toward integral embodiment will be conditioned—and thus helped or hindered—by the groups we belong to. In later chapters, and especially Part Three, I suggest certain principles by which to establish effective social support for a balanced and inclusive metanormal development.

[6.2]
THE GENETIC BASIS OF METANORMAL ATTRIBUTES

Studies such as those cited in the preceding section help us appreciate culture's formative influence upon individual development. We are social creatures at every level of our functioning, in each domain of experience. The most independent ascetic is shaped by postures he learns from his parents or peers, by emotions his community values, by concepts he receives from priests and philosophers. The fiercest hermit carries his society with him into the wilderness. This is either good news or bad news, of course, depending upon the quality of our culture's conditionings. We acquire both strengths and weaknesses from our families, friends, and teachers. Our communities reinforce some attributes that facilitate further growth, and others that hinder it.

How, then, does our genetic inheritance predispose us to high-level change? Do all of us, most of us, or just a few of us have the right genetic stuff for integral development? Given our limited understanding of human nature and the great differences between individuals, we cannot now answer these questions with great certainty, but we can make some preliminary observations.

First, no matter how many genetically transmitted talents or traits any one of us might have to support transformative practice (such as mental energy, emotional range, or physical stamina), no one is predisposed to have *all* the extraordinary attributes described in this book. If there are exceptions to this rule, they are unknown to me and the experts on human development I have consulted. Even the most gifted among us have limitations and disabilities.

Second, few of us can pinpoint all the talents for high-level change we possess, given our selective cultural conditioning. Anthropology and sociology have shown that each culture nurtures just some of our capacities while neglecting or suppressing others. As we have seen, this selectivity extends to modes of carriage, types of movement, patterns of emotion, styles of thought. And since that is the case, it is impossible to specify how each of us may be disposed to the various metanormal attributes noted in this book, some of which are not valued—or even recognized—in most communities today. For example, you may possess a telepathic empathy which none of your teachers has appreciated. During athletic performance, a young friend might experience ESP that none of his coaches recognizes. A family member might hear suprasensory music that he or she is afraid to describe. Indeed, the history of intelligence testing should make us cautious about making judgments regarding *anyone's* developmental possibilities. For more than half a century now, psychometricians have added new scales to mental tests as they have identified components of cognitive ability that had not until then been measured. Psychologist Howard Gardner, for example, has proposed that at least six major types of intelligence exist: linguistic, musical, logical-mathematical, spatial, kinesthetic, and personal.[17] Memory, too, has been found to be more complex than psychologists once believed. It has been shown, for example, that some people can remember things that occurred during the first months of their lives and that most of us can best recall certain images, perceptions, and events in particular mental states.[18] Given the recognition that human capacities are more complex than scientists once thought, it is reasonable to suppose that even more human abilities will be acknowledged in years to come.

Third, nonordinary states or processes—call them the graces of God, the workings of Buddha Mind, the powers of Tao, or agencies of emergent supernature—can lift us beyond ourselves, strengthening capacities we already possess or conferring new ones upon us. This book presents many examples of radical and unexpected self-exceeding. The type of love I call metanormal because it transcends ordinary needs is experienced by all sorts of people, some of whom are normally mean-spirited. Mystical exaltations have occurred among men and women who had not previously exhibited contemplative gifts. Astonishing agility has been exhibited by people whose athletic performance is usually undistinguished. Such phenomena suggest that each of us to some extent can realize our own evolutionary transcendence, our own exceeding of genetic endowment. In a manner analogous to the large jumps in development that occurred during earlier stages of evolution, we are sometimes lifted to new levels of excellence. The fact that such transcendence is not wholly predictable from what we know about genes and culture precludes our specifying the limits of anyone's potentials for growth.

Fourth, a number of surveys have shown that spiritual illuminations and other extraordinary experiences occur among large percentages of various subject groups (see appendix A.4). Such studies indicate that people of radically diverse talents, education, and temperament have luminous cognitions, emotions, and perceptions, often without benefit of creative discipline (see chapter 4.3). Given the results of such research, it is difficult to argue that metanormal capacities are limited to a few of us with special genetic gifts or fortunate cultural circumstances. To adopt a Buddhist metaphor, these surveys suggest that all of us can become "stream enterers" as we join the river of companionship, teaching, and discipline that flows toward extraordinary life. The assertions of religious teachers East and West that every human can be enlightened resonates with the evidence for human transformative capacity presented in this book. There is good reason to believe that we all can progress in transformative practice, and that we each have access to subliminal processes that help us. Our genetic endowments, shaped (for good or ill) by our respective cultures, can be developed by discipline and lifted by powers beyond our normal purview. In later chapters I discuss certain relations between nature, nurture, practice, and grace.

7

Philosophy, Religion, and Human Development

IN THIS CHAPTER, I briefly describe some ideas that help us, as well as some that do *not* help us, understand our transformative capacities. To help us understand and cultivate extraordinary functioning, such ideas must be refined or abandoned in light of our developing experience. They must be refashioned or rejected as we acquire new data from the natural and human sciences, psychical research, comparative religious studies, and other activities.

[7.1]
Metanormal Capacities and Grace as Features of Human Evolutionary Advance

Siddhis and Charisms

Most siddhis of Hindu-Buddhist lore, charisms of Catholic saints, and similar phenomena in other religions have been regarded in their respective traditions as inferior to enlightenment or union with God. Their inferior status is reflected in the Hindu adage "*moksha* before siddhi," liberation before power; the Roman Catholic doctrine that charisms, being *gratiae gratis datae*, gifts freely given, are not in themselves signs of virtue or sanctity; and the Zen Buddhist view of them as potential distractions from (or phenomena preparatory to) enlightenment. But from an evolutionary perspective, they can be seen in another way—as emergent features of human development, as

171

capacities inherent to the richer life that is available to us. Charisms and siddhis, when viewed in their rich diversity, signal a multileveled transformation of our entire nature. These dynamic by-products of religious practice can be seen not as hindrances to life's highest good, but as budding faculties of our greater humanity.

It is understandable, though, that most religious traditions have viewed many capacities of this kind with suspicion. First, some of them, such as metanormal kinesthesis or vitality, appear only transiently during contemplative practice and can seem alien or even perverse. Second, because some of them can stimulate egocentricity and various kinds of destructive activity, practical wisdom dictates their subordination to unitive awareness and loving service. And third, in earlier times they could not be seen in the context I am exploring because their host traditions were formed before the facts of evolution were discovered. When the Upanishads, Buddhist sutras, and Tao Te Ching were composed—indeed, when the first teachings of all the great contemplative schools were formulated—their authors did not have our modern knowledge of cosmic and biological development. Though the sacred traditions nurtured understanding of numerous extraordinary capacities, they could not comprehend the history of advances in structural complexity, behavioral repertoire, and awareness among living species. To the founders of most mystical philosophies, the world was not an arena in which livelier, more conscious forms appeared in the course of time, but a place marked essentially by suffering and death. For early Buddhists, the world apprehended through the ordinary senses was *samsara,* a grinding wheel of death and rebirth. For Neo-Platonists, it was the lowest level on the great scale of being. For many, if not most, Christian contemplatives, it was separated twice from God, by Creation and by the Fall.[1] Though all religious traditions have supremely life-affirming aspects, they have also produced powerful language about the world's uncertainty, its essential misery, its basic immunity to progress. Viewing the manifest world as a place to escape from, some of the greatest mystics have helped orient religious practice away from the integral development of human nature. "The body is a dung heap," wrote Thomas à Kempis. An illumined soul in the flesh, said Sri Ramakrishna, is like an elephant breaking out of a flimsy hut. For many ascetics to whom the body exemplifies the earth's transience and suffering (instead of its potential for extraordinary life), it seems logical to seek deliverance from embodiment, and sensible to think that most charisms or siddhis have no more value than any worldly capacity that supports our inferior state.

Such attitudes need not rule our thinking, however, for the discoveries of modern science give us new perspectives on the world and our human capacities. Our planet has been revealed in a new light, not as a static or cyclical world, but as an arena in which graduation upon graduation of species have occurred for several hundred million years. This stupendous advance suggests that humans might develop further. Evolution to date is a supreme inescapable gesture, pointing toward a mysterious future for living forms, leading us to suppose it could continue through further eons. Indeed, it has even exceeded its established laws and patterns. Because evolution has gone beyond its own bounds before—as, for example, when life arose from inorganic matter and humankind from its primate ancestors—and because we harbor transformative capacities such as those described in this book, it is not unreasonable to think that, in spite of our many liabilities, further progress, even a new kind of evolution, might be available to us.

The universe itself, as revealed to us by modern science, invites us to open our imagination and correlate our sense of human possibilities with evolution's dynamism. A vision of human advance embedded in the facts of cosmic development, it seems to me, gives new perspective to certain phenomena associated with religious life, including charisms and siddhis. For in order to realize a richer existence upon earth, we need capacities for creative interaction with our material existence. To interact with other people and our physical environment in the light of a higher consciousness, we need perceptual, communication, and movement abilities, volition, kinesthesis, and vitality, as well as the unitive awareness, love, and delight that are understandably celebrated in the sacred traditions. The Indian philosopher Sri Aurobindo wrote:

> We need not shun the siddhis and cannot shun them. There is a stage reached by the yogin when, unless he avoids all action in the world, he can no more avoid the use of the siddhis of power and knowledge than an ordinary man can avoid eating and breathing; for these things are the natural action of the consciousness to which he is rising, just as mental activity and physical motion are the natural action of man's ordinary life. All the ancient *rishis* used these powers, all great avatars and yogins have used them, nor is there any great man . . . who does not use them continually in an imperfect form, without knowing clearly what are these supreme faculties that he is enjoying.[2]

But to repeat: "siddhis of power and knowledge" such as Aurobindo referred to are often neglected by the very traditions in which they most dramatically arise. That is the case, in part, because a self-reinforcing suspicion of them arose as I have explained. In following the time-tested wisdom that a purging of egocentricity is necessary for enlightenment or union with God, many ascetics have spurned metanormal capacities that their discipline opened to them. Indeed, many yogis and monks have suppressed certain *normal* capacities. Ascetic excesses of some Christian desert fathers were widely emulated in the fourth and fifth centuries and into medieval times.[3] Sri Chaitanya (b. 1485), one of India's most beloved saints, spent the last years of his life in nearly continuous trance, setting an example that countless devotees have tried to imitate.[†] Though such extremes have not been the rule in religious life, they have existed in every tradition. In the great religions, there is a gamut of transformative activities ranging from those that drastically suppress mind and flesh to Zen-inspired martial arts (20.1), Taoist disciplines for bodily transformation (8.2, 21.5), and Hasidic celebrations of the holy in everyday life. These many types of discipline are represented in the graph shown as Figure 7.1.

In this figure, the diagonal line represents a development of human nature in which mind, the feelings, and flesh are *fundamental* to creative advance, not instrumental merely, nor detrimental, as they are deemed to be in approaches signified by lines near the vertical. On the diagonal line, siddhis of power and knowledge would have an important place as budding features of our greater nature. On it, our world-engaging capacities would be cultivated within a deepening spiritual realization and would be viewed as necessary to our development rather than as impediments to it.[‡]

But to be more precise, let us characterize integral practices as occupying a fairly broad portion of the spectrum in which individual lines of advance might fall. Given the extent of individual differences in all dimensions of human nature, many kinds of balanced progress can occur. This is signified in Figure 7.2.

[†] In Pondicherry, India, I watched a yogi in trance being fed by his followers. His eyeballs were rolled back so that only the whites of his eyes could be seen; he sat erect in the lotus position; and his face shone with ecstasy. According to the people around him, he had achieved *nirvikalpa samadhi*, and "would never return to this world."

[‡] The lines of this diagram, of course, are ideal representations. "Purely spiritual" transformation, even in highly ascetic yogas, always involves bodily change, while conversely, bodily developments conceived along strictly material lines could involve some enhancements of consciousness.

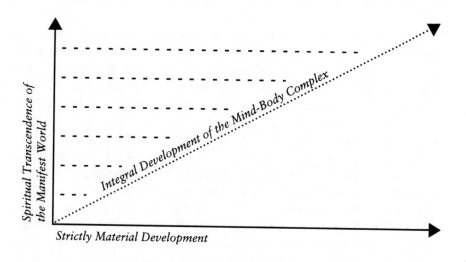

Figure 7.1. The vertical axis signifies the world-transcendence exemplified by yogis absorbed in trance, and the horizontal axis a strictly material program for bodily transformation (through genetic engineering, for example, or elaborate prosthetic devices). The hatched section between the diagonal and vertical axes represents the many directions religious practice has taken, from world-transcendence to integral transformation.

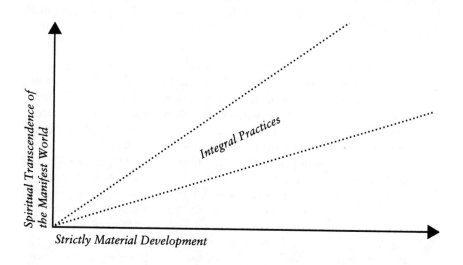

Figure 7.2.

Grace as a Mark of Evolutionary Transcendence

As I have suggested, doctrines of divine mercy or grace in the Western religions, of nonattainment in Buddhism, of surrender to God's will in certain Hindu sects, and of noninterference with the Tao refer to the same fact of human experience, namely that unitive awareness and other extraordinary capacities often appear to be given rather than earned, spontaneously revealed rather than attained through ego-centered effort (though dedicated practice sets the stage for their appearance and often brings them under conscious control). It is understandable that these similar ideas should arise in conjunction with quite different worldviews, since all of them refer to the same turn of human development—that is, the emergence of a greater nature whose dimensions are mysterious to us. Doctrines of grace, nonattainment, and noninterference with the Tao all point to a developing event on our planet, which (following Dobzhansky and Ayala) I have called *evolutionary transcendence*. In spite of their different metaphysical assumptions, every enduring religious tradition recognizes an ego-transcending process by which extraordinary life arises.

It has been argued, however, that we cannot equate the Jewish, Christian, or Muslim witness to Divine mercy or grace with the arising of Buddha-nature or surrender to an impersonal Tao. How can gifts from a personal God be reconciled with spiritual realization within an impersonal Order? In response to such objections I quote Philip Novak, a philosopher who for several years has practiced Buddhist meditation.

> In the midst of striving for apotheosis the aspirant is reminded that a ripeness beyond his control is all. The will to attend and the will to intend are kept delicately balanced by the knowledge that constraint can actually vitiate the inner work. Only when complemented by a surrender to a "higher" or more inclusive order can the effort of the personal will bear wholesome fruit. This is commonplace in Christianity. It is also a commonplace in the Zen tradition. There, stories abound of Masters undercutting disciples' strenuous efforts by reminding them . . . that ultimately there is nothing to do nor to attain. [Anyone] who has tried to sit still more or less constantly for sixteen hours a day at a Zen *sesshin* and yet amidst these agonizing efforts reads each day the passage from the Heart Sutra that asserts that there is "no path" and "nothing to attain"—this person knows well both legs of the paradox. Moreover, a strong case can be made for the proposition that Dogen [a principal founder of Japanese Zen] understood effortful Zen

practice as taking place in a Cosmos permeated by grace, in an Order, that is, in which surrender, or at least alignment, but not attainment, was the proper attitude.[4]

Commenting further on the equation of Christian and Buddhist understandings, Novak wrote:

Amid Dogen's exhortations to practice we find reminders to surrender to Buddhic activity. Dogen's sense of practice may be likened more closely to the reception of an omnipresent Gift than to the attainment of a distant goal. Indeed, this seems to be a central meaning of his doctrine of the identity of practice and enlightenment. The dynamism of Buddha-nature is all-pervading and the fulfillment of the bodhisattva vow depends not so much on our attaining to it, but on our not obstructing its continuous salvific activity. . . . The concept of special grace dispensed intermittently by a Supreme Being is, of course, totally foreign to Zen, but if grace goes largely unproclaimed in Zen, it is because the fish does not think to comment upon the ocean. Far closer to the truth than the notion that grace is absent is that it is always and everywhere present. One of Zen's laments is that we remain blind to the graceful immanence of the Buddha-nature, failing to draw on its power.[5]

The notion of special grace dispensed intermittently by a supernatural Deity, which Novak correctly says is foreign to Zen, is alien as well to a vision of human development embedded in the facts of evolution. Extraordinary life has its own patterns, its own regularities, which differ from those evident in ordinary functioning, but is not *arbitrarily* given by God. It is instead, like Dogen's Buddha-Nature, "always and everywhere present," everlastingly available, endlessly responsive to our aspirations for it.

Philosopher Marco Pallis also argued that doctrines of grace are implied in Buddhist thought.

The attractive influence of enlightenment experienced as providential and merciful (Buddhist) emanation from the luminous center, strikes on human consciousness in three ways, which might be described respectively as: 1.) invitation into enlightenment, 2.) companionship of enlightenment, and 3.) reminders of enlightenment. The first-named corresponds to (Christian) "conversion," the gift of faith. The second corresponds to man's being "in a state of grace," in virtue of which his apparent weakness is enabled to envisage tasks and surmount obstacles beyond ordinary human strength. The third coincides with the supplying of various "means of grace," that is to say, *upayas* consecrated by tradition—scriptural teachings, methods of meditation, initiatic rites, and the like.[6]

In an essay entitled "Is There Room for Grace in Buddhism?" Pallis described certain similar ways in which Christians, Buddhists, and Taoists are brought to spiritual fulfillment.

> The word "grace" corresponds to a whole dimension of spiritual experience; it is unthinkable that this should be absent from one of the great religions of the world. In fact, anyone who has lived in a traditionally Buddhist country knows that this dimension finds its expression there too, vehicled by the appropriate forms.
>
> In China the Taoists have always spoken of the "activity of heaven"; for us to speak of the "activity of enlightenment" is in no way far-fetched. This is in fact the function of grace, namely to condition men's homecoming to the center from start to finish. It is the very attraction of the center itself, revealed to us by various means, which provides the incentive to start on the way and the energy to face and overcome its many and various obstacles. Likewise grace is the welcoming hand into the center when man finds himself standing at long last on the brink of the great divide where all familiar human landmarks have disappeared.[7]

And the "welcoming hand into the center," which both Christianity and Buddhism bear witness to, is described in certain Hindu scriptures.[†] For example, the Bhagavad Gita's account of Krishna's relation with Arjuna has the three essential features that Pallis found in both Christian and Buddhist experience; namely, a call (or invitation) from Krishna to his human instrument, Krishna's subsequent companionship with Arjuna on the fields of Kurukshetra, and reminders (or skillful means) by which Arjuna could be enlightened in the midst of his suffering. The Koran propounds a similar doctrine. Its opening line, *Bismi 'Lahi'r-Rahmani 'r-Rahim*, refers at once to mercy as intrinsic to God, *Ar-Rahman*, and projected into creation, *Ar-Rahmin*. This formulation, which is central to Islamic teaching, implies that God's mercy in this world comes from His primordial essence. And belief in grace is expressed, too, in Jewish doctrines of God's mercy, Platonic affirmations that *daimones* inform and guide human souls,

---

[†] According to Sri Ramakrishna, " . . . the winds of God's trace are always blowing, but you must raise your sail" (Smith, H., 1983).

and belief in the Tao's beneficence. But such doctrines involve more than simple affirmations of gracious Power beyond the ordinary self. The sacred traditions bear a common witness to many aspects of grace, including those mentioned in the passages cited above. There is a considerable lore concerning the ways in which grace unfolds in human experience. A comparative study of such lore in different traditions would broaden our understanding of grace-laden human activity.

But if unitive awareness and self-existent delight are outcomes of evolutionary process and at the same time "not-attained," or freely given, as if they existed before any effort to realize them, we are presented with some philosophic dilemmas. How can such awareness and delight be "emergent" if they already exist? How can our limited, struggling efforts give rise to powers and knowings that seem fundamentally complete? In the next section, I present an answer to such questions proposed by philosophers such as Henry James, Sr., and Sri Aurobindo—namely, that evolution, with all its meandering, unfolds a luminous existence hidden or involved in this universe since its inception.

[7.2]
IDEAS THAT ILLUMINE THE RELATIONS BETWEEN EVOLUTION AND SUPERNATURE

To best appreciate our possibilities for extraordinary life, we need to embrace both the facts of evolution revealed by modern science and the witness of sacred traditions East and West. To do this, we must confront two fundamental realities: a universal development of (inorganic and living) forms that has lasted for several billion years, and transcendent orders of existence conceived or experienced for several millennia by people all over the earth. The possibility of integrating these two aspects of existence is fundamental to my inquiries here.

During the two centuries since "progress" became a prominent idea in the West, the philosophers Fichte, Schelling, Hegel, Solovyov, Berdyaev, Bergson, Whitehead, Samuel Alexander, C. Lloyd Morgan, Jean Gebser, Charles Hartshorne, Teilhard de Chardin, and Sri Aurobindo, among others, have tried to comprehend or explain the developing universe in relation to something ultimate, eternal, or everlasting. In various ways, each of them provides support for a simultaneous embrace of nature and supernature by linking the world's progress to

Geist, Deity, "the ever-present Origin," *Satchitananda,* or some other version of a world-transcending (yet immanent) Reality. The richness of their speculations shows what a promising field for philosophic inquiry appears when universal development is considered together with intuitions and experiences of a Supreme Principle or Divinity. Furthermore, their insights often illumine particular processes and specific human conditions that inhibit or facilitate metanormal capacities. Many insights of these philosophers suggest that the links between superordinary dimensions of existence and this world's developmental processes are ripe for new understanding.

Looking away from either aspect of existence, it seems to me, constitutes a supreme philosophic avoidance. Philosophers of ancient and medieval times should not be faulted because they did not know about evolution, but anyone now who attempts to build a comprehensive understanding of this world without contemplating its stupendous history is a self-blinded explorer. And conversely, no general theory of human development can in good faith overlook the enormous witness to mystical cognition and other metanormal abilities revealed by modern religious studies, psychical research, anthropological studies of shamanism, and other kinds of systematic inquiry into extraordinary experience. The evolving universe and supernature, however named, stand before us now as two inescapable facts. It is telling, I think, that so many great philosophers since the late eighteenth century have explored the relations between them.

This book draws upon several sets of ideas about evolution's relation to metanormal realities. Here I will describe five of them:

- □ notions of evolutionary novelty and emergence;
- □ concepts of evolutionary subsumption;
- □ perspectives that might be termed panpsychic, animistic, or nondualistic interactionism;
- □ proposals that new types of consciousness have developed at various times and places in the human race; and
- □ doctrines of involution-evolution.

I do not subscribe to all the concepts related to these five sets of ideas, but each set provides insight, language, or a philosophic stance that supports my speculations about human development. Like theories in science, each set stands or falls with the facts provided by scientific, parapsychological, contemplative, and other empirical methods.

Evolutionary Novelty and Emergence

Samuel Alexander, C. Lloyd Morgan, C. D. Broad, Joseph Needham, Michael Polanyi, and others have developed the idea that evolution produces emergent structures, processes, and laws (or habits) that had not previously existed.[8] According to most versions of this thesis, emergent items in the world cannot be explained or predicted from the conditions, events, or patterns they grew out of. They are fundamentally novel, not rearrangements of preexisting elements. They are qualitatively—not merely quantitatively—different from anything that existed before them.

The concept of emergence developed by the thinkers just noted implies the existence of levels; that is, portions of the world marked by qualities, forms, and regularities peculiar to them and emergent from other domains.[9] Emergent evolutionists have disagreed, however, about the number of levels existing in this universe. Morgan named four—psychophysical events, life, mind, and Spirit (or God); and Alexander five—space, time, matter, life, and Deity; while others have objected to such definite enumerations because countless gradations exist between inorganic matter, animal species, and humankind. According to philosopher T. A. Goudge, for example, "It is more in accord with the evidence to read life and mind as the final accumulative stages of a long series of minimal emergences rather than as abruptly occurring saltations."[10] But in spite of their differences, philosophers who have emphasized emergence and novelty in the world's development lend support to the notion of evolutionary transcendence described in chapter 3. Like biologists such as Dobzhansky and Ayala, they have helped to elucidate the fact that matter, life, and mind each operates with distinctive patterns. They help us view metanormal phenomena on their own terms, not in a reductionistic manner that obscures their significance for human development. Emergent evolutionists encourage us to oppose reductionisms that inhibit understanding of scientifically anomalous events, including those extraordinary experiences described in this book, that might signal a new evolutionary order.†

---

† Doctrines of emergence have been developed by numerous scientists who have not subscribed to evolutionary philosophies such as Alexander's or Morgan's. "[The] evolutionary version of the emergence doctrine," wrote Ernest Nagel, "is not entailed by the conception of emergence as irreducible hierarchical organization, and the two forms of the doctrine must be distinguished." See: Nagel 1979.

However, notions of emergence by themselves fail to account for certain knowings and powers that occur spontaneously or as a result of transformative practice. Countless saints and sages since antiquity have described a reality that appears to preexist any attempts to apprehend it. Blessings from this ego-transcending order appear, in Christian terms, to be *given by God*, or in Buddhist terms, to be *non-attained*. How do we reconcile experiences of eternal life with notions that such life is emergent? Later in this chapter, I describe a set of ideas that addresses this paradox, namely, doctrines of emanation and return spelled out or implied in Neo-Platonist, Vedantic, and other philosophies, and reformulated from an evolutionary perspective by modern thinkers such as Henry James, Sr., and Sri Aurobindo.

Evolutionary Subsumption

In the course of evolution, new levels build upon those that precede them, appropriating earlier processes in their unique activity. Thus life takes up inorganic elements, using them for its distinctive ends; and humans depend upon biological processes for a functioning more complex than that of their animal ancestors. Hegel saw an analogous process at work in psychosocial development, and used the German word *aufheben*—which suggests both annihilation and preservation—to describe the frequent subsumptions of cultural forms by their successors. In the dialectic of history, he claimed, earlier kinds of human behavior and consciousness (*Gestalten des Bewusstseins*) have been "lifted up" to higher levels.[11]

The Indian philosopher Sri Aurobindo also emphasized this aspect of evolutionary change. Emerging consciousness, which in Aurobindo's philosophy is essentially Divine, takes up the life form it inhabits

> to raise it up to a higher level, to give it higher values, to bring out of it higher potentialities. And this [it] does because evidently [it] does not intend to kill or destroy it, but, delight of existence being [its] eternal business and a harmony of various strains, not a sweet but monotonous melody the method of [its] music, [it] wishes to include the lower notes also and, by surcharging them with a deeper and finer significance, get more delight out of them. . . .[12]

In Aurobindo's view, to realize extraordinary life upon earth, we must embrace our diverse inheritance, including our biological processes and the inorganic elements that comprise them, rather than transcending them through ascetic discipline. In Part Three, I note some ways in which certain practices foreclose upon various faculties—for example, by neglecting imagination or intellect, suppressing potentially creative emotions, suppressing this or that virtue because it is not favored by the discipline's founder, or dampening our physical energies through bodily mortifications. However, by seeing that we can integrate the many dimensions of our human nature in a "harmony of various strains," as Aurobindo put it, and by appreciating the fact that such integration is analogous to the subsumptions evident in earlier stages of evolutionary progress, we can draw upon our rich heritage and cultivate the full range of our greater potentials.

Panpsychism and Animistic Interactionism

Like the emergent evolutionists noted above, the philosophers Henri Bergson, Alfred North Whitehead, and Charles Hartshorne have emphasized the novelty evident at all levels of the universe. Each in his own way has called into question both commonsense notions and philosophical doctrines that the fundamental units of the physical world are devoid of creativity.[13] Theologian David Griffin, who has developed certain ideas proposed by Whitehead and Hartshorne, wrote:

> For Whitehead, creativity is the ultimate reality of which all things are instances. This means that the basic things or entities are events, spatio-temporal processes of becoming. Whitehead's view not only generalized the Einsteinian notion of the convertibility of matter and energy, but also employed another basic Einsteinian notion, which is that space and time are inseparable. One cannot speak of space and time as separate, but only of space-time, or time-space. Whitehead's way of explaining this idea is that the things comprising the world are not essentially nontemporal things that just happen to be in time. They are essentially spatio-temporal events. They take—or make—time, as well as space, to occur. There is no "nature at an instant"—taking instant in the technical sense as having no duration or temporal extension. There are also no actualities devoid of spatial extension. All actualities are temporal-spatial events. Whitehead therefore says that all actual

entities are "actual occasions." No actual things simply endure passively. Each real thing is a spatio-temporal happening.

These actual occasions or events can be extremely brief, with hundreds, thousands, millions, or even billions within a second. Things that endure, such as electrons, atoms, molecules, cells, and minds, are each comprised of a series of brief events.

Moments in the life history of an electron, a cell, and a human being obviously differ immensely in terms of the forms they embody. But they all have one thing in common; each is an instance of creativity. Creativity is in this sense the ultimate reality, that which all actualities embody. All actual entities are thereby creative events.[14]

For Whitehead, Griffin, and others of similar persuasion, this view of physical things finds support from quantum theory, which suggests that subatomic events are not strictly determined. Yet these same subatomic events in large groups appear to be patterned, and at the molecular level are clearly structured. According to Whitehead, such patterning and structural integrity are mediated by an activity inherent in all spatio-temporal happenings, or actual occasions, namely *prehension*. Subatomic entities, cells, and humans alike *prehend* other spatio-temporal happenings, either including them in their own activity (by positive prehension) or excluding them (by negative prehension). Prehension does not necessarily involve sensory perception or consciousness, as it goes on continually in the physical elements and among humans at both conscious and unconscious levels. It does, however, involve contact with other entities (or occasions), and influence from and upon them.

Whitehead termed positive prehensions *feelings,* thus emphasizing their more-than-sensory-and-conceptual nature, while negative prehensions were said to *eliminate from feeling.*[15] Prehension has many kinds of subjective form, including "emotions, valuations, purposes, adversions, aversions, and consciousness."[16] Human self-awareness is simply one type of prehension. Through prehension, all entities constituting this universe continually contact and influence other entities, with some degree of creativity. The view held by Whitehead, Griffin, and others that the entire universe is a continually prehending process has been called *panpsychism,* or *panexperientialism,* in that it sees soul, experience, or subjectivity everywhere, even in the physical elements.[17]

This panpsychic or panexperiential view of the universe has led David Griffin to suggest a "nondualistic, or animistic interactionism" to account for telepathy, psychokinesis, and psychophysical transformations such as those described in this book. In this view, all consciousness, all soul, even a disembodied soul, interacts with other entities, whether inorganic, animal, human, or superhuman, through a prehension that does not always depend upon sensory processes, language, or the direct application of muscular force. In Griffin's words:

> Because the soul's perception of the world is not only or even primarily sensory perception, a soul separated from its physical body's sensory apparatus could still be capable of perception. While in the body, the soul's perception of the outside world by means of its bodily sensory organs presupposes the soul's perception of its body itself, especially its brain, and that perception is not sensory. Also, the present moment of the soul perceives its past moments, a type of perception that we call memory, and this perception is not sensory. The soul's direct perception of God, through which it learns of the reality of logical, moral, and aesthetic norms, as well as enjoying religious experience in the stricter sense, is not sensory experience.

In the animistic framework Griffin proposes, perception and causal influence are simply two aspects of the same event; that is,

> what is causal influence from the perspective of the cause is perception from the perspective of the effect. The soul is able consciously to influence the body's muscular system only because the bodily cells of this system perceive the conscious aims of the soul for its bodily parts; the soul unconsciously affects the body insofar as the bodily cells perceive its unconscious feelings. The body likewise influences the soul insofar as the soul perceives the feelings of its various bodily cells. To be perceived is to exert causal influence. If it is possible for souls in a discarnate state to perceive each other . . . it is therefore by definition possible for them to influence each other.
>
> This philosophical deduction from the animistic viewpoint is empirically supported by psychosomatic and parapsychological evidence. Psychosomatic studies show that the soul's influence on its body is not limited to the muscular system; the bodily cells in general seem to be responsive to the soul's aims and other feelings. Phenomena such as stigmata show how strong the soul's influence on its "involuntary" system can be.[18]

In chapters 11 through 23, I present a good deal of evidence that human will, imagery, emotions, somatic awareness, physiological processes, and bodily structures profoundly influence one another, and that all of them can respond to ego-transcendent agencies. The panpsychic or animistic interactionism implied in Whitehead's speculative philosophy and spelled out by Griffin gives us a picture of the universe strongly consonant with conclusions I draw from such evidence that we humans are capable of further psychophysical development. This picture of things indicates that our consciousness and volitions are not as separate from our bodies' physical processes as strictly mechanistic and dualistic worldviews suggest.

The Evolution of Consciousness

During the last 200 years, the idea that consciousness has advanced in the human race has been developed by Hegel, Bergson, Aurobindo, and other philosophers. The notion that mind is multileveled preceded Hegel, of course, having appeared, for example, in Plato's metaphor of the divided line (*The Republic* VI: 509–11); in Plotinus's distinction between *nous* and *dianoia*; and among Hindu and Buddhist thinkers since the time of the Upanishads. But modern philosophers such as Hegel and Bergson have emphasized the idea that new levels of consciousness appear in the course of human history, like other emergent properties of the universe. Even if they have roots in eternal or preexisting orders of existence, new kinds of knowing have become manifest for the first time *on earth* in particular societies and individuals, then spread to others by education or example.

According to Hegel, each stage of human development is canceled and preserved in the dialectical progress of history. In *The Phenomenology of Spirit* (1805), he traced this ongoing process from the slave of antiquity, who struggled successfully against nature's difficulties, to the stoic's establishment of freedom within himself independent of nature's demands, to the skeptic's development of freedom by dissipating restrictive categories of thought, to the Christian believer's discovery of freedom in a transcendent God, to the modern intellectual's appropriation of reason's highest principles. In this dialectic, successive forms of consciousness (*Gestalten des Bewusstseins*) subsume the forms that precede them.[19]

Henri Bergson, too, distinguished levels of consciousness that appeared over time, particularly *intellect* and *intuition*. Intuition, he believed, arose from the instinctive activity of animals, becoming disinterested and self-conscious in humankind. Like a painter's capacity for pure perception, it apprehends the world directly, yielding knowledge a philosopher needs for comprehensive understanding of the world and self. In his *Introduction to Metaphysics* (1903), Bergson emphasized the immediate, nonconceptual character of intuition, characterizing it as a direct participation in, or identification with, what is intuited. Directed at the external world, it becomes an act "by which one is transported into the interior of an object in order to coincide with what there is unique and consequently inexpressible about it." Turned toward the self, it becomes an immersion in the indivisible flow of consciousness, a grasp of pure becoming. Unlike the intellect, which remains detached from its objects, producing knowledge that is relative to some viewpoint, intuition enters into what it knows, dispensing with symbols.[20] Subsequently, Bergson modified this characterization of intuition. According to philosopher T. A. Goudge:

> He came to emphasize [its] cognitive character instead of its immediacy, and even spoke of it as a mode of thinking. As such, it is not a spontaneous flash of insight but an act that is engendered by mental effort. To achieve an intuition, we must turn our attention away from its natural concern with action. This act demands concentration of thought. Even when we are successful, the results are impermanent. Yet the intellect can effect a partial communication of the results by using "concrete ideas," supplemented by images. . . . Consequently, the knowledge attained by intuition is not altogether ineffable. Nor is it, in the strict sense, absolute, for intuition is a progressive activity that can widen and deepen its scope indefinitely. Its limits cannot be fixed a priori.[21]

The religious aspect of Bergson's thought became more pronounced toward the end of his life. He found a divine purpose in evolution, which could be known by intuition. *Elan vital,* the force that impels universal development, is sometimes communicated "in its entirety" to those mystics who achieve partial coincidence with the creative effort that is "of God, if it is not God Himself." This experience, however, does not end in passivity, but leads to intense activity that embodies God's love for the world. Mystics are impelled to advance the divine purpose by furthering the good of their fellows. The spirit of the mystics, Bergson believed, must become universal to ensure humanity's further development.

Aurobindo, too, argued that new types of consciousness arise in the course of human history, but his metapsychology was more elaborate than Hegel's or Bergson's. Like them, he distinguished more comprehensive from less comprehensive kinds of knowing, but described several levels of higher consciousness culminating in *Supermind,* in which the divine unity is expressed through diversity, individuals are harmonized with their universal Ground, and personal will is joined with the cosmic action. The hierarchical stages of mind leading to Supermind are typically expressed in particular kinds of extraordinary activity—*higher mind* in synoptic thought, *illumined mind* in mystical inspiration, *intuitive mind* in religious genius, and *overmind* in world-transforming action—while Supermind is yet to be embodied on this planet.

Aurobindo's spiritual realization was richly varied, it seems, embracing states of consciousness described in different religious traditions. Like the great Bengali mystic Sri Ramakrishna, he described his experiences of a personal and impersonal, transcendent and immanent, silent and dynamic Divinity, and like his Indian forerunner claimed that God's supernature has many dimensions. Given his varied contemplative experience and wide acquaintance with religious literature East and West, Aurobindo was bound to have a more elaborate spiritual psychology than Hegel or Bergson. Even if we remain agnostic (as I do) about his delineation of higher mental levels, we can accept the general idea behind it that consciousness is far more fluid and complex than most Western philosophers realize.[†]

Involution-Evolution

Hegel, Henry James, Sr., and Sri Aurobindo, among others, have advanced the idea that this world's unfoldment is based upon the implicit action, descent, or involution of a Supreme Principle or Divinity. Each of these philosophers believed that the progressive expression of higher forms or qualities is made possible by their secret existence or immanence in nature. In Hegel's conception, the *Geist* gradually reveals itself to itself through the long dialectic of history, recovering its fundamental completeness by a series of theses, antitheses, and syntheses in which one aspect of itself after another is subsumed (*aufgehoben*) in a higher fulfillment.

Henry James, Sr., who was overshadowed by his famous sons William and Henry, developed a synthesis of ethical, social, and

---

[†] Aurobindo described the levels between ordinary intellect and Supermind in *The Life Divine,* bk. 2, chap. 26, vol. 19 in *The Collected Works.* Less formal and more colorful descriptions of higher mental levels appear in some of Aurobindo's letters. See, for example: *Letters on Yoga,* pt. 1, vol. 11, pp. 233–377, in *The Collected Works.*

metaphysical thinking based largely upon Swedenborgian and Neo-Platonic ideas. For him, evolution was preceded by the involution of Divinity in this world. "Whatsoever creates a thing," he wrote, "gives it being, *in*-volves the thing, not the thing it. The Creator involves the creature; the creature *e*-volves the Creator."[22] James was led to this vision in part by Swedenborg. In his metaphysical work *Substance and Shadow* (1863), James wrote:

> In a word, according to Swedenborg, God creates us or gives us being only by thoroughly incarnating Himself in our nature; but inasmuch as this descent of the creator to creaturely limitations incidentally involves of course, on the part of the creature, the strictest inversion of the creative perfection . . . so it must necessarily provoke a corresponding ascending movement on God's part, giving us spiritual extrication from this infirmity. Otherwise creation would remain utterly inoperative save in a downward direction.
>
> Let us clearly understand then that the Divine operation in creation is made up of two movements: one . . . creative, giving us natural being or identity; the other . . . redemptive, which is a movement of glorification giving us the amplest individual or spiritual expansion out of that base root. The prior movement, the descending, statical, the properly creative one—gives us natural selfhood or consciousness, a consciousness of separation from God, of a power inhering in ourselves and independent of Him. The posterior movement—the ascending, dynamical, and properly redemptive one—gives us spiritual consciousness, a consciousness of union with God.[23]

But the "posterior, ascending, dynamical movement" was not limited to human consciousness. James proposed in some of his late works that inorganic nature, too, participates in the evolution of involved Divinity. Consciousness, he wrote,

> belongs to the mineral realm as truly, though not so distinctly, of course, as to the vegetable and animal. It is the most diffused or common form of consciousness, and therefore the least obvious to human apprehension.
>
> The mineral form then is the earliest or lowest evolution of the me. It is the me in an intensely inert state, in a passive state or state of rest simply. It is the me getting place or position first, in order to its subsequent experience of *growth* in the vegetable form, *motion* in the animal, and *action* in the human form.
>
> Nature is but the echo of the soul, and images nothing therefore of the Divine creation and providence which is not primarily impressed by the soul.[24]

Writing several decades later, Aurobindo articulated a doctrine of involution-evolution that resembles James's. Animal life, he wrote,

> is a living laboratory in which Nature has, it is said, worked out man. Man himself may well be a thinking and living laboratory in whom and with whose conscious co-operation she wills to manifest God. For if evolution is the progressive manifestation by Nature of that which slept or worked in her, involved, it is also the overt realization of that which she secretly is. We cannot, then, bid her pause at a given stage of her evolution, nor have we the right to condemn with the religionist as perverse and presumptuous or with the Rationalist as a disease or hallucination any intention she may evince or effort she may make to go beyond. If it be true that Spirit is involved in Matter and apparent Nature is secret God, then the manifestation of the divine in himself and the realization of God within and without are the highest and most legitimate aim possible to man upon earth.[25]

Though there are significant differences between their philosophies, both Aurobindo and James saw universal evolution arising from a previous involution of Divinity in nature. Both of them might be called evolutionary emanationists in that they regarded the manifest world to be an emanation of Divinity (or the One), but at the same time a dynamic process creatively seeking to reveal God in the physical world. James's biographer Frederic Young wrote: "To read Aurobindo's masterpiece, *The Life Divine,* is, to one who has read the senior James' works, to experience an indescribable feeling that Aurobindo and James must have corresponded and conversed with each other; so much spiritual kinship is there between the philosophies of these two thinkers!"[26] Both philosophers regarded "apparent Nature" to be "secret God," and saw the Supreme Reality emerging more fully in this world through the vicissitudes of time. Like philosophers since the late eighteenth century such as Hegel and Bergson, they "temporalized the great chain of being," to use the historian Arthur Lovejoy's phrase, conceiving the manifest world "not as the inventory but as the program of nature." In describing this temporalizing, Lovejoy wrote:

> An important group of the ruling ideas of the early eighteenth century—the conception of the Chain of Being, the principles of plenitude and continuity on which it rested, the optimism which it served to justify, the generally accepted biology—all were in accord with the supposedly Solomonic dictum . . . there not only is not, but there never

will be, anything new under the sun. The process of time brings no enrichment of the world's diversity; in a world which is the manifestation of eternal rationality, it could not conceivably do so. Yet it was in precisely the period when this implication of the old conception became most apparent that there began a reaction against it.

For one of the principal happenings in eighteenth-century thought was the temporalizing of the Chain of Being. The *plenum formarum* came to be conceived by some, not as the inventory but as the program of nature, which is being carried out gradually and exceedingly slowly in the cosmic history. While all the possibles demand realization, they are not accorded it all at once. Some have attained it in the past and have apparently since lost it; many are embodied in the kind of creatures which now exist; doubtless infinitely many more are destined to receive the gift of actual existence in the ages that are to come. It is only of the universe in its entire temporal span that the principle of plenitude holds good. The Demiurgus is not in a hurry; and his goodness is sufficiently exhibited if, soon or late, every Idea finds its manifestation in the sensible order.[27]

Until notions of progress and evolution became prominent in the West, the idea of emanation and return was usually embedded in worldviews that regarded the world to be a static (or cyclical) existence to which time adds nothing new.[28] That the emanationist vision has been wedded to both evolutionary and nonevolutionary cosmologies indicates its lasting appeal to the metaphysical imagination, and its resonance with an intuition prevalent in different times and cultures. The idea of divine emanation economically and beautifully reflects a realization reported by countless people since antiquity that they enjoy a secret contact, kinship, or identity with the founding Principle of this universe. Such realization, which gives philosophical doctrines of emanation-return (or involution-evolution) compelling justification and support, may be brief or long-lasting, spontaneous or the result of transformative practice. Philosophers and mystics of virtually every sacred tradition have expressed versions of it through parables, aphorisms, or metaphysical statements. Thus:

- □ "Before your parents were," asks a famous Zen *koan,* "what is your original face?" This celebrated line suggests that we enjoy an essential subjectivity or personhood that precedes our birth and outlasts death itself.
- □ In a famous Hindu parable, a tiger separated since birth from its mother is raised by sheep, believing itself to be one of them until

another tiger shows it its own reflection in a river. We are all tigers, the parable implies, all secretly God (or Brahman) though we think we are something else.

- The Platonic doctrine of *anamnesis*, or "recollection," which asserts that we can remember the Divine Ideas underlying sense impressions, is based upon the belief that humans have immortal souls that communed with those Ideas before assuming a mortal body. Though scholars have debated the extent to which Plato himself enjoyed mystical illumination, Platonist and Neo-Platonist thinkers have traditionally asserted that humans are secretly rooted in Divinity and can realize that fact through the practice of virtue, the pursuit of beauty, and philosophic inquiry (or dialectic). "God," wrote Plotinus, "is outside of none, present unperceived to all; we break away from Him, or rather from ourselves; what we turn from we cannot reach; astray ourselves, we cannot go in search of another; a child distraught will not recognize its father; to find ourselves is to know our source."[29] The metaphor of homecoming in this passage is expressed by the German poet Novalis in his famous line *immer nach hause*, "always homeward" to our secret Source.

- Though according to Christian dogma the human soul cannot enjoy identity with God, the Dominican priest Meister Eckhart wrote: "The knower and the known are one. Simple people imagine that they should see God, as if He stood there and they here. This is not so. God and I, we are one in knowledge." In similar fashion, Saint Catherine of Genoa claimed: "My Me is God, nor do I recognize any other Me except my God Himself."[30]

- And the Sufi saint Bayazid of Bistun wrote: "I went from God to God, until they cried from me in me, 'O thou I'!"

These lines from Buddhist, Hindu, Platonist, Christian, and Islamic traditions reflect an enduring realization, shared by countless people since ancient times, of a Reality ordinarily hidden but immediately recognized as our original face, our true identity (as a tiger among sheep), our immortal soul, our shared Ground with God, our secret at-oneness "with all the Gods."[31] Such realization supports and gives rise to doctrines of emanation and return. In its light, it is natural to see the manifest world either as a stage for the individual soul's return to its Source, or (like Aurobindo and James) as a universal evolutionary process by stages expressing its secret Divinity.

The involution-evolution idea helps account for certain long-ings, illuminations, and apparent remembrances of a primordial Superexistence; and helps explain the profound resonance between human volition, imagery, emotion, and flesh through which psychophysical transformations appear to be mediated. Our cells, by this account, respond to our thoughts and intentions (and to ego-transcendent agencies) because they arise from the same eternal or primordial Source. The involution-evolution idea resonates with my proposals that transformative practice (like evolution) can evoke our latent capacities for extraordinary life. According to this formulation, human nature can realize metanormal capacities because that is its predisposition. Because the Creator has descended into the creature, as the elder James put it, the redemptive movement of nature "gives us a consciousness of union with God." Or as Aurobindo wrote, "If apparent Nature is secret God, then . . . the realization of God within and without are the highest and most legitimate aim possible to [humanity]."

The involution-evolution doctrines articulated by Henry James, Sr., and Sri Aurobindo have much in common with the views of Alfred North Whitehead and Charles Hartshorne. Whitehead's *prehension*, for example, is of two types: physical prehension of actual entities, the spatio-temporal happenings that constitute universal process (see the discussion of animistic interactionism above); and conceptual prehension of eternal objects, which are reminiscent of the Platonic Forms. God, in his primordial nature, sustains eternal objects, which give to all actual occasions, or entities, their initial subjective aim. Or to put it another way, God attracts actual entities (including atoms, cells, and humans) to himself, though in his "consequent nature" he suffers with and is shaped by their activities.[32] Charles Hartshorne has developed much the same view. In their recognition of both eternal and developing aspects of Divinity, Whitehead and Hartshorne come close to the involution-evolution ideas proposed by Aurobindo and James. Indeed, all four thinkers can be deemed to be *panentheists* in that they acknowledge both transcendent and immanent aspects of Divinity. (*Panentheism* holds that the being of God includes and penetrates the whole universe, so that every part of it exists in him, but—as against *pantheism*—that his being is more than, and is not exhausted by, the universe.)[33] The British theologian John Robinson elaborated the panentheist doctrine in his *Exploration into God*. By his formulation, since we already exist in God we do not seek Him, but rather "explore into Him."[34]

The panentheist doctrine has become more and more compelling to me as I have studied reports of religious experience drawn from different cultures. People with different beliefs about God and human nature have since the beginnings of recorded history described their realization of a Reality that is both immanent and transcendent to this universe. Modern studies of secular and religious ecstasies by William James, Marghanita Laski, Raynor Johnson, Edward Robinson, and others have produced many detailed accounts of such experience.[35] Though it occurs among people with different cultural conditionings, such illumination typically involves an apprehension of something familiar, of a Reality secretly known, and frequently evokes statements such as "I've come home again," "this is where it all starts," or "this is who I really am."[†] And it frequently accompanies metanormal volition, vitality, and movement. The extraordinary powers described in this book are attributed, for example, to divine possession (13.1, 21.3), the presence of God (22), or the activity of "Emptiness (*sunyam*) in human hands and feet" (20.1). Virtually every type of metanormal attribute has been deemed to arise from a transcendent Reality fundamentally linked with ordinary human nature. For these reasons, it seems to me, the evidence for extraordinary functioning leads us toward some sort of panentheism.

In summary, these five sets of ideas, each of them developed in modern times by philosophers who tried to reconcile evolution with ultimate or eternal principles, resonate with my proposals here:

- First, anti-reductionist doctrines of emergence, with their emphasis on novelty and emergent levels of existence, because they provide insight and a philosophic stance to support the proposition that metanormal capacities are part of the world's creative advance.

- Second, the notion of subsumption, because it illuminates a central aspect of development in the cosmic, biological, and psychosocial domains, and because it supports the idea that transformative practice can take up all our parts to produce richer self-integrations.

---

† In the words of T. S. Eliot: "And the end of all our exploring/Will be to arrive where we started/And know the place for the first time" (*Four Quartets*). For a collection of parables, aphorisms, and metaphysical formulations that express this recognition, see: Metzner 1986, chap. 8: Returning to the Source.

- Third, panpsychic or animistic interactionism, because it integrates data from various scientific fields to suggest that our human capacity for development arises from a creativity present throughout the universe, and because it highlights evidence that entities operating at different levels, from subatomic events to human activity, prehend or contact one another in a mutually formative manner.

- Fourth, the concept that consciousness has advanced by stages in the human race, because it relates our understanding of individual growth to the long process of human history.

- Fifth, the involution-evolution idea because it so beautifully and economically relates different kinds of extraordinary human experience.

Each of these ideas—and many others, of course—can be developed to illumine our possibilities for creative advance. But let me emphasize the word *develop*. To serve practical life, including the integral disciplines proposed in this book, conceptual schemes must be refined in the light of our unfolding experience. They must be revised (or abandoned) in the wake of new findings provided by science, psychical research, comparative religious studies, transformative practice, or other activities. Though they provide orientation for our disciplines and life in general, we must remember that philosophic maps are not the territories they represent, and further, that they often hide or obscure many features of the things they are meant to depict. Thus they sometimes need to be supplemented by other maps and metaphors. Indeed, success in any field requires a willingness to entertain principles that might seem at first to be contradictory. Religious disciplines are assisted at times, for example, by notions of self-exceeding (and the heroic metaphors they inspire); at other times they require ideas of nonattainment, surrender to God, or self-acceptance. Because the extraordinary activities described in this book are produced by complex perceptual, cognitive, emotional, and physiological processes, we need a great variety of images to comprehend and describe them.[36]

Ideas That Impede Understanding of Metanormal Development

Remembering, then, that they are subject to revision, the five sets of ideas described above (as well as many others) help integrate theories and data that illumine our possibilities for growth. But there are other notions, some of them in vogue today, that do not assist such integration. Scientistic reductionisms that systematically exclude mystical truth claims and evidence for paranormal phenomena; mind-body dualisms that radically separate subjective experience from somatic process; and ascetic worldviews that denigrate the body's place in human fulfillment do not help us integrate insights concerning human development from the life sciences, psychical research, and comparative studies of religious experience. Nor do they help us conceive practices adequate to our greater capacities. To close this chapter, I want to discuss briefly two sets of ideas that inhibit some people's appreciation of human transformative capacity—namely, notions of the *Übermensch* stemming from Nietzsche, and implications in the thinking of Hegel and others that God uses humans as discardable means to an end, as mere steppingstones to higher forms of life.

Nietzsche's *Übermensch,* in philosopher Walter Kaufman's words, "made his public appearance" in *Thus Spoke Zarathustra.* The term itself had been used by Heinrich Muller (*Geistliche Erquickungsstunden,* 1664), and Goethe (in a poem *Zueignung* and in *Faust,* Part I, line 490); but Nietzsche gave it new meaning, which the English *superman* does not convey. The *Übermensch* is not a muscleman, of course, nor a tyrant, nor produced by natural selection. Nietzsche wrote in *Ecce Homo* (III.1) that only "scholarly oxen" could interpret his idea Darwinistically. The Overman, according to Kaufman, "has overcome his animal nature, organized the chaos of his passions, sublimated his impulses, and given style to his character."[37] The Overman's power comes from self-mastery, not dominance over others. His joy arises from a transcendent dimension of personhood (his true self), not from satisfaction of ordinary appetites. Nevertheless, various aspects and distortions of Nietzsche's proposals about the *Übermensch* have hindered discourse about human self-exceeding. The Nazis' appropriation of certain Nietzschean passages to justify their claims for a master race are foremost among these.

Commenting upon this misappropriation, Kaufman described Nietschze's lifelong opposition to anti-Semitism, his strong advocacy of racial mixture (to promote cultural vitality), his reverence for the Old Testament, his scorn for German nationalism, his disgust with most political functionaries, his belief that human advance depends upon self-cultivation rather than selective breeding, and his scorn for mass enthusiasms such as those Nazism would inspire. Nazi apologists such as Richard Oehler systematically misquoted Nietzsche, or took some of his words out of context, to support their contention that he anticipated Nazi claims about a German master race.[38] Kaufman wrote:

> Nietzsche's views are quite unequivocally opposed to those of the Nazis more so than those of almost any other prominent German of his time or before him . . . and these views are not temperamental antitheses but corollaries of his philosophy. Nietzsche was not more ambiguous in this respect than is the statement that the Nazis' way of citing him represents one of the darkest pages in the history of literary unscrupulousness.[39]

But clarifications of Nietzsche's views on race won't dispel a tendency among some people to associate conceptions of further human development with a power-hungry superman. Indeed, such an equation has hindered the ability of some people to think boldly about high-level change at all. Our thinking, however, does not have to be limited in this way. We are not obliged to equate extraordinary functioning with a master race or a narcissistic Overman, for the metanormal capacities discussed in this book seem to develop best in conjunction with love for other living things. The sensitivity, empathy, and love of life fostered by practices of the kind proposed here predispose us to care for the creatures of this world. There is a volition beyond drives for dominance, an ego-transcending identity based upon solidarity with others, a superabundant vitality that overflows to those in need.

Still, it is good, I believe, that worries about ambitious schemes for human growth are part of our intellectual climate, because extraordinary capacities can stimulate or support many kinds of destructive activity. The *Übermensch*, by his vivid presence in contemporary thought, reminds us that visions of human growth must be framed with wisdom and care.

Proposals that we can participate in a fundamental evolutionary advance are immediately associated by some people with notions that God (or some other superentity) uses humans as means to His (or Its) own ends, as mere instruments of a grand design. This reaction was dramatized by Kierkegaard, who felt that Hegel's vision of a world proceeding according to the Absolute's inexorable logic robbed each individual of significance, subjectivity, and sense of worth. And similar objections have been raised in response to other visionary philosophies, against Teilhard de Chardin's proclamations that the universe is moving relentlessly toward the Omega Point, for example, or against Marx's doctrine that the triumph of the proletariat is inevitable. But evidence for human transformative capacity, even when viewed in terms of a general evolutionary advance, does not necessarily imply that God, *Geist*, evolution, nature, the Dialectic, or any other entity is automatically implementing its grand design with indifference to individual choice or significance. Since evolution meanders more than it progresses, among animal species and humans alike, it is reasonable to suppose that evolutions in the metanormal domain are likely to be marked by ups and downs. The choice to cultivate our greater possibilities is ours, not God's. All high-level human change involves our subjectivity. There will be no further human development unless some of us work to realize it. Doctrines of automatic or necessary human advance have fallen into disrepute. Happily, objections to the inevitability of class struggle, the Absolute's dialectical progress, or God's unstoppable march toward the Omega Point have won the day. But such objections do not negate the fact that we harbor potentials for further development such as those described in this book.

Nevertheless, ideas such as involution-evolution call up strong objections in certain people. To philosophies that seem to make evil subordinate or instrumental to a march of history toward the Good, some of us respond with an instinctive "No!" With Ivan Karamazov, we decline the ticket of admission to a world in which innocent children suffer. In the face of life's monstrosities, we cannot believe in a secret Divinity that impels the human race toward a potentially glorious future. No intellectual argument by itself, however life-giving it may be, provides a sufficient answer to such objections. Our best response to evil consists in living so that we help reduce it. No ideas by themselves justify God's ways to us. This world's suffering requires transformative *acts* that promote a better life for the creatures of this planet.

Indeed, sweeping idealism can promote a contemplative view of human existence that works against direct engagement with actual problems. Most of us would agree that it is better to make improvements in the real world than it is to view life in Hegelian fashion, as if from a mountaintop. Finally, a worldview must be measured by its fruits, and philosophies that emphasize human limitation, perhaps, promote more good works than the visions of an Aurobindo or a Whitehead. Since my proposals here resonate with certain ideas of these two philosophers, how do they help lead us toward the common good?

Answers to that question are scattered through this book, but I will note a few of them here. First, the survey of extraordinary functioning in Parts One and Two can help us identify possibilities for growth that we might not otherwise recognize. In Part Three I build on that survey to suggest practices that would make us better able to serve our fellows. Furthermore, such an inventory indicates that there is potential creativity in certain human activities which at first sight seem strange or unattractive. In doing this, such an inventory can open up ways of growth that are closed by philosophies insensitive to our more radical capacities for transformation. That certain life-giving attributes emerge first as sickness or eccentricity is a recurring theme in this book. If we take a strictly reductionist or materialist view of human nature, we are not likely to see that certain upsetting or perverse-looking episodes have a tendency toward the release of capacities that can support activity for the sake of others. For example, the exceptional vitality associated with kundalini-type episodes or the *incendium amoris* (see chapter 5.4) often stimulates works of love and service. To know that such episodes might be first signs of creative capacities can help us integrate them into our everyday life.

The involution-evolution idea supports openness to life-giving capacities, then, but it can do more than that. By alerting us to the immanence of a supreme goodness in the world at large, it can—*without overlooking evil or suffering*—promote a general sense of hope, a faith in life, a readiness to see goodness in the things of this world. These virtues can help us to be creative and loving people, and thus better able to serve our fellows.

The view of human possibility developed here, with its involution-evolutionist perspective, also points us toward great adventures. If we actually harbor the potentials described in this book, we stand at the edge of an immense frontier. This frontier, conceivably, could attract our love of exploration, our need for new territories, our drive to exceed ourselves. In so doing it could help reduce certain evils caused by lack of creative outlet. I don't think it farfetched to suggest that much of our overconsumption, drug addiction, and need for deadly conflict result from energies that could be rechanneled. Integral practices of the kind proposed here can, I believe, give us new substitutes for violence, drugs, delinquency, and destruction of our natural surroundings. Cultivation of extraordinary capacities can make us more capable of enriching the world. By helping to promote such activity, the ideas reviewed in this chapter serve compassionate ends, rather than the disengaged contemplation of life against which many of us protest.

8

Metanormal Embodiment in Legend, Art, and Religious Doctrine

WHILE MOST RELIGIONS have supported body-denying attitudes and practices, they have also given rise to myths and doctrines of physical transfiguration. Some of these, it seems to me, point toward real possibilities. The Christian doctrine of glorification and Taoist legends about holy flesh, for example, might express premonitions of something that could actually happen. Though such legends and doctrines need not be taken literally, they might intuitively anticipate the extraordinary life humans could one day enjoy.

[8.1]
Christian Doctrines of the Glorified Body

That there will be a universal resurrection of the dead is a dogma framed in all major creeds of the Roman Catholic church. And with that belief, the church upholds another, formally defined by the Fourth Lateran Council (1215) in these terms: "all will rise with their own bodies which they now have so that they may receive according to their works, whether good or bad." The bodies of those whose works have been good will be glorified and will share the soul's beatific vision.[1] This doctrine has fascinated theologians since antiquity, inviting speculations from some of Christianity's greatest thinkers. According to religious historian Caroline Bynum:

Through the doctrinal controversies of the second to fifth centuries, the resurrection of the body was firmly established as an element of the Christian faith. Medieval councils confirmed this. The Fourth Lateran Council in 1215 required Cathars and other heretics to assent to the proposition that "all rise again with their own individual bodies, that is, the bodies which they now wear," and the Second Council of Lyon in 1274 reaffirmed the requirement.

Theologians of the high Middle Ages neither abandoned the doctrine nor ceased to discuss it. Several (e.g., Albert the Great and Giles of Rome) wrote treatises about it. Moreover, it came up again and again in quadlibetal disputes (i.e., disputes by university students and masters on topics of current interest), and it provided the occasion for debating certain key philosophical issues raised by Aristotle.[2]

Christian thinkers with quite different views of human nature have shared a fascination with ideas about bodily resurrection. Here I will quote Saint Paul, Origen, Thomas Aquinas, and the twentieth-century Catholic writer Romano Guardini to indicate the prominence of such ideas through the long course of Christian history.

The prime reference for most subsequent discussions of glorification is I Corinthians 15:39–44. This much-quoted passage reads:

> All flesh is not the same flesh: but there is one kind of flesh of men, another flesh of beasts, another of fishes, and another of birds.
>
> There are also celestial bodies, and bodies terrestrial: but the glory of the celestial is one, and the glory of the terrestrial is another.
>
> There is one glory of the sun, and another glory of the moon, and another glory of the stars: for one star differs from another star in glory.
>
> So also is the resurrection of the dead. It is sown in corruption; it is raised in incorruption:
>
> It is sown in dishonour; it is raised in glory: it is sown in weakness; it is raised in power:
>
> It is sown a natural body; it is raised a spiritual body. There is a natural body, and there is a spiritual body.

Christian theologians have frequently commented upon the four characteristics Saint Paul attributed to the glorified body, namely *impassibility* ("what is sown in corruption rises in incorruption"), *clarity* ("what is sown in dishonor rises in glory"), *agility* ("what is sown in weakness rises in power"), and *subtility* ("what is sown a natural body rises a spiritual body"). Thomas Aquinas described these four aspects of glorification in *Summa Contra Gentiles* 4.86. Following Aristotle, he regarded soul to be the body's "form," the agency that vivified and preserved human flesh. As a thought experiment, read the following passages while keeping in mind the extraordinary capacities described in chapter 5.

The glory and power of the soul elevated to the divine vision will add something more amply to the body united to itself. For this body will be entirely subject to the soul—the divine power will achieve this—not only in regard to its being, but also in regard to action, passion, movements, and bodily qualities.

Therefore, just as the soul which enjoys the divine vision will be filled with a kind of spiritual lightsomeness, so by a certain overflow from the soul to the body, the body will in its own way put on the lightsomeness of glory. Hence, the Apostle says: "It is sown in dishonor. It shall rise in glory" (I Cor. 15:43).

Moreover, the soul which will enjoy the divine vision, united to its ultimate end, will in all matters experience the fulfillment of desire. And since it is out of the soul's desire that the body is moved, the consequence will be the body's utter obedience to the spirit's slightest wish. Hence, the bodies of the blessed when they rise are going to have agility. This is what the Apostle says in the same place: "It is sown in weakness, it shall rise in power." For weakness is what we experience in a body found wanting in the strength to satisfy the desire of the soul in the movements and actions which the soul commands, and this weakness will be entirely taken away then, when power is overflowing into the body from a soul united to God. For this reason, also, Wisdom (3:7) says that the just "shall run to and fro like sparks among the reeds": this is not said because there is motion in them by reason of necessity—since they who have God want nothing but as an indication of their power.

They will not be able to suffer anything which is harmful to them. For this reason they will be incapable of suffering. Nonetheless, this incapability of suffering will not cut them off from the modification essential to sense knowledge, for they will use their senses for pleasure in the measure in which this is not incompatible with their state of incorruption. It is, then, to show their incapacity for suffering that the Apostle says: "It is sown in corruption, it shall rise in incorruption" (I Cor. 15:42).

Furthermore, the soul which is enjoying God will cleave to Him most perfectly, and will in its own fashion share in His goodness to the highest degree; and thus will the body be perfectly within the soul's dominion, and will share in what is the soul's very own characteristics so far as possible, in the perspicuity of sense knowledge, in the ordering of bodily appetite, and in the all-round perfection of nature; for a thing is the more perfect in nature, the more its matter is dominated by its form. And for this reason the Apostle says: "It is sown a natural body, it shall rise a spiritual body" (I Cor. 15:44). The body of the risen will be spiritual, indeed, but not because it is a spirit—as some have badly understood the point—whether in the sense of a spiritual substance, or in the sense of air or wind; it will be spiritual because it will be entirely subject to the spirit.[3]

Saint Thomas expressed a central Christian intuition when he said the risen body of the just "will be spiritual, but not because it is a spirit—as some have badly understood the point." No, it will truly be *body*. He passionately upheld the idea, in spite of its apparent absurdity, that our present bodies, the very flesh we have now, will be glorified if our works are good. It is not enough that the souls of the just will enjoy beatitude; their bodies must participate too. Indeed, it is against the nature of the soul to exist without its physical counterpart (*Summa Contra Gentiles* 2.68, 83; 4.79). Soul separate from body is imperfect, *in statu violento*. The resurrection is natural in that it reunites the two, though its cause is supernatural (*Summa Contra Gentiles* 4.81). Saint Thomas and other Christian thinkers felt an intuitive rightness in this vision, even though it contradicted the obvious facts of bodily decay. That intuitive rightness apparently weighed against conceptual difficulties such as these: If our bodies will be glorified centuries after their decomposition, from what material will they be reconstituted? Which stage of their development will become their template for eternity? And what about innocent children who die? Will they be children in heaven? According to Caroline Bynum:

> Hugh of St. Victor wondered whether we shall be able to open and close our eyes after the resurrection. Honorius (and Herrad of Landsberg who borrowed his discussion) queried what color we will be in heaven and whether we will wear clothes. Guibert of Nogent fulminated against the cults of the tooth of Christ and of the holy foreskin because they implied that Christ had not risen in total bodily perfection and that our resurrection might therefore be defective as well. Several theologians debated whether food taken in by the body during its lifetime would become part of that body and rise at the end.[4]

In spite of conceptual difficulties and apparent absurdities, such as Bynum noted, many Christian theologians maintained their belief in glorified embodiment. Though the doctrine of resurrection is em-

bedded in a prescientific worldview and is supported by plain superstition, it has persisted, conceivably, because intelligent and sensitive
thinkers have sensed the body's potentials for radical transformation.
Reframing Christian vision from the developmental perspective proposed here, doctrines of the Final Days might symbolize a new evolutionary domain, and "risen bodies of the just" might represent metanormal embodiment.

Origen, the third-century theologian held by many to be the
greatest Christian thinker of his day, wrote about the glorified body
with great sweep and imagination.[5] Though some of his teachings
were declared anathema and most of his writings have been lost (he
was reputed to have written more than a thousand books), his influential work *On First Principles* has come down to us, mainly in a Latin
translation by the fourth-century Christian translator Rufinus, but
also in fragments from the original Greek. As a thought experiment,
again, read the following passages while remembering the evidence
for extraordinary human capacities noted in chapter 5.

> For if bodies rise again, undoubtedly they rise again as a clothing for us,
> and if it is necessary, as it certainly is, for us to live in bodies, we ought
> to live in no other bodies but our own. And if it is true that they rise
> again and do so as "spiritual," there is no doubt that this means that
> they rise again from the dead with corruption banished and mortality
> laid aside; otherwise it would seem vain and useless for a man to rise
> from the dead in order to die over again. Finally, this can be the more
> clearly understood by carefully observing what is the quality of the
> "natural body" which, when sown in the earth, can reproduce the
> quality of a "spiritual body." For it is from the natural body that the
> very power and grace of the resurrection evokes the spiritual body,
> when it transforms it from dishonour to glory.
>
> Our flesh indeed is considered by the uneducated and by un-
> believers to perish so completely after death that nothing whatever of
> its substance is left. We, however, who believe in its resurrection, know
> that death only causes a change in it and that its substance certainly
> persists and is restored to life again at a definite time by the will of its
> Creator and once more undergoes a transformation.
>
> Into this condition, therefore, we must suppose that the entire
> substance of this body of ours will develop at the time when all things
> are restored and become one and when "God shall be all in all." We
> must not think, however, that it will happen all of a sudden, but gradu-
> ally and by degrees, during the lapse of infinite and immeasurable ages,
> seeing that the improvements and correction will be realised slowly
> and separately in each individual person.

For the faith of the Church does not accept the opinion derived from certain Greek philosophers, that besides this body which is composed of the four elements, there is a fifth body which is entirely other than and diverse from our present body; since we can neither produce from the holy scriptures the least suspicion of such an opinion, nor can its acceptance be allowed as a logical inference, particularly as the holy apostle clearly lays it down that no new bodies are to be given to those who rise from the dead, but that they are to receive the same ones which they possessed during life, only transformed from a worse to a better condition. For he says: "It is sown a natural body, it will rise a spiritual body," and "it is sown in corruption, it will rise in incorruption; it is sown in weakness, it will rise in power; it is sown in dishonour, it will rise in glory."

The whole argument, then, comes to this, that God has created two universal natures, a visible, that is, a bodily one, and an invisible one, which is incorporeal. These two natures each undergo their own different changes. The invisible, which is also the rational nature, is changed through the action of the mind and will by reason of the fact that it has been endowed with freedom of choice; and as a result of this it is found existing sometimes in the good and sometimes in its opposite. The bodily nature, however, admits of a change in substance, so that God the Artificer of all things, in whatever work of design or construction or restoration he may wish to engage, has at hand the service of this material for all purposes, and can transform and transfer it into whatever forms and species he desires, as the merits of things demand. It is to this, clearly, that the prophet points when he says, "God who makes and transforms all things."[6]

Origen anticipated proposals such as those in this book by his intuitive vision of our possibilities for extraordinary life; by his belief that material substance can undergo "every kind of transformation"; by his insistence that the glorified body would arise, however mysteriously, from our present body, not a "fifth body" such as that proposed by some Greek philosophers; and by his belief in a gradual development toward glorification during the "lapse of infinite and immeasurable ages." Though scholars still debate the sources of his thought, attributing them variously to Christian, Neo-Platonist, and Gnostic inspirations, we find a strong resonance in Origen with later Christian visions of the Glorified Body. Like Saint Paul before him and Thomas Aquinas a thousand years later, he anticipated the idea that human nature, including both mind and flesh, can realize a luminous existence.

That the doctrine of the Glorified Body remains central to some Christian thinkers today is evident in the following passages from the twentieth-century Catholic writer Romano Guardini.

To be man is to be spirit expressed and made active through the body. To be man is to be a bodily organism, subject to the operation of a personal spirit that bestows on that organism a form and capacity which of itself it is powerless to attain to; to be man is to occupy a specific place in history, in which the spirit with its dignity and responsibility takes its stand. Resurrection therefore signifies that the spiritual soul, true to its nature again becomes the soul of a body—indeed that only now is it fully liberated and empowered to inform the body. Resurrection signifies that the matter from which the soul has departed becomes once more personalized, spiritualized corporeality, that is, human body, though no longer limited by time and space, but become, as Paul says, a spiritual, a "pneumatic" body.

In all of these different kinds of existing things we find "body"— in the crystal, the apple tree, the horse and birds, and in the man I am addressing. At each stage physical matter is put at the service of a new principle that gives it not only fresh qualities and capacities, but indeed at every stage a new character. Stage by stage it overcomes inertia, weight, bondage, muteness, and gains lightness, space, height, freedom; its sphere of operation broadens and the operations themselves increase in importance. Both the power to act and the scope for action are enlarged.

Does this line of development stop with man as we know him? Our intuitive feeling tells us it must proceed further. Humanity is not a blind alley. The possibilities of what we call body are inexhaustible. A clear and direct indication of what they are is furnished by the rising stages the body reaches in man himself.

For the human body is not a finished, arrested form; it is ever in process of becoming. That a healthy body kept in condition by care and exercise is more "body" than a neglected one is self-evident. But does the body reach its greatest dignity and perfection in the face and carriage of a man who lives among noble objects and pursues a thinking life, or in the trained and healthy but anti-intellectual and superficial man? If the question surprises us, it is because we are in the habit of looking at the human body much as we look at the animal body, as nature merely. But the human body is definitely determined by spirit. The face of a man who is passionately searching for truth is not only more "spiritual" than that of the man with a dulled mind, it is also more of a face, that is to say, it is more genuinely, more intensively "body." And there is not only more "spirituality" in the bearing of a man with a free and generous heart than in that of a crude and selfish person, there is a more responsive body. With man begins a wholly new scale of development. The body as such becomes more animated, more vibrating, as it is more strongly informed by the life of the heart, mind, and spirit.

... What, then, will not be possible when eternity breaks into time and divine strength and holiness holds unrestricted sway, setting the spirit free in its absolute purity and power?[7]

[8.2]
LEGENDS ABOUT TAOIST IMMORTALS

The writer John Blofeld described a Chinese scholar who believed he would transform his flesh into a "shining, adamantine substance, weightless yet hard as jade." Within three years, the old man claimed, his transmutation would be complete. The contemplative master Hsu Yin, who in his late nineties could outwalk Blofeld, retained his youthful agility into his 120s. And according to the Chinese text *Histories of the Immortals*:

Tseng Ying, having attained the Way in his eighties, regained the glossy black hair of his younger days and such lightness of body that his arrows would hit a mark set up at a distance of a hundred paces. In a single day, he could walk several hundred *li* at a pace that made it difficult even for youngsters to keep up with him, and once he fasted for fifty days without being assailed by hunger.[8]

In his *Taoism: The Quest for Immortality*, Blofeld described the "yogic alchemy" by which immortal bodies are created. By the sublimation of *ching* (semen or psychophysical essence), *chi* (breath or vital force), and *shen* (mind or spirit), the adept forms a "spirit-child" or "spirit-body," the vehicle of everlasting life.[9] Though such teachings can be interpreted in various ways, they strongly suggest that some Taoists have believed humans can advance to higher forms of embodiment, in this flesh or a substance created within it.

[8.3]
SHAMANIC DISMEMBERMENT AND RESURRECTION

Shamanism exhibits certain strong similarities from culture to culture.[10] Rituals of symbolic dismemberment and resurrection, conversations with spirits, and trances in which the shaman journeys into other worlds have been observed among Siberians, Central Asians, North and South Americans, Africans, and peoples of the South Seas. These and other practices, I suggest, anticipate disciplines to produce metanormal embodiment. Consider, for example, the following activities described by Mircea Eliade.

> The ecstatic experience of dismemberment of the body followed by a renewal of the organs is also known to the Eskimo. They speak of an animal that wounds the candidate, tears him to pieces or devours him; then new flesh grows around his bones. . . . Usually these cases of spontaneous vocation are manifested, if not by a sickness, at least by an unusual accident (fight with a sea beast, fall under ice, etc.) that seriously injures the future shaman. But the majority of Eskimo shamans seek ecstatic initiation and in the course of it undergo many ordeals, sometimes very close to the Siberian and Central Asian shaman's dismemberment. In these cases there is a mystical experience of death and resurrection induced by contemplating one's own skeleton.
>
> Among the Warburton Ranges aborigines (West Australia) initiation takes place as follows. The aspirant enters a cave, and two totemic heroes (wildcat and emu) kill him, open his body, remove the organs, and replace them with magical substances. They also remove the shoulder bone and tibia, which they dry, and before putting them back, stuff them with the same substances. During this ordeal the aspirant is supervised by his initiatory master, who keeps fires lighted and observes his ecstatic experiences.[11]

Ilpailurkna, an aboriginal magician in Australia, told anthropologists B. Spencer and F. J. Gillen that

> when he was made into a medicine man, a very old doctor came one day and threw some of his *atnongara* stones at him with a spear-thrower. Some hit him on the chest, others went right through his head, from ear to ear, killing him. The old man then cut out all of his insides, intestines, liver, heart, lungs—everything in fact, and left him lying all night on the ground. In the morning the old man came and looked at him and placed some more *atnongara* stones inside his body and in his arms and legs, and covered over his face with leaves.
>
> Then he sang over him until his body was all swollen up. When this was so he provided him with a complete set of new inside parts, placed a lot more *atnongara* stones in him, and patted him on the head, which caused him to jump up alive. The old medicine man then made him drink water and eat meat containing *atnongara* stones. When he awoke he had no idea as to where he was, and said, "*Tju, tju, tju*—I

think I am lost." But when he looked round he saw the old medicine man standing beside him, and the old man said, "No, you are not lost; I killed you a long time ago." Ilpailurkna had completely forgotten who he was and all about his past life. After a time the old man led him back to his camp and showed it to him, and told him that the woman there was his *lubra*, for he had forgotten all about her. His coming back this way and his strange behavior at once showed the other natives that he had been made into a medicine man.[12]

According to Spencer and Gillen, the *atnongara* stones used in this initiation "are small crystalline structures which every medicine man is supposed to . . . produce at will from his body, through which it is believed that they are distributed. In fact, it is the possession of these stones which gives his virtue to the medicine man."[13]

Imagined dismemberment is practiced, too, in the Americas, Africa, and Indonesia, where according to Eliade "dream sickness or initiation ceremony [has the same central element]: death and symbolic resurrection of the neophyte, involving a cutting up of the body performed in various ways (dismemberment, gashing, opening the abdomen, etc.)."[14] Eskimos of Siberia, Greenland, and Labrador have also practiced rituals of death and rebirth. The great arctic explorer Knud Rasmussen described interviews with shamans in his *Intellectual Culture of the Iglulik Eskimos*, summarizing some of them in this passage.

> Though no shaman can explain to himself how and why, he can, by the power his brain derives from the supernatural, as it were by thought alone, divest his body of its flesh and blood, so that nothing remains but his bones. And he must then name all the parts of his body, mentioning every single bone by name; and in so doing, he must not use ordinary human speech, but only the special and sacred shaman's language which he has learned from his instructor. By thus seeing himself naked, altogether freed from the perishable and transient flesh and blood, he consecrates himself, in the sacred tongue of the shamans, to his great task, through that part of his body which will longest withstand the action of the sun, wind and weather, after he is dead.[15]

These shamanic initiations, I propose, reflect an aspiration for somatic transformation. Though we can dismiss them as products of a primitive mentality that reduces spiritual values to grossly physical terms, we can also see them as informed by intuitions that we may indeed be reborn, in both mind and body.

[8.4]
SUPERORDINARY POWERS IN FANTASY LITERATURE, CARTOONS, MOVIES, AND SCIENCE FICTION

Fantasy literature, cartoons, movies, and science fiction offer a rich imagery of human transformation. Such imagery varies in power and subtlety, ranging from the cartoon simplicity of Superman to the astronaut's mysterious rebirth in the stargate of Stanley Kubrick's *2001*, but in its entirety it reflects virtually all the metanormal capacities described in this book. In these images one finds hyperdimensional consciousness and luminous bodies, superordinary vitality and movement, mystical ecstasies and unitive awareness. Furthermore, many of the transformations that such fiction depicts are accomplished through metahuman agencies: in *Star Wars*, Luke Skywalker destroys the Death Star when he surrenders to the Force; Kubrick's astronaut is reborn through the immense and nameless intelligence that left a stargate on Jupiter's moon. And the subjects of such transmutation struggle through initiatory rites like shamans and mystics in real life. Luke Skywalker suffers under Yoda's guidance to become a vehicle of the Force. Kubrick's astronaut dies by stages to his old self as his mind and body mutate. In its depictions of transformative practice and process, in its suggestions of Divinity's grace, in its renderings of many extraordinary powers, popular visionary art today resonates with the picture of metanormality presented in this book. Using as a guide the capacities listed in chapters 4 and 5, for example, we find images of:

□ Metanormal perception in Ben Kenobe's sense of "disturbance in the Force" when Princess Leia's planet is destroyed by the Death Star; Ray's ability to see Shoeless Joe Jackson, though others without second sight cannot, in the movie *Field of Dreams*; and Jimmy Stewart's vision of the world's goodness in *It's a Wonderful Life*.

□ Metanormal kinesthesis in the protagonist of Paddy Chayevsky's book (and movie) *Altered States*, who relives evolution through profound contact with his own cells.

□ Metanormal communication abilities in Ben Kenobe's telepathic contacts with Luke Skywalker; and the telepathic powers of Jommy Cross, the "mutation after man," in A. E. Van Vogt's novel *Slan*.

□ Metanormal vitality in the luminous aliens of the movie *Cocoon*, who sustain themselves on Earth (for a while) without their customary sustenance; and the tireless energy of the superhero in Van Vogt's *Slan*.

□ Metanormal movement in the levitation produced by E. T. as his young companions fly on bicycles with him; the protagonist's ability to travel out of body in Jack London's *Star Traveler* and Van Vogt's *Supermind*; Ben Kenobe's postmortem materialization in ordinary space-time; and the walk by James Earl Jones into the land of departed baseball players during *Field of Dreams*.

□ Metanormal influence upon the environment in E. T.'s healing touch; Luke Skywalker's power to move imperial storm troopers from a distance; Yoda's ability to lift Luke's spacecraft without touching it; and Ray's creation of magical space by building a baseball diamond in *Field of Dreams*.

□ Metanormal delight in the ecstasies imparted to her ardent admirer by the beautiful alien of the movie *Cocoon* (after which the young man exclaims, "If this is foreplay, I'm a dead man!"); and the marriage of the mutated superwoman in Van Vogt's *Supermind* to Point Omega.

□ Metanormal cognition in the intuitive mind of homo intelligens, the successor of homo sapiens, in Lester del Rey's *Kindness;* the widely shared visions of a numinous spacecraft during the movie *Close Encounters of the Third Kind*; and Ray's vision in *Field of Dreams* that a baseball diamond will attract something great and marvelous.

□ Extraordinary volition in Luke Skywalker's power to destroy the Death Star by surrender to the Force, instead of reliance upon his airplane's computer.

□ Metanormal identity in the astronaut of *2001* as he dies to his ordinary mind and body so that he can incarnate the immense intelligence that beckons to the human race.

□ Extraordinary love in the numinous embrace of the woman protagonist by the mysterious underwater presence in the movie *The Abyss*; Michael Valentine Smith's sacrifice of his body for his fellows in Robert Heinlein's *Stranger in a Strange Land*; and Ben Kenobe's deliberate death at the hands of Darth Vader for the sake of the freedom fighters.

□ Metanormal embodiment in the shape-changing protagonist of Jack Williamson's *Darker Than You Think;* and the luminous aliens in *Cocoon,* who pass through physical barriers, levitate, live without food, and radiate ecstasy to one another and the humans they touch.

In a manner analogous to Jules Verne's anticipation of atomic power in 20,000 *Leagues Under the Sea,* certain cartoons, movies, and fantasy literature might express intuitions of capacities that are available to us. If, as Marshall McLuhan suggested, artists are a culture's "distant early warning system," the images just noted might prefigure luminous knowings and powers that can be realized by the human race.

9

EVOLUTION IN A
LARGER EARTH

IN A MUCH-QUOTED PASSAGE of his *Religio Medici,* English essayist Thomas Browne characterized man as "that great and true Amphibian whose nature is disposed to live, not only like other creatures in diverse elements, but in divided and distinguished worlds." We humans, in other words, exist on many levels at once, as subject and object, experience and electrochemical event. Like amphibians, we live in various domains, moving through dreams, reveries, and everyday consciousness, amid ideas, emotions, and sensory events, venturing at times to interior spaces that seem distant from our normal habitats. Here I draw upon Browne's metaphor to suggest that integral development can introduce us to a "larger Earth" rather than lifting us to separate planes of existence. Like amphibians, I propose, we can maintain access to our present milieus as we enter worlds that subsume them. While increasing our range of activity in a more spacious environment, we need not be separated from our present surroundings. The first living cells, after all, created new realms for organic development, and humans discover inner worlds to explore as their consciousness develops. Since animal species subsumed inorganic structures in their development, and since humans in turn subsume many features of animal functioning, we can suppose that metanormal life might evolve in analogous fashion. Rather than living exclusively in disembodied heavens or "separate realities," we can broaden our horizons in *this* world, conceivably, and realize new kinds of life in contact with more and more dimensions of the universe. Our further development, in short, would open the world we now perceive rather than disengage us from it.

Such possibilities have been pictured in fairytales and works of science fiction. The eldila, for example, live "hypersomatically" in C. S. Lewis's space trilogy, restructuring matter at will to enjoy their stupendous existence. In such stories, I suggest, a grand thought experiment is conducted, helping us conceive possibilities for extraordinary life. In Lewis's novel *Perelandra*, for example, the archons of Mars and Venus appear to the protagonist, Ransom, so that

> a tornado of sheer monstrosities seemed to be pouring over [him]. Darting pillars filled with eyes, lightning pulsations of flame, talons and beaks and billowy masses of what suggested snow, volleyed through cubes and heptagons into an infinite black void. "Stop it," he yelled, and the scene cleared. He gazed round on the field of lilies, and presently gave the eldila to understand that this kind of appearance was not suited to human sensations. "Look then on this," said the voices again. And he looked with some reluctance, and far off between the peaks on the other side of the little valley there came rolling wheels. There was nothing but that—concentric wheels moving with a rather sickening slowness one inside the other. There was nothing terrible about them if you could get used to their appalling size, but there was also nothing significant. He bade them to try yet a third time. And suddenly two human figures stood before him on the opposite side of the lake.
>
> They were taller than the Sorns, the giants whom he had met in Mars. They were perhaps thirty feet high. They were burning white like white-hot iron. The outline of their bodies when he looked at it steadily against the red landscape seemed to be faintly, swiftly undulating as though the permanence of their shape, like that of waterfalls or flames, coexisted with a rushing movement of the matter it contained. For a fraction of an inch inward from this outline the landscape was just visible through them: beyond that they were opaque.
>
> Whenever he looked straight at them they appeared to be rushing towards him with enormous speed: whenever his eyes took in their surroundings he realised that they were stationary. This may have been due in part to the fact that their long and sparkling hair stood out straight behind them as if in a great wind. But if there were a wind it was not made of air, for no petal of the flowers was shaken. They were not standing quite vertically in relation to the floor of the valley: but to Ransom it appeared (as it had appeared to me on Earth when I saw one) that the eldils were vertical. It was the valley—it was the whole world of Perelandra—which was aslant. He remembered the words of Oyarsa long ago in Mars, "I am not here in the same way you are here." It was borne in upon him that the creatures were really moving, though not moving in relation to him. This planet which inevitably seemed to him while he was in it an unmoving world was to them a thing moving through the heavens. In relation to their own celestial frame of reference they were rushing forward to keep abreast of the mountain valley. Had they stood still, they would have flashed past him too quickly for

him to see, doubly dropped behind by the planet's spin on its own axis and by its onward march around the Sun.

Their bodies, he said, were white. But a flush of diverse colours began at about the shoulders and streamed up the necks and flickered over face and head and stood out around the head like plumage or a halo. He told me he could in a sense remember these colours—that is, he would know them if he saw them again—but that he cannot by any effort call up a visual image of them nor give them any name. The very few people with whom he and I can discuss these matters all give the same explanation. We think that when creatures of the hypersomatic kind choose to "appear" to us, they are not in fact affecting our retina at all, but directly manipulating the relevant parts of our brain. If so, it is quite possible that they can produce there the sensations we should have if our eyes were capable of receiving those colours in the spectrum which are actually beyond their range.[1]

Integral development, conceivably, can by small steps introduce us to an existence analogous to that of Lewis's hypersomatic creatures. This possibility is suggested by the following phenomena.

Perceptions of Phantom Figures and Extraphysical Domains

Some people apprehend phantom figures or extraphysical worlds that seem to be more than figments of their imagination. Climbers Estcourt and Scott, explorer Ernest Shackleton, seafarer Joshua Slocum, and aviator Charles Lindbergh, for example, described disembodied entities that tried to communicate with them and that persisted for hours or days (5.1). Many people recount near-death episodes in which they perceive beings that seem to know what is happening in this world (10). Contemplative adepts have detailed their perceptions of extrasomatic creatures and vistas that seem not far removed from our normal environments (5.1). Certain mathematical and musical forms are perceived all at once, as if they already exist (5.8). During such experiences, the subject typically feels as if he or she is apprehending something real, something with its own objective status, that is somehow related to our physical universe. Are those somethings first glimpses of a "larger Earth"? To a frog with its simple eye, the world is a dim array of impressions in comparison to ours. Are we like frogs in relation to an environment we cannot yet perceive? Is some "otherworld" experience analogous to the vision of early amphibians?

Out-of-Body Experience, Traveling Clairvoyance, and Dematerialization

Instances of *bilocation* (or "traveling clairvoyance") are recorded in *Butler's Lives of the Saints*, and the Catholic church includes the phenomenon among its acknowledged charisms.[2] Taoist sages are said to travel in superordinary fashion with their "immortal, diamond forms."[3] According to certain Islamic texts, Iranian "men of light" move through the hyperdimensional worlds of *Hurqalya*, the "celestial Earth."[4] Does the movement reflected by these accounts involve configurations of matter, or dimensions of space-time, that science has not yet revealed? Such movement is dramatized in stories about UFOs and in works of science fiction.[†] Do legends about metanormal agility reflect the fact that it occasionally happens? Some philosophers have thought so.[5] And in recent years, some mathematicians and physicists have speculated that theories aiming to unify the fundamental forces in a space-time with more than four dimensions (sometimes called Kaluza-Klein theories) might account for paranormal phenomena.[‡] Physicist Saul-Paul Sirag has written, for example, that "we may view the body as a projection from the total hyperspace. Plato apparently had such a scheme in mind when he proposed in his cave allegory that we tend to identify ourselves with our three-dimensional bodies, just as the chained slaves identified themselves with their two-dimensional shadows."[6] In any case, hyperdimensional models of the universe, which are proliferating among mathematicians and physicists today, have a resonance with esoteric accounts of extraspatial worlds in which our familiar existence is embedded, and from which phantom figures, luminosities, odors of

---

[†] See for example Keith Thompson's *Angels and Aliens*, Addison-Wesley, 1991, especially chapter 16.

[‡] For an account of Kaluza-Klein theories of hyperdimensional space-time, see: Freedman & Van Nieuwenhuizen, 1985, The hidden dimensions of spacetime, *Scientific American* (Mar.), pp. 74–81. This article traces the development of hyperdimensional models from their origins in quantum theory and Einstein's general relativity to the proposals of Theodor Kaluza that space-time is embedded in a five-dimensional world, to contemporary superstring and supergravity theories. There is often a long time, Freedman and Van Nieuwenhuizen wrote,

> between the development of elegant theoretical ideas and the precise formulation of predictions that can be tested experimentally. It took 13 years, for example, to find the correct way of applying non-Abelian gauge theories to the unification of the fundamental forces. The current lack of clear indications that the ideas of Kaluza-Klein theory are experimentally correct does not necessarily indicate the ideas are wrong. They may simply require further theoretical work.

sanctity, and other extrasomatic phenomena materialize, and through which highly developed spirit-bodies move. Might our present movement abilities be analogous to those of early amphibians that had not learned to breathe or move freely on land?

Extraordinary Conditions of Energy and Matter

We can imagine an existence, then, with a range of perceptual events greater than ours, in which we might move with new agility and freedom. Such an existence, it seems, would involve superordinary energy states. Do the initiations of shamanic-religious lore involve such conditions? Contemplative transmission may be produced by a touch, a look, a smile, or an embrace, but can also be imparted at a distance without sensory cues (see chapter 5.3). The medium of such exchange is designated by words such as *spirit, prana,* or *ki,* but could be called with just as much precision some sort of hyperdimensional environment. It is conceivable that various joys and empowerments relayed from teacher to disciple, lover to lover, athlete to athlete, and parent to child involve processes and structures that neither science nor religion has adequately described. When sportspeople say they are "zoned," they seem to indicate a space beyond ordinary space that is intimately connected with both mind and body. Does the inexplicable radiance of a saintly face, the glow of certain athletes' skin, or the light in the eyes of a lover result from reorganizations of matter in some sort of hyperspace? Are these events made possible by multidimensional worlds such as those proposed by modern physicists? Might some instances of spiritual healing, or the streaming sensations that sometimes accompany athletic performance, or the dancing lights perceived around some saints be produced by energy patternings projected from a larger Earth? Does metanormal delight result from materializations of subatomic particles that facilitate opioid peptide production?

Testimonies to subtle bodies and subtle matter, which exist in virtually every sacred tradition, support speculation of this kind. The *ka* of Egyptian religious lore; the *sariras, koshas,* and *dehas* of Hindu-Buddhist yoga; the Greek *ochemata;* the *jism,* or astral body, of Iranian Neo-Platonism and Sufism; and the spirit-body of Taoist yogic alchemy (8.2) are said to be made of a matter different from ordinary flesh. Ensouled bodies made of this spirit-matter, it is said, can pass through physical barriers, materialize and dematerialize, and change shape or size. Such bodies are more responsive to emotion, thought, and volition than our present flesh. In the words of Henri Corbin, imagination is for them "comparable to what the hand is for the [physical] body."[7] Testimonies to subtle bodies, which exist both in religious lore and collections of psi phenomena made by modern psychical researchers (see appendix F), suggest that human flesh interacts with, and gives rise to, extraordinary kinds of matter.

Given the anecdotes, legends, and subjective reports just cited, it is at least conceivable that we could realize a metasomatic existence, coming ashore like amphibians into a world beyond our first habitat, transcending many patterns of ordinary human life. Returning to my diagrams from chapter 7, we can represent integral development as shown in Figure 9.1.

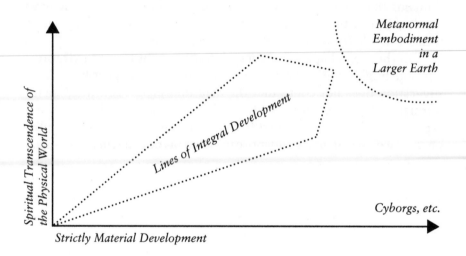

Figure 9.1.

As the diagram indicates, metanormal embodiment could be developed only along the right lines of development; i.e., vectors that integrate the cultivation of mind and body. Those who neglect human nature's active capacities—on a trajectory like the vertical line—would not grow into a "larger Earth." On the other hand, those who pursued purely physical development, through genetic manipulation or prosthetic devices, for example, would not mutate their living tissue and consciousness to superordinary states.

Different practices open different domains and cause us to leave different imprints upon our surroundings. It is conceivable, for example, that India's physical and cultural environments were injured by the otherworldly influence of Hindu philosophy. Spending much of their life in trance, their attention withdrawn from physical things, some of India's great yogis set a powerful example for others to follow. As role models for later generations, they contributed perhaps to India's neglect of material and social development, a neglect that was fortified by doctrines that this world is *maya*, illusion. But even if we don't blame asceticism for India's difficulties, the failure to cultivate our many-sided nature as we develop contemplative insight can cut us off from life upon earth. By valuing mystical cognition above our other metanormal capacities, or by withdrawing into ascetic solitude, we amputate a large part of our earth connection. Integral practices, however, would embrace both the universe we now perceive and any domains that subsume it. To support such practices, it might help to think of our further evolution in terms of a larger Earth rather than in terms of worlds or heavens separate from this one.

In the next chapter, I suggest that movement toward extraordinary life might continue after death in some sort of spirit-body. This suggestion does not contradict the idea that progress toward metanormal embodiment in this life might reveal a larger Earth. Now-separate worlds can be joined *both in this flesh and hereafter,* I propose, as we open to more and more dimensions of existence.

10

EVOLUTION AND
POSTMORTEM
SURVIVAL

SEVERAL PHENOMENA DISCUSSED in this book raise questions about postmortem survival. For example, do out-of-body states reveal a soul that can live after death? If certain mystical experiences reveal our metanormal identity, what sort of existence does that identity enjoy after the body's passing? Since we draw heavily upon the religious traditions for evidence of our greater capacities, shouldn't we accept their affirmations of an afterlife? I will respond to such questions, first, by noting that we can accept the fact that extrasensory and mystical experiences occur without subscribing to all beliefs associated with them. Indeed, the great variety of survival theories compels us to caution in considering claims about life after death. Hindu and Tibetan Buddhist yogis believe in reincarnation. Virtually all Jewish, Christian, and Moslem mystics have thought we live only one life on earth but exist after death in a disembodied condition. And some Taoist sages have claimed that just a few advanced spirits survive. Given these various beliefs, each attested to by religious adepts, there is good reason to remain agnostic about most pictures of postmortem existence.

Nevertheless, certain claims that bear upon the survival question are based upon firsthand experience common to people with different beliefs about the afterlife. It is indisputable, for example, that individuals in many eras and cultures—among them contemplative masters and philosophers who hold a central place in religious history—have enjoyed illuminations in which an eternal or immortal self was overwhelmingly apparent to them. In chapter 5.10, I cite mystics East and West who have described an identity that cannot die. It is also unarguable that people of all ages have out-of-body experiences during near-death episodes, sports competition, anesthesia, childbirth, high fever, or extreme pain. Such experience is detailed in every religious culture, and in medical journals, psychiatric case studies, and the annals of psychical research (see chapter 5.5 and appendix E). Nor can we deny that individuals with widely disparate beliefs about the afterlife have perceived what they took to be other worlds (and beings who inhabit them) during prayer, emotional crisis, and other kinds of experience. Though we are not compelled to believe that these experiences prove that a soul survives the body's demise, we must agree that such experience leads many people to affirm postmortem existence.

There are also less common experiences that provide indirect evidence that the soul survives death, among them cases of traveling clairvoyance in which the subject is apparently perceived as a phantom figure. In the Wilmot case cited in chapter 5.5, three people were involved: Mrs. Wilmot, who dreamed she stood by her distant husband; Mr. Wilmot, who dreamed his wife stood by his bunk; and Mr. Wilmot's cabin-mate, who claimed to see a woman in their cabin at the same time the Wilmots were dreaming. We could account for this episode by attributing the cabin-mate's vision to an illusion produced by his telepathic impressions of the other two parties, but it is possible as well that Mrs. Wilmot did indeed travel in some sort of spirit body that her husband perceived in a dream and his cabin-mate mistook for a fleshly figure. The Verity case cited in chapter 5.5 also belongs in this class of experience. And besides the better cases of projected phantoms described in the literature of psychical research, some religious anecdotes about this phenomenon are hard to dismiss entirely. If we strip away literal interpretations from stories of bilocation, discounting the idea that someone can be physically present in two places at once, we can give credence to some claims (in both Eastern and Western religions) that particular saints projected themselves paranormally.

There are also numerous accounts of communications with the dead. Studies of mediumship have been conducted by the British and American Societies for Psychical Research for more than a hundred years, and by other groups of dedicated investigators. I refer you again to Myers's *Human Personality* for representative studies of these, and to Alan Gauld's comprehensive survey of the field, *Mediumship and Survival.*[1] Several types of experience, in short, provide evidence for a soul or spirit-body that survives death, though they give rise to different beliefs about the afterlife. In the face of these different beliefs, thinkers such as William James, C. D. Broad, H. H. Price, and John Hick, as well as investigators of past-life memories such as Ian Stevenson, and contemporary students of near-death experience have approached the survival question with both critical distance and openness to the relevant data (see appendix E).[2]

If we suspend judgment, then, about the nature of life after death, what can we surmise about its possible relations to integral development? What connections might exist between extraordinary functioning and postmortem survival? In response to these questions, I suggest: first, that postmortem life might continue in some sort of body; second, if postmortem life happens in a world of spirit-bodies, progress toward metanormal embodiment could continue there; and third, if such afterworld progress can occur, cultivation of our various capacities in this life would facilitate their deployment and further development hereafter.

Postmortem Bodies

According to many accounts of extrasomatic experience, the soul or spirit has boundaries, unique form, a point of view, location (which can move) in some sort of environment, and continuity with its life upon earth. And according to some, the vehicle of survival even has a face and organs. In countless reports collected by contemporary near-death researchers, and in many accounts of mediums, mystics, and shamans who have traveled in trance, we find descriptions of a conscious self, or person, that inhabits particular locations in the spirit-world. It sometimes encounters other spirits, views marvelous or terrible soulscapes, hears unearthly music, touches other forms (and has intercourse with them), moves from place to place, and acts upon its environment. Such descriptions indicate that this traveling self, this soul, is a kind of body.[†] Indeed, there may be a fundamental equation between personhood and embodiment, in this world or any other. To be a *particular* person, after all, requires us to have distinctive fea-

---

[†] Emmanuel Swedenborg wrote:

> Manifold experience has witnessed to me that when a person crosses over from the natural into the spiritual world, which happens when he dies, he carries with him everything that is his, or everything belonging to his person, except his earthly body. For when a person enters the spiritual world, or life after death, he is in a body the way he was in this world. There seems to be no difference, since he does not feel or see any difference. Further, the spirit person enjoys every outward and inward sense he enjoyed in the world. As before, he sees, he hears and speaks, he smells and tastes; as before he feels the pressure when he is touched (Swedenborg 1960b, p. 358).

In *The Projection of the Astral Body*, Sylvan Muldoon and Herewood Carrington cite Prescott Hall's *Astral Projection*. During out-of-body experience, Hall wrote, the subject hears

> a hissing or whistling, as of escaping steam; single musical notes; musical phrases, generally new to the hearer; human tunes; the sound of a bell or bells, sometimes in harmony;

tures, and thus contours and boundaries, however hyperdimensional. We view the world from a certain standpoint, even if at the same time we perceive our unity with others. As embodied selves, we interact with our surroundings in distinctive ways, with sensibilities and histories that make us unique. As I suggested in chapter 5.10, saints and mystics of all cultures have exhibited well-formed and colorful features, both psychic and physical, *after* their spiritual illuminations. They have retained their unique identity and their idiosyncratic forms of expression. If they have realized a more distinctive embodiment in this life, it is conceivable that they would keep it hereafter.

That the soul has a postmortem body is an ancient idea reflected in terms such as the Egyptian *ka*, the Greek *ochema*, and the Sanskrit *kosha, deha,* and *sarira*, which represent vehicles of consciousness that can live after death. Different doctrines accompany the idea that we possess a subtle body (or bodies), too many to summarize here. Some were described by philosopher G. R. S. Mead in *The Doctrine of the Subtle Body in Western Tradition*,[3] but the most comprehensive discussion of them can be found in the magisterial four-volume work by the Dutch philosopher J. J. Poortman, *Vehicles of Consciousness: The Concept of Hylic Pluralism*.[4] A reading of Poortman's survey reveals the great diversity of evidence and arguments for the spirit-body to be found, for example, in the Rig Veda, Upanishads, Old Testament, pre-Socratic philosophers, Plato, Aristotle, certain Epicureans and Stoics, Plotinus, Porphyry, Iamblichus, Proclus, Origen, Saint Paul, many Gnostics, Augustine, Thomas Aquinas, Origen, Descartes, Paracelsus, Swedenborg, and works of several modern philosophers, novelists, and scientists. The great detail, beauty, and richness evident in descriptions of the subtle body, and the passionate affirmations of its existence by great saints and philosophers since antiquity, will impress most readers of Poortman's study.

Postmortem Progress

If postmortem life happens in a spirit-body, it is conceivable that the entire spectrum of metanormal capacities can develop after death. If

and metallic noises like the striking of an anvil. It is not unusual, at the beginning of a projection, to hear seemingly far-distant sounds which sound familiar to the subject. Often it is as if someone, far away, were calling in a musical voice. A very peculiar sensation is as if someone unseen were blowing in one's face. It is as if some unseen finger-tip touched one about the throat, mouth and nose, causing a "tickling" sensation (Muldoon & Carrington 1961).

spirit-bodies apprehend and act upon their extrasomatic environments, move about in some sort of spirit-space, and communicate
with others as so many accounts suggest, it is at least plausible that
they share the universal adventure of development, with its victories
and defeats, its forward and backward movements, its progress and
regress. And indeed, this idea is reflected in religious doctrine.
According to Christian and Moslem traditions, there are purgatories
in which we can come closer to God. According to many Hindu, Neo-
Platonist, and Buddhist teachings, we absorb the experience of our
past incarnation during the interval between it and our next. However
much these doctrines differ, they regard the afterworld to be a condition in which spirits suffer and grow.

Preparing for Postmortem Life

The preservation of continuous awareness into postmortem life is described in the Tibetan *Bardo Thodol*, or *Book of the Dead*.[5] It can also
be trained, according to some yoga schools, by the practice of conscious sleep in which one develops a constant witness, or lucid dreaming, that lasts through the night,[†] and by holding constant awareness
during strenuous vigils. Through contemplative practice, it is said, we
can learn to maintain waking consciousness through changes in
brain-state during which memory is normally obliterated. By retaining awareness during occasions in which we normally lose it, we can
prepare to hold it as we begin our postmortem existence. The training
of contemplative attention, in short, helps impart such attention to
the surviving spirit. But if there is indeed an afterlife, it must involve
more than awareness. Here I wish to propose that a wide range of capacities might develop after death. Through their cultivation in this
life, conceivably, they can be strengthened for a spirit-body's postmortem journey. For example:

---

† In a letter on yoga, Aurobindo wrote:

In ordinary sleep we do not become aware of the worlds within; the being seems sub-
merged in a deep subconscience. On the surface of this subconscience floats an obscure
layer in which dreams take place, as it seems to us, but, more correctly it may be said, are
recorded. When we go very deeply asleep, we have what appears to us as a dreamless
slumber; but, in fact, dreams are going on, but they are either too deep down to reach the
recording surface or are forgotten. Ordinary dreams are for the most part or seem to be
incoherent, because they are either woven by the subconscient out of deep-lying impressions left in it by our past inner and outer life, woven in a fantastic way which does not
easily yield any clue of meaning to the waking mind's remembrance, or are fragmentary
records, mostly distorted, of experiences which are going on behind the veil of sleep—

- Perception of events external to the individual, if broadened by transformative practice to include clairvoyance and awareness of extrasomatic luminosities, creatures, auditions, and tactile impressions, would conceivably carry over to the surviving entity. Since these perceptual abilities do not appear to depend upon physical receptors (see chapter 5.1), we can suppose that they would continue to function, extrasomatically, in the worlds of postmortem existence. And perception of the numinous could develop there too, since it finds beauty everywhere, in the "other worlds" of visionary trance as well as the physical universe.

- If internal clairvoyance does not depend upon physical senses, it presumably would not in an extrasomatic entity. If its extrasensory modalities can be strengthened in this life, they could conceivably be developed in the next.

- In chapter 5.3, I note three types of extrasensory communication: sustained telepathic exchange of particular emotions, thoughts, or intentions; telepathic transmission of spiritual illumination; and communal ecstasies that appear to be telepathically induced. Since none of these depends upon sensory cues, it is conceivable that they occur among spirit-bodies, or between spirit-bodies and people in this world. To the extent that we develop such abilities, we conceivably will be better able to interact with any creatures we meet hereafter, including those that might assist our further development.

- It is conceivable that the extraordinary vitality, or spiritual "warm-bloodedness," that enables contemplative adepts to withstand environmental extremes (5.4, 21.1–21.3, 22.6, 23.1)

very largely indeed these two elements get mixed up together. For, in fact, a large part of our consciousness in sleep does not get sunk into this subconscious state; it passes beyond the veil into other planes of being which are connected with our own inner planes, planes of supraphysical existence, worlds of a larger life, mind or psyche which are there behind and whose influences come to us without our knowledge. Occasionally we get a dream from these planes, or something more than a dream,—a dream experience which is a record direct or symbolic of what happens to us or around us there. As the inner consciousness grows by sadhana, these dream experiences increase in number, clearness, coherence, accuracy and after some growth of experience and consciousness, we can, if we observe, come to understand them and their significance to our inner life. Even we can by training become so conscious as to follow our own passage, usually veiled to our awareness and memory, through many realms and the process of the return to the waking state. At a certain pitch of this inner wakefulness this kind of sleep, a sleep of experiences, can replace the ordinary subconscious slumber (Sri Aurobindo, 1970, Letters on Yoga. Vol. 23 in The Collected Works, pp. 1023–24).

and survive for long periods without food (21.2, 21.7, 22.4, 23.1) carries over to postmortem existence, enabling those who possess it to master whatever environmental fluctuations or deprivations might occur there. There is considerable evidence that such vitality does not depend upon material sustenance. Since it does not depend upon food, drink, clothing, or other physical supports, it might develop as well in a spirit-body.

▢ In chapter 5.5, I propose that there are four kinds of metanormal movement ability: the extraordinary agility exhibited by certain shamans, martial artists, and athletes; levitation; traveling clairvoyance; and deliberate passage into other worlds. The latter two, if strengthened through transformative practice, would conceivably facilitate a spirit-body's capacity to move either from place to place in contact with the physical world or in the supraphysical dimensions of postmortem existence. There is a considerable lore, after all, about appearances in this world of departed saints and relatives, as well as their deliberate movement through worlds of the afterlife.

▢ Since telergy, telekinesis, and alteration by mental influence of some portion of space do not depend upon fleshly organs or limbs (see chapter 5.6), they could develop in a spirit-body. This supposition is supported by a considerable lore about influences upon the living from the dead, including the testimony of mediums and contemplatives that they touch and are touched by creatures in extrasomatic worlds.

▢ Through various transformative practices, we can learn to control pain and heighten pleasure, eliminating responses to internal or external stimuli that once caused us suffering (15.8, 17.2, 21.1–21.4, 23.1), and cultivate a self-existent delight that subsumes ordinary pains and joys (5.7). To the extent that we realize such mastery here, in a body given to many kinds of discomfort, we might incorporate it into a spirit-body. Having learned to enjoy impacts of this world once experienced as sources of suffering, our death-surviving form might master its response to the afterlife, finding its interactions there to be sources of recurring delight.

▢ People who have described out-of-body experience or communications with the dead have said that spirit-bodies enjoy cognitions like those experienced in this world. Shamans of stone-age cultures, mediums of modern times, and mystics of the world religions have claimed that souls can cognize their extrasomatic environments and grow in their knowledge of ego-transcendent realities. Furthermore, mystical illumination is said to develop in the afterlife. Given the fact that such illumination does not depend primarily upon sensory cues, it could flower during the soul's postmortem existence.

□ There is a type of willing that is felt to be in accord with a Power beyond ordinary life. At death, when physical bearings are lost, such volition could help to orient the spirit-body toward the extraordinary life described in contemplative and near-death literature. As the spirit-body becomes better oriented to its supraphysical existence, such volition might grow, deepening its alignment with a higher guidance and conferring ever greater self-mastery.

□ A profound subjectivity can arise through witness meditation. During contemplative practice in which thoughts and images are relinquished, it becomes apparent that the witnessing subject does not depend upon particular mental or physical objects. Since it does not depend upon particular things, whether psychic or material, this higher, deeper, or truest self (*purusha* or *mahatmya*) could develop beyond the grave.

□ Saints and mystics since ancient times have said that a love of God and other creatures which does not depend upon the satisfaction of ego-centered desires can develop in the afterlife. And it is conceivable, too, that all emotions and feeling-states flourish there, creating a hellish, mixed, or beatific existence. As on earth, evolution beyond the grave might stand still, regress, or progress in relation to metanormal embodiment.

□ There is much lore among mediums and contemplative adepts that spirit-bodies have articulated structures such as those noted in chapter 5.12. If they participate in the universal adventure of development, spirit-bodies might undergo changes to support their realization of metanormal capacities. If, as I have suggested, the universe is a body-building business, we can well imagine that such building occurs in the afterlife, and that progressive articulation of the mind-body complex in this life might facilitate such articulation hereafter.

Other Ideas About the Afterlife

We can, of course, make a great many further suppositions about postmortem existence based upon contemplative lore, psychical research, contemporary studies of near-death experience, and studies of so-called past-life memories. The English philosopher John Hick has done just that, examining theories about life after death in Judaism, Christianity, Hinduism, and Buddhism to find their respective strengths and weaknesses. According to Hick, belief in some sort of afterlife is not ruled out by any established scientific findings nor by any agreed-upon philosophical arguments. He suggests, however, that Eastern ideas about reincarnation, Western notions of an immortal ego, and Christian visions of eternal damnation are inadequate to both the data and several philosophic questions related to postmortem existence. Eastern and Western ideas of the afterlife point beyond themselves, toward a "common conception of the eschaton, the final and eternal state."[6] They convergently suggest, Hick proposes, that each soul, having come into existence with its human birth, lives through many lives in worlds other than ours, reincarnating not on this planet (or any other) but in ever higher spheres in which it can progressively realize beatific enjoyment of God and other souls.

Sri Aurobindo explored the survival question in a comprehensive manner like Hick's, joining his own contemplative realizations, his wide acquaintance with yogic lore, and examinations of various proposals about postmortem destiny. He was led by a combination of philosophic argument, religious testimony, and his own yoga experience to believe that each soul has opportunities through many lives to grow toward a beatific personhood. But unlike Hick, Aurobindo believed that succeeding lives occur in the physical world (with internatal periods on other planes of existence), and that the soul can progress toward a luminous embodiment involving the divinization of living matter.[†]

Aurobindo's and Hick's proposals about the soul's destiny show

---

[†] Aurobindo wrote:

A new life is not a taking up of the development exactly where it stopped in the last, it does not merely repeat and continue our past surface personality and formation of nature. There is an assimilation, a discarding and strengthening and rearrangement of the old characters and motives, a new ordering of the developments of the past and a selection for the purposes of the future without which the new start cannot be fruitful or carry forward the evolution. For each birth is a new start; it develops indeed from the past, but is not its mechanical continuation: rebirth is not a constant reiteration but a progression, it is the machinery of an evolutionary process. It is probable also that the integrating positive preparation would be carried out and the character of the new life would be decided by the soul itself in a resort to its native habitat, a plane of psychic repose, where it would draw all back into itself and await its new stage in the evolution. (Sri Aurobindo, 1970, *The Life Divine*. Vol. 19 in *The Collected Works*, pp. 802–803).

that similar philosophic approaches can lead to quite different views of the afterlife.[‡] And using the same data that Aurobindo and Hick employed, we can make yet other suppositions about postmortem existence. It is conceivable, for example, that a spirit-body might choose between incarnations on earth, or on other planets, or in supraphysical spheres, so that both Aurobindo and Hick are partly right in their proposals.[7] But no matter how life after death might proceed, there are good reasons to suppose that integral practices can facilitate our postmortem journey. As we realize metanormal capacities, conceivably we not only enjoy them in this life, but in worlds to come. Whatever the afterlife brings us, we can cultivate a unitive awareness of our Source, a love for others, and a self-existent delight that shines through all types of embodied existence.

[‡] It is more than coincidence, perhaps, that Aurobindo was born a Hindu (though he did not affiliate himself formally with any Hindu tradition) and Hick was born a Christian.

Evidence for Human Transformative Capacity

... but the sudden illumination—
We had the experience but missed the meaning,
And approach to the meaning restores the experience
In a different form, beyond any meaning
We can assign to happiness.

T. S. ELIOT

I have no doubt whatever that most people live, whether physically, intellectually or morally, in a very restricted circle of their potential being ... much like a man who, out of his whole bodily organism, should get into a habit of using and moving only his little finger. We all have reservoirs of life to draw upon, of which we do not dream.

WILLIAM JAMES

The part always has a tendency to unite with its whole in order to escape from its imperfections.

LEONARDO DA VINCI

IN PART TWO, I describe transformations of mind and body in contexts ranging from pathological dysfunction, to psychosomatic healing, to disciplines that make just a small demand upon their practitioners, to religious and other life-encompassing practices. I have organized this material to show:

- that significant alterations of consciousness and the flesh can occur, either in fragmented and dysfunctional, partially integrated, or exceptionally healthy and well-organized people; and
- that pyschophysical changes all along this spectrum of experience are caused in large part by certain identifiable activities, or transformative modalities, that can be incorporated into programs for human growth.

11

PSYCHOSOMATIC CHANGES IN ABNORMAL FUNCTIONING

IN THIS CHAPTER, I describe psychophysical changes that occur in hysterical stigmata, false pregnancy, and multiple-personality disorder. Although these changes typically involve disability and suffering, they dramatize transformative capacities that can, I believe, be cultivated to promote human growth.

[11.1]
HYSTERICAL STIGMATA

Hysterical stigmata resemble religious stigmata in that they represent with great specificity important images in the lives of their recipients, typically appear in conjunction with powerful emotions, and often occur among highly suggestible people. They differ from religious stigmata, however, in that they are rarely accompanied by ecstasy or other mystical phenomena. Their importance for this discussion lies in their demonstration of the body's malleability and its responsiveness to powerful volitions, emotions, and mental images, whether conscious or unconscious.

Psychiatrist Joseph Lifschutz treated a patient who exhibited physical changes of this kind. She "was a particularly sober and serious-minded person," he wrote, "and there was little question in my mind concerning the authenticity of her report."

When she was 13, [her] father scratched her down her back with his fingernails, leaving three long scars. These healed over in time. Four years later, at the age of 17, she had left home because of her father's brutality, and was living in the country with her brother. I am not sure how long she had not seen her father, but it is my impression it was many months, perhaps over a year. Somehow [he] found out where she was, and announced he would pay a visit. The patient reports now that as the time of the visit approached, her old back scars, which had been healed for four years, would redden and bleed. . . . This reddening and bleeding of the three old healed scars would recede spontaneously, but these episodes recurred several times, each with the anticipation of a visit from her father.[1†]

This patient's wounds resembled those of Catholic stigmatics in that they bled recurrently, though fear instead of joy accompanied their reactivation. Like the marks of Gemma Galgani, Anne Catherine Emmerich, and Padre Pio (22.3), they showed the great specificity with which the human body can dramatize highly charged issues. Psychoanalyst William Needles also described a series of stigmatic bleedings, in a man aged 31 who bled from his hands on three occasions, each time upon recalling situations that stirred up "Oedipal strivings, fantasies and guilt."[2] And psychotherapist Helene Deutsch described similar skin changes that appeared to express sexual impulses. "I have often seen cases," she wrote, "in which the patient's hand would swell up and become red whenever his associations led him to memories of repressed masturbation."[3]

A more elaborate set of stigmata was described by psychiatrist Ernest Hadley, whose patient, a 24-year-old man, bled from his left armpit during four- to five-day periods through at least seven regular monthly cycles. During one bleeding episode, there appeared "a quantity of serosanguineous material," Hadley wrote. "After washing the armpit, unbroken skin was revealed and there was no [skin discoloration]. Gradually, droplets of the bloody or hemoglobin-stained secretion reappeared." The fact that such bleeding occurred on something

---

† The psychoanalyst Sandor Ferenczi attributed some hysterical stigmata to a "materialization of the Oedipus fantasy which localized converted excitement masses at parts of the body which could be easily placed at the disposal of unconscious impulses" (Ferenczi 1926). From a similar psychoanalytic perspective, Otto Fenichel suggested that repressed thoughts find their substitute expression in a "material change of physical functions [through which] the afflicted organ unconsciously is used as a substitute for the genitals. This . . . may consist of objective changes within the tissues [such as] hyperemia and swelling representing erection; or it may be limited to abnormal sensations imitating genital sensations" (Fenichel 1945, p. 219).

like a menstrual cycle led Hadley to believe it represented his patient's identification with the female character. Menstruation symbolized both defense against sexual assault and female innocence, he suggested, for his patient had identified the armpit with the vagina since his childhood.[4†]

But psychogenic stigmata can also symbolize difficulties that are not primarily sexual. Psychiatrist Robert Moody, for example, described a man treated for attacks of somnambulism who exhibited indentations on his arms that resembled rope marks. These appeared when he was reliving an episode during which he had been tied in bed to inhibit his sleepwalking. In the British medical journal *Lancet*, Moody wrote:

> On the night of April 9, 1944, the patient was observed by the nurse on duty to be tossing and turning violently on his bed. He was holding his hands behind his back and appeared to be trying to free them from some imaginary constriction. [Afterward] the nurse noticed deep weals like rope marks on each arm, the patient being apparently unaware of their presence. Next day the marks were still visible and were observed by myself and others. The patient had only a vague recollection of what had happened the previous night. By the evening of April 11 the marks had disappeared, except for some residual subcutaneous haemorrhagic staining.
>
> On the night of April 11 the incident was abreacted under narcosis.
>
> I watched him writhing violently for at least three-quarters of an hour. After a few minutes weals appeared on both forearms; gradually these became indented; and finally some fresh haemorrhages appeared along their course.
>
> [Later] he gave a clear account of everything that had happened and related the incident to his experiences in hospital in India. Next morning the marks were still clearly visible and were photographed.[5]

The *Lancet* article included a photograph of the marks just described. Though it was conceivable, Moody wrote, that his patient might have tied a rope around himself to cause the first set of indentations, "on the second occasion strict observation made trickery impossible; so it is difficult to see how the occurrence can be explained in any way other than as a genuine psychosomatic phenomenon." Moody described other psychogenic bodily marks he had observed in his

---

† Hadley's description resembles accounts of Louise Lateau's stigmata (33.3). Like the wounds of the once-famous ecstatic, his patient's bleeding was not accompanied by bruises or scars.

therapeutic practice. During catharsis, a patient once buried by a bomb explosion exhibited a swelling on his left ankle where he had been struck, and a second on his head where he had been hit during the same incident. The abreaction of a merchant seaman who had fallen into an icy sea precipitated a localized inhibition of blood flow to his extremities. And a woman who relived a riding accident exhibited psychogenic bruising on her right side where she had fractured some ribs in her fall. Moody wrote:

> These phenomena show that when an experience in which there has been somatic as well as acute psychical trauma is dissociated from normal consciousness, the complex resulting from it may retain the living image, as it were, of the somatic and the psychical experience; and that, when such a complex is discharged into consciousness—even after remaining in a latent state for many years—the somatic component of the original experience may be re-expressed.[6]

In a letter to *Lancet* published in 1948, Moody described a female patient who had exhibited several psychogenic marks. These included a red mark on her shoulder that appeared as she relived a beating in which a whip had caught her on the very spot; a swelling of her right wrist after she recalled an accident in which the wrist had been fractured; red streaks on her legs that corresponded to wounds caused by yet another accident; and the appearance of a bruise that resembled the imprint of an elaborately carved stick her father had used to beat her. Moody also described an experiment he had conducted with the same patient in the presence of a fellow physician. After recalling an incident in which she had been struck across the hands, she exhibited red streaks where her father's whip had left bloody marks. Moody then encased her right hand in a firm plaster bandage, which he removed the following morning in the presence of his colleague to find "obvious bloodstains on the dressing immediately covering the weals." A third doctor also observed the patient's lesions shortly after her bandage was removed.[7]

Some stigmata reflect several traumas and conflicts at once. For example, a 27-year-old man exhibited weals along an elaborate tattoo of a dagger that resembled a female with tapering legs and a penis-like protuberance. As a number of aggressions, fears, guilts, and neurotic compliances, some of them related to the patient's transvestism, were brought to consciousness by psychotherapy, only the dagger tattoo among several on his body swelled. "The symptom," suggested therapists Norman Graff and Robert Wallerstein, "served to . . . resolve and thus lessen the danger of eruption of a whole complex of hostile, erotic, exhibitionistic impulses combined with self-punishing needs."[8] Photographs of this elaborate stigma that appeared in the journal *Psychosomatic Medicine* demonstrate the body's remarkable ability to express complex personal issues with intricate, highly specific psychogenic changes.

In 1955, Frank Gardner and Louis Diamond of Harvard University proposed that some people develop a sensitivity to their own red blood cells that causes bruising in response to trivial injury. They tested their hypothesis by injecting patients who suffered from recurrent spontaneous bruising with small amounts of their own blood, and found that pain and discoloration often appeared around the injected areas. Gardner and Diamond called this disorder autoerythrocyte sensitization.[9] But in reviewing 27 such cases that had been studied at Case Western Reserve University, Oscar Ratnoff and David Agle found that psychological trauma also contributed to the affliction.[10] Injections of saline solution and other substances produced the same kind of bruising in some of these patients that injections of their own blood had caused, and their complaints read "like a table of contents of a monograph on hysteria."[11] Many of these women exhibited dramatic self-display, wide mood swings, sexual frigidity, and demanding dependency. Several had undergone an extraordinary number of operations, ten being subjected to hysterectomy before they were forty years old. Their striking symptoms led Ratnoff and Agle to believe that emotional issues either caused or influenced their spontaneous bruising, even when they were reactive to their own blood.[†] Ratnoff wrote:

> How these emotional stresses were translated into physical symptomology has escaped us. But in four of five patients in whom hypnosis was successfully achieved, lesions resembling spontaneous bruises appeared at a suggested site. Further, we were able by hypnotic suggestion to inhibit the appearance of a positive skin test in a patient who had earlier responded to her own blood, and, conversely, to induce a positive reaction in a patient who had had repeated negative tests. Whatever the mechanism, then, suggestion appeared to be significant in the pathogenesis of the lesions of autoerythrocyte sensitization.

[†] H. Haxthausen tested eight patients exhibiting hysterical skin reactions to see if they had a purely physical reactivity to mechanical or chemical trauma, and found that none of them did. "In no single instance," he wrote, "with any one of the tests employed, did there occur a reaction to be distinguished qualitatively from corresponding responses in normal persons" (Haxthausen 1936).

[Is this affliction] a lay equivalent of [religious] stigmata? To ask this question is to delve into matters beyond our knowledge. But one of our patients, in a period of severe emotional distress, bled through an area on the . . . thigh about the size of a silver dollar, the blood oozing through [her] hair follicles. The bleeding stopped after a few minutes and no excoriations suggesting self-induction were found. Where mind and body meet, we all must recognize our ignorance.[12]

[11.2]
FALSE PREGNANCY

Pseudocyesis, a condition in which nonpregnant women exhibit signs of pregnancy, was known to doctors of antiquity. Around 300 B.C., Hippocrates described twelve women "who imagine they are pregnant seeing that the menses are suppressed and the matrices swollen."[13] Influenced perhaps by her father's strong desire for sons, Mary Tudor, the daughter of Henry VIII, twice had symptoms of pregnancy that culminated in false labor.[14] And Joanna Southcott, the nineteenth-century religious leader, believed herself at 64 to be pregnant with the second messiah and died shortly after she had a nonproductive labor.[15] In 1823, John Mason Good introduced the term *pseudocyesis* into the medical literature;[16] and in 1938 Flanders Dunbar distinguished the condition from hallucinatory pregnancy stemming from psychosis, pregnancy that was deliberately simulated, and pseudopregnancy caused by a tumor or other bodily defect.[17]

In 1937, G. D. Biven and M. P. Klinger reviewed 444 examples of the disorder. The females included in their review ranged in age from 5 to 79 years, though 73 percent of them were between 15 and 39, and 22 percent were menopausal. Overall, 182 had been truly pregnant at least once before they experienced a false pregnancy, and 307 were married; 43 percent had (false) pregnancy signs that lasted for nine months; and one 79-year-old mother of 10 had symptoms persisting for seven years. Twenty-three women of the 444 had the condition more than once, and one of them displayed signs of pregnancy every nine months from her marriage until her death.[18] In a later, less extensive case review, P. H. Fried and associates examined 27 examples of pseudocyesis.[19]

Reviewing the English-language medical literature in 1978, Jane Murray and Guy Abraham found 512 reports of false pregnancy, of which only 17 had been published since 1960.[20] Murray and Abra-

ham suggested that if this decline in case reporting reflected a true decline of the disorder, it might have resulted from social changes that
grant women satisfying roles other than motherhood and childbearing. Other researchers have proposed that fewer women develop the
condition today because there is more sophistication about medical
matters and because pregnancy tests have been made reliable, inexpensive, and more accessible.[21]

The pseudocyesis cases reviewed by Biven and Klinger and by
Fried involved the following symptoms, in the order of their frequency: (1) partial or complete disappearance of menses, usually lasting for nine months; (2) abdominal enlargement; (3) breast changes,
including swelling and tenderness; secretion of milk and colostrum,
pigmentation; and enlarged papillae; (4) sensations of "fetal movements"; (5) softening of the cervix and enlargement of the uterus; (6)
nausea and vomiting, sometimes with aberrations of appetite; and (7)
weight gain, usually greater than in pregnancy.[22]

Until the twentieth century, most doctors and physiologists attributed false pregnancy to simple physical causes such as air swallowing, excess fat, distension of the abdominal muscles, contraction of
the diaphragm, or retention of urine. The sensations of fetal movement described by many patients were attributed to contractions of
abdominal muscles or intestinal peristalsis. Since the early 1900s,
though, psychogenic explanations of false pregnancy have been proposed by several medical theorists. Among the psychological needs
and dynamics that might contribute to the development of pseudocyesis, these have been suggested: narcissistic self-involvement, by
which a woman liberates herself from undesirable burdens by unconsciously simulating pregnancy; the desire to stabilize marriage with a
husband who wants a child; the desire for affection, attention, and
care that pregnancy might bring from husband, parents, friends, or
doctors; a compensatory reaction against deep fears of pregnancy;
symbolic compensation for imagined or real loss; a desire for the
physical fullness and sense of completeness that pregnancy often confers; the alleviation of emotional emptiness and numbness; the overcoming of physical inadequacy through a swollen abdomen and
larger breasts; the desire for perfection or rebirth that a new child suggests; the wish to stop menstruation because it symbolizes disintegration and loss of boundaries; the need to prove one's identity as a
woman with reproductive abilities intact; a desire for power over husband, children, parents, friends, or circumstances, by forcing or stabilizing a marriage, perhaps, or by showing superiority over other

females; a need for self-punishment caused by forbidden thoughts or guilt-provoking deeds, particularly those associated with sexual experience; a need to experience pain and suffering; and the recurrent, compulsive need to prove oneself creative and fertile.[23] Although this list of contributing causes is not exhaustive, it suggests the range of factors that might help to precipitate the disorder. Long clinical experience has shown that many needs and conflicts contribute to pseudocyesis, though the commanding image of pregnancy rules them.

Some researchers have suggested ways in which psychological and physiological processes might interact to produce false pregnancy. E. J. Pawlowski and M. F. Pawlowski, for example, explained the condition in a manner that might be termed *somatopsychic*, proposing that minor bodily changes such as slight abdominal swelling or a missed period sometimes initiate in suggestible women the mistaken belief that they are pregnant. Such belief, the Pawlowskis argued, might then stimulate further signs of apparent pregnancy.[24]

A more complex psychosomatic model of the syndrome was proposed by Edward Brown and Peter Barglow, who suggested that depression upsets normal hormonal balance via the cortical and limbic systems so that ovulation and menstrual bleeding are inhibited. At the same time, increased prolactin resulting from this imbalance stimulates lactation, promotes amenorrhea, and increases the production of estrogen and progesterone to keep the uterus in a quasi-pregnant condition.[25] Brown and Barglow's theory is open to question, however, in part because hormonal profiles have varied significantly among women with pseudocyesis. Among six patients described by Monica Starkman and colleagues, for example, luteinizing hormone levels were elevated in two, normal in two, and low in two; follicle-stimulating hormone levels were low in two and normal in four; and prolactin was elevated in three and normal in three. These findings, Starkman and her colleagues wrote, showed that "no one common endocrine profile is seen in all patients with pseudocyesis."[26]

The study of pseudocyesis, in short, shows that the same type of bodily alteration can be caused by different psychological and physiological processes. In some cases, it seems, apparent signs of pregnancy initiate a mistaken belief that one is pregnant, which triggers further signs of pregnancy; while in other cases, conscious or unconscious images initiate the condition. I emphasize this observation here because other psychobiological changes, too, are produced in more than one way, along various psychological and physiological pathways.

Pseudocyesis in Males

Even if it were demonstrated that all women suffering from pseudocyesis had identical levels of circulating hormones, false pregnancy in men would show that similar bodily changes can be caused in different ways. Males, obviously, because they do not possess ovaries and other organs that produce signs of pregnancy, must give rise to symptoms of imminent childbirth by strange devices. A 33-year-old merchant seaman described by psychiatrist James Knight, for example, had expanded five inches at the waist even though he had gained little weight, suffered from no organic disorder, and exhibited no hormonal abnormalities. This patient, who according to Knight appeared to be "strong, healthy, and masculine," had suffered from morning nausea, apparent abdominal movements, and increased appetite following the onset of his stomach's swelling, all of which had convinced him that he was pregnant. His physical symptoms subsided as he gained insight into the needs and conflicts that had precipitated his strange condition.[27]

Psychiatrists Dwight Evans and Thomas Seely also described a case of male pseudocyesis, in a 40-year-old black man whose liver, thyroid, and blood chemistries were normal when he was hospitalized for depression and persecutory delusions. After eight days of treatment, the man said he wanted to have a child by impregnating his wife, in part because his five-year-old daughter "was no longer a baby." During his second week of hospitalization his abdomen began to swell, even though he gained no weight, and he experienced nausea and vomiting. On his fourteenth day of treatment, he felt something moving in his stomach, "like a baby." In the weeks that followed, he gained 16 pounds while his abdomen continued to grow. "Repeated physical examination," Evans and Seely wrote,

> revealed a protuberant, nontender abdomen with no organomegaly, palpable masses, or evidence of ascites [accumulation of fluid in the peritoneal cavity]. Repeated x-rays revealed no evidence of [intestinal obstruction]. Abdominal ultrasound revealed no abnormalities of the liver or pancreas. Gastroenterological consultation revealed no pathology to account for his abdominal protuberance.

After eight weeks of hospitalization, the man was discharged, though his abdomen was still swollen. Subsequently his mental state improved, and with lithium carbonate treatment his signs of pregnancy vanished.[28] Other cases of male pseudopregnancy have been reported by W. N. Evans and by G. J. Aronson.[29]

[11.3]
MULTIPLE PERSONALITY

Human transformative capacity is also dramatized, however perversely, in multiple-personality disorder (MPD), a condition in which different ego-states (which sometimes have no awareness of one another) alternately control mind and body. Changes from one personality to another, called *switching* by psychotherapists who deal with the disorder, can happen in a few seconds, though they more commonly take a few minutes or hours. As people with this disorder switch personality, their gestures, carriage, voice, mental set, emotions, and self-identity shift dramatically; their eye structure and vision sometimes change; and patterns of electrochemical activity in their brain are altered. Many MPD victims harbor personalities with radically different lifestyles, values, and religious beliefs. Many have at least one personality that takes itself to be less than 12 years old and at least one that belongs to the opposite sex.[30]

Psychiatrist Bennett Braun described the psychophysiological changes he had witnessed in various multiples. One of his patients, for example, was allergic to citrus juices in all of his personalities except one. "If this personality ate an orange," Braun wrote, "and remained in control of the body for a sufficient period of time to digest and metabolize [it], no ill effects [resulted]." Another patient, who was usually so allergic to cats that she itched and teared around them, could play with them for considerable periods of time in one of her ego-states, and even be scratched and licked, without any apparent allergic response.

Braun observed red spots approximately the diameter of a cigarette appear on a third patient's skin while she was controlled by a personality tortured by cigarette burning. Braun reported:

> Each time that personality returned, the spots returned. The same patient in another personality developed stripe marks across the lateral aspects of both arms, and some across the shoulders and back of the neck, all of which I was able to observe. They were reported to be the results of a whipping administered by the mother.[31]

Another patient of Braun's had psychogenic epilepsy in response to violence or the disclosure of abuse in childhood, exhibiting a powerful reaction to potentially disturbing issues common to multiples. Yet another handled pain by passing it from one personality to another, becoming comatose on one occasion with a peritonitis which she had denied. "When [she] awakened in the recovery room," Braun wrote, "she heard someone moaning and wondered about the patient in the next bed, only to find she was the only one in the room. She [the host personality] was hearing the pain being dealt with by her various alter personalities."[32]

Psychologist Scott Miller, at the University of Utah, had an ophthalmologist give standard optical tests to ten multiples in different states, and found that they experienced significant changes in visual acuity, in the shape and curvature of their eyes, and in their optical refraction. One woman with personalities aged 5, 17, and 35 had a childhood condition called "lazy eye" only in her 5-year-old state, while a male patient who had suffered an injury that made his left eye turn out exhibited the condition in just one of his personalities. The Miller study confirmed an observation made by various clinicians that multiples often have different eyeglasses for their different selves.[33]

There is also evidence that in achieving their remarkable changes, multiples experience profound alterations of brain activity. Although a study by Philip Coons and associates did not reveal inherent differences between the brains of multiples and those of normal persons,[34] psychiatrist Frank Putnam reported that 11 multiples exhibited greater variations in their EEG records than controls who were instructed to rehearse imaginary alternate personalities. "These results," Putnam wrote, "suggest that normal control subjects are not able to simply 'fake' this condition."[35] There is evidence, too, that some multiples experience abnormal changes in cerebral blood flow as they change personality.[36] The fact that some multiples switch handedness as they switch ego-state also suggests that they have a larger repertoire of brain behaviors than most normal people.[†]

Evidence such as Miller's, Braun's, and Putnam's that multiples experience extraordinary bodily changes is supported by a rich clinical lore that they recover more quickly than most people from various afflictions. It has been reported, for example, that they heal from burns with unusual speed, and physician Cornelia Wilbur has claimed that their aging process often seems to have slowed. This accumula-

---

[†] Psychiatrist Joel Brende reported marked fluctuations of electrodermal response in the hands of a multiple. "States of dissociation," he wrote, "marked the transition between the two alternate personalities, Jay and James, and were observed to be linked with electrodermal response desynchronization and presumed hemispheric dissociation. That is, each of the two cerebral hemispheres, hypothesized as 'housing' each of the two alternate personalities, seemed to function as if 'split' apart from the other" (Brende 1984).

tion of experimental and clinical evidence has supported the belief held by some physicians that dissociative states facilitate healing. Doctors who make this supposition cite cures by healers who place their subjects in trance and the positive results of operations conducted with hypnosis.

Most clinicians are convinced that virtually all multiples adopt their bizarre behavior in childhood to cope with extreme abuse, developing their various roles and psychophysical athleticism through a lifetime of practice. Frank Putnam, at the National Institute of Mental Health, compared the massive alterations of mood, cognitive style, and muscular-nervous-hormonal configurations in MPD victims to the dramatic changes evident in children, who switch emotions with great speed and flexibility. By entering states similar to hypnotic trance, and by dissociating from their painful experiences in a different ego-state, some chronically abused children distance themselves from memories of beating and other traumas. In becoming multiples, they carry the fluidity of childhood into adulthood. "It's adaptive for [abused] children to keep their states separate," Putnam told Daniel Goleman of the *New York Times,* "so that they can keep the awareness of abuse from their other selves. That way, [painful] feelings and memories don't flood them."[37] Psychiatry professor Eugene Bliss also suggested that multiples employ states like hypnotic trance to avoid disturbing memories of childhood abuse.

> Most are capable of posthypnotic amnesia and many can do automatic writing. Some are hypnotic virtuosos. This clinical impression has been supported by formal hypnotic testing. Most claimed to have never been hypnotized, but when they were questioned about past comparable experiences, many recalled them. They reported a multitude of hypnotic-like episodes that included almost every classical hypnotic feat.[38]

Bliss found that some of his patients could be induced through hypnosis to tell how they switched personalities. One, for example, said "she lies down, but can do it sitting up, concentrates very hard, clears her mind, blocks everything out, and then wishes for [another identity], but she isn't aware of what she is doing."[39] This patient, it appeared, employed great concentration to summon an alternate personality. She had a favored posture (lying down) and cognitive ritual ("blocking everything out"). Through long practice, she had perfected a method of self-alteration which had something in common with transformative disciplines discussed in subsequent chapters.

Like certain religious adepts (21 & 22), multiples alter their flesh and consciousness through highly focused intention, hypnosis-like trance, and the retention of childlike mutability, though they do so in a dissociated and psychologically destructive manner. Further studies of their self-transformative strategies, conceivably, could reveal more practices that they share—however perversely—with people who display well-integrated forms of extraordinary functioning.

12

PLACEBO EFFECTS

THE TERM PLACEBO refers to any component of therapy without specific effects that is used, deliberately or inadvertently, as a treatment or as a control in experimental studies.[1] A placebo by this definition may consist of a pill, injection, surgical procedure, psychotherapeutic format, or machine that does not have, or does not depend entirely upon, specific psychophysiological processes to produce particular effects. A *placebo effect* is the psychological or physiological outcome that a placebo produces.[†]

With the help of placebos, countless people have experienced alterations of mood and perception, improvement of their autonomic and motor functioning, or relief from disability and suffering. Like the hysterical stigmata and false pregnancies discussed in the previous chapter, psychosomatic changes induced by placebos dramatize our capacity for self-transformation.

[12.1]
PLACEBO RESEARCH

The placebo concept has a long history in Western medicine. The term itself, which is taken from the Latin verb meaning "I shall please" or "I shall serve," entered medical parlance in the late eighteenth century. Quincy's *Lexicon* of 1787 called it a "commonplace method of medicine"; Hooper's *Medical Dictionary* of 1811 defined it to mean any medicine "adapted more to please than benefit the patient."[2] But the idea behind the term was familiar to doctors of Greek antiquity,

---

[†] For a discussion of these definitions and their various connotations, see philosopher Adolf Grunbaum's Explication and Implications of the Placebo Concept, and Howard Brody's Placebo Effect: An Examination of Grunbaum's Definition, both of them chapters in *Placebo: Theory, Research, and Mechanisms* (White, L., et al. 1985).

who knew that their cures depended largely upon psychological factors such as their patients' expectation of therapeutic success. The idea that most prescriptions had morale-boosting properties as well as specific effects was part of ancient medical wisdom.

Placebo effects were not studied systematically, however, until the twentieth century. Since the early 1900s, hospitals, research centers, medical schools, and drug companies have found that dummy treatments relieve an amazing range of afflictions, produce toxic side effects, and catalyze alterations of mood and behavior. The dramatic changes induced by placebos in modern experiments have convinced many psychologists and medical people that human beings possess largely untapped capacities to balance and restore their own functioning.[3] For example, in a landmark study published in 1955, Harvard University's Henry Beecher reviewed fifteen double-blind experiments in which patients were given a placebo for postoperative pain, cough, angina pectoris discomfort, headache, seasickness, anxiety, drug-induced mood changes, or the common cold, and found that 35 percent of their 1,082 subjects had been satisfactorily relieved. Beecher wrote:

> The constancy of the placebo effect, in a fairly wide variety of conditions, including pain, nausea, and mood changes, suggests that a fundamental mechanism is operating in these several cases, one that surely deserves further study. Many "effective" drugs have power only a little greater. To separate out even fairly great true effects above those of a placebo is manifestly difficult or impossible on the basis of clinical impression.[4]

Psychiatrist Arthur Shapiro suggested that the history of medicine was largely the history of placebo effects. When illness kept him bedridden in 1956, he read through a 100-year file of medical journals to find recurring patterns of treatment. "Medical fads came and went," he wrote. "A treatment would look wonderful, produce marvelous results and then disappear. *Something* seemed to work."[5] Guided by this observation, medical people now generally believe that a placebo's effect is produced by various aspects of a medical procedure that do not have pharmacologically specific effects upon the patient's functioning. A treatment's reputation, the patient's expectations about it, the therapeutic setting, and the doctor's belief in his own diagnoses all contribute to medical success.[6] As several experiments have shown, even a capsule's size, shape, and color help determine its effectiveness.[7] No treatment, it seems, is without some placebo effect, whether it be healing or toxic. Pharmacologically active drugs such as insulin, penicillin, or Dilantin may become either more or less effective, depending upon their recipients' attitudes toward them and the circumstances in which they are given.

In a study published by *The New England Journal of Medicine*, Herbert Benson and David McCallie reviewed the history of treatments for angina pectoris. Summarizing their review, they wrote:

> Many types of therapy . . . have been advocated, only to be abandoned later. A partial list would include heart-muscle extract, pancreatic extract, various hormones, x-irradiation, anticoagulants, monoamine oxidase inhibitors, thyroidectomies, radioactive iodine, sympathectomies, various vitamins, choline, meprobamate, ligation of the internal mammary artery, epicardial abrasions and cobra venom. Since most of these are now known to have no specific physiologic effect in the treatment of angina pectoris, we can analyze the benefits [attributed to them] and assess the degree of influence of [their] placebo effect.

Nearly all of these treatments had been greeted with enthusiasm by the doctors who administered them and the patients who received them, but controlled tests of their effectiveness by skeptics who operated under circumstances that minimized their placebo effect eventually called them into question. As the number of negative findings increased or when another new therapy appeared, the original therapy was abandoned. Benson and McCallie wrote:

> Quantitatively, the pattern is consistent. The initial 70 to 90 percent effectiveness in the enthusiasts' reports [decreased] to 30 to 40 percent "base-line" placebo effectiveness in the skeptics' reports. This pattern was recognized by the nineteenth century French physician, Armand Trousseau, who allegedly stated, "You should treat as many patients as possible with the new drugs while they still have the power to heal."

Benson and McCallie illustrated their argument with histories of five abandoned treatments, and detailed the divergent results reported by enthusiasts and skeptics. They concluded:

> Even with inactive procedures, physicians in the past have consistently achieved marked symptomatic improvement in approximately 80 per-

cent of patients with angina pectoris. This remarkable efficacy should not be disregarded or ridiculed. After all, unlike most other forms of therapy, the placebo effect has withstood the test of time, and continues to be safe and inexpensive.[8]

Like psychogenic stigmata or the disappearance of allergies in multiple personality, placebo-induced cures and toxic effects demonstrate our capacity to mobilize latent transformative powers—for either sickness or health—without external devices. "The placebo is an emissary between the will to live and the body," wrote Norman Cousins. "But the emissary is expendable."[9]

[12.2]

PLACEBO-INDUCED RELIEF FROM PHYSICAL AFFLICTIONS

Angina Pectoris

Ligation of the internal mammary artery was used in the United States during the 1950s for the relief of angina pectoris in the belief that it improved coronary blood flow. Henry Beecher compared the results of enthusiastic and skeptical surgeons who performed the operation.[10] Four enthusiasts described in Beecher's paper operated on 213 patients, of whom 38 percent experienced complete relief, and 65 to 75 percent showed considerable improvement. However, studies by Dimond (1958) and L. A. Cobb (1959) led to the abandonment of the procedure. In these two studies, a skin incision was made in every case, but the internal mammary artery was ligated only in randomly selected patients. In Dimond's group, 100 percent of the nonligated and 76 percent of the ligated patients reported decreased need for nitroglycerine and increased exercise tolerance; and six months after their operation, five ligated and five nonligated patients in Cobb's group reported more than 40 percent subjective improvements.[11] "Both studies showed that ligation of the internal mammary artery was no better than a skin incision, and that such an incision could lead to a dramatic, sustained placebo effect," Benson and McCallie concluded in their review of angina treatments.

Benson and McCallie also summarized the results of the Vineberg procedure, in which the internal mammary artery was implanted into a 3- to 4-cm tunnel burrowed into the myocardium to improve coronary blood flow. "Although an improvement rate of 85 percent was reported," they wrote, "several investigations demonstrated that neither objective nor subjective measures of improvement correlated with patency of the implanted artery or establishment of collateral circulation. In none [of these investigations] did improvement correlate with angiographic evidence of revascularization." Yet some 10,000 to 15,000 operations of this kind were performed (with a mortality rate of approximately 5 percent) before the procedure was abandoned, so convincing were its results until skeptics published their studies of it. Several drugs used for angina treatment, too, became less effective when shown to be nonspecific.[12]

Warts

In 1934, physician Herman Allington reported an experiment in which 105 patients were given sulpharsphenamine, a drug used successfully in treating warts, and 120 a dose of distilled water that was colored to match it. The experiment was conducted double-blind—that is, the operators who gave the injections did not know if they were administering the drug or the placebo. Of those treated with sulpharsphenamine, 52.5 percent, against 47.6 percent of those receiving the placebo, achieved complete remission.[13] In an earlier study, reported in 1927, physician Bruno Bloch cured 44 percent of his patients who had suffered from one kind of wart and 88 percent who had suffered from another by suggestion alone, thus reinforcing the belief among many physicians that psychological factors play a large role in the onset and cure of the affliction.[14] After reviewing the research literature in 1959, psychiatrist Montague Ullman concluded that suggestion was the most important factor in the cure of warts, even when they were treated by X ray, drugs, or surgery.[†]

[†] Ullman 1959. In a later study, Ullman and Dudek analyzed the nature of the therapeutic relationship involved in such treatments, and found that 8 out of 15 patients who went into deep hypnotic trance were completely cured within four weeks compared to just 2 of 47 who could not be deeply hypnotized (Ullman & Dudek 1960).

For further research on the role of suggestion in the cure of warts, see: Dudek 1967 and Sulzberger & Wolf 1934.

Asthma

Much research has shown that placebo effects can either stimulate or relieve breathing difficulties among asthmatics. In one study, for example, 19 of 40 asthmatic subjects developed airway resistance after they inhaled saline solution that they believed to be allergenic, and 12 of those 19 developed full-blown wheezing and bronchial spasms. Three minutes after the inhalation of a saline-solution placebo, every subject's airway resistance returned to baseline levels.[15] In a second study by the same researchers, 15 of 29 asthmatic subjects developed bronchospasm after breathing saline solution which they believed to contain allergenic agents. "In light of [these] findings," the authors concluded, "a meaningful assessment of the precipitants of asthma and the treatment of asthmatic patients must necessarily include an appraisal of the role played by suggestion. The expectations of the patient may have a marked influence on the efficacy of any given therapeutic regimen."[16]

Pain

In a review of 11 double-blind studies conducted between 1959 and 1974, medical researcher Frederick Evans found that 36 percent of 908 subjects who received placebo medication achieved at least 50 percent relief from various kinds of pain. His results agree almost exactly with the 35 percent figure Henry Beecher found in his 1955 survey of 15 double-blind placebo experiments.[17]

Arthritis and Other Disorders

In a study of placebo effects among arthritic patients, about the same number benefited from placebo tablets as those who received conventional antiarthritic drugs. Furthermore, when patients who had not experienced relief were given placebo injections, they too felt their symptoms were alleviated. Among the benefits these subjects experienced were improvements in eating, sleeping, and elimination, as well as reductions in swelling.[18] Hay fever, coughing, headache, diabetes, peptic ulcer, seasickness, and the common cold, too, have been relieved or cured with placebos.[19]

[12.3]
MOODS AND BEHAVIOR INFLUENCED BY PLACEBOS

Anxiety

Both clinical and experimental research has shown that placebos facilitate pain reduction when they help lessen anxiety. In one study, for example, 14 subjects whose anxiety decreased after they took a placebo could endure experimentally induced pain for longer periods of time than they had before.[20] Frederick Evans, one of the experimenters, concluded that "these results and other recent studies suggest that one can anticipate a placebo-induced reduction in suffering when anxiety decreases."[21]

Depression

Like anxiety, depression is caused by organic disease or by affective disorders, and like anxiety it is sometimes relieved by placebos. In a double-blind study of 203 depressed patients, for example, subjects given placebos improved as much as patients in six groups given various antidepressant drugs.[22] These patients' symptoms included sadness, tearfulness, despondency, suicidal thoughts, loss of interest in self and environment, concentration difficulties, psychomotor retardation, sleep impairment, appetite disturbance, weight loss, and diminution of sexual desire. Studies by Louis Lasagna and other placebo researchers have produced similar results.[23]

Other Affective and Behavioral Changes

Various studies have shown that placebos can

- □ produce slowed or speeded pulse, observable calm or nervousness, and feelings of comfort or euphoria;
- □ help relieve insomnia and other sleep difficulties;
- □ induce emotional and perceptual changes among marijuana users that mimic experience with the drug; and
- □ help reduce obesity and urinary incontinence.[24]

Adverse Reactions to Placebos

Henry Beecher counted 35 toxic side effects of placebos in the fifteen studies he reviewed for his classic 1955 paper, among them nausea, dry mouth, heaviness, headache, concentration difficulties, drowsiness, fatigue, and unwanted sleep.[25] In a study of the drug mephenesin, Stewart Wolf and Ruth Pinsky found that placebos produced combinations of weakness, palpitation, nausea, rash, epigastric pain, diarrhea, urticaria, and swelling of the lips that mimicked known side effects of the drug in some of their subjects.[26] And a group of Mexican researchers reported that women given placebo contraceptives reported decreased or increased libido, headache, bloating in the lower abdomen, dizziness, lumbar pain, nervousness, dysmenorrhea, nausea, epigastric pain, anorexia, acne, blurred vision, and palpitations.[27] Studies such as these show that placebos can induce destructive as well as beneficial responses. Like psychotherapy or religious practice, they can act like a two-edged sword upon us.

[12.4]
PLACEBO EFFECTS AND TRANSFORMATIVE PRACTICE

The studies I have just reviewed, and others that a complete inventory of placebo research would include, show that afflictions can be relieved, that moods and perceptions can be altered, and that autonomic and motor functioning can be improved to some extent by the use of nonspecific treatments that promote their recipients' self-regulative powers. Such research, however, has provided little evidence to date that placebos can stimulate exceptional functioning of the kind produced by the conscious cultivation of particular capacities.[28] This is the case because we typically develop extraordinary attributes through activities that involve self-awareness, whereas a placebo's effectiveness generally depends upon its subject's *lack of awareness.*[29]

Nevertheless, placebo effects are important for this discussion because they dramatize our capacity for dramatic psychosomatic changes. Here I will briefly describe certain elements involved in both placebo effects and successful transformative practice.

Supportive Social Conditions

Positive social support generally facilitates placebo-induced healing and self-regulation. Several studies have shown, for example, that:

- Enlarging patient groups in which treatments are administered can improve responses to placebos, probably because the power of suggestion is increased by the greater number of participants.

- A placebo's effectiveness depends to a large extent upon the physician's interest in the patient involved, interest in the treatment, and concern about the treatment's results.

- A placebo's power is increased by experimental studies that impart a sense of interest and care to their subjects.

- Placebo effects in most treatments are increased when the treatment has a good reputation.[30]

Expectation, Positive Suggestions, and Faith

To demonstrate the role of suggestion in placebo effects, experimenters have administered the same dummy therapy with different descriptions of it. In one study, for example, volunteers received the same placebo on three separate occasions, each time with a magnetic tracer by which their stomach activity could be monitored. With the first administration, all subjects were told they were getting a drug that would cause strong gastric churning; with the second, that the drug would make their stomachs feel full and heavy; with the third, that the drug was a placebo. Though they took the same substance each time, these subjects showed changes in stomach activity that conformed to their expectations.[31] This study, and others, show that mental sets affect physiological processes with great specificity.

Physicians since antiquity have recognized the healing power of faith, evident in shrines such as Epidaurus and Lourdes (see chapter 13.1). Indeed, faith can be so strong that subjects suffering from upset stomachs have been relieved by *nausea-inducing* agents when told by doctors that such agents would help them.[32] Psychologists have identified several ways in which expectant faith facilitates healing. Richard Bootzin, for example, argued that patients with increased expectancies regarding their ability to cope tend to rehearse positive ideas, images, and moods. "The person receiving a supposedly effective therapy," Bootzin wrote, "may reduce the number of self-defeating thoughts and images in which he engages and may increase the frequency of [his] coping self-statements and positive images."[33] Psychologist William Plotkin, too, has suggested that faith is a distinctive process in the placebo effect:

As soon as the treatment is believed to have commenced persons who have faith in that treatment will begin, on the basis of that faith, to treat themselves as persons whose problems have been cured. The reason, again, is that if they do not lack faith in the treatment, there is no question in their minds as to whether it will succeed; if they see themselves as being in the process of being cured, and if they have the necessary competence to act accordingly, then we can say that they have already started to be persons who are cured—they are treating themselves that way and acting appropriately (not merely going through the motions).

. . . Coming to act as cured persons, patients may also strengthen those skills that are relevant to therapeutic improvement. Through practice and experience in acting as cured persons, they may improve at those behaviors that aid or express a cure. It should be noted that the person who has faith in a therapeutic procedure is even more likely to improve than one who merely has "positive expectancies," "conviction," or "hope": If an individual does not merely expect to be cured, but takes it for granted that he or she has been cured or is well on the road to a cure, then that individual is less likely to see things as evidence to the contrary, and, accordingly, is more likely to act in a manner consistent with and facilitative of cure.

. . . Consider a man who suffers from chronic muscle contraction headaches. If he is given a prescription of placebo pills, and if he has faith in the efficacy of the putative medicine, he will take it for granted that his headache problem has been cured and will treat himself accordingly. He may no longer expect to be stricken with headaches; he may cease to worry about being regularly incapacitated; he may experience a lifting of a tremendous burden, and may perhaps celebrate by spending his newly "won" time in pursuing various pleasurable and relaxing endeavors for which he previously felt ineligible; and he may no longer present himself to others (or to himself) as a headache sufferer, with a consequent reduction in social pressure to behave (and experience) accordingly.[34]

Expectant faith, in short, stimulates healing behavior, along with positive imagery and mood. Like Renaissance doctors who subscribed to the Hippocratic doctrine that elements of mind and body "tread in a ring," contemporary medical researchers have shown that cognition, affect, and behavior powerfully influence one another, for better or for worse.

Complex and Redundant Mediations

In 1978, it was reported in the British journal *Lancet* that some dental patients whose impacted molars were removed experienced significant reduction of pain when given placebos, whereas similar patients given naloxone, a substance that blocks the action of endorphins, did not. Furthermore, when the subjects who experienced pain relief were given naloxone, their pain increased to the levels experienced by those who did not respond to the placebo and by those who initially received the endorphin-blocker. Because naloxone undid the analgesia triggered by the placebo, the *Lancet* study authors concluded that endorphins help cause placebo pain relief.[35]

Subsequent studies, however, have suggested that the mediation of placebo pain relief is more complex. In 1983, Richard Gracely reported in *Nature* that placebos helped relieve pain in some patients suffering postoperative dental pain even though the endorphin receptors in their brains were blocked by naloxone. The results of this and other experiments suggest that naloxone binds preferentially to certain types of receptors, leaving others available to pain-relieving opioid peptides other than endorphins.[36] It is now evident that pain relief involves different kinds of physiologcal interaction. Indeed, ongoing medical research suggests that because our bodies are highly complex, many if not most somatic changes are mediated in multiple ways, by differing pathways or interaction chains, through various combinations of central nervous, immune, endocrine, or other activity. Our psychosomatic complexity, and the redundancy of many bodily processes, make it possible for transformative programs to work in various ways, overcoming obstacles such as blocked endorphin receptors, injured organs, or impaired nervous functioning. In an authoritative review of placebo research, Leonard White, Bernard Tursky, and Gary Schwartz noted, "There is no single placebo effect having a single mechanism and efficacy, but rather a multiplicity of effects, with differential efficacy and mechanisms."[37]

In summary, then, placebo effects depend upon supportive social conditions surrounding a treatment's administration; positive expectations (and accompanying suggestions) that the treatment will succeed; hopeful faith, which stimulates positive images, emotions, covert rehearsals, and behaviors supporting the treatment involved; and various physiological processes that can be mobilized for healing and growth. As we shall see, these factors operate in activities other than placebo-induced treatment, including the transformative disciplines discussed in this book.

13

SPIRITUAL HEALING

THE KINDS OF HEALING I discuss in this chapter have
been given various names, among them spiritual, shamanic, religious,
psychic, mental, charismatic, prayer, paranormal, and faith cure;
laying on of hands; magnetizing; and miracle.[1] Though such terms are
not strictly interchangeable, they generally refer to restorative activities that appear to depend in part upon agencies beyond normal
healing processes. If such activities involve a shaman, saint, or other
healer, they are attributed to something special in the healer, to his or
her healing gifts, closeness to God, superabundant vitality, or mana. If
they involve a temple or shrine, they are ascribed to the numinous
power the place possesses. And if they occur without healer or holy
place, they are said to result from contacts with a superordinary dimension of the self, disembodied entity, or Divinity itself.

Spiritual healing, of course, is produced largely by agencies or
conditions that operate in other kinds of therapy, among them: authority invested in the healer or restorative act, cultural support and
group suggestion, confident expectation and strong imagery of relief
(or at least unconscious predisposition to it), working through of underlying issues leading up to the cure, resolution of interpersonal
problems related to the suffering or disability involved, profound
arousal and/or relaxation, and altered states of consciousness that
concentrate curative processes and promote receptivity to them. But
such agencies and conditions, which have been illumined to some extent by psychology and medical science, are complemented in spiritual healing, it is proposed, by one or more of the following events:
transmissions of electrical, magnetic, or other forms of physical energy; infusions of vitality such as the *prana* or *chi* recognized by
Hindu, Buddhist, and Taoist yogas; psychokinetic influences; and
healing transactions involving supracosmic dimensions of existence.

Some psychologists and medical researchers, as well as most

healers themselves, believe that the different kinds of event or agency just noted operate simultaneously in spiritual healing. While shamans often pray for their subjects, thus activating or strengthening a spiritual unity with them, they sometimes transmit healing influence through touch. Many Christian saints, and healers such as Valentine Greatrakes, practiced such multileveled therapy (13.2); and pilgrims to therapeutic locations such as Lourdes, and persons who experience such healing without benefit of healer or place, also testify to extraordinary curative infusions with physical, vital, and transpersonal elements. Given the hierarchical structure of human nature, it is plausible that the cures discussed here involve many levels and kinds of therapeutic transaction.[2] The kinds of healing discussed in this chapter, then, differ from other types of therapy by their partial or total reliance upon vital energies such as *ki,* psi influence, holy places, deep regions of the self, or supracosmic agencies.[†] They are often accompanied by a sense of the numinous, of contact with a transcendent power and light. They sometimes occur in the midst of powerful emotions, or in trance, when perceptual constancies are suspended. They frequently lift their subject up, raising his or her "center of personal energy," as William James put it, establishing the self—however briefly—in a larger power and sufficiency. And they usually convey a clear sense that fundamental change has occurred, that "work is really done, that spiritual energy flows in and produces effects, psychological and material," to quote James again.[3] These characteristics of spiritual healing, and others, are discussed in the sections that follow.

[13.1]
HISTORICAL EXAMPLES OF SPIRITUAL HEALING

In ancient Egypt, healing rituals were immensely complex. The Ebers Papyrus describes invocations to the gods for relief of specific diseases. Physical therapies, some of them sophisticated by modern stan-

---

[†] People who turn to spiritual healing may or may not depend upon standard medical treatment. Many Christian Scientists, for example, will not use surgery or drugs, even if they are desperately ill, depending entirely upon prayer and God's grace. However, most sufferers who turn to healers, shrines, or prayer sensibly rely as well upon conventional medical support. At Lourdes, probably the most active center for spiritual healing in the world today, pilgrims are encouraged to seek both medical and religious help for their ailments. For a discussion of problems resulting from misguided ideas about spiritual healing, see: Wilber 1988.

dards, and religious healing were used side by side, such records show. The legendary Imhotep, said to have been chief physician, magician, and vizier to Zoser, pharaoh of the Third Dynasty, was revered in Egypt until the time of Alexander's conquest; and many temples founded in his name became seats of medical instruction.[4] That medicine, magic, and religion were joined in the person and cult of Imhotep reflects their integration in Egyptian culture. Magico-religious healing procedures were also elaborated in *Books of the Dead* and other treatises. Rites of purification and propitiation like those used by faith healers today were practiced regularly in Egypt for some three thousand years, from the times of the earliest pharaohs.[†]

The Greeks, too, joined physical and spiritual therapies, notably in some three hundred temples dedicated to Asklepios, the god of healing, that were built during the eighth and seventh centuries B.C. The most famous of these, located at Epidaurus, comprised a central shrine, baths, hostels, gymnasia, theatre, and stadium. In most Asklepia, the sick were cleansed by purificatory rites, then admitted to the porch of the temple, where they spent one or two days in prayer and meditation. Thus prepared, they slept in the central shrine, sometimes under the influence of narcotics, so that the god might appear to them in dreams for inspiration, healing, and guidance. Such treatment was often supplemented by baths, massage, gymnastics, and diet.

Hippocrates, the most celebrated physician of Greek antiquity, in discussing epilepsy (which he attributed to natural causes), described "the deity who purifies, sanctifies, and cleanses from sin."[5] Certain characters in Plato's dialogues also assert that medicine must treat flesh through the soul. "All good and evil," said Socrates in the *Charmides*, "originates in the soul, and overflows from thence, as if from the head into the eyes. And therefore if the head and body are to be well, you must begin by curing the soul; that is the first thing. Because they try to cure the body without the soul, the cure of many diseases is unknown to the physicians of Hellas."

Biblical Accounts of Healing

The first Book of Kings (17:20–24) describes Elijah's meeting with a mother whose son had died. Turning to God, the prophet asks:

[†] For descriptions of ancient Egyptian esotericism, see: Budge 1901 and Petrie 1909.

"O Lord my God, hast thou also brought evil upon the widow with whom I sojourn, by slaying her son?"

And he stretched himself upon the child three times, and cried unto the Lord, and said, "O Lord my God, I pray thee, let this child's soul come into him again."

And the Lord heard the voice of Elijah; and the soul of the child came into him again, and he revived.

In a similar story (2 Kings 4:18–37), the prophet Elisha resuscitates the Shunammite's son who had died suffering from pain in his head, and the Book of Isaiah (35:5–6) describes the coming king of righteousness who will show God's power:

Then the eyes of the blind shall be opened, and the ears of the deaf shall be unstopped.

Then shall the lame man leap as an hart, and the tongue of the dumb sing.

According to the four Gospels, Jesus healed:

- Four cases of blindness: Bartimaeus (Mark 10:46–52; Matthew 20:29–34; Luke 18:35–43); the blind man of Bethsaida (Mark 8:22–26); two blind men (Matthew 9:27–31); and the man blind from birth (John 9:1–34).
- Two cases of fever: Peter's mother-in-law (Matthew 8:14–15; Mark 1:30–31; Luke 4:38–39); and the official's son (John 4:46–54).
- Eleven lepers: a single leper (Matthew 8:1–4; Mark 1:40–45; Luke 5:12–16); and ten lepers (Luke 17:11–19).
- A deaf mute (Mark 7:31–37).
- A woman with a hemorrhage (Matthew 9:20–22; Mark 5:25–34; Luke 8:43–48).
- A man with a withered hand (Matthew 12:9–13; Mark 3:1–5; Luke 6:6–10).
- A man with dropsy (Luke 14:1–6).
- A centurion's servant with paralysis (Matthew 8:5–13; Luke 7:1–10).
- A wounded slave of the high priest (Luke 22:50–51).
- The sick of Capernaum (Matthew 8:16–17; Mark 1:32–34; Luke 4:40–41).

□ Many people from Tyre, Sidon, and Galilee (Matthew 12:15–21; Mark 3:7–12; Luke 6:17–19).

□ Those by the shore at Gennesaret (Matthew 14:34–36; Mark 6: 53–54).

□ The sick at Bethsaida (Matthew 14:14; Luke 9:1–11); and at Tiberius (John 6:2).

□ Crowds upon entering Judea (Matthew 19:2).[6]

Jesus laid his hands upon the sick (Mark 6:5), held their hands (Mark 1:31), or simply touched them (Matthew 8:3, Mark 1:41, Luke 5:13). He put his fingers into the ears of a deaf-mute, touched the man's tongue with saliva, and commanded that his senses be opened (Mark 7:33–34). He spit into the eyes of a blind man, then laid his hands upon him (Mark 8:23–25); and made clay with his spit to anoint the eyes of another (John 9:1–12). And sometimes when people appealed to him through intermediaries, he healed them from a distance. Like healers of many religious traditions, he responded to people in need with intuitive, spontaneous, dramatic certainty, claiming to be a vehicle of God's redemption. Though the gospel healing stories have been questioned by some biblical scholars, the methods of Jesus resemble those used by healers of many eras. Saint Francis of Assisi, Valentine Greatrakes, and countless others have laid their hands upon the sick, embraced them lovingly, spit on their wounds, or called to God in a commanding voice just as Jesus was said to have done, building a reputation by their manner and success that inspired confidence in their restorative powers. Healers since antiquity have shared certain methods and beliefs to which the gospel accounts conform.[†]

Healing in the Middle Ages

Shrines dedicated to saints and martyrs of the church, many of them housing relics of their patrons, became centers of therapy in the Middle Ages, and miracle lists compiled by medieval monks show that they catalyzed an immense range of cures.[7] Such lists, which the French medievalist Luchaire called "medical journals of the Middle Ages,"

[†] Some biblical scholars have suggested that Jesus was influenced by charismatic Jewish healers such as Hanina ben Dosa and by the legend of Elijah's cures so that his dramatic mission conformed to an established, though unconventional, tradition. See: Vermes 1978.

provide richly detailed evidence for the effectiveness of spiritual healing. Historian Ronald Finucane analyzed two such collections, one from the shrine of Saint Godric in northeastern England and another from the tomb of Thomas Becket in Canterbury, which include some six hundred cures between them, and found that they list nobles, knights, yeomen, ecclesiastics, craftsmen, and the poor. Blindness, deafness, paralysis, and other afflictions that were probably psychogenic; organic infirmities such as broken bones, epilepsy, leprosy, and gout; and overt mental disturbances were cured or relieved at both places, according to their miracle lists. "For many reasons . . . the two cults exhibited unique characteristics," Finucane wrote, "yet the behavior of pilgrims cured at both shrines was similar—indeed, was repeated at every other saint's shrine in medieval Europe—and can still be observed in some parts of Christendom today."[8]

After analyzing some 1,500 cures attributed to these and other medieval healing centers, Finucane reconstructed some typical rituals that pilgrims followed in their search for relief. Certain devotees, for example, were led to a particular place through dreams; others decided by casting lots or were sent there by a physician. Upon arrival at a shrine, many were purified through confession, while most made offerings and prayed to the presiding saint or martyr for their intercession. In all this they were usually accompanied by crowds of fellow sufferers, some of whom experienced cures that inspired new confidence in those around them. Like the sick at temples of Asklepios, some slept in the shrine's church, their vigils lasting for several days. For those who needed especially close contact with the saint's healing power, niches near his bones were sometimes provided. According to one medieval account:

> a blind girl was taken to Becket's shrine. Laying her head upon the martyr's tomb, she fell asleep. Waking up she rubbed her eyes and found that she could see. She turned to her mother, saying, "mother, I can see, though I couldn't before." Her mother replied, "Hush, and put your head back on the Sarcophagus." She did so, and after a little while arose wholly cured.[9]

Pilgrims' prayers were often accompanied by tears, moaning, swooning, vomiting, the cracking of bones, writhing, shrieks, or profuse bleeding, which induced an atmosphere that helped rouse everyone in the vicinity to new enthusiasm and emotional release. A deaf woman, for example, while standing in prayer at Becket's shrine, "felt a surge of pain, and within her head it seemed as if many twigs were being snapped into tiny bits. While she was thus afflicted she cried as if an interior infection had burst, a great deal of pus flowed from her ears, after which blood came out, and, after the blood, her hearing returned." And a blind woman led to Becket's shrine felt "as if she were engulfed in the raging heat of a furnace." Ripping the veil from her head and the garments at her breast, she fell to the floor and lay prone for an hour, after which, opening her eyes and getting up, she burst out, "I can see!"[10]

Some pilgrims were only partly relieved at a particular shrine before they were cured at another. Finucane described a blind woman whose sight was partly restored at Becket's tomb, who then went to Godric's tomb where, after spending the night, she recovered normal vision.[11] Like many pilgrims at Lourdes in the twentieth century, the people Finucane described were energized by strong expectation of relief and were then caught up by a shrine's highly charged atmosphere. Like many who are healed in similar circumstances today, they felt they had contacted a transcendent wholeness, a presence or energy beyond their ordinary functioning, that restored them in a manner that seemed miraculous.

Valentine Greatrakes

Valentine Greatrakes, the "Irish stroker," was born in 1629 of an English family settled in county Waterford. He wrote that, in 1662,

> I had an impulse, or a strange persuasion in my own mind (of which I am not able to give any rational account to another) which did very frequently suggest to me that there was bestowed on me the gift of curing the Kings-Evil. . . .Whether I were in private or public, sleeping or waking, still I had the same Impulse.[12]

The King's Evil was the name commonly given to scrofula, or tuberculosis of the cervical lymph nodes, an affliction that was sometimes amenable to cure by royal touch.[†] Greatrakes, however, did not limit his healing powers to this affliction. In the three years following

---

[†] Healing powers were ascribed to certain kings and queens of Europe, and were eventually regarded as a confirmation of the divine right to rule. During his exile in Holland, Charles II of England demonstrated his God-given right to govern by touching, or "stroking," thousands of sufferers; and after his restoration to the throne, he touched 22,982 of them, according to records of his reign (Crawfurd 1911. In Laver 1978).

his "Impulse," he cured people with various troubles in Waterford and nearby counties, and eventually attracted sufferers from England, where the Plague was raging. By the summer of 1665 his fame had spread to London, capturing the attention of Viscount Conway of Ragley Hall, whose wife had suffered incapacitating headaches since childhood. At the request of the viscount, Greatrakes went to Ragley Hall but was unable to help the viscountess. He healed hundreds of others, though, and was invited by the king himself to London, where he treated many thousands more, among them royalty, scientists, and men of letters. The English chemist and physicist Robert Boyle attested to seven of Greatrakes's cures that he had witnessed, though he attributed them to "a lucky commotion in the blood and spirits conducive to healing [stimulated] by exalted imagination and strong passions."[13] In spite of his successes, however, the Irish healer provoked a highly critical pamphlet by David Lloyd, a chaplain to Isaac Barrow, one of Britian's most prominent scientists. In response to Lloyd's attack, Greatrakes wrote an autobiography meant to establish his authenticity and sincerity, supporting it with testimonies to his cures signed by John Wilkins, a founder of the Royal Society; Ralph Cudworth and Benjamin Whichcote, the Cambridge Platonists; Andrew Marvell, the poet; several medical men; and other prominent figures. These documents and other records of the time indicate that Greatrakes made a great impression on the people of London.

Like celebrated healers of other eras, Greatrakes often responded to the sick in spontaneous, dramatic ways, stroking or massaging them, using a knife to lance boils upon occasion, offering some his urine to drink or rub on their wounds, spitting in the eyes of others. One witness of his ministrations wrote:

> When he stroked for Pains, he used nothing but his dry hand, if Ulcers or running Sores he would use Spittle on his Hand or Finger, but for the Evil if they came to him before it was broke, he stroked it, and ordered them to poultess it with boil'd Turneps, and so did every Day till it grew fit for lancing, he then lanc'd it and with his Fingers would squeeze out the Cores and Corruption.[14]

Though Greatrakes said he stroked with his hand "outside of the Cloathes only," not all witnesses concurred, one of them reporting that certain women "unlaced and untied their petticoats, and he followed [the pain] on their bare bodies till 'twas gone." Nevertheless, he always insisted that he was merely God's instrument, a claim that must have reinforced his effectiveness among the devout. "The resemblance in phenomena induced by [Greatrakes's] stroking to those elicited by Franz Anton Mesmer and his immediate successors nearly a century later are remarkable," wrote psychologist A. Bryan Laver.[15] Greatrakes's sufferers, for example, often exhibited convulsive crises like some of Mesmer's patients (15.1), as well as temporary anesthesias that could be successfully tested with pins and knives. Like Mesmer, the Irish healer sometimes drew pain out at the toe and threw it away like polluted water. And objects associated with the two men were sometimes used to accomplish cures when their owners were absent, like the bones of saints or other religious relics. By their reputations and vivid personalities, both Mesmer and Greatrakes built up an aura of belief that catalyzed the self-restorative powers of those they treated. Though Mesmer attributed his cures to infusions of magnetic fluid, and Greatrakes relied upon the grace of God, both stimulated many people to raise their "center of personal energy," as William James put it, "producing effects psychological and material."

New Thought and Christian Science

Faith healing has been nurtured in the United States by organizations affiliated with the New Thought movement inspired in large part by Phineas Quimby (1802–1866), a watchmaker who developed a mind cure influenced by theories of the French mesmerists. Of all Quimby's converts, the most influential was Mary Baker Eddy, who was cured by him from lifelong afflictions that were largely psychogenic. Mary Baker Eddy studied Quimby's teachings devoutly, then created her own brand of New Thought, which she formalized in *Science and Health,* published in 1875 (later editions were titled *Science and Health with Key to the Scriptures*). Until her death in 1910 she promoted her teaching through the Christian Science church, which spread throughout the United States and to other countries. "Matter is nothing beyond an image of mortal mind," she wrote. "Evil has no reality. It is simply a belief, an illusion of material sense. Nothing but God is real and eternal, and sickness is an error." The devout Christian Scientist typically treats health problems with prayer and such maxims, maintaining a steadfast denial of the "mistakes" that produce sickness. A trained practitioner may also help suffering individuals, perhaps at a distance, by helping to place their thoughts and emotions in rapport with the Divine Mind. Strict believers do not accept standard medical treatment beyond hygienic measures.

Christian Science places great emphasis upon suggestion and regards negative thought with "an almost superstitious abhorrence," as one writer put it.[16] Its denial of disease and misfortune has been criticized by physicians who argue that such an attitude discourages people from getting adequate care for conditions that are treatable by standard medicine, and by psychologists who argue that it encourages suppression of emotional issues that cannot be solved by suggestion alone. The religion's enormous success, however, indicates that relying primarily upon prayer and affirmations of goodness promotes well-being for many people.

American New Thought and Christian Science have given birth to many groups with similar beliefs, among them the Jewish Science of Rabbi Morris Lichtenstein. The following passages from his book *Jewish Science and Health* reveal Rabbi Lichtenstein's affinity with Phineas Quimby and Mary Baker Eddy:

> The Divine Mind communicates with the human mind through the imagination. A prayer, therefore, should be offered in the form of a mental image. Man must visualize the thing he desires, he must use his imaginative powers to form his petition in terms clearly outlined in his own mind. The profound concentration of attention and thought which this form of prayer requires fills also the heart with deep earnestness and devotion. Man must pray whole-heartedly as well as whole-mindedly; he must believe in his heart that his well-being depends completely upon his prayer.
>
> A prayer for health is offered in a manner akin to the prayer offered to any other need (and) consists of two parts: first, the visualization of divine giving and then of man's receiving; first the process of healing; second, the state of health restored through that process.
>
> If any particular organ of the body is affected, the affirmation must declare that health is saturating it, obliterating all defection and suffering. . . . This formula may be applied also when the ailment is in the lungs, digestive organs, liver, kidneys (or) any part of the body, care being taken to name the given part distinctly.[17]

[13.2]

SPIRITUAL HEALING IN THE TWENTIETH CENTURY

The authors of a study published by the *Journal of Nervous and Mental Disease* claimed to show that 43 pentacostal believers who experienced faith cures exhibited "denial, repression, projection, and disregard of reality" in their typical functioning, and that the primary function of such cures was "not to reduce symptomatology, but to reinforce a magical belief system that is consonant with one's subculture."[18] Criticisms of this kind, however, do not negate the good results of spiritual healing. Though denial and repression are evident among their members, healing cults often produce real cures. Religious healing has simply occurred too often, among too many kinds of people, to be dismissed as a stunted response to adversity. This fact is evident at Lourdes, which is visited annually by thousands of physicians—among them Catholics, Protestants, Jews, Moslems, Buddhists, and agnostics—many of whom testify to dramatic cures they witness there.

Healing at Lourdes

As part of its long-standing effort to establish the authenticity of religious phenomena, the Catholic church founded a Medical Bureau at Lourdes in 1883 (25 years after Bernadette Soubirous had her famous visions of the Virgin) to determine the nature of the cures that occurred at the famous shrine. Since 1954, the bureau has sent dossiers on those cases that seem scientifically inexplicable to the International Medical Committee of Lourdes, a group of physicians and medical researchers from several nations, where the cures in question are again examined. Saint John Dowling, a British physician who was a member of the International Committee, wrote in the *Journal of the Royal Society of Medicine:*

> If, after the initial scrutiny and follow-up, the Medical Bureau thinks that there is good evidence of an inexplicable cure, a dossier [on the cure] is sent to the [International Committee] which usually meets once a year in Paris. A preliminary examination of the data is made and if the members agree that the case is worth investigating they appoint one or two of their number to act as rapporteur. The rapporteur makes a thorough study of the case, usually seeing the patient himself, and presents the material in a detailed written dossier circulated to the members before the meeting at which they will take their decision.
>
> The report is then discussed critically at length under 18 headings, a vote being taken at each stage. In the first three stages the Committee considers the diagnosis and has to satisfy itself that a correct diagnosis has been made and proven by the production of the results of full physical examination, laboratory investigations, X-ray studies and endoscopy and biopsy where applicable: failure at this stage is com-

monly due to inadequate investigation or missing documents. At the next two stages the Committee must be satisfied that the disease was organic and serious without any significant degree of psychological overlay. Next it must make sure that the natural history of the disease precludes the possibility of spontaneous remission and that the medical treatment given cannot have effected the cure. . . . Then the evidence that the patient has indeed been cured is scrutinized and the Committee must be satisfied that both objective signs and subjective symptoms have disappeared and that investigations are normal. The suddenness and completeness of the cure are considered together with any sequelae. Finally, the adequacy of the length of follow-up is considered. After this detailed study, the question "Does the cure of this person constitute a phenomenon which is contrary to the observations and expectations of medical knowledge and scientifically inexplicable?" is put. A simple majority carries the case one way or the other.

The declaration by the [International Committee] that it considers the cure to be medically inexplicable does not make it a miracle because that is a matter for the Church, not doctors. The verdict is sent to the patient's bishop and if he thinks fit he appoints a Canonical Commission with its own medical advisers. If it reports favourably and the bishop accepts the report, he issues a decree declaring the case to be a miracle.[†]

The Medical Bureau at Lourdes and the International Committee have produced a wealth of material regarding spiritual healing. Their archives—which include X-ray pictures, photographs, records of biopsies, and summaries of other laboratory tests—are a valuable resource for studies of our regenerative capacities.[19] Prominent visitors to Lourdes have also provided persuasive testimony regarding cures at the shrine, as well as insights regarding their causes. Alexis Carrel, a Nobel Laureate in medicine and a medical director of the Rockefeller Institute, told the story of his first trip to Lourdes in 1903 with intimate detail.[20] Having gone there as a curious skeptic, he attached himself to a woman with tubercular peritonitis. While he watched, her sickly features changed so dramatically that he felt he might be "suffering an hallucination." Then, before his eyes, a large abdominal tumor which had filled her navel with pus disappeared within a few minutes. To Carrel's astonishment, the woman suddenly

---

[†] Dowling 1984. According to Theodore Mangiapan, director of the Medical Bureau for many years, the International Committee deemed more than 20 cures to be scientifically inexplicable between 1954 and 1990, but only some of these were declared to be miracles by the bishops of the dioceses in which they occurred.

seemed free of pain. Later that day, when he visited the woman at a local hospital, she was sitting up in bed, eyes gleaming, her cheeks full of color. Her abdomen seemed normal, showing no sign of the hopeless-looking tumor Carrel had seen just hours before. This and other cases convinced him that many of the cures at Lourdes were authentic, whether they were miracles or not, and could not be attributed solely to the relief of functional disorders. Various afflictions Carrel witnessed had seemed intractable before they were healed. As a rationalist and scientist, he said, he was forced by his experience at Lourdes to admit that human beings possessed mysterious capacities that science should explore as thoroughly as it did germs and new surgical procedures.[21] The following cases illustrate the range of organic disorders that have been healed at the shrine or by water taken from its springs.

Francis Pascal. Born a normal child, Pascal was afflicted with meningitis when he was three and was left blind and partly paralyzed. In August 1938, when he was four, he was instantly cured after two immersions at Lourdes. Members of the Bureau and other medical experts confirmed that both his blindness and paralysis were organic, not functional. The archbishop of Aix-en-Provence declared the cure to be miraculous in 1949, and Pascal lived to be a normal adult.[22]

Gerard Bailie. When he was two and a half years old, Bailie was striken with bilateral chorioretinitis and double optic atrophy, an eye disease that is normally incurable, and lost his sight after an operation. In September 1947, when he was six, his blindness was cured during a visit to Lourdes. Though his affliction was thought to be incurable because his optic nerves had atrophied, upon reexamination by the Medical Bureau, he could see objects clearly.

Delizia Cirolli. In 1976 Delizia, a child from Sicily, developed a painfully swollen right knee that was eventually diagnosed as a case of Ewing's tumor. Her parents refused to have her leg amputated, and her mother took her to Lourdes. But X rays taken the following month showed no improvement in the girl's condition, and the family prepared for her death. Friends and sympathizers continued to pray to the Virgin of Lourdes, however, and Delizia was regularly given water from the shrine. Then, some three months after the condition had appeared, it vanished. X rays showed repair of the bone that had metastasized. Four trips to the Medical Bureau in 1977, 1978, 1979, and 1980 showed that the cure was complete.[23]

Vittorio Micheli. In 1963 Micheli, of Scurele, Italy, experienced complete remission at Lourdes from a sarcoma that had destroyed part of his pelvic bones. X rays before and after his cure attached to a report by Professor Michel-Marie Salmon to the International Committee (which can be obtained from the Medical Bureau) show that Micheli's iliac bone was restored after suffering massive disintegration.[24] The reconstitution of Micheli's pelvis took several months, though at Lourdes he experienced sudden relief from pain and a sense that he was cured.

Serge Perrin. In 1970 Perrin, of Lion d'Angiers, was cured of organic hemiplegia with ocular lesions caused by cerebral circulatory defects. Doctors who examined him were convinced his afflictions were organic, and their diagnosis was confirmed by the International Committee. Perrin's cure was deemed scientifically inexplicable, in part because it was sudden and complete. According to a summary of his case prepared by the Medical Bureau:

> He was in a wheel-chair [at Lourdes], his wife and daughter beside him. The priest passed, putting Oil on his forehead and hands, then anointed his neighbour, then a woman. It was at this moment that he felt a strange warmth in his toes, which up to then were almost dead and felt numb. Then he realised that his feet were coming back to life, and the warmth spread through his feet. He thought it was caused by the hot water bottles, though normally he could not feel their heat. He lifted up his blanket and noticed that his right ankle was very swollen. Attributing this oedema to cardiac insufficiency, he thought this was the end. However, the heat spread to his legs, and he started to stir in his wheelchair. His wife, thinking he wished to urinate, asked him to be patient. He replied: "I do not know what has happened to me, but I have the impression that I will not need my sticks much longer, and that I could walk." The warming up of his body continued for approximately 10 minutes. He returned to the hospital which he had so recently left in a little voiture, pulled by his wife. On the way he lifted up his blanket and observed that his ankle was now normal. To get into the lift for the first floor, a wheel-chair was brought to him, and to the astonishment of the brancardier, who knew him well, he got up and sat down on his own chair without help. Back at his bed, he asked his wife to take him to the toilettes. She offered him his walking sticks, but he refused them: "I cannot, I will never walk with them again." He then got up and for the first time, without sticks or support crossed the ward, came back again to his bed, walking all the time unaided, and then, out of obedience, went down to the refectory in his wheel-chair. . . .

He was taken down to the Grotto and then to the Rosary Square for the Procession of the Blessed Sacrament. While waiting, he automatically took off his spectacles, and with his right eye covered, found that he could see the names of different bureaux under the arcades on the opposite side of the square with his left eye. He also saw the banners of the various pilgrimages. After the ceremony, he was taken to the station, and helped into his bunk in the ambulance coach, as on the outward journey. Dr. Sourice, his personal physician who knew him well, made a preliminary brief examination. In particular, he checked his visual fields and found that he had recovered his vision without any doubt at all. Mr. Perrin got up out of his couchette and went down the corridor to say good-bye to his family, who were standing in front of the coach. He even got down on to the platform, without help or use of his sticks, and got back up into the train again the same way a few minutes prior to departure.

The return journey was uneventful: he was given, quite intentionally, his last injection of Vincamine, and near Bordeaux, took one Binoctal sleeping tablet. At all times he was quite calm, manifesting no apparent emotion.

On Sunday, May 3rd, the day after his return, Dr. Sourice examined him more thoroughly and under better conditions. The neurological examination was quite normal. No motor lesion could be detected, his reflexes were brisk, but normal. Superficial and deep sensation were normal too. The cranial nerves and his vision were normal. The carotid and superficial temporal pulsations were normal and palpable.[25]

Among the cures at Lourdes, there have been complete remissions from ulcers with extensive gangrene, anterolateral spinal sclerosis (ALS), tuberculous, peritonitis, leg and abdominal tumors, dorsolumbar spondylitis, blindness of cerebral origin, bilateral optic atrophy, multiple sclerosis, sarcoma of the pelvis, and Budd-Chiari disease.[26]

Healing Among the Kung

Kia healing among the Kung hunter-gatherers of the Kalahari Desert promotes emotional release, physical reenergizing, social cohesion, and the realization of extraordinary capacities that give flavor and meaning to life. "A transformation of consciousness is at the core of the Kung experience of healing," wrote psychologist Richard Katz. "[It] comes only after a painful transition into an enhanced state of consciousness, [that] brings on a sense of connectedness between a spiritual healing power, the healers, and their community."[27] Kung healing requires the arousal, or "boiling," of *n/um*, a psychophysical energy or substance activated by ritual dancing, drumming, and chanting. Participants in this ritual confront the uncertainties and contradictions of their lives, resolve issues dividing them, and re-affirm their group's cohesion. Kung healers put *n/um* into people to "pull out" their sickness, frequently handle or walk on fire (some-times placing their head directly into the flames), see the insides of other people's bodies, observe scenes at a distance, and "travel to God's home." According to Katz, the Kung often say that *kia*, an ex-traordinary state of mind and body, gives them a stronger sense of identity, a sense of becoming more fully themselves. The superordi-nary sight associated with *kia* was described picturesquely by K'au Dau, a blind healer:

> God keeps my eyeballs in a little cloth bag. [He] plucked my eyeballs out and put them into the bag and then he tied the eyeballs to his belt and went up to heaven. And now, on the nights when I dance and the singing rises up, he comes down from heaven swinging the bag with the eyeballs above my head and then he lowers the eyeballs to my eye level, and as the singing gets strong, he puts the eyeballs into my sockets and they stay there and I cure. And then when the women stop singing and separate out, he removes the eyeballs, puts them back into the cloth bag and takes them up to heaven.[28]

Kung ritual dance rouses *n/um* through exertion and proximity to a fire. When it is strong, it is "heavy" and Kung healers "have a lot of it." Boiling *n/um* can be painful, even frightening, and sometimes destabilizes the healer's perception and behavior. Given its dangerous power, *n/um* must be controlled through practice and personal strength. Wi, an old and experienced healer, described Dau, who was still young in *kia:*

> What tells me that Dau isn't fully learned is the way he behaves. You see him staggering and running around. His eyes are rolling all over the place. If your eyes are rolling, you can't stare at sickness [with] steady eyes. . . . You need direct looking. Your thoughts don't whirl, the fire doesn't float above you, when you are seeing properly.[29]

Kia triggers extraordinary somatic awareness of the kind de-scribed in chapter 5.2. Wi, the old healer, said:

You see the *n/um* rising in other healers. You see the singing and the *n/um*, and you pick it up. As a healer in *kia,* you see everybody. You see that the insides of well people are fine. You see the insides of the one the spirits are trying to kill and you go there. Then you see the spirits and drive them away.[30]

[13.3]
Modern Research Involving Paranormal Aspects of Spiritual Healing

In the first laboratory psychokinesis studies, dice and other inanimate objects were used as targets. But in the 1940s and early 1950s a few researchers demonstrated PK effects upon living systems by mentally coaxing paramecia into particular quadrants of the microscopic field,[31] by mentally influencing the growth of seedlings,[32] and by affecting the rate of bacterial growth. Other studies have been conducted since then with healers, with subjects thought to be gifted in producing PK, and with ordinary subjects, to see if living organisms might be influenced paranormally. In most discussions of mental healing, these experiments have generally been grouped together on the assumption that mental healing is a form of PK. To analyze them more carefully, however, psychologist Jerry Solfvin divided them into two classes, one dealing primarily with healing and one focused primarily on PK, and separated each group of experiments according to the target materials used.[33] In this section, I draw heavily from Solfvin's listing. All the references listed below can be found in appendix D.1.

PK Studies with Living Target Materials

Barry (1968a, b) and Tedder & Monty (1981) reported successful PK experiments with fungus growth; Braud (1974) and Nash (1982) with bacteria; Haraldsson & Thorsteinsson (1973) and Nash & Nash (1967) with yeast; Metta (1972) with lepidopterous larvae; and Brier (1969) with the tracings of a polygraph that was connected to the leaves of a plant. Kief (1972) found no effects in attempts to mentally influence enzymes. Braud et al. (1979) reported effects on biochemical processes in blood cells after treatments by psychic healer Matthew Manning, but Deamer (1979) found no effects of Manning's treatments on blood cells, lipid monolayers, and proteins.

There are reports of PK effects, all significant in the predicted direction, on the behavior of a cat (Osis 1952), dogs (Bechterev 1948), mice (Gruber 1979a, b), rats (Extra 1971), ants (Duval 1971), baby chicks (Nash & Nash 1981), protozoa (Randall 1970), and paramecia (Richmond 1952). Gruber (1979a) reported a PK influence on human motor behaviors, and Braud (1978) on the galvanic skin responses of an unsuspecting subject.

Mental Healing Studies with Plant Life

Experiments with practicing mental healers have shown significant effects on yeast (Grad 1965), enzymes (Edge 1980; Smith 1968, 1973, 1977), and salmonella typhimurium (Rubik & Rauscher 1980). Snel (1980) and Kmetz (1981) have reported significant effects on in-vitro cancer cells; the effects, however, could have been at least partially due to handling artifacts.

On the other hand, MacDonald et al. (1976) found no effects on bacterial growth after treatments by a mental healer; and Knowles (1954) reported no success when he tried to influence the clotting time and sedimentation rate of human blood or induced skin burns in human patients.

Other studies have shown significant increases in the heights of plants that were watered with water or saline solution treated by healers: see, for example: Grad 1963, 1964; Loehr 1959; and Miller 1972.

Mental Healing Studies with Animals

The first published laboratory studies of spiritual healing with animals were conducted at McGill University in the 1950s and 1960s by Bernard Grad, who had already demonstrated significant increases in the yields and heights of plants whose seedlings were watered with solutions treated by a healer.[34] Working with Oskar Estebany, the same healer he had used in some of his plant experiments, Grad showed that mice fed an iodine-deprived diet with doses of thouracil to induce goiter, when treated by the laying on of hands, developed goiter more slowly than mice fed the same diet without healing treatments. In this series of experiments, Estebany could touch the mouse cages but not the mice directly. To rule out the effects of warmth, some of the control groups were treated by having heat tape wrapped around their cages, simulating the warming effects of Estebany's hands. Then, to remove himself even further from physical contact with the mice, Estebany treated cotton-wool pads that were placed in the cages of the target mice, while identical pads that were not treated were placed in the cages of the controls. Again, the treated group developed goiters more slowly than the control group. In a third experiment, goiters were allowed to develop for several weeks before the mice were separated randomly into target and control groups, then put on a normal diet. Though thyroid weight decreased in all the mice, it decreased more rapidly among those in the healer-treated group.[35]

In a later study, Grad had patches of skin removed from mice to form wounds whose size could be determined precisely. Measurements made during the eleventh and fourteenth days of this experiment showed that the treated group healed more rapidly than two control groups, one of which was placed in cages wrapped with heat tape to simulate hand warming, and one placed in ordinary cages.[36] In another experiment of this kind, which was performed with even stricter controls, Grad divided 300 wounded mice into groups that were either healed by Estebany, not treated at all, or placed in cages that were held periodically by medical students to simulate the healer's procedure. Again, Estebany's group showed smaller wounds (on the fifteenth and sixteenth days of the experiment) than the other two groups.[37] By now Grad had controlled his experiments enough to ensure that Estebany's success was not caused by stroking the mice or by warmth generated from his hands. As far as Grad could tell, only Estebany's healing intention differentiated the situations of targets from those of their controls.

Biologists G. K. and A. M. Watkins conducted several experiments to determine whether mental healing speeded the resuscitation of anesthetized mice. In their first trial, a healer was present in a room with his targets but was allowed no physical contact with them. Nevertheless, his mice recovered faster than controls rendered unconscious at the same time. In a second experiment, the healers' targets were hidden behind a screen, but once again they recovered faster than their controls. And in a third study, a number of healers viewed their mice through a one-way glass from outside the room in which they were placed. For the first four runs (of 24 trials each), during which the timers knew which mice were being healed, the targets recovered faster than those that were not acted upon at a distance. The next seven runs, however, were conducted with the timers blind to the mice being healed, and no differences were observed between the experimental and control groups. During these trials, the healers complained that the complicated procedure was distracting. For that reason, in the last four trials a compromise procedure was adopted that kept the experimenters blind. With this procedure, the target mice again recovered more quickly than their controls. Differences between the recovery times of targets and their controls in all the trials of this experiment, when lumped together, were significant, despite the seven runs that showed no effects.[38]

Subsequently, seven more series of this kind were conducted by the Watkinses and their colleagues. In one, a new timer and anesthetizer were used; in another, ordinary subjects replaced the gifted healers. All the experiments, except the one with nongifted healers, produced significant results. In the last two series, mice placed on the side of the table that had been acted upon by the healer before they were placed there recovered faster than their controls, even though the healer was absent. The "healed side of the table" continued to awaken the mice more quickly than the other half, suggesting a possible linger effect of PK and spiritual healing.[39]

Mental Healing with Humans

Stimulated by Grad's experiments on plants and animals, Dolores Krieger, an experienced nurse, suspected that humans exposed to healers might show increases of blood hemoglobin analogous to increases of chlorophyll in healer-treated plants, since hemoglobin plays a central role in several life-sustaining processes. Following her intuition, Krieger conducted three experiments on human subjects, measuring their hemoglobin levels before and after treatments by Oskar Estebany. Using matched subjects of various ages and both sexes in her first two studies, she found that the mean hemoglobin values of her experimental subjects exceeded their former values at the .01 level of confidence (meaning there is only one chance in a hundred that the excess could be caused merely by chance) after a laying on of hands by the healer. In a third experiment, the mean hemoglobin values of her healed subjects surpassed their former values at the .001 level of confidence (that is, at odds of one in a thousand), in spite of firmer control over her research design. These studies did not precisely control the times for post-healing hemoglobin measurement, yet similar responses have been noted in other experiments.[†] Krieger used such research findings and the practical lore of spiritual healing to develop an adjunct to patient care that she named Therapeutic Touch.[40] This approach is based on the concept that a healthy human body has an excess of energy that may be transferred to others for their benefit.

The results of a study conducted by Janet Quinn at St. Vincent's Medical Center in New York with 60 heart patients strongly suggest that the healer's state of mind is central to Therapeutic Touch. Quinn randomly divided a group of nurses, taught half of them the Krieger method for her experiment, and had the rest count backwards from 100 by sevens while performing the same hand motions. Independent observers could not distinguish the trained healers from their controls, and the patients were told only that the nurses were trying "to learn about the body through their hands." None of the patients, in other words, knew they were subjects in this experiment, yet those whose healers had been trained in Therapeutic Touch experienced less anxiety than those whose nurses merely counted backwards.[41]

Evidence that something more than hands-on contact operates in Therapeutic Touch was provided by a double-blind study conducted by Daniel Wirth. In this study, 44 healthy subjects were randomly placed in a treatment or control group, but none of them knew to which group they were assigned. Puncture wounds were then made in all subjects and measured on days 0, 8, and 16. A Therapeutic Touch practitioner treated the target subjects from a separate room. Results showed that treated subjects experienced a significant acceleration in the rate of their wound healing at day 8 as compared to nontreated subjects. Thirteen of 23 treated subjects versus none of 21 control subjects were completely healed by day 16. Placebo effects and the possible influences of suggestion were eliminated by isolating the subjects from the healer, by blinding them to the nature of the therapy, and by using an independent experimenter who did not know what therapy was employed.[42]

Joyce Goodrich, a student at the Union Graduate School of Cincinnati, Ohio, conducted a study of spiritual healing in which ques-

[†] Krieger 1975. An experiment by psychologist William Braud of the Mind Science Foundation of San Antonio, Texas, lent plausibility to Krieger's results. In Braud's study, 32 subjects attempted to protect human red blood cells in test tubes from hemolysis (disintegration and loss of hemoglobin) by mental intention and visualization, without any physical intervention. Rate of hemolysis was measured photometrically, and both subjects and experimenters were "blind" regarding critical aspects of the experiment. According to Braud, significant differences in rate of hemolysis between experimental (i.e., mentally "protected") and control blood samples were found in an extra-chance number of subjects ($p = 1.91 \times 10 - 5$). See: Braud 1988.

tionnaires regarding feelings, sensations, current health, and mood were given to both healers and clients before and after they participated in healing sessions. In their initial meetings, healers and clients participated in the Leshan Type One healing procedure, in which the healer imagines himself deeply connected with his patient in a spiritual light or other form of unity. In subsequent sessions, the patients were to sit quietly or meditate while their healers meditated upon them from a distance. Goodrich, however, had arranged for only half of the sessions to occur while healers and patients were concentrating at the same time. When the experiment was completed, the questionnaires, edited for time cues, were given to independent judges who tried to pick out which had been filled out before and after the synchronous sessions. The judges were able to do so beyond chance expectations.[43]

Working with the people Goodrich had used in her study, Shirley Winston examined the relationship between healer and client by assigning them to four degrees of contact. There was a trend toward more effective healing, she found, when the degree of contact was low, leading her to suggest that spiritual healing is a transpersonal event that can be hindered by interpersonal communications.[44] Winston's view, however, differed from that of F. W. Knowles, a physician who treated chronic pain in some of his patients by "willing it away." Knowles tested his therapy with experiments in which his patients sometimes did not know whether he was healing them mentally or not, and developed a theory that emphasized the simultaneous influence of normal and paranormal processes in the healer-patient relationship.[45] In his comprehensive review of mental healing experiments, Jerry Solfvin supported Knowles's suggestion that healing had both interpersonal and transpersonal elements.[46]

To test the effects of healing through intercessory prayer, physician Randolph Byrd randomly divided patients in a coronary care unit at San Francisco General Hospital into a target population of 192 and a control group of 201 patients. Neither patients, staff, doctors, nor Byrd himself knew which patients would be the targets of prayer; and the people who prayed were simply given their subject's first name, diagnosis, and general condition. Every subject was assigned three to seven intercessors, each of whom was asked to pray daily for their patient's rapid recovery, and Byrd collected information about all his subjects without knowing who were targets and who were controls. Congestive heart failure, cardiopulmonary arrest, and pneumonia occurred less frequently in the group receiving intercessory prayer than they did in the control group. Fewer patients in the prayer group required ventilatory support, antibiotics, or diuretics. In the prayer group, 85 percent were considered to have a "good hospital course" after entry versus 73 percent in the control group. And finally, a general analysis of the differences between targets and controls in their recovery from disease showed a high degree of statistical significance between the two groups.[47]

Studies of Influence at a Distance Involving Biofeedback from Subjects to Their Influencers

Thirteen studies involving 62 "influencers," 271 subjects, and 4 experimenters conducted by psychologist William Braud and anthropologist Marilyn Schlitz of the Mind Science Foundation in San Antonio, Texas, suggest that one person can influence another's skin activity at a distance by intending to do so, using various visualization exercises and biofeedback from their targets. In these 13 experiments, the "influencers" were presented with concurrent polygraph records of their distant subjects' electrodermal activity, which they tried either to quiet or to excite by employing one or more of the following strategies:

- imagery and self-regulation techniques to induce the intended condition (either relaxation or arousal) in *oneself*, while imaging and intending a corresponding change in the distant subject;
- imaging the *other person* in appropriate relaxing or activating settings;
- imaging the desired outcomes of the polygraph pen tracings— i.e., few and small pen deflections for calming periods, many and large pen deflections for activation periods.

Taken together, these 13 studies yielded significant evidence for the experimental effect ($p = .000023$, $z = 4.08$, and mean effect size $= 0.29$). The Braud-Schlitz research design was extremely well controlled and ruled out coincidence, common external stimuli (because influencers and their subjects were housed at separate locations), and common physiological rhythms. According to Braud and Schlitz, their results demonstrated

> reliable and relatively robust anomalous interactions between living systems at a distance. The effects may be interpreted as instances of an anomalous "causal" influence by one person directly upon the

physiological activity of another person, [or] an anomalous informational process combined with unconscious physiological self-regulation on the part of the influenced person.

Our experimental design guaranteed that the effect could not be attributed to conventional sensorimotor cues, common external stimuli, common internal rhythms, or chance coincidence. A number of additional potential artifacts can be mentioned here and can be effectively dismissed.

1. *The findings are the result of recording errors and motivated misreadings of polygraph records.* This explanation is rejected on the basis of blind-scoring of polygraph records and, later, by the use of completely automated assessment techniques and computer scoring of response activity.

2. *The subjects knew beforehand when influence attempts were to be made and "cooperated" by changing their own autonomic activity when appropriate.* This explanation may be rejected because the subjects were not told when or how many influence attempts would be made, nor was the experimenter aware of the influence/control epoch schedule until all preliminary interactions with the subject had been completed. Subjects did not know of the existence of, or have access to, the envelopes containing schedule information.

3. *Subjects could have become aware during the experimental sessions themselves of when influence epochs were in progress and could have altered their own physiological reactions during those periods.* This possibility was eliminated by isolating the subject from any such cues from the influencer. Subject and influencer were in separate, closed rooms at least 20 meters apart.

4. *Differences in autonomic activity between influence and control periods are due to systematic error—i.e., some progressive change in electrodermal activity over time.* This objection may be rejected. Progressive (time-based) errors could have been contributed by (a) changes in equipment sensitivity as the equipment warmed up, (b) changes in electrodermal activity due to adaptation or habituation to the experimental environment, or (c) changes in electrodermal activity due to polarization of the recording electrodes. Equipment was allowed to warm up for 15 to 20 minutes prior to the beginning of a session and therefore had become thermally stable before the experiment began. The use of electrodes with large surface areas, and the use of a constant-current electrodermal recording device reduced the possibility of polarization problems. The use of silver-silver chloride electrodes and partially conductive paste in other experiments further minimized a polarization problem. A special analysis of the data from Experiment 5 is relevant to the habituation question. Statistical evaluation of total electrodermal activity for the first halves *versus* the second halves of the sessions indi-

cated no evidence of an habituation effect. This absence of habituation could be attributed to the use of an adaptation period before the actual recording session began, and to the use of constantly changing auditory and visual stimulation of the subject (i.e., the use of the random tones and colored lights display). Thus, there was no progressive change in electrodermal activity due to any of the three possible processes mentioned above. However, even if a progressive change had occurred, the use of the ABBA counterbalanced design and the use of truly random influence/control sequencing in other experiments would have prevented this error from contributing differentially to influence versus control epochs.

5. *The findings are due to arbitrary selection of data.* This explanation may be rejected since total numbers of subjects and trials were prespecified, and the analyses reported include all recorded data.

6. *The results are due to fraud on the part of the subjects.* This explanation may be rejected. The subjects were unselected volunteers; it may be assumed that such subjects had no motive for trickery. However, even if a subject were motivated to cheat, such an opportunity was not present. Cheating would have required knowledge of a session's influence/control epoch sequence and of the precise starting time for the session, or the assistance of an accomplice. Both of these requirements were eliminated.

7. *The results are due to fraud on the part of the experimenters.* No experiment, however sophisticated, can ever be absolutely safe from experimenter fraud. Even if an experiment were controlled by an outside panel of disinterested persons, a hostile critic could still argue that collusion was involved. The imagined extent of such a conspiracy would be limited only by the imagination and degree of paranoia of the critic. We can only state that we used multiple-experimenter designs so that one experimenter's portion of the experiment served as a kind of control for another experimenter's portion. Only the successful replication of these findings by investigators in other laboratories would reduce experimenter fraud to a non-issue. We hope that this report will stimulte such replication attempts.

We conclude that our results cannot be attributed to any of the various potential artifacts or confounds mentioned above, and therefore are not spurious. Rather, the results reflect an anomalous psychophysical interaction between two individuals separated from one another in space.[48]

In addition to responding physiologically in a manner consistent with the imagery of the distant influencer, subjects in the Braud-Schlitz experiments often reported subjective responses that corresponded to the influencers' experience. For example, one subject reported a vivid impression of the influencer coming into his room, walking behind him, and vigorously shaking his chair. The impression was so strong that he found it difficult to believe that the event hadn't happened. In this session, the influencer had indeed employed such an image. And during another session, an experimenter remarked to an influencer that the electrodermal tracings of the subject reminded him of the German techno-pop instrumental musical group *Kraftwerk*. When the experimenter went to the subject's room at the end of the session, the subject's first comment was that early in the session, for some unknown reason, thoughts of the group *Kraftwerk* had come into her mind. The subject could not have overheard the experimenter's earlier comment to the influencer.[†]

Further evidence that extrasomatic influences operate in spiritual healing is provided by the pioneering Russian researcher L. L. Vasiliev, who reported that experimenters caused the muscles of blindfolded, deeply hypnotized subjects to contract by pointing at them;[49] by Graham Watkins, Anita Watkins, and Roger Wells, who placed light-shielded photographic films underneath mice during experiments with distant healing, and found that they were exposed after their experiments;[50] and by studies in which healers appear to change the quality of water or radiate ultraviolet light from their hands.[51]

The results of these experiments suggest that extrasensory influences might sometimes be at work in sports teams, orchestras, or jazz groups that are suddenly inspired; in mass healings at shrines or revivalist meetings; in hypnotic experiments during which most of the subjects enjoy special success; and during other group activities. It can be argued, of course, that the contagious aspects of group behavior are mediated entirely by sensory cues, but the experiments noted in this chapter provide further evidence that extrasensory influences also help to convey moods and volitions from one person to another.

[†] Parapsychologist Rex Stanford has proposed the concept of *conformance behavior* to account for telepathic interactions such as those evidenced in Braud and Schlitz's experiments. Such behavior involves "an anomalous interaction between physically isolated systems, and encompasses the phenomena of psychokinesis, [and] receptive psi." Braud has generalized Stanford's concept, suggesting that "under certain conditions a system possessing a greater degree of disorder (randomness, lability, noise, entropy) changes its organization so as to more closely match that of another system possessing less disorder, less entropy, greater structure." Braud cited the results of various experiments as examples of conformance behavior, among them changes of orientation among fish that were willed or visualized to be in a given position, and changes in the locomotor activity of gerbils that were similarly acted upon by intention and imagery from a distance. In these experiments, Braud wrote, humans, fish, and gerbils "seemed to be equally susceptible to conformance behavior" (Braud 1979).

[13.4]
SPIRITUAL HEALING AND TRANSFORMATIVE PRACTICE

Even if we don't comprehend all of its complexities,[†] spiritual healing indicates that mental intention can affect the flesh directly, through "whole-hearted, whole-minded" imagery and practiced attention to the structures or processes involved. It suggests that consciousness and living matter have a profound resonance with, and formative influence upon, each other. It shows that we can summon a wide range of energies for healing and growth without machines or other external devices, that the intention to help one's fellows might touch them through extrasomatic processes (even when they are unaware of it), that supportive cultures or groups can facilitate our general well-being. And it also strongly suggests that a principle, presence, or power beyond ordinary functioning can have profound effects upon us. As we shall see, there are good reasons to believe that all of these factors contribute to successful transformative practice.

[†] Taking such complexities into account, Larry Leshan has divided spiritual healing into two broad classes, one involving the transfer of energies from healer to patient, usually by the laying on of hands or similar rituals, and the other involving an imaginative joining by the healer with his subject in a spiritual unity. See: Leshan 1975.

14

Extraordinary Capacities of Disabled People

BLINDNESS, DEAFNESS, AND other disabilities lead many people to develop metanormal attributes. In this chapter, I briefly describe some men and women who by meeting severe afflictions with courage have discovered extraordinary depths of human understanding, new subtleties in sensory experience, and a profound joy within ordinary pleasure and suffering. Like the alterations of mind and flesh noted in other chapters, the remarkable capacities described here provide dramatic evidence for human transformative capacity.

Certain blind people can sense their location by sounds made with a walking stick, know the moon has risen and point to its location, determine the position and movement of clouds in the daylight sky, feel heat from distant objects, recognize the colors of fabrics by touch, or identify household objects by nontactile clues.[1] Such abilities, of course, for the most part involve the exceptional development of sensory capacities. Determining location by tapping a stick involves close attention to echoes discernible through ordinary hearing. The moon's position can be guessed through sensations of warmth that most of us can discriminate with training. Subtle differences of texture related to fabric colors are probably recognizable to anyone with normal tactile sensations. But it also seems that sightlessness, like deliberate sensory deprivation, can promote extrasensory capacities. Helen Keller, for example, exhibited knowledge at times that could not be fully explained in terms of ordinary sensory cues, and she enjoyed states of mind in her later life that resembled the illuminations of religious mystics.

Many deaf people, too, exhibit special powers. Some can follow another's conversation by holding a speaker's throat as he or she talks.

Others can appreciate songs by touching a turning phonograph record. A few enjoy choral or symphonic music in auditory hallucinations. The South African poet and novelist David Wright, who was deaf from the age of seven, described the "phantasmal voices" he heard as he watched a speaker's face. His deafness, he wrote,

> was made more difficult to perceive because from the very first my eyes had unconsciously begun to translate motion into sound. My mother spent most of the day beside me and I understood everything she said. Why not? Without knowing it I had been reading her mouth all my life. When she spoke I seemed to hear her voice. It was an illusion which persisted even after I knew it was an illusion. My father, my cousin, everyone I had known, retained phantasmal voices. That they were imaginary, the projections of habit and memory, did not come home to me until I had left the hospital. One day I was talking with my cousin and he, in a moment of inspiration, covered his mouth with his hand as he spoke. Silence! Once and for all I understood that when I could not see I could not hear.[2]

Commenting upon Wright's experience in his book *Seeing Voices*, the neurologist Oliver Sacks observed that there is a sensory "consensus" by which

> objects are heard, seen, felt, smelt, all at once, simultaneously; their sound, sight, smell, feel all go together. This correspondence is established by experience and association. This is not, normally, something we are conscious of, although we would be very startled if something didn't sound like it looked—if one of our senses gave a discrepant impression. But we may be made conscious, very suddenly and startlingly, of the senses' correspondence, if we are suddenly deprived of a sense, or gain one. Thus David Wright "heard" speech the moment he was deafened; an anosmic patient of mine "smelt" flowers whenever he saw them; and a patient described by Richard Gregory could at once read the time on a clock when he was given his sight (he had been blind from birth) by an eye operation: before that he had been used to feeling the hands of a watch with its watch-glass removed, but could make an instant "transmodal" transfer of this knowledge from the tactile to the visual, as soon as he was able to see.[3]

Helen Keller, who became deaf and blind at 19 months, utilized the sensory "consensus" noted by Sacks to develop extraordinary perceptual abilities. She derived pleasure from touching great works of art. As my fingertips trace line and curve, they discover the thought and emotion which the artist portrayed. I can feel hate, courage, and love, just as I can detect them in living faces.

The hands of those I meet are dumbly eloquent. The touch of some hands is an impertinence. I have met people so empty of joy, that when I clasped their finger tips, it seemed as if I were shaking hands with a northeast storm. Others have sunbeams in them, so that their grasp warms my heart. It may be only the clinging touch of a child's hand; but there is as much potential sunshine in it for me as there is in a loving glance for others.[4]

Keller developed extraordinary sensitivity by the age of eight. According to her teacher Anne Sullivan,

her whole body is so finely organized that she seems to use it as a medium for bringing herself into closer relations with her fellow creatures. She is able not only to distinguish with great accuracy the different undulations of the air and the vibrations of the floor made by various sounds and motions, and to recognize her friends and acquaintances the instant she touches their hands or clothing, but she also perceives the state of mind of those around her. It is impossible for anyone with whom Helen is conversing to be particularly happy or sad, and withhold the knowledge of this fact from her. She observes the slightest emphasis placed upon a word in conversation, and she discovers meaning in every change of position, and in the varied play of the muscles of the hand. She responds quickly to the gentle pressure of affection, the pat of approval, the jerk of impatience, the firm motion of command, and to the many other variations of the almost infinite language of the feelings.

Keller also appeared to know things at times that her friends and acquaintances could not account for in terms of sensory cues. Anne Sullivan wrote:

While making a visit at Brewster, Massachusetts, she one day accompanied my friend and me through the graveyard. She examined one stone after another, and seemed pleased when she could decipher a name. . . . When her attention was drawn to a marble slab inscribed with the name Florence in relief, she dropped upon the ground as though looking for something, then turned to me with a face full of trouble, and asked, "Where is poor little Florence?" I evaded the question, but she persisted. Turning to my friend, she asked, "Did you cry loud for poor little Florence?" Then she added: "I think she is very dead. Who put her in the big hole?" As she continued to ask these distressing questions, we left the cemetery. Florence was the daughter of my friend, and was a young lady at the time of her death; but Helen had been told nothing about her, nor did she even know that my friend had had a daughter. Helen had been given a bed and carriage for her dolls, which she had received and used like any other gift. On her return to the house after her visit to the cemetery, she ran to the closet where these toys were kept, and carried them to my friend, saying, "They are poor little Florence's." This was true, although we were at a loss to understand how she guessed it.[5]

In her old age, Keller wrote: "I dream of sensations, colors, odors, ideas, and things I cannot remember." She claimed that an extraordinary light was revealed to her in sleep: "and what a flash of glory it is! In sleep I never grope but walk freely." Helen Keller lived triumphantly in spite of her blindness and deafness. People in many parts of the world have been inspired by her creativity, her love of adventure, her great energy in the midst of extreme adversity. Her life suggests that extraordinary qualities can arise in us all despite our limitations. Indeed, her exceptional tactile, kinesthetic, and olfactory abilities; her great imagination and loving exuberance; and her contact with transcendent realities were probably catalyzed to some degree by her difficulties.

Jacques Lusseyran, who was accidentally blinded at eight when struck by another boy, developed remarkable sensitivity to other people, and a lasting sense of inner light. Born in Paris in 1924, he was 15 when the German occupation began. At 16 he organized an underground resistance movement called *Les Volontaires de la Liberté* with boys under the age of 21. Because he possessed remarkable intuitions about people's character, he was chosen by his young comrades to head the movement's recruiting effort. In a year, *Les Volontaires* grew to more than 600 members. Lusseyran was eventually arrested and spent 21 months in prison, 15 of them in Buchenwald. In 1963, he published a book about his early life, from which I have taken the following passages.

> Movement of the fingers was terribly important, and had to be uninterrupted because objects do not stand at a given point, fixed there, confined in one form. They are alive, even the stones. What is more, they vibrate and tremble. My fingers felt the pulsation distinctly, and if they failed to answer with a pulsation of their own, the fingers immediately

became helpless and lost their sense of touch. But when they went toward things, in sympathetic vibration with them, they recognized them right away.

Yet there was something still more important than movement, and that was pressure. If I put my hand on the table without pressing it, I knew the table was there, but knew nothing about it. To find out, my fingers had to bear down, and the amazing thing is that the pressure was answered by the table at once. Being blind I thought I should have to go out to meet things, but I found that they came to meet me instead.

With smell it was the same as it was with touch—like touch an obvious part of the loving substance of the universe. I began to guess what animals must feel like when they sniff the air. Like sound and shape, smell was more distinctive than I used to think it was. There were physical smells and moral ones.

Before I was ten years old I knew with absolute certainty that everything in the world was a sign of something else, ready to take its place if it should fall by the way. And this continuing miracle of healing I heard expressed fully in the Lord's Prayer I repeated at night before going to sleep. I was not afraid. Some people would say I had faith, and how should I not have it in the presence of the marvel which kept renewing itself? Inside me every sound, every scent, and every shape was forever changing into light, and light itself changing into color to make a kaleidoscope of my blindness.

As with the sense of touch, what came to me from objects was pressure, but pressure of a kind so new to me that at first I didn't think of calling it by that name. When I became really attentive and did not oppose my own pressure to my surroundings, then trees and rocks came to me and printed their shape upon me like fingers leaving their impression in wax.

This tendency of objects to project themselves beyond their physical limits produced sensations as definite as sight or hearing. I only needed a few years to grow accustomed to them, to tame them somewhat. Like all blind people, whether they know it or not, these are the senses I use when I walk by myself either outdoors or through a house.

I did not become a musician, and the reason was a strange one. I had no sooner made a sound on the A string, on D or G or C, than I no longer heard it. I looked at it. Tones, chords, melodies, rhythms, each was immediately transformed into pictures, curves, lines, shapes, landscapes, and most of all colors. Whenever I made the A string sound by itself with the bow, such a burst of light appeared before my eyes and lasted so long that often I had to stop playing.

At concerts, for me, the orchestra was like a painter. . . . If the violin came in by itself, I was suddenly filled with gold and fire, and with red so bright that I could not remember having seen it on any object. When it was the oboe's turn, a clear green ran all through me, so cool that I seemed to feel the breath of night.

There is nothing in the world which cannot be replaced with something else; sounds and colors are being exchanged endlessly, like the air we breathe and the life it gives us; nothing is ever isolated or lost . . . everything comes from God and returns to God along all the roadways of the world.[6]

Having inspired other young Frenchmen to form *Les Volontaires de la Liberté* in 1941, Lusseyran was chosen by the group's Central Committee to head its recruiting.

They claimed I had "the sense of human beings." I would hear more acutely and pay better attention. People would not easily deceive me. I [would] not forget names or places, addresses or telephone numbers. Every week I would report on the outlook without resorting to scraps of paper or lists. Everything written down, even in code, was a risk that none of us had the right to run.

In less than a year nearly six hundred boys took the road to the Boulevard Port-Royal. . . . If I could plumb these hearts and consciences—and I felt sure I could—it was because I was blind and for no other reason.

As each day went by, some of us had to get used to some strange phenomena. Since we had been in the Resistance our mental powers had grown stronger. Our memories were unbelievably agile. We read between the words and silences. Deeds which only two months before seemed impossible to us, putting walls or phantoms in our path, were now broken down into a dust of easy little tasks.

Reinforcements came every day. I was surprised to find I knew things they had not told me, surprised when I awoke in the morning feeling a sense of purpose strong and entirely new to me—one which I found out was shared, three hours later, by two or even ten of the comrades. The spirit of the Resistance was born, and was using me as its instrument.[7]

Lusseyran gave just a brief account of his prison experience. "I am not going to take you through Buchenwald," he wrote. "No one has ever been able to do it." He ended his book by summarizing two truths that ruled his life.

The first of these is that joy does not come from outside, for whatever happens to us, it is within. The second truth is that light does not come to us from without. Light is in us, even if we have no eyes.[8]

Largely to compensate for their blindness and deafness, Helen Keller and Jacques Lusseyran developed extraordinary sensory abilities and intuitions about other people, as well as contact with a spiritual presence and light. Met with a bravery and resilience like theirs, disabilities can help us realize our greater potentials. Like transformative practices, disabilities place constraints upon us that can help us know ourselves better and direct our best energies to creative ends. And this principle holds for difficulties other than blindness or deafness. Motorically handicapped people exhibit great endurance, flexibility, and strength during Special Olympics and other competitions. Oliver Sacks described the mathematical powers of idiot savants, and other impressive capacities exhibited by people with neurological disorders.[9] Isaac Newton discovered laws of gravitation in a highly agitated state that temporarily forced him into solitary life. Virginia Woolf wrote enduring stories while suffering auditory hallucinations. Van Gogh was driven to create by painful emotional turmoil.[10] Disabilities, in short, can lead us toward extraordinary life, revealing new depths of character, richer textures of the sensory world, and a radiance within ordinary joys and sufferings.

15

Mesmerism
and Hypnosis

SINCE FRANZ MESMER introduced his theory of animal magnetism in the 1770s, hypnotic suggestion has caused considerable controversy as a method for relieving affliction or enhancing unusual abilities. But in spite of its fluctuating reputation during the last two hundred years, it has been used by countless therapists and experimenters to alleviate pain, assist the healing of skin afflictions, relieve allergies, improve circulation, lower blood pressure, cure phobias and addictions, induce illusions and hallucinations, cause temperature changes in hands and feet, increase sensory acuity, improve cognitive performance of various kinds, assist memory recall, aid concentration in sports and musical performance, stimulate out-of-body sensations, induce experiences that appear to be paranormal, and produce exaltations that resemble mystical states. The clinical and educational usefulness of hypnosis has been confirmed by many carefully controlled experiments, and evidence for several of its effects is now overwhelming.[1]

[15.1]
History

Franz Anton Mesmer was born in a village near Lake Constance, Switzerland, in 1733. He studied medicine at the University of Vienna and took a medical degree there in 1766. Some mystery remains, however, about his early life and the origins of his famous theories. Paracelsus had prescribed the use of a magnet in the treatment of hysteria and had written that "man possesses a magnetic power by which he can attract certain effluvia of a good or evil quality in the same manner as a magnet will attract particles of Iron."[2] The Scottish physician William Maxwell had collected ancient beliefs about magnetic cures in his book *De Medicine Magnetica* published in 1679. And an English doctor, Richard Mead, had written a treatise titled *De Imperio Solis et Lunae in Corpora Humana*, which may have influenced Mesmer's thinking.[3] But whatever ideas may have influenced him, Mesmer in 1773 and 1774 relieved a 27-year-old patient, one Fräulein Oesterlin, of several afflictions by making her swallow a preparation containing iron and attaching three magnets to her body. This result, Mesmer claimed, was not primarily caused by the magnets but by a substance accumulated in his own body, which he called magnetic fluid or animal magnetism. The magnets, he proposed, only reinforced and directed its flow. Mesmer spent the rest of his life developing these ideas and healing methods.[4]

In some respects, Mesmer's therapy resembled methods used by healers since antiquity, but it was framed in pseudoscientific terms that fit the rationalism of eighteenth-century Europe. Mesmer conceived health, for example, as a harmonious distribution of magnetic fluid in the organism. This substance, he believed, permeated the universe. It could be affected by metal devices or hand passes, stored in the body, and influenced by the human will. Mesmer treated his patients by provoking convulsive crises in them through infusions of magnetic fluid, either from his own body or from physical devices constructed for that purpose. After such crises, he believed, the sufferer's balance and health would tend to be restored. Because many of Mesmer's patients were cured, his ideas were repeatedly verified. Like Valentine Greatrakes and European kings who healed by touch (13.1), he stimulated a self-fulfilling expectation of relief that made his therapy work. Though we now see that his cures were produced by the mobilization of his subjects' self-restorative powers, he believed they depended upon the lawful operations of animal magnetism.†

In Vienna Mesmer used magnetic plates contrived by the Jesuit astronomer Maximilian Hell to assist his therapy, but abandoned

---

† Mesmer's success depended upon his powers of suggestion but may also have resulted in part from the transference of energies from healer to subject of the sort discussed in chapter 13. Though he characterized hypnotic suggestion as "a successful appeal to the subliminal self," Frederick Myers felt Mesmer was partly right about the transference of a tangible influence from healer to subject. "To Mesmer," he wrote, "we owe the doctrine of a nervous influence or effluence passing from man to man,—a doctrine which, though it must assume a less exclusive importance than he assigned to it, cannot, in my view, be altogether ignored or denied" (Myers, F. W. H., 1954, vol. 1, p. 159).

them for the *baquet* when he moved to Paris in 1778. The latter device, a large oaken tub filled with bottles of water he had "magnetized," helped Mesmer evoke the extraordinary crises for which he grew famous. As one biographer wrote:

> The whole machine was a kind of travesty of the galvanic cell. To carry out the resemblance, [its] cover was pierced with holes, through which passed slender iron rods of varying lengths, which were jointed and movable, so that they could be readily applied to any part of the patient's body. Round this battery the patients were seated in a circle. . . . Further, a cord, attached at one end to the tub, was passed round the body of each of the sitters, so as to bind them all into a chain. Outside the first a second circle would frequently be formed, who would connect themselves together by holding hands. Mesmer, in a lilac robe, and his assistant operators—vigorous and handsome young men selected for the purpose—walked about the room, pointing their fingers or an iron rod held in their hands at the diseased parts.
>   The proceedings were enlivened throughout by music from a piano or other instrument.[5]

In 1784 Louis XVI appointed a commission from the Academy of Sciences and the Academy of Medicine to study Mesmer's methods. This commission, which included Benjamin Franklin, the astronomer Bailly, and the chemist Lavoisier, gave the following account of sessions at the *baquet:*

> The tableau presented by the patients is one of extreme diversity. Some are calm, composed, and feel nothing; others cough, spit, have slight pains, feel a glow locally or all over the body, accompanied by perspiration; others are shaken and tormented by convulsions. These convulsions are remarkable in their frequency, their duration, and their intensity. As soon as one attack begins others make their appearance. The Commission has seen them last for more than three hours . . . [they] are characterized by involuntary spasmodic movements of the limbs and of the whole body, by contractions of the throat, by spasms of the hypochondriac and epigastic regions; the eyes are wandering and distracted; there are piercing cries, tears, hiccoughs, and extravagant laughter. The convulsions are preceded and followed by a state of languor and reverie, by exhaustion and drowsiness. Any sudden noise causes the patients to start, and even a change in the music played on the piano has an effect—a lively tune agitates them afresh and renews the convulsions.[6]

Before serving on the royal commission, Franklin had approached mesmerism with both skepticism and good-natured tolerance. Some of his views are apparent in the following passage from a letter dated March 14, 1784, to a Monsieur de la Condamine:

> As to the animal magnetism, so much talked of, I am totally unacquainted with it, and I must doubt its existence till I can see or feel some effect of it. None of the cures said to be performed by it have fallen under my observations, and there being so many disorders which cure themselves, and such a disposition in mankind to deceive themselves and one another on these occasions, and living long has given me so frequent opportunities of seeing certain remedies cried up as curing everything, and yet soon after totally laid aside as useless, I cannot but fear that the expectation of great advantage from this new method of treating diseases will prove a delusion. The delusion may, however, in some cases be of use while it lasts. There are in every great, rich city a number of persons who are never in health, because they are fond of medicines, and always taking them whereby they derange the natural functions, and hurt their constitution. If these people can be persuaded to forbear their drugs, in expectation of being cured by only the physician's finger, or an iron rod pointing at them, they may possibly find good effects, though they mistake the cause.[7]

The behavior of Mesmer's patients resembled the swooning and groaning of medieval pilgrims at shrines such as Saint Godric's and Thomas Beckett's (13.1) or enthusiastic demonstrations at contemporary faith-healing meetings, though transfusions of so-called magnetic fluid, rather than the Holy Spirit, were deemed to be their cause. It is no wonder that scientists of the time, including most of those on the Royal Commission appointed in 1784, finding no evidence for a new kind of magnetism, attributed such behavior to the power of suggestion. With his commanding presence and reliance upon a special force, Mesmer resembled other spiritual healers, except that his theories were framed in the mechanistic language of the late eighteenth century. According to hypnosis researcher Ronald Shor, in Mesmer's day

> orthodox physicians had little to offer sufferers beyond impersonalized faith healing rationalized in the form of potions, bleeding, and purges. Mesmer's brand of faith healing on the other hand was a credo that enlisted enthusiastic belief and high expectancy of cure. . . . Mesmer did not fail to note that unless his patients were cooperative and sincerely wished to be cured they would not allow themselves to become receptive to the physician's healing influences. In other words, although his

physicalistic theoretical interpretations were incorrect, as a clinical practitioner Mesmer had an astute working knowledge of the importance of a favorable doctor-patient relationship.[8]

In 1784 one of Mesmer's disciples, Armand Chastenet, the Marquis de Puysegur, described his discovery of artificial somnambulism; later that year he elaborated his findings in the first edition of his *Memoires*. The Marquis had found that one of his magnetized patients, Victor Race, fell into a strange condition during which he had certain awarenesses that he lacked while awake. In this state, which Puysegur called a "perfect crisis," Race did not have convulsions such as those exhibited at the *baquet,* and spoke with a brighter intelligence than he usually possessed. His extraordinary functioning prompted Puysegur to induce a similar state in other patients. In this nonconvulsive crisis, it was widely reported, some people could diagnose their own illness, foresee its course, and prescribe treatments for it.[9] Following a long-standing tradition that kings and noblemen could heal those in their charge, the Marquis treated his peasants, servants, and neighbors with his own kind of mesmerism, eventually devising a procedure in which they held on to ropes hung from a magnetized elm tree while the magnetic fluid coursed through them. Many of these subjects fell into the same sort of trance that Puysegur had induced in Victor Race, partly because they were not influenced by stories of Mesmer's *baquet.* In such a state, they could speak lucidly, open their eyes, and walk about without difficulty, responding to their healer's commands and then forgetting their trance experiences when they returned to their normal condition. Working with subjects such as these, Puysegur and his followers discovered (or created) most of the major mesmeric-hypnotic phenomena acknowledged today, including the motor automatisms, catelepsy, amnesia, analgesia, hallucinations, posthypnotic phenomena, and individual differences in susceptibility.[10]

These first discoveries, however, were often entangled with excessive claims about powers that appeared in the somnambulistic state and with wild theories about the occult. These wilder assertions led the French Academy of Sciences to reject mesmerism during the 1820s and turned most scientists and doctors against it. But interest in it did not disappear completely. Despite the credulous excesses generated by mesmerism, exploration of its genuine phenomena continued within the existing framework of physical and psychological science. This stage of understanding was marked by three insights: first, that mesmeric-hypnotic phenomena were genuine and important; second, that they were essentially psychological; and third, that they were valid subjects of scientific research.[11] From this recognition, modern hypnosis studies were born.

The view that mesmerism was a psychological phenomenon that could be studied scientifically was pioneered by several men during the first half of the nineteenth century. The Portuguese priest Jose Faria, for example, rejected mesmeric ideas regarding magnetic fluid and claimed that lucid sleep, or somnambulism, was produced by the subject's confidence, expectations, and receptive attitude. In the 1840s James Braid, an English physician from Manchester, developed the concept of *monoideism*, or single-minded absorption in trance, and promulgated methods of hypnotic induction that dispensed with mesmeric passes. Braid also publicized the term *hypnotism*, which he coined in 1843, and thus helped many doctors and scientists take a fresh look at discredited mesmeric phenomena.

In Germany, however, mesmerism was developed in another direction for several decades, often by men of literary or scientific distinction and by universities that supported studies of animal magnetism.[12] Mesmer's theory of a universal fluid appealed to many German Romantics who conceived the universe to be a living organism unified by a spiritual presence; and the Marquis de Puysegur's discovery that magnetic somnambulism elicited paranormal powers permitted German philosophers to think the human mind could establish connections with the World Soul. For example, in a textbook on animal magnetism the mesmerist Carl Kluge described six degrees of magnetic sleep: the waking state; half-sleep; inner darkness, or full trance with outward insensitivity; inner clarity, which included extrasensory perceptions; self-contemplation, by which a subject could perceive the interior of his body as well as the insides of someone with whom he was in rapport; and universal clarity, the perception of things hidden in the past, the future, or at remote distances. Commenting on theories such as Kluge's, Henri Ellenberger wrote, "while the French were seeking extra-lucid *somnambules* as auxiliary subjects for medical practice, the Germans utilized them in an audacious attempt at experimental metaphysics."[13]

Mesmer had spoken of a sixth sense involving the magnetic fluid, and the Marquis de Puysegur had proposed that mesmeric trance produced clairvoyance and precognition. Inspired by such ideas, German intellectuals studied the unusual powers of extraordinary magnetic subjects. The poet Clemens Brentano, for example, broke with his former life to live near the German stigmatic Anne Catherine Emmerich (22.3) so that he could observe her ecstatic trances and record her visions, which unfolded each night according to the cycle of the liturgic year. With the material he recorded, he compiled two widely read books.[14] The German physician and poet Justinus Kerner, who first described botulism and was influential among both physicians and literary people, conducted a long study of the visionary Friedericke Hauffe, whom he had cured from convulsions by magnetic sleep. After her cure, Friedericke led a "bodiless life" in which her vital forces were sustained through daily magnetization. In trance she became a seeress, expressing herself in a solemn, musical voice while speaking High German. According to Kerner and other prominent witnesses, she appeared to see distant events, foretell the future, cause objects to move without physical contact, and speak ancient languages. Gradually she revealed a vision of magnetic circles that symbolized a hierarchy of spiritual conditions. Philosophers and theologians, among them Schelling, Eschenmayer, and Schleiermacher, consulted her about philosophical matters, and Kerner published a book, *Die Seherin von Prevorst*, that contained accounts of his experiments with her. Henri Ellenberger called Kerner's book, which was a prodigious success in Germany, "the first monograph devoted to an individual patient in the field of dynamic psychiatry."[†] Kerner's observations of Friedericke Hauffe prompted other reports of paranormal phenomena by physicians and laypeople, many of which were published by Kerner and his friends in the *Blatter von Prevorst* (1831–39) and the *Magikon* (1840–53), probably the first periodicals devoted to psychical research.

While Kluge, Kerner, and other Germans explored the metaphysical and paranormal aspects of mesmeric phenomena, some doctors continued to use hand passes and other methods of classical mesmerism to relieve a wide range of afflictions. Among these, two British doctors, John Elliotson and James Esdaile, were prominent in the 1840s and 1850s. Elliotson ran a hospital in London where he used magnetic sleep to anesthetize patients in surgery. He wrote a book about his work, and from 1843 to 1856 published a periodical, the *Zoist*, that described clinical applications of mesmerism as well as experiments with magnetic sleep on humans and animals.[15] While Elliotson was practicing mesmerism in London, Esdaile performed

---

[†] According to Ellenberger, Kerner "made great efforts to be objective, separating his observations from his experimentations and from philosophical interpretations, which he left to Eschenmayer." See: Ellenberger 1970, pp. 79–81.

several hundred operations in India using magnetic sleep as his anesthesia. In a preface to his book *Mesmerism in India,* he presented this list "showing the Number of painless Surgical Operations performed at Hooghly during the last eight months":

Arms amputated	1
Breast amputated	1
Tumor extracted from the upper jaw	1
Scirrhus testium extirpated	2
Penis amputated	2
Contracted knees straightened	3
Contracted arms straightened	3
Operations for cataract	3
Large tumor in the groin cut off	1
Operations for hydrocele	7
Operations for dropsy	2
Actual cautery applied to a sore	1
Muriatic acid applied to a sore	2
Unhealthy sores pared down	7
Abcesses opened	5
Sinus, six inches long, laid open	1
Heel flayed	1
End of thumb cut off	1
Teeth extracted	3
Gum cut away	1
Prepuce cut off	3
Piles cut off	1
Great toe nails cut out by the roots	5
Large tumor on leg removed	1
Scrotal tumors, weighing from 8 lb. to 80 lb.	14

As sleep and the absence of pain [Esdaile wrote] is the best condition of the body for promoting the resolution of inflammation by the powers of Nature, I have extinguished local inflammations by keeping the patients entranced till this was effected. I beg to state, for the satisfaction of those who have not yet a practical knowledge of the subject, that I have seen no bad consequences whatever arise from persons being operated on when in the mesmeric trance. Cases have occurred in which no pain has been felt subsequent to the operation even; the wounds healing in a few days . . . and in the rest, I have seen no indications of any injury being done to the constitution. On the contrary, it appears to me that less constitutional disturbance has followed than under ordinary circumstances. There has not been a death among the cases [I have] operated on.[16]

In his book *Natural and Mesmeric Clairvoyance*, Esdaile reprinted the following magazine account of an operation he had performed.

The woman was lying upon a charpoy (native cot), and one of the assistants was leaning over her head, mesmerising her. The charpoy was then removed and placed opposite to a window which was opened to admit light. I then looked attentively at the patient who, to all appearance, was in a most tranquil sleep.

Dr. Esdaile prepared for the operation, and I placed myself so as both to see what the doctor did, and to observe the countenance and movements (if any should take place) of the patient. The leg was taken off a little below the knee, and I was surprised at the small quantity of blood that flowed from the wound, being not more than two, or at the most three, tablespoonsful. The thigh and knee from whence the leg had been taken were, as well as every other part of her body, perfectly motionless; and the only evidence that existed that the doctor was not operating upon a corpse, was the heaving of her chest in respiration. She was not held, or tied down in any way, and during the whole operation not the least movement or change in her limbs, body, or countenance took place: she continued in the same apparently easy repose as at first, and I have no reason but to believe that she was at perfect ease.[17]

British doctors, journalists, and government officials observed Esdaile's work and helped make it famous, but their positive reports did not prevent hostility from the medical community. Indeed, both Esdaile and Elliotson complained about the violent opposition to them, some of it from the Royal Medical and Chirurgical Society.[18] But in spite of the skepticism and hostility they encountered, the two pioneering physicians continued to report their work in books and the *Zoist* for more than a decade, and published theories linking mesmerism to paranormal powers. Ether and chloroform, however, eventually supplanted mesmeric anesthesia. Doctors and scientists either rejected mesmerism completely or adopted ideas like Braid's that emphasized the purely psychological processes on which hypnotic phenomena depend. Commenting on the fact that mesmerism had lost its power to anesthetize as belief in it waned in London, Elliotson wrote:

I believe I was not wrong; I believe that in what I originally saw, mesmerism played the parts precisely that I claimed for it. It is a wicked error to suppose that I was a party to a deception, or to a whole series of deceptions, if you like; but I candidly say that mesmerism, at the present moment, has no power to remove pain. It is a mystery; it had power . . . but we are now in another cycle, and it seems to me that there are special periods only in which mesmeric phenomena can be induced.[19]

The same scorn and skepticism that Elliotson and Esdaile encountered kept most doctors who valued their reputations from working with mesmerism; and even Braid's hypnotism, which did not depend upon hand passes or theories about magnetic fluid, was generally disregarded by the medical profession. According to Ellenberger, "In the period from 1860 to 1880 . . . a physician working with these methods would irretrievably have compromised his scientific career and lost his medical practice."[20]

Hypnosis was given new scientific respectability, however, by the famous French neurologist Jean-Martin Charcot, who described his own studies of the hypnotic condition to the French Academy of Sciences in 1882. Because Charcot was generally considered to be the greatest neurologist of his time and had made important discoveries about both organic and functional nervous disorders, his theories on hypnosis were accepted by the same Academy of Sciences that upon various occasions had condemned mesmerism. Charcot's prestige gave hypnotism a new dignity, and his celebrated presentation to the Academy in 1882 triggered many studies of hypnotic phenomena, though he claimed incorrectly that the hypnotic condition passed through three definite stages (lethargy, catalepsy, and somnambulism) and that it was a diseased condition fundamentally linked with hysteria. In time, students of hypnosis realized that Charcot's subjects at the Salpetriere, the Paris hospital that housed his famous research, had been inadvertently cued to act as the master had predicted. Since many of his patients had been part of his hysteria research and were kept at the Salpetriere, they adopted the behaviors expected of them when they were hypnotized; and as their attending physicians began to observe this, Charcot's theory was debunked. Hypnosis itself, however, did not suffer its former disrepute.

Besides helping to establish hypnosis research, Charcot made other discoveries that would inform dynamic psychiatry. He analyzed differences between organic and functional paralyses, produced hypnotic paralyses experimentally, recognized the role of forgotten traumas in hysteria, and conceived nervous shock to be a hypnoid state that enabled its victim to fix his or her disability through auto-suggestion. He contrasted hysterical, posttraumatic, and hypnotic paralyses with organic disabilities, and distinguished dynamic amnesia, in which suppressed memories can be recovered by hypnosis, from organic amnesia in which memories are permanently lost.[21] These observations would profoundly influence Freud, who studied at the Salpetriere between October 1885 and February 1886. Charcot described dissociation in both normal and disturbed personality, the influence of repressed ideas on human functioning, and other psychological dynamics that bear either directly or indirectly on transformative disciplines; and his allegiance to strict observation and experimental method helped put hypnosis on a scientific footing.[22] Charcot's inaccurate theory would be corrected by the very research methods he championed.

Hypnotism was also promoted by a group of doctors, centered in the French city of Nancy, who developed a distinctive type of suggestive therapy. The Nancy school, inspired by Auguste Liebeault, a French country doctor, and led by Hippolyte Bernheim, a medical professor famous for his research on typhoid fever and other diseases, promulgated its methods through a network of physicians in Europe, Russia, and America. Albert Moll and Schrenck-Notzing in Germany, Krafft-Ebing in Austria, Bechterev in Russia, Milne Bramwell in England, Boris Sidis and Morton Prince in America, Auguste Forel in Switzerland, and Sigmund Freud were all influenced by the school's practices and theories.

Hypnotic therapy in the Nancy tradition included confident, authoritative suggestions of general well-being and symptom removal. Liebeault, a practical doctor, did not waste time with lengthy inductions, while Bernheim eventually made less and less use of hypnotism, contending that its effects could be obtained by suggestions in the waking state. The Liebeault-Bernheim approach, however, because it did not ordinarily involve diagnosis of a condition's causes and did not fully address dissociative aspects of consciousness, often produced superficial results. Nevertheless, it dramatized the therapeutic power of suggestion, helped hypnosis gain medical respectability, and inspired eminent psychiatrists such as Bechterev, Bramwell, Prince, and Freud.[23]

In the twentieth century, psychologists and medical people have built upon the pioneering work just noted. Some, such as Milton

Erickson, have discovered new induction methods and therapeutic applications of hypnosis, showing how diverse its uses and effects can be;[24] while others have developed the experimental approach championed by Charcot. During the 1930s, Clark Hull of Yale University developed quantitative analyses of hypnotic phenomena, and other investigators produced standarized procedures for rating hypnotic responsiveness in terms of observable behavioral criteria that minimized the role of the experimenter's judgment. In the 1950s and 1960s Ernest Hilgard and his colleagues at Stanford University developed this experimental work, in part through scales of hypnotic susceptibility that helped to simplify laboratory hypnosis research.

Experimental and clinical studies with hypnosis today continue the search for scientific explanation launched by Mesmer and incorporate the commitment to method emphasized by Charcot. In this they are assisted by the insights of dynamic psychiatry and the quantitative analyses made possible by statistical models. In achieving their solidity of method, though, most contemporary experimentalists have rejected the evidence for extrasensory events that hypnosis sometimes provides. While bringing new exactitude to hypnosis research, academic psychologists have lost interest in clairvoyant, telepathic, and quasi-mystical experiences such as those reported by Puysegur, Kluge, Kerner, Esdaile, Elliotson, and other pioneers of the field. Nevertheless, in recent decades a few researchers have explored the apparent connections between hypnotic and paranormal phenomena. In section 15.13 I review some of their findings.

[15.2]
SOME TYPICAL HYPNOTIC PHENOMENA

Amnesia

Posthypnotic amnesia, whether spontaneous or suggested, has been cited frequently as a mark of deep hypnosis since its discovery by the Marquis de Puysegur in the 1780s. When it occurs, suggestions, commands, or things learned within a hypnotic session are forgotten after the session is over, or earlier events in a trance may be forgotten while the subject remains hypnotized. Ordinarily, the amnesia is lifted by a prearranged signal. Some memory researchers regard this as a genuine memory disorder like ordinary clinical amnesias, while others regard it as a strategic social behavior similar to keeping secrets.[25]

Hypnotic amnesia dramatizes the malleability of memory and helps us see how memory can be manipulated, by oneself or others, in creative or destructive ways. Several researchers believe that people with multiple personality employ a kind of self-hypnosis to induce their various selves, and that they forget traumatic events or unacceptable conflicts as do hypnotic subjects given suggestions for amnesia (11.3).

Analgesia

In section 15.8 I describe several clinical accounts and experimental findings that indicate that hypnotic suggestion can alleviate pain and suffering.

Hypermnesia

Hypnotic hypermnesia, the recall of forgotten experiences, was observed by Puysegur and has been witnessed by many physicians during the past two centuries. Freud used it with some of his first patients to recover repressed memories before giving it up for dream interpretation and free association, and many psychotherapists still employ it. Experimental research, too, while showing that memory can be distorted by leading questions or the expectations of the hypnotist, has provided evidence that hypnosis can indeed facilitate recall of past events, forgotten languages, and other material.[26] Age regression, during which a subject acts as if he or she were reliving earlier experience, has also been an enduring feature of hypnosis.

Remembrance of forgotten material can also occur during other transformative activities—for example, when an athlete in extremis recalls a forgotten incident, or when a contemplative spontaneously reviews his life in deep meditation. The remembering of past events has a long history in religious practice and is central to many therapeutic disciplines. It can provide new perspective upon one's interpersonal and intrapsychic dynamics and can help to integrate dissociated parts of the self.

Feats of Dexterity and Strength

Hypnotic subjects sometimes exhibit great agility and strength during inductions intended to increase their confidence or deepen their trance. A hypnotized person, for example, might hold an arm outstretched for long periods, lie like a board between two chairs, or demonstrate extraordinary hand-eye coordination. Though such feats can be attributed to a desire to please the hypnotist or impress other onlookers, they also result in part from exceptional concentration and neuromuscular integration stimulated by hypnosis.

Hallucinations, Illusions, and Other Perceptual Effects

Several perceptual changes have been reported by hypnotized subjects, among them hallucinations, visual illusions, improved visual and auditory acuity, temporary color-blindness, tubular vision, optical blurring, improved reaction time during tachistoscopic tests, and partial or complete deafness.[27] All of these effects, whether enabling or disabling, reflect a capacity (among some subjects, at least) for perceptual deautomatization and restructuring, a capacity apparent among some meditators, artists, and creative athletes. In making this observation, I am aware that hypnosis researchers are still debating the nature of hypnotic illusions and hallucinations. There is good evidence, for example, that hypnotically induced color-blindness does not mimic congenital color-blindness; and Nicholas Spanos has provided evidence that casts doubt upon the authenticity of some hallucinations.[28] On the other hand, some hypnotized subjects do seem to have perceptual distortions that cannot be attributed to inattention or pretense. David Spiegel of Stanford University, for example, found that evoked cortical potentials are significantly altered by hypnotically induced hallucinations.[29]

Posthypnotic Effects

Posthypnotic effects, which are typically experienced by subjects amnesic to the suggestions that produced them, provide further evidence that significant behaviors, feelings, and cognitions can be shaped by human influences that operate outside ordinary waking consciousness. Because many of our impulses and inhibitions originate in the largely forgotten admonitions of parents, teachers, and other authorities, we can learn something about their production and removal from experiments with this class of hypnotic phenomena. In such experiments, hypnotized subjects are usually instructed to behave in a particular way without remembering their hypnotic instruction. A subject acting under such compulsion might perform mental or physical tasks without realizing he had been instructed to do so—in response to a certain signal perhaps, or spontaneously after a designated length of time. Or he might make subconscious calculations while awake, with no awareness that (or why) he is doing so. Edmund Gurney, for example, reported a number of experiments in which his awakened subjects automatically worked arithmetical sums while reading aloud, or wrote a second line to a couplet rhyming with a first line given during hypnotic trance. One of Gurney's subjects even multiplied complex numbers while repeating "God Save the Queen" aloud with every other word left out.[30]

Gurney described two experiments in which his subjects were instructed to perform a particular act at a given hour or after a designated number of minutes or days had passed. One subject was told that 39 days later,

at 9:30 P.M., he was to come and call on a gentleman resident in the house where I was lodging, with whom he had no acquaintance. He of course had no memory of this direction when awake. No reference was made to the command till March 19th, when he was suddenly asked, in the trance, how many days had elapsed since it was given. He *instantly* said sixteen, and added that there were twenty-three more to run, and that the day when he was due was Easter Monday. All these statements were correct. But the odd thing was that, on further questioning, he misdated both the day of the order and the day of fulfillment, respectively March 3rd and April 11th. This makes it tolerably clear that he did not originally arrive at the date of fulfillment by immediate reckoning from the date of command, and then fix it in his mind simply as a date. (Easter Monday, when so near as twenty-three days, might be arrived at in a moment by remarking the day of the week.) Moreover, if he made March 1st his *terminus a quo*, he ought to have said eighteen instead of sixteen, and would probably have had to pause to reckon. The reasonable interpretation of the result is surely that he was in some way actually counting the days as they passed.

In the next case, which occurred after the above remarks were written, I got an actual account of the process which singularly confirms them. P__LL was told, on March 26th, that on the one hundred and twenty-third day from then he was to put a blank sheet of paper in an envelope, and send it to a friend of mine whose name and residence he knew, but whom he had never seen. The subject was not referred to again till April 18th, when he was hypnotized and asked if he remembered anything in connection with this gentleman. He at once repeated the order, and said, "This is the twenty-third day; a hundred more."

s . "How do you know? Have you noted each day?"

P__LL . "No; it seemed natural."

s . "Have you thought of it often?"

P__LL . "It generally strikes me in the morning, early. Something seems to say to me, 'You've got to count.'"

s . "Does that happen every day?"

P__LL . "No, not every day—perhaps more like every other day. It goes from my mind; I never think of it during the day. I only know it's got to be done."

Questioned further, he made it clear that the interval between these impressions was never long enough to be doubtful. He "may not think of it for two or three days; then something seems to tell him." He was questioned again on April 20th, and at once said, "That's going on all right—twenty-five days"; and added, "Twenty-seven days." After he was woke on April 18th, I asked him if he knew the gentleman in question, or had been thinking about him. He was clearly surprised at the question, said he fancied he had once seen him in my room (which, however, was not the case), and that the idea of him had never since crossed his mind.[31]

Some subjects apparently count days as they pass (without waking awareness that they are doing so) in order to fulfill posthypnotic commands, but others employ a more mysterious means of estimating the passage of time. The British psychiatrist Milne Bramwell, for example, instructed some of his subjects to inscribe a cross on a piece of paper after various periods given in minutes. In one experiment a subject, Miss A., was told to inscribe a cross 24 hours and 2,880 minutes later, which she did correctly while amnesic to the hypnotic command. In subsequent experiments she did so after 4,417 minutes, 11,470 minutes, 10,070 minutes, and other equally demanding intervals. When questioned during hypnosis about her mental processes during these extraordinary feats, she said that when Bramwell's commands were made in hypnosis she did not calculate when they fell due, that she did not calculate them at any time afterward, that she had no recollection of them when awakened, that no memory of them ever arose in her waking state, that shortly before their fulfillment she experienced her fingers moving as if to grasp a pencil and perform the act of writing, that this impulse to write was immediately followed by the idea of making a cross and a recognition of the time, and that she never looked at a clock or watch until after she had written the figures.[32] In summarizing 55 experiments with this subject, Bramwell wrote:

. . . of these one, apparently, was either not carried out by Miss A., or unrecorded by me, while in another she mistook the original suggestion, but fulfilled it correctly in accordance with what she thought it had been. Forty-five were completely successful, i.e. not only did Miss A. write down the correct terminal time, but this was done, also, at the moment the experiment fell due. Eight were partially successful. In these the terminal time was correctly recorded in every instance, but there were minute differences, never exceeding five minutes, between the patient's correct estimate of when the suggestion fell due and the moment at which she carried it out. The proportion which these errors bear to their respective intervals varies between 1 to 2,028 and 1 to 21,420.

On twenty-four occasions Miss A. was asked to calculate (in hypnosis) when the suggestions fell due; she was wrong in the first nine instances, but in the remaining fifteen right in eleven and wrong in four. As the experiments advanced, not only the frequency, but also the extent of Miss A.'s errors in calculation decreased, and the answers were given much more rapidly. Sometimes the correct replies were almost instantaneous, and in these instances no conscious calculation could be traced. It is to be noted that Miss A.'s mistaken calculations had no effect on the correctness of her results.[33]

Frederic Myers noted:

The most fundamental difference between these cases and those of Gurney's seems to be that the method of estimation of time really used by Bramwell's subjects could not be discovered by questioning during hypnosis, but appeared to belong to some stratum of consciousness more profound than the hypnotic (suggesting an analogy with the "inspiration of genius"); whereas Gurney's subjects, when hypnotized again before a suggestion was fulfilled, were conscious of keeping count of the lapse of time.[34]

Experimental results of this kind by Gurney, Bramwell, Pierre Janet, and others led William James to write:

I cannot but think that the most important step forward that has occurred in psychology since I have been a student of that science is the discovery, first made in 1886, that, in certain subjects at least, there is not only the consciousness of the ordinary field, with its usual centre and margin, but an addition thereto in the shape of a set of memories, thoughts, and feelings which are extra-marginal and outside the primary consciousness altogether, but yet must be classed as conscious facts of some sort, able to reveal their presence by unmistakable signs.[35]

Studies of posthypnotic suggestion have been refined and more carefully controlled since the days of Gurney and Bramwell—for example, by Wesley Wells (1940), who concluded that both retention and amnesia for the commands given in his experiments were 100 percent complete, and that effective hypnotic suggestions endured for a year or more; by Griffith Edwards (1963), who found that one of his subjects retained a posthypnotic suggestion for 405 days; and by other researchers.[36]

It is important for this discussion to note that constraints as well as unusual behaviors can be induced by posthypnotic suggestion. A subject may be told that he *cannot* perform a certain physical act when his trance is lifted, or that he *cannot* make a particular calculation. Such inhibitions also operate in our everyday life, through direct suggestions or general cultural contraints we are not fully aware of. Some athletes have said, for example, that hypnosis revealed unconscious ideas that they could not achieve a particular standard of excellence; and many people in psychotherapy have loosened the hold of subconscious inhibitions induced by forgotten traumas or conditionings of which they were unaware. Given the power of posthypnotic suggestion in the relatively cool and inconsequential circumstances of laboratory experiments, it is no wonder that many transformative disciplines include the examination of our unconscious compulsions.[†]

But posthypnotic phenomena also demonstrate creative aspects of human volition. The persistence with which some of Gurney's and Bramwell's subjects subconsciously fulfilled their instructions resemble the tenacity of dedicated athletes, contemplatives, and other long-suffering participants in transformative activity, though the aim of their behavior was trivial. If nothing else, their compulsivity dramatizes our human predisposition for sustained and purposeful behavior, a predisposition that can be harnessed for ends great or small,

---

[†] Psychologist Charles Tart has compared hypnosis with the socialization process, arguing that enculturation has many features in common with the hypnotic inductions and behaviors we observe in the laboratory. Body movements are shaped to fit culture's norms, for example, in a way that resembles a hypnotist's induction of automatized movements in his subjects; while certain attitudes, fantasies, and aspirations are reinforced by suggestions repeatedly given from infancy by parents and teachers who act as "cultural hypnotists." A culture's power to induce hypnosis, however, is more pervasive, long-lasting, and effective than a single hypnotist's, Tart has argued, for it is not limited to a few sessions and may use innumerable punishments and rewards to strengthen its effects. Its subjects, for example, begin their inductions when they are too young to resist them. Nearly all members of a given culture share a "consensus trance," which is virtually unbreakable, so that they are continually rehypnotized by most of the people they encounter. See: Tart 1986.

creative or destructive. The fact that Bramwell's subject Miss A. kept calculating intervals of several thousand minutes during 55 separate experiments, with a power that Myers compared to the inspiration of genius, gives evidence of extraordinary will as well as cognition.

In Part Three I propose that volition can be trained with methods such as self-hypnosis that engage both our conscious and subliminal readiness for purposeful action. Physicians and contemplatives who have known that most people function in a dissociated way, with divided intentions that inhibit the enjoyment of life and the realization of metanormal capacities, have employed such methods for millennia.

[15.3]
HYPNOTIC INDUCTION

Various methods of hypnotic induction have been employed since Mesmer's day, often with great success until new theories or fashions lead to their rejection. Mesmer used the *baquet* to induce his convulsive crises; the Marquis de Puysegur roped his peasants to trees in order to elicit somnambulistic states; Elliotson and Esdaile made mesmeric passes to produce anesthesia for surgery; James Braid had his subjects maintain a "fixed upward and inward gaze" until they were hypnotized; Liebeault stared into his patients' eyes while he told them their symptoms were relieved; and Charcot sometimes slapped hysterics on the back, without any verbal suggestions, to produce temporary arm paralyses. Each of these prominent contributors to the field had his own ways of inducing hypnotic states. Today's theories and fashions, however, incline most hypnotists to use a conversational manner, relaxation, and calming images to induce a sleeplike trance.

But history teaches us that calming instructions are not the only way to induce a hypnotic condition. Arguing that most people today are unnecessarily biased by the notion that hypnosis resembles sleep, some contemporary researchers have induced hypnosis through strenuous activity. Arnold Ludwig and William Lyle, for example, induced hypnotic effects through spinning movements, knee bends, and verbal instructions designed to increase alertness, sensorimotor stimulation, and cognitive activity. The two researchers reported that their subjects were more suggestible, spontaneous, and active in hyperalert trance than they were after ordinary (relaxing) procedures.[37] Others have found that subjects hypnotized with suggestions to enhance the richness and intensity of their imagery exhibit more suggestibility than they do after sleeplike inductions;[38] that subjects hypnotized while pedaling a stationary bike exhibit approximately the same suggestibility as people hypnotized in the usual manner;[39] and that pedaling while listening to an induction tape produces more suggestibility than using the tape alone.[40] And Etzel Cardena discovered that somewhat different qualities of hypnotic experience are produced by inductions involving active pedaling, sitting on a machine that pedals automatically, or sitting in a chair.[41]

In short, several studies have shown that hypnotic experience can be produced by inductions involving intense physical and mental activity. Given the playing trance of certain athletes, the meditative mindsets of some martial artists, and the hypnosislike experience produced by religious drumming or whirling, this discovery should not surprise us. The success of active hypnotic inductions supports the general proposal in this book that consciousness, behavior, and physiological processes can be deliberately enhanced during physical activity as well as rest, in movement as well as immobile concentration.

Psychologist Theodore Barber, too, has questioned the cultural biases involved in contemporary notions of hypnotic induction, arguing that inductions are not necessary for suggestions to be successful. However, many experiments have shown that for most subjects inductions do indeed increase responsiveness to a hypnotist's commands. Though some gifted subjects respond well to waking suggestion, the responsiveness of many or perhaps most people is increased by hypnotic procedures.[†]

[15.4]
THE HYPNOTIC STATE

Whether hypnosis involves a distinctive state of consciousness has been debated for many years. For example, by classifying hypnosis as a form of role-enactment, psychologist Theodore Sarbin has frequently given the impression that there is something deceptive about the behavior of hypnotic subjects (though Sarbin has emphasized that hyp-

---

[†] Evans, F., 1963; and Weitzenhoffer & Sjoberg 1961. Charles Tart has compared responsive subjects to highly trained runners who can start workouts at faster speeds than people who are out of shape. Ordinary subjects, like runners who are not well-trained, Tart suggests, need the warm-up of induction procedures before they can approach a gifted subject's ability.

notic behavior is not sham and that much of it goes beyond a desire to please the hypnotist).[42] Psychologist Nicholas Spanos has argued that research on trance logic, hypnotic amnesia, and analgesia does not provide compelling evidence for a state theory of hypnosis.[43] And Theodore Barber has claimed that hypnotic trance is not fundamentally different from ordinary consciousness, frequently citing experimental evidence that suggestions without hypnotic inductions are just as effective as the same suggestions with them.[44] Ernest Hilgard and other researchers, however, have criticized studies such as those Barber cites because they indiscriminately combine highly hypnotizable and nonhypnotizable persons in their subject groups. In such experiments, nonhypnotizable subjects typically respond as well to waking suggestions as they do to hypnotic commands, but the hypnotizable subjects tend to drift into hypnosis during waking suggestions so that the statistical results that ensue are misleading.

Debates as to whether hypnosis is a state have also been criticized for being largely semantic. According to Hilgard, such arguments

> are largely irrelevant to the investigator, who is little likely to explain his findings by attributing them to a special "state" as causal. I have elsewhere used the analogy of sleep and dreams: the fact that dreams usually occur within sleep does not imply that [they] are to be satisfactorily explained by saying that sleep causes them. I prefer to define a *domain* of hypnosis, which then becomes a field of investigation.
>
> How can we define the hypnotic domain? A behavioral definition of hypnosis is best arrived at by pointing to the kinds of behavior shown by subjects highly responsive to hypnotic-like procedures, [that is] arm rigidities, loss of voluntary control, direct muscular responsiveness, hallucinations, analgesia, amnesia, response to post-hypnotic suggestion. It does not seem to matter very much what kind of induction is used, or no induction at all: these behaviors still cluster together, with the most hypnotically responsive yielding more, and the less responsive fewer of them.
>
> I can get along without a state concept, if it proves too troublesome, still investigating the domain of hypnotic-like behaviors. But for the most part that would be, for me, a mere circumlocution; even if state is merely a metaphor, it is a convenient one. But it is somewhat more than this; Orne, in a set of ingenious experiments has shown consistently that those truly hypnotized differ in their experience from those who simulate hypnosis.[45]

If hypnosis is not a distinctive state, as Barber has argued, but merely a type of role-enactment, people should be able to simulate it.

Martin Orne and his colleagues, however, in the experiments that Hilgard referred to above, found that truly hypnotized subjects experienced certain emotions, sensations, and cognitions that were not experienced by controls who tried to imitate hypnosis.[46] Though controls in such experiments are subject to different demands than subjects (Orne has called them simulators or quasi-controls), these studies indicate that the hypnotic state is something more than a metaphor.[47] Indeed, there is reason to believe that hypnosis involves more than one type of altered state. Nineteenth-century mesmerists such as Carl Kluge and Justinus Kerner (see 15.1), for example, claimed that magnetic sleep provided access to various dimensions of consciousness; and some studies in recent years have indicated that there are several distinct kinds of hypnotic state. I discuss this possibility in the following pages.

But it is not necessary to choose between state and role theories of hypnosis in either-or fashion. Psychologists associated with the state position have acknowledged the interpersonal and cultural factors shaping hypnotic behavior; researchers who have emphasized the sociocultural elements of hypnosis believe that state theories contribute to our understanding of hypnotized subjects.[48] Hypnosis, like all behavior and consciousness-altering activity, involves cultural conditioning as well as biologically rooted, state-dependent elements of human nature.

[15.5]
HYPNOTIC DEPTH

Psychologist Charles Tart has explored different levels and types of hypnotic state. In a study of very deep trance, for example, he drew upon hypnosis research, dynamic psychiatry, and religious studies to interpret the experience of a highly hypnotizable subject who had trance experiences that in some respects resembled contemplative illumination.[49] Tart's subject, William, described experiences of profoundly altered identity and "oneness with the universe."[50]

Inspired in part by Tart's study, other researchers have studied deep trance. Using electroencephalography and subject report, psychologist Spencer Sherman found that his subjects experienced feelings of "oneness with everything," episodes of profound mental quiet or emptiness, apprehensions of different reality levels, and sensations of great brightness. The profoundest hypnotic states produced in

Sherman's study were correlated with drastic reductions in EEG amplitude. Psychologist Brian Feldman conducted a similar study, in which the participants reported feelings of oneness with the environment, as well as a sense of awe and wonder. And using active induction methods with 12 subjects, psychologist Etzel Cardena induced feelings of oneness and merging with light. Furthermore, Cardena found that cluster analyses of phenomena reported at different levels of trance suggested that hypnosis includes several distinct modes of experiencing. The hypnotic state, Cardena suggested, is really a combination of states, or modalities, of experience.[51]

These findings by Tart, Sherman, Feldman, and Cardena, like the dicoveries of Kluge, Kerner, and other mesmerists, correspond with testimony in the shamanic and contemplative traditions that consciousness has many dimensions beyond ordinary waking awareness.

[15.6]
HYPNOTIC SUSCEPTIBILITY

The Assessment of Hypnotizability

Development of the Stanford Hypnotic Susceptibility Scales by Ernest Hilgard, A. M. Weitzenhoffer, and their colleagues in the late 1950s and early 1960s provided a strong catalyst for hypnosis research in the decades that followed. These tests, and others like them, enable researchers to assess the responsiveness of subjects in hypnotic experiments, to establish experimental and control groups matched for hypnotizability, to find nonsusceptible subjects who might simulate hypnosis, and to study the nature of susceptibility itself. They ordinarily provide specific suggestions for a number of hypnotic experiences (such as an extended arm getting heavy), which are scored in terms of behavioral criteria (for example, that an extended arm must drop 12 inches or more in less than 30 seconds). A subject's degree of susceptibility, or responsiveness, is scaled according to the number of hypnotic behaviors he or she exhibits and the number of typical hypnotic experiences he or she reports. Different versions of a given scale make test-retest studies possible, and similar psychometric instruments such as the Stanford Profile Scales of Hypnotic Susceptibility permit the assessment of individual strengths and weaknesses within the general domain of hypnosis—testing, for example, whether a subject can experience hallucinations, hypnotic dreams and regressions, amnesia, posthypnotic effects, or loss of motor control. Another assessment device, the Barber Suggestibility Scale, tests for hypnoticlike behaviors produced by suggestion without hypnotic inductions. The reliability and validity of scores produced by these various instruments have been established through decades of testing.

Studies using such scales have shown that hypnotizability varies greatly from person to person but that it remains relatively stable over time in a given subject. Studies at Stanford University have shown that the susceptibility scores of subjects tested when they were undergraduates were highly correlated (0.60) with the same subjects' scores 10 years later. Over short periods such correlations are higher.[52]

Correlates of Susceptibility

Josephine Hilgard, Theodore Barber, Patricia Bowers, and others have shown that highly responsive subjects have considerable capacity for imaginative absorption in daydreaming, movies, storytelling, acting, and reading fantasy literature. These hypnotic subjects typically begin to develop their imaginative proclivities in childhood.[53] Bowers has gathered evidence that hypnotizable subjects share several attributes with highly creative people, among them ease in experiencing mental images, ability to complete the gestalt of incomplete pictures, preference for tasks that stimulate fantasy, and ability to "regress in the service of the ego."[54] In making these comparisons, Bowers drew upon studies by Frank Barron, Donald MacKinnon, and their colleagues at the University of California's Institute of Personality Assessment and Research which suggest that eminent writers and architects have greater awareness of inner experience, less resort to repressive defenses, and greater independence of judgment than less creative people.[55] Bowers wrote:

> Certain types of creative people are able to allow the structure of a problem to affect associational processes without interposing conscious, volitional strategies. When the task is one requiring creative solutions such as writing portions of a novel, the person may experience solutions as simply "happening" or coming from "out of the blue," because the writer has not consciously determined the course of the solution.[56]

Someone who can make this focused surrender in hypnosis, Bowers suggested, can do so in creative tasks. Studies by Helen Crawford which show that hypnotizable subjects do better on gestalt closure tasks than nonsusceptible subjects support Bowers's suggestion. Crawford has argued that the imaginative involvement, capacity for absorption, and skill at gestalt closure that various researchers have found among highly susceptible subjects comprise a synthetic or holistic style of thinking that is central to creativity in general.[57]

Some studies suggest that hypnotizability is related to the hemispheric specialization of brain functioning. Ruben and Raquel Gur, for example, studied lateral eye movements that characterize styles of hemispheric functioning. Their experimental findings, wrote the Gurs, indicate that the ability to become hypnotized is mediated by the "nonverbal, holistic, synthetic, or 'appositional' hemisphere."[58] Further evidence for such a connection was provided by Kenneth Graham and Kevin Pernicano, who showed that hypnosis stimulated an illusion called the autokinetic effect, during which a stationary point of light located in dark surroundings appears to move. Subjects in the Graham-Pernicano experiment perceived the light moving to the left about 33 percent of the time when they were not hypnotized, and about 54 percent of the time when they were. These results, the experimenters wrote, support "a laterality theory . . . which views hypnosis in terms of a relative shift of cognitive functioning from the dominant (left) to the nondominant (right) hemisphere of the brain."[59] Supporting this supposition, studies by Harold Sackeim and associates showed that hypnotizable people tend to sit on the right side of classrooms.[60] In short, three sets of experimental results with hypnotizable subjects, involving lateral eye movements, the autokinetic effect, and patterns of classroom seating, all seem to reveal a tendency among hypnotizable people to process information in the nonverbal hemisphere of the brain.[61]

Styles of attention have also been correlated with hypnotic susceptibility. A. Tellegen and G. Atkinson, for example, reported a consistent relation between measures of hypnotizability and paper-and-pencil measures of capacity for absorption in the task at hand.[62] Charles Graham showed that highly susceptible subjects experience a marked narrowing of attention in the visual field when they enter hypnosis, but that nonsusceptible subjects continue to perceive their surroundings as they normally do after the same induction.[63] And David Van Nuys found a significant negative correlation ($-.42$) between the number of intrusive thoughts reported by subjects during a meditation exercise and their degree of hypnotizability.[64] These experiments on attentional style suggest that hypnotic susceptibility involves a capacity for unusually deep absorption.

With that possibility in mind, Charles Graham and Frederick Evans designed an experiment in which hypnotizable subjects were asked to generate random numbers. They should do this better than less responsive subjects, the two researchers believed, not only because they could focus their attention better but also because they could achieve the cognitive deautomatization such a task requires in the manner of creative people who surrender to fantasy and temporarily relinquish reality testing. Graham and Evans designed their experiment so that their subjects would feel the pressure of time, requiring them to produce a number from 1 to 10 each second for 100 seconds. The difficulty inherent in this task seems to arise from two sources. On a cognitive level,

> the individual [must] monitor his previous output in order to equalize probabilities and select alternatives under the pressure of producing a response every second for 100 seconds. On a different level, the task is somewhat analogous to the technique of free association, which also appears deceptively simple but can be very difficult for some individuals. The difficulty . . . seems to arise because being random is the antithesis of what we are taught to do throughout life. From birth, it seems, each of us is expected to become progressively more organized and predictable in our behavior, especially in our patterns of thought. To be successful at [this] task, however, would seem to require just the reverse: the controlled non-patterning of thought. It may also involve elements of memory distortion and imagery. Each of these aspects [of cognition] has also been proposed as common to the hypnotic situation.[65]

The Graham-Evans hypothesis was confirmed, as those subjects who could most successfully list numbers in a random sequence also tended to be more hypnotizable than other subjects in the experiment. Some of these more hypnotizable subjects used imagery (such as a row of ten lights, one of which brightened every second); while others disconnected their cognitive and verbal capacities, simply letting "the numbers pop out." Citing Arthur Deikman's proposal that deautomatization is central to meditation experience and various altered states, Graham and Evans suggested that success in their experiment required a capacity "to circumvent an established, automatized, cognitive organizational pattern." Their findings, they wrote, appeared to support Ernest Hilgard's claim that highly hypnotizable individuals are not afraid to relinquish reality testing or become deeply involved in subjective experiences.[66]

A study of selective attention reported by Robert Karlin supports Graham and Evans's results. Subjects in Karlin's experiment listened to two different tape recordings played simultaneously through a single sound source and were asked to rate their difficulty in understanding one of the two and to indicate what they could remember of it. A significant correlation was found between the subjects' success at remembering items in the target recordings and their degree of hypnotic susceptibility. "Taken together with the Graham and Evans (1977) study," Karlin wrote, "[these results suggest] that highly hypnotizable subjects are able to distribute attention on both perceptual and cognitive tasks in ways not available to less hypnotizable subjects."[67]

The attentional skill that the Graham-Evans and Karlin experiments reveal among hypnotizable subjects, which includes capacities for both sustained concentration and deautomatization of habitual cognitive patterns, resembles the deeply focused yet spontaneous responsivenesss exhibited by gifted athletes, dancers, martial artists, shamans, and Zen masters.[68] This combination of capacities, it seems to me, is a fundamental feature of extraordinary functioning and can be cultivated to facilitate the integral development I propose in this book. Correlates of hypnotizability such as perceptual flexibility, tolerance of ambiguity, capacity to relinquish reality testing, openness to strange experience, vivid and spontaneous imagery, and capacity for absorption in the task at hand, which are exhibited by creative people in many fields, can be developed to some extent by somatic disciplines (18), psychotherapy (17.1, 17.2), meditation (23.2), and other practices.

Fantasy-Prone Personality

Research psychologists Theodore Barber and Sheryl Wilson identified 27 women who were excellent hypnotic subjects and compared them to 25 women who were less hypnotically susceptible. The two psychologists found that with one exception the excellent hypnotic subjects had a profound fantasy life, that their fantasies were often hallucinatory, and that their involvement in fantasy contributed to their superb hypnotic performance. Apparently, these 26 subjects had developed their fantasy lives since childhood. Wilson and Barber wrote:

> Those who had played with dolls and toy animals believed that these objects were alive, had feelings, and had unique personalities. Consequently, they consistently treated their dolls and stuffed animals with

respect and consideration. For instance, a number of the fantasizers informed us that they slept each night with a different doll or stuffed animal so that none would have hurt feelings. Also, when they were not playing with them, they would leave them in a comfortable position such as lying in bed or sitting in a chair so that they could look out the window. If they were at all careless with or inadvertently "hurt" a doll or stuffed animal they would apologize. They also felt sorry for their dolls and toy animals when they left them home alone because they thought they would feel lonely.

When they were children, almost all of the fantasizers believed in fairies, leprechauns, elves, guardian angels, and other such things. The strength of these beliefs is illustrated by one subject's statement that, when she was finally convinced that Santa Claus was a fabrication, she could not understand why adults had tried to "make up" such a person when there were so many real beings around such as tree spirits and fairies. The strength of the fantasy-prone subjects' beliefs in such beings probably originates in their conviction that they have seen, heard, or even played with them. Many saw no more than "little legs disappearing around corners." However, for some, encounters with such beings were vivid and "as real as real"; for instance, one told us how as a child she would spend hours watching in fascination the little people who lived in her grandmother's cactus garden that adults kept insisting were not there. With few exceptions, their belief in elves, leprechauns, fairies, guardian angels, tree spirits, and other such creatures did not terminate during childhood; as adults, they either still believe in them or are not absolutely sure they really do not exist.

A small number of the fantasizers had either God or Jesus as an imaginary companion. Since these subjects were from broken homes or were abused children, their personal relationship with God or Jesus can be viewed as serving important needs. They had created an ideal companion who gave them unconditional love, approval, support, and advice in a world that otherwise was unstable, unsupportive, critical, and cruel.

With only two exceptions, the fantasy-prone subjects would pretend during childhood that they were someone else such as an orphan, a princess, an animal, a bird, or a fairytale character such as Cinderella or Snow White. They became so absorbed in these roles that they felt they actually became the character they were pretending to be. Their pretending often extended well beyond any delimited play period and continued into their daily life. For instance, one subject felt, not that she was pretending to be a bird, but that she was a bird pretending to be a girl. Similarly, another subject who continually pretended to be a princess told us that she actually felt that she was a princess pretending to be an ordinary child doing things ordinary children do, such as going to school, riding a bike, and so on. She saw her house as a castle complete with a real moat and drawbridge, and she told the children at her school that she was a princess and that she lived in a castle with a moat.[69]

The controls in this study had not experienced fantasies with nearly the same intensity as the fantasizers. Their imagery was typically less vivid, more constrained by reality testing, less frequent, and less enjoyable. Wilson and Barber's fantasy-prone subjects were normal in most respects, however, and could clearly differentiate their fantasies from ordinary reality. All of them, it seems, had learned to be secretive about their imaginative activities. The "princess" described above, for example,

> was accused of lying by her school friends when she brought them to see her fantasized castle (actually her middle-class home). She was shocked that they could not see the castle since it was real to her. Her secretiveness derives from that moment; henceforth, she never tried to share her fantasies and pretending with friends.[70]

According to Wilson and Barber, the following four factors contributed to their subjects' fantasy proneness:

> 1. At least 70% of the fantasizers recall being encouraged to fantasize by a significant adult [who] encouraged them to fantasize by one or more of the following means: (a) the adult read to the child or told the child fairytales or fantasy stories; (b) the adult praised the child for her make-believe and her fantasies; and (c) the adult treated the child's dolls and stuffed animals as if they were alive, thus encouraging the child to believe they were alive.
>
> 2. Sixteen of the fantasizers, in contrast to only one subject in the comparison group, perceived themselves as having been very lonely and isolated as children. These fantasizers felt they became deeply involved in fantasy to overcome their isolation and saw it as providing companionship and entertainment.
>
> 3. Nine of the fantasizers (and none of the subjects in the comparison group) reported that they had had a difficult or stressful early life. Their reports include: (a) serious physical abuse from a parent, a foster parent, or an older sibling (at times necessitating medical treatment), (b) a mother who had severe emotional problems, (c) a mother who deserted the family, (d) unstable living conditions such as living with various relatives and at several foster homes, and (e) various combinations or all of the above. All nine of the subjects in this group told us that they used fantasy to escape from their surroundings. Surprisingly, most of these had a secret hiding place (e.g., in a nearby

meadow or behind the sofa) where they would habitually "hide" in order to be undisturbed while they lived an entirely different life in fantasy.

4. At least nine fantasy-prone subjects had a special life situation that contributed to an extreme involvement in fantasy. By the time they were 2, 3, or 4 years of age, all nine subjects had begun intensive studies in piano, ballet, dramatic acting, or art; and six of the nine had begun studying two or more of these areas intensively. They utilized fantasy when carrying out these activities.[71]

As adults these women spent a large part of their waking hours exercising their fantasies—for example, while engaged in conversations that bored them, during routine tasks, during free time set aside for such activity, or while falling asleep. Their fantasies were often so vivid that imaginary smells, tastes, sounds, and tactile sensations had the immediacy and intensity of sensations caused by external stimuli. Seventy-five percent of these subjects said they had experienced orgasms produced solely by sexual fantasies, without physical stimulation; and 65 percent claimed they had experienced fantasies of hallucinatory intensity.[72]

Twenty-one of 22 fantasizers who were asked, in sharp contrast to just one of 11 control subjects, said they had been acutely aware of hallucinatory experiences since they were children because they found them enjoyable. Their memories of tactile sensations, smells, and sounds were more vivid and intense than those of the controls, which suggested to Wilson and Barber that the fantasizers' attention was generally more acute, not only to imagery but to external stimuli.[73] This attentional acuity, conceivably, made the memories of these fantasy-prone subjects especially vivid. The researchers wrote:

> Of the 26 fantasizers, 24 (and only 3 subjects in the comparison group) report many vivid memories of events that occurred prior to their third birthday. Of these, 8 report clear memories of events that occurred on or before their first birthday.
>
> When they try to remember their very early life, they appear to reexperience the thoughts, emotions, and feelings, including the feeling of being in their baby body, in the same way they did originally. For example, when one of our subjects was describing her memory of her first birthday, she vividly reexperienced her father singing "Happy Birthday," her mother standing behind her encouraging her to try to blow out the candle, and her sister sitting next to her. She could feel herself sitting in her high chair and could feel herself in her body as it was

at that time ("with my big, fat tummy"). Immediately following this vivid recall, she cried out to us, "Life was wonderful back then."

In addition to developing vivid memories for personal experiences, two of the fantasizers, apparently as a result of unusual circumstances in their early lives, also developed superior auditory memories for spoken material. By age 4 one of the fantasy-prone subjects appeared to read all her children's books perfectly out loud from cover to cover. However, she could not read at all. She would look at each page of the book and rehearse word for word the voice of the adult who originally read it to her. She retained this auditory eidetic imagery throughout her life and found it very useful.[74]

Twenty-five of these subjects got sick whenever they witnessed violence in television or movies. Nineteen reported physical symptoms related to prominent fantasies or memories, as for example when one thought (mistakenly) that she had eaten spoiled food. Thirteen of 22 who were asked said that they had experienced a false pregnancy with one or more of the following symptoms: loss of menstruation, enlargement of breasts, abdominal swelling, morning sickness, cravings for particular foods, and the impression of fetal movements. Two even went for abortions because their symptoms were so convincing. Such responsiveness led Barber and Wilson to suggest that these women possessed a remarkable psychosomatic plasticity by which their bodies conformed to their strongly held images.[75]

Ninety-two percent of these exceptional subjects claimed that their fantasies sometimes accompanied or triggered telepathy, clairvoyance, precognition, religious vision, or spiritual healing. Eighty-eight percent of them reported out-of-body states, compared to 8 percent of the control group, two of them during near-death experiences. And 50 percent had felt that "someone or something was using them to write a poem, song, or message."[76] In accounting for the particular combination of extraordinary capacities that their subjects possessed, Wilson and Barber hypothesized

that vivid sensory experiences, vivid memories, and vivid fantasies are causally interrelated as follows: individuals who focus on and vividly feel their sensory experiences have relatively vivid memories of their experiences; and individuals with vivid memories of their experiences are able to have relatively vivid fantasies because they can use their vivid memories as raw material from which they can creatively construct their fantasies.

Although the above hypotheses appear difficult to test, the relationship we have noted between sensory experiences, memories, and fantasies leads to predictions that are easily testable. For instance, we should be able to predict the vividness of individuals' fantasies by assessing the vividness of their sensory experiences or by asking them to recall specific personal events, for instance, their first day in school, the first time they smoked a cigarette, or the first time they became inebriated. The degree to which they have vivid sensory experiences and the degree to which they recall and reexperience earlier personal events vividly in all sense modalities should predict the vividness (hallucinatory qualities) of their fantasies (and correlated variables such as hypnotizability and psychic abilities).[77]

Barber and Wilson proposed that they had found a personality type which had not been described before, though some of their subjects' exceptional capacities had been studied separately. Hallucinatory ability in normal subjects, for example, had been noted in the nineteenth century by Francis Galton.[78] E. R. Jaensch had studied people with unusual powers of visualization whom he called *eidetikers*.[79] Vogt and Sultan had discussed the extraordinary visualizations of Nikola Tesla, who could construct intricate machines in his mind with photographic detail.[80] And the Russian psychologist A. R. Luria had written a classic study, *The Mind of a Mnemonist*, about a man whose images were so vivid they seemed to be real.[81] But these earlier studies, Wilson and Barber suggested, did not adequately relate the various aspects of fantasy-prone personality.

> To the best of our knowledge, the syndrome that we have uncovered, which includes involvement in fantasy, hallucinatory ability, vivid memories, hynotizability, and psychic abilities, has not been delineated previously as a unitary entity. However, each separate facet of the syndrome has been studied separately, and, at times, two or more facets have been interrelated.[82]

To explore the Wilson-Barber thesis, psychologists Steven Lynn and Judith Rhue compared 156 subjects with exceptional fantasy involvement to people with less active imaginations.

> General support was secured for Wilson and Barber's construct of fantasy-proneness. Fantasizers were found to differ from nonfantasizers and in many cases from medium-range subjects, on measures of

hypnotizability, imagination, waking suggestibility, hallucinatory ability, creativity, psychopathology, and childhood experiences.

> [However] individuals at the extreme end of the continuum of
> fantasy-proneness do not necessarily conform to a unitary personality
> type. Although we launched our research with the intent of studying a
> particular type of person, as we progressed, we were increasingly im-
> pressed by the diversity exhibited by the fantasizers we studied. . . .
> Fantasizers vary widely in their hypnotizability level, hallucinatory
> ability, developmental history, and psychological adjustment. Whereas
> some fantasizers are highly hypnotizable, others are not; whereas some
> fantasizers can hallucinate an object "as real as real," others are not
> able to generate convincing hallucinations; whereas some fantasizers
> report being abused as children, others paint a rosy picture of their
> childhood family environments; and whereas some fantasizers appear
> to be exemplars of healthy adjustment, other fantasizers' adjustment
> can be described as only marginal at best.[83]

Though Lynn and Rhue did not find a personality type so clearly
marked as Wilson and Barber had suggested, their findings indicate
that some people do indeed experience vivid, sometimes hallucinatory imagery on a regular basis. Taken as a whole, the investigations of
fantasy-proneness by Wilson, Barber, Lynn, and Rhue, like studies of
imaginative involvment by Josephine Hilgard and other hypnosis researchers, show that some people develop strong fantasy lives through
a lifetime of imagery practice. Whether or not there exists a definite
class or "unitary entity" of fantasy-proneness, the lifelong cultivation
of imagery reported by subjects in the studies just noted appears to
produce powerful imagination and memory, intense concentration,
openness to unusual experience, sensory acuity, and strong somatic
responsiveness to mental imagery. All of these capacities can facilitate
extraordinary functioning in general.

The Development of Hypnotic Responsiveness

A number of studies show that hypnotizability can be developed to
some degree. Michael Diamond reviewed several experiments in
which subjects became more hypnotically responsive through music,
silence, psychedelic drugs, biofeedback, sensory deprivation, a personal development program at Esalen Institute, training in hypnotic
behavior, operant conditioning, or the practice of relaxation. "Train-

ing in psychophysiological control, focused thinking, and imaginative present-centered involvement," Diamond wrote, "all appear to increase the ability to absorb oneself in hypnotic suggestions."[84] A subsequent review of hypnosis research by psychologist I. Wicramasekera supported Diamond's general proposals.[85] Indeed, during some of the studies Diamond reviewed, certain subjects developed capacities related to hypnotic responsiveness in a short period of time.[86] If people with limited training have altered some of their basic abilities during such experiments, it is no wonder that religious and other life-encompassing practices produce dramatic changes of consciousness and behavior.

[15.7]
SELF-HYPNOSIS

Besides inventing the term *hypnosis* and providing fruitful theories about hypnotic suggestion, James Braid was the first person to systematically explore self-hypnosis.[87] He used it for help with his own problems and was convinced that "patients can throw themselves into the nervous sleep, and manifest all the usual phenomena of mesmerism, through their own unaided efforts." Braid supported this contention with a series of experiments in which several subjects displayed hypnotic phenomena without the influence of a hypnotist.[88]

Since Braid's time, many physicians and psychotherapists have taught their patients self-hypnosis as a means of symptom relief, and the practice has been used as well by athletes, musicians, and others to enhance performance. Experimental studies have shown that through self-hypnosis highly susceptible subjects frequently experience more spontaneous and vivid imagery, more awareness of imagery detail, more levels of trance, and more diverse effects than they do with heterohypnosis. Most hypnotizable subjects, it seems, are responsive to self-hypnosis, though some have trouble with it. According to psychologist Lynn Johnson, who has systematically compared the two activities, self-hypnosis often promotes a spontaneous, expansive, and shifting mode of attention with unexpected awareness of internal events, while heterohypnosis by comparison produces a more tightly focused state with less imagery. In most cases, self-hypnotized subjects say that they are more creative and self-controlled than they seem in heterohypnosis.[89]

[15.8]
HYPNOTIC PAIN REDUCTION

Since the days of Mesmer's *baquet,* mesmerism and hypnosis have produced relief from suffering of many kinds. A substantial clinical literature has accumulated to show that hypnosis relieves pain caused by many kinds of affliction.[90]

Types of Pain and Emotional Distress That Hypnosis Has Relieved

In a series of articles published during the 1950s, physician Byron Butler reported the successful reduction of pain in cancer patients and gave a history of cancer treatment by hypnosis going back to 1890.[91] More recently, V. W. Cangello gave posthypnotic suggestions for pain reduction to 73 cancer patients and found that 30 of them reported excellent and 20 good results. His deeply hypnotizable patients generally experienced more relief than the others, though about half of his less susceptible subjects were also helped.[92] The results of these studies, wrote Ernest and Josephine Hilgard, "show a relationship between hypnotic responsiveness and success in pain reduction. The figure that commonly emerges—about 50 percent of the cases showing substantial improvement—is close to that reported by other clinicians."[93] Hypnosis has also been used to overcome apprehension and anxiety before surgery, assist the transition to convalescence, reduce the need for postoperative narcotics, alleviate nausea, and raise morale.[94]

Ernest and Josephine Hilgard compiled the following list of operations between 1955 and 1974 during which hypnotic pain reduction was used without chemical analgesics or anesthetics:[95]

Appendectomy	Tinterow 1960
Caesarean section	Kroger & DeLee 1957; Taugher 1958; and Tinterow 1960
Gastrostomy	Bonilla et al. 1961
Mammaplasty	Mason 1955
Breast tumor excision	Kroger 1963
Breast tissue excision	Van Dyke 1970

Skin grafting, debridement, etc.	Crasilneck et al. 1956; Tinterow 1960; and Finer & Nylen 1961
Cardiac surgery	Marmer 1959; and Tinterow 1960
Cardiac excision	Ruiz & Fernandez 1960
Fractures and dislocations	Goldie 1956; and Bernstein 1965
Cervical radium implantation	Crasilneck & Jenkins 1958
Curettage for endometritis	Taugher 1958
Vaginal hysterectomy	Tinterow 1960
Circumcision where phimosis present	Chong 1964
Prostate resection	Schwarcz 1965
Transurethral resection	Bowen 1973
Oophorectomy	Bartlett 1971
Hemorrhoidectomy	Tinterow 1960
Facial nerve repair	Crasilneck & Jenkins 1958
Thyroidectomy	Kroger 1959; Chong 1964; and Patton 1969
Ligation and stripping	Tinterow 1960
Removal of tack from child's nose	Bernstein 1965
Repair of lacerated chin in child	Bernstein 1965
Removal of fat mass from arm	Scott 1973

As you can see, even cardiac surgery and tumor extractions have been performed without drugs. Claims such as Esdaile's and Elliotson's that major surgery can be done painlessly with mesmeric sleep or hypnosis (15.1) have been confirmed again and again in recent decades.[96]

There is now a large clinical literature about hypnotic analgesia in childbirth. American physician R. V. August, for example, reported that 58 percent of 850 deliveries for which he was responsible required no medication at all, while 38 percent required mild analgesics such as Demerol, and only 4 percent (36 of the 850) a local or general anesthetic.[97] In a Soviet study, complete analgesia was produced by hypnosis in 29 percent of 501 deliveries, partial analgesia in 38 percent, doubtful results in 12 percent, and failure in 21 percent.[98] In a survey of 210 childbirths reported in the *British Medical Journal*, it was found that women taught autohypnosis suffered significantly less pain during labor than women who were taught relaxation and controlled breathing or given no training at all.[99] And in another American study, 16 of 22 women were relieved of pain during labor and delivery.[100]

Hypnosis was used to relieve pain during tooth extraction as early as 1837 but was generally supplanted by nitrous oxide and ether. It has, however, been used successfully since World War II in many kinds of dental operation.[101] Hypnosis has also been used to alleviate lower back pain, ulcer pain, phantom limb pain, intractable shoulder pain, postoperative discomfort from ophthalmic surgery, burn pain, and migraine headache.[102]

Pain Reduction as a Function of Hypnotizability

Several experiments have shown that capacity for hypnotic analgesia has about a 0.50 correlation with hypnotic responsiveness in general.[103] While nearly everyone can find some relief through the relaxation and distraction that hypnosis provides, most highly hypnotizable people experience more pain reduction through hypnosis than ordinary subjects do. There is good evidence, moreover, that hypnotic pain reduction involves more than placebo effects. In a study by McGlashan, Evans, and Orne, for example, hypnotizable subjects experienced significantly less distress with hypnotic analgesia than they did after taking a placebo that they believed to be a pain-killing drug, but nonresponsive subjects got about as much relief from the placebo as they did from hypnotic suggestion. The authors concluded that there are two components of hypnotic analgesia: first, a placebo effect; and second, "a distortion of perception specifically induced during deep hypnosis."[†] Commenting on these experimental findings, Ernest Hilgard wrote, "It is easy to see how, in the context of a practical clinical situation, results attributed to hypnosis might be produced by the placebo effect. In which case it might be concluded that hypnotic analgesia is independent of measured hypnotic responsiveness. The data . . . show how wrong such a conclusion would be."[104] Hypnotic analgesia is more than a placebo effect for most hypnotizable (as well as a few nonhypnotizable) subjects, and it is also more than a device for reducing anxiety.

[†] McGlashan et al. 1969. The nonsusceptible subjects in this study had a 0.76 correlation between their responses to hypnotic suggestion for pain reduction and a dummy drug they believed to be a powerful analgesic, while the hypnotizable subjects reported pain reduction exceeding the relief they derived from the same placebo.

Hypnotic Analgesia and Anxiety Reduction

In a study by Ronald Shor, subjects who simulated hypnosis by appearing to be relaxed were compared to a group of hypnotized subjects to see how they responded to electric shock. The simulation worked so well that judges could not tell the two groups apart, yet during the postexperimental inquiry, which emphasized honest reporting, the hypnotized subjects reported that they had felt no pain while the simulators said that they had. Relaxation by itself, while helpful in reducing discomfort, did not completely remove the pain the simulators felt.[105] Commenting on Shor's study, Ernest Hilgard wrote: "The conclusion [of this experiment] is clear: the analgesic effect of hypnotic suggestion is not to be confused with the relaxation or anxiety-reduction effect. [Relaxation] may indeed be important, as in the preparation for surgery, or during the recovery period, [yet] it is a mistake, in understanding what is happening, to equate the two effects."[106] Just as true hypnotic analgesia is more than a placebo effect, it is also more than a technique for anxiety reduction.[107]

Overt and Covert Pain

Ernest Hilgard and his associates at Stanford University, as well as other researchers, have developed methods through which subjects report pain perceived by a subliminal process that Hilgard has named the *hidden observer*. To elicit reports of concealed distress, an experimenter typically tells a subject that a part of him surveys somatic events like other regulators of bodily processes and that during hypnosis he will register pain by a prearranged signal even though he will not recall it in his waking consciousness. Under these circumstances, some hypnotizable subjects specify the degree of pain they feel even though it is out of awareness while they are hypnotized. Furthermore, these subjects commonly report that their pain is somewhat less than it would be in normal circumstances and that it does not cause significant discomfort. In subjects who report covert pain, Hilgard wrote, "it is evident that this aspect of [their] experience has been hidden by a kind of amnesic process." Such amnesia can be lifted, but differs from ordinary posthypnotic amnesia in that the deflection from memory in the overt report of pain occurs *before* pain is experienced. Amnesia usually involves the forgetting of something that was previously perceived or remembered.[108]

Conceivably, something like Hilgard's hidden observer operates for accomplished martial artists, athletes, and contemplatives. Such modulation, perhaps, enables people who deliberately embrace pain for the sake of their discipline to stop short of damaging themselves.

Pain Reduction Through Hypnosis and Waking Suggestion

Some hypnosis researchers have claimed that trance is not required for hypnotic pain reduction,[109] but Ernest Hilgard and others have criticized their arguments because they are based upon studies that indiscriminately combine highly hypnotizable and nonhypnotizable persons. In these poorly designed experiments, nonhypnotizable subjects typically respond as well to waking suggestions as they do to hypnotic ones, while hypnotizable subjects tend to drift into hypnosis during waking suggestions so that the ensuing statistical results are misleading. To get around this problem in research design, Hilgard and his associates used hypnotizable subjects as their own controls by comparing their responses to cold pressor pain (produced by placing one's hand in ice water) in three conditions: after suggestions of analgesia during the waking state, after the same suggestions in hypnotic trance, and in the waking state without such suggestions. In these responsive subjects, overt pain was indeed reduced by waking suggestion, but was reduced even more by analgesia induced by hypnotic trance.[110]

In other words, there are two components in hypnotic pain control. One involves diversion of attention, relaxation, and reduced anxiety and is available to all subjects, while the other involves an amnesialike process and is available only to highly hypnotizable subjects. These two components of hypnotic analgesia give rise to some of the disagreements between hypnosis experimenters and clinicians. Hilgard wrote:

Many practitioners of hypnotherapy believe that everyone is hypnotizable, while experimenters characteristically believe that only a small proportion of the population can be hypnotized sufficiently to yield substantial amnesia, hallucinations, and other evidences of profound hypnotic involvement. The two-component theory points out that therapeutic practices using hypnotic methods can be of benefit to nearly everyone through relaxation, anxiety-reduction, diversion of attention, and improved self-confidence, even though, in a careful assessment setting, some of those helped would be shown to be barely hypnotizable.[111]

Hypnotic Pain Reduction as Compliant Behavior

Some researchers, especially those who favor a role interpretation of hypnosis, have attributed hypnotic pain reduction to compliant behavior, arguing that a subject's behavior is largely based upon desire to please the hypnotist. However, psychiatrist David Spiegel and his associates at the Stanford University School of Medicine found that brain activity in response to electrical shock was modified in 10 hypnotizable subjects who received suggestions for pain reduction. "Highly hypnotizable subjects," Spiegel wrote, "but not those with low hypnotizability, showed changes in amplitude of event-related potentials consistent with hypnotic task demands. This study suggests that such sensory alterations are accompanied by an alteration in neuronal response to stimuli."[112]

That hypnotic analgesia involves something more than compliant behavior is evident, too, in the fact that some people control their own pain by self-suggestion without any hypnotist. Ernest Hilgard gave the following examples of self-induced pain reduction.

> One male student had an accident on a ski slope, resulting in a compound fracture of his leg. It took a long time for the rescue sled to be brought up the slopes in order to take him to the emergency hospital (but) he hypnotized himself and remained comfortable throughout, and the attendants at the hospital could not understand how someone with such a severe injury could arrive after protracted delay in an obviously relaxed and comfortable state. Another young man broke a bone in his foot when he was about to appear in a leading part for several performances in a college play requiring vigorous Mexican-style dancing. After having appropriate x-rays he discussed with his physician the possibility of permanent damage if he were to use the foot without a cast for the duration of the play. The physician agreed that the bone was not in a position in which placing stress on the foot would do any permanent damage, although putting weight on it would undoubtedly be very painful. Using what he had learned in the laboratory, he eliminated the pain hypnotically during each performance, and fulfilled all his obligations before having the foot placed in a cast until

the bone might heal. He reported only a single episode when he felt pain: one of the others in the play stepped on his foot; fortunately, he was able to recover and make it painless again. A young woman student had cut her knee seriously, the repair requiring 38 stitches. Because of an allergy to novocaine, she controlled the pain subjectively by blocking everything from her mind, concentrating on breathing, and picturing her head filled with something like foam rubber that would block sensation. This is compliant behavior, but compliant to her own demands for achieving comfort in the face of normally noxious stimulation.[113]

Physiological Mechanisms Involved in Hypnotic Pain Control

Ernest Hilgard and associates tested the possibility that hypnotic analgesia might act in the medial centers of the brain by releasing endorphins or similar substances. They did this by administering naloxone, an endorphin-blocker, to subjects given suggestions for pain reduction. Because their subjects' analgesia was not significantly affected by the drug, it seemed that hypnotic suggestion was mediated by mechanisms other than endorphins. Hilgard's results were supported by further experiments by David Spiegel and Albert Leonard in which naloxone failed to reverse the alleviation of chronic pain by hypnosis.[114] Results such as Hilgard's and Spiegel's indicate that physiological mechanisms mediating hypnotically induced pain reduction are quite complex. Indeed, some researchers doubt that all of them can ever be mapped, given the immense flexibility and redundancy of the human organism.

Hypnotic Pain Reduction and Contemplative Activity

Testimony by hypnotic subjects that they did not suffer from pain even though they "knew it was there" resembles the contemplative adage that "pain is an opinion," that it is simply information to which one can respond in various ways (5.7). In this respect, the awareness and self-control that sometimes accompanies hypnotic analgesia resembles the finely articulated consciousness and self-mastery produced by successful meditation.

[15.9]
THE RELIEF OF INJURY AND DISEASE
THROUGH HYPNOTIC SUGGESTION

Congenital Ichthyosiform Erythrodermia
(Fish-Skin Disease)

Fish-skin disease is characterized by thick, black, horny skin (often as hard as fingernails and numb for a depth of several millimeters) that cracks when bent so that it sometimes oozes a blood-stained serum. Though most medical texts once deemed it incurable, reports of its remission after hypnotic suggestion have appeared in the clinical literature. In a paper published by the *British Medical Journal* in 1952, English physician A. A. Mason described his successful treatment of the condition through suggestions that his patient's horny skin would fall off.[115] Another English doctor, C. A. S. Wink, reported similar results with two sisters, aged six and eight.[116] In 1966 C. Kidd reported a 90 percent improvement in one patient, and J. M. Schneck a 50 percent cure in another.[117]

Inspired by Mason's first report about his hypnotic treatment of fish-skin disease, Mullins, Murray, and Shapiro used the same approach with a similar congenital affliction, pachyonychia congenita, characterized by thickened skin and enlarged nails on the hands and feet. Their patient, an 11-year-old boy who needed crutches or crawled before they treated him, learned to walk with only slight impairment and stood without pain for the first time he could remember.[118]

Warts

In his book *Illustrations of the Influence of the Mind upon the Body in Health and Disease Designed to Elucidate the Action of the Imagination*, published in 1872, D. H. Tuke described a cure for warts that included rubbing the skin with beef stolen from a butcher's shop. The vivid imagery triggered by this dramatic treatment, Tuke suggested, contributed to the patient's cure.[119] A Zurich physician, B. Bloch, treated the affliction with another highly suggestive procedure, a "wart-killer" with noisy motor, flashing lights, and fake X ray. Of 179 patients, Bloch reported in 1927, 31 percent lost their warts after a single treatment with the fearsome machine.[120] Interpreting remissions produced by such procedures to be the result of strongly held images, many physicians have treated warts by hypnotic suggestion. In 1942 R. H. Rulison reviewed 921 cases, and in 1960 M. Ullman and S. Dudek noted many more in which warts were relieved or eliminated through hypnosis.[121] Reflecting upon studies such as these, essayist Lewis Thomas suggested that

> any mental apparatus that can reject a wart is something else again. This is not the sort of confused, disorderly process you would expect at the hands of the kind of unconscious you read about in books, at the edge of things making up dreams or getting mixed up on words, or having hysterics. Whatever, or whoever, is responsible for this has the accuracy and precision of a surgeon. There almost has to be a person in charge, running matters of meticulous detail beyond anyone's comprehension, a skilled engineer and manager, a chief executive officer, the head of the whole place. I never thought before that I possessed such a tenant. Or perhaps more accurately, such a landlord, since I would be, if this is so, nothing more than a lodger.
>
> Among other accomplishments, he must be a cell biologist of world class, capable of sorting through the various classes of lymphocytes, all with quite different functions that I do not understand, in order to mobilize the right ones and exclude the wrong ones for the task of tissue rejection.
>
> Some intelligence or other knows how to get rid of warts, and this is a disquieting thought. . . . It is also a wonderful problem, in need of solving. Just think what we would know if we had anything like a clear understanding of what goes on when a wart is hypnotized away. We would know the identity of the cellular and chemical participants in tissue rejection, conceivably with some added information about the way viruses create foreignness in cells. . . . Best of all, we would be finding out about a kind of superintelligence that exists in each of us, infinitely smarter and possessed of technical know-how far beyond our present understanding. It would be worth a War on Warts, a Conquest of Warts, a National Institute of Warts and All.[122]

Blood-flow Control During Injury, Surgery, and Disease

In 1975 Thomas Clawson and Richard Swade described several injuries and diseases in which blood flow was either increased, reduced, or stopped by hypnotic suggestion.[123] One dental surgeon, for example, inhibited bleeding in 75 hemophiliac patients during tooth extraction through suggestions that emphasized tranquillity; and another reduced blood loss in nine normal patients during dental surgery with a similar procedure.[124] Suggestion has also been used to control gastrointestinal bleeding and to alleviate hemophiliac bleeding after injury.[125]

Burns

Around 1887, the hypnosis researcher Delboeuf hypnotized a peasant girl, then suggested that her right arm was anesthetized and burned both of her arms with a red-hot iron. Her left arm, in which she had experienced pain, had a wound some three centimeters in diameter while her right arm had one less than one centimeter wide.[126] In 1917 J. A. Hadfield reported similar results in another experiment of this kind;[127] Chapman, Goodell, and Wolff performed yet another in 1959, adding suggestions that one of their subject's arms would be tender and sensitive. In the Chapman-Goodall-Wolff experiment, local vasodilation and skin temperature were greater and lasted longer in the arm for which suggestions of tenderness were given, and a histamine-like vasodilator was found in greater quantities there.[128] Such data led D. M. Ewin to use healing suggestions with burn victims. One of his patients was a man who had fallen into a pot of molten lead (950°C) up to his knee. After being given suggestions within half an hour of his injury that his burn was cooling, he developed burns much less severe than those usually produced by such an injury.[129] In later experiments, Moore and Kaplan used burned subjects as their own control, directing suggestions for healing to some of their burns while others were left to heal normally.[130] Ewin suggested that his patient's wounds were cooling in treating them less than half an hour after they occurred, while Moore and Kaplan used suggestions for warmth on the day following their subjects' injuries. Since cold compresses are normally applied immediately after burn injuries while heat is applied a day or two later, both Ewin and Moore-Kaplan hypnotically simulated the appropriate standard treatment.

Herpes Simplex, Psoriasis, and Contact Dermatitis

Herpes simplex and psoriasis have been relieved or eliminated through hypnotic suggestions of strong cell structure, healthy skin, hormonal balance, cleanliness, or cool sensations.[131] And rashes produced by various plants have been either reduced or induced by hypnosis. For example, 13 Japanese boys who were told that harmless leaves rubbed against their arms were from a tree to which they were sensitive showed some dermatitis; while 11 of the same group who were told that leaves of the offensive tree that were rubbed on their arms were harmless did not react with their accustomed itching or blisters.[132] These results, Theodore Barber wrote,

> showed that once a hypersensitive response to a contact allergen has been firmly established, the response can be reinstated by the feeling, thought, and belief that one is being stimulated by the [offending agent]. This, of course, has vast implications for immunology, since it indicates that at least some immune responses, such as the T-cell mediated skin response to an allergen, may be much more influenced by emotionally tinged feelings-thoughts-imaginings-beliefs than has heretofore been supposed.[†]

Musculoskeletal Disorders

John LeHew described the hypnotic relief of several musculoskeletal disorders, among them sprained backs, degenerative vertebral conditions, rheumatoid arthritis, bone fractures, bursitis, pulled muscles, and muscle spasms. The treatments Lehew described included hypnotic suggestions for relaxation, analgesia, muscle strengthening, serenity, and proper bone alignment.[133]

Asthma and Other Disorders

In the *British Medical Journal* of October 12, 1968, a subcommittee of the British Tuberculosis Association reported their favorable assessment of hypnotic treatment for asthma. The subcommittee accepted 252 patients for study, dividing them into one group given monthly hypnosis treatments and taught autohypnosis and another given exercises in progressive relaxation. Independent judges consid-

---

[†] Barber 1984, p. 72. Other allergic responses have also been altered by hypnotic suggestion, among them hay fever. And reactions to tuberculin have been especially dramatic among some hypnotizable subjects. Not only have they been eliminated, but they have been simultaneously inhibited and stimulated in subjects who were told one injection was harmless (when it contained tuberculin) and another real (when it was not). See: Mason & Black 1958; and Mason 1960.

ered the patients' asthma to be improved in 59 percent of the hypnosis group and 43 percent of the control group.[134] Many physicians and psychotherapists have successfully treated asthma with hypnosis, among them H. H. Diamond, who reported the complete cure of 40 asthmatic children; C. W. Moorefield, who described marked improvement in nine patients; and Z. Ben-Zvi and associates, who described their success in treating exercise-induced asthma attacks.[135] In these studies, suggestions were typically given for bronchial relaxation, general calm, and easy breathing, and were frequently reinforced by autohypnosis and self-suggestion.

Meir Gross reported that anorexia nervosa was relieved in 50 patients by hypnotic suggestions to reduce hyperactivity, correct distorted body images, increase kinesthetic awareness, eliminate feelings of inadequacy, reduce perfectionistic tendencies, and lower resistance to therapy.[136] In 1979 several articles reporting the treatment of obesity by hypnotic suggestion were reviewed in the *American Journal of Clinical Hypnosis*.[137]

Using hypnotic suggestion, Harold Crasilneck successfully treated patients suffering from impotency. He wrote:

> Based on an approximate total of 1,875 males treated for psychogenic impotency during the past 29 years, of which 600 were treated in a cooperative, empirical study, it is my opinion that hypnosis should be considered as a primary treatment [when] cases of psychogenic impotency do not respond to a pharmacological regimen.[138]

Hypnosis has also been used to relieve or eliminate phobias. See, for example, the April 1981 issue of the *American Journal of Clinical Hypnosis*.[139]

[15.10]
CHANGES IN BODILY STRUCTURE AND PROCESS PRODUCED BY HYPNOTIC SUGGESTION

Breast Development

A number of researchers and clinicians have produced breast enlargement in women through hypnosis, usually by promoting blood flow in their subjects' breasts through images of sunshine upon them, or the reinstatement of tender, swelling feelings typical of puberty, or the induction of pulsing, tingling sensations associated with sexual arousal. Some 70 participants in five hypnosis experiments experienced an average bust increase of 1.5 inches, while some women's measurements increased by 3 inches or more. In most of these studies, measurements were carefully made, with consideration of weight changes and the menstrual cycle. As might be expected, highly hypnotizable subjects and those with strong motivation or expectation of success generally experienced the most dramatic results. Some women in these studies had to work through sexual inhibitions that had affected them since puberty.[140]

Inflammation, Weals, Blisters, and Bruises

Hypnotic suggestions have produced inflammations, bruises, and blisters. During one experiment for example, a woman responded to suggestions that she was burned with a bright red inflammation covering most of her hand. Her hypnotic suggestions, it seems, revived memories of a previous hand burn caused by an accident in her kitchen, and reinstated the painful sensations she felt at the time.[141] In another experiment a woman's face, shoulders, and arms turned beet red in response to suggestions that she was sitting on a sunny beach.[142]

Hypnotic suggestions can also elicit weals and blisters. J. A. Hadfield, for example, in 1917 suggested to one of his subjects that a red-hot iron was being applied to his arm, then had the arm bandaged and observed continuously for 24 hours until it was unwrapped in the presence of three physicians. A blister had begun to form where Hadfield had suggested it would, and grew larger as time went on.[143] Montague Ullman observed a complete blister form on a soldier's hand in response to suggestions that the hand had been grazed by a molten shell fragment.[144] And in a study reported by a Russian researcher, a highly susceptible female subject exhibited redness, swelling, then a blister following suggestions that a coin placed on her arm was hot.[145] Reviews by F. A. Pattie (1941), A. M. Weitzenhoffer (1953), Theodore Barber (1961), G. L. Paul (1963), and Leon Chertok (1981) have described several hundred instances of hypnotically suggested weals or blisters.[146] According to Barber, most hypnotically induced inflammations occur in subjects who have formed strong relations with their hypnotists, are fantasy prone, or exhibit the cutaneous lability that underlies nervous hives and dermatitis.[147] Such subjects, Barber suggested, can induce the localized dilation of capillaries that occurs in response to actual burning.

Bruises have also been produced by hypnosis. David Agle, Oscar Ratnoff, and Marvin Wasman, for example, induced ecchymotic lesions (skin discolorations due to extravasation of blood) at designated points in four patients through suggestion. These subjects, who suffered from autoerythrocyte sensitization (a reactivity to their own extravasated red cells that causes spontaneous skin marks) exhibited a propensity to express psychological problems through bodily symptoms.[148]

In summary, the subjects of many experiments have exhibited marks on their skin induced by suggestion. Like religious stigmata (22.3) or psychogenic lesions triggered by therapeutic catharsis (11.1), hypnotically induced blisters and bruises demonstrate the specificity with which strong mental images can influence bodily structures and processes.[†]

Skin Temperature Changes

A Polish researcher described 38 hypnotized subjects who exhibited skin temperature increases on their fingers and toes that averaged 2.7°C (with a 63 percent increase of capillary blood flow) after suggestions for warmth.[149] In a study of self-hypnosis and biofeedback, subjects from 5 to 15 years old raised the temperatures of their index fingers up to 3.7°F and lowered them as much as 8.8°F.[150] And during a demonstration of hypnotic control described by J. A. Hadfield, a subject who had raised the temperature in both hands to 95°F by vigorous exercise experienced a drop to 68°F in his right palm in response to hypnotic suggestion while the temperature in his left palm remained at 94°F.[151] Many bodily changes induced by hypnotic sug-

---

[†] Since bruises can be induced through hypnotic suggestion, it is reasonable to suppose that they can also be reduced by some sort of cognitive control. The fact that some religious stigmatics have achieved rapid remission from their wounds through prayer supports that supposition (22.3). Indeed, some people believe they acquire a resistance to bruising. George Leonard, who is chief instructor of an aikido school in California, wrote in a personal communication to me:

> I have noted that when students first begin practicing aikido, they frequently show bruises from being grasped firmly by the wrists and from being struck on the arms as they block incoming blows. After as little as a few weeks' practice, they can withstand grasps or blows of the same or even considerably more force and show no signs of bruising. Since aikido involves little in the way of upper body conditioning, muscular buildup does not appear to explain this phenomenon.

gestion are mediated in large part by changes in blood flow. "We can hypothesize," Theodore Barber wrote, "that believed-in suggestions, which are incorporated into ongoing cognitions, affect blood supply in localized areas, and the altered blood supply, in turn, plays a role in producing phenomena such as localized inflammations, the cure of warts, the enlargement of mammary glands, the production of blisters and bruises, and the reduction of burns and bleeding."[152]

Modulation of Gastric Acid Secretion

In two studies with healthy, highly hypnotizable subjects, Kenneth Klein and David Spiegel found that hypnotic suggestion can modulate gastric acid production. In a first study, 13 women and 15 men were hypnotized after their basal acid secretion was measured, and were then told to imagine they were eating a series of delicious meals. Their mean acid output increased by 89 percent. In a second study by the same experimenters, 7 women and 10 men underwent two sessions of gastric analysis in random order, once with no hypnosis and once under hypnotic instruction to relax. Compared to the no-hypnosis session, the subjects exhibited a mean reduction of 39 percent in their basal acid output under the influence of hypnotic suggestions that diverted their thoughts from hunger. "We have shown," Klein and Spiegel wrote, "that by using different hypnotic suggestions to induce intense mental imagery it is possible to either promote or inhibit gastric acid secretion."[153]

[15.11]
OTHER RESULTS OF HYPNOTIC SUGGESTION

Perceptual Alterations

Theodore Barber reviewed experiments by Milton Erickson and others that produced hypnotically induced blindness or color-blindness; and studies of visual distortion are noted in the comprehensive reviews of hypnosis research by Ernest Hilgard in 1965 and 1975 and by John Kihlstrom in 1985.[154] But hypnotic suggestion has helped improve vision as well as to inhibit or distort it. M. V. Kline and A. M. Weitzenhoffer reported that nearsighted individuals experienced improvements in their eyesight during hypnosis; Milton Erickson reported similar improvements in subjects regressed to an age when they did not wear glasses; Gerald Davison and Lawrence Singleton accidentally discovered that one of their subjects experienced improved vision during hypnosis; Charles Graham and Herschel Leibowitz found that several susceptible subjects who were nearsighted exhibited improvements of vision after direct and posthypnotic suggestions; and Sheehan, Smith, and Forrest produced improvements of monocular visual acuity by suggestions in a group of myopic students.[155]

Like vision, hearing can be distorted by hypnosis. The sensation of hearing mosquitoes buzzing, for example, has been induced in thousands of subjects given the Stanford Hypnotic Susceptibility Scales, which use auditory hallucinations of mosquitoes as a standard measure of hypnotizability. And tests of magnitude estimation have shown clear changes in auditory sensitivity following suggestions for deafness.[156] Few laboratory studies have explored the use of hypnosis to *improve* hearing, however, although suggestions for deafness have helped reduce speech errors.[157]

Jerome Schneck described several tactile and kinesthetic sensations experienced by one of his patients in hypnotherapy, among them feelings that his head was turned away from Schneck during therapy, that his arms "were not there," that he was weightless, that his head was about to burst, that he was rigid, that he was walking on a tightrope, that his arms were numb and lifeless, that he was dizzy, that he was rising above the earth, that he was pressed flat, and that he was paralyzed. Schneck interpreted these experiences to be symbolic somatizations of various emotional issues his patient was confronting at the time.[158] Wallace found that hypnotically induced anesthesia disrupted perceptual adaptation to errors produced by distorting prisms; Nash, Lynn, and Stanley produced out-of-body sensations in a number of susceptible subjects; Brenman, Gill, and Hacker found changes in body awareness to be the most frequent spontaneous changes they observed in hypnosis; and Josephine Hilgard reported that body-image distortions are sometimes unintended sequelae of hypnotic trance.[159]

Improvements of Cognition and Learning

Theodore Barber reviewed several studies in which hypnotic suggestion improved concentration, study habits, and retention. G. S. Blum and associates found that hypnotic suggestion can affect learning and memory by influencing anxiety, cognitive arousal, pleasure, and attention. Helen Crawford and Steven Allen reported that hypnotically responsive subjects during hypnosis improved their performance on a visual memory discrimination task. And G. H. Bower and colleagues found that hypnotically suggested mood states could, under some conditions, induce state-dependent enhancements of memory.[160] (*State-dependent learning* and *state-dependent retrieval* are terms used by psychologists to denote learning or remembering that depends upon particular configurations of emotion, imagery, and physiological process. Considerable research suggests that things learned or experienced in a particular state can best be retrieved in the same or a similar state.)

Improvements of Physical Performance

Reports of hypnotism's usefulness in sports are supported by laboratory studies which have shown that hypnosis can help some people improve their strength and motor skills. J. A. Hadfield and W. R. Wells, for example, found that subjects given hypnotic suggestions for increased strength increased their grip on a hand dynamometer; while seven subjects in a study by M. Ikai and A. H. Steinhaus exhibited greater forearm strength after suggestions that they would produce more power. N. C. Nicholson found that seven subjects could lift three kilograms of weight with the flexor muscles of their middle fingers at the rate of 30 lifts per minute for at least 20 minutes after they were given hypnotic suggestions; and Williams replicated Nicholson's experiment with a single subject who lifted a weight of 4 kilograms with his middle finger 80 times a minute for more than half an hour, and with five subjects who lifted 5 kilograms 80 times a minute. Other studies have shown that motivational suggestions, given after induction or in the waking state, increase strength, endurance, and coordination in many subjects using jump-and-reach tests, arm rigidity tests, hanging-by-the-hands tests, arm dynamometers, back dynamometers, leg dynamometers, bicycle ergometers, barbell-pressing tests, weight-holding tests, and the "human plank" test in which the body is rigidly stretched between two bars while supporting a weight.[161]

Kenneth Callen suggested that long-distance running naturally induces a kind of autohypnosis that facilitates a runner's enjoyment and performance. More than half of 424 runners who completed a questionnaire concerning their running habits said they experienced vivid imagery and trancelike states with increased receptivity to internal events, which are hallmarks of autohypnosis. Running itself is a form of induction, Callen argued, in that it involves deep rhythmical breathing and frequently induces imagery, eye fixation, and obsessive repetition of a sound or phrase.[162]

[15.12]
SOME NEGATIVE EFFECTS OF HYPNOSIS

Hypnosis, of course, can have negative outcomes. Used for entertainment, whether informally or during stage performances, it can help induce people to say or do things about which they are embarrassed later, or attempt strenuous feats that cause physical injury, or go into a trance that leaves them disoriented. Indeed, people near psychological breakdown have been led into emotional crisis during ostensibly harmless hypnosis demonstrations. In his book *Hypnosis Complications,* therapist Frank MacHovec cited several reports of negative hypnosis outcomes, most of them produced by stage hypnotists or untrained people.[163] Hypnotically assisted psychological treatment can also produce undesirable outcomes, though most therapists who have worked with hypnosis believe that interpretations of behavior and other therapeutic modalities—rather than hypnosis itself—are usually the primary causes of negative hypnosis results during therapy.[164]

In laboratory experiments or psychological treatments conducted by trained practitioners, however, when hypnotized subjects approach potentially disturbing issues they usually drift into harmless fantasy, fall asleep, or awake from whatever trance they have entered. Indeed, careful studies have shown that responsibly conduced hypnosis experiments produce few bad results. In a study reported in 1968, for example, more than 100 subjects had fewer negative changes in their scores on the Minnesota Multiphasic Personality Inventory, fewer self-referrals to the counseling center at which their study was conducted, and less need for medical attention in general within 90 days of their hypnotic sessions than a group of matched controls who were not hypnotized.[165] In 1974, psychiatrist Josephine Hilgard reported that among 120 university students who were hypnotized, only 19 had short-term and 18 longer-term aftereffects. Among the former, one experienced headache, one a stiff neck, eight drowsiness or sleep, eight cognitive distortion or confusion, and one anxiety, but all 19 felt normal an hour later. Among those with longer aftereffects, three had headaches, one dizziness, one a stiff arm, seven drowsiness or sleep, two cognitive distortion or confusion, and four dreams that seemed related to their hypnotic experience, but none of their reactions appeared serious, half clearing up within a few hours and the rest within a few days.[166] And in 1979, William Coe and Klazina Ryken reported that a group of university students had no more negative experiences from hypnosis than they did after a brief verbal learning experiment, an examination, attending class, or "college life in general." Indeed, the two researchers rated the students' exam to be the most stressful among these events, while the students tended to rate the hypnosis sessions as the most pleasant.[167]

Even hypnotically assisted psychotherapy produces relatively few reports of negative outcome. A 1961 survey by Ernest and Josephine Hilgard and M. R. Newman of journal articles about such therapy, and representative discussions of the subject by Meares (1961), Pulver (1963), Kost (1965), and West and Deckert (1965), show that among skilled therapists few serious problems arise. "But [such articles] all warn that therapy should be done by those trained in therapy," Josephine Hilgard wrote, "and not by those trained in hypnotic techniques alone."[168]

Hypnosis can be misused, too, in law trials and police investigations when employed to facilitate the memory of victims and witnesses. A commission convened by the American Medical Association in 1985 cautioned against the systematic use of hypnosis to assist recollection, first because hypnotic recall can result in confabulation as well as clearer memory, and second because it can promote an artificial sense of certainty among hypnotized subjects about material they mistakenly believe they have recalled. Given the uncertainties involved in hypnotically assisted recall, New York, Arizona, New Jersey, and other states confine testimony to prehypnosis memories; and California forbids testimony by a hypnotized witness or victim on the premise that they may present confabulated material, or become invulnerable to cross-examination because their mistaken convictions are strengthened.[169]

[15.13]
PARANORMAL EXPERIENCE AND QUASI-MYSTICAL STATES INDUCED BY HYPNOTIC SUGGESTION

Authoritative 10-year reviews of hypnosis research by Ernest Hilgard and John Kihlstrom that were published in 1965, 1975, and 1985 by the *Annual Review of Psychology* contained no references to studies involving extrasensory perception and just two or three to studies that dealt with transpersonal states.[170] This neglect reveals a general bias among contemporary academic psychologists against parapsychology, since several studies relating hypnotic and extrasensory phenomena had been conducted between the 1950s and early 1980s, of which 14 were reviewed in the *American Journal of Clinical Hypnosis* in 1969 by R. L. Van De Castle and 25 in the *Journal of the American Society for Psychical Research* in 1984 by Ephraim Schechter.[171] Hilgard and Kihlstrom's failure to mention any of the studies that Van De Castle and Schechter reviewed reflects the general effort by experimental researchers (and many clinicians) to win scientific acceptance for hypnosis by freeing it from association with mysticism and occultism. Hypnosis researchers, rather than parapsychologists, have kept the two fields separated in recent decades.

But such caution stands in vivid contrast to attitudes held by great hypnosis researchers of the past. Mesmer, for example, claimed that some magnetized persons could paranormally perceive past and future.[172] The Marquis de Puysegur described subjects who apparently diagnosed the afflictions of others by clairvoyance, or verbalized his unspoken thoughts.[173] James Esdaile described subjects whom he hypnotized from a distance and a magnetized woman who could identify the number and suit of playing cards placed on her stomach.[174] And as noted in 15.1, Carl Kluge, Justinus Kerner, and other German mesmerists studied paranormal capacities produced by magnetic sleep.

Distinguished researchers from other fields, too, have studied extrasensory experience in hypnotized subjects. Henry Sidgwick, a prominent professor of moral philosophy at Cambridge University (and a founder of the Society for Psychical Research) with Edmund Gurney described a minister's wife who could correctly identify tastes in her hypnotist's mouth, as if by telepathy.[175] Though he knew it might damage his reputation to do so, Freud described apparent telepathic interactions with some of his patients (17.3). And Pierre Janet, a principal founder of dynamic psychiatry, described a remarkable series of experiments in which a Dr. Gibert of Le Havre apparently hypnotized his subject Leonie from a distance. During 1885 and 1886, Janet and Gibert conducted 25 experiments with telepathic hypnosis, 19 of which they deemed to be successful (see appendix A.7). Frederic Myers, Charcot, and other prominent figures observed these proceedings and subsequently described various instances in which Leonie performed specific tasks that Gibert telepathically commanded her to perform (as, for example, when she walked to Gibert's house while some of her distinguished observers watched from behind lampposts and bushes).[176] Janet even suggested upon one occasion that Leonie travel clairvoyantly while in hypnotic trance to report on the famous French physiologist Claude Richet, and was astonished when she correctly reported that Richet's laboratory was on fire.[177]

Sir William Barrett, the prominent English physicist, reported his own experiment with hypnotically induced clairvoyance, during which an Irish peasant girl accurately described an optician's shop in London with which Barrett was familiar, reporting details such as a clock hanging over the building's entrance.[178] However, the British Association for the Advancement of Science refused in 1876 to consider Barrett's account of the experiment until Alfred Russell Wallace, the co-discoverer of natural selection, intervened on his behalf. When Barrett reported his experiment to an anthropology subsection of the association, much of his audience left the room. The general skepticism among scientists about hypnosis and paranormal phenomena reflected by that event was captured in this statement by the physicist Hermann Helmholtz:

> I cannot believe it. Neither the testimony of all of the Fellows of the Royal Society, nor even the evidence of my own senses would lead me to believe in the transmission of thought from one person to another independently of the recognized channels of sensation. It is clearly impossible.[†]

William James, however, held a different view. In a discussion of Frederic Myers's concept of the subliminal mind, he wrote:

> The subliminal region being thus established as an actuality, the next question is as to its farther limits. . . . My subliminal, for instance, has my ordinary consciousness for one of its environments, but has it additional environments on the remoter side? Has it direct relations of intercourse, for example, with the consciousness, subliminal or supraliminal, of other men? Some of the phenomena of hypnotism or mesmerism suggest that this is actually the case. I refer to the reports (several of them irreproachably recorded) of hypnotism at a distance,

---

† Heywood 1961. A similar sentiment was expressed in an article published in *Science* by G. R. Price, who wrote: "Not 1000 experiments with ten million trials and by 100 separate investigators giving total odds against change of 10 to the one-thousandth to 1" could make him accept ESP." See: Price, G. R., 1955.

of obedience to unspoken orders, and of "community of sensation" between hypnotizer and subject.

The system of faculty of a subject under hypnosis is quite different from his waking system of faculty. While portions of the usual waking system are inhibited, other portions are sometimes supernormally energized in hypnosis, producing not only hallucinations, but after-results in the way of sense discrimination and control of organic function, to which the waking consciousness is unable to attain.[179]

Suggestion at a distance, which Janet and Gibert had helped to publicize through their experiments with Leonie in Le Havre, was studied, too, by the Russian physiologist L. L. Vasiliev. Using several measures of physical relaxation to indicate the onset of hypnotic trance, Vasiliev found that subjects could be induced to become either more drowsy or alert through telepathic influence. When Vasiliev placed some of his agents in a structure made of lead slabs and his subjects in a Faraday chamber constructed of sheet iron to screen out electromagnetic influences that might mediate suggestion at a distance, he still got positive results. After many experiments of this kind over the course of several years, he rejected the idea that extrasensory communication was carried in the electromagnetic spectrum. "Telepathic transmission," he wrote, "is accomplished by some kind of energy or factor so far unknown to us."[180]

Though he eventually rejected hypnosis as a means of facilitating ESP, J. B. Rhine conducted experiments with hypnotic suggestion during the early 1930s. In his first book detailing work at the Duke Parapsychology Laboratory, he described a study in which 30 hypnotized subjects carried out 530 trials with single digits as target stimuli, 340 trials during which the same subjects raised their right or left hand while their agents were looking at the words *right* or *left,* and 245 trials in which they tried to pick which octant of a circle their experimenters were looking at. Rhine's results in this series of experiments were positive at a level 25 to 1 beyond chance.[181] Rhine said this study did not evoke significant ESP ability, but his results can be interpreted to show that hypnosis did in fact facilitate telepathy or clairvoyance.

The supposition that hypnotic suggestion can facilitate paranormal capacities was strongly supported by two reviews of experiments that compared ESP scores during hypnosis with scores in the waking state. The first, by R. L. Van De Castle, reviewed 14 studies in which hypnosis was used to improve ESP. Van De Castle calculated that these experiments produced cumulative evidence of ESP at a probability level of 10^{-10}, or *one in 10 billion*.[182] The second review, published in 1984 by Ephraim Schechter, examined 25 studies reported by 1984 (in 20 papers by 12 researchers or research teams from 10 laboratories). Seven of these showed significantly stronger ESP results after hypnotic induction; nine produced positive but statistically nonsignificant results. No study produced a negative outcome. Schecter calculated the probability that seven studies would produce significant results by chance to be only 34 in a million, and the likelihood that 16 of the experiments would produce stronger performance after induction by chance to be only 6 in 10,000. After analyzing each of these studies for experimental flaws, Schechter estimated that the correlation between a score he devised for potential design problems and the studies' success rating was far from statistical significance. There was no apparent systematic relationship, he concluded, between the experiments' inadequacies and their positive outcomes. "It seems that the difference between ESP performances in the [hypnotic] induction and control conditions is a reliable effect," he wrote. "It occurs more often than would be expected by chance and it does not appear to be due to faulty ESP testing or experimental design."[183]

These experimental results are supported by many clinical reports that paranormal experience occurs during hypnosis. For example, 24 percent of 200 respondents to a questionnaire sent to every member of the Australian Society of Hypnosis indicated that they had experienced ESP while using hypnosis. The survey's author Lorna Channon wrote that, as a confirmed skeptic concerning paranormal phenomena, she had not anticipated this level of positive response to the single question that asked whether hypnosis had produced extrasensory communication for the questionnaire respondents. Channon suggested that hypnotists who fall into partial trance themselves while inducing hypnosis might open themselves more fully to extrasensory messages, partly because they achieve greater rapport with their subjects and feel a deeper empathy for them.[†]

Quasi-mystical experience, too, has been evoked by hypnosis in contemporary experiments. Bernard Aaronson, for example, guided trained subjects into a detachment from sensory experience, a relinquishment of ego-identification, and a discarding of ordinary mental categories that produced experiences resembling mystical states.[184]

[†] Channon 1984. The survey question regarding ESP read: "While using hypnosis, have you ever had any experiences which have suggested that hypnosis produces a kind of extrasensory communication between you and your patient?"

Informed in part by Aaronson's methods, Paul Sacerdote induced quasi-mystical experience to relieve subjects from pain.[185] And as noted in 15.5, Charles Tart has described transpersonal states that occurred in one of his hypnotic subjects.[186]

[15.14]
HYPNOTIC PHENOMENA AND
TRANSFORMATIVE PRACTICE

Mesmerism and hypnosis have not typically produced the illumination and creativity evident, for example, in contemplative practice. This is the case because usually they have been conceived as therapeutic activities rather than programs for high-level change, even though the Marquis de Puysegur, Kluge, Kerner, Esdaile, Myers, James, and others have described their metanormal products. Nearly everyone who has experienced hypnosis has done so episodically, for the alleviation of some affliction, or less frequently for the improvement of a particular skill. Hypnotic suggestion by itself is not a life-encompassing practice such as Christian mystical prayer, Buddhist yoga, or shamanism, nor has it been typically used as a long-term discipline in the manner of some martial arts.

Nevertheless, mesmerism and hypnotism have stimulated experiences and capacities that play an important role in transformative practice, among them creative absorption, perceptual flexibility, hypermnesia, exceptional physiological control, psychosomatic plasticity, and access to subliminal levels of consciousness. A mysterious intelligence often comes into play during hypnotic suggestion. As Lewis Thomas suggested, this is not the sort of confused, disorderly intelligence you would expect from a Freudian unconscious. "There almost has to be a person in charge," Thomas wrote, "running matters of meticulous detail beyond anyone's comprehension, a skilled engineer and manager, a chief executive officer . . . a cell biologist of world class."[187] This, perhaps, is the most fundamental insight we gain from hypnosis, this glimpse of a superior intelligence within us that can drastically alter our perceptions and thinking, restore healthy functioning, and enhance our most basic capacities. From the wealth of mesmeric and hypnotic phenomena we learn that human functioning can be developed in dramatic ways by evoking something that resembles Frederic Myers's "subliminal mind."

Among other things, hypnosis research teaches us that

- deep absorption can stimulate various kinds of exceptional functioning;
- anxiety, pain, and physiological processes can be modified through focused intention;
- certain capacities that promote creativity in general are correlated with hypnotic responsiveness, among them perceptual flexibility, tolerance of ambiguity, the ability to relinquish reality testing, openness to strange experience, vivid and spontaneous imagery, the ability for gestalt closure, and holistic or synthetic thinking;
- hypnotic responsiveness and its psychophysical correlates can be cultivated;
- hypnotic virtuosos develop their exceptional skills by years of practice (usually beginning in childhood);
- such practice can produce a psychosomatic flexibility through which the body can be creatively altered; and
- social contexts (such as those evident in the interpersonal dynamics of hypnotic process) enhance or inhibit many human capacities.

16

BIOFEEDBACK
TRAINING

BIOFEEDBACK TRAINING IS a method for learning psychosomatic self-regulation for the promotion of health, the improvement of particular skills, or the attainment of desired mental states. During such training, activity in some body part is detected by sensitive instruments and fed back to a subject through sounds or visual displays so that the subject can learn to modify it. Functioning once considered to be inaccessible to voluntary control is now commonly modified with the help of instrumented feedback. Though the cognitive and somatic processes that make such control possible are not completely understood, the widespread success of biofeedback training has shown that most, if not all, people can improve their powers of self-regulation. Echoing the belief of many researchers, Alyce and Elmer Green of the Menninger Foundation wrote: "It may be possible to bring under some degree of voluntary control any physiological process that can continuously be monitored, amplified, and displayed."[1] Furthermore, numerous studies have shown that self-regulation skills acquired through biofeedback training can be retained after instrumented feedback is dispensed with. By 1990, for example, more than 2,000 subjects at the Menninger Foundation had learned to modify various bodily processes through a combination of feedback, Autogenic Training (see 18.3), and visualization so that their new self-control did not depend upon machines. In 16.6 I discuss some aspects of biofeedback training that bear upon transformative practice in general.

[16.1]
PRECURSORS OF CONTEMPORARY BIOFEEDBACK TRAINING

Many instruments that facilitate self-regulation can be thought of as feedback devices. A rattle or doll, for example, can help an infant learn motor skills. A ball's flight provides information that helps a child perfect a throwing motion. A tippy canoe gives instant feedback to someone practicing footwork and balance. Such devices—and countless others—are antecedents of modern biofeedback machines. Instruments to facilitate physiological self-mastery became especially ingenious, however, during the late nineteenth and early twentieth centuries. To help lip readers, for example, Alexander Graham Bell used the phonautograph invented by Frenchman Leon Scott. This instrument consisted of a diaphragm stretched across the narrow end of a cone. When the user spoke into the cone, a brush traced wave forms on smoked glass. Bell was also interested in the manometric flame of Rudolph Koenig. In this device, a membrane was stretched across a hole in a gas pipe so that a voice near it produced movement in a flame from the pipe, and the flame was reflected from mirrors mounted on a revolving wheel. These reflections exhibited characteristic light patterns, with which deaf people could experiment until they made the sounds they wanted.[2] The phonautograph and manometric flame were precursors of the sonagram, which is used in the study of speech characteristics; and Bell's pioneering work helped establish a basis for contemporary training of the deaf.[3]

Instrumented biofeedback was pioneered as well by O. Mowrer and W. Mowrer (1938), who used an alarm system triggered by urine to stop children from bed-wetting; by Lawrence Kubie (1943), who fed back amplified breathing sounds to induce "hypnogogic reverie"; and by Edmund Jacobson (1939), who inserted electrodes into arm flexor muscles to provide feedback for muscle relaxation.

Other clinicians and experimentalists laid the groundwork for biofeedback training through studies of noninstrumented self-regulation. The physiologist J. R. Tarchanoff (1885) described a male subject who could accelerate his heart rate at will: such control, Tarchanoff suggested, was mediated by some unknown center in the brain. Other researchers described a medical student who could increase his heart rate by 27 beats per minute, a 29-year-old male who could initiate tachycardia by "slight abrupt muscular efforts," and two subjects who voluntarily accelerated their pulse.[4] Though these studies consisted mainly of anecdotal accounts, they helped stimulate a suspicion among some psychologists and medical researchers that heart rate could be deliberately modified.

Investigations of exceptional muscular control were also conducted from time to time during the late nineteenth and early twentieth centuries. Schafer, Canney, and Tunstall (1886) measured the rate of human muscle contraction in response to volitional impulses. J. H. Bair (1901) trained students to wiggle their ears by having them jolt their retrahent muscle with a small electric shock and then simulate the resulting spasms by flexing their jaw and face muscles. R. S. Woodworth (1901) explored the voluntary control of movement force. And beginning in the 1920s, Edmund Jacobson developed his landmark studies of progressive muscular relaxation (see 18.6 and appendix H). These pioneering investigators helped encourage the widespread research on muscular self-regulation that emerged after 1960.

Other self-regulation methods, too, anticipated biofeedback training. The Alexander Technique and Autogenic Training were developed before instrumented biofeedback was popular, and their practitioners promoted the idea that kinesthetic sensitivity and control of both muscular and autonomic processes can be trained (18.2, 18.3). At the same time, Hindu, Buddhist, and Sufi groups in the West helped create a climate in which biofeedback could achieve its sudden celebrity. In the 1960s especially, Eastern religions became popular in Europe and America and informed some of the early biofeedback studies.

[16.2]
CONTEMPORARY BIOFEEDBACK TRAINING

In spite of the work just noted, before the 1960s most physiologists and research psychologists believed that autonomic functions could not be controlled voluntarily like muscular and skeletal activity.[5] According to most experimentalists, the self-control that some subjects seemed to exhibit was either caused indirectly through processes under voluntary control or was the result of placebo effects. Given this widespread belief, demonstrations that heart-rate control in paralyzed rats could be conditioned gained instant celebrity.

During the late 1960s, the widely respected psychologist Neal Miller, with L. V. DiCara and other associates, appeared to show that the heart rate of rats temporarily immobilized by the drug curare could be instrumentally conditioned by electrical brain stimulation. Miller and his associates also thought they had shown that conditioned heartbeat control in curarized rats could be transferred to the nonparalyzed state.[6] Inspired in part by Miller's work, other research-

ers reported similar results, providing support for the idea that autonomic systems could be controlled just as yogic adepts and a few
pioneering experimenters had claimed. Miller eventually found it difficult to replicate his first results and came to have serious doubts
about them, but he had encouraged new attempts to prove that control of autonomic functions was possible in humans.[7]

Indeed, Miller's work with curarized rats caused a sensation
among experimentalists, in part because it appeared to disprove the
long-standing theory that autonomic responses could be modified
only by classical conditioning;[†] and in part because it dramatically
supported the growing belief that autonomic processes could be
brought under conscious control. John Basmajian had shown that
normal subjects could consciously activate single motor units, and
had reported his work in *Science* in 1963. Johann Stoyva, T. H. Budzynski, and others had developed techniques with which the elec-
tromyograph could be used to control muscle tension. Joe Kamiya had
found in the late 1950s that subjects could control brain activity registered on the EEG. And H. D. Kimmel had used operant conditioning
to modify the galvanic skin response, a measure of electrical conduc-
tivity on the skin related to emotional arousal.[8]

As a result of these studies, which were being reported in both
scientific journals and the popular press, a historic shift of belief
among physiologists and experimental psychologists began to occur
in the late 1960s, away from a conviction that the autonomic nervous
system was relatively inaccessible toward a recognition that it could
be deliberately modified. In spite of extravagant claims about biofeedback training and the controversy they caused, this fundamental

---

[†] Miller's research helped break down a tenet of learning theory that responses controlled by
the autonomic nervous system (ANS) were learned by different means than those controlled by
the central nervous system (CNS). ANS learning, it was thought, occurred, often in a random
fashion, through classical conditioning, while CNS responses could be operantly or instrumentally conditioned. In classical conditioning, a conditioned stimulus such as a bell is associated
with an unconditioned stimulus such as food, so that a particular response such as a dog's saliva-
tion is elicited. The dog in this case responds to the experimenter's manipulations automatically,
involuntarily, and reflexively. In operant conditioning, on the other hand, the subject is rewarded
after a desired behavior, with food or praise, for example, so that he tends to act the same way
again. The belief that autonomic processes could not be operantly conditioned arose partly by
historical accident because B. F. Skinner and his behaviorist followers, most of whom were psychologists, focused on the operant model in studying learning processes. Their experimental
findings encouraged a legend (which came to be taken as fact) that the processes of operant conditioning applied to external behaviors mediated by skeletal muscles but not to autonomic
behaviors.

change in outlook has supported all subsequent work with self-regulation.[9]

Growing acknowledgment that autonomic processes could be voluntarily modified stimulated the formation of self-regulation training programs in many parts of the world. In 1970 the Biofeedback Research Society was founded and was subsequently expanded to reflect the widespread clinical, research, and educational interest in self-regulation throughout the United States. Subsequently called the Biofeedback Society of America, it began publishing the journal *Biofeedback and Self-Regulation* in 1975. A bibliography of the literature by Francine Butler, published in 1978, included more than 2,300 citations; and in 1983 John Basmajian guessed that "in its short life [the field] has been documented in some ten thousand professional articles, and its practitioners and scientists number in the many thousands around the world." Though other experts in the field give different estimates of its size, the biofeedback literature is immense.[10]

[16.3]
VOLUNTARY CONTROL OF MUSCLE ACTIVITY

Voluntary Control of Single Motor Units

In 1934 Olive Smith described the conscious control of individual motor unit potentials. D. B. Lindsley (1935) confirmed Smith's findings and showed that some subjects could relax a muscle so completely that no active units could be found. Others continued such research on fine motor control, and John Basmajian built upon it to show that normal subjects could be trained to isolate and consciously activate separate motor units (i.e., individual neurons and the muscle fibers they innervate).[11] Basmajian's demonstration of this remarkably specific self-control has since been repeated by other researchers and has provided a powerful stimulus for biofeedback research in general. Single motor unit (SMU) control has now been trained in many muscle sites.[12] Furthermore, SMU firing patterns are now altered in different ways. Their frequency can be changed, for example, or their rhythms varied. Basmajian has called such variations special effects, naming some drumrolls or gallops, and has dramatized them by amplifying their signals through loudspeakers.

This marvelous specificity of self-regulation has been developed and maintained by many people. In one study, for example, 13 normal subjects learned to control SMUs from three sites in the right trapezius muscle and then to maintain such control without feedback. Four of these subjects were able to maintain such control while drinking water from a glass, thus activating the trapezius as a whole while simultaneously suppressing activity of SMUs in separate parts of the muscle.[13]

To sense the internal systems involved in single motor unit control, to learn their nonverbal "language," a subject must correlate his weak and unstable kinesthetic feedback with visual or auditory displays. Researchers A. J. Lloyd and J. T. Shurley suggested that the biofeedback approach, viewed from a simple cybernetic analysis, is a technique that amplifies a weak proprioceptive signal in a relatively noisy system. But instead of amplifying the signal, they proposed, one might reduce noise in the system. Lloyd and Shurley did just that by training SMU control in a quiet, darkened chamber; they showed that this was a much more efficient way to do such training.[14] Their findings support the long-held belief that yogis and other contemplatives achieve their exceptional self-control in part by limiting sensory input so that they can better apprehend kinesthetic signals.[15]

Voluntary Control of Gross Muscle Activity

Since the 1960s, subjects in many biofeedback studies have extended their control of muscle activity.[16] Conditions that have been cured or relieved through biofeedback include temporomandibular joint dysfunction, which is characterized by pain in the face and jaw, ringing in the ears, teeth grinding during sleep, difficulty swallowing, fatigue, and other symptoms; and orofacial dyskinesia, a disorder characterized by uncoordinated movements of the face, jaw, tongue, and neck.[17] Tension headache has been reduced or eliminated through relaxation of the frontalis muscle.[18] And several researchers have successfully treated torticollis, a muscular contracture in which the head is twisted to one side.[19] Instrumented feedback has also helped numerous subjects find relief from cerebral palsy symptoms, compulsive subvocalization, esophageal dysfunction, excessive nasality, involuntary spasms of the eyelids, trouble opening the eyes following psychological trauma, muscle paralysis caused by cerebrovascular accidents, and difficulty playing musical instruments.[20]

[16.4]
VOLUNTARY CONTROL OF THE BRAIN'S ELECTRICAL ACTIVITY

Since the late nineteenth century, neuroscientists, physiologists, and psychologists have studied electrical activity of the mammalian nervous system, at first with galvanometers, kymographs, and other simple electromechanical devices, and now with electroencephalographs. In 1875 Richard Caton, a British physiologist, recorded electrical activity in the cerebral cortex of rabbits, cats, and monkeys. In 1890 Adolph Beck of Poland recorded oscillations from the human brain. And in 1924 Hans Berger published a landmark description of human EEG readings that grew partly from research to find physical means by which telepathic impressions are mediated. Berger assigned the Greek letters delta, theta, alpha, and beta to particular wave forms and spectra of electrical activity recorded by the electroencephalograph.[21]

Alpha-Wave Training

At the University of Chicago in the late 1950s, research psychologist Joe Kamiya found that some people could control their alpha rhythms with the help of EEG feedback. Kamiya's work aroused considerable interest among his fellow psychologists even before he published his results, and it helped stimulate popular curiosity about biofeedback after he published articles about it in 1968, 1969, and 1970.[22] Kamiya reported that his subjects had learned to increase and decrease the frequency of their alpha waves (8–12 cycles per second), often in alternating five-minute epochs; and he showed that visualization, mental arithmetic, and other focused intellectual operations helped suppress alpha activity. He also drew attention to the possible importance of a drift effect by which the percentage of alpha waves in a subject's EEG tends to increase spontaneously in the experimental situation. Experiments by Barbara Brown at the Veterans Administration Hospital in Sepulveda, California, which seemed to show that subjects could modify alpha frequency and amplitude by manipulating their feeling states, also helped promote the idea that electrical activity in the brain could be modified.

These first studies stimulated public enthusiasm for alpha training, in part because they led many people to think that alpha waves were correlated with mystical states. But they also raised many questions. Erik Peper and T. B. Mulholland, for example, showed that subjects could decrease alpha with EEG feedback but could not increase it beyond levels produced while at rest with their eyes closed; and their results were supported by other investigators.[23] These results led Peper to suggest that certain motor impulses block alpha, and that the relaxation of such impulses permits alpha activity to return to its normal resting levels. In a paper published in 1971 he argued that these alpha-suppressing impulses include corrective commands to the oculomotor system, which suppress alpha measured at the occipital region of the brain.[24] In subsequent studies Peper and Mulholland showed that elevating the eyes or blurring a target while tracking it was correlated with greater alpha abundance; and other researchers demonstrated that increased eye convergence led to a decrease in alpha, while the reverse was the case when the eyes diverge.[25] Furthermore, direct feedback from an electroculogram, which records subtle eye movements, has been as effective as EEG feedback in helping some subjects produce such changes in brain activity. Far vision, it seems, enhances alpha, while near vision suppresses it. Given this correlation, it is likely that the "soft eyes" practiced in aikido, the relaxed gaze of *zazen*, and certain hypnotic suggestions involve the relaxation of alpha-blocking eye convergence.

Leading biofeedback researchers do not typically claim that alpha frequencies are reliable indices of mystical experience. Kamiya, for example, has steadfastly maintained that the well-being associated with increased alpha activity should not be equated with satori or other religious states, even though Akira Kasamatsu and Tomio Hirai had found that alpha waves were abundant in the EEG records of Zen masters and their advanced students (23.1). As a subject in some of his experiments during 1965, I can testify to Kamiya's refusal to equate control of alpha waves with mystical illumination. Though he was aware of Kasamatsu and Hirai's studies, he would not make facile comparisons between satori and the experiences of biofeedback subjects. According to psychologist William Plotkin, who carefully reviewed biofeedback studies involving the EEG, the so-called alpha experience is caused in large part by sensory deprivation, sustained alertness, introspective sensitization, perceived success at the feedback task, and myths surrounding alpha training.[26] Plotkin wrote:

> Biofeedback training is not merely a form of manipulation of human physiology; it is a complex social-behavioral interaction in which not merely physiology but attitudes, expectations, motivations, attention, alertness, and understandings are directly and indirectly influenced independently of any contingencies between physiology and feedback.[27]

A highly regarded study by D. H. Walsh supports Plotkin's view, in that some of its subjects experienced no specific state during their increased alpha production and others reported unpleasant experiences.[28] The fact that brain waves recorded by the EEG are complex entities produced by many interactions of the central nervous system also makes it unlikely that any precise correlation can be made between them and a particular subject's experience. According to research psychologist Joel Lubar:

> The EEG recorded with scalp electrodes in humans represents the tip of an iceberg. We often make the mistake of thinking that EEG demarcates unitary processes that define behavioral states. In recording the EEG from a particular location using bipolar electrodes we really see the difference between the electrical potentials at the two points where the electrodes are placed. This potential difference in turn contains the following elements: dendritic activity from the upper cortical layers, direct current shifts occurring across large regions of the cortex, occasional evoked or event potentials elicited by sensory stimuli during the time the recording is taken, and relays of activity representing activation and inhibition from thalamocortical pools and pacemakers which are in turn influenced by relays from the brainstem reticular formation. This is significant because, when we teach an individual to alter a particular EEG component such as alpha rhythm, beta spindles or sensorimotor rhythm, we do not know which underlying processes are most related to the acquisition of the task. In some cases, it may be changes in dendritic potentials. In other cases it may be direct activity from neuronal pools under the recording electrodes, and/or it might be activity of thalamic nuclei which are only remotely affecting the overlying cortical activity.
>
> It may also be that conditioning in a particular EEG pattern in one particular individual might represent different processes than it would in another individual. That is, a particular pattern of EEG activity may come about as a result of different kinds of subcortical and brainstem activity in different individuals. This would help to explain why the subjective reports associated with alpha EEG biofeedback conditioning in humans are quite variable.[29]

Theta-Wave Training

Theta waves (4–7 cycles per second), which can be recorded from many parts of the cortex, are especially abundant during drowsiness and sleep and are associated with daydreaming, imagery, and creative visualization. They have been voluntarily modified with the help of feedback in studies by Elmer and Alyce Green, Barbara Brown, Joel Lubar, and others. The Greens and their associates at the Menninger Foundation, for example, have trained subjects to augment theta activity for the enhancement of creativity, for the induction of hypnogogic imagery during psychotherapy, and to catalyze paranormal experience; and Lubar has used theta feedback in treatments for sleep disorders. Subjects of such therapy, Lubar claimed, reported that disturbing thoughts during drowsiness were replaced by images of pleasant colors, forms, and places.[30]

Beta-Wave Training

Psychophysiologist D. E. Sheer suggested that a narrow frequency band centered at 40 cycles per second (cps) reflects a focused arousal associated with learning processes. He based his assertion upon studies in which he found 40 cps activity in the visual cortex of cats during visual discrimination learning and in children who were mastering tasks involving short-term memory. Studies conducted in the 1950s by Das and Gastaut (23.1) also led him to these conclusions. Sheer's research suggests that 40 cps training might help some people improve memory consolidation and relieve learning disabilities. Lending support to Sheer's suggestions, B. L. Bird and associates trained normal subjects to increase or decrease 40 cps activity and in further experiments showed that their subjects could produce such activity without feedback.[31]

Brain-Wave Asymmetry Training

Erik Peper found that some people can learn to produce different amounts of alpha in each brain hemisphere simultaneously. In a study reported in 1971, his subjects exhibited such control after they were provided with auditory feedback whenever alpha asymmetry was present. Suspicious that his results were confounded by his subjects' initial alpha asymmetry, which may have been amplified when their total alpha activity increased, he conducted a second study using separate tones to index the presence of symmetry and asymmetry, but got equivocal results.[32] Subsequently, however, other researchers demonstrated that alpha activity could be controlled in one hemisphere alone; that a 14-year-old boy could increase alpha in his left hemisphere while increasing beta or theta responses in his right; that male and female subjects could suppress alpha in both hemispheres or suppress it in one while enhancing it in the other; and that certain subjects could increase alpha at one site while decreasing it at another in the same hemisphere.[33]

[16.5]
VOLUNTARY CONTROL OF OTHER BODILY FUNCTIONS

Heart Rate

After Tarchanoff published his studies of heart control in 1885 (see 16.1), a few physicians and physiologists described subjects who could modify their heartbeat. In the 1960s such observations helped stimulate numerous conditioning experiments in which the subjects were not informed that their heart rates were being altered.[34] However, psychologist Peter Lang and associates showed that subjects with true feedback reduced their heart-rate variability more than those who were told that such feedback was a tracking task. Lang's studies helped discredit belief that a subject's knowledge that his heart was being conditioned would interfere with the heart's experimental control.[35] Stimulated by Lang's study and the growing interest in biofeedback generally, several researchers began to study heart-rate self-regulation.[36] Lang and his associates also showed that, while instructions alone induced subjects to alter their heart rates, feedback helped them augment such control and maintain it when no feedback was given. Subsequently, other researchers confirmed Lang's finding.[37]

Heart malfunctioning has also been modified—and sometimes eliminated—through biofeedback training. Sinus tachycardias, Wolff-Parkinson-White syndrome, and fixed atrial fibrillation have been controlled by patients given beat-to-beat feedback during laboratory sessions and at home; and patients with premature ventricular contractions have learned to reduce the prevalence of their dysfunctional beats.[38]

Blood Pressure

In the 1970s and 1980s, studies of blood-pressure modification were refined and controlled for several confounding effects, and their cumulative results showed that:

- with the help of feedback or relaxation training, many subjects can modify their systolic pressure;
- systolic pressure is easier for most subjects to control than diastolic pressure;
- diastolic pressure may be easier for most subjects to *raise* than systolic; and
- subjects trained to lower their blood pressure with both feedback and relaxation training can sustain marked reductions of their diastolic pressures for more than one year.

Johann Stoyva showed that relaxation disciplines can complement biofeedback in helping people acquire such control.[39]

Electrodermal Activity

Electrodermal activity is measured by passing a small constant current between two points on the skin (often on the palm of the hand) and determining the resistance to it while it is in a steady state or changes in resistance to it that occur spontaneously or in response to internal or external stimuli. Experimental psychologists have studied such activity in large part because it reflects stress and anxiety, and much experimental evidence now strongly suggests that electrodermal activity can be operantly conditioned.[40] But subjects of operant conditioning research are typically not told which bodily processes are being conditioned. To paraphrase Elmer Green, in such experiments "the cerebral cortex of the experimenter conditions the autonomic systems of the subjects."

However, several studies conducted since the 1960s have shown that subjects can modify their electrodermal activity using either feedback or exercises such as recalling stressful situations. Some researchers found, for example, that actors trained in the Method school of acting could increase or decrease their skin resistance better than those who were not Method-trained.[41] Such studies have also shown that subjects vary greatly in their ability to discriminate changes in their electrodermal activity as well as in their ability to modify it. The voluntary control of skin activity, it seems, is acquired more effectively by some subjects through relaxation, meditation, or imagery than through instrumented feedback.[42]

Peripheral Temperature and Circulation

Changes in hand temperature and peripheral blood flow have also been brought under some measure of control with the help of feedback in several clinics and laboratories—to facilitate relaxation, to treat migraine headache, or to relieve other disorders.[43]

Gastrointestinal Functioning

Through biofeedback training, many people have learned to modify their gastrointestinal secretion and motility. Such control, which involves glands and smooth muscles that are usually regulated by the autonomic nervous system, can be therapeutic. Patients with reflux esophagitis have learned to increase their lower esophageal sphincter contraction, for example, thus providing a barrier against reflux. People with fecal incontinence resulting from neuromuscular impairment have learned to control their anal sphincter. And patients suffering from stomach acidity, ulcers, or irritable bowels have learned to suppress their abnormal smooth-muscle responses and acid secretions.[44]

[16.6]
Aspects of Biofeedback Training That Bear upon Transformative Practice in General

In spite of the practical success of biofeedback training, experimentalists and clinicians have argued about its effectiveness and possible mediations. Because certain issues framed by this debate have a bearing upon transformative practice in general, I will briefly review them here and provide references to fuller discussions of them.

The Quality of Training Required to Make Biofeedback Effective

Among the several thousand biofeedback studies published in scientific journals, many have appeared to show that biofeedback has not worked with a particular subject group, or has not relieved a particular affliction, or has not helped its subjects master a particular skill. According to Alyce and Elmer Green, 15 of the first 16 published studies using instrumented feedback for control of hypertension either failed completely or had positive results without clinical value. The Greens wrote:

> These research failures were generally well designed from a conditioning point of view, used excellent statistical methods, but were *clinically* defective. Specifically, successful clinicians do not use methods in which the patient is prevented from (or not aided in) developing self-awareness. The research designs and protocols used in [these] investigations satisfied those who favor control of subjects by instrumental conditioning, but . . . not those who favor freedom from conditioning, increased consciousness, and success *by the patient* in developing skill in self-regulation.
>
> One professor, for instance, demonstrated that biofeedback for control of heart rate is of no value whatsoever. He used a large number of subjects (183 college sophomores), processed the data with impeccable statistics, but provided his subjects with only *six 5-minute training sessions!* With an equally large "n," with the same statistics, and with six 5-minute training sessions, we could demonstrate that hearing (auditory feedback) is of no value whatsoever in learning to play the piccolo.
>
> The single early research study which reported success at followup in training hypertension patients (Patel 1973 & 1975a) had a clinical design in which home practice of *shavasan*, a traditional yogic practice for deep relaxation and bodily self awareness, was used in conjunction with laboratory feedback of galvanic skin response.
>
> In contrast to the above-mentioned conditioning studies (in which consciousness was ignored), our own work in the field of hypertension uses a variety of self-regulation techniques, including thermal biofeedback (emphasizing increased blood flow in the feet), striate muscular relaxation, breathing exercises, and Autogenic Training which includes visualization. We stress self-awareness, blood flow control, blood pressure reduction, and medication reduction. In our first series of patients, 6 out of 7 who had been using prescription medications for hypertension control (from six months to twenty years), were able to regulate

their blood pressures and at the same time reduce their medication levels to *zero*. Later, additional hypertensive patients were trained, many in groups, with similar positive results.

Our medication-free patients attributed maintenance of their success over a long period of time (the longest follow-up being 8 years) to being able to maintain the self-awareness and self-control which developed during their original training.[45]

Biofeedback researchers Solomon Steiner and W. Dince recommended that to achieve sound results in laboratory experiments

the length of training on the biofeedback task must be increased so that it becomes more consistent with existing clinical practice. While 8 to 25 hours of training have proven necessary for effective and lasting symptom alleviation . . . most biofeedback studies provide the subject with less than 4 hours. Most efficacy studies to date have provided less than 50% of the training time required for a good clinical outcome.[46]

In his presidential address to the Biofeedback Society of America in 1983, psychologist Steven Fahrion made a similar set of suggestions. Researchers, he recommended, should demonstrate that their subjects have learned the self-regulatory task involved.[47] Fahrion, Steiner, Dince, and the Greens all suggested that, to be effective, biofeedback training must last long enough and be designed well enough to provide its subjects with reliable self-regulation skills. Good design here requires that experimenters give subjects a rationale for their procedures, as well as a program that aims to promote their awareness and self-mastery. Operant conditioning, which neglects the subject's awareness and volition, should be replaced by experimental models that emphasize voluntary control of physiological processes. The Greens have argued, too, against those who object that an experimenter's high hopes for a treatment merely invoke the placebo effect. It is an error, they wrote,

not to recognize that the placebo effect is merely a subdivision of the general "self-regulation effect." When visualization and expectancy are activated through the use of a placebo, the effect almost invariably fades as time passes because it is attached to an inactive medical factor coupled with fading visualization and expectancy. But when visualization and expectancy are attached to one's own volitional effort, then a true skill can be developed, embodying genuine self-regulation of the autonomic nervous system. . . . For training to be most effective, the patient must thoroughly understand that self-regulation is psychological, neurological, and practical.[48]

These suggestions about biofeedback studies apply to all transformative disciplines in that they highlight the need for sufficient practice, motivation, and self-awareness to achieve self-mastery. Few of us can realize lasting beneficial changes of any kind without great commitment and effort.

The Role of Awareness and Volition in Self-Regulation

Researchers frequently have trouble providing their subjects with precise instructions for controlling autonomic functions, and subjects who have learned to control such functions often have trouble describing how they did so. Difficulties in talking about biofeedback experience concern some researchers but should not surprise anyone acquainted with nonverbal activity. The acquisition of all motor and sensory skills involves subtleties that escape detection, as well as features that don't have commonly accepted names. With practice, however, one can discriminate fleeting sensations and images associated with somatic processes and use them as cues for controlling such processes without instrumented feedback.

But operant conditioners have cited the fact that some subjects cannot detect any physiological awareness associated with their newfound control to justify their belief that awareness and volition do not play central roles in the extension of self-control to autonomic processes. Several researchers, however, after long-term research with many subjects, have found that awareness can indeed develop around subtle feats of self-control such as a single nerve cell's firing.[49] Internal cues associated with particular physiological functions are often reported by subjects well trained in self-regulation; and such cues can be substituted for the instrumented biofeedback with which they were associated. "The machine's information," wrote the Greens, "[is then] replaced by direct body awareness."[50] Furthermore, somatic awareness does not require verbal descriptions. If a subject apprehends the subtle, fluctuating, and sometimes evanescent sensations associated with changes in his brain waves, he need not find terms to describe them in order to control their activity. Such verbal activity, in fact, can make self-control more difficult (or impossible), as it would if one were descending a stairway three steps at a time or performing any other task that demands strict attention.

Given the role of somatic awareness and volition in acquired self-regulation, several researchers have conceived biofeedback training to be a kind of skills learning. Psychologist Peter Lang has argued that it is more analogous to learning to play darts or tennis than to an animal being operantly conditioned.[51] In Lang's view, feedback should be regarded as information rather than reinforcement. Drawing from the literature of motor skills learning, David Shapiro and Richard Surwit suggested that

> while reinforcement is most effective the closer it occurs in time to the preceding response, feedback operates best the closer it is to the next response. That is, in the feedback information conception, the delay from feedback to the next response is more crucial than the delay from the response to feedback (reinforcer).[52]

In operant conditioning, moreover, the schedule of reinforcement is crucial to how quickly learning takes place and how long lasting it turns out to be, while motor skills learning is affected more by the content of feedback than by its schedule. Feedback, Shapiro and Surwit argued, contains information that provides response-correcting properties as well as confirmation of correct performance. Such information may be reinforcing when it gives immediate encouragement and reminds the subject of rewards to come, but it is more than that. In this sense, the skills learning model of biofeedback training includes reinforcement theory while adding elements to it.

But Shapiro and Surwit have also suggested that operant conditioning can be used to maintain the behaviors that biofeedback facilitates. New self-regulation skills, they suggested, can be placed under reinforcing social contingencies once they are acquired.[53]

The Signal-to-Noise Ratio in Biofeedback Training

As noted in 16.3, Lloyd and Shurley found that sensory isolation can facilitate biofeedback training. Their experimental design was prompted by the notion that biofeedback could be viewed from a simple cybernetic analysis of signal-to-noise ratio. By reducing sensory distractions, they proposed, experimenters enhance their subjects' attention to feedback displays and to proprioceptive signals. The success of the Lloyd-Shurley experiments supports the ancient observation that mental quietude and solitude promote yogic self-control. Indeed, the first admonition of Patanjali's Sutras, India's most celebrated yoga scripture, is *yogas citta-vritti nirodhah,* "through yoga the mind's subtle turnings cease." By silencing the noise, by calming the turbulence that interferes with kinesthetic or sensory perception, we come to greater self-knowledge and mastery. In Part Three, I propose that inner quietude and the heightened sensitivity it brings are fundamental for transformative practice.

Individual Differences in Developing Self-Regulation Skills

The research picture is still sketchy regarding individual differences in learning self-regulation skills.[54] This is the case, perhaps, because few researchers have given their subjects sufficient self-regulation practice to uncover significant differences between them. Elmer Green observed in a letter to me, for example, that good "somatizers" who quickly turn personal issues into vivid bodily responses are generally superior subjects in biofeedback programs. If more clinicians and experimentalists assessed their subjects carefully, he suggested, they would recognize that fact more clearly. Green's description of good biofeedback subjects as somatizers resembles Theodore Barber's characterization of hypnotic virtuosos (15.6).

It is also possible that people differ in the raw strength of their volition, so that subjects with high internal awareness may not be able to increase their self-control as easily as individuals with less awareness but stronger will. And it is also conceivable that individuals vary in their ability to coordinate awareness with volition. This is conjecture, but does not seem farfetched if we consider self-regulation to be a complex skill involving several components.

The Variety of Cognitive and Somatic Mechanisms That Mediate Self-Regulation

Somatic functioning can be modified in various ways. All of us can alter our heart rate, for example, even when sitting still, through simple muscle movements in the chest or by changing the speed and depth of our breathing; and some people can increase and decrease it even without these activities.[55] Electrical activity in the brain can also be changed by different maneuvers: by movements of the eyes, for example, especially their convergence, which decreases alpha-wave activity, or their divergence, which increases it; by inducing changes in heart rate; or by cognitive exercises such as visualization and unfocused attention.[56] In short, self-regulation skills, like other capacities, are mediated in more than one way and can therefore be developed through different kinds of practice.

Specific and General Control of Cognitive and Somatic Processes

Some biofeedback researchers propose that it is simpler to control integrative physiological networks than to modify the more specific processes they include.[57] One such integrative network involves the ergotropic and trophotropic systems.[58] (Activation of the ergotropic system occurs during anger, excitement, fear, and other forms of arousal, while the trophotropic is activated by relaxation.) Somatic responses that result from stimulation of the two systems are immensely complex, involving striate muscles, smooth muscles, hormones, opioid peptides, the brain, and autonomic nervous system. Although no one has shown that biofeedback training can cope with all of these in a coordinated fashion,[59] meditation and somatic disciplines such as Autogenic Training, by stimulating the relaxation response, modify several at once.

At the same time, however, certain highly specific physiological processes can be modified for beneficial results. For example, tension headaches can be relieved by controlling the frontalis muscle, and migraine can be alleviated by redistribution of blood flow through hands and feet. Specific responses can also be modified to promote extraordinary functioning, for example, when ecstatic trance is induced by the slowing of breath. In short, both cognitive and somatic, and both normal and extraordinary functioning can be improved by controlling either general systems or specific organs and muscles. Autogenic Training includes both exercises for general relaxation and organ-specific formulae to modify specific functions. Jacobsen's progressive relaxation uses both general and specific methods for self-control (18.3, 18.6). Like these disciplines, integral practices can employ interventions at various levels of the organism's complex hierarchy, from a single cell to generalized responses mediated by the central nervous system.

17

PSYCHOTHERAPY AND
IMAGERY PRACTICE

[17.1]
THE VARIETIES OF PSYCHOTHERAPY

Many types of psychotherapy are practiced today.[1] Some, like Jungian therapy, are departures from psychoanalysis; others are based on recent refinements of Freudian theory; still others have roots in behaviorism, cognitive psychology, and other kinds of psychological inquiry. These many therapies include countless techniques and address different problems or levels of development. *Structure-building* methods, for example, aim to strengthen the underdeveloped self by reinforcing self-reliant initiatives and by revealing the clients' tendencies to over-identify with parents, mentors, and peers.[2] *Uncovering* techniques such as free association, on the other hand, are designed to lift repression in the already developed self. *Role clarification* (in Transactional Analysis and family therapy, for example) is used to illuminate hidden agendas, dishonest transactions, and confused self-representations.[3] *Cognitive restructuring* is designed to correct limiting or destructive beliefs.[4]

Methods such as these are meant to alleviate suffering and overcome psychological disabilities. But some therapies address more than pathology or marked dysfunction. The existential approaches of Ludwig Binswanger, Medard Boss, Rollo May, and James Bugental embrace fundamental concerns with life-meaning, authenticity, and death, and in so doing support autonomy and self-reflection.[5] Jungian therapy aims to enrich consciousness and promote individuation. Transpersonal therapies encourage spiritual insight, joining elements of medical therapy and meditation practice in comprehensive programs for personal growth.[6]

As therapeutic experience has developed in the twentieth century, sophistication has developed (among some therapists at least) about the need for proper diagnosis in the face of complex needs, difficulties, and possibilities for personal growth. Competent psychotherapists, for example, will not usually emphasize uncovering techniques with people who have poor impulse control or underdeveloped egos, while experienced transpersonal therapists might not recommend meditation for depressed or obsessional clients. A number of psychiatrists have matched types of psychological problems with their appropriate therapeutic approaches;[7] and the American Psychiatric Association publishes a Diagnostic and Statistical Manual that distinguishes psychological afflictions and suggests the best treatments for them.[8] Philosopher Ken Wilber has made a brave attempt to correlate a wide spectrum of difficulties—from psychosis to miscarriages of contemplative development—with their suitable therapies.[9]

With their different aims and methods, modern psychotherapists have claimed many kinds of therapeutic success. Critics of the field often ask, however, whether such claims are justified. Does this or that therapy actually produce the positive results its advocates advertise? Answers depend, of course, on the particular school or therapist under scrutiny: some methods and some therapists have better records than others, and many therapeutic successes result from nothing more than placebo effects. Nevertheless, it is clear that many people have benefited from psychotherapy. Some readers, I suspect, have enjoyed good results themselves; and there can be little doubt that all the positive outcomes reported in reputable journals have been realized upon occasion. This holds true, it seems to me, even though there is considerable disagreement about the efficacy of particular approaches.

In the decades since psychiatrist H. J. Eysenck argued in a much-cited article that psychotherapy produces no more positive change than normal life experience, much research has shown that therapy can alleviate suffering and promote various kinds of growth.[†] For example, a landmark survey of 474 outcome studies (involving some 25,000 clients) indicated that the average person in psychotherapy was better off at the end of treatment than 80 percent of control subjects. This study embraced many kinds of therapy, estimated the magnitude of treatment effects, and examined complex correlations between client and therapist attributes, techniques, and actual outcomes. Though it drew upon some studies that were poorly controlled, and

† Eysenck claimed that outcome studies available at the time "did not support the hypothesis that psychotherapy facilitates recovery from neurotic disorder" (Eysenck 1952).

forced divergent methods together for statistical analysis, it showed that therapy does indeed produce more positive than negative results.[10] Even if we have doubts about this or that method, and sensibly distrust particular therapists, it is unreasonable to deny that many people have changed for the good during psychological treatment. Indeed, psychotherapy produces many kinds of positive result, from alleviation of anxiety to improved communication skills to life-enhancing philosophic insight. Many people find relief from depression, for example, or discover new ways to show affection, or overcome vacillations of will. Some receive insight into a chronic sense of failure or guidance in probing the very meaning of life. It is probably impossible to enumerate every kind of assistance psychotherapy provides today. A complex culture of human development has arisen under its banner, with elements of medical treatment, emotional education, and moral training. That is why most therapists use more than one approach, even if they give first allegiance to a particular school.[†] But no social historian has adequately described the field's many-sided development nor the great range of its accomplishments. Most laypeople, and many academic psychologists, are unaware how broad the field is, identifying the whole of it with psychoanalysis, behaviorism, or some other school. The richness of contemporary psychotherapy is generally unappreciated by social scientists and the general public alike.

On the other hand, psychological treatment is not always successful. It can, for example, promote some capacities at the expense of others. Certain therapists fail to appreciate the mutual entailment of the virtues, encouraging honesty without much attention to kindness, courage without prudence, or self-assertiveness without empathy. Most therapists have their limitations, and all therapeutic approaches have weaknesses as well as strengths. Nevertheless, psychotherapy can facilitate ethical-emotional education, psychic repair, philosophic insight, and even new joy and purpose in life. The field has produced insights and methods that are crucial to programs for human growth. The integral development proposed in this book depends upon healthy emotions, self-awareness, and the coordination of conflicting

---

[†] Responding to a survey in 1973, 55 percent of the members of the American Psychological Association's Division of Clinical Psychology said they drew upon more than one orientation or method (Garfield & Kurtz 1974 & 1976). Such eclecticism, along with the fact that most therapies have nonspecific as well as specific effects, makes it impossible to limit particular outcomes to particular approaches. For example, a course of behavior modification for a specific phobia can foster self-confidence and improved self-esteem, while a Jungian analysis to promote understanding of the deep self might relieve a once-intractable phobia.

impulses, all of which can be promoted in ways developed by modern therapists. In chapter 26.1, I note some psychotherapeutic methods that facilitate particular virtues and extraordinary attributes.

[17.2]
IMAGERY PRACTICE FOR HEALING AND GROWTH

Many clinical and experimental studies have shown that imagery practice can facilitate relief from various afflictions, among them depression, anxiety, insomnia, obesity, sexual problems, chronic pain, phobias, psychosomatic illnesses, cancer, and other diseases.[11] Furthermore, imagery practices have been developed to promote particular virtues, skills, and capacities. Anees Sheikh and Charles Jordan have written a useful summary of such disciplines, from which I have drawn the following list.[12]

Janet. In the 1890s French psychiatrist and philosopher Pierre Janet pioneered the use of imagery in therapy, in part by showing that one image could be substituted for another in the ideation of his hysterical patients.[13]

Freud. In his early clinical work, Freud evoked images in his patients by pressing their foreheads, but he abandoned the method around 1900. Nevertheless, some Freudian psychoanalytic procedures help stimulate imagery, among them reclining in a restful attitude, mild sensory deprivation produced by looking at the ceiling, attention to dreams and transference fantasies, use of free association, and emphasis on early memories.

Jung. Jung regarded mental imagery to be a fundamental, ongoing, creative aspect of the psyche that could be attended to for cure and personal growth. His *active imagination* method, which has influenced many psychotherapists, brings imagery into focus so that it might enrich awareness and guide the individuation process.[14] Ordinary fantasies, Jung maintained, are merely personal inventions operating on the surface of things, while active imagination arises from the psyche's larger purposes and mythic structures. In some of his writings Jung claimed that active imagination was superior to dreams in quickening psychological development.

Binet. Psychologist Alfred Binet, famous for his intelligence tests, encouraged his patients to converse with their visual images in a state he called provoked introspection. Like Janet, he believed that psychological material that emerged during such introspection revealed his clients' diverse though largely unconscious subpersonalities.

Happich. In Germany during the 1920s, Carl Happich developed Binet's therapeutic approach by inducing emergent images through muscular relaxation, attention to breath, and meditation. He described a meditative zone in which productions that matured in the unconscious became visible to the mind's eye. To stimulate imagery, he had his patients imagine specific scenes such as a chapel, a meadow, or a mountain.[15]

Caslant. In France, Eugene Caslant encouraged his subjects to ascend or descend in imaginal space. He aimed in such exercises to free his subjects from habitual mental restraints, thus helping them generate imagery that could facilitate extrasensory experience and personal growth.[16]

Post–World War II European clinicians. Since World War II, several influential European clinicians have employed imagery as their primary therapeutic modality. Robert Desoille gave impetus to such work through his *le reve eveille dirige,* or directed daydream, a method derived from Caslant, Jung, and Freud. Andre Virel and R. Fretigny, who were influenced by Desoille, developed oneirodrama, or dream drama, which employs fantasies in narrative form; and Hans Carl Leuner worked with a method, called guided affective imagery, which he traced to Freud's early therapy. All of these therapists used extended fantasies to obtain information about their subjects' motivations, conflicts, perceptual distortions, self-perceptions, and early memories.

North American behavior therapists. From 1920 to 1960, largely because the influence of behaviorism was so strong, relatively little clinical or experimental work on mental imagery was conducted in North America, and not a single book on the topic was published there. But sensory deprivation studies, sleep research, and work with hallucinogenic drugs in the late 1950s and early 1960s revived interest in subjective experience among experimentalists, while clinicians of various schools developed a new appreciation of imagery's usefulness in therapy. In an ironic twist of history, behavior therapists were among the first clinicians to make imagery's use respectable. For example:

- *Systematic desensitization,* credited in large part to Joseph Wolpe, gradually introduces visualizations of disturbing stimuli while the subject is relaxed, so that objects and situations they represent become less threatening.[17]

- *Implosion, or flooding,* exposes a subject to his or her fears through intense visualizations of threatening objects (such as snakes) or situations (such as speaking in public) so that the fears can be extinguished without aversive consequences.[18] Unlike systematic desensitization, implosion often arouses a high degree of anxiety, in part by leading the subject (imaginatively) beyond his worst fears, into situations he would rarely if ever encounter in ordinary circumstances.

- *Covert conditioning,* developed by Joseph Cautela, includes a number of techniques that are the covert analogues of operant and social-learning procedures, among them covert positive reinforcement (imagining rewards for particular behaviors), covert negative reinforcement (imagining painful consequences of particular behaviors), covert extinction (imagining some unwanted trait or behavior disappearing), and covert modeling (imagining a desired behavior).[19]

Eugene Gendlin. Gendlin and his colleagues have developed *experiential focusing* as a therapy and means of personal growth. In this approach, an image often emerges to move the subject from a general, intuitive sense of a problem to its fundamental core. According to Gendlin, once an image is sufficiently engaged it changes, with a characteristic sense of release, and its meaning opens up. Therapists who use this method claim that such opening is typically accompanied by the client's new understanding of personal issues or by creative solutions to the problems at hand.[20]

Reyher and Storr. Joseph Reyher, Joseph Shorr, and other psychoanalytically oriented therapists have incorporated imagery as a primary modality in their clinical work. Reyher has used classical free association, but with an emphasis on images rather than words. It is puzzling to clients, for example, when anxiety-provoking images appear to be innocent. "Their curiosity thus piqued," Reyher wrote, "they are invited to revisualize [them]. Anxiety and resistance is [thus] intensified and symptoms are exacerbated as underlying strivings become depicted with increasing clarity." Reyher has called this process *emer-*

gent uncovering, maintaining that a patient can derive significant insight from it even without a therapist's interpretation.[21]

Joseph Shorr has employed more active intervention than Reyher in his *psycho-imagination therapy,* though like Reyher he has based his work largely upon the evocation and interpretation of imagery. In his group therapy, for example, all the participants imaginatively visualize a particular member, who then visualizes the others; or all the participants might visualize the therapist in response to his images of them. In the therapeutic situation, Shorr has encouraged his patients to distinguish their own view of self from those attributed to them by significant others.[22]

Eidetic psychotherapy. Developed by Anees Sheikh, Akhter Ahsen, and their associates, eidetic psychotherapy emphasizes the personality's multiplicity of states rather than its unity, and it relies upon the elicitation and manipulation of tridimensional eidetic images. These are conceived to be semipermanent representations of formative events in one's past consisting of a visual part (image), a set of bodily feelings and tensions (somatic pattern), and a cognitive or experiential meaning. The eidetic image, in this conception, "displays certain lawful tendencies toward change."[23] Eidetic therapists try to revive tridimensional images, which are usually fragmented or distorted, so that they might move toward wholesome expression and promote personality development.[24]

Gestalt Therapy. Fritz Perls, the primary inventor of Gestalt Therapy, helped popularize imagery-based psychotherapy. Gestalt therapists commonly employ various forms of psychodrama in which their clients act out images of their subpersonalities or various segments of their dreams, fantasies, and life situations so that they might incorporate dissociated parts of themselves into their conscious functioning.[25]

Humanistic and transpersonal psychology. Several therapists and educators affiliated with humanistic and transpersonal psychology, among them Ira Progoff, Robert Gerard, Robert Masters, and Jean Houston, have developed ways to use imagery for healing and personal growth.[26]

By removing certain obstacles to creative functioning and by promoting particular virtues and traits, the imagery practices listed here can contribute to programs for balanced human growth. In chapter 26.1, I note certain ways in which they can be incorporated into integral disciplines.

[17.3]
PARANORMAL EVENTS AND TRANSPERSONAL PERSPECTIVES IN PSYCHOTHERAPY

Freud's Acceptance of Telepathy

In 1921 psychoanalyst Wilhelm Stekel published a book about telepathy in dreams;[27] and Freud himself concluded, after decades of intellectual conflict, that telepathy was a fact, publicly acknowledging his conclusion in 1925 in an essay prepared for his *Collected Writings*. Both Stekel and Freud suggested that telepathically acquired information was filtered and shaped in the same largely unconscious ways in which sensorily acquired information is mediated.[†] Freud's British colleague and future biographer Ernest Jones was so alarmed by the master's acknowledgment of telepathy that he sent a circular letter to fellow analysts in Britain and Europe, warning that

> this month's "Psyche" has a leading article entitled "The Conversion of Freud," and also a second one on the subject containing the following passage: "A few years ago the analysis of dreams must have seemed to many adherents of the Viennese school to be developing into a not altogether inexact science. . . . But today the wild men are once more not far from the fold—for if Telepathy be accepted the possibility of a definite oneiric aetiology recedes some decades, if not centuries, into the future." Much wilder articles to the same effect have appeared in the popular Press.[28]

Freud responded to Jones's protest with his own circular to colleagues and friends, asserting that

> our friend Jones seems to me to be too unhappy about the sensation that my conversion to telepathy has made in English periodicals. He will recollect how near to such a conversion I came in the communica-

---

† In an earlier paper on dreams and telepathy, read by Karl Abraham before the Vienna Psychoanalytic Society and published in the March 1922 issue of *Imago,* 8:1–22, Freud had not affirmed a belief that telepathy actually happened. "You will learn nothing from this paper," he wrote, "about the enigma of telepathy; indeed, you will not even gather whether I believe in the existence of 'telepathy' or not" (see: Freud 1955). In his 1925 essay, though, Freud indicated his belief that telepathy was a fact of human experience (see: Freud 1964). Some of the observations upon which Freud based his conclusions regarding telepathy may be found in: Jones, E., 1957, vol. 3, chap. 14; in the two essays cited above; and in Devereux 1953, chaps. 3–8.

tion I had the occasion to make during our Harz travels. Considerations of external policy since that time held me back long enough, but finally one must show one's color and need bother about the scandal this time as little as on earlier, perhaps still more important occasions.[29]

Jones then wrote to Freud directly:

You are doubtless right, as usual, when you say that I am too much oppressed by the telepathy matter, for in time we shall overcome the resistance it evokes just as we do all others. But you are lucky to live in a country where "Christian Science" together with all forms of so-called "psychical research" mingled with hocus-pocus and palmistry do not prevail as they do here to heighten opposition to all psychology. Two books were written here recently trying to discredit psycho-analysis on this ground alone. You also forget sometimes in what a special position you are personally. When many things pass under the name of psycho-analysis our answer to inquirers is "psycho-analysis is Freud," so now the statement that psycho-analysis leads logically to telepathy, etc., is more difficult to meet. In private political opinions you might be a Bolshevist, but you would not help the spread of psycho-analysis by announcing it. So when "considerations of external policy" kept you silent before I do not know how the situation should have changed in this respect.[30]

Again Freud answered Jones:

I am extremely sorry that my utterance about telepathy should have plunged you into fresh difficulties. But it is really hard not to offend English susceptibilities. . . . I have no prospect of pacifying public opinion in England, but I should like at least to explain to you my apparent inconsistency in the matter of telepathy. You remember how I had already at the time of our Harz travels expressed a favorable prejudice towards telepathy. But there seemed no need to do so publicly, my own conviction was not very strong, and the diplomatic consideration of guarding psychoanalysis from any approach to occultism easily gained the upper hand. Now the revising of *The Interpretation of Dreams* for the Collected Edition was a spur to reconsider the problem of telepathy. Moreover my own experiences through tests I made with Ferenczi and my daughter won such a convincing force for me that the diplomatic considerations on the other side had to give way. I was once more faced with a case where on a reduced scale I had to repeat the great experiment of my life: namely, to proclaim a conviction without taking into account any echo from the outer world. So then it was unavoidable. When anyone adduces my fall into sin, just answer him calmly that conversion to telepathy is my private affair like my Jewishness, my passion for smoking and many other things, and that the theme of telepathy is in essence alien to psychoanalysis.[31]

Upon receiving the letter just quoted, Jones apparently resigned himself to Freud's decision. "There was no more to be said," he wrote. Freud was going on record that he believed in telepathy even though his admission might impede the acceptance of psychoanalysis by the medical community. The analysts associated with Freud, concerned to achieve scientific respectability for their already controversial theories, realized that to acknowledge telepathy as a fact was a step of great consequence. "It would mean admitting the essential claim of the occultists," wrote Jones, "that mental processes can be independent of the human body."[32] Freud understood this well, and made the point in print. *"Dans ces cas pareils, ce n'est que le premier pas qui coute. Das weitere findet sich,"* he wrote about the acceptance of telepathy. "It is only the first step that counts. The rest follows." Jones wrote:

In the years before the Great War, I had several talks with Freud on occultism and kindred topics. He was fond, especially after midnight, of regaling me with strange or uncanny experiences with patients, characteristically about misfortunes or deaths supervening many years after a wish or prediction. He had a particular relish for such stories and was evidently impressed by their more mysterious aspects. When I would protest at some of the taller stories, Freud was wont to reply with his favorite quotation: "There are more things in heaven and earth than are dreamed of in your philosophy." Some of the incidents sounded like mere coincidences, others like the obscure workings of unconscious motives. When they were concerned with clairvoyant visions of episodes at a distance, or visitations from departed spirits, I ventured to reprove him for his inclination to accept occult beliefs on flimsy evidence. His reply was: "I don't like it at all myself, but there is some truth in it," both sides of his nature coming to expression in a short sentence. I then asked him where such beliefs could halt: if one could believe in mental processes floating in the air, one could go on to a belief in angels. He closed the discussion at this point (about three in the morning!) with the remark: "Quite so, even *der liebe Gott.*" This was said in a jocular tone as if agreeing with my *reductio ad absurdum* and with a quizzical look as if he were pleased at shocking me. But there was something searching also in the glance, and I went away not entirely happy lest there be some more serious undertone as well.[33]

Jung's Theories of Synchronicity and Transpersonal Archetypes

That accepting telepathy would open the door to occult and mystical phenomena, as Jones worried, even to *der liebe Gott,* was made evident by the life and work of Carl Jung. The famous psychiatrist's theories of synchronicity, or meaningful coincidence, and transpersonal archetypes, and his sympathetic attention to religious experience extended psychotherapy well beyond conventional medical treatment. Reading Jung's autobiography *Memories, Dreams, and Reflections* and other accounts of his development, one observes the inrush of extraordinary phenomena that Jones seemed to fear.[34] While the first psychoanalysts were trying to show that many or most religious beliefs were projections of unconscious processes or products of longings for omnipotent parents, Jung moved beyond such concerns to chart the autonomous psyche, an immeasurable world with its own objective reality.

Like Freud, in his published writings Jung did not subscribe to common beliefs about the spirit world (though he seemed to do so in some of his letters), but he widened the boundaries of psychotherapy to include many kinds of extrasensory events. Claiming that psi phenomena arise from a biologically rooted collective unconscious, he helped legitimize their acceptance by psychotherapists of many persuasions. Jung's beliefs about such matters are reflected in the following excerpts from a letter he wrote in 1960 to A. D. Cornell, then president of the Cambridge University Society for Psychical Research.

> Paranormal psychic phenomena have interested me all my life. Usually, as I have said, they occur in acute psychological states (emotionality, depression, shock, etc.), or, more frequently, with individuals characterized by a peculiar or pathological personality structure, where the threshold to the collective unconscious is habitually lowered. People with a creative genius also belong to this type.
>
> Just as in physics we cannot observe nuclear processes directly, so there can be no direct observation of the contents of the collective unconscious. In both cases their actual nature can be inferred only from their effects—just as the trajectory of a nuclear particle in a (cloud) chamber can be traced only by observing the condensation trail that follows its movement and thus makes it visible.
>
> In practice we observe the archetypal "traces" primarily in dreams, where they become perceptible as psychic forms. But this is not

the only way they reach perception: they can appear objectively and concretely in the form of physical facts just as well. In this case the observation is not an endopsychic perception (fantasy, intuition, vision, hallucination, etc.) but a real outer object which behaves as if it were motivated or evoked by, or as if it were expressing, a thought corresponding to the archetype.

The archetype is not evoked by a conscious act of the will; experience shows that it is activated, independently of the will, in a psychic situation that needs compensating by an archetype. One might even speak of a spontaneous archetypal intervention. The language of religion calls these happenings "God's will"—quite correctly insofar as this refers to the peculiar behavior of the archetype, its spontaneity and its functional relation to the actual situation.

I must add, however, that I have observed and also partially analysed people who seemed to possess a supranormal faculty and were able to make use of it at will. But the apparently supranormal faculty consisted in their already being in, or voluntarily putting themselves into, a state corresponding to an archetypal constellation—a state of numinous possession in which synchronistic phenomena become possible and even, to some extent, probable. This faculty was clearly coupled with a religious attitude which enabled them to give suitable expression to their sense of the ego's subordination to the archetype.[35]

Dynamic Psychiatry and Psi

In 1947 a group of psychiatrists, most of them with psychoanalytic training, formed a Medical Section of the American Society for Psychical Research to explore paranormal phenomena from the perspective of dynamic psychiatry. Jan Ehrenwald, Jule Eisenbud, Joost Meerloo, Montague Ullman, and G. Pederson-Krag were members of this group. Like Frederic Myers and other researchers I frequently cite in this book, they collected much evidence for paranormal functioning, and their published data strongly suggest that telepathy, clairvoyance, and even psychokinesis occur in therapy, sometimes dramatically.[36] Some of them have argued, moreover, that many extranormal events escape the notice of parapsychologists who are unacquainted with the largely unconscious associational patterns and defensive maneuvers revealed by psychoanalysis. Because psi is often mediated by subliminal processes, much of it can best be understood with psychoanalytic insights and methods. The picture of psi-conditioned behavior that has emerged from certain data provided by psychotherapists since

Stekel and Freud suggests that paranormal capacities help us satisfy our basic needs. Such data indicate that psi can serve both our destructive and constructive impulses. Eisenbud, Ehrenwald, and their colleagues have provided evidence that paranormal events are woven into all of our functioning, whether hateful or loving. They have made a strong case, it seems to me, that psi-conditioned activity is part of our everyday life. Eisenbud wrote:

> In 1921 Wilhelm Stekel reported a number of patients' telepathic dreams which indicated that the dreamers did not concern themselves mainly with catastrophic events of major proportions but with the everyday experiences out of which dreams are ordinarily fashioned. The motives behind the paranormal acquisition of information showing up in these dreams, Stekel reported, arose not chiefly out of empathic concern over the misfortunes of loved ones but more often out of hate, discord and jealousy. In the following year Sigmund Freud showed, in a penetrating analysis of a presumptively telepathic dream sent him by a correspondent, that the laws by which the dreamer attempted to cope with a conflictful situation by his use of telepathically acquired information were the same as those that governed other areas of his unconscious mental life.
>
> Both Stekel and Freud extended their observations over the next few years. Stekel, besides emphasizing the type of everyday emotional linkage in telepathic interactions, showed that information telepathically received need not be consciously apprehended as such: for example, a woman might experience a painful bodily symptom rather than a veridical vision at the precise moment that her husband was betraying her. Freud showed that fortune-tellers were apt to pick up the repressed, unconsciously entertained wishes of their clients and that these wishes could be murderous. He also convincingly showed that a patient in treatment was sensitive to material of concern to the analyst, and that the reaction which was evident in the patient's associations—for instance jealousy—could be related dynamically to this material, just as if the patient had acquired conscious knowledge of what was taking place in the analyst's mind.
>
> Other investigators soon extended these observations. When a clinically based system of dream analysis was brought to bear, the everyday dreams of people were seen to incorporate, more frequently than heretofore had been imagined, information paranormally received. This information was demonstrably woven into the texture of the dreams, not only in the same manner as normally acquired information but also for the same purposes—chiefly the attempt to resolve conflict situations by the magical if only fantasied manipulation of the internal and external environment. As in ordinary dream elaboration, the

paranormally acquired information might be utilized in a symbolic manner, for example, the element fire to symbolize sexual arousal, the element water to symbolize birth. Sometimes elements such as newspaper headlines, or radio or telephone communication, would signify that telepathy or clairvoyance was being utilized in the dream.

In addition, data from waking life showed that paranormal processes are capable of conditioning not only all areas of thought and awareness but also our deepest levels of bodily functioning. Clinical data confirmed Stekel's observation that almost any type of symptom, from disturbances of the musculoskeletal system to disturbances of the cardiovascular-respiratory system, could be related to external events presumptively apprehended in a paranormal manner.[37]

Like Eisenbud, psychiatrist Jan Ehrenwald argued that psycho-analytic insights help show how psi operates in therapy. In his *New Dimensions of Deep Analysis,* first published in 1954, he presented case studies in which telepathy between him and his patients appeared to have a significant role. Summarizing these, he wrote:

A favourable psychological setting (such as psychotherapy) may create a more or less sustained telepathic susceptibility to highly complex motivational systems prevailing in the mind of a potential agent. [The] unintentional effect of preconscious motivations and attitudes . . . is here described as telepathic leakage and it is precisely the unintentional, preconscious nature of such leakage which makes it a matter of prime importance [for therapy]. As long as the therapist remains unaware of its operation, telepathic leakage may easily influence the result of his analytic inquiry and thus become a potential source of error vitiating his conclusions. True telepathic leakage goes beyond cultural conditioning. For instance, Jung's patients, at certain periods of his work, [had dreams that] contained grandiose allusions to ancient Egyptian or Assyrian or Hindu texts. At other times they seemed to bear out some of the fanciful beliefs of medieval astrologers and alchemists. No Freudian or Adlerian analyst seems to have ever recorded material of this kind. Yet I, for myself, have noticed peculiar changes in the dream imagery used by my patients at various periods of my analytic work. The changes seemed to be largely conditioned by my own preoccupation with the dream material encountered in the records of various analytic schools.

One could object at this point that all these discrepancies simply reflect changes within the general outlook of both patient and therapist, reinforced by their community of interests and intellectual preoccupations. This is a perfectly legitimate objection—but it only holds true for cases in which the transmission of specific material from the therapist to the patient can be explained in terms of overt verbal communication. It is not applicable to such observations as form the main topic of the present study.[38]

Eisenbud's and Ehrenwald's confident assertions about the pervasiveness of psi in doctor-patient relations seem a long step from Freud's cautious acceptance of telepathy 30 years before. By the 1950s, several psychiatrists had come to believe that telepathy plays a central role in both therapy and everyday life. Indeed, failure to appreciate the ubiquitous influence of paranormal phenomena, Eisenbud argued, could impede the therapist's understanding of the patient. He wrote:

> It is possible that the vast depths of the dream can never be plumbed without the systematic and wide ranging application of the psi hypothesis. However, the most fruitful field for the application of the psi hypothesis to human behavior is not necessarily in the area of dreams but in the ordinary transactions of daily life. These can be investigated under a variety of circumstances, the best point of departure being the analytic hour. The psi-conditioned interplay between the analyst and the patient in this affect-charged situation provides a highly concentrated and sharply focused sample of what presumably goes on in the variegated interactions between people in other areas.[39]

According to Eisenbud, Ehrenwald, and other psychotherapists, the paranormal interplay that happens in therapy occurs as well in everyday life. "Psi-conditioned events," Eisenbud wrote, "are to be looked for not just here, or there, or under just certain conditions, but everywhere, throughout the warp and woof of our existence."[40] In appendix A.5, I have listed several books that include psychoanalytic studies of paranormal interaction. This material can help us understand how our latent capacities for extraordinary life, as well as ordinary functioning, can be distorted by largely unconscious needs, drives, and conflicts.

Transpersonal Psychology

The therapeutic engagement with extrasensory experience that Stekel, Freud, and Jung initiated has been developed, too, by a loosely affiliated group of transpersonal psychologists in the Americas, Europe,

Japan, and Australia. No single figure governs their thinking, nor does any unified approach command their allegiance, but they frequently turn to Jung, Roberto Assagioli, Abraham Maslow, and William James, and to Eastern and Western contemplative practices for ideas about personal growth. The transpersonal orientation has not been supported by much experimental research but has been developed to some degree in articles published in the *Journal of Transpersonal Psychology* since 1969 and during meetings of the Association for Transpersonal Psychology.[41] Anthony Sutich, a principal founder of the *Journal* and Association, summarized some major assumptions underlying the transpersonal position:

> Transpersonal therapy is concerned with the psychological processes related to the realization, or making real, of states such as "illumination," "mystical union," "transcendence," and "cosmic unity." It is also concerned with the psychological conditions or psychodynamic processes which directly or indirectly form barriers to these transpersonal realizations. In both the past and the present, individuals have necessarily had different relationships to their impulses towards ultimate states and emotional growth; they have also varied in their levels of development in these at different times in their life cycles.[42]

Philosopher Ken Wilber has proposed a model of human growth that integrates a transpersonal perspective with the developmental schemes of Piaget, Maslow, Jane Loevinger, Margaret Mahler, and other psychologists, the moral development theories of Lawrence Kohlberg, and numerous contemplative teachings. In arguing that contemporary developmental theories can be joined with religious insights, Wilber represents the views of most transpersonal therapists.[43]

Many therapists with a transpersonal orientation have been influenced by the Italian psychiatrist Roberto Assagioli, who from the 1920s until he died in 1974 developed a comprehensive discipline for healing and growth that he called Psychosynthesis. Assagioli's approach draws upon psychoanalysis and the work of therapists such as Desoille, Leuner, and Jung, and it employs meditative techniques from both Eastern and Western contemplative traditions. It aims to shift the center of personality from normal consciousness to its fundamental core, which Assagioli sometimes called the true or higher self. The methods used in Psychosynthesis include symbolic visualization and guided daydreaming to illuminate unconscious issues or open up new dimensions of consciousness; active imagination to establish contact with the psyche's deep purposes and dynamics; and concentration upon positive symbols to reinforce desirable capacities.[44] In its integration of healing procedures with cultivation of metanormal capacities, and in its engagement with many aspects of human functioning, Psychosynthesis points the way toward the integral practices I discuss in Part Three.[45]

18

SOMATIC DISCIPLINES

THE FIELD OF SOMATIC EDUCATION

Somatic education is a term that is applied to several disciplines developed during the twentieth century in Europe and America. In recent years the field comprised by these methods has been called Somatics by American philosopher Thomas Hanna and *Somatotherapie* by French physicians and educators.[1] Philosopher Don Johnson, who more than anyone else has stimulated understanding of the field's basic principles, has distinguished somatic education from osteopathy, chiropractic, and standard medical practices that aim primarily or exclusively for symptom relief. In Johnson's words:

> Somatics is legitimately characterized as a field because its many methods share a common focus on the relationships between the body and cognition, emotion, volition, and other dimensions of the self. While mainstream medicine, orthopedics, physical therapy, chiropractic, and osteopathy treat the body as an independent entity, somatic practices explore the body in relationship to an individual's entire experience. Within that general unity of the field, a particular somatic method can be defined by its concentration on one or more bodily systems. Rolfing, for example, explores how the organization of connective tissue affects thought, perception, and emotion. The methods developed by F. M. Alexander and Moshe Feldenkrais focus on the neuro-muscular-skeletal systems. In a very different mode, the family of therapies that derive from Wilhelm Reich explore the individual's autonomic impulses and peristalsis, while the sensory awareness work of Elsa Gindler and Charlotte Selver involve a radical examination of perceptual abilities.
>
> A second characteristic which unites the field is a shared assumption that therapy or healing derives from fundamental transformations of experience and the cultivation of new capacities. Since they are primarily educators, practitioners of Rolfing, the Feldenkrais Method,

Sensory Awareness, and other somatic disciplines characteristically do not promise medical therapy. However, they share the assumption that transformations of bodily experience can increase one's self-healing capacities and facilitate symptom reduction.[2]

In this chapter I describe seven prominent somatic disciplines. Each of them, I believe, can contribute to the integral practices discussed in Part Three.

[18.2]
THE ALEXANDER TECHNIQUE

History and Method

Frederick Matthias Alexander, an Australian actor born in 1869, developed his educational approach while curing himself from a disconcerting loss of voice. When doctors could not help him, he began to observe his own muscular tensions, sometimes with mirrors, and found that his disability was caused by a tendency to pull his head backwards and downwards.[3] In alleviating this condition, Alexander came to believe that all human movements, no matter how small, involve the entire person. In his words,

> The term psycho-physical is used throughout my works to indicate the impossibility of separating "physical" and "mental" operations in our conception of the human organism. . . . Hence I use the term psycho-physical activity to indicate all human manifestations, and psycho-physical mechanism to indicate the instrument which makes these manifestations possible.[4]

Alexander's discovery that postural contortions had caused his own distress led him to develop educational methods that promote kinesthetic awareness and inhibit responses that prevent optimal functioning. Though his approach ultimately involves the entire body, it usually begins by focusing attention on the relations between head and torso. Alexander instructions typically elicit relaxation of the neck so that the head moves forward and up and the torso effortlessly lengthens and widens. But Alexander insisted that his method was more than a series of exercises, since optimal functioning requires

conscious self-control and a developing sensory-kinesthetic awareness instead of accumulated, subconscious habit. As such control and
self-awareness develop, body alignment improves, the head typically
moves up from the top of the spine, and the back lengthens to relieve
abnormal pressures upon it. But these health-promoting changes are
only part of the constructive self-control a practitioner of this method
can acquire. Frank Jones, a classics scholar who became an Alexander
practitioner, reflected the founder's philosophy when he wrote that
the Alexander Technique

> teaches you how to bring more practical intelligence into what you are
> already doing; how to eliminate stereotyped responses; how to deal
> with habit and change. It leaves you free to choose your own goal but
> gives you a better use of yourself while you work at it. Alexander dis-
> covered a method for expanding consciousness to take in inhibition as
> well as excitation and thus obtain a better integration of the reflex and
> voluntary elements in a response pattern. The procedure makes any
> movement or activity smoother and easier, and is strongly reinforcing.[5]

When it is successful, the Alexander Technique promotes both
excitatory and inhibitory self-control, self-awareness, and spontaneity, mastery through surrender, and a rewarding sense of lightness
and freedom. Alexander influenced other somatic disciplines, including those of Wilhelm Reich and Moshe Feldenkrais, as well as many
psychotherapies. His influence has been furthered by some of his
famous pupils, including Aldous Huxley (who modeled a central
character in *Eyeless in Gaza* upon Alexander), George Bernard Shaw,
Raymond Dart, John Dewey, Sir Stafford Cripps, Nikolaas Tinbergen,
and Charles Sherrington. Dewey wrote prefaces to three of Alexander's books and advocated the Alexander Technique to his colleagues for decades; and Tinbergen devoted half of his acceptance
speech to Alexander upon receiving the Nobel Prize in 1973 for his
work in ethology.

Research

Alexander's approach has rarely been explored in experimental studies, partly because the founder himself resisted attempts to assess his
work scientifically. Like many somatic educators since, Alexander
feared that most scientists would not comprehend the subtleties of his
method. His reluctance to work with experimentalists frustrated

John Dewey, who tried for many years to win acceptance for Alexander among educators and the general public.[†] Frank Jones, however, undertook experiments such as Dewey proposed. In several studies at Tufts University, he examined subjects of the Alexander Technique with electromyography, X rays, and multiple-image photography, as well as through subjective report, providing a well-written and detailed account of his results in a book, *Body Awareness in Action*. Summarizing these experiments, Jones concluded that:

- The reflex response of the organism to gravity is a fundamental feedback mechanism which integrates other reflex systems.

- Under civilized conditions this mechanism is commonly interfered with by habitual, learned responses which disturb the tonic relation between head, neck and trunk.

- When this interference is perceived kinesthetically, it can be inhibited. By this means the antigravity response is facilitated and its integrative effect on the organism is restored.[6]

[18.3]
AUTOGENIC TRAINING

History

Johannes Schultz, a German neurologist, developed Autogenic Training after studying the work of physiologist Oskar Vogt in Berlin at the turn of the century. While exploring the mechanisms of sleep and hypnosis, Vogt had observed that intelligent and critically minded subjects could induce autosuggestive states. Inspired by Vogt's discoveries, Schultz undertook his own studies of psychophysiological mechanisms involved in hypnosis. During these first experiments, his subjects consistently reported sensations of heaviness in their extremities, usually followed by sensations of warmth. Schultz concluded from such reports that muscular relaxation, which produced the feeling of heaviness, and vasorelaxation, which produced the sensations of warmth, were basic factors in producing hypnosis. This conclusion led him to wonder if "the psychophysiologic mechanisms

---

[†] Dewey wrote to Frank Jones that Alexander "was never able to undertake [experimental studies of his work] because of early obstinate prejudices" (Jones, F. P., 1979, p. 105).

which are responsible for inducing heaviness and warmth could be mobilized by autosuggestions, and if thus a state of amplified relaxation similar to the hypnotic state would result."[7]

Further clinical and experimental studies led Schultz in 1932 to publish the first edition of his *Das Autogene Training*, in which he described his methods and theory. In the years that followed, psychotherapists and physicians in Europe and America began to use Schultz's approach in their clinical practice, and many experimental psychologists tested its effectiveness in treating various afflictions. By the 1980s Autogenic Training had helped stimulate the development of biofeedback training (see chapter 16), as well as some behavioral and imagery-based therapies. It has also informed contemporary meditation research. When Schultz published the first English-language edition of his book (*Autogenic Training*) with his student and colleague Wolfgang Luthe in 1959, he could draw upon more than 600 published studies involving his theories and methods.

Method

Because many subjects had reported heaviness and warmth in their extremities during hypnosis, Schultz began reinforcing these sensations in his hypnotic inductions. To deepen his subjects' relaxation, he added instructions that they slow their heart rate and respiration; and because warm baths and cool compresses had tranquilizing effects on excited patients, he added further suggestions that they imagine warmth in their abdomen and coolness on their forehead. "These six physiologically oriented steps: heaviness and warmth in the extremities, regulation of cardiac activity and respiration, abdominal warmth and cooling of the forehead," he wrote, "[are] the core of Autogenic Training."[8] Eventually, Schultz added a number of precise verbal instructions and training postures to his six basic suggestions to form what he termed the autogenic standard exercises.

The autogenic standard exercises are most effective, Schultz claimed, when practiced in quiet surroundings. To this end, Schultz advised his subjects to sit or lie down in a quiet place, loosen restrictive clothing, and close their eyes. Then the six standard exercises should be practiced one by one, at first for some 30 to 60 seconds each. Whether a trainer gives the six instructions or the subject practices them on his or her own, the verbal formulae should be gradually shortened so that they induce their effects more quickly. Eventually, the subject need only say something such as "extremities warm and heavy," or simply "warm and heavy." Though Schultz provided standardized procedures, practitioners have adapted his exercises in various ways to their training or living situations. With practice, pronounced relaxation can be produced by recitation of autogenic phrases even when a subject is physically active. Detailed instructions in the method, accompanied by brief descriptions of experiences they elicit, are included in *Autogenic Training*.[9]

According to Schultz, the six standard exercises provide the basis for further self-instructions designed to promote healing or particular capacities. These more advanced training techniques include organ-specific formulae, intentional formulae, and the following seven meditative exercises.

1. the induction of spontaneously experienced colors;
2. the induction of particular colors that can trigger specific images or emotions;
3. the visualization of concrete objects;
4. meditation upon abstract qualities such as justice, freedom, or happiness, using visual, auditory, tactile, kinesthetic, gustatory, and olfactory images;
5. the experience of selected feeling-states;
6. the visualization of other persons; and
7. the elicitation of specific information from the unconscious.[10]

Organ-specific formulae address particular somatic disorders. To alleviate hay fever or bronchial asthma, for example, a subject relaxes using the standard exercises, then repeats a phrase such as "my eyes are cool" or "my eyelids are cool and numb" until he or she feels relief. Intentional formulae are designed primarily to modify mental processes—for example, by suggesting that particular environmental conditions need not cause an asthma attack, or by affirming a positive attitude toward school examinations, or by reinforcing the will to stop smoking.[11]

Theory

Schultz and Luthe did not claim to understand all the physiological mechanisms responsible for the changes that Autogenic Training produces. However, their research indicated to them that such changes occur partly because the central nervous system is quieted by reduction of afferent (incoming) stimuli. Such quieting, they believed, facilitates the deactivation of ergotropic mechanisms (which produce the fight-or-flight response) and the promotion of trophotropic functions (which cause general relaxation). Schultz and Luthe based their theory of self-regulation upon the model of trophotropic-ergotropic responses of parasympathetic and sympathetic nervous stimulation developed by Walter Hess.

Several studies have shown that meditation evokes physiological responses like those produced by Autogenic Training. Herbert Benson has attributed these similar results, which he named the relaxation response, to a strengthening of the trophotropic response described by Hess (23.2), and has argued that quiet meditation, Autogenic Training, and other relaxation methods offset "the harmful effects of the inappropriate elicitation of the fight-or-flight response."

Applications

Schultz and Luthe could report in 1959 that their approach had helped alleviate bronchial asthma and tuberculosis; disorders of the gastrointestinal tract; "cardiac neuroses" and functional problems of the heart; angina pectoris, high blood pressure, and other circulatory afflictions; disorders of the endocrine system, the urogenital system, and pregnancy; diseases of the eyes and skin; epilepsy; stuttering; alcoholism; and problems with sleep.[12] Unlike many others who make such claims, Schultz, Luthe, and their followers stimulated a large body of work by responsible clinicians and experimentalists, much of which was reported in leading scientific journals. The 604 books and articles listed in the 1959 edition of *Autogenic Training* indicate the range of such research.[13]

[18.4]
THE FELDENKRAIS METHOD

History

Moshe Feldenkrais, born in 1904, was prompted to develop his somatic discipline in part by his own disability, a knee malfunction caused by athletic injury. Originally trained as a physicist, he studied anatomy, physiology, and psychology in order to rehabilitate himself without surgery, and in doing so he developed both a philosophy of life and his educational method. Like Alexander, he came to believe that humans can replace automatic, uncultivated behavior with a highly articulated, freely determined, self-aware, and spontaneous functioning. During the 1970s he developed a large following in Europe and the United States, and for many years until his death in 1984 he taught his philosophy and methods to students from many parts of the world at the Feldenkrais Institute in Tel Aviv.

Method

Most people's posture and coordination are considerably impaired during infancy and childhood, Feldenkrais claimed, so that their general functioning is limited for most of their lives.

> We tend to stop learning when we have mastered sufficient skills to attain our immediate objective. Thus, for instance, we improve our speech until we can make ourselves understood. But any person who wishes to speak with the clarity of an actor discovers that he must study speech for several years in order to achieve anything approaching his maximum potential. An intricate process of limiting ability accustoms [us] to make do with [a small part of our] potential.[14]

This stunting causes a general physical atrophy, Feldenkrais believed, which is accompanied by a loss of spontaneity, a shrunken self-image, inhibitions against further learning, underdeveloped sensory and kinesthetic awareness, and unnecessary limitations upon emotional and mental capacities.[15] The Feldenkrais Method aims to restore atrophied abilities and stimulate further growth by simultaneously cultivating sensation, emotion, thinking, and motor activity, though it is based primarily on physical movement. For Feldenkrais, movement affords an especially efficient starting point for transformative practice. Through the initiation of new motor patterns, one can open up and cultivate other aspects of human functioning.

The Feldenkrais Method includes two general components. The first, called functional integration, consists of hands-on body work tailored to the particular needs of the patient or student. A trained instructor helps the client become aware of the body's limiting patterns by means of physical manipulations, simple touch, and instruction. It is a gentle, sensitive method (when properly done) that can foster increased awareness of sensations, emotions, and thoughts while introducing unfamiliar kinds of movement. By altering motor patterns, it helps broaden behavioral repertoire. Like the Alexander Technique, it can produce a feeling of freedom, lightness, and balance accompanied by new spontaneity and pleasure.

The second kind of Feldenkrais work, generally called awareness through movement, is conducted in groups and includes exercises in which limbs, head, neck, and torso are moved in unaccustomed ways. Following Feldenkrais's dictum that learning happens best when it is pleasurable, instructors frequently tell their students not to do anything that hurts as they attempt unfamiliar types of movement. Each exercise is performed slowly and as pleasantly as possible, without strain, so that each participant might sense the smallest changes in tonus and alignment. Admonitions to relax and go slowly are helpful, Feldenkrais believed, because they help the body relinquish its habitual tendency to operate in hurtful, self-defeating ways. Such exercises increase flexibility, vitality, self-awareness, and pleasure, he argued, because they circumvent habitual involuntary movements developed since infancy. They break up stultifying patterns of behavior and begin to reveal new possibilities for action and feeling. With practice, they tend to have a contagious effect on the rest of one's activities, stimulating innovative ideas and richer emotions. By the time he died, Feldenkrais had invented more than a thousand exercises to promote awareness and new movement patterns.[16]

[18.5]
ROLFING

History

Rolfing, or Structural Integration, is a system of physical manipulations that aims to align and integrate the body so that it might function more fully. The discipline's inventor, Ida Rolf, was born in 1896 and trained in biological chemistry while working at the Rockefeller Institute in New York. Her methods were informed by her studies with Pierre Bernard, an American yoga teacher, and by the work of Amy Cochran, an osteopath Rolf met in the 1930s. Though Rolf had trained others in her method, she did not develop a significant following until the late 1960s, when she started to teach at the Esalen Institute in Big Sur, California. During the 1970s the Rolf Institute was founded in Boulder, Colorado, to train practitioners and set standards for the practice of Structural Integration.

Theory and Method

The Rolfing practitioner manipulates the client's body, sometimes painfully. Myofascia, the connective tissues enveloping the muscles, are the primary target in this procedure. With hands, forearms, or elbows, a Rolfer tries to tease or force these tissues apart so that the muscles they enclose can function more freely in a well-articulated, integrated fashion. According to Rolf, "the basic law of Rolfing is that you add structure to the body. In so doing you are demanding a change in function."[17] In his book *The Protean Body*, Don Johnson described some functional changes he had experienced or observed during Rolfing sessions, both as a recipient and practitioner.

Once, when Ida had done a session with me, spending most of the hour loosening and restructuring my sacrum, I walked out onto a dock in the Gulf of Mexico. . . . I experienced being a part of the sea breeze, the movement of the water and the fish, the light rays cast by the sun, the colors of the palms and tropical flowers. I had no sense of past or future. It was not a particularly blissful experience: it was terrifying. It was the kind of experience I'd invested a lot of energy in avoiding.

The psychotic quality of this experience came from my fear, which itself was rooted in my flesh. Over the years, as my body has become more integrated, the same kind of experience is available without the terror.

My goal in this session was to balance [Ray's] shoulders and rib cage. Originally, his whole torso looked as if someone had twisted it slightly clockwise around his center. His eleventh and twelfth ribs, especially on the left side, were so distorted that in the early sessions I could barely locate them. The compression in his upper three ribs and the myofascial structures associated with the shoulder joints made it almost impossible for him to allow me to touch them. But this time, I found the lower ribs had, because of our earlier work, shifted to where I could easily work with them to coax them into a normal position. He allowed me to work deeply in the armpit structures, stretching the pectoralis minor and inducing more movement in the uppermost ribs. I also lengthened the tissues in his arms, with special attention to his right wrist, which he had badly damaged in an acrobatic accident.

While I was working on his left forearm, he recalled having driven a nail through his right lower leg. At the end of the session, his ribs and shoulders were fuller and more balanced. The new position of his thorax gave more freedom to his pelvis. [Photographs of him] again demonstrated significant change.[18]

Powerful images and emotions are frequently triggered by Rolfing sessions, while forgotten traumas may surface with surprising force. Though Rolf generally maintained that her practitioners should not try to be psychotherapists, some Rolfers discuss their clients' feelings and insights as they arise during Rolfing sessions, and some psychologists use Structural Integration as an adjunct in their therapeutic work.

Research

Though Rolfing has produced a large clinical lore, it has not been studied extensively in the laboratory. In 1973, however, psychologist Julian Silverman, then Esalen Institute's research director, and kinesthesiologist Valerie Hunt, then working at UCLA's Movement Behavior Lab, found that Structural Integration produced striking results in 11 male subjects. Working with different teams to assess the effects produced by 10 Rolfing sessions, Silverman and Hunt found that their separate experimental results produced a coherent picture of change in their subjects. Silverman's study showed that all the participants in the experiment exhibited increased amplitudes in their averaged evoked EEG responses after Rolfing. This finding was suggestive, Silverman wrote, because other researchers had found that large evoked EEG amplitudes are generally correlated with "an open, more receptive orientation to external stimuli."[19] Furthermore, the subjects' EEG responses generally showed less variability after Rolfing, which suggested that these subjects had learned to organize sensory stimulation efficiently. And the "slope" of some subjects' EEG wave forms after Rolfing became more like those of people who are highly responsive to subtle stimulation but who can effectively modulate strong inputs. These changes in brain activity were consistent with the subjects' verbal reports, which indicated that they experienced more openness and receptivity to sensory input after their Rolfing was completed.

When their EEG results were combined with measures of blood

chemistry and analyzed statistically, the subjects fell into three groups.[20] Full-body photographs of them all, grouped according to Silverman's statistical analysis, were presented to Emmett Hutchins, an experienced Rolf practitioner. Without knowing anything about the experimental results that led to Silverman's grouping, Hutchins gave the following descriptions of the bodies he saw in the photographs.

> The greatest degree of [muscular] balance was achieved with the subjects in Group 1. These subjects can easily be recognized as those with the most change toward normal.
>
> Subjects in Group 2 belong to the soft-bodied type in which the intrinsic musculature is hypertense while the outer muscles are soft and toneless. Although the Structural Integration processing produced much improvement in this inner-outer balance, [the subjects remained] basically unchanged. Further Rolf work is indicated for this group.
>
> Group 3 is the reverse of Group 2. These subjects can be viewed as having a soft core structure and a hard outer sleeve structure. As in Group 2, much of the inner/outer imbalance still remains even after 10 hours of Structural Integration and further treatments are indicated.[21]

These cluster descriptions, Silverman wrote, were consistent with the results of biochemical and electrophysiological measures. The subjects whom Hutchins had judged to have the greatest degree of balance between intrinsic and extrinsic muscles, resembled

> a laboratory prototype of an open, receptive, efficient sensory information processor. These individuals exhibited the most marked reduction slopes on the EEG evoked response procedure. They exhibited the lowest amplitude at the high intensity and the least variability on the EEG procedure. This pattern suggests efficient sensory modulation. They exhibited [EEG readings] associated with high sensitivity, low CPK [a muscle enzyme that is elevated during certain stress conditions] . . . and the lowest SGOT [which also indicates lower levels of stress]. Their pattern of response to low and high intensity stimulation is considered to be an adaptive one. It indicates both sensitivity to stimulation and a capacity to modulate strong stimulation efficiently. Personality data, available on half of the subjects in the study, also were in accord with this formulation. It is important to note that the physical differentiations made by Emmett Hutchins were anything but obvious to a number of the scientists on the research team. Rather, these distinctions appeared to be the result of a specialized way of viewing the body that Structural Integration practitioners cultivate.[22]

Valerie Hunt and a team of associates independently tested Silverman's subjects with telemetric electromyography before and after their Rolfing sessions. These recordings, Hunt wrote, supported the hypothesis that Rolfing induces "alterations of energy release and frequency of muscle depolarization leading to improved motor capacity. [To produce such changes in their subjects' myograms] some form of reorganization must have taken place in the central nervous system, possibly a functional alteration in the neural chain of command exercised in volitionary coordinated motor functioning." Rolfed subjects, for example, used less energy while throwing balls and when they were lying down. "These . . . findings," Hunt wrote, "indicated that after [their Structural Integration sessions] the subjects were at greater rest during inactivity, and increased their mechanical efficiency when vigorously moving."[23] This simultaneous articulation of perceptual and motor functioning, both Silverman and Hunt concluded, suggested that Rolfing produces interdependent improvements in muscular and central nervous system activity.[24]

Hunt and Silverman concluded that their separate findings indicated, among other things, that: different kinds of physiques have different responses to Rolfing; such differences in response can be measured; and trained body workers have insights about the body that can be verified in the laboratory. Clinical skill and experimental science can be joined, these studies indicate, to improve our understanding of somatic education.[†]

The Misuse of Rolfing

Though Rolfing has beneficial effects, some people have questioned the rigidity with which it is sometimes practiced. Don Johnson, who is both a practitioner and a theorist in the field, has criticized Rolf herself for promoting a restrictive, sometimes damaging, idea of body alignment. Johnson has called this Procrustean tendency somatic Platonism because it is based on the belief that all bodies are healthy to

---

[†] In another study, Hunt examined state-trait anxiety, brain activity, muscle activity, and electrical fields in 24 Rolfed subjects who were compared with 24 controls matched for posture, body structure, amount of physical activity, height, weight, and age. Hunt found that Rolfing decreased her subjects' anxiety, increased the dominance of their right hemisphere brain activity during tasks thought to require it, improved some of their muscular functioning, increased the efficiency of their motor performance in general, and affected the flow of electrical currents between various points in their bodies (Hunt et al. 1977).

the degree that they conform to a universal ideal of posture and spinal configuration. A similar rigidity has also been exhibited by Alexander, many Reichians, and some bioenergetics practitioners. Johnson wrote:

> I suspect that working with an idealistic model is a major factor in producing the painful experiences which are common to a number of somatic therapies. If I work on your foot, for example, with the idea that I know before entering your connective tissue where your ankle belongs, I will tend to force the ankle toward the "norm." But the direction in which I've chosen to move the foot may be in conflict with the organism's innate sense of rightness, and the conflict produces pain, often interpreted as resistance on the part of the client. [Thus] the client's experience of pain may signal, not his or her resistance to change, but a conflict between his or her bodily wisdom and the goals of the therapist.[25]

But somatic disciplines are not the only practices subject to abuse of this kind. The imposition of restrictive norms upon mind and body has frequently limited or distorted psychotherapeutic and religious practice. *Programs for creative human change need to be based on a clear perception of each individual's psychophysical uniqueness.* I will return to this observation in subsequent chapters.

[18.6]
PROGRESSIVE RELAXATION

History and Theory

The American physician Edmund Jacobson (1888–1983) developed a method for self-regulation that he called Progressive Relaxation and pioneered experimental research on the voluntary control of somatic functioning. His work, like Johannes Schultz's with Autogenic Training, provides an exemplary model for the development of somatic and other transformative disciplines in that it joined clinical and experimental studies.

Jacobson was led to his experimental work when he realized that favorable patient reports did not in themselves constitute scientific conclusions about his methods. "I was repeatedly led to ask," he wrote, "whether I was really securing a greater degree of relaxation than [other doctors]. This question led to laboratory investigations to which [Progressive Relaxation] largely owes its growth."[26] Jacobson's first experiments, on the inhibition of sensation, were reported in *Psychological Review* in 1911 and were followed by other studies of neuromuscular self-control that were published until the 1960s. The representative list of Jacobson's theoretical and experimental papers presented in appendix H suggests the range, variety, and pioneering nature of his work.

In some of his first experimental studies, which were conducted at Harvard beginning in 1908, Jacobson showed that muscular tensions exhibited by knitted brows or postural rigidity predisposed his subjects to experience a pronounced startle response. Relaxed subjects, on the other hand, did not react to potentially jarring stimulation with such strong contractions. These observations suggested to Jacobson that muscles already contracted were more likely to contract further in response to an unexpected stimulus. The truth of this explanation, he wrote,

> could be tested by producing a heightened contraction or tonus, whereupon we could test whether sudden noise would occasion an extent of start varying in amplitude with the tonus. Accordingly the subject was directed to stiffen the muscles of the arms, legs, head and trunk; that is, to sit still while partially contracting his muscles. In other tests he was directed to relax his muscles as completely as possible. An intermediate condition was also directed, in which the muscles were held only slightly tense.
> . . . The contraction seemed in these preliminary tests very much greater with extreme muscular tenseness than with mild; while with relaxation, there was no start and no shock, and the sound seemed to lose its irritating character. In this way it was suggested that in addition to the factors mentioned by other writers, the involuntary start depends upon the previous neuromuscular state; and that, if a start characteristically occurs when there is interruption of thought-processes or of attention, this is because the mental activity is accompanied by skeletal muscle contractions.[27]

Jacobson linked his findings on the startle response with research into the effects of strong stimuli upon pulse, respiration, blood pressure, cerebrospinal fluid pressure, bladder musculature, and contractions of the colon and esophagus. Such research indicates that viscera do not respond to outside stimuli as strongly while relaxed as they do when tense. When relaxation becomes a ready habit, Jacobson argued, large muscles and viscera alike react more calmly to external stimulation.

> Individuals who start tend to explain their reaction as due to the strong character of the stimulus. However, according to our tests, a violent reaction along with subjective disturbance seems to depend upon the preceding state of general muscular tonus. Some individuals do not visibly start at all as a rule. . . . These observations [suggest] that all subjective irritation or distress might be reduced if the individual were to become sufficiently relaxed; and this hypothesis remains today a beacon-light for further experiments and observations.[28]

In later studies at Cornell University, Jacobson found that subjects he trained exhibited relaxation that could be observed by others. These studies, he wrote, "suggest that what is commonly called *effort* in our subjective experience consists in part of readily observable contractions of skeletal muscles."[29] In the trained subject,

> a distinctly different type of reaction to the environment becomes possible. His responses become less determined by fixed associations or habitual conditioning. His emotions become subject to a certain measure of control. He can work with less neuromuscular tension. He can be taught to distinguish between issue and attitude—between the problems which he faces and his own neuromuscular actions. Thus [his] efforts are freed to operate in the real interest of [his] organism and there can be greater efficiency, which implies less pathological wear and tear. This is because effort on the part of the organism . . . evidently corresponds with the usage of any instrument. Notwithstanding the amazing tissue restoration characteristic of the organism, excessive effort tends in the long run toward organic disruption or senile change.[30]

Jacobson's work also indicated that mental events unrelated to external stimuli are often accompanied by muscular contractions. In experiments at the University of Chicago in 1922–23, he observed the facial expressions and postural changes exhibited by subjects engaged in mental imagery. These studies showed that mental images are usually accompanied by subtle eye movements and small muscular contractions of the forehead, tongue, torso, and limbs. Like Jacobson's previous experiments with the startle response, these showed that unnecessary movements can be diminished, or even eliminated, through relaxation. These and subsequent findings, which were published in the *American Journal of Physiology* and other journals during the early 1930s, anticipated subsequent biofeedback research. Jacobson's paper on the voluntary control of the esophagus (1925), which has often been cited in the literature on self-regulation, and his experimental studies of muscular relaxation, influenced many clinicians and experimentalists who developed biofeedback training in the 1960s and 1970s.[31]

Like contemplative practice and other transformative disciplines, Progressive Relaxation is based upon the idea that human distress is largely self-activated. Instead of attributing our problems to outside stimuli, Jacobson argued, we can learn to control ourselves so that we act more calmly and efficiently in all circumstances, in spite of potentially noxious inputs. Taking increasing responsibility for all of our functioning, we can respond to the environment's demands with an even temper. Jacobson developed this basic idea into a general philosophy of life, much as Alexander, Reich, and other therapists have done with their clinical insights. We are peculiarly suited for self-regulation, he maintained, because our nervous system is so widely and finely deployed through our entire body, and because we have a fundamental capacity to alter its functioning through self-awareness and conscious volition. "Effective effort-control," he wrote, "means efficiency in every occupation."[32]

Method

Jacobson distinguished general relaxation of the entire body from differential relaxation of various body parts, though both are achieved by the same basic procedures in the practice of Progressive Relaxation. First, a subject learns to distinguish the various sensations that arise from the muscles—usually by contracting, then relaxing, particular muscle groups. This basic procedure can be followed while lying down, sitting, standing, or moving, so that the subject gains a progressive mastery of more and more muscular tissue during rest or activity. Jacobson described some of his methods in *Progressive Relaxation*.

A minimum of words is to pass between the physician and the patient, who is to learn from concrete experience rather than from discussion. While lying down, the individual flexes his left arm steadily, avoiding unnecessary contraction of other groups . . . and reports whether he notes the sensation from contraction of the biceps group. To strengthen the sensation, the physician offers passive resistance. Some patients perceive at once, while others require many repetitions, particularly the agitated and unobserving types.

> When the sensation is clearly perceived while the individual is flexing the biceps, his attention may be sharply called to the issue by saying to him, "This is *you* doing! What we wish is simply the reverse of this, namely, not doing!" As he then relaxes, he begins to learn clearly what it is *not to do*. . . . After he has relaxed his arm for several minutes to illustrate this point, he is requested to contract it, and then to relax again. This time it is called to his notice that his act of relaxation involved no effort: he did not have to contract his arm or any other part in order to relax.
> If these matters evidently have been assimilated, the individual is made to contract the biceps again and then to let it go. He is requested to let the part go further and further every minute. "Whatever it is that you do or do not do when you begin to relax, that you are to continue on and on, past the point where the part seems to you perfectly relaxed!" This instruction, if clearly illustrated, conveys to him the meaning of progressive relaxation in terms of an immediate experience. If he now lies quietly with eyes closed and seems to be set to relax, he may be left alone.

> Sensations of tenseness are readily overlooked because of their relative faintness. These sensations are sometimes "unconscious," in the sense that they are commonly overlooked. Without doubt, in this sense, "unconscious" experiences can be relaxed away. Another distinction to be taught is between the experiences of moving and static tensions, that is, changing muscular contractions versus tonic states. Flexion of the arm provides an instance of a moving tension, while holding it rigid illustrates a static tension.[33]

In *Progressive Relaxation,* Jacobson outlined methods for the differential relaxation of various muscle groups in different postures and physical actions. Relaxation of this kind, he wrote, is well known in the arts. Voice teachers, he wrote, cultivate

> relaxation of the muscles of the throat, larynx and respiration. Singers learn that a loud tone is not required in order to be heard in the back row of an auditorium with good acoustics.
> In aesthetic and ballet dancing, relaxation plays a conspicuous role. The individual who holds himself rigid in these arts fails in his effects. A particular exercise is repeated until grace is attained. This

means that those muscles alone are used which are needed for the act, and that no excess tension appears in them.

Differential relaxation accordingly means a minimum of tensions in the muscles requisite for an act along with the relaxation of other muscles. A large variety of instances can be found in daily life. The speaker with trained voice does not tire even after prolonged effort if he keeps his throat differentially relaxed. The billiard player spoils the delicate shot if he is generally too tense. The golf or tennis player must learn to mingle a certain relaxation with the strokes in order to be successful. The restless or emotional student finds it difficult to concentrate. The excited salesman fails to impress his prospective client. The clever acrobat produces an impression of grace and of ease by relaxing the muscles he does not require. The comedian often makes his laughable effects depend upon extreme relaxation of certain parts while others are active or held rigid. It seems safe to say that every learning process depends upon the acquisition of certain tensions with concomitant relaxations. Psychological textbooks commonly illustrate the early learning process with the child at the piano who squirms and shifts, perhaps even protruding the tongue, as the notes are first studied. As skill is acquired, these tensions disappear: in the present terms, a certain degree of differential relaxation sets in.[34]

Jacobson developed procedures to cultivate self-control during specific behaviors. Such training, he wrote, was designed to eliminate those secondary activities that do not require muscular contractions during the task at hand. While reading, for example,

a noise may be followed by looking up and turning the head in that direction. Any distracting sensory stimulus may be followed by such secondary activity. Very often the average individual while reading or otherwise engaged has an undercurrent of distracting thought-processes in the form of worries, reflections, irrelevant recollections, intentions to do this or that thing, and very often even songs or strains of music are silently but almost incessantly repeated.[35]

With the help of Progressive Relaxation, we can cultivate freedom from subtle movements and the distracting thoughts or feelings that accompany them, which interfere with any activity. In summarizing his approach, Jacobson wrote:

[This] point of view has seemed to throw light upon the physiology and psychology of normal behavior, and consequently also of pathological behavior, opening a wide field for investigation. I have offered evidence

that proprioceptive sensations from muscular contraction are important elements in the "stream of consciousness"; that as these diminish
with advancing relaxation, not alone kinesthetic but also visual and auditory images become fewer, until for recurrent brief periods, at least,
mental activity is so to speak shut off. Corresponding to the excess of
voluntary and reflex muscular activities occurring in nervous hypertension, it was submitted that external observation of the individual reveals . . . contractions of skeletal and visceral musculature. These
muscular states are of clinical no less than of laboratory interest. Thus
a trifold correspondence (which is at large an identity) is found to exist
between the muscular states, the proprioceptive sensory impulses, and
the conscious processes of the individual, apparently yielding to us a
particularly intimate and detailed insight into matters concerning the
relation of "the mind and the body."[36]

[18.7]
THE SENSORY AWARENESS WORK OF
ELSA GINDLER AND CHARLOTTE SELVER

History and Philosophy

In 1910 Elsa Gindler, a young physical educator in Berlin, developed
tuberculosis. Because treatment at a sanitorium was beyond her
means, she looked for ways to promote her body's natural restorative
powers. In her search for health, she practiced the sensitive attention
to breathing, posture, and movement that formed the basis of her later work. "In this practice," her colleague Elfriede Hengstenberg
wrote, "she came into a state where she was no longer disturbed by her
own thoughts and worries. She came to see that calm in the physical
field *(gelassenheit)* is equivalent to trust in the psychic field. This was
her discovery, and it was basic to all her subsequent research."[37]
Gelassenheit, a word that captures much of the attitude implicit in
Gindler's work, may be translated into English as "calmness," but it
also means more than that. *Lassen,* which connotes "allowing" or
"letting go," implies a trust of gravity, and in Gindler's usage suggests
that by sensing the earth's pull upon us we can experience a reassuring
self-confidence.

To heal herself, Gindler tried to rest the parts of her lungs infected by tuberculosis. She did this by cultivating awareness of her

throat, diaphragm, rib cage, and stomach, thus permitting them to arrive at an optimal tonus. This practice led her to the same insight Alexander and Jacobson achieved, namely that kinesthetic awareness can facilitate motor, visceral, and other functioning. In the words of Thomas Hanna, who practiced several somatic disciplines:

> Elsa Gindler took seriously a phenomenon with which human beings have always been familiar: that directing one's attention to some part of one's body is an action that immediately affects that bodily part. If, at this instant, you continue to hold this book exactly as you are doing, without changing it, and you notice the relative tension or comfort of your hands and fingers, it is like turning a spotlight on [them]. Whatever sensations you were feeling a moment ago in your hands are minor compared with what you feel now. Your awareness highlights what is actually happening in your hands, and if your hands were unconsciously overtensed or uncomfortably positioned, you may notice that you now automatically adjust your hands so that they hold the book more easily.
>
> Awareness of kinesthetic feedback from our neural system predisposes the muscular system to adjust itself to more efficient functioning . . . [then] the changed motor patterns cycle back and give new sensory impressions, which again readjust the muscular coordination, and so on. It is a constant feedback loop that never ceases, from birth until death.[38]

In healing herself, Gindler came to believe that she had found a new approach to living. In the words of Charles Brooks, who helped develop her work in America, Gindler's intuitive self-restoration took her far beyond therapy.[39] For 50 years, until her death in 1961, she taught her method to many devoted students, calling it simply *Arbeit am Menschen* (work on the human being). Though most of her written materials were destroyed in Berlin when a Nazi firebombed the building she lived in, her essay "Gymnastik for Everyone" was translated into English and published in the journal *Somatics*.[40] In the United States, Charlotte Selver and Carola Speads developed Gindler's approach, which Selver named Sensory Awareness. In the 1940s Selver introduced her work to the psychoanalyst Erich Fromm and to Fritz Perls, who incorporated some of it into his Gestalt Therapy. In 1956 she met the philosopher Alan Watts, who conducted seminars with her in New York and California; and in 1963 she gave the first of many programs at Esalen Institute that introduced her methods to a broader audience.

Method

The Gindler-Selver approach resists easy formulations. It cannot be captured by a series of instructions or a particular set of exercises, because it proceeds informally through activities adapted to the person and circumstances involved. According to Don Johnson, it exemplifies the principle of "pure exploration." It encourages kinesthetic sensitivity, play, and openness for their own sake before symptom removal or the development of particular sensory or motor skills. Like all educational methods that are mainly nonverbal, it must be experienced to be well understood. Nevertheless, the following passages from Charles Brooks's *Sensory Awareness* convey something of its flavor.

It is unlikely that we are conscious of anything more than the vaguest sensation of the structure and function of the feet, even if we know the names of the bones and anatomical divisions and have a picture of it all in our mind from the anatomy book.

One thing we can easily do to make good for this lack is to sit down and explore our feet directly. With our own hands we may go deeply into them, discovering and enlivening the many joints and ligaments of which a foot consists. How far and deep must one go to follow the identity of a given toe until it becomes lost in the interior? What can we feel of the architecture of the arch? How does the heel seem to our palm and our fingers, in its aspect as bone and its aspect as padding?

Of course, we may equally explore the foot of a partner. This in a new group, can create some tense moments. For who has held the foot of a stranger in his hand and worked with it dispassionately?

We take time to feel where the floor is. How do we relate to it? Many people now feel that they are relating to the floor. They no longer stand on their feet but on something which they feel really supports them from below. The feet feel flexible and alive, not stood on but free to explore what they touch, as the hands a moment ago were exploring them.

Now the leader may ask, "Do you allow the connection with the floor upward into you?" And a little later, "Do you allow it through your knees?" Afterward, a number of people may well report that they found their knees were locked. When they gave up this locking, readjustments could be felt taking place in the ankles or the pelvis or higher.

We may proceed upward in all sorts of ways in this question of allowing a fuller, more organic connection with what we stand on. . . . Very often breathing changes, as one release triggers another, or perhaps sets up a constriction somewhere else.

Still standing, let us bring our hands gently to resting on the top of our heads. Through palms and fingers we can feel, if we are sensitive, not only our hair but also the temperature and perhaps the animation of living tissue underneath.[41]

[18.8]
REICHIAN THERAPY

History and Theory

Psychiatrist Wilhelm Reich (1897–1957) conducted the Vienna Seminar for Psychoanalytic Therapy from 1924 until 1930, and some of the papers he wrote at the time have been recommended ever since for students of the analytic method.[42] He founded sex-hygiene clinics in Vienna, Berlin, and other German cities, and developed theories about the destructiveness of authoritarian social structures, which he outlined in *The Mass Psychology of Fascism* (1933) and other writings. Departing from orthodox Freudian theory, Reich rejected symptom analysis in favor of a character analysis in which the entire personality structure was examined. Unless the characterological basis of psychological problems was modified, he believed, similar problems were likely to appear in their place.[43] His views on character analysis anticipated, and often inspired, other psychotherapies, and his insights about the social determinents of neurosis have influenced family counseling, group therapy, and industrial psychology, either directly or indirectly.

Reich also broke from most psychoanalysts with his theory of the orgasm. Stated simply, this theory holds that a patient is not fully cured until he or she achieves deep and healthy gratification in sexual intercourse. "Orgastic potency," Reich believed, heightens the enjoyment of sex while stabilizing the whole organism. Such potency includes the "the capacity for surrender to the flow of energy without any inhibitions; the capacity for complete discharge of all dammed-up sexual excitation through involuntary pleasurable contractions free of anxiety and displeasure and unaccompanied by phantasies."[44] Various rigidities and malformations of personality, which Reich called the character armor, prevent orgastic potency and the development of a mature, or genital, character that is free of blocked instinctual energy. Character analysis aims to eliminate such armoring not by focusing upon single symptoms or forgotten childhood traumas, but by directly confronting the patient's habitual responses and attitudes. In analysis of this kind, Reich maintained, "the patient no longer talked about his hatred, he felt it; he could not escape it as long as his armor was being correctly taken apart."[45] Character armor varies from patient to patient. Its layers, Reich wrote, might be "compared to geological or archeological strata which, similarly, are solidified history. A conflict which has been active at a certain period of life always leaves its trace in the character, in the form of a rigidity."[46]

In the 1930s Reich began to focus his therapy upon the chronic muscular contractions, or body armor, that anchored character armor. Posture, gesture, facial expression, and other behavior, he came to believe, express and maintain personality structure. It is a mistake to regard emotions as "mere events in the mind," for they are rooted in somatic processes. Reich criticized the tendency to regard theoretical terms such as *id* or *superego* as actual entities. By examining the physical patterns produced by a person's experiences, however, one can see how all of the person's functioning is shaped. Exerpts from two case studies presented in *Character Analysis* reflect Reich's concern for bodily expression in his therapy. One patient, he wrote,

> had a pleasing external appearance, was of medium height, had a reserved countenance, was serious and somewhat arrogant. His measured, noble stride caught one's attention—it took him a good while to come through the door and walk across the room to the couch. It was evident that he avoided—or concealed—any haste or excitement. His speech was well phrased and balanced, soft and eloquent. Occasionally, he would interject an emphatic, staccato "Yes!", at the same time stretching both arms forward, then passing one hand across his brow. He lay on the couch with crossed legs, very much at ease. There was very little if any change in his composure and refinement, even when very ticklish and otherwise narcissistic subjects were discussed. When, after several days of analysis, he discussed his relationship to his dearly loved mother, he quite obviously accentuated his noble pose in an effort to master the excitement which seized him. I told him that there was no need to be embarrassed and I urged him to express his feelings freely— but to no avail. He maintained his patrician bearing and refined manner of speech. One day, indeed, when tears welled up in his eyes and his voice was clearly choked, he raised his handkerchief to dry his eyes with the same dignified composure.
>
> Soon after the obvious excitement had subsided, I asked him what impression his analytic situation had made on him. He replied that it was all very interesting but it did not affect him very deeply—the tears had simply "escaped" him; it had been very embarrassing. An

explanation of the necessity and fruitfulness of such excitement was to no avail. His resistance increased perceptibly; his communications became superficial; his attitude, on the other hand, grew more and more pronounced, i.e., more noble, more composed, more reserved.

Perhaps it was merely an insignificant coincidence that one day the term "aristocrat" occurred to me for his behavior. I told him that he was playing the role of an English lord and that the reasons for this could be traced back to his adolescence and childhood. The contemporary defensive function of his "aristocratic manner" was also explained to him. He then produced the most important element of his family story: as a child he had never believed that he could be the son of the small insignificant Jewish merchant who was his father; he must be, he thought, of English descent. As a small boy he had heard that his grandmother had had an affair with a real English lord and he thought of his mother as having English blood in her veins. In his dreams about the future, the fantasy of some day going to England as an ambassador played a leading role.

As we probed consistently into his "lordly" behavior, we found that it was closely related to a second character trait, a tendency to deride his fellow men and the malicious joy he derived from seeing them come to grief. The analysis of this character trait offered considerable difficulty. He expressed his contempt and derision in a grand manner, as from a throne. At the same time, however, this served to gratify his especially intense sadistic impulses. To be sure, he had already talked about the many sadistic fantasies he had had as an adolescent. But he had merely talked about them. It wasn't until we began to ferret them out in their contemporary anchoring, in the tendency to ridicule, that he began to experience them. The lordly quality in his behavior was a protection against the excessive extension of his ridicule into sadistic activity. The sadistic fantasies were not repressed; they were gratified in ridiculing others and warded off in the aristocratic pose. Thus, his arrogant nature was structured exactly as a symptom: it served as a defense against and, at the same time, gratification of an instinctual drive.[47]

Method

Reich called his treatment of body armor *vegetotherapy* because it dealt directly with the body's vegetative or peripheral nervous systems, but he insisted that it did not replace character analysis. It was, he said, "character analysis in the realm of the body." David Boadella, who has practiced a therapy heavily influenced by Reich, described the ideas and methods of vegetotherapy in his book *Wilhelm Reich:*

The tensions of the body can be viewed as a series of constrictions, the function of which is to limit movement, breathing and feeling. Reich later described a number of body segments, each with its characteristic patterns of blockage. In the upper half of the head, for example, Reich found that many neurotics had a particularly tense scalp and forehead, which was often associated with a tendency to headaches. He distinguished between headaches due to frontal tensions, such as a chronic raising of the eyebrows and contraction of the muscles of the forehead, and occipital headaches due to tension in the muscles of the neck. When these tensions were dissolved he found that the frontal tensions corresponded to the bodily expression of anxious anticipation. In sudden fright one instinctively opens the eyes wide and tenses the scalp muscles. . . . A patient may look at the therapist with studied seriousness or with an anxious shifting glance; he may have a superior stare or a worried frown; he may have the typical "far-away" look of the schizoid person. These differing expressions reflect how the individual meets the world. They contain, in locked-up form, their own story of how the early relationships with parents and siblings were experienced. The tension patterns of the body contain the history of their origin.

Patients came to therapy with many kinds of smiles and mouth expressions. They had fixed sardonic grins or downturned mouths of despair. The compulsive character was saddled with his stiff upper lip. There were tight jaws, weak chins, gaunt cheeks, all expressing the use which the patient had learnt to make of his facial muscles. The healthy child or adult has muscles which can express the full range of emotions according to the requirements of the occasion. His face is mobile and adaptable. The tense person is restricted and limited to a narrow range of facial expressions which he acquired in order to cope with stress situations. He cannot consciously alter these expressions easily. Reich found that fundamental change only occurred when the emotions held down behind the facial tensions could be released.

The main emotions held down in the throat area are the noisy expressions of shouting, sobbing and screaming. In most cultures children are not supposed to make too much noise. Screaming and sobbing are disturbing emotions for many adults to have to face. But what else is an infant to do in intolerable situations of stress? It can only learn to swallow down its anger, to choke over its sorrows. When these held-down feelings were elicited in their original strength by Reich's vegetotherapeutic techniques, patients regularly experienced a sense of "clearing" in the head and a feeling of unity between the head and trunk, which had formerly been lacking.

Anger held back in the neck was connected with the muscles of the shoulders and, indeed, large areas of the back. Reich was at first astonished to discover how much rage was contained in people's backs. . . . In the "military posture" one had the replacement of organic rhythm with a machine-like breathing and posture. There was an obvious relationship between this bodily attitude and the emotional attitude that reached its most developed expression in the ideology of Fascism.

When Reich first began to influence the bodily tensions he continued using purely character-analytic methods; that is, he painstakingly described the patient's bodily expression to him, or imitated it himself in order to make the patient more and more aware of the detailed manner in which he used different parts of his body to suppress vital feelings. He would encourage his patients first to intensify a particular tension deliberately, in order to help their awareness of it. By intensifying it he was often able to elicit in an acute form the emotion which had been bound by the chronic form of the tension. Only then could the tension be properly abandoned. Increasingly, however, Reich began to use his hands directly on the bodies of his patients in order to work directly on the tense muscle knots. He was at great pains to point out that this was quite different from a physiological massage or manipulation, since it was necessary to understand the role of each tension in the total armouring of the patient. In vegetotherapy he was guided always by the emotional function of the tensions. If this was not grasped, mechanical pressure on muscle groups had only very superficial effects.[48]

While Reich's original character analysis was designed to facilitate orgasmic potency and healthy self-regulation, vegetotherapy was meant to complement it by promoting "vegetative liveliness" in the entire organism. Such liveliness is characterized by streaming sensations produced by muscular release. According to Boadella:

As patients gave up more and more of the bodily armour and their breathing became much freer, the capacity to surrender to spontaneous and involuntary movements increased greatly. Little by little sensations of warmth, of prickling in the skin, and of shuddering movements in the limbs and trunks began to integrate themselves into a convulsive reflex movement of the whole body in which there was a clonic involuntary flexion and extension of the spine. Looked at as a whole the body appeared to be expanding and contracting in a pulsatile manner. Because of its close resemblance to the clonic movements of orgasm, Reich termed this the "orgasm reflex."[49]

Reich's Biological and Cosmic Theories

Reich found support for his vegetotherapy in biological research that indicated to him that all living substance was animated by alternating periods of contraction and dilation with accompanying buildups and discharges of electric current.[50] In his later years Reich connected the psychophysical processes he observed in his patients with a universal life-force he called orgone. As Reich eventually conceived it, the orgone extended through all space, like the ether, and was drawn upon by all living things, forming a field around them that could be perceived by sensitive people. Like the Hindu *prana* or the Chinese *chi,* it pervaded the organism and could be transmitted directly through the hands or sexual intercourse.

Taken as a whole, it seems to me, Reich's cosmic theories comprise a quasi-mystical vision of the universe translated into a materialistic framework. Thus, Reich perceived all things to be unified by a tangible biophysical energy, the orgone, and he espoused a way of liberation (through orgastic potency and contact with the orgone's streaming). Like mystics since the dawn of history, he saw the world suffused with a single presence and wanted others to know its ecstasy, but he reduced the hierarchies of heaven and earth to fit his limited philosophy.

And like certain religious reformers, he advocated a human ideal with both liberating and tyrannical features. For his doctrines of orgastic potency and mature, or genital, character do not embrace every kind of human growth. Indeed, many people have exhibited the love and joy Reich sought in his therapy who were virgins all their lives or at some point renounced sexual activity altogether. Reich taught us much about the relations between cognitive, emotional, and somatic experience; he discovered certain ways in which we embody psychic issues; and his work has informed somatic educators for more than half a century; but he did not appreciate (or recognize) all our potentials for abundant life.

[18.9]
SOMATIC DISCIPLINES AND INTEGRAL PRACTICES

There are somatic disciplines besides those described in this chapter. The pioneering methods of Emily Conrad D'aoud and Bonnie Bainbridge Cohen, for example, have attracted large followings. Both are now used to relieve many kinds of disability and to cultivate sensory, kinesthetic, emotional, and cognitive functioning. The seven approaches discussed here, however, while not comprising the entire field of Somatics, have either been central to its development or have had a great influence upon it for several decades. Though just two of them (Autogenic Training and Progressive Relaxation) have been studied thoroughly with experimental controls, each has produced many types of healing and growth. Among them, they include methods that promote:

- sensory and kinesthetic awareness
- control of autonomic processes
- efficient modulation of sensory input
- sensorimotor coordination
- the articulation and coordination of particular muscle groups
- grace and efficiency of posture, carriage, and movement
- new patterns of movement
- flexibility of facial and gestural expression
- general relaxation as well as the relaxation of particular body parts during complex behaviors
- recuperation from stress
- vitality
- awareness and control of emotions and mental processes
- sensory, kinesthetic, emotional, and intellectual pleasure

In producing these outcomes, somatic disciplines can contribute to balanced programs for growth. By helping to articulate and coordinate various bodily and psychological processes, they can give rise to extraordinary functioning and strengthen the integral practices proposed in Part Three.

19

ADVENTURE
AND SPORT

THE WORLDS OF physical adventure and sport drama-
tize our capacity for self-exceeding. The immense variety of challenges athletes embrace on land, air, and water; the increasing number
of new games and contests (and the growing complexity of many old
ones); the exquisite analysis of technique and physiology used to
achieve high-level performance; the wide range of physiques cultivated for competitions of many kinds; and the pain, risk, and sacrifice
many people accept for the sake of sport all demonstrate our human
capacity and drive to achieve new levels and kinds of functioning.
Considering sport in its immense complexity, one might imagine that
the modern world has unwittingly made it a vast laboratory to experiment with our bodily powers.

Athletes and adventurers, moreover, often experience paranormal events, altered states, and ecstatic moments bordering on the
mystical. That such experience comes unbidden into the lives of many
sportspeople, affecting them deeply and challenging their beliefs
about themselves, dramatizes the fact that flesh and consciousness
tend to coevolve during the practice of strenuous disciplines. The fact
that spiritual moods occur spontaneously in many athletes indicates
that disciplines for the body sometimes catalyze depths of the mind,
even in people who have little or no understanding of such experience. The mind frequently opens in sport, suffusing bone and muscle
with its latent energies, whether or not the athlete can describe what is
happening to him. Indeed, I have come to believe that sport sometimes becomes a Western yoga of sorts, an earthy form of transformative practice. Through a contagion we are hardly aware of, athletes in
top form awaken a secret sense that we harbor capacities for extraordinary life.

Though modern sportspeople rarely have a philosophy to account for their illuminations or self-transcending feats, the connections between such experience and sport have been recognized in several cultures. In certain Amerindian tribes, for example, ritual running developed from religious ceremony.[1] Kathakali and other forms of Hindu dance join elements of yoga, athletics, and religious devotion. The temples of Asklepios combined physical exercise with devotion to God. In chapter 19.5 I discuss certain correspondences between athletic and religious practice.

Organized sport has a long history. The Olympic Games began in 776 B.C. (or earlier according to some accounts), and so captured the Hellenic world's imagination that the Greek city-states observed a one-month peace at the time of their celebration. By the sixteenth century, Europeans played field games, court games, rink games, course games, target games, table games, and street games. Today, increasing leisure and affluence permit the greatest expansion of athletics in history. Demanding activities such as hang gliding and windsurfing have achieved worldwide followings in a few years, while esoteric games such as underwater hockey attract many participants. Such exotic sports, as well as established ones, require highly specialized training that often depends on scientific research and new technologies. Professional football players, for example, have their weight and muscle size specified for them, and decathletes develop physiques that combine strengths of the shot-putter, sprinter, and vaulter. Some body builders have made their training an art: Frank Zane, for example, redesigned his physique for successive Mr. Olympia contests, winning three world championships with three distinct versions of himself.[†]

Women have extended this transformational enterprise. Their records have improved more dramatically than men's wherever women have begun to compete in numbers, notably in swimming, long-distance running, and track events; and they now participate successfully in dangerous sports such as mountaineering, skydiving, and hang gliding. Even more remarkable, perhaps, is the growth of athletic achievement among middle-aged and elderly men and women. Marks for older age groups are improving constantly in swimming and running.

[†] In several conversations with me, Zane said he had planned the physique he wanted for each of the three Mr. Olympia contests he won. The differences in his musculature from contest to contest are apparent in photographs.

But sport has developed without its participants perceiving the evolutionary possibilities I am exploring. Few can read all the dials in this vast laboratory, particularly those that give readings on subjective states. But the adventure of athletic discovery moves on—producing new records, revealing abilities no one had observed before, including more and more people in its compass. Sport will continue to amaze and empower us, suggesting new frontiers for human growth.

[19.1]
THE PROGRESSION OF MARKS AMONG ATHLETES

The inexorable advance of athletic records provides dramatic evidence that the human body has great potential for several kinds of development. The improvement of marks for men and women of all ages in swimming, distance running, and track and field events is unarguable. It is evident in world, national, and Olympic marks, in age records for either 10-year, 5-year, or 1-year groups, and in the growing number of world-class athletes. In 1982, for example, 150 women surpassed the marathon time that had stood as the world record in 1973, so sudden was the flowering of marathon talent. In 1980, more than 100 men broke the marathon world record, 2:15:17, that stood in 1962. In 1990, 170 men broke 2:15.[†]

Roger Bannister broke the four-minute barrier for the mile in 1954. By 1990, 595 men had run the mile in under four minutes, and during 1990, 62 men accomplished the feat.

Age records are falling even more dramatically. In 1985, Portugal's Carlos Lopes, who was 38 at the time, ran the marathon in 2:07:12, setting a new world record. When he won the Olympic marathon in 1984, Lopes defeated 37 men who were at least 10 years younger than he was. In 1981, 20 years after he set 14 world or national swimming records, Lance Larson, 41, exceeded his previous best times at several distances. Britain's Joyce Smith improved the marathon record for women over 40 to 2:29:57, a mark almost 10 minutes better than the previous record (Micky Gorman's 2:39:11), in races from 1979 through 1981. Here is a list of other remarkable age records, which was provided by Peter Mundle, the world's leading compiler of running records for people over 40 years old.

[†] These marks were provided by Jeff Hollobaugh of *Track & Field News*.

□ When he was 95, American Herb Kirk ran 800 meters in 6 minutes, 3 seconds.

□ When he was 90, American Buell Crane threw the discus 67 feet, 6 inches, and the hammer 55 feet, 8 inches.

□ When he was 87 years old, West Germany's Josef Galia ran 1,500 meters in 7:29:4 and the mile in 8:04:07.

Marks are improving more slowly in certain athletic events than in others. In men's sprints, for example, improvements have come less frequently in recent years than they have in distance events, though times for the fiftieth and hundredth places are better. Does this indicate that for these events the human body is approaching an absolute limit? This question is much debated, for apparent barriers like the four-minute mile have fallen with such consistency that we are led to suppose there will be breakthroughs in events that appear to offer little chance for improvement. Nevertheless, the body offers more opportunities for development along certain lines than it does along others. The explosive improvement of women's records in running and swimming demonstrates this obvious fact: there was more undeveloped potential available to women than there was to men. Men's distance running records also have improved markedly in recent decades, partly because African runners have had more opportunity to compete internationally than ever before.

In addition, abilities that were once thought to be extraordinary are now commonly exhibited by athletes in many parts of the world. Countless European, African, and American basketball players, for example, can dribble, jump, and shoot with more skill than the greatest stars of the 1940s. It wasn't until the 1940s, after all, that Stanford University's Hank Lusetti made the one-handed shot famous; and few players before World War II could slamdunk the ball. Such increase in skill is especially apparent in sports such as gymnastics and figure skating, where competitions are decided by judges. In these sports, advances in technique and achievement are as impressive as the new records in swimming and running (see 19.3).

Extraordinary physical functioning is also evoked by recently invented or once-esoteric sports. For example: spelunkers climb into caves through openings only slightly larger than the circumference of a climber's head, frequently at risk to their lives; underwater hockey players stay under water for two minutes or more while struggling against their opponents to move a puck along the bottom of a swimming pool; sailboarders do 360-degree flips on their boards without breaking their sail masts or falling; and climbers ascend ever more dangerous mountains and rock faces, overcoming obstacles once thought to be nontraversable. In each of these sports, and in many others that are growing in popularity today, previous standards of excellence are continually surpassed. Record-breaking of some sort occurs in every one of them, and each enables its participants to develop extraordinary attributes of mind and body.

The passion of sportspeople to exceed themselves reflects a drive for self-surpassing that all of us share to some extent. That drive, it seems to me, could find an outlet in the development of the metanormal capacities discussed in this book.

[19.2]
PHYSICAL FITNESS AND HEALTH RESEARCH

Modern epidemiological, clinical, and experimental studies give us new understanding of the physiological processes and particular practices that facilitate health. These studies show that men and women of all ages can improve their vitality and mental state through intelligently designed exercise programs. Never before in human history has the relationship between physical fitness and health been so thoroughly studied or well understood. Discussions about fitness and exercise sometimes suffer from a confusion of terms, however, so I will define certain words that are central to this discussion. *Physical activity* will refer to any bodily movement produced by skeletal muscles. The term *exercise,* on the other hand, will refer to physical activity that is planned, structured, and purposive—for play, sport, or the attainment of health. *Health* in this discussion will mean freedom from disease and a general vitality in work and leisure; while *physical fitness* will refer to a composite of agility, balance, coordination, speed, power, and reaction time as well as more health-related attributes such as cardiorespiratory endurance, muscular endurance, strength, body composition, and flexibility. Because the various components of fitness do not always vary in concert, I use terms such as *cardiorespiratory fitness* or *degree of flexibility* to signify development along particular lines.[2]

History

Health research supports the enduring insight that exercise promotes both psychological and physical well-being. The Greek ideal *mens sana in corpore sano,* a sound mind in a sound body, which has informed the teachings of doctors since antiquity, was articulated by Plato in the *Timaeus* (88):

> We should not move the body without the soul or the soul without the body, and thus they will be on their guard against each other, and be healthy and well-balanced. And therefore the mathematician or anyone else whose thoughts are much absorbed in some intellectual pursuit must allow his body also to have due exercise, and practice gymnastic; and he who is careful to fashion the body, should in turn impart to the soul its proper motions, and should cultivate music and philosophy, if he would deserve to be called truly fair and good.

In *Laws* 7, Plato recommended exercise through gymnastic training for the children of his ideal society:

> Education has two branches—one of gymnastic, which is concerned with the body, and the other of music, which is designed for the improvement of the soul. And gymnastic has also two branches—dancing and wrestling; and one sort of dancing imitates musical recitation, and aims at preserving dignity and freedom, the other aims at producing health, agility, and beauty in the limbs and parts of the body, giving the proper flexion and extension to each of them, a harmonious motion being diffused everywhere, and forming suitable accompaniment to the dance.
> . . . The art of wrestling erect and keeping free the neck and hands and sides, working with energy and constancy, with a composed strength, and for the sake of health—these are always useful, and are not to be neglected, but to be enjoined alike on masters and scholars.

Hippocrates of Cos, long celebrated as the patron saint of Western medicine, believed it was impossible to treat a patient without considering the entire body. In the *Corpus Hippocraticum,* a collection of ancient writings attributed to the famous doctor, physical activity is hailed as a cure for many ailments.[3] And exercise routines were prescribed at the healing centers of Greek antiquity. One Aristides described the exercises he practiced at Pergamum:

We were ordered to do many paradoxical things; among those which I recall there is a race which I had to run barefoot in winter-time, and again horse riding, the hardest of undertakings, and I also recall an exercise of the following kind: when the harbor waves were swollen by the wind and ships were in distress, I had to sail across to the opposite side, eating honey and acorns from an oak tree, and vomit; then complete purification is achieved. All these things were done while [my] inflammation was at its peak.[4]

At Epidaurus, festival games called the Asclepieia were held to promote the principle of *mens sana in corpore sano;* and similar games called the Nicephoria were held at Pergamum for several centuries.[5] The basis that physical culture provides for psychological health has been recognized, too, in Eastern cultures. In Patanjali's Yoga Sutras, meditative postures (*asanas*) and breathing exercises (*pranayamas*) complement mental discipline, and martial arts such as tai chi are often part of Taoist training.

But in spite of ancient wisdom about the beneficial effects of exercise, sustained scientific investigation of the effects of physical activity upon health did not begin until the 1950s. In 1953 J. Morris, of the London School of Hygiene and Tropical Medicine, and his associates reported that active bus conductors in London, who continually climbed the stairs of their buses, had less heart disease than did drivers, who sat at the wheel through the day.[6] Because the rising incidence of cardiovascular disability was causing worldwide concern, the Morris study prompted other researchers to assess the effects of occupation on health. Letter carriers were compared with mail clerks, American farmers with sedentary townsmen, Italian railroad workers with inactive office workers.[7] Researchers studied the bodily condition and work habits of Israeli kibbutzim laborers, and the incidence of heart disease among San Francisco longshoremen.[8] Though these first studies strongly suggested that regular physical activity provided protection against cardiovascular and other disease, other surveys of work habits and health produced ambiguous results because they did not account for nonwork activity, different physical requirements within occupations, or ethnic-cultural differences among their target populations.[9] Since the 1970s, however, epidemiological research has grown more sophisticated, and long-term investigations of large groups such as the San Francisco longshoremen study, the Framingham study, and the Harvard alumni study now provide convincing evidence that regular physical activity and exercise lessen the risk of cardiovascular and other diseases. In 1979 the U.S. Public Health Service declared that physical fitness was central to the improvement of public health; and to further the epidemiologic study of physical activity, the U.S. Centers for Disease Control created a Behavioral Epidemiology and Evaluation Branch in 1983. A workshop on the public-health aspects of physical activity conducted by the Centers for Disease Control in September 1984 provided a landmark recognition of physical fitness research by the American medical community.[10]

As the medical community's interest in exercise has grown, public participation in fitness-promoting activities has increased in many nations. In North America, Europe, Australia, Japan, and Russia, many millions of people have begun to hike, run, swim, or cycle on a regular basis. Indeed, most medical researchers now agree that such activity has contributed to the marked reduction in death from cardiovascular disease among Americans since the late 1960s.[11]

Epidemiological and Longitudinal Studies of Health and Exercise

Nationwide surveys of leisure-time physical activity, six of them in the United States and two in Canada, conducted between 1972 and 1983 indicated that approximately 20 percent of the two countries' populations exercised with an intensity and frequency generally recommended for cardiovascular health, while another 40 percent, though less active, exercised enough to receive some health benefit.[12] The Gallup Poll supported these survey results when it found that regular physical activity among its respondents had increased from 24 percent to 59 percent between 1961 and 1984, a jump of 246 percent.[13] Considerable data from community studies also suggest that physical activity has increased substantially in Canada and the United States.[14]

The San Francisco longshoremen study, which was initiated in 1951, tracked the work habits and fatalities of 3,686 San Francisco longshoremen for 22 years. Physical measurements of each man's exercise were made during on-the-job situations; changes in assignment were checked annually; and mortality rates were based on official death certificates. When cargo handlers were compared to more sedentary foremen and clerks, it was found that the death rate from coronary heart disease for men who expended more than 8,500 kilocalories per week was only about half (0.56) the rate for men with less energetic duties. A preliminary study showed that the participants' leisure activity was of little importance in these calculations.[15]

The Harvard alumni study has yielded considerable evidence that physical activity provides protection from cardiorespiratory afflictions, diabetes, and other diseases. An analysis of the study's data published by Ralph Paffenbarger and other medical researchers in 1984 showed that physically active alumni had a lower incidence of heart disease, stroke, respiratory disease, cancer, and deaths from all causes than those who were sedentary.[16] In a subsequent article, Paffenbarger and associates reported that the death rates of the study's participants declined steadily as the energy they expended increased, up to a level of 3,500 kilocalories per week (beyond which death rates increased slightly.) Exercise lowered death rates among the participants even if they had hypertension, smoked regularly, were overweight, or had a parent who died before the age of 65. Smokers, however, improved their chances for longer life by giving up smoking.[17]

The Framingham study, a carefully constructed investigation of various habits and physical characteristics related to incidence of heart and blood-vessel disease, has shown (among other things) that the percentage of high-density lipoprotein cholesterol (HDL-C) in relation to total cholesterol is inversely correlated with coronary afflictions in both men and women. Since this percentage is increased by regular endurance exercise, the Framingham results provide further (indirect) evidence that such exercise promotes health and longevity.[18]

The Lipid Research Clinics Mortality Study followed 4,276 men for an average of 8.5 years. The study's results strongly suggest that a lower level of physical fitness is associated with a higher risk of death from cardiovascular disease.[19]

The *Aerobic Center Longitudinal Study* followed 10,224 men and 3,120 women for an average period of slightly more than 8 years, for a total of 110,482 person-years of observation. Age-adjusted, all-cause mortality rates were 64.0 per 10,000 person-years in the least fit men, but only 18.6 in the most fit men. There were 39.5 deaths per 10,000 person-years for the least fit women, but only 8.5 among those who were most fit. This strong correlation between physical fitness and mortality rates remained after statistical adjustments were made for age, smoking habits, cholesterol levels, systolic blood pressure, fasting blood-glucose levels, parental history of coronary heart disease, and follow-up intervals. The study's authors concluded that physical fitness tends to lengthen life, in large part by reducing cancer and cardiovascular disorders.[20]

It has been argued that regular physical activity is associated with freedom from heart disease, hypertension, diabetes mellitus, and osteoporosis because people with such disabilities *cannot* exercise as much as healthy people. The research results described above might then reflect a selective rather than a protective factor. Repeated examinations of these studies, however, have shown that physical activity does indeed offer genuine protection from the conditions just noted. For example, analysis of four longitudinal studies with a combined population of more than 40,000 people showed that: (1) each study began with a healthy population, thus minimizing the potential confounding that would result from the initial inclusion of many sick people; (2) other potentially confounding factors such as age, smoking, obesity, hypertension, and family history were accounted for in their data analyses; (3) constitutional differences could not account for the correlation of exercise and health, at least in the Harvard study, because physical activity in adult life rather than college athletics predicted lower risk of heart disease; and (4) excluding cases of heart disease that occurred early in the follow-up period or accounting for changes in job classification did not alter the relationship between activity and freedom from cardiac disease.[21]

Epidemiological investigations of health habits strongly indicate that exercise has both a direct effect in curbing cardiovascular disease and an indirect effect through its reduction of blood pressure, obesity, and other factors that contribute to sickness in general. It should be pointed out, however, that while vigorous exercise is protective in the long run, it involves some short-term risk. Paul Thompson, a heart-disease researcher affiliated with Brown University, found in a study of deaths that occurred during jogging that the rate of death for joggers in Rhode Island—one death per year for every 7,620 joggers—was seven times higher than the estimated death rate from coronary heart disease during more sedentary activities. This increased risk during exercise, however, was powerfully compensated for by the long-term protection running afforded.

Thompson wrote in the *New England Journal of Medicine* that although "vigorous physical exercise both protects against and provokes sudden cardiac death, any short-term risk of exercise is out-weighed by [its] long-term beneficial effects."[22] In their review of heart disease studies, Siscovick and associates (1985) concluded that although "the risk of primary cardiac arrest is transiently increased during the act of vigorous exercise, men who are habitually vigorous have a lower overall risk of primary cardiac arrest than sedentary men."[23] Commenting upon the balance of risk and protection pro-duced by running and jogging, Thompson told Jane Brody of the *New York Times* that "if a person has only an hour to spend and wants to know the safest way to spend it, I'd say sit down—or better yet, lie down—and do nothing. But if a person has a lifetime to spend, it is not safest to spend that hour sitting; he'd be much better off exercising."[24]

Beneficial Physical Changes Produced by Exercise

In making the following list of changes resulting from exercise, I want to highlight the fact that transformative activities that have stood the test of time typically produce several beneficial results at once. Like the relaxation response produced by meditation, fitness through ex-ercise involves a large return on psychological and physical invest-ment. In chapter 24.2, I discuss complex, all-at-once responses that facilitate creative human change, and suggest that their employment makes transformative disciplines more efficient.

Enlarged and strengthened heart muscle. Regular exercise often en-larges and strengthens the heart. This kind of heart enlargement dif-fers from that caused by disease. It does not result from resistance to blood flow or some other maladaptive functioning, but rather from structural changes that give the heart greater strength and efficiency. Indeed, the average heart volume of a well-trained athlete is generally about 25 percent larger than that of a sedentary person matched for gender, age, and other features. Though its dimensions or configura-tions depend on the sport involved, an athlete's heart typically ex-hibits increased left ventricular thickness and mass, an enlarged ven-tricular cavity, thickening of individual myofibrils, thickening of the septum, and a larger number of contractile filaments within the mus-cle fibers.[25] Though such cardiac strengthening must be maintained through exercise, it provides protection from degenerative disease while it lasts.[26]

Increased stroke volume. World-class endurance athletes have cardiac outputs of 35 to 40 liters of blood per minute, while sedentary but healthy people have much less.[27] Since cardiac output determines how much oxygen, nutrients, and waste can be circulated by the blood, high output is correlated with physical efficiency, strength, and endurance. But endurance athletes achieve high cardiac output without raising their heart rates beyond those of comparable sedentary types because their stroke volume—the amount of blood pumped per stroke—is greatly increased. In one study, for example, the cardiac output of Olympic cross-country skiers increased almost eight times above rest to 40 liters a minute with a stroke volume of 210 milliliters per beat, which was nearly twice the capacity of healthy, sedentary people of similar age.[28]

Lower resting heart rate (bradycardia). Because their hearts are more powerful and efficient than those of untrained persons, endurance athletes usually have lower resting heart rates than people matched for age and other characteristics.[29] Many studies have shown, for example, that their rates can be as low as 35 to 40 beats per minute.[30] Furthermore, an endurance athlete's heart accelerates more slowly than a poorly trained heart during exercise, in large part because it pumps more efficiently.[31]

Increased blood and hemoglobin levels. Plasma volume and total hemoglobin tend to increase with endurance training. This adaptation, it seems, promotes circulatory and thermoregulatory mechanisms that assist oxygen delivery during exercise.[32]

Improved venous return. Compression and relaxation of the veins, as well as the one-way action of the venous valves, provide a "milking" action in the heart. This mechanical compression of the veins in the exercising muscle facilitates cardiac output and is the combined result of an increase in venous return, compression of the intraabdominal veins by the abdominal wall, and the action of the thoracic respiratory pump. As cardiac output increases with endurance training, venous tone also improves in both working and nonworking muscles.[33]

Increased VO2Max. The point at which oxygen consumption plateaus and shows no further increases, called maximal oxygen uptake or VO2Max, is generally regarded to be the most accurate single measure of the cardiorespiratory system's capacity for energy transfer.[34]

Endurance training can increase this vital capacity some 15 to 30 percent in the first three months of training and as much as 50 percent over a two-year period.[35] Indeed, exercise training has been shown in reliable studies to produce increases in maximal oxygen uptake as high as 93 percent.[36]

Improved circulation. Physically fit individuals perform submaximal exercise with less cardiac output than their untrained counterparts, partly because exercise promotes a muscle cell's ability to extract and utilize oxygen. Consequently, less regional blood flow is required to meet the muscle's oxygen needs, and free blood can be delivered to nonworking tissues.[37] Athletic training also increases capillary density, which shortens diffusion distance for oxygen, and facilitates blood profusion in the muscles.[38] Repetitious vasodilation during exercise also evokes a training effect on the arterioles, increasing their ability to dilate.[39]

Decreased blood pressure. Clinical and experimental studies have shown that both systolic and diastolic blood pressure can be lowered by endurance exercise. Such lowering has been observed in people with normal blood pressure and in those with hypertension.[40] Indeed, among patients with coronary artery disease and borderline hypertension, the effect of exercise training on blood pressure is sometimes dramatic.[41] The fact that vigorous exercise dilates the blood vessels has led some researchers to think that it produces a chronic vasodilation that reduces vascular resistance to blood flow.[42] Several epidemiological studies, too, indicate that physically active people have significantly lower systolic and diastolic blood pressure at rest than their less active, age-matched peers, and thus a lower risk of cardiovascular disease.[43]

Greater bone mass. People who maintain an active lifestyle have stronger bones than sedentary people, even into the seventh and eighth decades of life.[44] Several studies of postmenopausal women have shown a strong correlation between exercise and bone density.[45] Exercise of a weight-bearing nature, such as walking, running, or dancing, has the most beneficial effect in this regard, in part because it modifies bone metabolism at the point of stress. Leg bones of cross-country runners, for example, or the bones in athletes' throwing arms, have greater mineral content than less active bones.[46]

Decreased degeneration of joints and ligaments. Exercise stimulates a proliferation of connective tissue and satellite cells around individual muscle fibers and improves the structural and functional integrity of tendons and ligaments.[47] These changes provide some protection from joint and muscle injury.

Increased muscular strength. Muscles grow in response to exercise by enlargement of individual muscle fibers. Within the muscle cells, myofibrils thicken and increase in number, and protein breakdown decreases.[48] Habitual physical exercise helps muscle tissue retain protein and thus prolongs strength in older people.[49] Endurance training also increases myoglobin content in skeletal muscle, so that the quantity of oxygen within the muscle cells is increased.[50]

Improved reaction time. Age-related slowing of reaction time can be prevented to some degree by regular physical activity. One study found, for example, that active people in their sixties had reaction times as good as inactive people in their twenties.[51]

Increased ability to utilize fats and carbohydrates. Regular exercise increases the trained muscle's ability to mobilize and oxidize fats and carbohydrates because it promotes increased blood flow within the muscle, activity of fat-mobilizing and fat-metabolizing enzymes, efficient burning of free fatty acids, oxidative capacity of the mitochondria, and glycogen storage within the trained muscle.[52]

Decreased body fat. Among overweight people, regular endurance training causes a reduction in body weight accompanied by a decrease in body fat, in part because exercise mobilizes hormones that help reduce fat deposits throughout the body.[53]

Normalized blood lipid levels. Coronary heart disease varies inversely with the level of high-density lipoprotein (HDL) cholesterol, while there is a positive correlation between the frequency of the disease and low-density lipoprotein cholesterol (LDL) concentrations. Therefore, the ratio of total serum cholesterol to HDL is a good predictor of coronary heart disease. There is good evidence that endurance exercise has a beneficial effect on levels of these blood lipids, raising HDL while lowering LDL.[54]

Improved mobilization of lactic acid. Lactic acid, which increases with exercise, is a valuable source of chemical energy that is retained in the body during heavy physical exercise. Several studies have shown that after trained athletes perform strenuous short-term exercise, their blood lactic acid levels are 20 to 30 percent higher than those of untrained subjects.[55]

Improved hormonal balance. Intelligent exercise affects the secretion of the following hormones to improve endocrine balance:[56]

- Growth hormone, which stimulates tissue growth and helps turn fatty acids to energy.
- Prolactin, which helps mobilize fatty acids.
- Endorphins, which block pain and promote the feeling of an "exercise high."
- Vasopressin, which controls water excretion by the kidneys.
- Cortisol, which promotes use of fatty acids and protein catabolism and helps conserve blood sugar.
- Epinephrine and norepinephrine, which facilitate activity of the sympathetic nervous system, increase cardiac output, regulate blood vessels, and increase glycogen catabolism and fatty-acid release.
- Thyroxine T3 and T4, which stimlate metabolic rate and help regulate cell growth.
- Insulin, which promotes carbohydrate transport into cells, decreases blood glucose, and promotes fatty-acid and amino-acid transport.

Increased bloodclot-dissolving enzymes and inhibition of platelet aggregation. Several studies have shown that moderate exercise protects against heart attacks by increasing enzymes that help dissolve blood-clots and help inhibit platelet aggregations in the blood.[†]

[†] Ekelund, 1988.
Rauramaa, R. et. al. Effects of mild physical exercise on serum lipoprotein and metabolites of arachidonic acid. *British Medical Journal,* 1984. 288: 603–06. Rauramaa, R. et. al. Inhibition of platelet aggregation by moderate-intensity physical exercise: a randomized clinical trial in overweight men. *Circulation,* 1986. 74:939–44. Williams, R. et. al. Physical conditioning augments the fibrinolytic response to venous occlusion in healthy adults. *New England Journal of Medicine,* 1980. 302:987–91.

Strengthened immune system. Several studies have shown that moderate exercise promotes immune system activity that facilitates recovery from stress and helps ward off disease. For example:

- Seventeen healthy subjects, eight of them women between 21 and 39 years old, and nine of them women more than 65 years old, exhibited significant increases of natural killer-cell activity after bicycle exercise tests.[57]
- Components of the immune system called cytokines (including interleukin-1, interleukin-2, and tumor necrosis factor) are produced by exercise. There are strong reasons to believe that moderate exercise helps maintain these cytokines at levels that strengthen immunity to disease, but that chronically strenuous and exhausting exercise produces an imbalance in their production that depletes the immune system.[58]
- Aerobic training increases components of the immune system called helper/inducer (CD4) cells in people infected with the AIDS virus.[59]
- Regular exercise helps counteract day-to-day depressive disorders that suppress resistance to cancer. Exercise does this in part by stimulating endorphins that modulate various components of the immune system.[60]

But here a warning must be added. Overtraining can suppress immunity to infection. Indeed, studies from several exercise laboratories have shown that after a single exhausting exercise session there is temporary immune depression, with marked changes in the number and functional capacities of lymphocytes. These changes, which can last for several hours, have been observed in both well-conditioned athletes and untrained people.[61]

Reversal of coronary heart disease. A sustained regimen of low-fat vegetarian diet, nonsmoking, stress-management training, group therapy, meditation, and moderate exercise produced reversal of coronary heart disease in a group of patients studied by physician Dean Ornish and his colleagues. After one year of this regimen, 28 experimental subjects were compared with 20 patients in a usual-care group using coronary angiography, which reveals the size of arterial lesions. Overall, 82 percent of the experimental-group patients had reduced lesions in their blood vessels, while most members of the usual-care group did not.[†]

Improved resistence to cancer. In the Harvard alumni study, the cancer rate was significantly lower in those participants who burned more than 2,000 calories per week through physical activity than in those who burned less than 500.[62] In another study, female nonexercisers had almost twice the incidence of breast cancer and reproductive system cancer as former athletes;[63] and a 1986 Swedish survey showed an inverse relationship between job activity and colon cancer.[64]

Improved Mental Functioning and Psychological Health

Among the positive mental and emotional results attributed to regular physical activity, these have been reported in the medical literature: improved academic performance, self-confidence, emotional stability, independence, cognitive functioning, memory, mood, perception, body image, self-control, sexual satisfaction, and work efficiency; and decreased alcohol abuse, anger, anxiety, confusion, depression, dysmenorrhea, headaches, hostility, phobias, stress response, psychotic behavior, tension, and work errors. While claims for such results are based largely upon anecdote, opinion, or methodologically limited studies, there is also good experimental evidence that regular physical activity does indeed enhance certain aspects of mental and emotional functioning. Several well-regarded studies, for example, have shown that exercise can relieve mild to moderate depression and anxiety, promote self-confidence, and stimulate a general sense of well-being.[65]

In 1987 researchers at the U.S. National Institute of Mental Health evaluated clinical findings and other evidence linking mental health and exercise, and issued a consensus statement that physical fitness produces feelings of well-being and helps reduce anxiety, depression, and neuroticism. The statement concluded: "Current clinical opinion holds that exercise has beneficial emotional effects across all ages and in both sexes."[66]

[†] Ornish, Dean et. al. Can lifestyle changes reverse coronary heart disease? *The Lancet,* July 21, 1990. 336: 129–35.

Negative Outcomes of Exercise

Like other potentially beneficial activities, physical exercise is some-times destructive, producing (or contributing to) inadequate concern for job or marriage, excessive fatigue, overcompetitiveness, unwhole-some preoccupation with diet, obsession with body image, extreme self-centeredness, or physical injuries. Like any other kind of self-cultivation, physical training requires intelligence, balance, and good judgment to produce good results.[67]

[19.3]
THE ELABORATION OF SPORT

Gymnastics

At the 1972 Olympic Games, Olga Korbut helped create a worldwide interest in gymnastics, stimulating an enormous audience for Nadia Comaneci and other young prodigies who followed her. But few of Korbut's admirers appreciate how many gymnasts and coaches pre-pared the way for *her* through centuries of innovation. The charis-matic Russian gymnast inherited an enormous repertoire of maneu-vers, a highly developed set of training methods, and the example of many virtuosos. Like all great traditions—in science, art, and other sports—gymnastics gives its practitioners a complex achievement to draw upon while it makes increasing demands upon them. Indeed, some of the activities included in modern gymnastics were developed in the ancient world—in Crete, for example, where acrobats leaped over running bulls; in Greek gymnasiums, where exercise routines prepared athletes for wrestling and racing; and in Roman military schools, where wooden horses were employed to teach mounting skills. In late medieval and Renaissance Europe, gymnastic skills were perpetuated. Balancing routines were popular in the fourteenth cen-tury, in part because European noblemen admired the performances of Saracen girls in the court of Frederick II, and Rabelais described wandering acrobats of the sixteenth century who placed ropes be-tween trees "and there did swing by the hands touching nothing."

Gymnastics was given systematic formulation in 1793 by the German educator Johann Guts Muths. From his theories, gymnastics began its development into the sport we see today. Two men especially,

Johann Friederich Jahn (1778–1852) and Pehr Henrik Ling (1766–1839), carried his program forward. Inspired by Guts Muths, Jahn opened an outdoor gymnastics center in 1811 on the outskirts of Berlin; and in the years that followed he founded gymnastic clubs, promoted three pieces of apparatus now in standard use—the parallel bars, rings, and high bar—and developed several exercise routines. Ling, too, founded a gymnasium after reading Guts Muths, but was also influenced by the work of the Danish educator Franz Nachtegall. Adapting Nachtegall's work for his native Sweden, Ling devised the beam and wall bars, founded an institute for training gymnastic teachers, and established exercise principles for the Swedish army. His programs put more emphasis on the free flow of movement than Jahn's and did not require the same muscular force as the German's.

During the nineteenth century, the work of Jahn and Ling was developed by educators, athletes, and military men in Europe and America who organized gymnastics programs in clubs, schools, and military training programs. In 1881 representatives of several European gymnastic societies formed an international association for the sport. It was called the European Gymnastics Federation until 1921, when its name was changed to the International Gymnastics Federation, or FIG. The FIG is the oldest of all international sports federations.

Gymnastics was one of nine sports included in the first modern Olympics in 1896, but its format has been changed from one Olympiad to the next. Among the maneuvers introduced to the sport, there were:

- Italian Alberto Braglia's swing with straight arms around the high bar in 1912.
- Hungarian gymnast Gaki Meszaros's split on the high beam in 1934.
- Czechoslovak Alois Hudec's inverted crucifix on the rings (an upside-down hand balance with arms outstretched).
- Swiss gymnast Josef Stalder's "shoot" on the high bar, a clear circle from the handstand position passing through straddle support (which he performed in the London Games of 1948).
- Italian Franco Menichelli's 1960s innovations in the floor exercises, including expressive tumbling routines and maneuvers such as the Arabian high dive forward roll.
- Soviet gymnast Albert Azaryan's Iron Cross on the rings in 1960.

- Soviet gymnast Polina Astakhova's held lever sit, which consists of a lift from a straddled support on the balance beam to a straight-armed handstand and then a slow levered move to pass straight-legged on either side of the hands so that the body ends up in a 90-degree position with the hands between the knees.
- Yugoslav Miroslav Cerar's high shear on the pommel horse, which consisted of high leg kicks as he moved from one end of the horse to the other.
- Japanese Haruhiro Yamashita's handspring on the vault with a pike (a levering of the trunk and extended legs so that they make an angle of 90 degrees or less) at the 1962 World Championships.
- Soviet Mikhail Voronin's dramatic swinging routines on the rings.
- Soviet Sergei Diomidov's 360-degree turn on the parallel bars in the handstand position, a maneuver now named after him.
- Mikhail Voronin's high-piked front vault on the high bar, after which he recaught the bar on the side opposite the one he jumped from (now called the Voronin Hop).
- Japanese Mitsuo Tsukahara's 1970 vault with a half-turn of the entire body onto the horse and a one-and-a-half back somersault before landing (now called the Tsukahara Vault).
- Tsukahara's somersault from the high bar during the 1972 Olympics.
- Olga Korbut's back somersault on the beam and her flip from the top asymmetrical bar, both parts of the routine with which she delighted a worldwide television audience at the Munich Olympics.
- Soviet Nelli Kim's 540-degree whirl in the floor exercises during competitions in 1974.
- Romanian Nadia Comaneci's front aerial walkover with a half-twist on the beam and her dismount from the asymmetrical bars with a high twist and back somersault, in 1975; and her dramatic routine on the asymmetrical bars during the 1976 Olympics, which won the first perfect score of 10 in Olympic gymnastic history.
- Soviet Elena Mukhina's Korbut flip from the high asymmetrical bar with an added full twist and back somersault dismount, in 1977; and her full twist in a standing somersault to recatch the top asymmetrical bar, in 1978.

□ American Kurt Thomas's flare (now named after him), a flashy series of swinging leg movements performed on the floor or on the horse, in 1978.

□ Nadia Comaneci's midair somersault to start an asymmetrical bars routine, in 1980.

□ Yelena Davydova's 1980 double somersault during the floor exercise.

Other maneuvers have been added to gymnastics since 1980, but enough have been listed here to suggest the sport's great complexity. To achieve world eminence now, gymnasts must innovate as well as show mastery of established routines. They must demonstrate a great variety of skills as well as dramatic flare. To do this, they must train for many years with regimens more challenging by far than those required of former gymnasts. To be an all-around champion, a gymnast must develop strength, flexibility, agility, coordination, balance, and cardiopulmonary conditioning. Like accomplished decathletes and ballet dancers, skilled gymnasts today possess an enormous number of physical capacities.

Aerial Sport

Aerial sport expresses the human drive for transcendence of earthly restraints. In balloons, airplanes, parachutes, and spacecraft, or falling freely through space, increasing numbers of men and women have explored the possibilities of flight. In doing so, many of them have discovered new freedoms of movement and awareness. In his book *The Ultimate Athlete,* George Leonard wrote:

> When we fly, we add one dimension to our life; [and] we add it all at once. We approach this moment with apprehension, and once the line is crossed we are filled with a sensation of strangeness. Yet, often, there is also a sense of recognition. It is as if we have entered a familiar place. The exhilaration of flight is not just sensory. It is also the exhilaration of return, of coming home to a freedom we have longed for and once known.[68]

Other flyers have noted the simultaneous perception of strangeness and familiarity in aerial sport. Describing his sensations during free fall, physiological researcher Harry Armstrong reported that his thinking and perception seemed normal, contrary to expectations they would be blurred or obliterated. His breathing was undisturbed. The pressure of air on his body, he wrote, felt like "a very gentle, evenly distributed, generalized, superficial pressure on the surface of [my] body towards the earth. The nearest possible similar earthly experience is that of being lowered gently into a great bed of down."[69] Skydivers often experience this sensation. Before they open their parachutes, some of them turn and glide, join with others in elaborate formations, pass batons, or make balletic maneuvers.[70] And at times, the joy of flight opens into something beyond earthly horizons. In chapter 5.1, I cited Charles Lindbergh's account of such experience.[71]

Like Lindbergh, astronaut Russell Schweickart described the wonder of flying. In spite of the bodily violation he felt during his Apollo Nine adventure, he felt an immediate pleasure in the sensations of zero gravity.[72] The epiphanies that accompany free movement in three dimensions, and the strangely familiar nature of flight, have kindled the belief among pilots such as Lindbergh and Schweickart that we might enter a freer existence that is our secret birthright. The sudden familiarity, the sense of something remembered in a new dimension, is characteristic of much transformative experience, whether it takes place in alien physical domains such as outer space or in altered states produced by disciplined meditation. George Leonard and other flyers have believed that humans are built for adventurous journey, that we are by nature boundary-crossing, environment-transcending animals. Conceivably, the hunger for new experience that people exhibit in aerial sport might empower a crossing into richer embodiment, into that multidimensional existence I am exploring in this book.

Human flight, which had long been pictured in myth and art, was not realized until the late eighteenth century. Some had tried before that—among them an Arab, Armen Firman, who made wings and jumped from a tower in 852, and a Benedictine monk of Malmesbury Abbey who broke both legs during a similar attempt in 1020. But as far as we can tell, no one jumped safely from a great height until a Frenchman, Louis Sebastien Lenormand, parachuted from the tower of Montpellier observatory on the day after Christmas in 1783.[73] A large balloon had been launched that year by Joseph and Etienne Montgolfier, causing a sensation in Europe and inspiring ascents with animals; and on November 21, five weeks before Lenormand's first parachute jump, Pilatre de Rozier, curator of France's Royal Museum of Natural History, and the Marquis d'Arlandes had risen some 500 feet over the Bois de Bologne before a cheering crowd. Ballooning became a popular sport, and two Frenchmen fought a duel in balloons, firing at each other with blunderbusses above the Tuileries Gardens.

Parachute jumping developed rapidly after Andres Garnerin made a series of jumps in the early 1800s from heights up to 8,000 feet. Inspired by Garnerin's success, engineers devised several types of parachute and glider wing. There were countless injuries and deaths during these aerial experiments, but ballooning became an international sport and parachuting a form of entertainment. Borrowing acts from circus gymnastics, men jumped from balloons on trapezes attached to inflated canopies, sometimes hanging by their legs or their teeth. Attempts to fly were made with propeller-driven balloons and with winged machines, none of which were more thorough than the German inventor Otto Lilienthal's experiments with gliders during the 1880s and '90s. Lilienthal made more than 2,000 glides in safety (few of them longer than 20 seconds); developed numerous wing designs and air pressure tables; and published a book, *Bird Flight as the Basis of the Art of Flying*, before he was killed in a glider crash in 1896.

The airplane gave aerial sport an even greater impetus. Orville and Wilbur Wrights' first flight took place in December 1903, and in 1909 Louis Bleriot flew across the English Channel. In 1911 Calbraith Rogers took his *Vin Pix* from New York to Pasadena, California, in 3½ days flying time spread over 49 days. In the years that followed, altitude and distance records were continually broken; planes were invented that could loop and dive and roll. The air seemed an ever friendlier place.

As machine-powered flight developed, parachuting developed in concert with it. The first ripcord device was used in 1908. An American Army colonel, Albert Berry, made the first jump from a plane in 1912, and another American, Les Irvin, showed that free fall was possible in 1919, disproving a long-standing belief that it would knock a jumper unconscious. Subsequently, more and more jumpers "made mayonnaise," falling out of control before they opened their chutes. Some did cutaways or double jumps in which they opened a chute, cut it loose, then opened a second chute. (Charles Lindbergh did a cutaway in 1922, on his first jump from a plane.) Others tried for altitude marks, among them American Joe Crane, who set the first widely acknowledged world record, at 20,300 feet, in 1924. A few descended from planes with cloth wings attached to their arms and legs, extending their time in free fall while they made gliding loops and spirals. In 1942 Russian air force Major Boris Kharankhonov set a new altitude record of 40,813 feet. Skydiving as we know it only began, however, after the French parachutist Leo Valentin invented a technique, now called the Valentin position, in which the jumper extends his arms and legs to maintain stability. Since Valentin developed this maneuver, jumpers have invented countless stunts and extended their time in free fall. By many accounts, the long periods they spend in midair sometimes produce altered states of mind and quasi-mystical experience.

Eventually, parachuting and skydiving competitions were established in Europe, Russia, and the United States. The first world championships were held in Yugoslavia in 1951, the second in France in 1954, the third near Moscow in 1956. These and other contests include both accuracy and free-fall competition. In accuracy jumping, parachutists leave their planes at heights between 600 and 2,000 meters, aiming for a target with a bull's-eye that until recently was 10 centimeters (approximately four inches) in diameter. Jumpers have gained such control that they can land on the bull's-eye of their landing targets from heights of 15,000 feet or more. In 1972, at the world championships in the United States, a Czech skydiver had nine dead-center strikes in a row. At Yuma, Arizona, in 1978, Dwight Reynolds scored 105 consecutive dead-centers in daylight over a period of several days, and Bill Wenger and Phil Munden of the U.S. Army's Golden Knights each scored 43 consecutive dead-centers at night. Such accuracy prompted a change in the size of the bull's-eye from 10 to 5 centimeters.

In style or free-fall contests, competitors perform a series of aerial gymnastics for which they are awarded points by panels of judges. These contests include free-fall relative work in which two or more jumpers hold hands, pass batons, or perform synchronized movements. In 1965, eight jumpers formed the first eight-way star, and other jumpers subsequently developed static geometric forms such as squares and diamonds. In the late 1960s, skydivers started performing established maneuvers at greater speed, sometimes completing two or more static group formations during a single drop; and eventually jumpers learned to synchronize their maneuvers without contact. In the skydiving community, records are under constant assault. The altitude mark, for example, is now more than 20 miles, while static formations have included 75 members or more. National teams have more depth and all-around skill than before, and there are more virtuosos among women and older jumpers.

Hang gliding too has developed in recent years. Hang-glider pilots frequently break records for duration of flight, height of ascent, and distance covered. There are also competitions for speed and acrobatic maneuvers such as sailing slalom-like around balloons. In 1976, American Bob McCaffrey made a hang-glider descent of 31,600 feet from a balloon over the Mojave Desert.

New Sports

While established sports have been given new complexity and beauty, many new ones have been invented. Among them are the following.

Underwater hockey, or *octopush,* in which skin-diving players use a modified hockey stick to strike a puck against an opposing team's pool wall. The pool in which the game is played must be at least six feet deep, and the puck generally remains at the bottom. This can be a very rough game, and its contestants must have great lung capacity.

Freestyle skiing, which produces many injuries, and includes acrobatic stunts such as:

- □ sitting back and then lying down on the skis while moving
- □ touching one's back with the rear tips of the skis while the front tips are pointed down
- □ moving backward down a slope in the herringbone position, making a series of 360-degree turns
- □ jumping with one ski forward and up and the other backward and down before bringing them into alignment
- □ double aerial turns through 360 degrees
- □ bringing both skis up behind the skier to form a cross
- □ lifting a ski and placing it near the tail of the other, then transferring one's weight to it
- □ crouching on one ski while the other is extended to the side off the ground
- □ turning 360 degrees supported by ski poles
- □ rising to the tips of the skis with one's heels crossed

Each sport I have discussed in this section makes enormous demands upon those who pursue it. Each evokes extraordinary functioning and dramatizes the human passion for self-exceeding.

[19.4]
ELEMENTS OF SPORT THAT CONTRIBUTE
TO ATHLETIC SUCCESS

Certain things that promote athletic success can facilitate psycho-physical development in general. The technological innovations, improved medical knowledge and training techniques, new insights about human psychodynamics, social support, growing numbers of participants, and inspiration of small training groups that help produce the great beauty of sport—all of these can contribute to the integral practices discussed in this book.

Technological Innovation

Improved running shoes, synthetic track and jumping surfaces, safer landing pits that allow high jumpers and pole-vaulters more extension and exposure of their bodies, and other improvements of equipment and athletic arenas have contributed to the improvement of marks in track and field. Maneuverable surfboards and lightweight climbing gear enhance the abilities of skydivers, surfers, and mountaineers. Portable monitoring devices enable coaches and doctors to keep a close watch on athletes in training. And training procedures involving videotapes of desirable performance have helped many athletes improve their skills. Such innovation could contribute to any physical discipline that was part of an integral practice.

Improved Medical Knowledge and Training Techniques

As nutrition has improved and motivations for athletic success have increased in many parts of the world, scientific understanding of the body has grown enormously. New knowledge about the effects of stress, the interactions of hormones and exercise, the elements of cardiopulmonary fitness, and other aspects of physical functioning has led to more effective training regimens. Never before have coaches known so much about the body's response to various work loads and athletic maneuvers. Never before could athletes train so efficiently without breaking down. Strenuous year-round training was unusual in most sports until recent decades, but higher levels of excellence require it, and increased medical understanding makes it possible. The growing alliance of medical knowledge and sports promises even more improvements of training methods in the years to come.

Among the many lessons learned about athletic training, one should be specially noted here: namely, that great athletes can inadvertently set a bad example. Al Oerter, for example, won four Olympic championships in the discus—more in a single event than any other Olympic track and field competitor—with a flawed technique that was subsequently imitated by many throwers. The science of biomechanics eventually revealed Oerter's flaws and helped him improve his marks when he was more than 40 years old, after he had inspired others with his defective motion. We can learn something here about other kinds of training. Because the frontiers of human development are revealed by different kinds of people, including some with marked peculiarities and notable flaws, it is a great mistake to focus exclusively on single examples of extraordinary functioning for our constant guidance. We need to be careful about the exemplars we choose in creating a transformative practice.

New Insights About Human Psychodynamics

Psychology offers insights that sports training programs can draw on to promote athletic success. For example, several studies have shown that visualization combined with physical training can enhance motor skills,[74] and nationwide physical education programs in Sweden have introduced relaxation, concentration, visualization, and meditation training to several thousand athletes from different sports. Like the physical training methods outlined above, psychological approaches can be adapted to disciplines other than sports, including programs for integral development.

Social Support

The news media have promoted (and benefited from) sports, bringing constant attention to athletic events. By conservative estimate, more than 500 magazines devoted to particular sports were published around the world in 1983. Schools, churches, clubs, and youth leagues are also motivators through their teams and education programs. Such cultural support facilitates the great athletic accomplishment we witness today. Furthermore, many international competitions have arisen, providing great incentives for athletes. The motivating power of the Olympic Games is especially apparent in the improvement of marks during the months leading up to them, when many world-class athletes achieve their highest levels of skill and conditioning. The worldwide love of sports makes the Olympics possible, while the Olympics increase people's commitment to sports. As is usually the case, a culture's institutions give their valued activities special prominence.

Growing Numbers of Participants

Young people are turning to sports in increasing numbers. Because the destructive effects of World War II (which contributed to the relative lack of world-class athletic marks in the 1940s and '50s) have not been repeated, the talent pool grows in Russia, Europe, Japan, and America. And there is no telling how much athletic genius might be released in the Third World, given the recent emergence of world-class peformers in the People's Republic of China and Africa. Would there be an equivalent increase of metanormal capacities in general if they were better understood and more widely cultivated?

The Inspiration of Small Training Groups

When Roger Bannister broke the four-minute barrier in the mile in 1954, he was paced by his comrades Chris Chataway and Chris Brasher. The three had trained together before the race, and their strategy had been planned in advance. Brasher paced the first two laps and Chataway the third, letting Bannister pass in good position to go under four minutes. Bannister said he wouldn't have run that historic 3:59:4 without the support and inspiration of his friends. The first sub-four-minute-mile was a team affair. "We had done it—the three of us!" Bannister wrote. "We shared a place where no man had ventured—secure for all time, however fast men might run miles in the future. We had done it where we wanted, when we wanted, how we wanted."[75] Like Bannister's group, John Walker, the 1976 Olympic 1,500-meter champion and first miler to break 3:50, Rod Dixon, the great middle-distance runner, and Dick Quax, who held the world record for 5,000 meters, trained together for several years. All of them reached new peaks of performance while working out together in New Zealand during the 1970s.

In these two groups, the great example each person provided for the others gave their training the inspirational quality of championship events. In such cohorts, legendary contests have developed during ordinary workouts, lifting their participants to greater excellence. Percy Cerutty, who coached the great Australian miler Herb Elliott, deliberately cultivated this kind of élan. At his training facility in Portsea, Australia, he attracted young racers with his Spartan regimen, philosophic inspiration, and great enthusiasm for sport, building up a charge, it was said, that helped lift Elliott and others to greatness. Groups analogous to Cerutty's and Bannister's, it seems to me, could serve the development of extraordinary functioning in general.

[19.5]
Sport as Transformative Practice

In this and other chapters I have noted several kinds of extraordinary experience reported by sportspeople, including exalted states of mind and metanormal movement abilities. Here I will describe certain elements of sport that help to evoke such experience.

Regular practice of particular physical movements with the intent to improve them. Practice entrains countless physical and psychological processes to produce new skills, not only in sport but in every field of endeavor. This is the case even if the practitioner knows little or nothing about the specific changes such learning entails. And more than that, practice invites answering graces. As I have noted in previous chapters, disciplined activities sometimes give rise to spontaneous metanormalities. At times, the part of our nature most stretched gives birth to an extraordinary version of itself, as for example when a mother's long-suffering love gives rise to a sense of oneness between her and her child. And so it is in the heat of sport: inspired jumping can give rise to apparent levitation, disciplined ball-handling to unearthly hand-eye coordination, practiced footwork to superordinary agility. Where capacities are stretched to their limits, metanormalities tend to appear, despite the expectations or desires of their recipients.

Sustained and focused attention. Every sport requires concentration and freedom from distraction, qualities fundamental as well to contemplative devotion. Athletic skill demands unbroken attention to the environment, to the objects and other people involved in a sporting event, and to kinesthetic sensations. Success is correlated with constant presence in the moment. Indeed, concentration can produce a state of mind graced by extraordinary clarity and focus.[76] British golfer Tony Jacklin said, for example:

> When I'm in this state, this cocoon of concentration, I'm fully in the present, not moving out of it. I'm aware of every half inch of my swing. . . . I'm absolutely engaged, involved in what I'm doing at that particular moment. That's the important thing. That's the difficult state to arrive at. It comes and it goes, and the pure fact that you go out on the first tee of a tournament and say, "I must concentrate today," is no good. It won't work. It has to already be there.[77]

Many sportspeople have described "the zone," a condition beyond their normal functioning. Describing such a condition to me, quarterback John Brodie said:

> Often in the heat and excitement of a game, a player's perception and coordination will improve dramatically. At times, and with increasing frequency now, I experience a kind of clarity that I've never seen adequately described in a football story.[78]

As they try to describe such experience, athletes sometimes begin to use metaphors similar to those used in religious writing. Listening to such accounts, I have come to believe that athletic feats can mirror contemplative graces.

Imagery rehearsal. Jack Nicklaus described how imagination can contribute to successful golf shots:

> I never hit a shot, not even in practice, without having a very sharp, in-focus picture of it in my head. It's like a color movie. First I "see" the ball where I want it to finish, nice and white and sitting up high on the bright green grass. Then the scene quickly changes and I "see" the ball going there: its path, trajectory, and shape, even its behavior on landing. Then there is a fadeout, and the next scene shows me making the kind of swing that will turn the previous images into reality.[79]

Olympic high jumper Dwight Stones also used visualization to improve his performance. Tennis star Chris Evert mentally practiced her matches by anticipating her opponents' strokes and visualizing her own attack.[80] And some body builders use imagery to enhance their muscular development. Both Arnold Schwarzenegger and Frank Zane, for example, held images of the body they desired while training to win world championships. "A pump when I picture the muscle I want," Schwarzenegger told me, "is worth ten with my mind drifting." Both he and Zane have said they deliberately planned their physique, incorporating the symmetry, definition, and contours they wanted into their mental pictures of it. Their training, both of them told me, depended to a large extent upon mental practice.

After studying the imagery used by successful athletes, sports psychologist Richard Suinn distinguished it from lazy reverie or mere reflection upon past events. Such imagery, he wrote,

> is more than visual. It is also tactile, auditory, emotional, and muscular. One swimmer reported that the scene in her mind changed from black and white to color as soon as she dove mentally into a pool, and she could feel the coldness of the water. A skier who qualified for the U.S. Alpine ski team experienced the same "irritability" that she felt during actual races when she mentally practiced being in the starting gate. Without fail, athletes feel their muscles in action as they rehearse their sport.[81]

Imagery of this kind, according to Suinn, catalyzes the reintegration of physical performance. It is a well-controlled copy of experience, a sort of body-thinking similar to the powerful illusions of nocturnal dreaming.[82] In Part Three, I note ways in which such imagery can facilitate extraordinary functioning in general.

Relinquishment of limiting cognitive, volitional, emotional, and sensorimotor patterns. To some extent, sportspeople must acquire (or open to) another nature, which is why some of them say that in sport they are reborn (though the old self typically returns when the sporting event is over). Even the most commonplace sports require abandonment of habitual responses. Runners, for example, must resist the impulse to quit if they exercise lungs, heart, and legs beyond their usual limits. The urge to slow down is evident to everyone in races and hard training sessions. But the compensating pleasure of exceeding a limit, the glow of fitness, and the satisfaction that comes from overcoming resistance, come into play as well. Many sportspeople testify to both the pains and joys of self-surpassing, to the freedom that emerges in the midst of discomfort, to the second energy that comes with transcendence of ordinary patterns. By enduring the naysaying voices inside them, athletes discover new capacities. Breakdown for buildup—whether physical, emotional, or cognitive—is the rule in sport, as it is in all transformative practice.

Practiced detachment from results. Sport inevitably requires detachment from immediate results, even in the poorest loser. Without some ability to transcend failure, sports participants cannot finish most games or contests. Detachment produced by the ups and downs of training and competition, by the strange twists of certain games, by the fluctuations of winning and losing often leads to a marked equanimity. Many who persevere in athletics learn that there is an interior freedom and grace beyond their sport's uncertain results.

Long-term commitment, which sport often commands, facilitates all the elements of discipline noted above. No other activity requires more dedication to somatic alteration, more willingness to restructure physique and physical behavior.

Sport's boundedness. Johan Huizinga, in his classic study of play, *Homo Ludens,* described the role of boundaries in sport.

> Into an imperfect world and into the confusion of life [those boundaries] bring a temporary, a limited perfection. Play demands order absolute and supreme. The least deviation from it "spoils the game," robs it of character and makes it worthless. The profound affinity between play and order is perhaps the reason why play seems to lie to such a large extent in the field of aesthetics. Play has a tendency to be beautiful. It may be that this aesthetic factor is identical with the impulse to create orderly form, which animates play in all its aspects. The words we use to denote the elements of play belong for the most part to aesthetics, terms with which we try to describe the effects of beauty: tension, poise, balance, contrast, variation, solution, resolution, etc. Play casts a spell over us; it is "enchanting," "captivating." It is invested with the noblest qualities we are capable of perceiving in things: rhythm and harmony.[83]

A site such as Fenway Park, an Olympic stadium, or the Old Course at St. Andrews quickens the spirit, concentrates energies, connects us with heroes past and future. And even without a manmade arena, sport can enchant time and place. A mountain to be climbed, an ocean to be crossed, a stretch of countryside to be raced upon can provide special significance simply by being designated the field of adventure. The spatial and temporal boundedness of sport helps order and sublimate our energies, focus our mind, and release extraordinary capacities.

New integrations of mind and body. Volition, self-awareness, imagination, emotions, the senses, and motor control are involved in surpassing performance. To be successful, sports training must harmonize countless psychological and physical systems. It needs to establish new kinds of ordered functioning, yet allow spontaneous action. It must take our repertoires apart and reassemble them with new power and beauty.

The elements of sport I have just described are present in other transformative practices, but in the context of sports they evoke a greater range of physical abilities than that produced by any other family of disciplines. In the service of integral practices, they can, I believe, evoke a still greater range of capacities.

20

THE MARTIAL ARTS

AS MUCH AS any transformative practice that commands a significant following today, certain martial arts facilitate a many-sided, integral development of human nature. At their best they simultaneously promote moral sensitivity, athletic abilities, and a degree of unitive awareness. Some, such as aikido, are superior to modern sports in their reliance upon spiritual principles, and superior to quiet meditation in their cultivation of stillness in action. In chapter 20.1 I discuss some elements that make martial arts comprehensive programs for personal growth. In chapter 20.2 I note a few scientific studies of them. And in chapter 20.3, I illustrate their transformative power by describing the influence of aikido on the great Japanese baseball player Sadaharu Oh.

[20.1]

TRANSFORMATIVE ELEMENTS OF THE MARTIAL ARTS

The Basis of the Martial Arts in a Spiritual Ground

For many centuries, martial artists have made their training a *way* (in Japanese a *do*, derived from the Chinese *tao*) to realize extraordinary capacities. Most ideas that inform such training come from Buddhist, Taoist, Confucian, Shintoist, and shamanic teachings of the Far East, some of which trace their lineages to India. In the sixth century A.D., the monk Bodhidharma (Tamo in Chinese), who brought China the form of Indian Buddhism that evolved into Japanese Zen, introduced a series of self-defense exercises to the monks of the Shaolin Temple in Honan Province. The temple boxing that evolved from Bodhidharma's exercises had a formative influence on later fighting arts in China, Japan, and Okinawa, among them tai chi chuan, which joined some forms of Shaolin boxing with Taoist breathing exercises and medical lore. According to one legend, the Taoist monk Chang San-feng of the Yuan Dynasty (A.D. 1279–1368) created the 13 basic postures of tai chi to reflect the I Ching's eight trigrams and Chinese cosmology's five elements, but the true founder of the discipline is probably unknown. Modern tai chi may have originated with one Wang Tsung-yueh of Shansi Province who introduced it in Honan during the Chien-lung period (1736–1795) of the Ching Dynasty.[1]

Some practitioners of the Chinese martial arts classify approaches based on spiritual principles as members of the Internal System, in order to distinguish them from purely muscular exercises. Buddhist and Taoist meditation training that influenced the Internal System used several methods common to Indian yoga, among them cultivation of the vital energy *chi* (called *ki* in Japanese and *prana* in Sanskrit), eye fixation, breathing exercises, and controlled diets. Tai chi, however, gives spiritual awareness a fuller physical expression than most forms of yoga. According to a famous Taoist precept, "Meditation in activity is a hundred, a thousand, a million times superior to meditation in repose. The stillness in stillness is not the real stillness; only when there is stillness in movement does the universal rhythm manifest."

In Japan, another stream of martial arts was influenced by the culture of *bushido* (the way of the warrior), which drew on Confucian moral teaching, Shinto ritual, and Zen meditation. In the relatively peaceful Tokugawa era of the seventeenth, eighteenth, and nineteenth centuries, many samurai transformed their *bujutsu*, or fighting arts, into *budo*, fighting ways that aimed for spiritual development as well as physical prowess. "*Budo* are less combatively oriented and lack the practical element inherent in the [*bujutsu*]," wrote martial-arts historians Donn Draeger and Robert Smith. "The principle rather than the resultant technique is emphasized, and in some cases they have deviated so far from the [*bujutsu*] forms from which they sprang that they have lost all utility in practical combat."[2] The traditional Japanese *bujutsu* are combat forms whose names usually include the suffix -*jutsu*, but when they were converted into more spiritual practices they were given the suffix -*do*. Thus Jigoro Kano derived judo from jujutsu (and founded its first institute in Tokyo in 1882); and Morehei Uyeshiba (1883–1969) derived his aikido largely from aikijutusu, to cultivate *ki*, the cosmic energy, and *ai*, the harmonious blending of self with the universe. In disciplines such as aikido that are a *do* instead of a *jutsu*, the practitioner aspires to shift his functioning from the ordinary ego to his deeper nature. This transformation was described by the Japanese Buddhist scholar Daisetz Suzuki, who quoted the great swordsman Yagyu Tajima.

> The Mind (*kokoro*) is Emptiness (*ku, k'ung, sunyata*) itself, but out of this Emptiness an infinity of acts is produced: in hands it grasps, in feet it walks, in eyes it sees . . . it is indeed very difficult to have this experience because we cannot get it from mere learning, from the mere listening to others talk about it. Swordsmanship consists in personally going through this experience. When this is done, one's words are sincerity itself, one's behavior comes right out of the Original Mind emptied of all ego-centered contents.[3]

In his treatise on the sword, Yagyu Tajima quoted an old Japanese poem.

> *It is mind that deludes Mind,*
> *For there is no other mind,*
> *O Mind, do not let yourself*
> *Be misled by mind.*

Commenting upon these lines, Suzuki wrote:

> [Yagyu Tajima] distinguishes two kinds of mind, true or absolute and false or relative. The one is the subject of psychological studies, while the other is Reality, which constitutes the basis of all realities. In the poem quoted, "mind" is the false one and "Mind" is the true one. The true one is to be protected from the false one in order to preserve its purity and freedom.
>
> Zen in this case generally uses the term *kufu* (*kung-fu* in Chinese) which is synonymous with "discipline" or "training" (*shungyo; hsiu-hsing*). *Kufu* means "employing oneself assiduously to discover the way to the objective."[4]

For Morehei Uyeshiba, aikido was such a *kufu* because it could bring the Mind beyond mind into physical action. "Putting Aiki into action," he wrote, "we must constantly bring forth power that is without limits. The soul of the universe possesses a power capable of resolving all things regardless of their nature."[5] As a young man, Uyeshiba studied with several teachers, most notably Takeda Sokaku, one of Japan's most famous martial artists, and Onisaburo Deguchi, the founder of the latter-day Japanese religion *Omoto-kyo*. Like many martial artists before him, Uyeshiba combined several fighting arts and meditative practices in his own *budo*. Though he was famous for his warrior skills in prewar Japan, he retired to the countryside in 1942, deeply disturbed by Japanese militarism. Until his death in 1969, he spent much of his life in contemplative prayer while developing aikido into a practice that embodied ethical and spiritual principles.

An American writer and student of aikido, John Stevens, has described Uyeshiba's extraordinary powers in a book, *Abundant Peace*; and several movies reveal the master's great coordination and agility. The testimonies of people who knew him, collected in numerous books and articles since the 1930s, provide strong evidence that Uyeshiba possessed telepathic and other extrasensory abilities. Having talked at length with some of his students, and having studied the books and movies just noted, I find it conceivable that Uyeshiba realized most of the principal metanormal capacities described in chapter 5, including extraordinary movement abilities, mystical cognition, and a personal radiance that inspired countless people (among them several political leaders of postwar Japan). As he grew older, Uyeshiba embodied his developing realizations in the repertoires of aikido, so that it could more and more fully express the love and power latent in human nature.[6]

Martial Arts and the Development of Ki

The Japanese term *ki,* derived from the Chinese *chi,* can be translated as vital energy or force. Like the Sanskrit *prana* or Greek *pneuma,* it is conceived to be a subtle breath, spirit, or energy that pervades the universe. According to the Chinese philosopher Chuang-tzu, by relinquishing psychic turbulence through "mind-fasting" one can hear and see with *chi* to apprehend "the emptiness (*kyo, hsu*) in which infinite possibilities are harbored."[7] The practitioner of martial arts can either manipulate *ki* or surrender to it. By imagining heaviness, for example, or by visualizing his extended arm as a stream of energy, the martial artist can mobilize *ki* to change his weight or to make his arm grow stronger. In these and other practices, *ki* is summoned to accomplish specific tasks or to develop particular capacities.[8] On the other hand, it can be fostered as Chuang-tzu suggested by a "mind-fasting" through which the practitioner relinquishes thoughts or feelings that impede his deeper nature. In aikido, tai chi, and other martial arts, *ki* is channeled through a point below the navel called *hara* in Japanese and *tantien* in Chinese. Numerous exercises are designed to mobilize a centered awareness there.[9]

Like other martial-arts masters, the great Japanese swordsman Musashi told his students not to concentrate on their opponent's sword or particular movements. "Having mastery of the Way you will see the weight of his spirit," he wrote. "If you fix your eyes on details and neglect important things, your spirit will become bewildered, and victory will escape you."[10] Aikido expert George Leonard described "soft eyes," a visual relaxation that promotes perception of *ki* and permits action in a complex field. "In aikido," he wrote, "when you are attacked by four people simultaneously, you don't have time to see everything in sharp focus, but you must see movement and relationship clearly. [Soft eyes provide] a way of seeing everything at once, being part of everything."[11]

According to martial-arts lore, *ki* operates in many ways, flowing through an arm to strengthen it, massing itself in the body to increase weight, or evanescing to make one lighter. It can also be gathered in a single spot, like a focused beam of light. In some schools such focusing, called *kime* (*noi cun* in Chinese), is demonstrated by breaking wood blocks with blows that reputedly stop short of their target. By *kime*, one can penetrate armor, injure an opponent by delayed effect, and perform other prodigies.[12] The voice, too, can channel *ki*, especially by the *kiai*, or spirit-shout, directed from the *hara*. Through breathing exercises derived from Taoist and Indian yoga, the lungs, diaphragm, and larynx can be trained to produce *kiais* that knock down opponents or break physical objects. Like the *kime*, the *kiai* concentrates *ki* to restructure the vibrations that constitute both living and inanimate matter. Some adepts, it is said, can deliver it in silence.

But while there is a considerable lore about *ki* in the martial-arts traditions, mainstream science has not provided much evidence for it. Perhaps the term *ki* is simply a metaphor that is useful for martial-arts practice, or a symbol representing physical energies that physics or medical science can explain. George Leonard noted that

> the body radiates several forms of energy that can be measured by the instruments of Western science. Each of us is surrounded by an aura, if you will, of radiant heat; this heat may be perceived several inches from the skin by a sensitive hand, and from much greater distances by thermistor and infrared sensors. We are surrounded by what anthropologist Edward T. Hall terms an "olfactory bubble"; individuals of some cultures . . . feel uncomfortable when talking to someone they can't smell. There is also an electromagnetic field, associated with the pulsing of the heart, in and around the body; highly sensitive instruments have measured this field at a distance of several inches. In addition, the body

is surrounded by a cloud of ionized sweat that can be measured by electrostatic indicators. We might also bear in mind that we trail a cloud of warmed air, water vapor, carbon dioxide, bacteria, and viruses from our breathing, and that all this material, which has circulated through a most intimate cavity within our bodies, is very rapidly intermingled with that of all the others who share our breathing space.

All in all, we are not nearly so separate and skin-encapsulated as we are generally led to believe. In the 1930's, psychologist Kurt Lewin theorized that people exist within a psychological "life space" and that they interact with the outside world by means of this permeable and malleable field rather than by direct contact. It becomes clear after only a moment's thought that we are by no means imprisoned within our skins. Our interactions with the world are multiple and various. That we exist as intermingling fields, that we possess many ways of sensing one another at a distance, is not really remarkable.[13]

Perhaps *ki* is nothing more than the fields Leonard noted. Or maybe it is simply an abstraction that helps one develop skills such as "soft eyes," *kime*, and the *kiai*. On the other hand, it might be the sort of energy martial artists take it to be. Countless martial-arts students have had their abilities to perceive or use *ki* confirmed—or disconfirmed—by fellow students and teachers; for example, by breaking (or not breaking) objects with a *kime*, or throwing opponents without touching them. Though apparent prodigies such as these can be explained in terms of standard physics or social suggestion, it seems to me there is enough evidence for *ki* to warrant our belief that it might exist. Such evidence can be found, for example, in the journal *Subtle Energies*, which publishes clinical and experimental research involving scientifically inexplicable energies that interact with human psychological and somatic processes.[14]

The Cultivation of Sensory and Extrasensory Awareness

The ability to sense fine shadings of atmosphere, the faint rustling of leaves and grass, the murmuring of insects, or the distant sounds of human movement is cultivated in martial-arts training through breathing exercises, solitary retreat, meditation, and other practices that quiet the mind and focus the senses. "See first with the mind, then with the eyes, and finally with the torso and limbs," Yagyu Tajima wrote.[15] When the mind is freed from distracting thoughts, the senses will function with new scope and clarity.

And as we achieve this mental stillness, we can develop extrasensory capacities. "A sort of telepathy seems to develop in some swordsmen," wrote Daisetz Suzuki. "Yagyu Tajima was said to have had this 'sixth sense'." The great swordsman, for example, felt a murderous air in his garden when an attendant imagined he could successfully attack him;[16] and several of Morehei Uyeshiba's students described the master's ability to sense would-be attackers who tried to surprise him in his sleep.[17] The development of this so-called sixth sense is still a central element in much martial-arts training. In his California *aikidojo*, for example, George Leonard helps his students find magnetic north blindfolded, locate people hidden outside the building, and experience other types of extraordinary sensitivity.[†]

The Inculcation of Moral Values

As warfare decreased in Japan during the Tokugawa era, some samurai turned to revelry and crime instead of adopting the culture of *bushido*, while others became roaming fighters, or *ronin*, offering their skills for pay. Amoral professionalism of this kind was institutionalized in the fighting art of *ninjutsu*, which included methods of harassment, killing, and intelligence gathering. Practitioners, the *ninja*, were typically trained by secret groups, usually in the remote Iga and Koga areas of Japan, which were protected from enemies and travelers. Their members were divided into three classes: the *jonin*, who drew up contracts; the *chunin*, or clan leaders; and the *genin*, to whom missions were assigned. *Ninja* of *genin* status were considered the lowest stratum of Japanese society and if captured were often tortured and maimed. *Ninjas* developed methods of disguise, escape, poisoning, burning, seduction, secret entry, and weapons use. Many were highly skilled athletes who could scale commanding heights and cover a hundred miles or more a day on foot. Their dramatization in modern films as black-uniformed killers with rope ladders and poisoned darts reflects their ruthlessness, cunning, and amorality.[18]

But in spite of the *ninjas'* mercenary spirit and some *ronins'* unprincipled behavior, many martial arts have embodied a profound and beautiful ethics. The Taoist values of harmony and yielding, for

---

[†] A few scientific studies provide support for some of Leonard's results though they do not provide evidence for ESP. Sorbonne physics professor Yves Rocard, for example, showed that untrained subjects could sense changes in a local magnetic field. See: Rocard 1964.

example, were reflected in fighting schools that taught meditation to promote childlike innocence, pliable surrender, and freedom born of nonattachment.[19] Mencius, the celebrated Confucian philosopher, claimed he had cultivated *hao jan chih chi,* an "immensely great, immensely strong *chi,*" in contrast to a lower kind of vitality. *Chi,* he said, ultimately depends upon the accumulation of righteous deeds and arises from inherent goodness rather than mechanical exercises. Though *i,* or will, directs *chi,* neither can be fully realized without right living. The cultivation of *i* and *chi* are complemented in Taoist practice by *wu wei,* disciplined flow and surrender to deep nature, and *tzu jan,* disciplined spontaneity. The doctrine of *wu wei* leads its practitioners to refrain from contention, to remain silent and aloof, while *tzu jan* prompts them to respond naturally and spontaneously to attacking forces. The Internal System of Chinese boxing has its foundations in these four Taoist values, *i, chi, wu wei,* and *tzu jan,* and all four are expressed in aikido, tai chi, and other disciplines today.[20]

In Japan, Zen Buddhism influenced the martial arts as Taoism and Confucianism had in China. Zen's emphasis on self-reliance, spontaneity, equanimity, and indifference to death appealed to many samurai, and its teachings eventually cohered with Confucian morals and the Shintoists' reverence for life to help form the culture of *bushido.* This transformation of warrior virtues into an ethic of self-development was hastened during the Tokugawa Shogunate (1603–1867).[†] In the warrior ethics of the Tokugawa period, the ancient code of loyalty to one's lord and one's calling was preserved but was framed in terms derived from Confucianism. The Japanese writer Nitobe's book *Bushido* listed the culture's seven distinctive virtues:

- justice, which precludes all deceit or dishonesty
- courage, both moral and physical, based upon serenity, experience, and vigilance
- benevolence, which includes magnanimity, love, and sympathy (*bushi no nasake,* "the tenderness of a warrior," shaped all of *bushido*'s precepts)
- politeness, to refine body and soul

[†] The writings of Kamo Mabuchi (1697–1769), Motoori Morinaga (1730–1801), and Hirata Atsutane (1776–1843); the Confucian teachings of Yamaga Soko (1622–1685); the *Hagakure* of Tsunemoto Yamamoto (1649–1716); and the seventeenth-century *Primer of Bushido* by Daidoji Usan helped give birth to the *bushido* concept.

- veracity (the expression *bushi no ichi-gon* means a warrior's word does not need written pledges to support it)
- honor
- loyalty

Ritual suicide by disembowelment, known as *seppuku* or *hara kiri,* was important in the code of *bushido* because it showed that the warrior held certain values to be more important than life, and was a legal institution from the beginning of the thirteenth century. "Based on the Confucian principle that 'man should not live under the same heaven with the murderer of his leader-lord-father'," wrote Draeger and Smith, "*seppuku* embraced a 'mental physiology' according to which the 'seat of the soul' rested in the lower part of the abdominal region. By opening the soul, the warrior would expose his purity for all to see."[21]

Though many martial arts today do not emphasize moral values, some still embody virtues derived from the nobler forms of *bushido* and other warrior traditions. The entire repertoire of Morehei Uyeshiba's aikido, for example, incorporates a well-defined ethics of harmony, love, and reconciliation. Aikido's ethics are expressed in its physical repertoire, so that its practitioners build them into their muscles and nervous system. When moral values are embedded in the body's movements, they have a firmer basis than they would in good intentions alone. "Understand aikido first as *budo* and then as a way of service to construct the World Family," Uyeshiba wrote. "True *budo* is the loving protection of all beings. Reconciliation means to allow the completion of everyone's mission. True *budo* is a way of love." This ethic is based on Uyeshiba's belief that the soul can progress in concert with the deepest universal forces. "In aikido," he said, "we do not train to become powerful or to throw down some opponent. Rather, we train in hopes of being of some use, however small our role may be, in the task of bringing peace to mankind around the world. In this hope we become one with the Universal."[22]

Uyeshiba gave form to such ideals in the training, strategy, and movements of aikido.[†] Typically as a contest begins—on a training mat or in the streets—the person under attack, called the *nage,* remains calmly immobile until he turns the attacker's momentum,

---

[†] Aikido's terms reflect its purely defensive nature, as the defender is called a *nage,* from a Japanese word meaning throw, while the attacker is called the *uke,* from a word that connotes falling.

whirling him to the ground or twisting one of his limbs to control him. When correctly performed, aikido's moves are flowing and circular, looking more like a dance than a fight. In a *rondori,* or mass attack, the *nage* moves quietly between attempted tackles and blows, throwing his opponents by joining with their movement. Practiced by a master, aikido joins the serenity of contemplative awareness with inspired action. Indeed, its basic strategy cannot succeed without a measure of grace under pressure, for it requires the *nage* to step near his attacker to redirect his momentum. In this, aikido gives expression to the traditional Zen Buddhist-samurai virtue of serene spontaneity in trying circumstances and to the Taoist value of harmony. Furthermore, harmony, serenity, spontaneity, and reconciliation can be imparted to a *nage's* antagonist. During the *tenchi nage,* or "heaven and earth throw," for example, the *nage* raises one arm and lowers the other, splitting his attacker "between heaven and earth" so that he must achieve his own reconciliation through falling. Other moral values embedded in aikido include balance, which is cultivated by whirling steps and rhythmic breathing; sensitivity, trained by "soft eyes" and the apprehension of *ki*; and psychological strength, which develops through the rough-and-tumble of training. But like all sophisticated, time-tested disciplines, aikido embodies more values than its philosophers have named and more virtues than its adepts are fully aware of.

Martial Arts and the Cultivation of Psychological and Physical Beauty

In the beauty they express, some martial arts rival the finest forms of dance. Disciplines such as *kendo* (swordsmanship), *kyudo* (bowmanship), and aikido, with their elegant dojos (schools) and uniforms and their graceful rituals, reflect the emphasis that Chinese and Japanese cultures place on aesthetic values in everyday life. Tai chi and aikido movements, for example, are graceful because they are based on principles of harmony and disciplined spontaneity, and masterful because they embody faithful practice. The influence of such virtues is supported, moreover, by the philosophies that helped give the martial arts birth, with their emphasis on a primordial unity, cosmic vitality, and human capacity for transcendence. Everywhere and always, such philosophies hold, beauty can be expressed, even in the smallest act, even in discord or suffering.

The Complex Behavioral Repertoires of the Martial Arts

The martial arts derive much of their transformative power from the richness of their repertoires. The sheer number of movements they demand helps their practitioners develop coordination, balance, agility, flexibility, strength, endurance, and speed of hand and foot. They incorporate countless discoveries about our psychophysical potential, which is apparent in the fact that many of them use methods that modern somatic disciplines have only recently discovered. (Tai chi, for example, promotes alignment and extension of the neck like the Alexander Technique, and cultivates the soft, pleasurable stretching of the Feldenkrais Method; see 18.2 and 18.4.) The martial arts today embody an immense tacit knowledge about human physiology and movement, much of it derived from Oriental medicine and religious practice.[23]

[20.2]
SCIENTIFIC STUDIES OF THE MARTIAL ARTS

Scientists have shown that the hand of a karate expert can develop a peak velocity of 10 to 14 meters per second and exert a force of more than 675 pounds (or 3,000 newtons).[24] If the hand is positioned properly, it can easily withstand the resulting counterforce so that humans can break wood and concrete blocks with ease. The rupture modulus of bone, a measure of its breaking point, "is more than 40 times greater than that of concrete," researcher Michael Feld and his associates wrote in *Scientific American*. "If a cylinder of [bone] two centimeters in diameter and six centimeters long were simply supported at its ends, it could withstand a force exerted at its center in excess of 25,000 newtons. Such a force is eight times greater than the force concrete exerts on the hand in a karate strike. [But] the hand can actually withstand forces much larger than 25,000 newtons, because it is not a single piece of bone but a network of bones connected by visco-elastic tissues."† K. Hirata, F. B. Theiss, and H. T. Yura have explained other principles of physics and anatomy involved in the extraordinary striking power generated by some martial artists.[25]

---

† Feld et al. 1979. Another *Scientific American* article described some basic throws in judo and aikido to show how the physics of forces can be used against an opponent. See: Walker 1980.

Phillip Rasch and Eugene O'Connell found that a group of karate students were significantly stronger in their legs per pound of body weight than intercollegiate wrestlers with whom they were matched for age and athletic experience. The researchers guessed that their subjects' leg strength was generally superior to the wrestlers' because "a considerable number of different kicking techniques are practiced during [their] training sessions, and [their] legs are vigorously exercised through their full range of motion in several different ways."[26] Rasch and William Pierson estimated that several karate practitioners they studied had significantly faster reaction times than a matched group of wrestlers, if the results of their tests were adjusted to reflect their subjects' age differences.[27] Another researcher reported that, on the whole, a group of 42 karate students had faster reaction times than 42 tennis players of comparable athletic ability.[28]

Research psychologists observed a Korean karate expert who calmly controlled his emotions, his bleeding, and other physiological responses as his forearm was pierced with a bicycle spike from which a 25-pound weight was suspended. The subject's extraordinary control was revealed by recordings from five EEG leads, by measures of galvanic skin response, by measures of blood-flow velocity, and by the experimenters' observations.[29] This experiment is an exception to most scientific studies of martial artists in that it examined a highly skilled subject rather than beginners or intermediate students. Most martial-arts research, like most experimental studies of biofeedback, meditation, and other transformative activities, has been limited by its subjects' lack of experience.

In 1981 George Leonard studied improvements in physical functioning among his aikido students in Mill Valley, California. These six elements of Leonard's approach were especially significant for the people he studied:

- □ centering in the *hara*, which helped a 54-year-old woman decrease her heart rate in tense situations
- □ imaging energy flow, which one participant used to eliminate a finger deformity that her doctor had said was caused by incurable arthritis
- □ focused surrender to goals the participants had stated, which helped a 34-year-old man achieve marked flexibility in a surgically repaired knee
- □ breathing from the lower abdomen

459

- visualization
- heightened awareness of goals stimulated by the participants' written affirmations

"Although scientific controls were not used in this experiment," Leonard wrote, "a written self-assessment form indicated that its participants were surprisingly successful in changing their body in significant ways." Among the changes Leonard's subjects reported were an improvement in an EKG reading, new ability (among several students) to raise hand temperature, and increased muscle tone.

In 1982 Leonard conducted a second study with 26 students at his aikido school, using low gradient affirmations for bodily changes such as weight loss that could be realized through "straightforward and unremarkable actions"; medium gradient affirmations to relieve asthma attacks or produce other effects that would be "difficult to achieve but fairly easy to explain in Western scientific terms"; and high gradient affirmations for changes that would be hard to achieve *and* hard to explain scientifically.[30] Leonard's class had been meeting twice a week for three months when his subjects began their experiment, and during that time Leonard had introduced them to the fundamental elements of his approach, helping them remain balanced under increasing amounts of pressure. With its many uncontrollable variables, the project proved nothing statistically, nor was it meant to, but some of its participants reported dramatic results. One woman, for example, experienced significant relief from a deteriorating congenital heart condition. Her remission was confirmed by medical tests at the University of California Medical Center in San Francisco. At the time of this writing, she is free from heart trouble and has been told by her doctors that she could run a marathon. Another participant scheduled for arthroscopic surgery rehabilitated her badly injured knee so that she could practice aikido without any trouble. And other students significantly increased their flexibility, agility, strength, and cardiopulmonary endurance.

[20.3]

THE MARTIAL ARTS IN MODERN SPORT

Sadaharu Oh, Japan's famous baseball player, hit 868 career home runs to surpass Hank Aaron's American record, and won 15 professional home-run titles in Japan during a 22-year career. He also helped the Tokyo Giants win many national championships, including nine in a row from 1965 through 1973. But he might not have achieved his great success without special training with Hiroshi Arakawa, a baseball instructor. Oh has told his story simply and eloquently with writer David Falkner.[31]

Though he had been a high-school star as a left-handed pitcher, Oh was assigned to first base as a professional because he was a powerful hitter. But during his first three years with the Tokyo Giants, he did not fulfill his great promise and often drank to excess in Tokyo's Ginza district. Because Oh was struggling, the Giants' manager hired Arakawa to work with him in 1962. Arakawa extracted a promise from him that he would stop drinking and smoking, and during their first months of training introduced him to Morehei Uyeshiba, who offered them insights from aikido. Uyeshiba taught Oh about *ma*, the "psychic time and space" in which a contest occurs, and other aikido principles. But these first lessons did not have an immediate effect. Not until Arakawa made Oh adopt an unusual one-legged batting stance did he show marked improvement in his hitting. This dramatic change in style helped focus Oh's aikido training. The athlete wrote:

> I had reached a point where Aikido had become absolutely necessary rather than merely complementary to what I did. Without Aikido, I would not learn to stand on one foot, I would not "understand" it.
>
> One of the first things a student of Aikido learns is to become conscious of his "one point." This is an energy or spirit-center in the body located about two fingers below the navel. While many martial arts make use of this center, it is essential in the practice of Aikido, [which] requires tremendous balance and agility, neither of which are possible unless you are perfectly centered. So much of our early work was getting me to pose simply with the one point in mind. I would get up on my one foot and cock my bat, all the while remaining conscious of this energy center in my lower abdomen. I discovered that if I located my energy in this part of my body I was better balanced than if I located it elsewhere. If I located my energy in my chest, for example, I found that I was too emotional. I also learned that energy located in the upper part of the body tends to make one top-heavy. Balance and a steady mind are thus associated with the one point.[32]

Besides centering, Oh learned other things through aikido, among them awareness of *ki* and the power of waiting.

As long as I had [a] hitch in my swing, I could not begin to think of using ki in my batting. But posing on one foot, the goal of using ki did not seem so far-fetched—if I could learn to steady myself enough.

Earlier in the season, when we had simply been trying to overcome my hitching habit, Arakawa-san had had yet another discussion with Ueshiba Sensei about the problem. The Sensei, not being much of a baseball fan, had cut the talk short.

"Look," he said, "the ball comes flying in whether you like it or not, doesn't it? Then all you can do is wait for it to come to you. To wait, this is the traditional Japanese style. Wait. Teach him to wait."[33]

During the 1962 season, Arakawa incorporated concentration, *ki*, centering, balance, and waiting into Oh's baseball technique, so that he would achieve the "Body of a Rock" described by Musashi, the legendary Japanese swordsman. Oh wrote, "The Body of a Rock! The image entered my mind as simply as a bird alighting on a branch. The goal of perfecting what was in my body seemed entirely natural." Sometimes Oh practiced in front of a mirror, visualizing the many kinds of pitches he would face. To strengthen Oh's form, Arakawa had him imagine that his body was a gymnast's bar that could bear immense pressure without breaking. Oh practiced his stance with this image in mind before mirrors, in locker rooms, or at Arakawa's home, with a bat, a wooden pole, or other instruments until his blisters callused over. But his form was still imperfect. With his teacher's help, he realized that his upper body and bat position also needed to be reorganized. Only after strenuous months of practice could he balance his entire body so that his power was fully concentrated.

And what had been accomplished? Our training enabled me to hit thirty-eight home runs, twenty-eight of them coming after July 1. I raised by batting average to .272 and my RBI total to eighty-five, both career highs. Most important, I won the home-run and RBI titles for the Central League that year. [But] I received no particular praise from the Master of the Arakawa School that year. I accepted that. I knew he had his reasons.

"Think of it this way, Oh," he said to me. "Gain and loss are opposite sides of the same coin. It is best to forget them both."

It turned out that Arakawa-san had all the while been making his own plans. These had little to do with my having won a title or two. His mind was already in the future.[34]

Thus began a new stage in Oh's training based upon the *budo* of swordsmanship as well as aikido. At a dojo for the sword in Tokyo, Oh learned to channel *ki* through his bat as if each stroke "were a matter of life and death." He learned *metsuke*, a way of seeing as if with two sets of eyes, which Musashi had called a "distanced view of close things." With such ability, one could read a pitcher's intentions as well as his motion.[35] Oh also learned about the geometry of physical action, about the triangle made by arms and legs when a sword or bat is extended, and the power of circular movement in the pivot of the hips. "If the position of the driving elbow in a swordsman—or batter—is too far from the body," he wrote, "it is not possible to form a good triangle with the torso."[36] Day after day Oh practiced, and during the 1963 season he won his second home-run title and raised his batting average to .305. He wrote:

> But I was hungry for more. Sometime in that season, I woke to find that my heart was on fire. It did not matter if I went to Ginza or put away a whole bottle of scotch and a few beers in a night, I always woke clear-headed and on fire for baseball. I had reached the point where I simply lived to hit. How can I say it without sounding foolish? I craved hitting a baseball in the way a samurai craved following the Way of the Sword. It was my life.[37]

Oh hit 55 home runs in the 1964 season and raised his average to .320, still studying with Arakawa. When opposing teams shifted their fielders to the right side of the diamond, where he usually pulled the ball, he simply hit through them or over them. He had become the greatest and most feared hitter in Japanese baseball. When he retired 16 years later at the age of 40, he had broken more records and won more acclaim than any other Japanese player.

Other sportspeople have employed martial arts to promote their athletic performance.[38] Like Sadaharu Oh, they have testified that martial-arts exercises enhanced their physical skills, improved their mental states, and catalyzed various kinds of extraordinary functioning. But such training has influenced more than the world of sport. In recent decades, people in many walks of life have found new vitality, balance, and strength through practices such as aikido. The growing popularity of martial arts in many parts of the world today reflects a general attraction to disciplines that reveal the greater possibilities of mind and body.

21

Extraordinary Capacities of Religious Adepts

THE TRANSFORMATIVE POWER of religious practice has long been celebrated in mythic accounts of shamans, yogis, saints, and sages. Such myths, of course, do not constitute evidence for metanormal capacities but conceivably contain intuitions of attributes that humans might eventually develop (see chapter 8). In this chapter, however, I focus not upon myth but upon biographies, travelers' reports, anthropological studies, and other accounts that provide plausible testimony to extraordinary functioning produced by religious practice.

[21.1]
The Yogic Death of Calanus

In his *Anabasis of Alexander,* probably antiquity's most authoritative account of Alexander's adventures in Asia, Arrian described the yogic suicide of Calanus, an Indian sage who had joined the Macedonians' expedition. The holy man "grew enfeebled in Persia," the Greek historian wrote,

> and yet he would not submit to the ordinary way of life of an invalid, but said to Alexander that he was glad to make an end as he was. Alexander, however, argued with him at some length; but perceiving that Calanus would not give in, but would choose some other way of death, if one should not yield to him, ordered, as Calanus desired, that a pyre should be built for him.

Distributing gifts he had received from Alexander and chanting the names of Indian gods,

> Calanus mounted the pyre and lay down with solemnity, in the sight of all the host. As for Alexander, he felt this spectacle to be unseemly, with one for whom he had an affection; as for the rest, they felt nothing but astonishment to see that Calanus flinched not one whit in the flames. And when the pyre was lit . . . the trumpets sounded, as Alexander had ordered, and the whole army raised the cry which they raise when entering battle, and the elephants trumpeted their shrill war-cry, in honour of Calanus.[†]

[21.2]
THE CHRISTIAN DESERT FATHERS

During the third and fourth centuries A.D., many Christian contemplatives in the Middle East became famous for their holiness and ascetic heroism. According to the Abbe Duchesne's history of the early church, Macarius of Alexandria "could never hear of any feat of asceticism without at once trying to surpass it." When some monks went without sleep, Macarius kept himself awake for 20 nights. For an entire Lent he stood upright, eating nothing but cabbage once a week. And according to the historian W. E. Lecky, Macarius slept in a marsh for six months, constantly exposing his body to poisonous insects.[1] For some 40 years, the monk Bessarion never lay down while he slept; and Pachomius, the famous founder of monasteries, did not lay down for 50 years. Simeon of Stylites, it was said, often went without food for the full Lenten season. In 422, at Kalat Seman in northern Syria, he lived on a six-foot column, then built taller ones, until he made his permanent home on a pillar some 60 feet high. Though the circumference of its platform was little more than three feet (with a railing to keep him from falling), Simeon lived on the column without interruption for 30 years. From this prominent vantage point he preached sermons, condemned the unfaithful, performed cures, played ecclesiastical politics, shamed moneylenders into reducing their interest charges, and intimidated countless pilgrims.[2] His example helped create a fashion of ascetic pillar hermits that lasted for 12 centuries and persisted in secularized form into the twentieth century.[3]

[†] Arrian, bk. 8, 3. According to the eminent historian P. A. Brunt, "Arrian unquestionably provides us with the best evidence we have for Alexander. A comparison with other ancient accounts of Alexander shows clearly enough that his is best" (Brunt 1976, pp. xvi–xvii).

Besides heroic demonstrations of fortitude and self-mastery, some of these contemplative monks exhibited extrasensory powers. According to the *Historia Monachorum,* an account of the desert fathers written in the early fifth century A.D., John of Lycopolis predicted future events, successfully advising the emperor Theodosius and a Roman general about their military ventures. The anonymous author of the *Historia Monachorum* also noted the monk Eulogius' ability to read hearts and minds, Apollo's precognitions that strangers were about to visit his cell, and Theon's gift of prophecy. The desert fathers were known, too, for their healing powers. John of Lycopolis blessed oil for a senator's wife, who used it to cure her blindness from cataracts. Abba Or performed cures "without ceasing, so that monks flocked to him from all sides, gathering round him in their thousands." Elias, Apollo, John the Hermit, Macarius, and Amoun were visited by great crowds in search of healing. Many of these anchorites were also noted for their robust health. Saint Anthony, the most famous and influential of them all, lived until he was 105 (251–356). Abba Or and John of Lycopolis were about 90 years old when the author of the *Historia Monachorum* visited them in 394–95. And Elias was 100, though some desert fathers said no one remembered when he had entered his mountain retreat. Regular habits and joyous serenity, inspired in part by the "huge silence" of the desert, contributed to the longevity of these remarkable ascetics.[4]

Many accounts of the desert fathers are derived from scholarly churchmen noted for good judgment and critical sense, among them Saint Jerome, the author of the Latin Bible; Rufinus of Aquileia, who made a Latin translation of the *Historia Monachorum*; Athanasius, bishop of Alexandria, whose account of his friend Saint Anthony was widely influential in the Greek-speaking world of the early Church; and Evagrius of Antioch, who translated Athanasius' *Life of Saint Anthony* into Latin. Reports of the desert monks by men such as these, through their internal consistency and accord with other records of the day, lend plausibility to stories about the holy fathers' transcendence of pain and discomfort, their small intake of food and sleepless vigils, their longevity, serenity, and healing powers, and their contagious exuberance.

From the time of Saint Anthony, these joyous anchorites inspired thousands to follow their example, so that the desert came to be populated with both hermits and immense monastic communities that included people from the highest strata of Roman society. The love of God was in the air, and visitors from many parts of the world flocked to see the radiant and life-giving ascetics. Saint Jerome reflected this general excitement when he wrote to Rufinus in 375, "I hear that you are penetrating the hidden places of Egypt, visiting the bands of monks and going the round of heaven's family on earth. At last the full weight of truth has burst upon me: Rufinus is in Nitria and has reached the blessed Macarius!" These early Christian contemplatives, this "family of heaven on earth," had captured the imagination of the Roman world.[5]

[21.3]
SHAMANIC POWERS

Mircea Eliade drew upon studies by numerous anthropologists to characterize the essential elements of shamanism. He wrote:

> Shamans for all their apparent likeness to epileptics and hysterics, show proof of a more than normal nervous constitution; they achieve a degree of concentration beyond the capacity of the profane; they sustain exhausting efforts; they control their ecstatic movements. . . . Preparation for his future work leads the neophyte to strengthen his body and perfect his intellectual qualities. Mytchyll, a Yakut shaman known to Sieroszewski, though an old man, during a performance outdid the youngest by the height of his leaps and the energy of his gestures. He became animated, bubbled over with intelligence and vitality. He gashed himself with a knife, swallowed sticks, and ate burning coals.[6]

Shamanic virtuosity was evident, too, in a young Labrador Eskimo who remained for five days and nights in the sea so that he was given the title *angalok*, his people's equivalent of shaman. Eliade equated the power to survive so long in icy water with the Tibetan *tumo* (see 5.4), the ability of Manchu initiates to swim under ice in freezing seas, and the Rig Veda's *tapas*. The mastery of internal heat noted by Eliade, which has been attributed to yogis, Eskimo *angaloks*, and Tibetan lamas, has also been observed among shamans of the Solomon Islands, Sumatra, the Malay archipelago, and various North American Indian tribes.[7] Pavioso shamans of North America, for example, put burning embers in their mouths and touch red-hot irons.[8] Kung bushmen dance ecstatically in the flames of their campfires.[9] Araucanian shamans of Chile have walked barefoot on fire without burning their limbs or garments.[10]

Another kind of invulnerability was described by the Oglala Sioux medicine man Black Elk. Speaking to writer John Neihardt, he described the battle between Indians and white men at Wounded Knee, South Dakota, on December 29, 1890.

> They were fighting right there, and a Lakota cried to me: "Black Elk, this is the kind of a day in which to do something great!" I answered: "How" (signifying assent).
>
> Then I got off my horse and rubbed earth on myself, to show the Powers that I was nothing without their help. Then I took my rifle, got on my horse and galloped up to the top of the hill. Right below me the soldiers were shooting, and my people called out to me not to go down there; that there were some good shots among the soldiers and I should get killed for nothing.
>
> But I remembered my great vision, the part where the geese of the north appeared. I depended upon their power. Stretching out my arms with my gun in the right hand, like a goose soaring when it flies low to turn in a change of weather, I made the sound the geese make— br-r-r-p, br-r-r-p, br-r-r-p; and, doing this, I charged. The soldiers saw, and began shooting fast at me. I kept right on with my buckskin running, shot in their faces when I was near, then swung wide and rode back up the hill.
>
> All this time the bullets were buzzing around me but I was not touched. I was not even afraid. It was like being in a dream about shooting. But just as I had reached the very top of the hill, suddenly it was like waking up, and I was afraid. I felt something strike my belt as though someone had hit me there with the back of an ax. I nearly fell out of my saddle, but I managed to hold on, and rode over the hill.[11]

Several anthropologists have watched shamans perform ritual surgeries. Waldemar Bogoras, for example, observed a Chuckchee shamaness open her son's abdomen with a knife, thrust her hand into the wound, then close it without sutures so that Bogoras could not detect any scar afterward. Another Chuckchee shaman, after drumming to inspire freedom from pain, cut his own abdomen open for others to see.[12] Bogoras also reported the following incident. As a Chuckchee shaman drummed and sang to induce a trance in which he would descend to the underworld, voices were heard from several directions, some of them seeming to come from a distance. Apparently possessed by a spirit, the shaman spoke in falsetto while the tent shook and pieces of wood flew through the air. Another student of Siberian shamanism, Sergei Shirokogoroff, reported similar phenomena he had witnessed among the Tungus, including apparent spirit-voices and telekinesis.[†]

In the stone-age cultures of Siberia, the Americas, and Central Asia, the shaman's ecstatic freedom is expressed in myths of his magical flight and descents to the underworld. Besides expressing the soul's autonomy, the shaman's reputed ability to move through various cosmic zones, I propose, prefigures a superordinary mobility we can cultivate through transformative practice (see 5.5). Like Catholic doctrines of the Glorified Body, stories about shamanic journeys might anticipate a latent human capacity to produce or move through various configurations of matter (see chapters 8 and 9).

[21.4]
INDIAN RUNNING

In his book *Indian Running*, Peter Nabokov wrote:

> Indian runners relied on powers beyond their own abilities to help them run for war, hunting, and sport. To dodge, maintain long distances, spurt for shorter ones, to breathe correctly and transcend oneself called for a relationship with strengths and skills which were the property of animals, trails, stars and elements. Without their tutelage and beneficence one's potential could never be realized.[13]

In a contest chronicled by Westerners since the sixteenth century, Indians of Eastern Brazil compete in teams that carry logs weighing from 100 to 200 pounds. The runners are painted, with each tribe displaying favorite styles. The Kraho Indians claim that their ancestors, the sun and moon, invented these contests for themselves, passing them on to their human children.[14] The ethnographer Curt Nimuendaju described one race of this kind.

> Four men from each team lift their own log to the first racer's shoulder, and he immediately dashes off in the direction of the village, followed

---

[†] Shirokogoroff 1935. In Eliade 1964, p. 243. The exceptional stamina, agility, self-control, and healing abilities of many shamans, as well as the paranormal phenomena they sometimes appear to stimulate, are frequently inspired by ecstatic trance. As Eliade and other anthropologists have observed, such trance is not epileptic in that it can be deliberately induced. But its convulsiveness often seems epileptoid, resembling the violent raptures of Saint Teresa of Avila and the seizures of dervishes. See: Eliade 1964, p. 4.

and surrounded by a tumultuous troop of his fellows. The first choice is always one of the best runners, in order to have an encouraging start. When the two log bearers are running side by side, each has for his lane that side of the track on which his team had its log lying at the start of the race.

At once a mad chase starts in. Yelling and inciting the racers to greater efforts, blowing trumpets and ocarinas, the Indians with their waving grass ornaments bound deerlike to the right and left of the log-bearers' path, leaping over tufts of grass and low steppe bushes. After a distance of about a hundred and fifty meters, a fellow member runs up to the log bearer, who without stopping in his course dexterously twists his body around so as to transfer his load to his mate's shoulder and the race continues without the least interruption. Thus it goes on madly down the slopes of the hills in the torrid sunshine of the shadeless steppe, across brooks and uphill again, in the burning, loose sand. And now we come to the feature that remains incomprehensible to the Neo-brazilian and leads to his constantly ascribing ulterior motives to this Indian game: The victor and the others who have desperately exerted themselves to the bitter end receive not a word of praise, nor are the losers and outstripped runners subject to the least censure; there are neither triumphant nor disgruntled faces. The sport is an end in itself, not the means to satisfy personal or group vanity. Not a trace of jealousy or animosity is to be detected between the teams. Each participant has done his best because he likes to do so in a log race. Who turns out to be the victor or loser makes as little difference as who has eaten most at a banquet.[15]

Ceremonial runs among the Papago and Pima on the Gulf of California have sometimes lasted for eight hours. "For years men trained in anticipation of this experience," Nabokov wrote. "[They] made offerings to the ocean, [and] ran twenty miles to a headland and back, straining for a life-guiding vision."[16] Anthropologist Ruth Underhill described the visionary experience of these Indian runners. One of them, she wrote, saw white cranes that seemed to be setting his pace and realized he was destined to become a great kick-ball racer. Another saw a mountain revolving before him, and composed a song about it that won him prominence among his fellows. And a third heard a call from the "sea shaman," then retired to a cave for several years and turned himself into a sorcerer.[17]

These 40-mile runs of the Pima, however, are exceeded by the Tarahumara Indians of Mexico. Karl Kernberger, who photographed Tarahumara running events for several years, described one of their kick-ball races that lasted for two days and two nights. "They didn't seem to be pushing themselves, or crossing any pain threshold," he told Nabokov. "They just seemed to be doing what the Tarahumara have to do."[18] These prodigious runs promote social cohesion, provide outlets for aggression, and stimulate the health of their participants; but they do more than that. A drive for transcendence impels some Indian runners. "These ancient American traditions help to dignify our hunger to [participate in] what Mircea Eliade calls 'the cosmic totality'," Nabokov wrote. " I suspect that in some way all runners share the yearning expressed in the Navajo chant":

The mountain, I become part of it . . .
The herbs, the fir tree,
I become part of it.
The morning mists,
The clouds, the gathering waters,
I become part of it.
The sun that sweeps across the earth,
I become part of it.
The wilderness, the dew drops, the pollen . . .
I become part of it.

[21.5]
TAOIST MONKS AND MAGICIANS

The Buddhist scholar John Blofeld described a tribal Thai magician who was seated facing a shrine while chanting with enormous power. "Now and then," Blofeld wrote, "an extraordinary, indeed an awful, thing would happen. With a frightful scream, he would shoot into the air and land back upon the bench with [great] force. Such a movement by a seated man whose legs never once straightened for the jump was . . . uncanny. . . ."[19] Blofeld also described fire walking, flesh-piercing, control of bleeding, and exhibitions of great strength and agility by Taoist masters and their students.[20] The abilities revealed in these demonstrations, he wrote, were developed through meditation, breathing exercises, yogic postures, and other contemplative methods, many of them part of a Taoist program to achieve physical immortality (see 8.2).[21]

[21.6]
JEWISH MYSTICS

For most Jewish mystics, the greatest adventures of spirit occur in the common tasks of life, when *zohar*, the splendor of existence, becomes manifest. "Any event can awaken our higher faculties," wrote Edward Hoffman, a student of Kabbalism and Hasidism, "for the transcendent can be found in everything and everywhere."[22] But in spite of its focus upon the goodness of everyday life, Jewish mystical literature describes metanormal abilities similar to the charisms of Catholic saints and the siddhis of Indian yoga. Hasidism's founder, the Baal Shem Tov, for example, was said to have actualized a supplicant's wishes in a manner that demonstrated his extraordinary powers of mind over matter, predicted future events, and read the history of persons and objects; and the Rabbi Dov Baer of Mezritch (the Maggid), as well as the Rabbi Zalman of Liady, founder of the Lubavitcher sect flourishing today, frequently exhibited extrasensory abilities.[23]

[21.7]
THE YOGIC BURIALS OF HARIDAS

In a book entitled *Observations on Trance or Human Hibernation* published in 1850, James Braid, the pioneering physician who coined the term *hypnosis,* described an experiment that typified several scientifically inspired studies of religious adepts that were conducted during the nineteenth century.[24] He based his account on a report by Sir Claude Wade, then British Resident at the court of the Maharaja Runjeet Singh of Lahore, detailing the burial of a yogi named Haridas in 1837. To impress doubtful witnesses, the yogi had asked the Maharaja to supervise his burial for six weeks. Being skeptical about feats of this kind, the Maharaja took extraordinary precautions against fraud. According to Wade's report:

> On the approach of the appointed time, according to invitation, I accompanied Runjeet Singh to the spot where the fakir was buried. It was a square of the gardens, adjoining the palace at Lahore, with an open verandah all round, having an enclosed room in the centre. On arriving there, Runjeet Singh, who was attended on the occasion by the whole of his court, dismounting from his elephant, asked me to join him in examining the building to satisfy himself that it was closed as he had left

it. We did so; there had been a door on each of the four sides of the room, three of which were perfectly closed with brick and mortar, the fourth had a strong door, which was also closed with mud up to the padlock, which was sealed with the private seal of Runjeet Singh in his own presence, when the Fakir was interred. Indeed, the exterior of the building presented no aperture by which air could be admitted, or any communication held by which food could be conveyed to the fakir. I may also add, that the walls closing the doorway bore no mark whatever of having been recently disturbed or removed.

Runjeet Singh recognized the seal as the one which he had affixed, and as he was as sceptical as a European could be of the success of such an enterprise,—to guard as far as possible against any collusion,—he had placed two companies from his own personal escort near the building, from which four sentries were furnished and relieved every two hours, night and day, to guard the building from intrusion. At the same time, he ordered one of the principal officers of his Court to visit the place occasionally, and to report the result of his inspection to him, while he himself, or his Minister, kept the seal which closed the hole of the padlock, and the latter received the report, morning and evening, from the officer of the guard.

After our examination, we seated ourselves in the verandah opposite the door, while some of Runjeet Singh's people dug away the mud wall, and one of his officers broke the seal and opened the padlock. When the door was thrown open, nothing but a dark room was to be seen. Runjeet Singh and myself then entered it, in company with the servant of the fakir, and a light being brought, we descended about three feet below the floor of the room, into a sort of cell, where a wooden box, about four feet long by three broad, with a sloping roof, containing the fakir, was placed upright, the door of which had also a padlock and seal similar to that on the outside. On opening it we saw a figure enclosed in a bag of white linen, fastened by a string over the head—on the exposure of which a grand salute was fired, and the surrounding multitude came crowding to the door to see the spectacle. After they had gratified their curiosity, the fakir's servant, putting his arms into the box, took the figure out, and closing the door, placed it with its back against it, exactly as the fakir had been squatted (like a Hindu idol) in the box itself.

Runjeet Singh and myself then descended into the cell, which was so small that we were only able to sit on the ground in front of the body, and so close to it as to touch it with our hands and knees.

The servant then began pouring warm water over the figure; but as my object was to see if any fraudulent practices could be detected, I proposed to Runjeet Singh to tear open the bag, and have a perfect view of the body before any means of resuscitation were employed. I accordingly did so; and may here remark, that the bag, when first seen by us, looked mildewed, as if it had been buried some time. The legs and arms of the body were shrivelled and stiff, the face full, the head reclining on the shoulder like that of a corpse. I then called to the medical gentleman who was attending me to come down and inspect the body, which he did, but could discover no pulsation in the heart, the temples, or the arm. There was, however, a heat about the region of the brain, which no other part of the body exhibited.

The servant then recommended bathing him with hot water, and we gradually relaxed his arms and legs from the rigid state in which they were contracted, Runjeet Singh taking his right and I his left leg, and by friction restoring them to their proper action; during which time the servant placed a hot wheaten cake, about an inch thick, on the top of his head,—a process which he twice or thrice renewed. He then pulled out of his nostrils and ears the wax and cotton with which they were stopped; and after great exertion opened his mouth by inserting the point of a knife between his teeth, and, while holding his jaws open with his left hand, drew the tongue forward with his right,—in the course of which the tongue flew back several times to its curved position upwards, in which it had originally been, so as to close the gullet.

He then rubbed his eyelids with ghee for some seconds, until he succeeded in opening them, when the eyes appeared quite motionless and glazed. After the cake had been applied for the third time to the top of his head, the body was violently convulsed, the nostrils became inflated, when respiration ensued, and the limbs began to assume a natural fulness; but the pulsation was still faintly perceptible. The servant then put some of the ghee on his tongue and made him swallow it. A few minutes afterwards the eyeballs became dilated, and recovered their natural color, when the fakir, recognizing Runjeet Singh sitting close to him, articulated in a low sepulchral tone, scarcely audible, "Do you believe me now?" Runjeet Singh replied in the affirmative, and invested the fakir with a pearl necklace and superb pair of gold bracelets, and pieces of silk and muslin, and shawls, forming what is called a *khelat;* such as is usually conferred by the Princes of India on persons of distinction.

From the time of the box being opened, to the recovery of the voice, not more than half an hour could have elapsed; and in another half-hour the fakir talked with myself and those about him freely, though feebly, like a sick person; and we then left him, convinced that there had been no fraud or collusion in the exhibition we had witnessed.[25]

Wade's careful account was confirmed by Johann Martin Honigberger, a German physician who in 1839 was told by the English general Ventura and other credible witnesses about Haridas' interment. In a book entitled *Fruchte aus dem Morgenlande,* published in 1851, Honigberger told a story almost identical to Braid's, adding certain details that Wade's report had omitted.[26] According to Honigberger, Haridas had severed the ligament of his tongue while training for his strange profession, and in preparing for his 40-day burial had cleaned his stomach with a long strip of cloth and his bowels with enemas. Before his burial, his ears, rectum, and nostrils were sealed with wax, and he was wrapped in a linen cloth that was also sealed. He was then laid into a chest, which the Maharaja locked, and lowered into the enclosure that Wade described. Upon his resuscitation, Honigberger wrote, the yogi's attendant blew air into his throat and ears so that the wax plugs in his nostrils were loudly ejected. Having been sealed up in this manner, Haridas had survived with virtually no ventilation.

During a subsequent experiment by the Maharaja, Haridas' box was buried again and

> earth was turned [upon it] and trodden down, so as completely to surround and cover [it]; a crop of barley was sown over it; and a constant guard maintained on the spot. Moreover, twice during the period of interment, Runjeet Singh had the body dug up, when it was found to be exactly in the same position as when interred, and in a state of apparently entirely suspended animation. At the expiration of this prolonged interment, the fakir recovered under the usual treatment.[27]

[21.8]
THE BODILY TRANSFORMATIONS
OF SRI RAMAKRISHNA

The nineteenth-century Indian ecstatic Sri Ramakrishna Paramahansa (1836–1886) worshiped God as his master, child, mother, friend, and lover. He practiced several Tantric disciplines, realizing the different ecstasies they were intended to promote, and entered *nirvikalpa samadhi* upon meeting the Vedantic teacher Totapuri, remaining in trance for a full six months. In the course of his spiritual realizations, he had visions of Krishna, Christ, and Mohammed that led him to declare that the contemplative paths of every religion were fulfilled in a single Divinity. No other modern mystic has dramatized so many kinds of religious experience.[28]

During his ascetic practices, Ramakrishna exhibited remarkable bodily changes. While worshiping Rama as his devotee Hanuman, the monkey chieftain of the Ramayana, his movements resembled those of a monkey. "His eyes became restless," wrote the Vedantic scholar Swami Nikhilananda. "He lived on fruits and roots. With a cloth tied around his waist, a portion of it hanging in the form of a tail, he jumped from place to place instead of walking."[29] In his biography of Ramakrishna, novelist Christopher Isherwood paraphrased the saint's own description of his strange behavior: "I didn't do this of my own accord; it happened of itself. And the most marvelous thing was— the lower end of my spine lengthened, nearly an inch! Later, when I stopped practicing this kind of devotion, it gradually went back to its normal size."[†]

Ramakrishna also reported kundalini experiences, some of them accompanied by visions of chakras and other features of Indian esoteric anatomy. In his ecstasies he shed tears, trembled convulsively, bled through his pores, felt his joints loosening or locking, perspired heavily, and felt burning sensations.[30] Like other ecstatics, Ramakrishna possessed a powerful imagination that helped produce his exalted states and physical changes; and like Theodore Barber's exceptional hypnotic subjects (15.6), he was prey to suggestions from his environment. Upon one occasion, for example, clots of blood appeared on his lips after a cousin said he would bleed from his mouth. The suggestion took root strongly, but Ramakrishna was restored when a sadhu prompted him to stop his bleeding by exercising self-control.[31] But the famous saint would not use his self-transformative powers to try to save himself from cancer. When a friend suggested he remove his illness through yogic concentration, he rebuked him, asking how he could withdraw his mind from God and turn it to "this worthless cage of flesh."[32] Like many ascetics, he did not deem his body's restoration to be a worthy cause.

The accounts in this chapter describe only a small fraction of the metanormal capacities attributed to religious figures. Calanus's serene composure during his death by burning, the desert fathers' healing

---

[†] Isherwood 1959, p. 71. Isherwood's paraphrasing of Ramakrishna's description is almost identical to one in an earlier biography that drew upon firsthand accounts by the saint's closest disciples; see: *Life of Sri Ramakrishna* 1924, p. 82.

abilities, shamanic mastery of internal heat, Black Elk's invulnerability in battle, the Chuckchee healer's control of pain during his self-mutilation, the Thai magician's extraordinary movement, Haridas' metabolic control during his 40-day interment, and Ramakrishna's remarkable bodily changes reveal the transformative power of religious life.

22

THE CHARISMS OF
CATHOLIC SAINTS
AND MYSTICS

CATHOLIC SAINTS AND MYSTICS have exhibited psy-chophysical changes as dramatic as those found in any other religious tradition. Moreover, such phenomena have been subjected to much skeptical appraisal through canonization proceedings and investigations by journalists and medical researchers. Surveying the enormous literature about Catholic religious life produced in the past 200 years, I have been impressed by the sheer volume of medical reports, ecclesiastical reviews, and investigative journalists' accounts it contains. Taken as a whole, studies of Roman Catholic sanctity provide a unique body of evidence for human transformative capacities.

[22.1]
CATHOLIC CANONIZATION PROCEDURES

In this chapter I describe several extraordinary phenomena, or charisms, attributed to Catholic religious figures. My primary sources for this discussion are biographies of saints that include eyewitness reports from canonization proceedings; *Butler's Lives of the Saints*, as edited by Herbert Thurston and Donald Attwater; books on the physical phenomena of mysticism by Herbert Thurston, Caroline Bynum, Antoine Imbert-Gourbeyre, and other writers; and, indirectly, scholarly appraisals of Catholic hagiography contained in the Bollandists' *Acta Sanctorum* and *Analecta Bollandiana*.[1] Before commencing this review, I will briefly describe certain church procedures for examining the lives of its heroic figures.

In judging evidence from Roman Catholic sources about metanormal phenomena, we should appreciate the thoroughness with which such phenomena have been investigated by church authorities, medical researchers, and historians determined to uncover pious fraud or temper uncritical belief. To this description of Catholic investigative procedures I will add a note about the scholarship of the Bollandists and Father Herbert Thurston. Thurston's balanced, highly informed examinations of the physical prodigies associated with Catholic sanctity are a primary source for the material presented here.

Historical Background

Until well into the Middle Ages, canonization was usually carried out in each region by one or more bishops and their parishioners to confirm and to celebrate the sanctity of someone venerated for holiness or martyrdom. In the eleventh or twelfth century, however, papal canonization started to come into effect and was gradually extended to the entire church. In 1234 Gregory IX decreed that such papal procedure was the only legitimate inquiry into a saint's life and reputed miracles; and in 1588 Pope Sixtus V entrusted all canonization proceedings to the Congregation of Rites, which included cardinals, bishops, and other prelates chosen to represent the entire church. The Congregation adopted a stable practice, and in 1642 Urban VIII ordered the publication in one volume of all the decrees and subsequent interpretations issued on the canonization of saints during his pontificate.[2] In the following century, Benedict XIV produced his famous treatise *De Servorum Dei Beatificatione et Beatorum Canonizatione*, which defined the elements of canonization, clarified the concept of heroic virtue, and established criteria by which to judge the authenticity of miracles. The canonization process as it now operates has been summarized in a preface to *The Book of Saints* by the Benedictine Monks of Saint Augustine's Abbey, Ramsgate.

> [It] is exceedingly complex, [consisting] in the first place of a thorough investigation into all the particulars that can be ascertained of the life and death of the alleged Saint. . . . In this, as in all subsequent stages of the procedure, every witness is examined under oath and in the presence of a trained Church lawyer, who is obligated to urge all the objections he can think of, and who is at liberty not only to cross-examine the witnesses put forward but to call any number of others he pleases in order to rebut their testimony. Supposing the judgment of the Court of First Instance to be favourable, the case goes for retrial to a higher tribunal. In these proceedings not only are witnesses called to testify to individual facts, but particular stress is put upon the popular verdict concerning the alleged Saint, that is, upon the repute in which he was

held by those who may have had dealings with him or had opportunities of forming an opinion about him. Depositions of all kinds must be gathered together with as little delay as possible, and duly sworn to; but in order to guard against mere enthusiasm playing any part in the matter, at one stage of the proceedings a surcease of at least ten years is enjoined.

Usually, the procedure in causes of canonisation takes many years to complete; for there are numerous hearings and rehearings to be allowed for. A first stage is that of "beatification," which is reached on proof of extraordinary holiness of life and of two miracles. In modern procedure this is rarely reached within fifty years of the death of the Saint. At Beatification, permission is given for local veneration. For Canonisation proper, proof of two more miracles wrought since Beatification is demanded. The Servant of God is then enrolled in the Canon of Saints, his or her name being inserted in the Roman Martyrology or official catalogue of Saints proposed to the veneration of the Universal Church.[3]

Because the entire course of canonization now takes decades or even centuries, phenomena such as stigmata and luminosity have been discussed in detail by many people who claimed to have observed them, and by doctors, physiologists, physicists, and others who might ascertain their authenticity and causes. Because testimony is presented under oath, and because lies or exaggerations are considered by the church to be sins, there is a powerful constraint upon Catholic experts and witnesses to tell the truth in these proceedings. Furthermore, any testimony can be called into question by the *Promotor Fidei,* or Devil's Advocate, who challenges the evidence at hand Canonization records since the sixteenth century contain many carefully examined accounts of extraordinary functioning given at different times and places. Because they have been carefully recorded, these separate accounts can be compared and tested with a thoroughness that cannot be applied to the orally transmitted legends of other faiths. Descriptions of a particular saint's abstinence from food, for example, might vary so much from witness to witness that all of them are called into question, either by the Congregation of Rites or by those of us who study them now. We do not have to depend upon the church's verdict regarding saintly attributes in making our own assessments of them.

Roman Catholic canonization records provide compelling evidence for several types of metanormal capacity, and they undoubtedly contain clues about extraordinary functioning that no one has yet pursued. Someday, perhaps, their immense store of firsthand reports will be searched more thoroughly for insights about psychosomatic transformation.

Rules of the Church for Assessing the Marvels of Sanctity

Benedict XIV (formerly Prosper Lambertini), who was *Promotor Fidei* before he was pope, formulated rules for assessing miraculous cures, visions, and other religious phenomena. At his request, the Jesuit Emmanuel de Azevedo made a doctrinal synthesis of the many decrees by which the criteria for judging miraculous events had been defined.[4] The rules of evidence spelled out by Benedict XIV and Father Azevedo have been refined in the light of modern science. These rules distinguish physical and mental agencies that mediate unusual cures. They discriminate between different classes of charism and describe the role of imagination in sickness and health. They establish several types of vision and distinguish authentic mystical experience from false enthusiasms. During the past 200 years, they have been adapted to modern explanations of cures that once seemed miraculous, to psychiatric insights about the hysterical symptoms exhibited by some ecstatics, and to discoveries of psychical research that bear upon extranormal phenomena. With them, the Congregation of Rites can study potential saints using insights and methods of contemporary science while drawing on the church's long experience with many kinds of religious experience. The evidence for extraordinary functioning provided by canonization proceedings has been tested for exaggeration, fraud, and delusion, even if it has been interpreted by church authorities in ways we do not agree with. Considered in its entirety, this body of evidence has been winnowed more thoroughly than has the anecdotal material provided by other religious traditions.[†]

Herbert Thurston and the Bollandists

Herbert Thurston was born in London in 1856, entered the Jesuit order in 1874, and joined the staff of a Catholic journal, *The Month*,

---

[†] Besides the many investigative groups that serve in canonization proceedings, there are commissions and bureaus of inquiry in the church to examine phenomena put forward as miraculous. The Bureau Medicale, which was established in 1883 to study cures at Lourdes, for example, has examined many thousands of cases since its founding (see 13.2).

in 1894, where he worked until his death in 1939. A leading scholar of Catholic history, liturgy, and sanctity, he contributed more than 760 articles to scholarly journals and popular magazines, as well as 180 pieces to the Catholic encyclopedia. He helped carry out a major revision of *Butler's Lives of the Saints* and was a student of psychical research and psychiatry, especially as they contributed to the understanding of religious mystics. Most important for this discussion, however, is the fact that he studied the physical phenomena of Christian contemplative life with great care and persistence. Some of his articles were gathered into a single volume, *The Physical Phenomena of Mysticism,* by the Jesuit religious scholar Joseph Crehan. I have found this book to be the most useful guide in the English language to material from canonization proceedings, psychical research studies, and psychiatric case histories related to the psychophysical transformations of Catholic saints and mystics. The eminent religious scholar Hippolyte Delehaye, president of the Bollandists for many years until his death in 1941, called Thurston the leading scholar of hagiographical literature of his time.[5] A memoir of Thurston's life by Joseph Crehan includes a list of his publications that reveals the great range of his scholarship.[6]

Father Thurston often drew on the critical studies of the Bollandists, a group of Jesuits organized by Jean Bolland (1596–1665) to gather and analyze documents bearing on the lives and cults of the saints.[7] The Bollandists' *Acta Sanctorum* contains much scholarly analysis of Catholic sanctity; and their review *Analecta Bollandiana* includes studies of hagiographical and historical issues related to their ongoing research. Since the seventeenth century, these Jesuit scholars have separated fact from fiction in traditional accounts of Christian heroes, sometimes stirring up calls by offended churchmen that their work be deemed heretical.

[22.2]

CHARISMATIC PHENOMENA RECOGNIZED BY CATHOLIC AUTHORITIES

Though they are not considered by the church to be signs in themselves of virtue or sanctity, several extraordinary phenomena called charisms frequently accompany religious devotion. According to *The New Catholic Encyclopedia,* they include:

- *Visions*, the perception of normally invisible objects.
- *Locutions*, interior illuminations by means of words or statements, sometimes accompanied by a vision and seeming to proceed from the object represented.
- *Reading of hearts*, telepathic knowledge of secret thoughts or mood without sensory cues.
- *Incendium amoris*, burning sensations in the body without apparent cause. These include interior heat, usually a sensation around the heart, which gradually extends to other parts of the body; intense ardors (when the heat becomes unbearable and cold applications must be used); and material burning that scorches clothing or blisters the skin.
- *Stigmata*, the spontaneous appearance of wounds and bleeding that resemble the wounds of Christ.
- *Tears of blood and bloody sweat (hematidrosis)*, the effusion of blood from the eyes, as in weeping, or from pores of the skin.
- *Exchange of hearts*, the appearance of a pronounced ridge of flesh on a finger, representing a ring designating mystical marriage with Christ.
- *Bilocation*, the simultaneous presence of a material body in two distinct places at once.
- *Agility*, the instantaneous movement of a material body from one place to another without passing through the intervening space.
- *Levitation*, elevation of the human body above the ground without visible cause and its suspension in the air without natural support. It may also appear in the form of ecstatic flight or ecstatic walk.
- *Compenetration of bodies*, when one material body appears to pass through another.
- *Bodily incombustibility*, the ability of bodies to withstand the natural laws of combustibility.
- *Bodily elongation or shrinking.*
- *Inedia*, abstinence from all nourishment for great lengths of time.
- *Mystical aureoles and illuminations*, radiance from the body, especially during ecstasy or contemplation, which is considered to be an anticipation of the Glorified Body.
- *Blood prodigies, bodily incorruptibility, and absence of rigor mortis* in human cadavers.[8]

There is more evidence for some of these phenomena than for others, of course. There can be no reasonable doubt, for example, that many saints have exhibited stigmata, while the case for levitation is far less certain, and bilocation—if taken to mean the appearance of the same flesh (rather than a projected phantom) in two places at once—is impossible. In this chapter, I discuss only those phenomena that have been frequently described by competent witnesses in different periods of Christian history.

[22.3]

STIGMATA

Stigmata that correspond to Christ's crucifixion wounds had appeared on the bodies of at least 60 people, most of them women, by the 1930s, according to Herbert Thurston's conservative estimate; and many more have since been reported in medical journals and popular magazines. A far greater number of Christian believers, moreover, have experienced Christ's wounds vicariously without exhibiting visible marks.[9] Most typically, visible stigmata have consisted of bruises, welts, cicatrices, or bleeding wounds on the hands, feet, and side, but stigmatics have also exhibited apparent abrasions or punctures on the head representing the imprint made by Christ's crown of thorns, crosses on the back or chest, and lacerations representing Christ's scourging. As we shall see, these marks sometimes resemble objects or ideas peculiar to the stigmatic's experience. Some stigmata bleed every Friday or on particular Fridays; some bleed every day. And their skin texture varies, from reddened epidermis and blood blisters to wounds that require bandaging.[†] A few ecstatics have had nine or ten such marks on their body at once, but most have had fewer, and some have had just two or three.

Saint Francis of Assisi

Whether or not Saint Francis of Assisi was the first stigmatic, his example made an immense impression upon his contemporaries.[10] In the decades following his death, other Christian religious, virtually all of

---

[†] Many observers have pointed out that few stigmata have become infected in the manner of ordinary wounds. In this they resemble the psychogenic bodily marks discussed in 11.1.

them women, exhibited wounds like his. Since every stigmatic about whose wounds we can be certain came after him, Saint Francis might be said to have pioneered this form of bodily transformation. Indeed, news of the saint's stigmata, which his followers withheld from the public until his death, caused an immediate sensation. Soon after Saint Francis's passing, his confidant Brother Elias wrote to the Provincial of France:

> I announce to you great joy, even a new miracle. . . . For, a long while before his death, our Father and Brother appeared crucified, bearing in his body the five wounds which are verily the Stigmata of the Christ; for his hands and feet had as it were piercings made by nails fixed in from above and below, which laid open the scars and had the black appearance of nails; while his side appeared to have been lanced, and blood often trickled therefrom.[11]

Priests and storytellers throughout Europe expressed amazement about Saint Francis's stigmata, which were confirmed again and again by his closest companions. The following, for example, was written in the hand of Brother Leo, the saint's closest confidant, beside a blessing that Saint Francis himself had inscribed over his own name.

> The Blessed Francis, two years before his death, kept a Lent in the hermitage of the Alverna. . . . And the hand of the Lord was laid upon him. After the vision and speech he had of a seraph, and the impression in his body of the Stigmata of Christ, he made these Praises which are written on the other side of the sheet, and with his own hand he wrote them out, giving thanks to God for the favor that had been conferred on him.[12]

"The authenticity of this priceless memorial of the saint is now, practically speaking, uncontested," wrote Herbert Thurston, "and we cannot too strongly emphasize the fact that Brother Leo himself in this autograph note not only vouches for the reality of the stigmata . . . but also expressly assigns their appearance to the forty days retreat on Mount Alverna two years before [Saint Francis's] death."[13] Thomas of Celano, who was commissioned by Pope Gregory IX to write a life of Saint Francis, described the saint's stigmata as follows:

> And while [Francis] continued without any clear perception of its meaning (i.e. the vision of the seraph), and the strangeness of the vision

was perplexing his breast, marks of nails began to appear in his hands and feet, such as he had seen a little while before in the Man crucified who had stood over him. His hands and feet seemed pierced in the midst by nails, the heads of the nails appearing in the inner part of the hands and in the upper part of the feet and their points over against them. Now these marks were round on the inner side of the hands and elongated on the outer side, and certain small pieces of flesh were seen like the ends of nails bent and driven back, projecting from the rest of the flesh. So also the marks of nails were imprinted in his feet, and raised above the rest of the flesh. Moreover his right side, as it had been pierced by a lance, was overlaid with a scar, and often shed forth blood so that his tunic and drawers were many times sprinkled with the sacred blood.[14]

While accounts such as this one were almost certainly embellished, they share certain striking features with modern cases that have been closely observed and sometimes photographed. Lumps that looked like nailheads, for example, appeared on the hands of Domenica Lazzari, a stigmatic who lived in the Tyrol during the 1830s and 1840s.[15] Though there are differences between Thomas of Celano's descriptions of the "nails" in Saint Francis's hands and accounts of Domenica Lazzari's wounds, the two sets of stigmata were clearly similar. And it is absolutely certain that other modern stigmatics, among them Padre Pio and Theresa Neumann, have exhibited dark, raised cicatrices when their wounds were not bleeding. Since the early nineteenth century, many kinds of stigmata have been carefully documented by skeptical medical researchers. Each stigmatic described in the following sections lived after 1800 and was studied carefully for many years.

Anne Catherine Emmerich

Born in 1774 near Koesfeld, Westphalia, Anne Catherine Emmerich, an Augustinian nun, began to exhibit bleeding wounds in 1812. Like other Catholic stigmatics, she had raptures in which she vicariously experienced Christ's passion. Her diocesan, the vicar general of Munster, Clement August von Droste, assigned ecclesiastics to observe her over a period of several years, partly to satisfy his own curiosity and partly to quiet the sensation her stigmata had caused among believers and skeptics alike. Because the European Enlightenment had produced widespread skepticism about phenomena connected with religious ecstasy, church leaders were intent to preserve the integrity of their faith by debunking fraud and by separating pathology from sanctity.

The reports of ecclesiastics and doctors who examined Anne Emmerich make a convincing case for the authenticity of her stigmata, which included a Y-shaped mark resembling a crucifix at Koesfeld to which she had prayed as a child.[16] The following passages are from notes by Vicar General von Droste, who with two physicians and another churchman examined the stigmatic in the city of Dulmen on April 20, 1813.

> I examined the blood-crust of the left hand with a magnifying glass and found it very thin and a little rugose, or plaited like the epidermis when seen under a lens.
>
> The cross on the breast did not bleed, but appeared of a pale red color caused by the hood under the epidermis. I examined also the lines forming a cross, as well as the skin around them, and I could distinctly see that they did not break the skin. The epidermis over the lines and the skin surrounding them to some distance was unbroken and, through the glass, appeared as if peeling off a little.
>
> The wound on the right side was not bleeding, but the upper part of it was encrusted with dried blood, as might be produced by extravasated blood just below the epidermis.
>
> The cross on the breast was red with blood. I washed the upper part and examined it again. Had the skin been broken I should certainly have remarked it. I think there was near the cross a short streak which seemed to be a depression filled with blood.
>
> Her hair being very thick, it was impossible to examine the punctures around her head. She consented to have it cut close, though not so close, however, as to allow the blood instantaneously to soak her headdress and pillow. She requested this for the sake of neatness. The blood being washed away, a number of fine bloody marks could be seen with the naked eye scattered irregularly over the forehead and extending from the middle of it almost to the top of the head.[17]

After three visits to examine the celebrated stigmatic, the vicar general wrote to the Commissary General of French Police, who had jurisdiction in Dulmen:

> Sister Emmerich desires only to be forgotten by the world that she may be free for spiritual things which alone interest her. She asks nothing, she accepts nothing, she desires not to be spoken of, and I trust the public will soon forget her. I cannot discover the slightest shadow of imposture in her case, yet I shall continue to observe her closely.[18]

Anne Emmerich was one of the first ecstatics scrutinized by people with a modern outlook, and her wounds are well documented. Given the large number of doctors and skeptical churchmen who observed her, many of them suspecting a pious fraud, it seems likely that her stigmata appeared as she said they did, without external manipulation.

Gemma Galgani

Gemma Galgani was born in 1878 and died in Lucca in 1903. The following account of her ecstasies and stigmata, written by her confessor and biographer Father Germano di S. Stanislao, was confirmed by others who observed her.

> From this day forward the phenomenon continued to repeat itself on the same day every week, namely on Thursday evening about eight o'clock and continued until three o'clock on Friday afternoon. No preparation preceded it; no sense of pain or impression in those parts of the body were affected by it; nothing announced its approach except the recollection of spirit that preceded the ecstasy. Scarcely had this come as a forerunner than red marks showed themselves on the backs and palms of both hands; and under the epidermis a rent in the flesh was seen to open by degrees; this was oblong on the backs of the hands and irregularly round in the palms.
>
> Sometimes the laceration appeared to be only on the surface; at other times it was scarcely perceptible with the naked eye; but as a rule it was very deep, and seemed to pass through the hand—the openings on both sides reaching the other. I say seemed to pass, because those cavities were full of blood, partly flowing and partly congealed, and when the blood ceased to flow they closed immediately, so that it was not easy to sound them without a probe. . . . In her feet, besides the wounds being large and livid around the edges, their size in an inverse sense differed from those of her hands; that is, there was a larger diameter on the instep and a smaller one on the sole; furthermore, the wound in the instep of the right foot was as large as that in the sole of the left. Thus it must certainly have been with our Savior, supposing that both His Sacred Feet were fixed to the Cross with only one nail.[19]

According to Herbert Thurston, apparent lash marks on Gemma Galgani's body closely corresponded in size and in position with the wounds depicted in a big crucifix before which she was accustomed to pray. When Anne Catherine Emmerich was first marked with a cross on her breast, [it] was a Y-shaped cross, reproducing the form of a crucifix at Koesfeld to which she had great devotion in her childhood. All these things seem to point to an auto-suggested effect rather than to the operation of an external cause whatever its nature.[20]

Louise Lateau

Louise Lateau, a Belgian woman born in Bois d'Haine in 1850, exhibited the stigmata from 1868 until her death in 1883. Gerald Molloy, a theologian and rector of University College, Dublin, who observed her at firsthand, wrote that when a quantity of blood had exuded from her wounds, spectators wiped it away so that her

> stigmata were then more distinctly seen. They are oval marks of a bright red hue, appearing on the back and palm of each hand about the centre. Speaking roughly, each stigma is about an inch in length and somewhat more than half an inch in breadth. There is no wound properly so-called, but the blood seemed to force its way through the unbroken skin. In a very short time, sufficient blood had flowed again to gratify the devotion of other pilgrims, who applied their handkerchiefs as had been done before, until all the blood had been wiped away a second time. This process was repeated several times during the course of our visit.[21]

According to her family and friends, Louise Lateau bled through her wounds on every Friday but two after she was stigmatized on April 24, 1868, until her death on August 25, 1883. During a typical week, she began to feel burning sensations in her stigmata on Tuesday. At the same time her skin grew hot and dry, her pulse spasmodic and rapid, while shooting pains gathered in her heart, forming "a regular current" between it and her extremities. Then blisters appeared, rising on the swollen scars of her wounds. According to a Dr. Lefebvre, professor of general pathology at the Catholic University of Louvain:

> On each of the reddish surfaces of the hands and feet a blister is seen to appear and gradually rise; when it is fully developed it forms a rounded, hemispheric protrusion on the surface of the skin; its base is of the same dimensions as the reddish surface on which it rests, that is,

about two and a half centimetres long and one and a half wide. . . .
This blister is full of limpid serum. The zone of skin surrounding the
blister is not the seat of any turgescence or reddening.[22]

These wounds usually started to bleed on Thursday night between midnight and one in the morning, as Louise Lateau passed
through ecstasies of the Passion. Medical tests showed the blood they
produced to be normal except for an abundance of white corpuscles
and a high proportion of serum.[23] On Saturday, the stigmata were
usually dry and a little shiny, according to Professor Lefebvre, with
scales of dried blood that dropped away.

A commission of inquiry was organized by Cardinal Deschamps,
archbishop of Malines, in concert with the bishop of Tournai and
several medical people to test the authenticity and nature of Louise
Lateau's stigmata; and a second major investigation was initiated by
Belgium's Royal Academy of Medicine in 1874 following a widely
publicized claim at a scientific congress in Breslau that the stigmatic's
wounds were fraudulent. The Royal Academy's study lasted until October 1876, when the academy formally admitted that the facts of the
case were genuine and that it could not provide an adequate explanation of them. In the course of such investigations, Louise Lateau was
sometimes observed day and night by skeptical witnesses. On December 16, 1868, leather gloves were placed on each of her hands, then
tightly sealed, and a shoe was fixed in similar fashion to one of her
feet. A day later, after confirming that the three seals had not been
broken or tampered with, a Dr. Spiltoir de Marchienne, in the presence of eight witnesses, took off the gloves and shoe. Blood was flowing freely from the stigmata on both palms, while blisters had appeared on the backs of both hands and her covered foot.[24] Professor
Lefebvre conducted a further experiment. He wrote:

> On this day blood flowed abundantly from all the stigmata, especially
> on the dorsal face of the left hand; the dermis, exposed over a length of
> two and a half centimetres, bled without interruption.
>
> On the same surface of this hand I applied liquid ammonia on a
> rounded surface about two centimetres in diameter, alongside the
> bleeding stigma. At the same time I was careful to leave a strip of sound
> skin, so that the two wounds would not touch at the edges, and the phe-
> nomena deriving from them would remain distinct. Twelve minutes af-
> ter the application of the ammonia, a fine circular blister formed, filled
> with transparent serum. . . . I opened the epidermis and removed the

shreds, so as to expose the surface of the dermis over a circular area of more than two cms. diameter. Thus here were now two wounds side by side, situated on the same anatomical constitution. We observed carefully: the stigmatic surface continued to bleed, and when I left the girl at 2 P.M. the flow was still continuing, and there was no indication that it would soon stop. The artificial stigma did not yield a single drop of blood. I observed it for two and a half hours; a colourless serum sweated from it for about half an hour, then the surface dried up. I rubbed it with a rough cloth: the serum which soaked this cloth during the rubbing was slightly pink, but when I ceased the friction not another drop of blood flowed.[25]

Marie-Julie Jahenny

Marie-Julie Jahenny, a Breton peasant girl, exhibited the stigmata for the first time in 1873, some five years after the appearance of Louise Lateau's wounds. Like the Belgian ecstatic, she was studied thoroughly for many years by churchmen and medical people. Antoine Imbert-Gourbeyre, for 36 years a professor at the Medical School of Clermont-Ferrand, and the author of a comprehensive work on stigmatics, observed her for more than 20 years, partly because he was asked to by her diocesan, the bishop of Nantes, and partly to satisfy a curiosity that led him to study Lateau and other ecstatics. In 1894 he summarized the history of Jahenny's stigmata.

On the 21st March, 1873, she received the marks of the five wounds; the crown of thorns followed on Oct. 5th; on the 25th of November appeared an imprint on the left shoulder, and on the 6th of December the dorsal stigmata in hands and feet. On Jan. 12th, 1874, her wrists showed marks corresponding to those which the cords must have produced when our Savior's arms were bound, and on the same day a sort of emblematic pattern developed in front of her heart. By Jan. 14th stripes had appeared on her ankles, legs and forearms in memory of the scourging, and a few days afterwards there were two weals on her side. On the 20th of February a stigmatic ring was seen on the fourth finger of the right hand in token of her mystic espousals; later on there appeared various inscriptions on the breast, and finally on Dec. 7th, 1875 the words O Crux Ave with a cross and a flower.[26]

Describing the incident of December 7, 1875, Imbert-Gourbeyre wrote:

Marie-Julie announced a month beforehand, and several times over, that she was shortly to receive a new stigmatization, and that a cross and a flower with the words O *Crux Ave* were to be impressed upon her breast. More than a week before the events occurred she named the precise day: it was to be the 7th of December. The day before this her breast was examined, when it was ascertained that the emblems spoken of had not yet made their appearance. On the morrow, before the ecstasy came on, she offered to submit to another examination, but this was considered unnecessary; she had the right to expect that we should take her word for it. Soon after she passed into a state of trance, and while this wonderful device was developing, her family and the witnesses who were present were able to testify to the incomparable fragrance which exuded from her body and made itself perceptible through her clothes. When the ecstasy was over, the cross, the flower and the inscription could be clearly seen upon her breast.[27]

In 1894, nearly 20 years later, the French doctor wrote that "the flower and the inscription are visible still."

While expressing reservations about Jahenny's interest in the impression she was making, which contrasted with most stigmatics' unwillingness to court notoriety, Thurston argued for the authenticity of her various marks. "She seems to have lived in obscurity for more than twenty years," he wrote, "and to have retained during that time the good opinion of ecclesiastic authority. Further, it is difficult to believe that a physician of some standing in the medical world could have been imposed upon for all that time by a peasant girl in a remote country village who had simply tatooed herself with a needle or had recourse to some other obvious trick. . . . In any case, we may admit that patterns and inscriptions, though evanescent in their nature, have been produced at the Salpetriere upon the flesh of hysterical subjects by simple suggestion. This is the 'dermographisme' of T. Bathelemy and other French medical scientists."[28] As Thurston suggested, if dermographism, or skin-writing, can be produced through hypnotic suggestion, sacred words on the skin might be caused by an ecstatic's passionately held images of them.

Padre Pio

Born Francesco Forgione in 1887 at Pietrelcina in southeastern Italy, Padre Pio joined the Capuchin order in 1903. After experiencing the pain of Christ's wounds intermittently for several years, he exhibited permanent stigmata on his hands, feet, and side from September 20, 1918, until his death in 1968. He became an international celebrity, not only for his stigmata but for his vivid personality, sanctity, and reputed healing powers. Photographs of Padre Pio's wounds taken between the 1920s and his death confirmed descriptions of them in numerous medical reports and ecclesiastical studies.[29] The following chronology suggests the immense impact Padre Pio made on those around him, and the exceptional scrutiny he was subject to.

In 1907 he took his final vows, and was ordained a priest on August 10, 1910, celebrating his first mass the next day at his home church in Pietrelcina. Four weeks later he showed puncture wounds in his hands to a parish priest, but they disappeared after his prayers for their removal. In 1916 he entered a monastery, Santa Maria delle Grazie in San Giovanni Rotondo, there to spend the rest of his life. On August 5, 1918, he saw an apparition of a celestial person who hurled a spear at him. On September 20 he had the same vision, and found visible stigmata on his hands, feet, and side. Nine days later his fellow friars learned of his wounds when they discovered his blood-stained bedding.

By 1919, word of Padre Pio's stigmata had spread around the world, and pilgrims began to visit his monastery. In May he was examined by a Dr. Luigi Romanelli at the request of the Capuchin order; and in July the Holy Office sent Professor Amico Bignami, an agnostic pathologist of the Roman University, to study his wounds. In a formal report, Professor Bignami described superficial scars on the monk's hands and feet, plus a cross upon his left breast. These marks were extremely sensitive, but the professor did not consider them to have been artificially produced. He characterized them instead as "a necrosis of the epidermis of neurotic origin." He considered their symmetrical arrangement to be caused by "unconscious suggestion."[30]

In the fall of 1919, the General Curia of the Capuchins asked Dr. Giorgio Festa to examine the now celebrated monk; and like Drs. Bignami and Romanelli, he asserted that Padre Pio's wounds were authentic.[31] By now, several medical people had observed them oozing blood without external manipulation. Commenting on reports by Bignami and Festa, Herbert Thurston wrote that he knew of no other satisfactory example of a male stigmatic since the days of Saint Francis.[32]

In 1920 Pope Benedict XV sent his own physician, Professor Bastianelli, to San Giovanni Rotondo, and once again the authenticity of Padre Pio's stigmata was confirmed. In 1922, after the ascension of

Pope Pius XI, the Holy Office began yet another investigation. Padre Pio was forced to offer mass at a different and unannounced hour each day and was forbidden to bless crowds, show the stigmata, or answer letters from laypersons. He was also deprived of his spiritual advisor, Father Benedetto. All these moves were taken by the church to control the enthusiasm of Padre Pio's admirers and to guard against fraud of any kind by him or his close companions. But in May 1923, a transfer proposed by Church authorities was blocked by the angry demonstrations of citizens of San Giovanni Rotondo, and new orders from Rome permitted the monk to stay at Santa Maria delle Grazie.

On July 24, 1924, the Holy Office officially proclaimed uncertainty about the supernatural origin of Padre Pio's stigmata, partly to quiet the furor caused by his admirers and partly to deemphasize the role of such phenomena in a virtuous life. In 1925, as a result of Padre Pio's appeals, the Hospital of Saint Francis was established in San Giovanni Rotondo. In 1931 the Vatican informed Padre Raffaele, father guardian of Santa Maria delle Grazie, that the public could not have access to Padre Pio because the Holy Office would again investigate his wounds. His priestly rights were not restored until 1933. Between 1933 and 1957, however, Padre Pio's ministry flourished. The saintly priest promoted prayer groups around the world and developed his hospital, which was funded in part by a grant from the United Nations Relief Agency. In the later years of his life, several hundred—sometimes several thousand—people gathered each day to be near him. On May 5, 1957, Pius XII named him director for life of the third order at Santa Maria delle Grazie; and John XXIII made his first apostolic blessing for Padre Pio's work. Nevertheless, Monsignor Carlo Maccari came to San Giovanni Rotondo in 1960 to conduct yet another investigation of the famous monk.

Padre Pio died on September 23, 1968. Noting the church's long-standing caution about Padre Pio's stigmata and sanctity, Thurston wrote:

> the Roman authorities, guided by the experience of many centuries, are wisely disdainful of abnormal favours of the psychophysical order in which hysteria and other pathological causes, or even fraudulent simulation, may at any time play a part. The Church never canonizes any of her children in their lifetime, and even after death she does not accept such manifestations, however well-grounded may be the belief in their supernatural origin, as the sole or principal foundation for her favourable judgement.[33]

Theresa Neumann

Theresa Neumann of Konnersreuth, whose father was a farmer and part-time tailor, exhibited the stigmata from 1926 until her death in 1962. She was studied by churchmen and medical people for some 35 years in the relentless manner that characterized the scrutiny of Padre Pio. Like the Italian friar, she became the focus of both immense devotion and relentless inquiry. Unlike Padre Pio, though, she suffered from physical afflictions and "holy anorexia" in the manner of bedridden "victim souls" such as Louis Lateau. She is reputed by some to have lived on little more than Communion bread for the last 30 years of her life.

In the years preceding the appearance of her stigmata, she was relieved of several neurotic afflictions, including temporary blindness, paralysis of her back, and apparent appendicitis. But as she got well she predicted that she would eventually suffer "in ways no doctor could cope with." During the Lenten season of 1926, she received five of the stigmata, first on her side, then on her hands and feet, sometimes in the midst of religious visions. In November 1926 she began to feel Christ's crown of thorns, and two weeks later eight pronounced wounds on her head started bleeding. On Good Friday, 1927, the wounds on her hands and feet appeared to work their way through to the palms and soles, while her eyes bled until they were swollen shut and encrusted. In 1929 a lesion appeared on her shoulder in the place she imagined Christ's wound to be from carrying the cross to Golgotha. For the rest of her life she suffered intermittently from one or more of these stigmata. The consistency and volume of testimony regarding Theresa Neumann's wounds establishes their authenticity beyond reasonable doubt. If she sometimes manipulated her stigmata, as some skeptics maintained, she also frequently experienced spontaneous bleeding, according to numerous witnesses.[34]

Less-Celebrated Stigmatics

Marie Rose Ferron, born in 1902 near Quebec, moved with her Catholic family to Woonsocket, Rhode Island, in 1925. Having been a sickly child, she was bedridden and partially paralyzed from the time she moved until her death in 1936. Like other stigmatics, she ate little besides Communion bread and was often absorbed in prayer. In 1926 marks representing the wounds of Christ's flagellation appeared on her arms; in Lent of 1927 stigmata formed on her hands and feet; and in January 1928 punctures signifying Christ's crown of thorns began to bleed on her forehead. Though all these marks disappeared in 1931, a number of devotees were attracted by her extraordinary afflictions and deep spirituality. Many Catholics still call her a saint.[35]

Arthur Otto Moock, a resident of Hamburg, Germany, exhibited wounds in his hands, feet, and side that bled profusely every four weeks or so from 1933 until 1956. Moock, who was not a Catholic, claimed he was not particularly religious, but he suffered greatly from his affliction. At his request several doctors tried to cure him, but without success. According to a priest who took interest in him,

> [He] apparently lacks the deep interior life and strong faith of other stigmatics. He is hardly a willing ascetic [and] has not tried to profit materially or spiritually from his condition by passing it off as supernatural or submitting religiously to it. He has discouraged the procession of the pious that have come to him by being utterly candid about his unusual illness and unorthodox beliefs.[36]

The *Southern Medical Journal* for November 1980 included an account of a 23-year-old Mexican-American female stigmatic. The sixth of 12 children, she had been raised a Roman Catholic and claimed to have seen Christ placing a crown of thorns on her head in visions and dreams. Following her father's example, she quit the church to join a local Pentecostal sect, where she "experienced the Holy Spirit" and began praying constantly, sometimes asking to receive the stigmata. Wounds began to form on her hands some six months after she was married, while she was singing in church. Subsequently, new wounds appeared on her feet, head, and back. Though they opened on a few occasions, they usually bled through unbroken skin. According to the two physicians who studied her, the woman gave birth to a female child who subsequently exhibited stigmata on her hands, feet, and head. In their report, the two doctors described the results of various tests—among them studies of blood, EEG, pulse, respiration, and galvanic skin response—to which they submitted the mother. Their firsthand observations convinced them that neither the mother's nor child's stigmata were produced by external manipulations.[37]

In the *Archives of General Psychiatry* for February 1974, a pediatrician Loretta Early and a psychiatrist Joseph Lifschutz described a 10-year-old black Baptist girl who exhibited stigmata over a three-week period preceding Easter Sunday 1972. "The closest possible scrutiny," they wrote, "made it highly unlikely that [her] lesions were self-induced." The child was intensely religious, came from a large, lower-middle-class family in Oakland, California, and appeared to be physically normal. The two doctors wrote:

> No history was elicited for the patient and her family of prolonged bleeding, easy bruising, spontaneous bleeding, or psychiatric disorder. She had always been in excellent health and never had a serious illness or accident.
>
> The initial physical examination by [Dr. Early] revealed about 1 ml. of dried blood in the patient's left palm, the bleeding having recurred at school about ten minutes earlier. When the blood was washed away there was no cutaneous nor mucosal membrane lesion. Over the next five days the bleeding sites were examined using a five and ten power magnifying lens, revealing normal skin.
>
> She bled from the right palm on the fourth day, from the dorsum of the left foot on the sixth day, from the dorsum of the right foot and the right thorax on the seventh day, and from the middle of her forehead on the 14th day, seven days before Easter Sunday. For a total of 19 days, various persons and the patient reported bleeding from these sites, usually one to five times daily, but with the frequency decreasing to once every two days. [These bleedings] were observed by her school teachers, the school nurse, nurse's assistant, physician, and on one occasion other hospital staff.[38]

According to Early and Lifschutz, their subject experienced auditory hallucinations, usually during bedtime prayers, which consisted of simple statements such as "your prayers will be answered." Her dreams frequently included biblical events, some involving Christ. In the week before her bleeding began, she had read a book and had watched a television movie about the Crucifixion. She denied having knowledge of stigmata other than Christ's, but identified strongly with Saint Francis of Assisi when told about him.

Stigmata and Extraordinary Functioning

These recent cases and others like them reported in medical journals and the press confirm Thurston's opinion that bodily marks resembling Christ's wounds are exhibited more frequently than is commonly realized, even among people who are not mystically gifted or especially virtuous.[†] Many kinds of people, it seems, experience religious stigmata. Some have suffered from neurotic afflictions, while others have been active and hearty like Padre Pio, who lived to 81 and was said to have drunk one to three tumblers of wine at lunch and a bottle of beer each evening.[39] Some have been chronically bedridden, while others have served the sick, counseled pilgrims, or ministered to the poor. With all their differences, however, stigmatics as a group teach us something about extraordinary functioning in general. They show us, for example, that:

Pioneers can inspire novel forms of psychophysical transformation. Once news of Saint Francis's wounds spread through Europe, other people exhibited them too. According to Imbert-Gourbeyre's survey, there were records of at least another 31 stigmatics by the end of the thirteenth century; and since then many more have experienced similar wounds in spite of frequent admonitions by the church that heroic asceticism is to be admired but not emulated. There is a parallel here with other kinds of extraordinary human experience. When Roger Bannister broke the four-minute barrier in the mile, John Landy did it five weeks later, and many soon followed them. After Nijinsky perfected the ballon, he made such buoyant leaps not only thinkable but also mandatory for male ballet dancers. New kinds of physical capacity—even some that seem impossible—begin to spread after someone dramatizes them.

Compelling images help produce specific bodily changes. Ecstatics in non-Christian religions rarely if ever suffer stigmata that resemble the wounds of Christ. Every tradition has its physical expressions of choice, as it were, such as the stigmata of Roman Catholicism, the (stigmatic) battle wounds of Mohammed in Islam,[40] or the control of involuntary functions in hatha yoga.

A passionate desire to embody an ideal can facilitate great changes in cognitive, emotional, and physical functioning. Most stigmatics have wanted to share Christ's ordeals, often praying that they might experience his suffering. Indeed, some have intensified their identification with Christ by deliberately wounding themselves. According to her

---

[†] Another stigmatic, Ethel Chapman, an English member of the Anglican church who lived from 1921 until 1980, is described in Harrison, T., 1981; and a Polish Catholic woman living in Australia who bled from her eyelids during Friday ecstasies is described in Whitlock & Hynes 1978.

biographer, Lukardis, a nun of Oberweimar who died in 1309, "would strike violently the place of the wounds in each palm; delivering another fierce blow in the same spot, the tip of her finger seeming somehow to be pointed like a nail." She persisted in this practice for two years, until she had a vision during which a beautiful stigmatized youth pressed her right hand against his, saying, "I wish you to suffer with me." In that instant, a wound formed in her right hand, and in the weeks that followed, her left hand and feet exhibited stigmata. Her wounds then bled regularly on Fridays without external manipulation. Lukardis's behavior suggests that an autonomous process of stigmatization can be triggered by purely physical means as well as by imagery and ardent aspiration.[†]

Mixed motives in religious practice can produce mixed results. If yearning for union with God is partly motivated by guilt or unconscious aggressions, it might produce forms of suffering such as those exhibited by Louise Lateau and the other "victim souls" noted in this chapter.[‡] Internal conflicts and unassimilated traumas cannot by themselves produce—and may eventually subvert—metanormal capacities. Though they help drive some ecstatics toward ascetic discipline, morbid motives are poor substitutes for self-awareness, balanced practice, and healthy intentions in achieving new levels and kinds of functioning.

Deeply absorbed states of mind can facilitate significant bodily changes. Lukardis of Oberweimar received her first wound in trance, if we can believe her biographer. Saint Francis was stigmatized during his vision of a wounded seraph. And in more recent times, Louise Lateau, Marie-Julie Jahenny, Gemma Galgani, Theresa Neumann, and Padre Pio experienced their first stigmata in rapturous states. Similarly, bodily changes produced by athletic and martial-arts training, hypnosis, and other contemporary practices are facilitated by

---

[†] Thurston 1952, pp. 41–43. It is possible, too, that some stigmatics have manipulated old wounds to preserve them, either to maintain their symbolic nearness to God or to keep their devoted following. Theresa Neumann was accused of doing just that by some of her investigators. Evidence of such manipulation, though, does not disprove the authentic spontaneity of an ecstatic's original stigmata. Even her worst critics did not call Theresa Neumann a total fraud, as her wounds sometimes bled spontaneously in the presence of many witnesses.

[‡] Psychiatrist Richard Lord suggested that neurotic motives play a role in the formation of some religious stigmata, among them a desire to avoid menstruation by suffering periodic wounding, an urge to punish oneself for masturbatory wishes, and a longing to identify with a nonsexual lover (Lord 1957).

absorbed states of mind (15.6, 19.5, 20.1, 20.3). Images, it seems, often work upon the flesh most effectively in states of deep mental absorption.[†]

Certain transformative processes have their own momentum and dynamics. The bodily changes experienced by certain stigmatics resemble the kundalini of Hindu-Buddhist lore in their dramatic, sometimes sudden, unexpected appearance and in their seemingly autonomous progression. In spite of their great differences, some kundalini experiences and Christian stigmatizations seem to have a direction and purpose of their own once they have started, regardless of protests and struggles against them.

Various Hindu and Buddhist schools, for example, maintain that human nature must be balanced to assimilate kundalini, which has a largely autonomous dynamism that will either reorganize the body at a higher level of functioning or damage it in various ways.[41] Similarly, stigmatization among Christian ecstatics seems to occur with its own momentum and typical progression, impressing those who witness it by the suffering it causes and attendant phenomena such as the *incendium amoris* (see 22.6).[‡] Both stigmatization and kundalini are analogous to human birth in that they cause a tearing and buffeting of the body as they produce a new form of life.[42]

[†] Dr. Alfred Lechler described his production of stigmata by suggestive hypnosis in a pamphlet *Das Ratsel von Konnersreuth* published in Elberfeld in 1933 (cited by Thurston 1952, pp. 204, 207–208). Lechler's subject was an Austrian peasant girl. When she saw a movie of Christ's passion that left her with pain in her hands and feet, Lechler placed her in trance and suggested that she had been pierced by nails in the manner of the Crucifixion. After several hypnotic sessions, bleeding wounds appeared on her extremities, which Lechler photographed. Thurston wrote:

> Subsequently, [Lechler] induced a condition in which tears of blood streamed freely from the eyes and in which bleeding punctures appeared on the forehead corresponding to the crown of thorns. There also supervened an inflamed condition of the shoulder caused by her imaginary carrying of the cross. [Lechler's] photographs have every appearance of being a trustworthy record of the results obtained.

[‡] Thurston observed that stigmata appear in certain typical patterns, often in one extremity first, then the others, then on the head as the crown of thorns or on a finger as a token of espousal. Accounts of religious stigmata by medieval chroniclers have great value, he wrote,

> because on the one hand they were written down at a period when no tradition could yet have been formed as to the course of development followed by the phenomena, and on the other because they never attained any notable publicity. Indeed, it may be said that these details have only been printed and given to the world within quite recent times. Nevertheless, they accurately agree in their more general features with the description furnished by numerous witnesses of the manifestations common among the more remarkable stigmatics of the seventeenth and eighteenth centuries, and still frequently to be met with in our own day (Thurston 1952, p. 62).

Anne Emmerich, Louise Lateau, Gemma Galgani, and Theresa Neumann were swept beyond their own expectations and the reinforcements of their family and friends, often against their will and protestations. At times, it appeared they were carried by a process with a mind of its own. Cultural influences determined some elements of their experience but did not prepare them entirely, it seems, for the overwhelming, rapturous, surprising, sometimes devastating aspects of it. When ecstatics assert that such experience is given by God, they express a fundamental truth in the language of Christian faith, namely that they are filled with powers, knowings, and joys from a something beyond ordinary consciousness and cultural conditioning. However:

Stigmatics typically participate, whether consciously or unconsciously, in their transformations. Marie-Julie Jahenny correctly predicted that she would receive a new stigmatization consisting of a cross, a flower, and the words O Crux Ave upon her breast.[43] Like hypnotized subjects who exhibit skin-writing, sometimes at the precise moment their hypnotist specifies, it is conceivable that Jahenny unconsciously produced her stigmata on the day she predicted, even though she believed they were given by God. Self-control is sometimes evident, too, in the reductions of stigmata through prayer. Anne Emmerich, Gemma Galgani, Theresa Neumann, and Padre Pio experienced such remissions. If they subliminally influence their wounds' formation, we can suppose that some stigmatics also effect their reduction. Thurston wrote:

> This matter of concealment seems in a vast number of instances to have had a decisive influence upon the appearance, disappearance, and development of the stigmata. Many of the holiest of such mystics, finding that the wounds upon their hands could not be hidden and that they attracted general respect and veneration, prayed earnestly to God to be freed from such a snare to their humility. They asked that they might still share the pain which their crucified Savior felt in His hands and feet, without any external manifestation of their privileged condition. As a result we find that in a considerable number of cases the woundmarks disappeared within a short time of their first infliction and before the attention of anyone, except perhaps the confessor, the superior, or some trusted confidant had been directed to the matter.[44]

The idiosyncratic forms of some stigmata also indicate their recipients' subliminal production of them. As we have seen, the Y-shaped cross on the breast of Anne Emmerich resembled a prominent cross before which she had prayed as a child. Marks on Gemma Galgani's shoulders corresponded in size and position to the lash-marks on a crucifix familiar to her. While most stigmata correspond to images prevalent throughout Western Christendom, Anne Emmerich's cross and Gemma Galgani's wounds were not only culture specific but also locality specific: they resembled images peculiar to the two stigmatics' immediate environments. The psychophysical transformations associated with ecstatic experience, it seems, have universal, cultural, and completely personal features.

Some Christian ecstatics, in short, demonstrate a precise though subliminal self-control, exhibiting transformative self-suggestion in the guise of God's intercession. Given our contemporary knowledge about hypnotic responsiveness, we can suppose that most stigmatics are profoundly suggestible and would be exceptional hypnotic subjects: Theodore Barber's "fantasy-prone personalities," who were hypnotic virtuosos (15.6), exhibited a psychosomatic plasticity like the saints and mystics described here. And there are correspondences, too, between certain stigmatics and people who suffer from multiple-personality disorder. Theresa Neumann, for example, often passed into a condition that one of her observers called "the state of exalted repose," during which she assumed a strange voice and unusual postures to pronounce upon various subjects. Elizabeth of Herkenrode, Domenica Lazzari, Costante Maria Castreca, and Beatrice Mary of Jesus sometimes spoke or behaved in ways they could not remember later.[45] Like multiples today, these ecstatics employed trance to release capacities they normally suppressed, allowing themselves to do or feel things their dominant self-images would not allow (11.3). While attributing their trances and charisms to the grace of God, they participated to some degree, at least at subliminal levels, in their remarkable but sometimes grotesque transformations.

[22.4]
INEDIA

Long abstinence from food and drink, or inedia, is a principal Catholic charism and has been attributed to countless men and women since Christian antiquity. Saint Lidwina of Schiedam (d. 1433), it is alleged, ate nothing for 28 years; Venerable Domenica dal Paradiso (d. 1553) for 20 years; Blessed Nicholas Von Flue (d. 1487) for 19 years; Blessed Elizabeth von Reute (d. 1420) for 15 years; and Louise Lateau (d. 1883) for 12 years.[46] The historians Caroline Bynum and Rudolph Bell have documented such claims, as well as the intermittent bingeing, or bulemia, that typically accompanies heroic fasting. Bell, for example, traced the stages by which Saint Catherine of Siena gave up normal sustenance. In her late teens she lived—intermittently—on bread, water, and raw vegetables. At about age 23 she gave up bread, surviving on Communion wafers, cold water, and bitter herbs that she either sucked and spit out or swallowed and vomited. In January 1380, when she was about 33 years old, she abstained from water for a month in expiation for a crisis of the church in Italy.[47] Her biographers described her restless energy and sleeplessness, which increased as she ate less and less. According to Bynum, "One might interpret as binge-eating her pattern of long fasts followed by eating several times a day or her practice of forcing herself first to eat and then to vomit."[48]

Bynum and Bell have described other "holy anorexics" who intermittently fasted and feasted.[49] Inedia, it seems, has typically been punctuated by eating binges. Nevertheless, it is indisputable that many Catholic men and women have gone without food, and sometimes without drink as well, for long periods, sometimes without loss of energy. This fact has been established by several ecclesiastical and medical groups that have studied heroic fasting with care, sometimes submitting their subjects to around-the-clock observation and chemical analysis of their urine, blood, and vomit. Here I will describe three women who were studied in this manner.

Louise Lateau, the Belgian stigmatic described above, did manual labor for a few years after her stigmatization, but lost strength and appetite as her ecstatic states developed. According to her family and confessors, she could not digest anything without acute suffering after March 30, 1871, often vomiting when she was forced to eat. As we have seen, the Belgian Academy of Medicine had formed a commission to study her stigmata, but by 1876 it was also involved in acrimonious debates about the reality of her fasting.[50] The academy did not dispute claims that she went without food, however, as no evidence of fraud could be produced against her.

More recently, Theresa Neumann was alleged to have gone without food except for Communion bread for periods of several years, prompting doctors and priests to study her carefully and cross-examine her confidants. In 1927 the bishop of Regensburg appointed a commission to observe her. According to Thurston's account:

Four nursing sisters of Mallersdorf were chosen for the purpose, and a very strict code of regulations was drafted, to the observance of which they were required to bind themselves by oath. Relieving each other by pairs, two of the four were to be continually on duty night and day, never allowing the girl during the prescribed fortnight of observation to be out of their sight even for the shortest interval. Her weight, temperature, pulse, etc. were to be frequently taken. All excreta, whether in the process of natural relief, or by the flow of blood from the stigmata, or by vomiting, etc. were to be preserved, weighed, and subsequently submitted for analysis. Her room, clothes, bed, etc. were subjected to a thorough search, and she was always to be under close observation in her intercourse with her parents, family and all other persons. It cannot be questioned that these precautions were strictly necessary if any conclusion was to be reached which would be respected by those—mainly non-Catholics—who declared her to be a vulgar impostor.

The fortnight's observation of Theresa Neumann has proved to the satisfaction of all unprejudiced persons that she did not during that period take either food or drink. What is even more striking, the pronounced loss of weight which occurred during the Friday ecstasies was in each case made good during the two or three days which followed. On Wednesday, July 13, 1927, the day before the period of observation began, Theresa weighed 55 kilograms (= 121 lbs.); on Saturday, July 16, she weighed 51 kilograms (= 112 lbs.). On Wednesday, July 20, 54 kilograms (= 119 lbs.) were recorded, but this again had fallen by the following Saturday to 52 kilograms (= 115 lbs.), though on the Thursday, the final day, it stood once more at 55 kilograms, just as before the experiment. The extreme range of loss and gain was therefore about 8 lbs. It is curious that on two occasions within the fortnight (the 15th and 22nd) there is record of natural relief to the amount of half a litre. There was also on the two Fridays some vomiting, not very considerable in amount, which seems to have been due to the blood from her eyes or forehead [of her stigmata] running into her mouth. No trace of food was discoverable in the matter thus ejected.[51]

Alexandrina da Costa (1904–1955) lived in the town of Balasar near Oporto, Portugal. Crippled at 14 when she jumped from a window to escape a sexual assault, she was bedridden for much of her life and developed a passionate religious devotion. On Fridays she experienced Christ's crucifixion in trance, often rising from bed to kneel or prostrate herself, overcoming her paralysis to some degree. As she grew older, her raptures, healing influence, and apparent sanctity attracted pilgrims from Europe and the Americas. According to her confidants and confessors, she ate and drank nothing but the bread and wine of daily Communion for the last 13 years of her life; and like Louise Lateau and Theresa Neumann, she was studied by skeptical ecclesiastical and medical groups. At the conclusion of one investigation, the directing physician, a Dr. Gomez de Araujo of the Royal Academy of Medicine in Madrid, issued a formal statement in which he said, "It is absolutely certain that during forty days of being bedridden in [our] hospital, the sick woman did not eat or drink." Nurses and attending physicians also testified to Alexandrina's total abstinence from food and drink. Dr. Araujo's medical report was accompanied by a certificate with this declaration.

> We the undersigned, Dr. C. A. di Lima, Professor of the Faculty of Medicine of Oporto and Dr. E. A. D. de Azevedo, doctor graduate of the same Faculty, having examined Alexandrina Maria da Costa, aged 39, born and resident at Balasar, testify that the bedridden woman, from 10 June to 20 July 1943, remained in the sector for infantile paralysis at the Hospital of Foce del Duro, under the direction of Dr. Araujo and under the day and night surveillance by impartial persons desirous of discovering the truth of her fast. Her abstinence from solids and liquids was *absolute* during all that time. We testify also that she retained her weight; (that) her temperature, breathing, blood pressure, pulse and blood were normal while her mental faculties were constant and lucid; and she had not, during these forty days, any natural necessities.
>
> The examination of the blood, made three weeks after her arrival in the hospital, is attached to this certificate and from it one sees how, considering the aforesaid abstinence from solids and liquids, science has no explanation. For the sake of truth, we have prepared this certificate which we sign. Oporto, 26 July, 1943.[52]

Conceivably, the body has access to superordinary energies that can be triggered by religious passion. Though most people who have fasted for long periods have been sedentary or bedridden, some have been physically active. Heroic fasting suggests that the body can reconstitute its elements in extraordinary ways, dramatically altering its habitual physiological activity.

[22.5]
LUMINOUS PHENOMENA

Whereas stigmata that resemble the wounds of Christ have rarely appeared among members of non-Christian cultures (or even among Eastern Orthodox Christians), other phenomena associated with religious devotion are reported in most religions. Among these are the luminosities exhibited by Christian saints, Sufis, Taoist sages, Hindu yogis, and Buddhist mystics. Stories of saintly radiance are told in virtually all sacred traditions. Thurston wrote:

> There are so many stories of holy priests who lit up a dark cell or a whole chapel by the light which streamed from them or upon them, that I am strongly inclined to adhere to the more literal interpretation [of them]. For example, we read of the fourteenth-century Carthusian, John Tornerius, then at the Grande Chartreuse near Grenoble, that when he did not arrive in time to celebrate his first Mass, the sacristan went to his cell to fetch him, and found that the little room was radiant with light which seemed to be diffused all round the good Father. Similarly, in the process of beatification of the Franciscan Observant, Blessed Thomas da Cori, witnesses stated that the whole church on a dark morning was lit up by the radiance which glowed in the Father's countenance. Further, we learn from what is seemingly the earliest account preserved to us of Blessed Giles of Assisi, that in the night time on one occasion "so great a light shone round him that the light of the moon was wholly eclipsed thereby." So again, that the house of Blessed Aleidis of Scarbeke seemed to be on fire when she, with a radiant countenance, was praying within; or . . . that the cell of St. Lewis Bertrand "appeared as if the whole room was illuminated with the most powerful lamps."[53]

The Blessed Bernardino Realini; Father Francis Suarez, the famous theologian; Saint Lidwina of Schiedam; Saint Philip Neri; and other Catholic religious figures were said to have shone with a similar radiance.[54] More recently, luminous phenomena were attributed to Theresa Neumann. On May 17, 1927, for example, she had an ecstatic vision of Saint Therese of Lisieux. According to a biographer, Albert Schimberg:

> All at once a number of bystanders exclaimed that the stigma of the left hand shone brightly. Father Naber, who did not see the phenomenon, had asked the Konnersreuth village teacher to take a picture of Therese. When the photographic plate was developed, there appeared on it a bright, strong light, an aura as it were, about the left-hand stigma.[55]

Pope Benedict XIV, in establishing criteria for sanctity, drew upon the witness of many doctors and scholars when he wrote:

It seems to be a fact that there are natural flames which at times visibly encircle the human head, and also that from a man's whole person fire may on occasion radiate naturally, not, however, like a flame which streams upwards, but rather in the form of sparks which are given off all round; further, that some people become resplendent with a blaze of light, though this is not inherent in themselves, but attaches rather to their clothes, or the staff or to the spear which they are carrying.[56]

Commenting on Pope Benedict's discussion, Herbert Thurston wrote:

[He] is somewhat chary about admitting isolated cases of such effulgence to be incontestably miraculous, though in view of the recognized holiness of such servants of God as St. Philip Neri, St. Charles Borromeo, St. Ignatius of Loyola, St. Francis of Sales and many more, he does not dispute that the brilliant light which was seen on occasion to surround them when preaching, or when offering the holy sacrifice, was of supernatural origin. It is unquestionably true, as he tells us, that there are hundreds of such examples to be found in our hagiographical records, and although a great number of these rest upon quite insufficient testimony, there are others which cannot lightly be set aside.[57]

As noted in chapter 5.1, scintillae, soul-sparks, and other kinds of physical radiance have been attributed to Sufis and other Islamic mystics, to Neo-Platonist contemplatives, and to adepts of other religions. Luminosities are described with enough consistency from culture to culture to suggest that they actually happen.[58] Like inedia and the *incendium amoris* (see below), the luminous phenomena of Catholic sanctity give evidence that human nature has access to energies that are as yet inexplicable to mainstream science.

[22.6]

INCENDIUM AMORIS

In chapter 5.4, I compare the *incendium amoris* of Catholic sanctity with manifestations of extraordinary vitality in other cultures, among them Tibetan *tumo*, Hindu-Buddhist kundalini phenomena, and the "boiling *n/um*" of the Kalahari bushmen. This charism involves great bodily heat associated with ecstatic devotion. Saint Catherine of Genoa, for example, was said to have warmed the things she touched in extraordinary fashion.[59] The Venerable Serafina di Dio, a Carmelite nun of Capri who died in 1699, warmed those near her, "even in winter time."[60] The Dominican nun Sister Maria Villani of Naples, who died in 1670 at the age of 86, had to drink three gallons of water a day, it was said, to cool her internal fires.[61] And Orsola Benincasa, a sixteenth-century Italian ecstatic, sometimes needed basins of cold water to relieve her ecstasy.[62] Each of these women had a contagious vital force that impressed friends and confessors. Like Saint Philip Neri, who did not need a shirt in the wintertime and vitalized others by his presence (5.4), they had a prodigious warm-bloodedness that was triggered by their passion and discipline.

[22.7]
ODORS OF SANCTITY AND
HOLY BODY FLUIDS

Since the first centuries of Christian history, people have claimed that the bodies of certain martyrs and saints have an extraordinary fragrance. It was said that Saint Polycarp, who was martyred in 155; Saint Simeon Stylites, the fifth-century anchorite; Saint Guthlac, an Anglo-Saxon hermit; and other notable figures of the early church filled the air with sweetness, sometimes imbuing entire buildings with the smell of the "herb ambrosia."[63] Saint Teresa of Avila, a skeptic about many religious claims, believed that holy living could produce an odor of sanctity. In her *Book of Foundations,* she described a famous Spanish ascetic, Catalina de Cardona:

> All the nuns assured me that there was such a fragrance about her, like that of relics, that it clung even to her habit and her girdle, which, as they had taken them from her and given her fresh ones, she left behind. The fragrance was so sweet that it moved them to praise Our Lord. And the nearer they came to her, the sweeter it was, though her dress was of such a kind that, in that heat, which was very severe, one would have expected its odour to be offensive.[64]

Saint Catherine de Ricci was also known for her perfumelike fragrance. During her canonization proceedings, wrote Herbert Thurston, some 20 or 30 nuns in her convent at Prato swore under oath "to the strange odor which was especially noticeable in the chamber of death, although some of them had also perceived a similar perfume clinging to her on certain occasions in her lifetime. Some of [them] described it as resembling the scent of . . . a species of violet."[65] And

Saint Veronica Giuliani was said to have exuded such a scent from her stigmata. Quoting depositions from her canonization proceedings, a biographer wrote:

> When the wounds were open, they emitted so delicious a fragrance throughout the whole of the convent that this alone was sufficient to inform the nuns whenever the stigmata had been renewed, and on several occasions the religious were convinced by ocular demonstration that they had not been deceived. When the bandages which had been applied to these mysterious wounds were put away, they communicated the same sweet perfume to everything near them. The fact is attested by her confidant, the Blessed Florida Ceoli.[66]

Such fragrances, it has been alleged, sometimes spread from one part of the body to another, and from things the holy person touches to other objects. For example, upon the finger of Sister Giovanna Maria della Croce of Roveredo (d. 1673) there appeared a stigmatic ring signifying her marriage to Christ, which according to several witnesses emitted a powerful perfume. A biographer wrote:

> Thus it happened that Sister Mary Ursula having touched that finger in the holy nun's first illness, her hand for several days afterwards retained an exquisite fragrance. This scent was particularly perceptible when Giovanna Maria was ill, because she could not then take any precautions to disguise it. From her finger the perfume extended gradually to the whole hand and then to her body, and communicated itself to all the objects which she touched. . . . It was more powerful when she came back from Communion. It exuded not only from her body but also from her clothes long after she had ceased to wear them, from her straw mattress and from the objects in her room. It spread through the whole house and betrayed her comings and goings and her every movement.[67]

A similar fragrance was attributed to Saint Maria Francesca delle Cinque Piaghe, a Franciscan nun who died at Naples in 1791; to Agnes of Jesus, a Dominican prioress of Langeac who died in 1634; to the Blessed Maria degli Angeli, a Carmelite nun who died at Turin in 1717; and to Sister Mary of Jesus Crucified, a Carmelite nun of Pau, who died at Bethlehem in 1878.[68] More recently, Padre Pio was reputed to cause mysterious fragrances around people who had seen or touched him.[69] Assessing accounts regarding odors of sanctity, Thurston wrote:

> Now while, of course, we are bound to recognize that one or other of these descriptions may owe something to the fervent imagination of a

single reporter, writing possibly under deep emotional stimulus, still the accord among these witnesses, so widely separated in place and time, is not a little remarkable, and, what is perhaps more striking, there is a consensus of testimony as to the occurrence of similar manifestations in recent centuries which cannot be ignored . . . what lends a certain confirmation to the accounts referred to is the occurrence of facts of a similar nature among spiritualistic phenomena.[70]

Thurston quoted a detailed account by Stainton Moses, the English medium, of odors that appeared spontaneously during his seances. Moses wrote:

> It is only now that we are beginning to understand the phenomena of mediumship, which showed themselves among the monks, nuns, and recluses of the middle ages. They were in many cases powerful mediums [and] the odor of sanctity became a well-known occurrence among them. . . . It was a phenomenon of mediumship which was rife then, and which exists now, perhaps more frequently than we know.

Odors of sanctity, like other charisms, are called *gratiae gratis datae* ("things freely bestowed"), because they are not typically sought or expected and because they appear to be given by agencies beyond the self. In certain cases, they arise without any apparent source and are attributed to a distant or deceased saint, angels, or God.[71] In chapter 9, I mention speculations by contemporary physicists that certain extraordinary phenomena might be projections from a hyperdimensional space-time in which this world is embedded. Might some odors of sanctity be accounted for in this manner? Might they be materializations from a "larger Earth"?

Odors of sanctity are sometimes associated with another charism of Catholic sanctity, the production of holy body fluids. In medieval times it was widely believed that some women exuded a special blood, oil, milk, or saliva. Lidwina of Schiedam, who was famous for her inedia (see above), was said to produce such effluvia.[72] According to Caroline Bynum:

> The theme of exuding appears in several Italian *vitae*, as it does in those from the Low Countries and Germany. Flemish women were more apt to exude milk, Italian and German women oil or manna; but in *vitae* from all regions the theme is clear. Rita of Cascia, for example, who closed her body to ordinary eating, developed a permanently running sore on her forehead (supposedly made by the crown of thorns); after death she exuded a sweet odor. . . . The body of Rose of Viterbo (d. ca. 1252) was found at exhumation to have produced "manna like sweet

smelling oil." Whether oil, milk, or manna, the precious substance exuded by the woman's body was usually seen as curing and feeding.
Moreover, the exuding of extraordinary fluids was accompanied by extraordinary closure. A woman whose body was communicated to others was usually a woman who herself ate only the special food of
God—the eucharist.[73]

Men or women who exuded holy fluids, sometimes called *myroblutai*, or unguent spouters, were described by Greek-speaking
members of the early church. Extraordinary exudations from the
bodies of saints have been reported for almost two millennia.

[22.8]
INCORRUPTION

The cadavers of many Catholic religious have partially or entirely resisted decay, sometimes for centuries. Thurston listed the following
phenomena associated with such incorruption: a fragrance perceived
near the body of the deceased, which sometimes persists for months
or years; bleeding of the cadaver weeks, months, or even years after
death; the exuding of an oily, often fragrant fluid from some cadavers;
and, less frequently, the persistence of warmth in the cadaver after life
has ceased.[74] Reports of these phenomena appeared in the fourth century and continue to the present day.

Drawing on careful studies of saintly cadavers, Thurston and
Joan Carroll Cruz have written extensively about incorruption.[75] As
they point out, the cadavers of ordinary people are sometimes preserved by freezing or other natural causes. Only the high proportion
of well-preserved corpses left by Catholic saints, and their frequent
flexibility, fragrance, and bleeding, make the phenomenon notable.
To study incorruption more carefully, Thurston chose the evidence
contained in the canonization proceedings for those saints who lived
between 1400 and 1900 whose commemoration days are celebrated
by the entire church. By limiting himself to these 42 saints among the
several thousand recognized by Catholic communities, he could draw
upon records more complete than those for obscure religious figures.

Confining ourselves to the saints who have lived within the last five centuries, we find that forty-two of these are included in the Roman Calendar and are honoured in Mass and Office by all priests who follow the

Roman rite. To these forty-two I have confined my little census. This limitation has the conspicuous advantage that we may be quite sure that the mere fact of the body remaining incorrupt has in none of these cases led to the inclusion of the saint in the Calendar. Each one of these forty-two has been chosen as remarkable in some other way, either as the founder of a religious order or as a typical missionary or as a pattern of charity or innocence, etc.[76]

Thurston's inventory is in appendix K.1. Commenting on it, Thurston wrote:

As I have indicated by the letters A, B and C, our data may be divided into three classes. In no less than twenty-two cases (marked A) . . . there is good evidence that the body of the saint was found incorrupt after an interval of time which in normal individuals almost invariably sees the development either of an advanced stage of decomposition or of complete decay. There are seven more cases marked B, in which we have indications of the occurrence of unusual phenomena of a somewhat similar character. Finally, even in the C class, where little or nothing out of the common is recorded, the negative testimony we possess is not always conclusive. . . . No doubt we all have heard or read many stories concerning the discovery of human remains in an incorrupt state when coffins have accidentally been broken open or new graves have been dug. But the proportion of such instances to the thousands of skeletons which under certain systems of burial are constantly being cleared away from cemeteries to make room for other occupants, is extraordinarily small.

If it be contended that the abstemiousness with regard to food and drink characteristic of all such ascetics may profoundly modify the conditions of normal metabolism and tend to eliminate certain classes of microbes which are most active in the process of putrefaction, we may reply that the very poor are of dire necessity abstemious, while no observations point in their case to any similar immunity. Moreover, it ought to follow that when famine reigns in the land, the corpses of its victims should be proof against the agents of corruption, but no recorded experience seems to bear this out.

Like most charisms, incorruptibility is not limited to Roman Catholics. Swami Paramahansa Yogananda, who taught yoga in the West from the early 1920s until his death in 1952, was by many accounts an attractive and spiritually experienced teacher. According to the following notarized statement by Harry Rowe, mortuary director at Forest Lawn Memorial Park in Los Angeles:

Officials of Forest Lawn viewed the body of Paramahansa Yogananda an hour after his death on March 7, 1952. The body was then taken to his home on Mount Washington in Los Angeles, where many friends gathered to [see it].

For protection of the public health, embalming is desirable if a dead body is to be exposed for several days to public view. Embalming of the body of Paramahansa Yogananda took place twenty-four hours after his demise. In normal room-temperature, the enzyme action of the intestines of deceased persons causes distention of the tissues in the abdominal region about six hours after death. Such distention did not occur at any time in the case of Paramahansa Yogananda. When our Mortuary received his body for embalming, it presented no signs of physical deterioration and no putrefactive odor—two very unusual absences when a death has occurred twenty-four hours earlier.

Paramahansa Yogananda's body was embalmed on the night of March 8th, with that quantity of fluid which is customarily used in any body of similar size. No unusual treatment was given.

In cases of persons that are embalmed and exhibited to friends for a period of two or three weeks, it is necessary, to insure presentability, for the embalmer to apply, on the face and hands of the deceased, a creamy pore-sealing emulsion that temporarily prevents the outward appearance of mold. In Paramahansa Yogananda's case, however, no emulsions were used. They were superfluous, inasmuch as his tissues underwent no visible transformations.

The physical appearance of Paramahansa Yogananda on March 27th, just before the bronze cover of the casket was put into position, was the same as it had been on March 7th. He looked on March 27th as fresh and as unravaged by decay as he had looked on the night of his death. On March 27th there was no reason to say that his body had suffered any visible physical disintegration at all. For these reasons we state again that the case of Paramahansa Yogananda is unique in our experience.[77]

Describing incorruptibles, Thurston wrote:

Among the Greeks, before the Schism of the Eastern Church, [incorruptibles] were known as *myroblutai* (unguent spouters). One [was] St. Walburga, an Englishwoman by birth [who] became Abbess of Heidenheim, where she died in 779. Although there is no record of her body having been preserved entire, still from her bones . . . an oily fluid has trickled for more than a thousand years, and the phenomenon continues to this day.[78]

The relics of Saint Gerard Majella, who died in 1756, also exuded a mysterious substance. In a life of the saint, we read:

> The ecclesiastical authorities having ordered that all the relics of the servant of God should be officially examined, Gerard's tomb was opened for the first time on June 26th, 1856. It was then noticed that a mysterious oil oozed forth in such abundance from the brain and bones as to fill up more than one basin. . . . On Oct. 11th the body of the saint was again examined in the presence of two doctors. They found the bones more or less damp, but as this could be attributed to the humidity of the soil, it attracted but little attention. They were dried with all due care, and then placed in a chest lined with white silk. Four hours later the chest having been opened, it was discovered that a kind of white oil, shedding a sweet fragrance, was coming forth anew from the holy relics, and resting like drops of dew upon the silk lining.[79]

And during the disinterment of Saint Hugh of Lincoln in 1887:

> The body of the holy prelate, although it had been deposited there for well nigh eighty years, was found incorrupt and almost unchanged. As soon the Archbishop laid his hand on the glorious head of the Saint, it separated from the shoulders, leaving the neck fresh and red just as if death had been recent.
>
> In the tomb where the body had rested there was found a great quantity of pure oil. . . . On the following morning in the course of the ceremony it happened that the Bishop of Lincoln took up the head of St. Hugh and held it for a while reverently before him. As he did this an abundance of the same pure oil flowed from the jaw over the Bishop's hands, and this notwithstanding that the venerable head had been carefully washed a few hours before and had been found quite dry in the morning. The oil only ceased to flow when the Bishop had placed his previous burden upon the silver dish upon which this relic was to be borne in procession.[80]

Thurston and Cruz recount other stories of saintly cadavers that reputedly bled or exuded fragrant oils, pointing to the consistency and long history of such accounts in Christian tradition. The corpse of Saint Catherine of Bologna, for example, bled profusely three months after death; the corpse of Saint Peter Regelatus, a Franciscan who died in 1456, was exhumed in 1492 and bled profusely when it was cut; and dozens of other saints' cadavers produced fresh blood many years after their burial.[81] Like the emission of fragrant oil, blood prodigies of this kind have been noted from the first centuries of Christian history.

The practice of embalming has shown that most bodies buried like those in Thurston's inventory soon putrefy. According to a standard textbook, *The Principles and Practices of Embalming* by Clarence Strub, green and purple staining typically appears across the abdomen two or three days after death, and within three weeks organs and cavities begin to burst so that disfiguration is extreme. Four weeks after death, a corpse buried in the ground has usually given way to "slimy liquefaction and disruption of all soft tissues."[82] Modern medicine has clearly specified the patterns of postmortem decomposition, and thus (indirectly) confirms the extraordinary nature of incorruption. But this phenomenon has not attracted the attention of medical science. Like other unusual phenomena noted in this book, incorruption has not been systematically studied. This neglect is regrettable, given the fact that such study might yield important discoveries about the body's transformative capacities. Nevertheless, we can speculate about its causes. Arguably, incorruption provides more evidence that consciousness tends to conform the flesh to its particular qualities. If chronic anger and fear can damage physiological processes, and happiness can improve the immune system, the radiant vitality evident in most saints whose bodies did not decompose might have a preserving effect. Conceivably, unitive states of mind impart a powerful integrity to the bodies of incorruptibles. As they contemplate the divine beauty, certain saints might embody it to some degree through the lasting suppleness of their muscle and skin. As it apprehends eternity, human consciousness might transmit a self-sustaining permanence to the form it inhabits.[†]

[22.9]
BODILY ELONGATION

"One cannot help suspecting," Herbert Thurston wrote, "that the witnesses who gave evidence in processes of beatification may sometimes have been expecting the manifestations they reported because they were well aware that such things often occurred in the lives of

---

[†] In her book *Holy Feast and Holy Fast,* historian Caroline Bynum reviewed the testimony of monks, nuns, priests, and hagiographers that living saints, too, have exuded oil, milk, and blood after they had "closed off their bodies" through abstinence from food and drink. Bynum 1987, pp. 122–23, 145–46. See also: Bell 1985.

saints. Any indication which seemed to point to stigmatization, or elevation above the ground in prayer, or celestial radiance, or emanations of perfume was likely to be interpreted without discussion as something unquestionably miraculous." For that reason, Thurston regarded certain idiosyncratic bodily changes associated with mystical passion to be especially significant. "It is a matter worthy of notice," he wrote, "when we find some phenomenon recorded of a holy person which is not likely to have suggested itself to religious observers as a mark of sanctity, but which does at the same time hold a recognized place among the manifestations which psychical researchers have recorded in recent times." This is the case with so-called bodily elongations.[83] Drawing from proceedings for the beatification of Sister Veronica Laparelli, a nun who died in 1620 at the age of 83, Thurston quoted a sister Margherita Cortonesi.

> On one occasion, among others, when [Sister Veronica] being in a trance state was reciting her Office alternately with some invisible being, she was observed gradually to stretch out until the length of her throat seemed to be out of all proportion in such a way that she was altogether much taller than usual. We, noticing this strange occurrence, looked to see if she was raised from the ground, but this, so far as our eyes could tell, was not the case. So, to make sure, we took a yard-measure and measured her height, and afterwards when she had come to herself we measured her again, and she was at least a span (ten inches or more) shorter. This we have seen with our own eyes, all of us nuns who were in the chapel.[84]

In 1629, a Donna Hortenzia Ghini stated under oath that

> Sister Lisabetta Pancrazi, formerly a nun in the same convent, told me that on one occasion, seeing that the said Sister Veronica when in ecstasy seemed taller than in her normal state, took a yard-measure and measured her height, and that after the said Sister Veronica came to herself she measured her again with the said yard-measure, and she found that she was half an arm's length shorter.[85]

The *Promotor Fidei* in these proceedings remarked that this elongation was not only improbable but served no purpose of edification or utility. It could not benefit Sister Veronica, he said, and would cause repulsion and alarm rather than devotion in the beholder. Among other religious who allegedly exhibited elongations, the Capuchiness Abbess Costante Maria Castreca was said to have grown a considerable height from the ground during a religious ecstasy; the Venerable Domenica dal Paradiso grew taller in trance, according to her spiritual director and other confidants; and the Dominican nun Stefana Quinzani stretched her left arm considerably beyond its normal length during her reenactments of Christ's Passion.[86] Because such phenomena were not thought to be marks of holiness, they were noted simply because they were grotesque and unusual.

But it seems that increases in height or reach are not the only kind of bodily distortion visited upon Catholic mystics. Marie-Julie Jahenny, the stigmatic noted above, was sometimes distended in trance so that her "body was compressed, her limbs shortened, and her tongue swollen." During one of her ecstasies, her whole frame formed a ball, and her shoulders convulsed so that they formed right angles to her collarbone. These movements were followed by a dilation of the right side of her body from armpit to hip. After observing the episode, Antoine Imbert-Gourbeyre, a pathologist of long experience, said that it did not resemble any condition he was familiar with.[87]

I include such phenomena in this discussion because they indicate the body's responsiveness to altered states of mind. When consciousness is released from some of its ordinary constraints, whether in ecstasies or dissociated states, ligaments and muscles are sometimes liberated too. Though such release can facilitate grotesque behaviors, it also permits creative movement. Many athletes and dancers, for example, claim that their inspired performances are made possible by states of profound mental freedom (see 5.5).†

[22.10]
LEVITATION

There is overwhelming evidence that bleeding stigmata have appeared on the bodies of some ecstatics, and that some religious have gone without any food except Communion bread for many weeks or months. But other charisms are far less plausible. One of these, levitation of the human body, has a long history in Christian lore, however, and has been said to occur among adepts of other religions. There is

---

† Curiously, a modern practice to lengthen the spine is called *elongation*. Originally systematized for dancers by Joseph Pilates in the 1920s, it has roots in tai chi, ballet, and various therapies for somatic alignment, and some of its exercises resemble the odd or grotesque positions called bodily elongations by Catholic authorities on the contemplative life. See *Self* (the popular magazine) September 1991, pages 152–57.

just enough testimony to it by credible witnesses to warrant its inclusion in this inventory, though the case for its actual occurrence is weaker by far than the one for stigmata and inedia.

Saint Teresa of Avila, the famous reformer of the Carmelite order whose writings on mysticism are among the most influential in Christian history, had this to say about levitation in chapter 20 of her autobiography:

> Though rapture brings us delight, the weakness of our nature at first makes us afraid of it, and we need to be resolute and courageous in soul. . . . Occasionally I have been able to make some resistance but at the cost of great exhaustion, for I would feel weary afterwards as though I had been fighting with a powerful giant. At other times, resistance has been impossible: my soul has been borne away, and indeed as a rule my head also, without my being able to prevent it: sometimes my whole body has been affected to the point of being raised up from the ground.[88]

Saint Teresa described her efforts to resist such ecstasies and their physical effects.

> It seemed when I tried to make some resistance, as if a great force beneath my feet lifted me up. I confess that it threw me into great fear, very great indeed at first; for in seeing one's body thus lifted up from the earth, though the spirit draws it upwards after itself (and that with great sweetness, if unresisted), the senses are not lost; at least I was so much myself as to be able to see that I was being lifted up. . . . After the rapture was over, I have to say that my body seemed frequently to be buoyant, as if all weight had departed from it, so much so that now and then I scarcely knew that my feet touched the ground.[89]

There is no doubt that these words were written by Saint Teresa herself: they appear in the facsimile edition of her handwritten copy, exactly as it was submitted to the Inquisition in her lifetime. Furthermore, several witnesses said that they saw Teresa rise from the ground. In a deposition during the saint's beatification proceedings, for example, Sister Anne of the Incarnation stated that

> I was in the choir waiting for the bell to ring when our holy Mother entered and knelt down. . . . As I was looking on, she was raised about half a yard from the ground without her feet touching it. At this I was terrified and she, for her part, was trembling all over. So I moved to

where she was and I put my hands under her feet, over which I remained weeping for something like half an hour while the ecstasy lasted. Then suddenly she sank down and rested on her feet and turning her head round to me she asked me who I was and whether I had been there all the while. I said yes, and then she ordered me under obedience to say nothing of what I had seen, and I have in fact said nothing until the present moment.[90]

According to her biographer and friend Bishop Yepes, Saint Teresa once resisted a rapture during Communion by grabbing the bars of a grill as she rose in the air, crying out for deliverance from her ecstasy. Mother Maria Baptista, a Carmelite nun, said that she saw her raised from the ground on two occasions. And the *Acta Sanctorum* cites 10 separate depositions from the saint's canonization proceedings of witnesses who described similar incidents.[91]

Sister Maria Villani, a famous Dominican nun of the seventeenth century, described her own levitation:

On one occasion when I was in my cell I was conscious of a new experience. I felt myself seized and ravished out of my senses, so powerfully that I found myself lifted up completely by the very soles of my feet, just as the magnet draws up a fragment of iron, but with a gentleness that was marvelous and most delightful. At first I felt much fear, but afterwards I remained in the greatest possible contentment and joy of spirit. Though I was quite beside myself, still, in spite of that, I knew that I was raised some distance above the earth, my whole body being suspended for a considerable space of time. Down to last Christmas eve (1618) this happened to me on five different occasions.[92]

Of the many stories about this implausible phenomenon, the most dramatic concern Saint Joseph of Copertino, a seventeenth-century Franciscan monk. Accounts of Saint Joseph, who was said to have been observed while levitated on more than a hundred occasions, provides the most widely discussed evidence for levitation on record, at least among Christian ecstatics. His case is especially interesting because it was analyzed by Prosper Lambertini, later Pope Benedict XIV, while he was *Promotor Fidei*. Lambertini, who was deeply influenced by the European Enlightenment, established criteria still used today by which to judge the authenticity of religious phenomena (see 22.1). According to Thurston, the skeptical churchman's analyses, or "animadversions" upon the evidence, submitted to the Congregation of Rites on Joseph's behalf were thorough and searching. Any doubts

Lambertini may have entertained about Saint Joseph were apparently dispelled, however, for he published the decree of the saint's beatification in 1753, when he was pope, and wrote the following passage in his classic work on canonization.

> While I discharged the office of Promotor of the Faith, the cause of the venerable Servant of God, Joseph of Copertino came up for discussion in the Congregation of Rites, which after my retirement was brought to a favorable conclusion, and in this eye-witnesses of unchallengeable integrity gave evidence of the famous upliftings from the ground and prolonged flights of the aforesaid Servant of God when rapt in ecstasy.[93]

Thurston wrote:

> There can be no doubt that Benedict XIV, a critically minded man who knew the value of evidence and who had studied the original depositions as probably no one else had studied them, believed that the witnesses of St. Joseph's levitations had really seen what they professed to have seen. It is also certain . . . that these witnesses made their depositions upon oath at Osimo, Assisi, and other places in 1665–1666, only two years after the saint's death.[94]

Thurston described other cases of apparent levitation in *The Physical Phenomena of Mysticism,* and assembled a list of levitated ecstatics, which can be found in appendix K.2.

As noted in chapter 5.5, the Transcendental Meditation Society has conducted levitation training courses for many years with no certain success, and Herbert Benson could not find Tibetan lamas who were able to rise from the ground.[95] We have no experimental evidence or photographic records to prove that levitation has actually happened. Still, there is testimony to it by several Catholic saints, by many nuns and priests, and by laypeople such as those examined during the canonization proceedings for Saint Joseph of Copertino. There exist, it seems to me, enough compelling anecdotes about the phenomenon to warrant our considering it a human possibility. Indeed, we can speculate that there is a spectrum of physical movement involving various degrees of levitation. Certain athletes and dancers conceivably, though they do not levitate, are moved to some extent by the same forces that upon occasion lift saints from the ground.

[22.11]
TELEKINESIS

The word *telekinesis* is defined by the *Oxford English Dictionary* as "movement of or in a body, alleged to occur at a distance from, and without material connection with, the motive cause or agent." The phenomenon has been attested to by credible psychical researchers, by shamans and mystics in many cultures, and by athletes and martial artists. Here again, Catholic sanctity provides evidence for a metanormal capacity that has been testified to in different circumstances. For example, the Cure d'Ars, a French priest of the nineteenth century famous for his saintly integrity, described the inexplicable movement of a Communion wafer from his hand to the mouth of a communicant (see 5.6).[96] And a similar incident was described by Raymund of Capua, a general of the Dominican order beatified by Leo XIII in 1899. Raymund was confessor to the celebrated Italian ecstatic Saint Catherine of Sienna and frequently gave her Communion. He described the following events in a long, meticulous account preserved in the *Acta Sanctorum*, April, Volume 3, pp. 940–42. Thurston summarized the incidents as follows.

> In order to gratify her intense desire of receiving Communion, [Raymund of Capua] put on his vestments and proceeded to celebrate the Holy Sacrifice. When he turned round to give the general absolution before Communion he saw her face radiant and transformed with light. He was almost overpowered at the spectacle, and on once more facing towards the altar in order to take up the sacred particle he apostrophized It mentally, saying "Come O Lord, to Thy spouse." "The thought," he goes on, "had hardly framed itself in my mind when, before I touched It, the Sacred Host, as I clearly perceived, moved forward of Itself, the distance of three inches or more, coming close to the paten which I was holding in my hand." Whether it then leaped on to the paten, Raymund tells us he is unable to say.
>
> . . . On another occasion he had been waiting to begin Mass until the Saint, who was ill, could come to the church. Finally receiving a message that she was unable to communicate, he concluded that she had not left her home, and he thereupon offered the Holy Sacrifice believing, erroneously, that she was not present. She was, however, in point of fact, at the extreme end of the church in a place in which he could not, or did not, see her.
>
> "After the consecration and the *Pater Noster*," he wrote, "I proceeded according to the rubrics to divide the Host. At the first fraction,

the Sacred Host, instead of separating into two portions, divided into three, two large and one small, which seemed to me about the length of a bean but not so wide. This particle which I attentively observed, appeared to fall on the corporal by the side of the chalice about which I had broken the Host; I clearly saw it descend towards the altar, but I could not afterwards distinguish it on the corporal. Presuming that it was the whiteness of the corporal which prevented my discerning the particle, I broke off another, and after saying the *Agnus Dei*, consumed the Sacred Host. As soon as my right hand was at liberty, I felt on the corporal for the particle on the spot it had fallen; but I found nothing."

Raymund then gives an account of his profound distress and the fruitless search he instituted on every part of the altar and the floor. As he was giving up the quest in despair, he was interrupted by a visitor desirous of having an immediate interview with St. Catherine.

They found the Saint kneeling with some of her companions at the far end of the building remote from the altar. St. Catherine was in an ecstasy, but, as the need seemed urgent, Raymund persuaded her companions to try to rouse her.

"Mother, I verily believe that it was you who took that consecrated particle," Raymund said.

"Nay, Father," she replied, "do not accuse me of that; it was not I but *Another*; all I can tell you is, you will never find [the particle] again . . . it was brought me by our Divine Lord Himself."[97]

Thurston described other incidents of apparent telekinesis involving Sister Domenica dal Paradiso, a Florentine nun who founded a convent in the sixteenth century; Monsieur Olier, the saintly seventeenth-century founder of Saint Sulpice; and Saint Maria Francesca delle Cinque Piaghe, a nineteenth-century Italian ecstatic. More recently, one of Theresa Neumann's biographers, Fritz Gerlich, claimed that he saw a Communion wafer move inexplicably as it was handed to the stigmatic.[98] According to an account of Theresa's life by the Reverend Charles Carty, several photographs show "without a doubt the Host floating from the fingers of the priest on the way to her."[99] Similar incidents are recounted in biographies of Theresa by Archbishop Jozef Teodorowitz and Kaplan Fahsel.[100]

Like levitation, telekinesis has not been demonstrated in controlled experiments nor documented in motion pictures. Still, there exist many compelling anecdotes about it. Like spiritual healing and other types of telergy, it can be taken to be a metanormal version of abilities to affect the environment directly (see chapter 5.6).

[22.12]
EXTRASENSORY POWERS

In reviewing *Butler's Lives of the Saints* for evidence of extrasensory phenomena, parapsychologist Rhea White found that 676, or 29 percent, of the 2,532 listed saints had reputedly experienced one or more paranormal events.[101] White counted 31 cases of clairvoyance in Butler's collection, 24 cases of telepathy, and 20 instances of "discernment of spirits," the ability to read hearts telepathically. Jean Baptiste-Vianney, the famous Cure d'Ars (*Butler's Lives* 3:290), Veronica of Binasco (*Butler's Lives* 1:313), and Mary of Oignies (*Butler's Lives* 2:625) could learn about the past in an extrasensory manner. In 52 cases, a general statement is made that the saint was endowed with gifts of prophecy: Saint Severinus, for example, left the city of Astura after predicting correctly that it would be destroyed by invading Huns; Daniel the Stylite foretold a large fire in Constantinople; and Saint Mary Magdalen de Pazzi prophesied that Alexander de Medici would be pope. Fifty-five saints were said to have prophesized specific events, and 32 the approximate date of their death.[102]

White suggested that saintly ESP often appears to be under voluntary control. Religious practice seems to facilitate such experience, she wrote, as several saints "become psychic only later in life, after they had been contemplatives for a considerable length of time."[103] White's observation corresponds with a long-standing testimony in other religious traditions that extrasensory powers can be cultivated. Yogis, Zen masters, shamans, and Sufis, it is said, can learn to use telepathy and clairvoyance at will.

[22.13]
IMMOBILITY OF THE LIMBS AND DIMINUTION
OF THE SENSES IN ECSTATIC RAPTURE

That cataleptic trance has been regarded by Catholic authorities to be a chief characteristic of mystical union is evident in Augustin Poulain's *The Graces of Interior Prayer*, perhaps the most thorough account of Catholic contemplative experience produced in the twentieth century. According to Father Poulain, "the eleventh character of the mystic union," namely its action upon the body, is exercised in four ways.

1. The senses cease to act, or they convey a confused knowledge only. As the cessation of action on the part of the senses is complete or *almost* complete, the ecstasy itself is called *complete* or *incomplete*.

2. As a general rule, the limbs become immovable . . .

3. Respiration is almost arrested . . .

4. The *vital heat* seems to disappear, a coldness sets in at the extremities of the limbs.[104]

To support this characterization of mystical union, Father Poulain presented statements from Catholic mystics and saints about the physiological features of ecstasy. For example, Saint Teresa of Avila wrote:

> *When the rapture is at its highest,* [the soul] neither sees nor hears, nor perceives (*Life*, ch. 20).

> Upon ecstasy in a slight degree: (a) The soul . . . is, as it were, utterly fainting away in a kind of trance; . . . it cannot even move the hands without great pain. (b) The eyes . . . if they are open . . . are as if they saw nothing; nor is reading possible,—the very letters seem strange and cannot be distinguished,—the letters indeed are visible, but as the understanding furnishes no help, all reading is impracticable . . . (c) The ears hear; but what is heard is not comprehended. (d) It is useless to try to speak. . . . (e) If the soul is making a meditation on any subject, the memory of it is lost at once, just as if it had never been thought of. If it reads, what is read is not remembered nor dwelt upon; neither is it otherwise with vocal prayer.

In the words of Father Francescus Suarez (1540–1617), a Jesuit theologian and authority on contemplative prayer: "Ecstatics sometimes seem to have neither pulse nor heart-beat. And further it is with difficulty that any remnant of vital heat is detected in them; they have the appearance of death."

And according to Jean Baptise Scaramelli (1687–1752), who was the Jesuit author of *Directorio Mistico*, which Poulain called "the clearest and most complete treatise on mysticism existing":

> the beating of the heart is very feeble and the respiration is so slight that it is very difficult to distinguish it, as is clearly deduced from numerous experiments made with great care upon ecstatic persons . . . the imagination remains drowsy, without calling up any images, and it is the same with the sensitive appetite; this occurs even in the case of the full union.[105]

These passages reflect a widespread Roman Catholic assumption that mystical union inevitably produces sensorimotor immobility. Such assumptions have probably influenced retiring, highly introverted mystics such as Louise Lateau, Anne Emmerich, and Alexandrina of Portugal. Orsola Benincasa, for example, began having trances when she was 10. According to Thurston, these were preceded by violent palpitations that shook her entire frame.

> Those standing by, we are told, could hear the child's heart thumping against her ribs. When the trance itself came on, she stood like a marble statue, impervious to every form of external disturbance. They pricked her with needles and lancets, they pulled her hair, they pinched her and shook her, they scorched her with a naked flame, but though she felt the effects of this rough treatment afterwards, it produced no impression at the time. In her ecstasies at a somewhat more advanced age, we find the most extraordinary statements made regarding her immobility. Twenty men could not shift her position, and although this may be an exaggeration, quite a number of witnesses who gave evidence in the Beatification Process declared that they had often tried to bend her limbs, or to move her, or to take from her something she held in her hand, but were unable to do so. . . . Many of those who had known her longest describe her life as an almost continual ecstasy. She is said on some occasions to have lost consciousness eight times in the course of a single hour.[106]

Many Catholic authorities on the contemplative life share an assumption that the highest forms of mystical union come when the body is immobile, when "it is utterly fainting away," as Saint Teresa put it, when "the doors of the senses are closed against their will, in order that they might have more abundantly the fruition of our Lord." But many contemplatives have different assumptions. In the Bhagavad Gita, Krishna tells Arjuna that he can realize God through action in this world; and karma yogins generally believe that illlumined activity, or waking samadhi, is the greatest spiritual condition. According to a famous Taoist saying, "Meditation in activity is a hundred, a thousand, a million times superior to meditation in repose." And Roman Catholicism, of course, is notable for great exemplars of inspired good works such as Saint Francis of Assisi and Saint Francis Xavier. Indeed, many Catholic men and women have affirmed bodily life by their passionate asceticism. In the words of Caroline Bynum:

Medieval asceticism should not be understood as rooted in dualism, in a radical sense of spirit opposed to or entrapped by body. The extravagant penitential practices of the thirteenth to the fifteenth century, the cultivation of pain and patience, the literalism of *imitatio crucis* are . . . not primarily an attempt to escape from body. They are not the products of an epistemology or psychology or theology that sees soul struggling against its opposite, matter. Therefore they are not—as historians have often suggested—a world-denying, self-hating, decadent response of a society wracked by plague, famine, heresy, war, and ecclesiastical corruption. Rather, late medieval asceticism was an effort to plumb and to realize all the possibilities of the flesh. It was a profound expression of the doctrine of the Incarnation: the doctrine that Christ, by becoming human, saves *all* that the human being is. . . . [Medieval saints] were not rebelling against or torturing their flesh out of guilt over its capabilities so much as using the possibilities of its full sensual and affective range to soar ever closer to God. [107]

23

SCIENTIFIC STUDIES
OF CONTEMPLATIVE
EXPERIENCE

AS WE HAVE SEEN in chapters 13.2 and 22.1, the Catholic church has used scientific procedures during investigations of spiritual healing and contemplative phenomena. Since the early nineteenth century especially, stigmata and other charisms have been examined by churchmen, doctors, and medical researchers who were determined to maintain their objectivity. Critical distance has been evident, too, among students of yoga and shamanism. The stern tests of yogic burial conducted by the Maharaja Runjeet Singh (21.7), for example, and studies of Siberian shamanism by Russian anthropologists (21.3) exemplify the spirit of science. The scientific study of religious adepts can be traced to the early 1800s. In this chapter, I describe the instrumented examination of yogis and Zen monks that began in the 1930s and review the systematic meditation research that has burgeoned since the 1970s.

[23.1]
STUDIES OF YOGIS AND ZEN BUDDHIST MONKS

Studies of Indian Yogis

In 1931 Kovoor Behanan, an Indian graduate student in psychology at Yale, was awarded a Sterling Fellowship to study yoga. Supported in this research by Walter Miles, an eminent professor of psychology,

Behanan wrote a book about yoga that described instrumented studies of his own yogic breathing. During 72 days of experiments at Yale, he found that one breathing exercise, or *pranayama*, increased his oxygen consumption by 24.5 percent, a second by 18.5 percent, and a third by 12 percent.[1] This study helped stimulate interest in meditation research by showing that yoga's physiological effects could be examined in the laboratory.[2] Unlike many tales by travelers to the East, Behanan's straightforward, well-observed account of his laboratory research was free of exaggeration and mystification.

Behanan also studied Indian yogis. He was guided in this work by Swami Kuvalayananda, who promoted yoga research at a center for meditation practice he founded in the 1920s at Lonavla, a hill station near Bombay. Kuvalayananda developed a system of physical culture that included *asanas* and *pranayamas*, and he established a yogic therapy for many afflictions. His work was supported by several Indian states, two provincial governments of British India, Indian health agencies, and American foundations. For many years, the results of his laboratory research were published in a quarterly journal, *Yoga-Mimamsa*, which also provided instruction on postures, breathing exercises, and other disciplines. Many people interested in yoga research visited Lonavla, among them psychologists Basu Bagchi of the University of Michigan Medical Center and M. A. Wenger of UCLA, who gave new impetus to meditation studies in the 1950s. From the 1920s into the 1960s, Swami Kuvalayananda did much to promote the scientific study of yoga.

In 1935 a French cardiologist, Therese Brosse, took an electrocardiograph to India and studied yogis who said they could stop their heart. According to Brosse's published report, readings produced by a single EKG lead and pulse recordings indicated that the heart potentials and pulse of one of her subjects decreased almost to zero, where they stayed for several seconds.[3] Her finding was criticized, though, by Wenger, Bagchi, and B. K. Anand in their later, more thorough studies of yogic adepts (see below). Brosse also studied a yogi who was buried for 10 hours, and described other examples of self-control she had witnessed. Like Behanan and Swami Kuvalayananda, she helped promote the idea that yogic feats could be studied with scientific instruments.

The instrumented study of yogic functioning was expanded by Bagchi, Wenger, and Anand, who was then chairman of the Department of Physiology at the All-India Institute of Medical Sciences in

Delhi. Their landmark studies during the late 1950s were reported in American scientific journals, and with Akira Kasamatsu and Tomio Hirai's studies of Zen masters in Japan (see below) gave new momentum to meditation research. For five months in 1957, Bagchi and Wenger traveled through India with an eight-channel electroencephalograph and accessory instruments to record respiration, skin temperature, skin conductance, and finger blood-volume changes. During their trip they established experiments in Calcutta, Madras, Lonavla, and New Delhi, and conducted further tests in homes and a mountain retreat.[4] Among the subjects they examined, one could perspire from his forehead upon command in his freezing Himalayan retreat; a second could regurgitate at will to cleanse himself.[5] Three others altered their heartbeats so that they could not be heard with a stethoscope, though EKG and plethysmographic records showed that their hearts were active and their pulses had not disappeared.[†] In tests to compare relaxation in a supine position with seated meditation, Bagchi and Wenger found that four yoga students had faster heart rates, lower finger temperatures, greater palmar sweating, and higher blood pressure during meditation, though their respiration rates were reduced. Five yogis given similar tests exhibited even faster heart rates, lower finger temperatures, greater palmar conductance, and higher blood pressures during meditation than the students, though their breathing was slower. Such differences suggested that for these yogis meditation was an active rather than a passive process.[6]

Bagchi and Wenger also studied the effects of breathing exercises and found that some of their subjects, especially experienced ones, could produce bidirectional changes in every autonomic variable the experimenters measured. Though the two psychologists found that their subjects exhibited some dramatic physiological changes, they were cautious in drawing conclusions about yogic claims in general. "Direct voluntary control of autonomic functions

---

[†] Wenger & Bagchi 1961. Wenger, Bagchi, and Anand guessed that these three subjects used the Valsalva maneuver, consisting of strong abdominal contractions and breath arrest, to reduce venous return to the heart. "With little blood to pump the heart," they wrote, "sounds are diminished . . . and the palpable radial pulse seems to disappear. High amplification finger plethysmography continued to show pulse waves, however; and the electrocardiograph showed heart [contractions]." During such breath retention, moreover, their subjects' hearts changed position so that the potentials in one of their EKG leads decreased, which led Wenger and Bagchi to suggest that Brosse's earlier demonstration of complete heart cessation might have resulted from her use of a single EKG lead that lost its potentials when her subject's heart position shifted.

is probably rare among yogis," they wrote. "When such control is claimed, intervening voluntary mechanisms are usually employed." They made this qualification, however: "We have met many dedicated yogis who described experiences to us that few Western scientists have heard of and none has investigated. It is possible that the mere presence of a foreigner precludes optimum results."[7]

Other researchers have confirmed the discovery by Bagchi and Wenger that some subjects exhibit more than one pattern of physiological activity during their yogic practices. N. N. Das and H. Gastaut studied seven Indian yogis, who registered no muscular electrical activity during periods of complete immobility though their heart rates accelerated in almost perfect parallel with accelerations of their brain waves during moments of ecstasy. The most accomplished among these seven subjects, moreover, exhibited "progressive and very spectacular modifications" in their EEG records during their deepest meditations, including recurrent beta rhythms of 18–20 cycles per second in the Rolandic area of the brain, a generalized fast activity of small amplitude as high as 40–45 cycles per second with occasional amplitudes reaching 30 to 50 microvolts, and the reappearance of slower alpha waves after samadhi, or ecstasy, ended. In summarizing their study, Das and Gastaut concluded that

> the modifications [we] recorded during very deep meditation are much more dramatic than those known up till now, which leads us to suppose that western subjects are far from being able to attain the yogi state of mental concentration.
>
> It is probable that this supreme concentration of attention . . . is responsible for the perfect insensibility of the yogi during *samadhi*; this insensibility, accompanied by immobility and pallor often led people to describe this state as sleep, lethargy, anesthesia or coma. The electroencephalographic evidence here described contradicts such opinions and suggests that a state of intense generalized cortical stimulation is sufficient to explain such states without having to invoke associated processes of diffuse or local inhibition.[8]

Das and Gastaut's conclusion does not contradict the widespread findings of subsequent meditation studies that many or most meditators experience the trophotropic or relaxation response described by E. Gellhorn, W. Kiely, Herbert Benson, and other researchers.[9] Most subjects in meditation studies do not experience yogic ecstasy and so do not exhibit the cortical excitement that Das and Gastaut observed. Furthermore, different kinds of religious practice produce different types of experience accompanied by different types of physiological change. Kasamatsu and Hirai's Zen masters, for example, exhibited high-amplitude alpha and theta waves—not beta waves—during their deepest meditations (see below).

Further evidence that contemplative practice produces different physiological profiles was provided by B. K. Anand, G. S. Chhina, and Baldev Singh, who found that four yogis exhibited persistent alpha activity with increased amplitude during trance. These four yogis exhibited no alpha-wave blocking when they were bombarded with loud banging, strong lights, and other sensory stimuli, and two of them showed persistent alpha activity while holding their hands in ice-cold water for 45 to 55 minutes.[10] The yogis in this experiment exhibited physiological differences during meditation from at least two other groups of accomplished meditators. They did not exhibit alpha blocking in response to strong stimuli, in contrast to the Zen masters studied by Kasamatsu and Hirai (see below). Nor did they exhibit the beta waves that appeared on the EEGs of Das and Gastaut's subjects. The difference from the Zen masters probably resulted from a basic difference in focus between the two groups, the yogis having withdrawn their attention from external stimuli, whereas the Zen masters remained aware of their external environment. Their difference from Das and Gastaut's yogis, on the other hand, might have been due to differences between their meditation styles, the conditions of the experiments, or the qualities of their experience. The strong stimuli Anand gave his subjects, for example, may well have prevented the more ecstatic absorptions experienced by Das and Gastaut's yogis. The published reports of the Das-Gastaut and Anand-Chhina-Singh experiments do not provide enough detail to fully explain their different results, but they remind us that there are different kinds of contemplative experience. Roland Fischer, Julian Davidson, and other researchers have proposed some ways in which internal states might be correlated with different physiological profiles.[11]

In a study published in 1958, the Indian researchers G. G. Satyanarayanamurthi and B. P. Shastry described a yogi whose heart kept beating even though his radial pulse could not be felt and his heart could not be heard with a stethoscope for 30 seconds. This yogi's EKG showed no abnormalities, moreover, and finger plethysmography showed that his pulse was present though greatly reduced. The two researchers claimed that fluoroscopy conducted while the yogi was lying down showed that for several 30-second periods his heart's beating was just a "flicker along the left border below the pulmonary conus and in the apical segment of the left ventrical." They concluded that he achieved this control through the Valsalva maneuver.[†]

Elmer and Alyce Green, with their colleagues at the Menninger Foundation in Topeka, Kansas, also observed exhibitions of yogic heart control. Their subject Swami Rama, while sitting perfectly still, produced an atrial flutter of 306 beats per minute that lasted for 16 seconds. During a fibrillation of this kind, a section of the heart oscillates rapidly while its chambers do not fill and its valves do not work properly, but Swami Rama gave no sign that the maneuver caused him any pain or heart damage. The swami also produced an 11°F difference between the left and right sides of his right palm. While he did this, the left side of his palm turned pink and the right side gray.[12]

Yogis frequently use abdominal contractions to slow their heart rate rather than intervening more directly through the central nervous system. Curiously, though, an earlier study had examined a man with no yogic training at all who could stop his heart without such maneuvers, simply by relaxing and "allowing everything to stop." By this procedure, he could induce a gradual slowing of his pulse until he started to faint, at which point he would take a deep breath. When EKG tests showed that his heartbeat did indeed disappear, the doctor who examined him concluded that the man's cardiac arrest was "induced through some mechanism which, although under voluntary control, is not known to the patient himself. Careful observation did not reveal any breath-holding or Valsalva maneuver. Apparently the patient simply abolished all sympathetic tone by complete mental and physical relaxation."[13]

Like heart stopping, the live burial of yogis has excited the inter-

---

[†] Satyanarayanamurthi & Shastry 1958. Anand and Chhina, again, investigated three yogis who said they could stop their hearts. They found that to accomplish this, all three increased their intrathoracic pressure by forceful abdominal contractions with closed glottis after inspiration or expiration. Like Bagchi and Wenger, they discovered that their subjects' heartbeats could not be detected with a stethoscope after such a maneuver and that their arterial pulse could not be felt, though EKGs showed that their hearts were contracting normally with a deviation of axis to the right when the subjects held their breath after inspiration, and a deviation to the left after expiration. Furthermore, X-ray examinations showed that each subject's heart became narrower in transverse diameter and somewhat tubular while he was trying to stop it. The three yogis "could not stop . . . their heart beats," Anand and Chhina wrote, "[but] they greatly decreased their cardiac output by decreasing venous return [and] the decrease in cardiac output is responsible for the imperceptible arterial pulse. This practice of yogis is identical with the Valsalva maneuver." Like Bagchi and Wenger, they suggested that Brosse's experiment had been flawed because she had used a single EKG lead with her subject.

est of several researchers. A physician, Rustom Jal Vakil, published an account in the *Lancet* of such a confinement that was witnessed by some 10,000 people near Bombay in February 1950. According to Vakil, an emaciated sadhu named Ramdasji sat cross-legged in a subterranean 216-cubic-foot cubicle and remained there for 62 hours. His pulse remained steady at 80 beats per minute; his blood pressure was 112/78; and his respiratory rate fluctuated from 8 to 10 breaths per minute. Though he had some scratches and cuts, Vakil wrote, Ramdasji appeared "none the worse for his grueling experience."[14]

In June 1956, a more closely observed study of yogic confinement was conducted under the auspices of the All-India Institute of Mental Health in Bangalore with a hatha yogi, Krishna Iyengar. J. Hoenig, a psychiatrist from the University of Manchester, witnessed the experiment and described it in a review of yoga research published in 1968.[15] According to Hoenig's report, a pit some two by three by four feet was dug on the institute's grounds and covered with wire meshing, a rubber sheet, and cotton carpet. An electrode junction box connected to an EEG and an EKG was placed in the pit along with instruments to measure temperatures and concentrations of gas. The yogi was confined for nine hours. When he was released he immediately walked about the grounds, according to Hoenig's firsthand account, and demonstrated athletic feats including a headstand with his legs in the lotus position. The percentage of carbon dioxide in the air in his enclosure, which was 1.34 percent at the beginning of the experiment, was only 3.8 percent at the end, lower than would normally be expected. Iyengar's heart rate gradually slowed from 100 to 40 beats a minute in recurring 20- to 25-minute cycles, but his EKG record did not register any other abnormality and the cycles did not coincide with his breathing or brain-wave patterns. The yogi's EEG showed a normal waking record for the full nine hours, characterized by a stable alpha rhythm of 50 microvolts with no evidence of sleep or interference caused by physical movement. From these records, the experimenters concluded that their subject lay motionless and wide awake, without the active cognition that would have reduced or eliminated his alpha rhythm. Iyengar said he had maintained the *shavasana,* or corpse pose, using *ujjaya* breathing while remembering the names of God. He was surprised that his heart had speeded and slowed, and could not explain why it had done so. It beat normally, however, after the experiment.

Because the earthen pits used in most yogic confinements leak oxygen and carbon dioxide, Anand, Chhina, and Singh tested a yogi named Ramanand in an airtight glass and metal box, once for 8 hours and again for 10 hours. The yogi's average oxygen use during the first experiment decreased from the basal rate of 19.5 liters per hour to 12.2, and during the second experiment to 13.3 liters per hour. His carbon dioxide output went down during both experiments. Ramanand, moreover, did not exhibit any rapid breathing or speeded heart rate as the oxygen in his box diminished and carbon dioxide increased. "Sri Ramanand Yogi could reduce his oxygen intake and carbon dioxide output to levels significantly lower than his requirements under basal conditions," Anand and his colleagues wrote. "It appears from this study that [he] could voluntarily reduce his basal metabolic rate on both occasions he went into the box."[†]

During a remarkable experiment reported by L. K. Kothari and associates, a yogi was buried for eight days in an earthen pit and connected by leads to an EKG in a nearby laboratory. After the pit was boarded up, the subject's heart rate sometimes went as high as 250 beats per minute, until a straight line appeared on the EKG tracing when the yogi had been in the pit for 29 hours. There had been no slowing of his heart immediately before the straight line appeared, nor any sign of electrical disturbance, but the experimenters proceeded with certainty that their subject had not died. Suspecting that their EKG leads had been deliberately or accidentally disconnected, they checked their machine and continued to monitor its tracings. To their astonishment, it started to register electrical activity some seven days later, about a half hour before the yogi's scheduled disinterment. "After some initial disturbance," they wrote, "a normal configuration appeared. The [speeded heart rate] was again there but there was no

---

[†] Anand et al. 1961. A second study with an airtight box reported by P. V. Karambelkar and associates compared the reactions of an accomplished yogi, a yoga student, and two controls during confinements ranging from 12 to 18 hours. The box used in this experiment was closely monitored for oxygen and carbon dioxide content, having been thoroughly tested for leakage, and the subjects were attached to an EKG, a respiratory strain gauge belt, an EEG, a blood-pressure recording device, and a measure of their galvanic skin response. Each subject stayed in the box until its CO_2 level caused him discomfort. The yogi remained for 18 hours, until the air he was breathing reached 7.7 percent CO_2, while the other three stayed from 12 ½ to 13 ¾ hours, when their CO_2 levels reached 6.6 to 7.2 percent. The yogi stayed longer, the authors suggested, because he was habituated to such situations. But their yoga student, not their professional yogi, showed the least reduction in oxygen consumption as his CO_2 levels increased. He could withstand higher levels of CO_2, the authors argued, because for three years he had practiced the *kumbhaka* or breath-holding exercise of *pranayama,* which had trained his body to function with the increased alveolar CO_2 the exercise produces. Subsequently, the professional yogi increased his *pranayama* practice and exhibited improved adaptation to CO_2 (Karambelkar, Vinekar, & Bhole 1968; and Bhole et al. 1967).

other abnormality." When the pit was opened, the yogi was found sitting in the same posture he had started in, but in a stuporous condition. In accounting for his remarkable EKG record, the experimenters argued that a disconnection of the EKG lead would have produced obvious markings on the tracings in their laboratory, as they found when they tried to simulate ways in which the yogi might have tinkered with it. Furthermore, the yogi was ignorant about such machines, and the pit was completely dark. If the machine had malfunctioned in some way they could not ascertain, it seemed an extraordinary coincidence that it started again just a half hour before their subject's scheduled release. Apparently, the yogi was operating with some kind of internal clock that did not depend upon the daily cycles of light and darkness, for the most likely cause of the straight line on his EKG tracing was a dramatic decrease in his heart's activity. Kothari and his colleagues finally could not account for this remarkable cardiac record.[16]

Studies of Zen Buddist Monks

In a study that attracted much attention among meditation and biofeedback researchers during the 1960s, Akira Kasamatsu and Tomio Hirai, physicians at the University of Tokyo, studied the EEG changes exhibited during meditation by Zen teachers and their disciples (48 in all) from Soto and Rinzai centers in Japan. For experimental control, they studied the EEGs of 22 subjects with no meditation experience. They made EEG recordings; recorded their subjects' pulse rates, respiration, and galvanic skin response; and tested their responses to sensory stimuli during meditation. The recordings on the Zen monks were made during a weeklong retreat, or *sesshin*, at a Zendo, except for a few tests at the experimenters' laboratory. The Zen teachers and their most experienced students exhibited a typical progression of brain-wave activity during meditation, which Kasamatsu and Hirai divided into these four stages.

Stage I: characterized by the appearance of alpha waves in spite of opened eyes.

Stage II: characterized by an increase in amplitude of persistent alpha waves.

Stage III: characterized by a decrease in alpha frequency.

Stage IV: characterized by the appearance of rhythmical theta trains.[17]

Not all four stages were evident in every Zen practitioner, nor in any of the controls, but a strong correlation existed between the number of stages a given student exhibited and that student's length of time in Zen training. This correlation was supported by a Zen teacher's evaluation of each student's proficiency. The teacher ranked the students in three levels, without seeing their EEG records, and his rankings correlated well with Kasamatsu and Hirai's assessment of their EEGs.

The Kasamatsu-Hirai study also revealed significant differences between four Zen masters and four control subjects in their response to repetitive click stimuli. Like the Zen masters, the controls exhibited a blocking of alpha when a click sound first occurred, but they gradually became habituated to such stimuli so that their brain-wave activity no longer responded when a click was made. The Zen masters, however, did not become habituated, but continued to exhibit blocking as long as the stimuli continued. This finding indicates that Zen practice promotes a serene, alert awareness that is consistently responsive to both external and internal stimuli.[18]

Difficulties of Research with Religious Adepts

Though people testified under oath before the Congregation of Rites that they had seen Saint Teresa of Avila or Saint Joseph of Copertino defy gravity (22.10), no scientific studies have recorded instances of levitation. There are at least three possible reasons for this lack of evidence. First, of course, it may be that levitation has never happened. Second, the contemplative traditions may have lost their power to evoke the phenomenon. Third, levitation may only occur during rare and spontaneous ecstasies that cannot be programmed to meet the requirements of a scientific experiment. Superordinary lifting from the ground, if it in fact occurs, would require an improbable set of circumstances, which a scientist would be lucky to witness. Levitation, like other holy powers, would have to be caught "in the wild." In a laboratory, with wires attached to his head and a thermometer up his rectum, a yogi or lama is unlikely to exhibit a capacity that is rare in any case. As I have said in other chapters, in studies of extraordinary functioning there is a trade-off between robust results and scientific precision. Uninhibited by recording machines and safety rules, for example, the Maharaja Runjeet Singh could bury Haridas for 40 days

(21.7). More recent studies of yogic confinement, however, have been constrained by procedural controls and humane considerations.

Furthermore, there is often a disjunction between a scientist's attitude toward exceptional powers and an adept's ideas about them. Elmer Green, for example, described differences he had with the healer Jack Schwartz in interpreting Schwartz's intuitive diagnosis of illness.

> The question is: Are the auras he sees always radiatory patterns of energy from the human body . . . or are they automatic mental projections of one kind or another that are used psychologically to interpret a "knowing"? Sometimes when we "know" something in this way we tend to "see" it in the same way that we see a memory.[19]

Green was sympathetic to Schwartz, however, realizing that a scientist's constant doubt can inhibit or destroy a psychic's intuitions. This fundamental difference between scientists and psychics, Green wrote,

> need not cause problems if each takes time to understand the framework in which the other necessarily operates. If the psychic tries to pull apart every perception in order to find out if it is incorrect, so as to better determine the "truth," what is most likely to be pulled apart is the faculty of "seeing." The talent for perceiving might well fade away. On the other hand, if scientists stopped trying to find alternate explanations for the facts, they might get lost in a maze of [incoherent] ideas. For both scientists and mystics, however, the area of facts—rather than interpretations—is common ground. Excluding the opinions of fanatics, most of the arguments that we are aware of between the two camps have revolved around interpretations. Because psychics almost always have idiosyncratic factors in their frames of reference, scientists often do not understand them. And psychics do not understand what seems to them to be a destructive attitude on the part of scientists.[20]

Sympathy between scientists and adepts was evident in Swami Kuvalayananda's projects noted above, and in other experimenter-subject teams described in the preceding pages. Even the stern mutual challenge between Haridas and Maharaja Runjeet Singh exhibited an exemplary, if somewhat perverse, cooperation. Productive study of extraordinary functioning requires understanding between accomplished subjects and imaginative experimenters.

CONTEMPORARY MEDITATION RESEARCH

Meditation research increased dramatically during the 1970s and 1980s, particularly in the United States. This burgeoning research effort was stimulated in part by the studies of yogis and Zen masters noted in the previous section, and in part by the publication of landmark studies by Herbert Benson and Keith Wallace in *Science,* the *American Journal of Physiology,* and *Scientific American* between 1970 and 1972.[21] The Transcendental Meditation Society supported much of this work, though its enthusiastic claims and advertising efforts caused doubts among some researchers about the highly favorable outcomes in studies it sponsored.[22] These doubts led to further research, which has either contradicted, tempered, or confirmed the TM-sponsored claims. Since the early 1970s, more than a thousand studies of meditation have been reported in English-language journals, books, and graduate theses. Steven Donovan and I listed some 1,300 of these produced between 1931 and 1990 in a monograph, *The Physical and Psychological Effects of Meditation,* published by Esalen Institute's Study of Exceptional Functioning. The range of outcomes included in this research has grown considerably since the studies of yogis and Zen masters by Bagchi, Wenger, Kasamatsu, and Hirai. Cardiovascular, cortical, hormonal, and metabolic changes, several behavioral effects, and alterations of consciousness resulting from meditation have been explored in recent years. The medical instrumentation, psychological tests, and methods of analysis used in such experiments have been improved, and the range of subject populations has been enlarged to include different kinds of subject groups. This growth in sophistication of method is gradually improving our scientific understanding of meditation in ways that complement the insights contained in the traditional contemplative literature. However, the overall picture of meditation results produced by modern research is still uneven. Some effects have appeared consistently, but others have not.

The apparent inconsistencies of meditation's effects can be accounted for in various ways. Some physiological processes, perhaps, are unaffected by meditation, no matter how proficient or experienced the meditator might be; or perhaps they are affected to an insignificant degree. For some changes, such as amino acid concentrations in the blood, there has not been enough research to establish a consistent picture, partly because there has not been as much interest in them as in meditation's effects on blood pressure, heart rate, and other indices that have an obvious bearing on health. Taking blood samples during meditation, moreover, is harder to accomplish than recording blood pressure or skin responses.

Individual differences also present a special problem for understanding the results of meditation studies, because subject populations have included people of both sexes, all ages, various levels of education, and different kinds of social background. Many subjects have been college students with no previous meditation experience; others have been recent converts to religious groups; but only a few have been highly skilled in spiritual practice. The incentives to concentrate during experimental sessions have also varied; some subjects have wanted success for religious or other reasons, while others seem not to have been well motivated. And differences between meditation styles also complicate the results of such research. Though most studies have used some type of quiet concentration, some have used active methods such as rapid breathing. Julian Davidson, Roland Fischer, and others have distinguished between two classes of meditation— relaxing and excitatory—associating their effects with the trophotropic and ergotropic conditions of the central nervous system modeled by Gellhorn and Kiely.[23]

The results of meditation research are accumulating now in the manner of scientific knowledge generally, forming a publicly accessible body of empirical data that can serve generations to come. Unfortunately, however, these data are derived mainly from meditation beginners, and taken as a whole do not reflect the richness of experience described in traditional contemplative teachings. They are also limited by the conventional scientific insistence that results be repeatable. Certain important experiences occur only rarely in meditation, and a science that disregards them loses much of its empirical results. For these reasons, contemporary research does not illumine the full range of experience described in the contemplative scriptures and the oral traditions from which they come. Modern studies give us only a first picture of meditation's foothills, with a few glimpses of its peaks. Still, what they give us corresponds in several ways with traditional accounts. In appendix C, I list meditation outcomes for which there is good experimental evidence.

Transformative
Practices

Wine in ferment is a beggar suing for our ferment; Heaven
in revolution is a beggar suing for our consciousness.
Wine was intoxicated with us, not we with it; the body
came into being from us, not we from it.
We are as bees, and bodies as the honeycomb. We have
made the body, cell by cell, like wax.
RUMI—*The Mathnawi*

Ishavasyam Idam sarvam.
All this is for habitation by the Lord.
ISHA UPANISHAD

METANORMAL CAPACITIES TYPICALLY emerge from
normal abilities that have been developed through life-encompassing
practice. Some mystics and saints, for example, have loved God with
such passion that they realized a joy that did not depend on the
satisfaction of ordinary needs or desires. Mozart had a musical imagination so exquisitely trained that he sometimes heard symphonies
completely formed in his mind. Certain religious ascetics develop
such fortitude that they endure great adversities with self-existent delight (21.2, 21.3, & 23.1). But dramatic transformations of mind and
flesh are evident, too, in certain pathologies, in healing produced by
placebos and prayer, in many incidents of everyday life, and during

practices that make only a partial claim on their practitioners' commitment.

Taken in its entirety, the material presented in this book suggests that human nature harbors extraordinary attributes that may appear in sickness, healing, or programs for growth, either spontaneously or through formal discipline. While such attributes require long-term cultivation for their fullest development, they frequently appear to be freely given, sometimes when we do not seek or expect them. In the following chapters, I discuss their cultivation through various disciplines, with attention to our multiple roots in animal life, human culture, and ego-transcending dimensions of mind and body.

24

THE BASES OF TRANSFORMATIVE PRACTICE

IN ESTABLISHING TRANSFORMATIVE PRACTICES, we depend upon inherited bodily processes. We can cultivate somatic awareness and control, for example, because nerve cells that evolved from analogous structures in the earliest vertebrates are deployed throughout our bodies.[1] Relaxation exercises are effective because we possess a parasympathetic system that developed during the long course of mammalian evolution.[2] We can become creatively absorbed in work, perhaps, because we have inherited capacities for catalepsy, analgesia, and selective amnesia that facilitate escape and hunting. In short, self-regulation skills, regenerative relaxation, and performance trance, like other kinds of creative functioning, are based on capacities that developed among our animal forebears. And while transformative practices draw upon our animal inheritance, they also employ uniquely human activities. The imagination we use to enjoy books can be cultivated to induce metanormal cognitions or to facilitate extraordinary physical skills. The self-reflection we sometimes practice when confronted by difficulty can be deepened by means of sustained meditation. Transformative disciplines use both inherited and socially acquired attributes to improve many kinds of functioning.

But the very activities that facilitate growth can also inhibit it. Imagination and self-reflection, for example, if they enlist destructive images and patterns of thought, can impede our further development. Indeed, we are simultaneously supported and held fast by the mind-body complex we inhabit, with its functioning so exquisitely adapted to life on earth. That abiding tension of our animal-human nature, its simultaneous resistance to life-threatening change and capacity for dramatic restructuring, must be skillfully negotiated in transformative practice. We must respect the basic regularities of our intricate makeup as we begin to refashion them, remembering that breakdown must serve buildup (either psychic or physical), making sure we do not injure ourselves as we develop new capacities. Such negotiation can be difficult, of course, and often seems impossible, but it is sometimes assisted by mysterious agencies that seem to originate beyond the ordinary self. These spontaneous empowerments—call them the graces of God or the workings of Buddha-mind—answer our calls for a greater life and confirm our aspirations for it.

Transformative disciplines, then, rely on inherited, socially acquired, and ego-transcending activities, and these all play roles in the psychophysical alterations described in this book. Here I will emphasize some that are especially important for integral development. I will refer to their deliberate employment as *transformative moves* or *modalities*.

[24.1]
TRANSFORMATIVE MODALITIES IN SICKNESS, HEALING, AND GROWTH

Hysterical stigmata and false pregnancy demonstrate the great specificity with which the flesh can be shaped by emotionally laden mental pictures accompanied by strong conscious or unconscious intentions (11.1 & 11.2). Both pathologies reveal the precision with which highly charged images can shape somatic processes. And in multiple personality we also observe commonly shared processes employed in the service of abnormal functioning. Multiples, it seems, perfect their different selves through the lifetime practice of different roles, and in so doing develop extraordinary psychophysical plasticity (11.3).

Placebo effects and spiritual healing, too, depend on suggestive imagery, as well as expectation of success, the deliberate or covert practice of healing affirmations, health-mimicking behaviors, and confidence in a physician, procedure, religious figure, or other healing influence (12 & 13). Like the other psychophysical changes discussed in Part Two, both placebo-induced and faith cures are catalyzed by processes that operate in everyday life. In chapter 14, I describe the remarkable results of such processes among people with severe disabilities.

➔

The practices reviewed in chapter 15 through 19 also rely on common modalities of change. Hypnotic suggestion, for example, depends on the same kind of responsiveness to significant imagery that is evident in hysterical stigmata, the flexible role assumption involved in multiple personality, the expectation of success that sustains placebo effects, and the rapport or focused surrender that facilitates spiritual healing. As noted in chapter 15.6, the same absorption we experience while reading a good book promotes hypnotic alterations of consciousness and behavior. By deliberately inducing a creative trance (with the help of a hypnotist), we can partially integrate our fragmented volitions, so that we act with more strength and decisiveness, feel with more depth, think with new clarity and precision. Harnessing capacities that ordinarily operate in haphazard fashion, we can use hypnotic attention to facilitate various kinds of extraordinary functioning.

In biofeedback training, we observe a transition from the largely dissociated processes that produce hysterical stigmata or multiple personality, and the marked dependence on others involved in hypnosis, to a more conscious, self-reliant practice. With the help of instruments, a subject receives feedback from autonomic processes, which he or she then learns to modify. Cultivating the kinesthetic awareness and deliberate control of autonomic processes that virtually everyone uses upon occasion, most people can learn to raise or lower their blood pressure, change their brain waves, alter the flow of gastric acid, or modify other physiological functions (16).

Mental imagery, again, becomes a transformative modality in psychotherapy. In conjunction with catharsis, focused intention, and the acceptance (or reowning) of dissociated functioning, such imagery facilitates healing and growth by helping to restructure ruling attitudes, by providing access to repressed or previously unnoticed psychological processes, by helping to modify perception, by facilitating skill rehearsal, and by mediating awareness of bodily structures (17.1 & 17.2).

Somatic disciplines, too, employ many common capacities. Autogenic Training enlists the trophotropic response to facilitate deep relaxation and cultivates the precise attention to kinesthetic sensations we are sometimes impelled to by sickness or crisis (18.3). The Feldenkrais Method stimulates muscles and ligaments in the manner of children's learning and employs pleasurable stretching movements familiar to everyone (18.4). Progressive Relaxation enlists our natural capacities for self-regulation through simple contraction and relaxation of particular muscles, by focusing attention on specific body parts, and by articulating our capacity for selective control (18.6 & appendix H). The Sensory Awareness work of Elsa Gindler and Charlotte Selver employs close attention to sitting, standing, and walking, directing awareness to the simplest acts, extending our natural kinesthetic sense by childlike exercises (18.7). And Reichian therapies use the catharsis, trembling, and psychosomatic release that help most of us discharge tension (18.8).

Athletic regimens strengthen common running, jumping, and throwing skills, increase manual dexterity, deepen cardiopulmonary stamina, and enhance general coordination (19). They also draw upon our natural playfulness. The physical improvisations all children exhibit are used in sport and are lifted at times to brilliance. Sport depends on qualities of heart and mind, such as courage, alertness, and self-control, that exist to some degree in everyone.

➔

In chapters 20 through 23, I turned from disciplines that make partial claims upon their practitioners to the life-encompassing programs of certain martial arts, shamanism, yoga, and contemplative discipline. At their best, these promote more virtues and capacities, and embrace more human nature, than do hypnosis, biofeedback, psychotherapy, or sport. The integral practices proposed in this book would emulate their depth and inclusiveness and incorporate the impulse toward wholeness they frequently exhibit.

[24.2]
THE ROLE OF GENERAL, ALL-AT-ONCE RESPONSES IN TRANSFORMATIVE PRACTICE

Among the human endowments on which transformative practices are based, some can be cultivated to produce especially fruitful outcomes. For example, quieting meditation enlists the trophotropic response to produce coordinated changes in brain activity, heart rate, blood pressure, breath count, lactate production, and adrenal secretions (23.2). In producing this well-integrated outcome, apparently simple procedures such as counting breaths produce complex and creative results. Similarly, hypnotic and performance trance seem to be mediated by a general response inherited from our mammalian ances-

tors. The absorption, analgesia, and selective amnesia that characterize states of deep creative absorption may be functionally analogous
to—and derived from—the freezing behaviors, temporary freedom
from pain, and blindness to difficulty we observe among animals in
the wild. The nearly cataleptic focus of attention, the ability to endure
discomfort, and the forgetfulness of pain evident in some meditators,
athletes, hypnotized subjects, and people involved in creative work
come into play all at once, by a kind of reflex. And a similar appropriation of integrated process is evident in sport and the martial arts
when a challenge triggers adrenaline flow, quickens heart rate, and
produces new energy for exceptional performance. In the trophotropic response; in the coordinated appearance of absorption, analgesia, and selective amnesia; and in the fight-or-flight response,
various processes come into play synergistically, producing several
psychophysical changes at once.

I emphasize this fact in order to suggest that all transformative
moves have some of this all-at-once character, by which they produce
creative change economically. The cultivation of imagery, for example, facilitates new sensorimotor skills or extraordinary states of consciousness by the recruitment of countless psychological and somatic
processes. By means of such recruitment, countless cells, all of them
interlocked in the service of homeostasis, are led in tandem by mental
images. Similarly, the practice of different roles or subpersonalities,
the repetition of healing affirmations, sustained expectation of success, concentration, employment of kinesthetic awareness to modify
autonomic functioning, deliberate recall of unnoticed or dissociated
processes, focused intention, surrender to ego-transcending powers,
as well as other transformative modalities can trigger highly complex
reactions in us. Like good business deals or scientific theories, they
can yield great return on investment.

The many-sided effect of transformative modalities has been described by George Leonard, who has developed a discipline he calls
Leonard Energy Training (LET). The core elements of Leonard's approach include exercises for balancing and centering. These, Leonard
wrote,

> draw the subject's attention to various parts of the body, especially the
> center of the abdomen, in order to encourage a posture and inner aware-
> ness that is equal and even—right and left, top and bottom, front and
> back. These exercises are accomplished under varying circumstances:
> standing and sitting, eyes open and closed, in linear and circular move-

ment, under physical or psychological pressure from a partner. In some of the exercises, the subject is forced off balance, then taught to return quickly to a balanced and centered state. Most of the time, the gaze is "soft"; that is, the eyes are open but not fixed on any particular point.

These exercises produce physiological results similar to those produced by other relaxation procedures, including warming of the extremities, deepening and slowing of respiration, slowing of pulse, and increase in EEG alpha waves. Subjects say that they feel both calm and alert. They also report an increase in visual acuity and depth of field, an increased ability to discern fine color gradations and to perceive patterns of movement. Peripheral vision is measurably enhanced. Kinesthetic awareness tends to spread, so that it includes head, neck, torso, arms, and legs. Movement becomes more graceful, coordinated, efficient, and spontaneous. Complex and creative maneuvers are made without premeditation. In the balanced and centered state, most subjects are more physically stable than they are when their attention is focused intently upon a particular body part. This is demonstrated by having partners attempt to push them off center.

The balanced and centered state influences psychological as well as physical processes, producing in most subjects an increased mental stability and clarity, especially in pressure situations. Perceptions of a partner's feelings are enhanced, and subjects often report a general increase of sensitivity to other peoples' mood and intentions. Indeed, there is some evidence that telepathic and clairvoyant abilities are activated. During experiments in our *aikidojo*, students trained in centering and balancing have located magnetic north while blindfolded, located people hidden outside the building, and exhibited other types of extraordinary sensitivity.[3]

In this brief account, Leonard notes more than 20 outcomes of LET practice, each of which involves many psychosomatic changes. Here again, simple exercises produce complex, well-coordinated, creative results.

[24.3]
HIGH-LEVEL CHANGE INVOLVING EGO-TRANSCENDING AGENCIES

Some high-level changes, it seems, are produced by agencies other than those evident in animal or ordinary human activity. As I have suggested, certain awarenesses and behaviors are attributed to powers ouside the ordinary self, not only because they radically exceed our normal capacities but also because they seem well formed (or fundamentally complete) before their appearance in us, and because we cannot adequately account for them in terms of familiar human processes. Mystical illuminations, artistic or scientific epiphanies, inspired movements in sport, and other extraordinary experiences seem to arise in part from a life beyond the life to which we are accustomed.

These ego-surpassing experiences have inspired metaphysical doctrines such as divine emanation and return, or panentheism, that posit a Supreme Person or Principle that is both immanent in and transcendent to this world (7.2). Versions of these ideas have arisen among Vedantic, Buddhist, Taoist, Platonist, Kabbalistic, Christian, and Islamic mystics, and are reflected in dicta such as the Upanishads' "*Tat Tvam Asi,*" thou art that, the Kabbalists' "as above, so below," and the Christian "not my will, but Thine be done." According to mystics East and West, religious practice reveals human nature's basic, if hidden, connection or identity with a transcendent order. And as a corollary to this proposition there is usually a second—namely, that we are graced at times by extraordinary incursions of that transcendent order. Besides enjoying an essential connection or identity with it, we are sometimes graced by special inflowings of its substance and power. Frederic Myers approached this idea from the perspective of his own research with paranormal phenomena and subliminal workings of the mind.

In ancient and modern times, in East and West, among Pagans, Buddhists, Brahmins, Mohamedans, Christians, Infidels—everywhere it has seemed possible for men and women, by a certain stress of soul, to become in great measure superior to pain, and often to renew vitality with a success for which medical science cannot account. The true meaning of this far-reaching and multiplex power of self-suggestion is one of the standing puzzles—one of the growing puzzles—alike of biology and of psychology. Without pretending to solve it, I have nevertheless stated and defined it in a manner which may now serve to bring it into relation with an even wider range of phenomena. For I have spoken of it as a fluctuation in the intensity of the draft which each man's life makes upon the Unseen. I have urged that while our life is maintained by continual inflow from the World-soul, that inflow may vary in abundance or energy in correspondence with variations in the attitude of our own minds. So soon as this definition is made, we see that every form of self-suggestion falls within the limits which we have assigned to supplication. The supplication of the Lourdes pilgrims, the

adoring contemplation of the Christian Scientists, the inward concentration of the self-suggesters, the trustful anticipation of the hypnotized subject—all these are mere shades of the same mood of mind—of
the mountain-moving faith which can in actual fact draw fresh life
from the Infinite.[4]

Artistic inspirations, certain feats that amaze us in sports, the astonishing vitality evident in *tumo*, "boiling *n/um*," kundalini experiences, and the *incendium amoris* (5.4)—indeed, all the extraordinary
activities noted in this book—have been attributed to inflows such as
Myers described, whether these are ascribed to God, the gods, Tao, or
Buddha-mind. As we have seen, inspiration is not limited to the contemplative life. Metanormal expressions of all our capacities have
been attributed to an affinity we enjoy with higher powers. This affinity, it can be argued, is the predisposing condition for all particular
mediations of high-level change, the metamediation of all extraordinary functioning.[†] William James was led to such a view by his own
studies of religious experience. The further limits of human nature,
he wrote,

> plunge into an altogether other dimension of existence from the sens-
> ible and merely "understandable" world. Name it the mystical region,
> or the supernatural region, whichever you choose. So far as our ideal
> impulses originate in this region (and most of them do originate in it,
> for we find them possessing us in a way for which we cannot articu-
> lately account), we belong to it in a more intimate sense than that in
> which we belong to the visible world, for we belong in the most intimate
> sense wherever our ideals belong. Yet the unseen region in question is
> not merely ideal, for it produces effects in this world. When we com-
> mune with it, work is actually done upon our finite personality, for we
> are turned into new men, and consequences in the way of conduct fol-
> low in the natural world upon our regenerative change. But that which
> produces effects within another reality must be termed a reality itself,
> so I feel as if we had no philosophic excuse for calling the unseen or
> mystical world unreal. . . . God is real since he produces real effects.
>
> If asked just where the differences in fact which are due to God's

---

[†] But such resonance or affinity can also serve misguided and destructive development. Imagery, for example, can draw upon our latent powers to serve pathology as well as growth. That is
why images of desired outcomes must be modified in the course of transformative practice, why
they have to change, too, in the negotiations between our present condition and emergent
capacities.

existence come in, I should have to say that in general I have no hypothesis to offer beyond what the phenomenon of "prayerful communion," especially when certain kinds of incursion from the subconscious region take part in it, immediately suggests. The appearance is that in this phenomenon something ideal, which in one sense is part of ourselves and in another sense is not ourselves, actually exerts an influence, raises our centre of personal energy, and produces regenerative effects unattainable in other ways. If, then, there be a wider world of being than that of our every-day consciousness, if in it there be forces whose effects on us are intermittent, if one facilitating condition of the effects be the openness of the "subliminal" door, we have the elements of a theory to which the phenomena of religious life lend plausibility. I am so impressed by the importance of these phenomena that I adopt the hypothesis which they so naturally suggest. At these places at least, I say, it would seem as though transmundane energies, God, if you will, produced immediate effects within the natural world to which the rest of our experience belongs.[5]

In this book, I have described many kinds of activity that indicate that something beyond the ordinary self "actually exerts an influence, raises our centre of personal energy, and produces regenerative effects unattainable in other ways." When the Japanese swordsmaster Yagyu Tajima said an infinity of acts is produced from Emptiness, *sunyata*, so that "in hands it grasps, in feet it walks, in eyes it sees" (20.1), he referred to a Something such as James described that is in one sense part of ourselves and in another sense not. Physical action as well as contemplation, bodily structure as well as inspired cognition, can bring forth "fresh life from the infinite."

The developmental scheme proposed here links the access to transcendent things that Myers and James described with the development of our various cognitive, vital, and somatic capacities, and projects a future in which humans might realize extraordinary life upon earth. This perspective forces certain valuations of all transformative practices. Hypnosis, psychotherapy, or sports, for example, will not by themselves facilitate the integral transformations I am exploring because they are not designed to do so, nor will a religious asceticism that inhibits our physical abilities. None of these disciplines will sustain a coevolution of our various attributes with illumined consciousness. Only practices that enhance our psychological and somatic functioning while making special "drafts upon the Unseen" are likely to facilitate a balanced growth of our greater capacities.

[24.4]
THE IMPULSE TOWARD WHOLENESS IN ANCIENT
AND MODERN TRANSFORMATIVE PRACTICES

The human impulse toward wholeness is evident in both ancient and modern transformative disciplines. Here I briefly describe five that promote many-sided growth—one rooted in stone-age culture, two others more than a thousand years old, and two conceived in the twentieth century.

Shamanism includes practices that aim for integral transformation. By their ecstatic flights to the gods and subsequent falls to the under-world, certain stone-age shamans embrace both supernormal and subnormal dimensions of existence. By climbing the World-Tree, they imaginatively explore nature's many levels. By their ritual dismember-ment, they attempt their own psychophysical restructuring. The Mex-ican *brujo* Ramon Silva Medina climbed a waterfall with extraordi-nary agility, according to anthropologist Peter Furst, to dramatize the power he received from the gods;[6] ethnographer Curt Nimuendaju observed the inspired running of Brazilian Indians during games in-herited from the sun and moon;[7] Mircea Eliade described a Labrador Eskimo who stayed in the sea for five days and nights to win the title of *angalok* (21.3); and other anthropologists have chronicled astonish-ing physical feats by shamans in Sibera, Africa, and Australia. As I said in chapter 8.3, some forms of shamanism exhibit an aspiration for metanormal embodiment.

Patanjali's Yoga Sutras also promote many-sided development. They outline a practice with eight steps, the famous *ashtanga,* or eight-limbed yoga, consisting of (1) *yama* and (2) *niyama* exercises to de-velop sound morals and basic life habits; (3) *asanas,* yogic postures to enhance psychophysical functioning; (4) *pranayamas,* breathing ex-ercises to promote vitality; (5) *pratyahara,* the withdrawal of atten-tion from outward things; (6) *dharana,* concentration; (7) *dhyana,* contemplation; and (8) *samadhi,* union with the object of attention. These eight dimensions of practice include physical, emotional, ethi-cal, cognitive, and volitional training and evoke many kinds of meta-normal capacity.[8]

Zen Buddhist training, too, exemplifies balanced practice. It joins physical and cognitive disciplines and has deeply influenced the arts, daily aesthetics, and other aspects of Japanese culture. As much as any contemplative practice, it graces the things of this world, for example in Japanese tea ceremony, flower arrangement, temple architecture, gardening, archery, and rituals for the dead. In its power to illuminate work, play, and life's great transitions, Zen Buddhism sets a powerful example for integral practices. In the goodness it brings to the routines of everyday life, in its enhancement of the simplest gestures, it shows how responsive common activities can be to transformative discipline.[9]

Roberto Assagioli's Psychosynthesis draws on many psychiatric and religious schools (17.3). It cultivates a wide range of capacities and is insightful about contemporary forms of pathology in relation to the spiritual quest. More than most therapies, it aims to facilitate the growth of consciousness and moral sensibility beyond the healing of personal conflict; and more than most religious disciplines, it incorporates the discoveries of modern psychiatry. As much as any practice that commands a significant following today, it joins ancient and modern insights regarding normal and metanormal functioning. With all these strengths, however, it does not typically involve the systematic cultivation of kinesthetic or movement abilities. Assagioli drew little, if at all, from the martial arts, hatha yoga, or somatic disciplines; and most Psychosynthesis practitioners have not emphasized the body's potentials for dynamic activity and restructuring.[10]

Sri Aurobindo's Integral yoga, however, aims to produce radical somatic transformation. Like Assagioli, Aurobindo joined evolutionary perspectives with insights from the contemplative traditions. There is much practical lore in Aurobindo's letters on yoga, along with a comprehensive psychology of mystical experience.[11] The different but complementary experiences of transcendent and immanent, silent and dynamic, personal and impersonal aspects of Divinity are emphasized in his practice. As much as any philosopher, Aurobindo spells out the complexities of psychophysical development.

But with its richness and grand perspectives, Aurobindo's discipline suffers from certain limitations. It has not been deeply informed by modern depth psychology or scientific study of the body, and it relies too much, I believe, on Aurobindo's doctrine that Supermind, the dynamic aspect of the Absolute, will transform earthly life by a "universal descent." What Aurobindo meant exactly by this epochal event has puzzled many who are otherwise sympathetic to his vision.

Nevertheless, Aurobindo's synoptic metaphysics, his psychology of superordinary states, his doctrine of bodily transformation, and many practical aspects of his yoga point the way to integral practices for our time.

Other disciplines could be included in this list, of course, since virtually every sacred tradition has aimed to promote some version of human wholeness. Indeed, as we consider the great variety of transformative practices that have appeared in the last three thousand years, we might imagine that the human race has been zeroing in on an optimal set of paths toward its greater realization. In a long, meandering process typical of evolutionary process in general, humans have approximated—though not yet achieved—a breakthrough analogous to our emergence from hominids.

25

ELEMENTS OF TRANSFORMATIVE PRACTICE

[25.1]
SHORTCOMINGS OF TRANSFORMATIVE PRACTICE

Several psychotherapeutic, somatic, and martial-arts practices, and most religious disciplines, aim to promote some version of personal wholeness. Few if any, however, cultivate all of our capacities for extraordinary life or address the full range of our greater possibilities.

But even on their own terms, transformative disciplines often fail. Psychotherapists, for example, sometimes foster obsessive concern with small corners of the self, misguided indulgence of destructive emotions, or misreadings of personal dynamics that subvert healing and growth. As Gestalt therapist Fritz Perls remarked, therapy frequently becomes a defense against self-understanding and integration. Similarly, Rolfers and Reichian therapists can inadvertently promote inflexible images of the body that they impose (however painfully) on their trusting clients, sometimes with destructive effect. Somatic educator Don Johnson has described this perverse activity (18.5). Sport also produces its share of casualties, both spiritual and physical; and for all their stated moral concern, some religious groups give rise to cruel, even monstrous, activities.[1]

No discipline is immune to excess or lack of wisdom. All programs for human betterment can be undermined by ignorance, incompetence, or moral perversity. Indeed, all programs of self-cultivation are affected by inherited or culturally induced disabilities that afflict human activity in general. Frederic Myers, William James, Carl Jung, Roberto Assagioli, Abraham Maslow, James Hillman, and others have described many ways in which common human afflictions either trigger or prevent human growth;[2] and philosopher Ken Wilber has noted pathologies and developmental failures that occur at different stages of life, from infancy to the higher reaches of contemplative development.[3] Here I want to emphasize four destructive effects of religious and therapeutic disciplines that are particularly important for the formulation of integral practices.

A practice can reinforce limiting traits, preventing their removal or transformation. Ashrams and monasteries sometimes attract people who have trouble in their relations with others, then divorce them permanently from significant human contact; and ascetic contemplative practice often permits cruelty toward self and others. As Saint John of the Cross, Sri Aurobindo, and other religious teachers have warned us, spiritual disciplines can conceal and exacerbate all sorts of personal weakness.[4] Psychotherapy, too, can become a defense against creative change, as for example when someone rationalizes impulse-ridden behavior as the expression of dammed-up emotion, the release of body armor, or the expression of authenticity. People are frequently attracted to activities that preserve and disguise their shortcomings.

A practice can support limiting beliefs, giving them greater power in the life of an individual or culture. Some yogis, for example, are led by their ascetic philosophy to see the world as illusion, and are then confirmed in their otherworldly beliefs by the perceptions of other people's suffering that their discipline helps to promote. And psychotherapy, too, has this self-validating power—for example, when it rationalizes egocentricity in the name of psychological liberation.

A practice can subvert balanced growth by emphasizing some virtues at the expense of others. Every transformative discipline—indeed, every culture or group—tends to promote certain virtues while neglecting or suppressing others.[†] Psychotherapists who emphasize openness or honesty, for example, sometimes fail to encourage empathy and kindness. Martial-arts teachers who reward power before sensitivity can produce more brutality (and injuries) than those who give first priority to kinesthetic and interpersonal awareness.

A practice can limit integral development when it focuses on partial though authentic experiences of superordinary reality. Any metanormal cognition, by dominating the experience of its recipient, can

---

† For a discussion of different sets of the virtues that have arisen at different periods of Western history, see: McIntyre 1984.

strengthen particular beliefs about the world or human nature. Most mystics have been ruled by some archetype or dimension of the Transcendent and have been reinforced in its particular truth by their temperament, their training, and the beliefs they inherited with their practice and culture.[†]

When practices reinforce destructive traits, thus preventing their removal or transformation, when they support one-sided beliefs or limited sets of virtues, and when they are strengthened in such activity by the genuine realizations they induce, they become powerful obstacles to many-sided development, exerting a tyranny that prevents our deepest fulfillment. There are antidotes to such restriction, however, in the emerging understanding of pathogenic disciplines noted above, in the knowledge that there are many kinds of metanormal experience (including all those noted in chapters 4 and 5), and in a philosophy that emphasizes the balanced unfoldment of many virtues and capacities. With these safeguards, many disciplines discussed in this book can contribute to an integral practice. The religious traditions give us ways to develop metanormal volition and cognition, and bear witness to the ethics their cultivation requires. Modern depth psychologies, and the affective education they inform, complement the emotional disciplines nurtured by religious traditions and add new dimensions to them. Somatic training and sports provide methods to develop the body, and some martial arts show us how to join spiritual, ethical, and physical development. In this chapter I suggest some ways in which we can draw on all of these practices. But let me start with a few words about criteria for growth and regress.

There exist many standards by which to judge human development along particular lines, but only some of these apply universally. For example, athletic skills assessed with a clock or tape measure are acknowledged around the world, but the authenticity of mystical

---

[†] William James, for example, distinguished "healthy-minded" mystics who typically emphasize the world's goodness, from religiously minded "sick souls" who see evil and darkness everywhere. The two sensibilities, James argued, give rise to different philosophies (James, W., 1902, lectures 4 & 5: The Religion of Healthy-Mindedness, lectures 6 & 7: The Sick Soul). Philosopher Walter Stace distinguished introvertive from extrovertive mysticism. Introvertive mystics, Stace suggested, tend to perceive God within the self, while extrovertive mystics tend to see him pervading the universe (Stace 1987). Evelyn Underhill, Rudolph Otto, Marghanita Laski, and Sri Aurobindo have also distinguished various kinds of mystical realization and have examined ways in which different religious experiences engender different worldviews. See: Underhill, E., 1961, chap. 4: Mysticism and Theology, pp. 95–124; Otto 1960, pp. 57–105; Laski 1990, pp. 89–102; and Sri Aurobindo, *The Collected Works*, vol. 22, sec. 2: Integral Yoga and Other Paths, sec. 5: Planes and Parts of the Being, vol. 23, sec. 2: Synthetic Method & Integral Yoga. See also: Sri Ramakrishna 1969, chaps. 39 & 44.

experience is judged according to different criteria in different religious traditions. The features of successive contemplative stages outlined in Theravada Buddhism's famous text, the Visuddhimagga, for example, differ from those in the great summary of Catholic mystical practice, Poulain's *Graces of Interior Prayer*. Because they are based on somewhat different aims and different experiences of transcendent realities, the marks of contemplative progress in different traditions cannot be exactly equated (though they can be used in tests of a contemplative's experience within their host community: see 2.1). And there are differences, too, among the criteria of psychological maturation proposed by personality theorists. The various schemes of development put forth by Erik Erikson, Margaret Mahler, Jane Loevinger, Abraham Maslow, and others are still being refined; all of them have been criticized by eminent psychologists; and none has been universally accepted.

Recognizing the lack of consensus among developmental psychologists today about the marks of maturation, and considering the different signs of progress accepted by various contemplative schools, it is difficult to establish detailed criteria of advance toward metanormal functioning. For integral practices, we can make commonsense judgments about health and sickness, or goodness and badness, in relation to moral and emotional growth; we can draw on the few widely accepted marks of development modern psychologists do agree upon; and we can learn something from separate contemplative traditions about the signs of metanormal cognition. But to attempt more is a risky venture. We simply do not know enough yet to set up systematic growth criteria for the integral development proposed here. Though we can describe various kinds of extraordinary functioning, we will have difficulty specifying clearly marked steps toward their unfoldment. Integral practices depend largely on adventurous exploration.

[25.2]

THE INTERDEPENDENCE OF
TRANSFORMATIVE PRACTICES

Stoic philosophers of antiquity held that every virtue requires other virtues to complete it. They believed, for example, that courage without moderation is a bestial imitation of the real virtue, and that prudence without justice is not prudence. This view was put forth by

Plato in *Gorgias* (507c), by Aristotle in his *Nicomachean Ethics* (1144b32ff), and by earlier Greek thinkers such as Xenocrates, though the Stoics seem to have been the first to state it formally.[5] The Stoics' term for this mutual entailment of the virtues, *antakolouthia,* was used by philosophers of many persuasions in ancient times, including Platonists and Aristotelians. That courage, moderation, justice, wisdom, and other prime character traits depend to some extent on one or more of the others was a fundamental tenet of much Greek philosophy.

The same idea, adapted to include the many virtues and traits we now value, might help us understand the complexities of transformative practice. To develop awareness, for example, we need a measure of courage. To achieve lasting control of autonomic processes, we need sensitivity to somatic activity. The essential elements of practice discussed in this book always require others. Indeed, the interdependence exhibited by transformative methods is evident in the difficulty of classifying them. It is sometimes hard to differentiate them because they are so closely interwoven. Helpful understanding of others requires both empathy and detachment. Meditation requires both concentration and relaxation. Strengthening of will sometimes involves a yielding of ordinary volitions. Each potentially transformative method, each result of transformative practice, involves others. Their mutual entailment reflects the psychological integration Greek philosophers represented by the doctrine of *antakolouthia.* We must grant that interdependence, it seems to me, even though (1) extraordinary virtues and traits can exist side by side with great emotional, intellectual, or moral deficiencies; (2) different physiological systems, though interdependent, operate with considerable autonomy; and (3) much human functioning is affected by dissociated motives, attitudes, and feelings. Though our many psychological and somatic processes operate with a certain (sometimes large) degree of independence, they affect one another either directly or indirectly.

Psychologists Carl Jung and James Hillman, philosopher James Ogilvy, mythologist Joseph Campbell, and others have explored the polytheism of human nature, showing that our inner life is many-dimensional, multilayered, and teeming with presences of various kinds.[6] This rich complexity, they have argued, is inescapable. If we do not accommodate the many powers in which we are secretly rooted, they come to us anyway, as physical sickness, depression, obsession, or unexpected epiphanies that disrupt our everyday functioning. We are led, gradually or suddenly, sometimes successfully and sometimes not, to live in many dimensions at once. In his book *The Myth of Analysis*, Hillman claimed that

> each cosmos which each God brings does not exclude another; neither the archetypal structures of consciousness nor their ways of being in the world are mutually exclusive. Rather, they require one another, as the Gods call upon one another for help. They supplement and complement. Moreover, their interdependence is given with their nature. Jung said at Eranos in 1934: "The fact is that the single archetypes are not isolated . . . but are in a state of contamination, of the most complete, mutual interpenetration and interfusion." In his statement Jung voiced the Neoplatonic tradition. As Wind says, "The mutual entailment of the gods was a genuine Platonic lesson." For Ficino, "it is a mistake to worship one god alone." For Schiller "Belonging to one God only, any single cosmos, any single way of being in the world, is itself a kind of hubris."[7]

An analogous idea was developed by the Indian seers Ramakrishna, Vivekananda, and Aurobindo, who described their experiences of the personal and impersonal, transcendent and immanent, silent and dynamic aspects of Divinity. The transcendent order is immensely complex, the three claimed, and thus cannot be revealed through narrow ideas or practices.[8] With mystics such as Rumi and Kabir, they bore witness to the richness of our developing humanity. Indeed, the testimony of many religious adepts suggests that the interdependence of transformative methods reflects the integrated complexity of our emerging capacities. To recognize and embrace the complexity of metanormal experience, we need to expand most notions of it. In chapter 7.1, I criticize narrow ideas about grace, arguing that they limit receptivity to some of our greater possibilities. If our philosophic frameworks exclude particular attributes (whether normal or metanormal), those attributes are likely to be neglected or suppressed. By limiting ourselves to just part of the spectrum of grace, we will not open to all the life that awaits us.

But there is another source of our *antakolouthia*. Since transformative practice has multiple roots in our animal and human ancestry, the interdependence of its methods reflects the integrated complexity of ordinary functioning. For whenever we alter a mental, emotional, or physical process, we affect our entire organism, which is governed by homeostasis and informed by feedback from all its parts. That is why proven disciplines cultivate balance as they provoke dehabituating changes. Arousal is complemented by relaxation in (successful) yoga. Painful recognitions are supported by self-acceptance during (successful) psychotherapy. The release of certain muscles is accompanied by their realignment in (successful) somatic training.[†]

Normal functioning and our latent supernature intersect during high-level change, but we can foster or subvert their integration by the kinds of practices we choose. As I have said, certain flaws in our discipline can lead us away from integral transformation. Practices can be used to ends for which they were not designed; they can reinforce limiting traits or beliefs; they can give certain kinds of spiritual realization destructive sway over all the others. To avoid such dangers, we need to be wise mediators between our normal functioning and the metanormal activities emergent in us. The analogy that systems theorists make between organisms and societies, enlarged to include ego-transcendent realities, might help us think about such mediation.

Picturing our many structures and processes to be members of a single society, for example, we see that they operate with multiple controls in a multitudinous hierarchy, for the most part without our awareness or guidance. They know how to do this through programs inherited from our animal and human ancestors and by further training in this lifetime. Conducting their business in working harmony with other members of our organism-as-society, they accept a certain amount of in-fighting among competing factions, seek help from other groups when they need it, warn each other of dangers when necessary, and take all sorts of initiatives to ensure our survival. Now if

---

† The following studies suggest other ways in which various disciplines can complement one another.

  □ Meditation researcher Herbert Benson and his colleagues showed that meditation reduces oxygen consumption during fixed-intensity tasks such as running on a treadmill; and sports psychologist Richard Suinn found that distance runners significantly improved their performance while "running relaxed." These experimental results suggest that deliberate relaxation improves performance in aerobic sports by reducing muscular stress and improving the body's general efficiency (Benson et al. 1978; and Suinn 1976).

  □ Research psychologists R. J. Davidson and Gary Schwartz have presented evidence that meditation and aerobic sports complement one another in reducing anxiety. Meditation, they suggested, reduces cognitive anxiety, which has psychological causes, while aerobic sports reduce somatic anxiety, which is physically stimulated (Davidson & Schwartz 1984).

  □ By reducing the distractions of sensory stimuli so that kinesthetic impressions are more easily apprehended, sensory deprivation can facilitate biofeedback training (Yates 1980, pp. 79–80).

  □ By selectively relaxing unnecessary tension during movement, Progressive Relaxation has improved performance in dance and sport (Jacobson 1974, pp. 47–57).

we want to introduce a transformative program into this complex hierarchy, we can do so clumsily, stamping out entire working groups (by ascetic practices, for example, that diminish particular senses), or isolating certain groups from the whole by the persistent denial of feedback from them, or failing to recognize or surrender to agencies beyond ordinary functioning. If we operate in this way, we will suffer accordingly, from self-mutilation, dissociation, or lack of help from our secret resources. Though we might institute new kinds of functioning, we will cut ourselves off from many of our creative attributes.

On the other hand, we can introduce our transformative program with balance and skill, gradually leading our working groups toward new routines without stamping them out, maintaining intramural communication, opening to help from metanormal agencies (which seem to have their own processes, their own kinds of energy, their own abilities to enhance our present activities). Thus, we can preserve our rich society, enhance its internal cohesion, and bring its multitudinous hierarchy into resonance with new sources of harmony, goodness, and power.[†]

But perhaps I have worked this metaphor enough to illuminate my central suggestion, namely that creative practices draw upon our entire organism, sensitively guiding its various processes toward new efficiencies, enhancing contact among them, bringing them into resonance with metanormal activities. To do this, our practices must promote perceptual, kinesthetic, communication, and movement abilities; vitality; cognition; volition; command of pain and pleasure; love; and bodily structures. All of this involves social creativity, as none of us can develop without considerable help from our fellows. Indeed, we need many virtues and traits that help produce good societies in general, among them charity, courage, forgiveness, and balance. In chapter 26.1, I list some of these virtues and practices to cultivate them. Comprehensive disciplines, in short, need all our parts, all our processes, and creative social support.

[†] Analogies between organisms (or minds) and societies have been developed at length, though to support different theories, in Miller, J. G., 1978; in Minsky 1986; and in Bateson 1975. Using a metaphor about learning we might adapt to my analogy above, Minsky wrote:

> Human minds don't merely learn new ways to reach old goals: they also learn new ways to learn new goals. If we did that without constraint, we'd soon fall prey to accidents— both in the world and in the mind. At the simplest levels, we need protection against accidents like learning not to breathe, at higher levels we must not acquire lethal goals like learning to suppress our other goals entirely—the way that certain saints and mystics do (Minsky 1986).

[25.3]
THE COMPARATIVE STUDY OF
TRANSFORMATIVE PRACTICES

Among the advantages we enjoy today in creating integral practices is the proliferation of disciplines for cognitive, emotional, and bodily development. General semantics, linguistics, and related disciplines, most of them influenced directly or indirectly by the analytic philosophy that has flourished in British and American universities, give us new understanding of mental process. Never before have the foibles of thought, the good and bad habits of mind, the means of clear intellectual activity, been so thoroughly examined. Any comparative study of transformative practice, and any such practice itself, needs the lessons such disciplines offer. The education of emotions, too, has developed in recent times. Modern depth psychology has increased our understanding of repressed or dissociated feelings, unconscious motivations, and psychodynamics in general, while offering new approaches to health and exceptional functioning. By their insights into the effects of unconscious volitions, and by their discoveries about culture's formative influence on each person's makeup, the human sciences complement the transpersonal perspectives embedded in the religious traditions. (Note that some Christian orders and Eastern meditation schools employ psychologists to counsel their members.) Since Freud, the modern West has produced a yoga of the emotions that can support other transformative disciplines. Contemporary psychotherapy, and the affective education it informs, give us many ways to cultivate our relationships, volitions, and feelings to enhance integral practices.

At the same time, medical science, contemporary sports, and somatic education give us the basis for a physical training with unprecedented variety, richness, and robustness. Never before have so many athletic abilities been cultivated, nor have so many people tried to stretch their physical limits in so many ways, nor has human physiology been so thoroughly understood. As I have suggested, modern sports and the attendant fields of sports medicine and sports psychology constitute a vast laboratory for bodily transformation (see chapter 19). The discoveries by athletes and their trainers of optimal methods for superior performance; the growing lore among somatic educators about sensory, kinesthetic, and motor skills training (see chapter 18); and the developing insights about bodily functioning provided by medical science can assist any practice oriented toward metanormal embodiment. The cognitive, affective, and physical aspects of human functioning, in short, can be improved by numerous discoveries that few, if any previous culture enjoyed. These discoveries and the transformative disciplines they inform could comprise a yoga of yogas, as it were, to embrace our many capacities.

But there are also difficulties for integral practices in those countries that otherwise provide enough leisure and wealth to support them. There isn't much dialogue, for example, between the various organizations that promote athletic, therapeutic, and religious disciplines. The gulf between such disciplines, and the distance between them and academia, impede the cooperative study of high-level change. Furthermore, contemplative practice is not vibrant in Europe, the Americas, or most nations today. Indeed, much of its lore has been lost. The distractions of modern life, the widespread academic distrust of educational programs that embrace metanormal phenomena, and the lack of philosophic support for the cultivation of capacities that do not seem useful to society's immediate needs have produced a social climate not wholly propitious for the enterprise I am proposing. In many ways, integral practices have to work against the cultural grain. Nevertheless, they can incorporate the discoveries noted above, and they will inevitably be pursued by adventurous people. Like outer space, possibilities for extraordinary life beckon, and some of us will accept their challenge.

Given these opportunities and obstacles, then, how might we organize integral practices? One way to begin, I propose, is to compare those transformative methods that promote particular kinds of healing or growth. Theravada Buddhism's *vipassana,* Samkhya yoga, Zen Buddhist *zazen,* Psychosynthesis, and Gestalt Therapy, for example, rely on the noninterfering observation of thoughts, emotions, and sensations. These five practices, three of them ancient and two of them modern, use some form of witness meditation. Similarly, many yoga systems, martial arts, and somatic disciplines rely on slow stretching movements to articulate the functioning of particular muscle groups; while many therapies and contemplative schools employ visualization. A comparative study of transformative practices would reveal many methods that facilitate specific kinds of development.

That such methods and their results are analogous to one another may not be apparent, however, when they are embedded in different traditions. For example, both Theravada Buddhism's *vipassana* and Samkhya practice depend upon noninterfering self-observation, yet these similar disciplines are characterized in different ways. Theravada Buddhism with its doctrine of *anatta* emphasizes the illusory nature of ordinary selfhood; whereas Samkhya philosophy posits a *purusha,* or observing self, that is liberated by witness meditation from *prakriti,* the observable world. In *vipassana,* the ego-sense gives way to emptiness; whereas in Samkhya yoga *purusha* realizes its own essence more clearly. But in spite of their different supporting philosophies, the two meditation practices require the same close attention to internal processes, and both produce a sense of freedom, mastery, and delight. Furthermore, they closely resemble the choiceless awareness encouraged by Zen Buddhism, Psychosynthesis, Gestalt Therapy, and other programs for human growth. In psychotherapy, too, the same method or result is sometimes characterized in different ways. The cathartic recall of traumatic events, for example, might be termed "contact with dissociated parts of the self" by a Gestalt therapist or the "lifting of repression" by a psychoanalyst. Similarly, a successful hypnotic induction might be attributed to role compliance or a distinctive cognitive state by different hypnosis researchers (15.4). In contemplative practice, psychotherapy, and hypnosis, as well as in other disciplines, similar methods and outcomes are frequently interpreted in ways that obscure their similarity.

A comparative analysis of transformative activities also reveals functional analogies between different kinds of practice. Rolfing, for example, promotes an articulation of bodily movement that is analogous to the articulation of awareness catalyzed by witness meditation. In this somatic discipline, adhering myofascia are separated so that the client's muscles might operate more freely, while in analogous fashion, witness meditation tends to loosen congested mental contents. In these two practices, physical or mental structures are teased apart so that the entire organism might function more freely, with improved articulation of its various parts. Their methods are not identical, but functionally analogous.[†] Given the immense number of approaches to human development, it would clarify our understanding of them if we

---

[†] Psychologist James Hillman has described a similar articulation of psychic functioning through the personification of various experiences or complexes as gods or archetypes. By the heightened awareness of inner complexity that such personification produces,

> we become internally more separated, we become aware of distinct parts. Even should unity of personality be an aim, "only separated things unite," as we learn from the old alchemical psychologists. Separation comes first . . . [it] offers internal detachment, as if there were now more interior space for movement and for placing events, where before there was a conglomerate adhesion of parts or a monolithic identification with each and all (Hillman 1975, p. 31).

were to find that some of them depend on certain widely used methods. It would be useful to see whether we could break their molecular structures down into commonly shared elements. If we could thus identify the most effective methods for achieving particular kinds of change, we would be better able to formulate integral practices.

The methods that constitute therapeutic, somatic, athletic, and religious disciplines cannot be exactly mapped, since they change from place to place and over time, but a comparative analysis of them would be useful in several ways. We might see, for example, which have been used in different cultures, and which are culture-bound. By comparing their stated aims and outcomes we might find which human failings they do or do not address, and discover where some are weak and others strong in relation to particular human attributes. Indeed, we might uncover dimensions of transformative practice that are little known today and incorporate once-esoteric insight into a publicly available body of knowledge. To begin such a study, the most enduring and prominent practices from ancient and modern times might be listed and their methods identified. These methods could then be grouped according to their essential modalities and results. The psychological and somatic processes they affect, the capacities they facilitate, the virtues they promote, the cultural norms, expectations, and belief systems in which they have been embedded could be compared and analyzed. Such an enterprise would be a fertile adjunct to many fields of inquiry, including philosophy, psychology, and medical science.

But there is a simpler way to begin this comparative study, namely to list the capacities, virtues, and traits that balanced development involves, then identify methods that promote them. I will attempt this approach in the chapter that follows. Since, however, my purpose is simply to be suggestive, I will frame practice outcomes in quite general terms and will not attempt a comprehensive inventory of methods to realize them.

Carl Jung, too, wrote at length about the dual process of *solve et coagulo,* separation and coagulation, by which the European alchemists symbolically projected the articulation of psychic process into their distillation of the physical elements.

The alchemist saw the essence of his art in separation and analysis on the one hand and synthesis and consolidation on the other. For him there was first of all an initial state in which opposite tendencies or forces were in conflict; secondly, there was the great question of a procedure which would be capable of bringing the hostile elements and qualities, once they were separated, back to unity again (Jung 1970b, p. xiv).

26

INTEGRAL PRACTICES

[26.1]
TYPES OF CHANGE NEEDED FOR INTEGRAL DEVELOPMENT AND PRACTICES THAT PROMOTE THEM

Here I will list types of change needed for integral development in two groups, the first related to the 12 sets of developing attributes listed in chapters 4 and 5, and the second related to virtues or traits that are necessary for many-sided human growth. Though practice outcomes could be grouped in other ways, this inventory exemplifies one way in which transformative practices can be compared. Since all our psychological and somatic processes are interdependent to some degree, it is impossible to separate practice outcomes and methods into neat categories.

There is another reason why we cannot map practice outcomes with great exactitude; namely, that desired capacities can be cultivated directly—for example, when concentration is improved by focusing upon a single object; or indirectly—for example, when tightly focused concentration loosens perceptual patterning. The countless ways in which desired outcomes are practiced directly include: strengthening of empathy by imaginatively embracing another's condition; exercise of courage in fear-provoking situations; cultivation of an image by close attention to it; deliberate relinquishment of mental events to strengthen detachment; separation of myofascia by Rolfing to improve the articulation of muscular activity; and deliberate relaxation of particular muscles. The innumerable ways desired outcomes can be cultivated indirectly include: facilitation of the relaxation response by intense recitation of a mantram; elicitation of high arousal by quiet meditation;[†] evocation of free-flowing

---

[†] Roland Fischer, Julian Davidson, and other researchers have described swings between the trophotropic and ergotropic responses induced by religious practice, as for example when ecstasy arises from deep spiritual repose (Fischer 1971; and Davidson, J. M., 1976).

imagery through attention to a single point; strengthening of will by surrender to the demands of a difficult regimen; and the cultivation of serenity by strenuous physical activity.

Practice-Outcomes Related to the 12 Sets of Attributes Listed in Chapters 4 and 5

Perception of external events. The perception of ordinary sensory stimuli can be cultivated through:

1. Hypnotic suggestions for visual, auditory, tactile, gustatory, or olfactory range and acuity.
2. Somatic disciplines such as Sensory Awareness that promote sensitive hearing, taste, or touch.
3. Fitness training and somatic education that promote general sensory alertness.
4. Martial-arts training for visual and auditory range and discrimination.
5. Skill-specific training for sensory acuity and range, as in sports or wine tasting (see 5.1).
6. Meditation that quiets mental interference with sensory experience, stimulates synesthesias, and gives rise to perceptions of numinous beauty in the world at large.

There is also evidence that clairvoyant perception of physical objects, or remote viewing, and perceptions of extrasomatic events can be cultivated through procedures described by religious adepts and modern psychical researchers (see appendices A.1 and A.2, and the subsection on metanormal clairvoyance in chapter 5.1).

Somatic awareness and self-regulation can be cultivated through:

1. Biofeedback training.
2. The practice of visual, auditory, gustatory, or olfactory imagery that conveys information about particular body parts.
3. Disciplined observation of kinesthetic and tactile sensations during physical movement or repose.
4. The deliberate elicitation and modification of kinesthetic experience, as in Sensory Awareness, the Feldenkrais Method, or Reichian therapy.

5. Elicitation of the trophotropic response, as in Autogenic Training and quiet meditation.

6. Close attention to kinesthetic and tactile impressions during any of the practices noted in other parts of this inventory.

Communication abilities involve language, gesture, facial expression, touch, and extrasomatic interactions, which can be cultivated through:

1. Somatic disciplines that increase the range, subtlety, and liveliness of expression by relieving chronic tensions, increasing somatic awareness, and articulating the activity of various body parts.

2. Psychotherapy and affective education that facilitate sensitive, straightforward, and creative relationships by:
   a. promoting awareness of one's habitual attitudes, motives, and impulses
   b. increasing empathy through role-playing
   c. eliciting interpersonal activity for reflection upon it
   d. encouraging new forms of personal interaction, as in affective education, psychodrama, or family therapy

3. Sustained self-reflection that reveals strictures on expressive abilities, deepens empathy, and helps one reach out to others with greater understanding and skill.

4. Prayer, meditation, and other religious practices that produce sustained rapport among friends or student and teacher.

5. Acts of service and charity that promote helping skills and strengthen bonds with other people.

Vitality can be increased through:

1. Psychotherapy that lifts repressions, resolves internal conflicts, and unblocks defenses against strong feeling.

2. Somatic disciplines that reduce chronic tensions, promote regenerative relaxation, and make available energetic reserves.

3. Athletic training that improves blood circulation, metabolic efficiency, and general fitness so that more energy is available for mental or physical activity.

4. Martial arts that promote mental alertness, emotional balance in stressful circumstances, and general somatic efficiency.

5. Meditation or other religious practices that reduce draining emotions, unify conflicting volitions, and promote access to the subliminal depths of mind and body.

All the practices just noted can increase their practitioners' vitality, and sometimes trigger energetic mutations that resemble the *tumo*, kundalini, *incendium amoris*, "boiling *n/um*," and shamanic fortitude described in chapters 5.4, 13.2, 21.2–4, and 22.6.

Movement abilities. The body's movement abilities depend on several physical capacities, among them:

1. Balance, which can be developed through aikido, tai chi, and other martial arts; dance; and sports such as rock climbing, gymnastics, ice skating, and surfing.
2. Cardiopulmonary endurance, which can be developed through running, swimming, rowing, cross-country skiing, race-walking, cycling, and other endurance sports.
3. Agility, which can be developed through aikido, karate, and other martial arts; dance; and sports such as rock climbing, basketball, and soccer.
4. Flexibility, which can be developed through various stretching routines, modern dance, hatha yoga, tai chi, aikido, and somatic disciplines such as the Feldenkrais Method.
5. Muscular strength and power, which can be developed through weightlifting, calisthenics, and physical labor.
6. Coordination, which can be developed through sports, dance, and the martial arts, as well as by somatic disciplines such as Rolfing that articulate the functioning of separate muscles and ligaments.
7. Speed, which can be developed by cultivating all the bodily capacities noted above, and by the speed training developed by modern athletes and coaches.

The cultivation of all these capacities can be assisted by hypnotic suggestions, imagery-practice and mental discipline developed by sports psychologists (see chapters 15.10, 15.11, 17.2, and 20.3), and by certain elements of sport that contribute to general athletic success (see chapter 19.4). There also exist practices to develop out-of-body experience and entrance into "other worlds," which I have called types of metanormal movement (see chapter 5.5 and appendix E).

Abilities to alter the environment directly. In chapter 5.6, I described two ways in which humans can influence their environments; namely, through hand-eye skills and through extrasomatic influences (psycho-kinesis). Hand-eye skills can be developed through:

1. Somatic disciplines (such as Rolfing that articulates the muscles and ligaments of the hand, or Sensory Awareness) that promote sensitive, tender, and loving touch.
2. Sports such as tennis, baseball, and golf.
3. Martial arts such as karate, kendo (swordsmanship), and kyudo (archery).

Extrasomatic influence upon the environment can be developed through:

4. Prayer, spiritual healing exercises, and other practices in which telergy is exercised (see chapter 13).
5. Martial arts, shamanic, yogic, and contemplative disciplines that promote telekinetic abilities (see 20.1 and 22.11).
6. Long meditation practice that alters some portion of physical space (5.6 and 13.1).

Pain and pleasure. Pain can be controlled or eliminated, pleasure increased, and need-transcending joy induced through:

1. Behavior modification or other therapies by which chronic fears are overcome, desires for approval are transcended, and other pain-engendering responses are creatively modified (17.2).
2. Hypnosis, yoga, and other disciplines during which once-painful feats are accomplished without discomfort (15.8, 15.9).
3. Sustained fantasies, during which intensely pleasurable feelings are evoked for long periods (15.6).
4. Strenuous sports or adventure, during which blows are received without pain.
5. Meditation and other religious practices by which painful stimuli are turned into sources of pleasure, and once-boring routines are enjoyed.

Cognition. Transformative practice can develop cognition by bringing new material into its purview or by articulating and strengthening its various processes. I have not made separate listings, though, for methods that extend, articulate, or strengthen cognition because they would be nearly identical. In other words, almost every practice that increases cognition promotes its extension, articulation, and strengthening to some degree. In witness meditation, for example, one might remember long-forgotten events or uncover repressed emotions (thus extending one's cognitive range), while long-familiar trains of thought, which have surfaced countless times before, appear with new articulation. Strong and highly articulated concentration, on the other hand, which can be thought of as a spotlight in relation to the floodlight of witness meditation, often triggers the release of dissociated material for subsequent self-integration. Furthermore, there are many kinds of cognitive skill that can be developed by skill-specific training. Mathematical abilities, for example, can be fostered by the study of mathematics, logical reasoning by the practice of logic, critical thinking by training in analytic philosophy, imagination by creative writing, and contemplative knowing by meditation or prayer. Several practices also facilitate cognitive activity in general. For example:

1. The resolving of psychological conflicts that impede imagination or analytic thought, as in good psychotherapy.
2. The recall of repressed or habitually unnoticed imagery; for example, by emotional catharsis or witness meditation, so that such imagery enriches mental process.
3. The reduction of inhibition to unusual ideas, imagery, or associative process; for example, by psychotherapy, meditation, or philosophic reflection that makes them philosophically and morally acceptable.
4. Strengthening concentration.
5. Exercising unfamiliar types of knowing; for example, through:
   a. concentration on evocative ideas, visual images, sounds, or other stimuli
   b. intensely imagining new worlds suggested by fantastical literature, contemplative writings, dreams, or altered states of mind
   c. establishing contact with ego-transcendent realities by
      i. imagining such realities with concentrated attention until tangible contact with them is established
      ii. prayerful communion with them
      iii. surrendering to their activity

    iv. noninterfering self-observation that deepens an awareness more fundamental than particular mental contents

    v. deliberately emptying the mind so that its fundamental essence is directly experienced

6. Integration of analytic, holistic, and imaginative thought by the study of philosophy, myth, artistic works, or religious symbols.

Volition. There are many kinds of volition, many of them biologically rooted and all of them shaped by social conditioning. Some are dissociated from ordinary waking consciousness; some are pathologically compulsive; some are easily accessible to deliberate modification and control. Indeed, we are multivolitioned and driven by many kinds of willing. But we can cultivate and harmonize these various volitions by:

1. Noninterfering observation of them.
2. Focused recall of those that are repressed or unnoticed through:
   a. guided imagery
   b. catharsis in meditation or psychotherapy
   c. exaggerated exercise of them, as in psychodrama or Gestalt Therapy
3. Articulating them through:
   a. the practice of particular physical acts
   b. the control of autonomic processes and sensory activity, as in biofeedback training and hatha yoga
   c. the practice of complex social skills; for example, during group therapy, psychodrama, and conflict-resolution programs
   d. the practice of complex cognitive skills
4. Strengthening them through:
   a. action against habits and impulses, as in ascetic renunciation
   b. acceptance of pain; for example, in strenuous athletic training
   c. facing fears, as in certain behavior therapies or high-risk sports
   d. deliberate repetition of mental or physical acts, as in prayer, meditation, or endurance sports
   e. suggestions and exhortations to encourage difficult activities, as in sports or religious retreats

5. Integrating them through:
   a. resolution of internal conflicts, so that one's aims can be pursued more wholeheartedly
   b. resolution of moral dilemmas that impede creative activity in general
6. Joining them with ego-transcendent agencies during inspired moments in sports, meditation, prayer, artistic work, charitable service, or other activities.

Individuation and sense of self. A subjectivity that perceives its fundamental oneness with others can be cultivated through:

1. Witness meditation in action or repose during which thoughts, imagery, emotions, sensations, and volitions are noted and relinquished.
2. Communion with a transcendent Presence, Power, or Principle that produces an identity beyond one's ordinary sense of self.
3. Action dedicated to a transcendent principle in which self-regarding concerns for results are relinquished.

The realization of a subjectivity, or personhood, beyond the ordinary self provides a strong basis for the development of all human attributes. The psychological freedom and security that such experience imparts facilitates the growth of cognitive, emotional, and behavioral capacities, and thus promotes individual uniqueness.

Love has many elements, among them delight in others for their own sake, and:

1. Empathy, which can be developed by:
   a. experiencing actual situations that others experience
   b. imaginatively entering another's experience during role-playing, intimate conversation, or solitary reflection
   c. extending the range and depth of emotions through:
      i. cathartic recall of suppressed or forgotten feelings
      ii. noninterfering observation of them
      iii. concentration upon visual, auditory, or other imagery that evokes them
      iv. reframing beliefs and attitudes to permit their cultivation

  v. articulating them in therapy, play, or artistic expression
  vi. strengthening them through practiced fantasies and expressive routines
2. Desire that others thrive, which can be strengthened by:
   a. self-examination or therapy that reduces competitiveness and needs for dominance
   b. practices that promote one's own integration, well-being, and sense of personal security
   c. philosphical reflection that reveals the similarity or identity of one's own and another's highest ends
3. A well-being that overflows to others, which can be cultivated through all the practices noted here and by mutual self-disclosure.

Bodily structures and processes can be improved by:

1. Physical manipulations that articulate the structure of particular muscles, ligaments, and tendons, as in Rolfing.
2. Vibratory movements that break down chronic tensions, as in Reichian therapy.
3. Stretching routines, weightlifting, and other exercises that articulate or strengthen particular muscles, tendons, and ligaments.
4. Routines that develop specific components of physical conditioning, such as interval training for cardiopulmonary endurance, agility exercises, and flexibility training.
5. Training voluntary control of particular autonomic processes, as with biofeedback.
6. Exercises such as meditation that facilitate the Relaxation Response.
7. Mental imagery of desired somatic changes.
8. Comprehensive practices such as tai chi and aikido that help integrate the entire body.

Virtues and Traits Necessary for Integral Development

All the capacities noted above depend on the exercise of many virtues and traits, all of which can be developed through practice. For example:

Honesty can be cultivated through:

1. Disciplined self-reflection or small group interactions that promote self-awareness.
2. The practice of truth-telling.
3. Feedback from others about the effects of one's dishonesty and truth-telling.

Creativity can be nurtured by the cultivation of:

1. Openness to experience, willingness to temporarily relinquish reality testing, nondefensiveness in relation to strong ideas or feelings, perceptual flexibility, and tolerance of ambiguity, which can be developed through:
   a. small group processes that provide safe circumstances for unusual self-disclosure
   b. psychotherapy or meditation that provokes the recovery of repressed or dissociated psychological material
   c. hetero- or self-suggestions that fortify acceptance of potentially threatening thoughts, perceptions, feelings, and impulses
   d. arguments or exhortations that encourage dispassionate examination of disturbing stimuli
   e. regular exposure to fantasy, science fiction, or other writing that recontextualizes ordinary experience
   f. exposure to foreign cultures
   g. meditation, religious ritual, and other practices that promote altered states of consciousness.
2. Independence of judgment, which can be developed through:
   a. disciplined practice of critical thinking
   b. therapy, self-reflection, and other practices that clarify one's compulsive attachment to particular concepts, beliefs, and expectations
   c. meditation that strengthens personal security by revealing a spiritual ground or identity beyond particular ideas
3. Regression in the service of the ego, triggered by sustained witness meditation, psychotherapy, immersion in imaginative literature, or surrealistic reveries during which barriers to unusual thoughts, feelings, perceptions, or behaviors are lowered.
4. Willingness to take risks, which can be developed by exercising courage in:

a. safe therapeutic settings
b. brainstorming exercises that prohibit inhibiting criticism
c. nonstructured situations in which one is not penalized for failure and is positively reinforced for success.
5. Unusual capacities for intuition, which can be cultivated by hypnosis, imagery practice, or sustained meditation that triggers perceptions about people or things that can be checked for their truth or falsity.

Courage can be developed through:

1. Controlling fear by:
   a. self-reflection, therapy, meditation, and other practices that reveal hidden causes of fear so that they may be clearly addressed
   b. exposure to stimuli of gradually increasing threat, as in some behavior-modification procedures
   c. direct exposure to threatening situations with a resolve fortified by self-reflection and support from others
   d. catharsis resulting from practices such as those noted above, which helps alleviate anxiety.
2. Rebound after setback or defeat, which can be strengthened by:
   a. group support and encouragement
   b. meditation, self-reflection, and imagery practice that produce emotional buoyancy and relinquishment of self-defeating mental processes
   c. physical exercise that induces a sense of well-being
   d. hypnotic or waking suggestions to reframe one's ideas about failure
   e. acceptance of psychological or physical blows as gifts, as in aikido and Leonard Energy Training[1]
   f. dedication to a long-term practice
3. Initiative in the face of difficulty, which can be strengthened by attempting things one is normally afraid to do.

Balance, stability, and calm in the midst of turbulent or threatening circumstances can be developed through:

1. Physical exercise that reduces free-floating anxiety while promoting steadiness of mind and body.

2. Therapy that reduces internal conflicts, anxiety, and vulnerabilities to particular challenges.
3. Martial-arts training, dance, and sports that promote grace under pressure.
4. Visualization of a personal or impersonal Divinity, higher self, or other supreme principle.
5. Witness meditation.
6. Contemplative prayer.

Resilience can be developed through:

1. Reduction of habitual muscular tensions by somatic disciplines or biofeedback training.
2. Self-reflection that produces hopeful, though realistic, perspectives on life in general.
3. Relinquishment of chronically negative attitudes in psychotherapy or witness meditation.
4. Practiced contact with the self-existent delight revealed by witness meditation, contemplative prayer, and other religious exercises.

This brief inventory could be greatly extended, of course, but is sufficient to show that we can learn something about transformative practices by a comparative analysis of them. The listing above, for example, reveals the interdependence of capacities that constitute particular traits or virtues. Honesty, creativity, courage, balance, and resilience, like other prime character traits, have more than one component and might best be characterized as virtue families or attribute clusters.

Furthermore, a listing of this kind suggests that many disciplines promote attributes beyond those to which they are primarily addressed. Athletic regimens that have as their goal improved cardiopulmonary fitness also can improve mood and emotional stability. Religious exercises that aim for contemplative illumination can produce great physical stamina. Indeed, the frequent appearance of certain practices in this inventory shows that they enhance many kinds of capacity. Like the general, all-at-once responses described in previous chapters, they accomplish many changes in economical fashion. Even if, like fitness training, they ordinarily serve limited ends, they can also facilitate integral practices. A comparative study of transformative disciplines suggests that the cultivation of particular virtues or capacities typically gives rise to others. Caring for other people tends to produce a joy in their good that lifts us beyond everyday needs and desires. Loving service and friendship, in short, have a tendency toward metanormal delight. Similarly, intellectual curiosity can lead us to investigate a particular problem with such passion that we inadvertently relinquish habitual modes of thought, inhibitions upon our imagination, or limiting beliefs, so that new dimensions of consciousness open to us. Or the simple courage to face new self-perceptions can evoke awareness of an ego-transcending identity, a personhood beyond one's ordinary constructions of self. The cultivation of any virtue or capacity can facilitate more than one creative attribute.

[26.2]
SOME GUIDING PRINCIPLES FOR INTEGRAL PRACTICES AND INSTITUTIONS THAT SUPPORT THEM

They must be suited to each practitioner's makeup. All transformative disciplines must be adapted to their individual practitioners, even in groups with narrow ideas about human development. The most dogmatic martial-arts or contemplative teachers must bend their methods to some extent to accommodate the different capacities of those in their charge. But the need for flexibility is especially pronounced if our aim is integral transformation. A multidimensional approach requires methods adapted to each of its practitioner's shortcomings, strengths, and stage of growth. For that reason there can be no single or "right" kind of integral discipline with a universally applicable and strictly specified set of techniques. If experience has shown that flexibility is necessary to accomplish narrow ends, it emphatically teaches us that a many-sided development of human nature requires rich, diverse, and adaptable practices.

They promote a simultaneous development of our various faculties. Because they involve all dimensions of human activity, integral practices develop through a complex series of outcomes, each of them produced or nurtured by specific methods such as those listed above. Such methods embrace the entire mind-body complex, illuminating volitions and cognitive processes, enriching emotion and relations with others, promoting many virtues, and giving the body new power and beauty.

They generally require several mentors, rather than a single guru. Since integral practices engage our various capacities, they depend upon teachers of many kinds, including meditation masters, somatic and psychological educators, athletic trainers, and instructors in cognitive skills. People turning today to diverse programs for growth reflect a general interest in many-sided development as well as a general dissatisfaction with various shortcomings of particular religious or therapeutic approaches. Though such eclecticism can encourage shallowness or lack of commitment, it often comes from a drive toward wholeness and a hunger for many-sided fulfillment.

They require a strong and developing autonomy. Only by assuming responsibility for our own development can we learn from our mistakes, sharpen our discrimination, and establish an identity strong enough to sustain the difficulties of high-level change. If we give such responsibility to a guru or group, we may limit our discipline and overlook our particular strengths and shortcomings. If we do not measure guidance from others by our own experience and discernment, we can diminish our unique resources and weaken our personhood. Only through healthy ego-strength can we transcend ourselves.

And even when we surrender to agencies beyond our normal functioning, we must hold to a final reference within. That reference is gradually transformed into an identity beyond the ordinary constructions of self, but it remains an identity nevertheless, not the clone of another creature.

They are facilitated by personal traits that promote creativity in general. Frank Barron, Henry Murray, and other psychologists have identified personality traits commonly possessed by creative people, among them:

- tolerance of ambiguity
- openess to experience (including altered states)
- willingness to temporarily set aside reality testing
- nondefensiveness in relation to strong feelings or unusual ideas
- independence of judgment
- attraction to complexity and asymmetry
- capacity for creative regression
- capacity for right-brained, or nondominant, hemispheric functioning that mediates holistic thinking
- flexibility of perception and ego boundaries
- the ability to create order from disorder and to see similarities in the dissimilar

- □ the capacity to perceive obscure figures or gestalts
- □ a flair for rendering novel forms from complex stimuli
- □ unusual capacities for intuition
- □ psychological risk-taking
- □ emotional sensitivity
- □ a strong drive to find pattern and meaning
- □ feelings of oneness with others[2]

Though these traits have generally been identified among artists, scientists, and others who produce works separate from themselves, they also facilitate practices that develop human nature itself. They help to catalyze new richness of awareness and broader repertoires of behavior. They help bring new enjoyment, beauty, and creativity to our relationships and into our everyday life.

These traits are evident, too, in hypnotic responsiveness, which may be drawn upon to promote certain kinds of exceptional functioning (see chapter 15.6 and appendix G),[3] in the capacity for transformative imagery,[4] in meditation,[5] and in sports (witness the unpredictable moves of great basketball players, the surprising ring strategies of inspired boxers, the novel movements of great gymnasts and dancers). The development of metanormal capacities requires receptivity to unusual experience and a willingness to substitute novel for habitual responses. The realization of new states and behaviors often necessitates a temporary suspension of reality testing. To explore the terra incognita of our latent capacities, we must love adventure, complexity, and strange territory. Integral practices reorder elements of body and mind as if they were artistic materials, into new forms of power and beauty, and for that reason require those personality traits that promote creativity in general.[†]

[†] In a formulation I find useful for understanding certain aspects of high-level change, Arthur Koestler proposed that creation in science, art, and humor involves the combination of two or more frames of reference that seem incompatible. "I have coined the term 'bisociation'," he wrote, "in order to make a distinction between the routine skills of thinking on a single 'plane', as it were, and the creative act, which . . . always operates on more than one plane. The former may be called single-minded, the latter a double-minded transitory state of unstable equilibrium where the balance of both emotion and thought is disturbed" (Koestler 1964). Like the artistic and scientific acts of creation that Koestler described, convulsive crises in spiritual healing, psychotherapy, athletic training, and religious practice join habitually incompatible elements of mind and body, at least temporarily, to produce new experience and performance. Like all creative acts, they bisociate elements of human experience in novel combinations.

Though they encourage individual autonomy, they require surrender at times to transformative agencies beyond ordinary functioning. In the ego-transcending love and knowing at the heart of integral development, indeed in all the metanormal activities described here, a power or presence beyond our familiar faculties sometimes inspires us. This superordinary activity alters our sense of autonomy, annihilating and fulfilling it at once as it improves our various capacities.

They require patience and the love of practice for its own sake. In his book *Mastery*, George Leonard describes some ways in which human growth is subverted by unwillingness to persevere in self-discipline.[6] There is an antidote to such failure, though, in the recognition that improvements usually come in spurts after long periods of little or no advance. By enjoying the long plateaus of the learning curve and practice for its own sake, we are more likely to persevere in transformative disciplines than if we constantly demand immediate results.

They utilize inherited all-at-once responses, or psychosomatic compliance for high-level change. Meditation elicits many beneficial changes (23.2). Similarly, a simple running program produces several dozen kinds of hormonal, cardiopulmonary, and muscular improvement (19.2). As I have suggested, general all-at-once responses can facilitate transformative practices, making them more efficient and comprehensive at once. Like a good business deal or fertile scientific theory, they produce great return on investment. Paraphrasing Freud, we might call these simultaneous and manifold changes instances of "psychosomatic compliance for high-level change."

Indeed, every significant human change involves a huge number of psychological and somatic events, so that distinguishing general from specific response can be misleading. Nevertheless, the distinction can help us think about transformative practice. Even if it is only a relative distinction, the difference between general and specific responses can help us to improve our strategies for integral development.

They utilize the manifold changes catalyzed by images and altered states. As noted in previous chapters, images and altered states can facilitate complex developments of human functioning. They can activate intricate webs of psychophysical events for healing or growth, help integrate different body parts, and lift various capacities to extraordinary levels.[7] In doing this they resemble all-at-once responses derived from our animal ancestors, but whereas inherited responses act in well-established ways, images and altered states can overrule homeostatic reflexes to produce truly novel abilities. In their more radical effects, they resemble the subsumptive processes evident in much evolutionary advance.

They enlist more than one mediation to achieve particular outcomes. Swami Rama, a yogi whom Elmer Green studied in his laboratory at the Menninger Foundation, could alter his heart rate at will. The swami could do this in at least three ways: by the Valsalva maneuver, by direct control of the vagus nerve, or by "immediate influence upon the cardiac muscle."[8] In this versatility, he exemplified a principle evident in other kinds of self-mastery, namely that particular behaviors can be modified through various physiological pathways, by different processes, using more than one type of mediation.

They surpass limits by negotiation rather than force. Any significant improvement in physical functioning requires a reordering of bodily processes. Distance runners must alter their metabolism as they reach new levels of fitness; body builders must break down muscle fibers to rebuild their physique; gymnasts must stretch tendons to improve their flexibility. And qualities of heart and mind, too, are restructured by transformative practice. The psychic exhaustion experienced by athletes and martial artists, the initiatory sickness of aspiring shamans, the dark night of the soul described by Christian mystics, the pain of Zen meditation retreats, and the Sufis' annihilation in God all result to some extent from breakdowns that facilitate self-surpassing.[9] But such breakdowns can be too severe. To succeed, transformative disciplines must strike a fine balance between deficient and excessive effort, respecting limits while they aim to surpass them. For that reason, successful athletes and realized contemplatives alike are frequently connoisseurs of their own condition. A distance runner may feel his poor training, for example, in the missing glow after a half-hearted workout, or—conversely—in the taut restlessness of excessive fatigue; while an experienced meditator knows the difference between firmly held mindfulness and the jaundiced aura of driven concentration. Discipline is negotiated best with a skillful, as well as a determined, will.

Such negotiation requires experience, because limits change according to circumstance and the organism's fluctuating disposition. Accomplished athletes, martial artists, shamans, and contemplatives must be existentialists, as it were, centered in the now as they move forward in practice, accommodating their efforts to their psychological and physical conditions of the moment. Because the target of transformative practice is always moving, disciplinary skill requires adaptability, patience, and a willingness to stay on the long plateaus of the learning curve as negotiations with limits to growth continue.

In this, of course, limits stretch as well as fluctuate. Our first constraints, whether psychological or somatic, can yield to skillful practice. At the same time, our aspirations must bend to accommodate homeostasis. Self-cultivation involves alterations of our ideals as well as our present makeup. Such accommodation is evident, for example, in the use of imagery for skill acquisition. Though imagery practice can improve performance, the spontaneity required by a scrambling quarterback or driving point-guard can be interrupted by images inappropriate to the situation at hand. The delicacy required for a tricky golf shot can be impeded by obsessive mental pictures. Imagery, like muscles, must conform to requirements of the moment. And a similar malleability is needed, too, in those rarer activities I have termed metanormal. In deep meditation, we sometimes apprehend images of extraordinary presences, partly visual, partly kinesthetic, or even auditory, which have to give way as something still greater happens. Or in aikido, images vanish as our body moves in ways we had not imagined. In all dimensions of growth, in all aspects of our functioning, inspired images must yield at times to greater inspirations. Even our most luminous mental pictures are subject to the negotiations of transformative practice.

Thus in practice we look two ways at once, toward those elements wanting change and toward our ever-developing intimations of a greater activity in us. By simultaneously embracing our ideals and actualities, both of which continually develop, we can begin to effect their marriage. And this holds for the philosophies and mythologies that attend our aspirations for growth. The imagery contained in this book, for example, must develop in the light of further experience if it is to have any usefulness for speculation, research, or practice.

They depend upon improvisation. Any successful discipline remakes us to some extent. It alters habits, eliminates outworn routines, changes psychic and physical structures. This is especially true for practices that simultaneously embrace our cognitive, affective, and physical parts while respecting our unique makeup. Although some disciplines with relatively limited aims have well-marked stages, methods, and results, integral practices make their way in new territories. Though they can draw upon experience from many fields, they have no detailed maps for all their activities, or few clearly established lines of progression. For that reason they depend upon a certain amount of trial and error, the love of adventure, and improvisation.

They utilize images of unity. Meditation on symbols of oneness are part of both ancient and modern transformative practices. Because our various parts tend to mirror one another, such practices facilitate internal unity and our sense of solidarity with others.[10]

They require (and facilitate) conscious transitions between different states of consciousness. The cultivation of continuous awareness in sleep has been taught in various yogic traditions. Sri Aurobindo wrote:

> It is even possible to become wholly conscious in sleep from beginning to end or over large stretches of our dream-experience; then we are aware of ourselves passing from state after state of consciousness to a brief period of dreamless rest, which is the true restorer of the energies of the waking nature, and then returning by the same way to the waking consciousness.[11]

It is also possible to promote conscious transitions between the ordinary waking state and altered consciousness other than sleep. Some Indian Tantric schools, for example, have permitted the use of alcohol as part of their discipline, encouraging their followers to maintain self-awareness while intoxicated. Dervishes hold their inner eye steady while dancing at the edge of exhaustion. Gurdjieff taught self-observation during manual labor, intense exercise, or other conditions where attention typically falters. The Tibetan *Book of the Dead* gives instructions for the maintenance of consciousness through stages of death. By extending unbroken awareness through all our activities, such practices open connections between our dissociated parts, give us more command of habitual behaviors, promote access to our subliminal depths, and deepen realization of a subjectivity beyond all mental or physical happenings.

They depend on a developing awareness that transcends psychological and somatic functioning. Such awareness can disidentify itself from all images, thoughts, impulses, feelings, and sensations by noninterfering self-observation. It comes into its own, so to speak, through a choice-less awareness in action or repose that relinquishes attachment to par-ticular events, either psychic or physical. It provides a growing free-dom from mental, emotional, and physical habits, as well as brief moments of joy and recognitions of a deeper freedom still. With cul-tivation, it opens onto larger vistas and becomes a boundless subjec-tivity that realizes unity everywhere.

This fundamental awareness is its own reward and one of the supreme results of transformative practice. It is also necessary for practical success in both the early and later stages of integral develop-ment. Only its regenerative freedom can mediate the shocks, disin-tegrations, readjustments, and restructurings required for high-level change. Only its comprehensive embrace is spacious enough, clear enough, and constant enough to sustain a radical reconstruction of the human organism, so well adapted to life on Earth. Only an aware-ness beyond our particular parts can outlast our homeostasis.

The Sanskrit language of illumined states is suggestive in this re-gard. *Nirvikalpa samadhi,* for example, a supreme unitive state in which our fundamental identity shines, has no (*nir*) parts (*kalpas*). Of its own fundamental necessity, it transcends formations that arise within it. With its extraordinary perspective, this essential awareness sustains negotiation of internal conflicts more effectively than limited parts of the mind. And even when we resist or reject particular im-pulses and reflexive behaviors, it helps balance our feedback-laced systems. This, of course, is a secret of the best modern therapy and the ancient practice of witness meditation. From this undivided, all-inclusive, buoyant awareness, we can bring our disparate elements into contact and eventual integration.

They orient all our capacities and somatic processes toward the ex-traordinary life arising in us. The 12 sets of attributes described in chapters 4 and 5, which can be cultivated in ways such as those listed in this chapter, give rise to extraordinary versions of themselves. Inte-gral practices orient us toward these emergent attributes so that the full spectrum of grace can operate in us. To do this, they must be sus-tained by a philosophy that embraces our many parts, by an aspiration for many-sided development, and by surrender to an existence greater than the one with which most of us are presently familiar. They place us on a path toward extraordinary life, which, I believe, includes types of love, joy, and embodiment beyond our present abilities to conceive.

FREQUENTLY
USED TERMS

Charism. A term used by Roman Catholic authorities on the religious life to signify special experiences or abilities associated with a life of prayer. Among these are luminosity, "odors of sanctity," telekinesis, and stigmata. For a discussion of such phenomena in Catholic contemplative practice, see chapter 22.

Extraordinary, or metanormal, functioning. Human functioning that in some respect radically surpasses the functioning typical of most people living today. There are many kinds and degrees of such functioning, and at times it appears to be mediated by agencies or principles beyond those apparent in normal experience. In chapter 3, I amplify this definition. *Extraordinary* and *metanormal,* as I use the words here, are synonymous, and I use them interchangeably throughout this book. The word *metanormal* was suggested to me by George Leonard.

Extrasomatic, or extrasensory, experience. Experience that does not depend upon physical processes.

Grace. The process by which unitive knowing, self-transcending love, and other extraordinary capacities emerge in us, during which such capacities appear to be freely given rather than earned, spontaneously revealed rather than attained through ego-centered effort. I amplify this definition in chapter 7.1.

Human nature. As I use the term in this book, it refers to the mind-body complex as a whole, including its somatic, vital, affective, cognitive, volitional, extrasensory, and transpersonal dimensions.

Imagery. Quasi-sensory experience that occurs in the absence of stimuli that produce its genuine sensory counterparts. (For a discussion of various definitions and conceptual confusions associated with the term, see Richardson 1983). Such experience can be visual, auditory, tactile, kinesthetic, gustatory, or olfactory, either faint or vivid, noticed or unnoticed, easy or difficult to apprehend. Evidence provided by anthropology, psychotherapy, experimental psychology, and common human experience strongly suggests that the capacity for imagery is species wide and that human adults have a continual, night-and-day stream of it, to which they may or may not attend (see Pope & Singer 1978).

Integral practice. A discipline to cultivate the physical, vital, affective, cognitive, volitional, and transpersonal dimensions of human functioning in an integrated way. In chapter 26 I discuss some guiding principles for such discipline.

Kundalini. A Sanskrit term for a special power of the mind-body complex that can be cultivated through the practice of yoga. Its awakening is experienced in many ways, some of which have been described in yoga scriptures and some by contemporary psychologists. For a brief discussion of kundalini-type experiences, see chapter 5.4.

Metanormal. See *extraordinary, or metanormal, functioning.*

Paranormal. A term used in reference to certain phenomena that in some way exceed the limits of what is deemed possible on current scientific assumptions. Most of these phenomena (such as telepathy and psychokinesis) involve action at a distance (either temporal or spatial) between minds and other minds or minds and matter.

By the distinction I make in this book, paranormal interactions that occur among animals or in largely unconscious ways during everyday human life are not *metanormal.* (Such interactions have been described by Freud and other psychoanalysts; see 17.3.) Capacities for conscious paranormal functioning (such as deliberate clairvoyance) comprise a sub-set of those capacities I call metanormal (see 5.1).

Psi. A term proposed by B. P. Wiesner and R. H. Thouless, which can be used as either a noun or adjective, to signify paranormal processes and causation. *Psi-gamma* stands for paranormal cognition, *psi-kappa* for paranormal action. *Expressive psi,* a term used by David

Griffin, is synonymous with psi-kappa, *receptive psi* with psi-gamma. The term *psi* is meant to suggest that various paranormal phenomena are aspects of a single process, which in its active or expressive (psi-kappa) mode is called psychokinesis, telekinesis, or telergy, and in its receptive or cognitive (psi-gamma) mode is called telepathy, clairvoyance, or extrasensory perception (ESP).

Siddhi. A Sanskrit term, roughly synonymous with the Roman Catholic term *charism,* for special human capacities that, although they occur spontaneously outside the context of formal discipline, more typically arise as byproducts of transformative practice. Among such capacities are mystical cognitions, clairvoyance, and extraordinary physical abilities.

Synoptic, or integral, empiricism. The acquisition and verification of data from mainstream science, psychical research, comparative studies of religious experience, and other fields that are ordinarily kept apart by scientists and scholars. This approach utilizes experience acquired through sensory, kinesthetic, and extrasensory modalities. In chapter 2.1, I amplify this definition.

Transformative practice. A complex and coherent set of activities that produces positive changes in a person or group. The practices to which I most commonly refer in using the term are either religious, yogic, shamanic, athletic, somatic, therapeutic, or martial arts–related, but other disciplines, too, as well as professions and common activities such as parenting or marriage are so termed in certain chapters.

In most cases, frequently used foreign-language terms such as siddhi and kundalini will not be italicized.

Appendix A

[A.1]
Eminent Accounts of Parapsychology
and Psychical Research

Broad, C. D. 1953. *Religion, Philosophy and Psychical Research.* Harcourt, Brace.

Edge, H., et al. 1986. *Foundations of Parapsychology.* Routledge & Kegan Paul.

Gurney, E., F. W. H. Myers, & F. Podmore. 1886, 1970. *Phantasms of the Living.* Vols. 1 & 2. Scholars' Fascimiles & Reprints.

James, W. 1986. *Essays in Psychical Research.* In *The Works of William James.* Harvard University Press.

Murphy, G. 1961. *Challenge of Psychical Research.* Harper.

Myers, F. W. H. 1903, 1954. *Human Personality and Its Survival of Bodily Death.* Vols. 1 & 2. Longmans, Green.

Proceedings and Journal of the American Society for Psychical Research. American Society for Psychical Research, New York.

Proceedings and Journal of the British Society for Psychical Research. Society for Psychical Research, London.

White R. & L. Dale. 1973. *Parapsychology: Sources of Information.* Scarecrow.

White, R. 1990. *Parapsychology: New Sources of Information.* Scarecrow.

[A.2]
Highly Regarded Parapsychology Experiments

Nils Wiklund of Stockholm's Karolinska Institute asked several past presidents of the Parapsychological Association (which is based in the United States with members from various nations) to list the best parapsychology experiments. Seven of the eight groups of experiments were most frequently nominated in response to Wiklund's request, of which seven were also chosen by members (not presidents) of the same association. One set, however, has since been called into question by parapsychologists. The sets of experiments that were chosen in both surveys, minus the questioned set, are listed here, each in chronological order. See: Wiklund, N. 1984. On the assessment of evidence for psi. *European Journal of Parapsychology,* 5:245–59.

(1)

Schmidt, H. 1969. Precognition of a quantum process. *Journal of Parapsychology,* 33:99–108.

Schmidt, H. 1969. Clairvoyance tests with a machine. *Journal of Parapsychology,* 33:300–306.

Schmidt, H. 1973. PK tests with a high-speed random number generator. *Journal of Parapsychology*, 37:105–19.

(2)

Pratt, J., et al. 1968. Identification of concealed randomized objects through acquired response habits of stimulus and word association. *Nature*, 220:89–91.

Blom, J., & J. Pratt. 1969. A second confirmatory ESP experiment with Pavel Stepanek as a "borrowed" subject. *Journal of the American Society for Psychical Research*, 62:28–45. See also letter by Pratt in *J.A.S.P.R.* 1969, 63:207–209.

Pratt, J. G. 1973. A decade of research with a selected ESP subject: An overview and reappraisal of the work with Pavel Stepanek. *Proceedings of the American Society for Psychical Research*, 30:1–78.

Keil, H. H. J. 1977. Pavel Stepanek and the focussing effect. Research letter of the Parapsychology Laboratory, University of Utrecht (Oct.), no. 8.

(3)

Honorton, C. 1977. Psi and internal attention states. In B. Wolman (ed.), *Handbook of Parapsychology*, pt. 5, chap. 1.

(4)

Schouten, S., & E. F. Kelley. 1978. On the experiment of Brugmans, Heymans, and Weinberg. *European Journal of Parapsychology*, 2:247–90.

(5)

Rhine, J., & J. Pratt. 1954. A review of the Pearce-Pratt distance series of ESP tests. *Journal of Parapsychology*, 18:165–77.

Hansel, C. 1980. *ESP and Parapsychology. A Critical Re-Evaluation*. Chap. 10. Prometheus.

(6)

Pratt, J., & J. Woodruff. 1939. Size of stimulus symbols in extrasensory perception. *Journal of Parapsychology*, 3:121–58.

Pratt, J. 1976. New evidence supporting the ESP interpretation of the Pratt-Woodruff experiment. *Journal of Parapsychology*, 40:217–27.

Hansel, C. 1980. *ESP and Parapsychology. A Critical Re-Evaluation*. Chap. 11. Prometheus.

[A.3]

STUDIES OF SPONTANEOUS PARANORMAL PHENOMENA

Besterman, T. 1932. The psychology of testimony in relation to paraphysical phenomena: Report of an experiment. *Proceedings of the Society for Psychical Research* (May), 40:365–87.

Bridge, A. 1970. *Moments of Knowing: Personal Experiences in the Realm of Extrasensory Perception*. McGraw-Hill.

Dale, L., R. White, & G. Murphy. 1962. A selection of cases from a recent survey of spontaneous ESP phenomena. *Journal of the American Society for Psychical Research*, 56:1:3–47.

Green, C. 1960. Analysis of spontaneous cases. *Proceedings of the Society for Psychical Research,* 53:97–117.

Haight, J. 1979. Spontaneous psi cases: A survey and preliminary study of ESP, attitude and personality relationships. *Journal of Parapsychology* (Sept.), 43:179–204.

MacKenzie, A. 1987. *The Seen and the Unseen.* Weidenfeld & Nicolson.

Osis, K., & E. Haraldsson. 1986. *At the Hour of Death.* Hastings.

Prince, W. F. 1963. *Noted Witnesses for Psychic Occurrences.* University Books.

Rhine, L. E. 1961. *Hidden Channels of the Mind.* Macmillan.

Rhine, L. E. 1967. *ESP in Life & Lab: Tracing Hidden Channels.* Macmillan.

Schwartz, E. K. 1952. The psychodynamics of spontaneous psi experiences. *Journal of the American Society for Psychical Research,* 46:3–10.

Schwarz, B. E. 1980. *Psychic-Nexus: Psychic Phenomena in Psychiatry and Everyday Life.* Van Nostrand Reinhold.

Stanford, R. G. 1973. Psi in everyday life. *ASPR Newsletter,* 16:1–2.

Stevenson, I. 1966. Announcement re an unusual collection of spontaneous case material available to investigation and scholars. *Journal of American Society for Psychical Research,* 60:178–79.

Stevenson, I. 1970. *Telepathic Impressions: A Review and Report of Thirty-Five New Cases.* University Press of Virginia.

Stevenson, I., J. Palmer, & R. Stanford. 1977. An authenticity rating scale for reports of spontaneous cases. *Journal of the American Society for Psychical Research,* 71:273–88.

Tyrell, G. N. 1962. *Apparitions.* Macmillan.

West, D. J. 1948. The investigation of spontaneous cases. *Proceedings of the Society for Psychical Research,* 48:264–300.

White, R., & R. Anderson. 1990. *Psychic Experiences: A Bibliography.* Parapsychology Sources of Information Center.

White, R. 1981. Saintly Psi: A study of spontaneous ESP in saints. *Journal of Religion and Psychical Research,* 4:157–67.

[A.4]
STUDIES CONCERNING THE PREVALENCE OF PARANORMAL PHENOMENA

Evans, C. 1973. Parapsychology—What the questionnaire revealed. *New Scientist,* 517:209.

Gallup, G. 1982. *Adventures in Immortality.* McGraw-Hill.

Gallup Poll. 1985. Press release, May 14.

Greeley, A. 1975. *The Sociology of the Paranormal: A Reconnaissance.* Sage Publications.

Greeley, A. 1987. Mysticism goes mainstream. *American Health* (Jan./Feb.), pp. 47–49.

Hay, D. 1979. Religious experience amongst a group of postgraduate students: A qualitative study. *Journal for the Scientific Study of Religion,* 18:164–82.

Hay, D. 1982. *Exploring Inner Space: Scientists and Religious Experience.* Penguin.

McCready, W., & A. Greeley. 1976. *The Ultimate Values of the American Population.* Sage Publications.

Palmer, J. 1979. A community mail survey of psychic experiences. *Journal of the American Society for Psychical Research,* 73:221–51.

Sidgwick, E. 1984. Report on *Census of Hallucinations Proceedings of the Society for Psychical Research,* 10:25–422.

Thomas, L., & P. Cooper. 1978. Measurement and incidence of mystical experience: An exploratory study. *Journal for the Scientific Study of Religion,* 17:433–43.

Thomas, L., & P. Cooper. 1980. Incidence and psychological correlates of intense spiritual experiences. *Journal of Transpersonal Psychology,* 12:1:75–85.

West, D. J. 1948. A mass-observation questionnaire on hallucinations. *Journal of the Society for Psychical Research,* 34:187–98.

[A.5]
PSYCHIATRIC AND PSYCHOLOGICAL STUDIES
OF PARANORMAL PHENOMENA

Avriol, B. 1988. Telepathy occurrences and psychoanalytic practice. *Revue Francaise de Psychotronic* (April–June), 1:19–20.

Devereux, G., ed. 1953. *Psychoanalysis and the Occult.* International Universities Press.

Enrenwald, J. 1940. Psychopathological aspects of telepathy. *Proceedings of the Society for Psychical Research,* 46:224–44.

Ehrenwald, J. 1942. Telepathy in dreams. *British Journal of Medical Psychology,* 19:313–23.

Ehrenwald, J. 1944. Telepathy in the psychoanalytic situation. *British Journal of Medical Psychology,* 20:51–62.

Ehrenwald, J. 1948a. Neurobiological aspects of telepathy. *Journal of the American Society of Psychical Research,* 42:132–41.

Ehrenwald, J. 1948b. *Telepathy and Medical Psychology.* W. W. Norton.

Ehrenwald, J. 1949. Quest for psychics and psychical phenomena in psychiatric studies of personality. *Psychiatric Quarterly,* 23:236–47.

Ehrenwald, J. 1950a. Presumptively telepathic incidents during analysis. *Psychiatric Quarterly,* 24:726–43.

Ehrenwald, J. 1950b. Psychotherapy and the telepathy hypothesis. *American Journal of Psychotherapy,* 4:51–79.

Ehrenwald, J. 1954. Telepathy and the child–parent relationship. *Journal of the American Society of Psychical Research,* 48:43–55.

Ehrenwald, J. 1956. Telepathy: Concepts, criteria, and consequences. *Psychiatric Quarterly,* 30:425–49.

Ehrenwald, J. 1966–67. Why Psi? *Psychoanalytic Review,* 53:647–63.

Ehrenwald, J. 1975. *New Dimensions of Deep Analysis.* Arno.

Eisenbud, J. 1954. Behavioral correspondences to normally unpredictable future events. *Psychoanalytic Quarterly,* 23:205–33, 355–87.

Eisenbud, J. 1949. Psychiatric contribution to parapsychology: A review. *Journal of Parapsychology,* 13:247–62.

Eisenbud, J. 1955. On the use of the Psi hypothesis in psychoanalysis. *International Journal of Psychoanalysis*, 36:1–5.

Eisenbud, J. 1967. *The World of Ted Serios*. Morrow.

Eisenbud, J. 1970. *Psi and Psychoanalysis*. Grune & Stratton.

Eisenbud, J. 1982. *Paranormal Foreknowledge: Problems and Perplexities*. Human Sciences Press.

Eisenbud, J. 1983. *Parapsychology and the Unconscious*. North Atlantic Books.

Fodor, N. 1945. The psychoanalytic approach to the problems of occultism. *Journal of Clinical Psychopathology*, 7:65–87.

Fodor, N. 1971. *Freud, Jung and Occultism*. University Books.

Freud, S. Dreams and Telepathy. Vol. 18. In the *Standard Edition of the Complete Psychological Works of Sigmund Freud*. Hogarth.

Freud, S. Dreams and Occultism. Vol. 22. In the *Complete Psychological Works*. Hogarth.

Jung, C. 1963. *Memories, Dreams, and Reflections*. Ed. A. Jaffe, trans. R. & C. Winston. Pantheon.

Meerloo, J. A. 1948–51. Papers read to the Medical Section, American Society of Psychical Research.

Meerloo, J. A. 1949. Telepathy as a form of archaic communication. *Psychiatric Quarterly*, 23:691–704.

Meerloo, J. A. 1950. *Patterns of Panic*. International Universities Press.

Meerloo, J. A. 1953. *Communication and Conversation*. International Universities Press.

Pederson-Krag, G. 1947. Telepathy and repression. *Psychoanalytic Quarterly*, 6:61–68.

Schwarz, B. E. 1969. Synchronicity and telepathy. *Psychoanalytic Quarterly*, 56:44–56.

Stekel, W. 1921. *Der Telepathische Traum*. Johannes Baum Verlag.

Ullman, M. 1948. Communication. *Journal of the American Society of Psychical Research*, 42.

Ullman, M. 1952. On the nature of resistence to psi experiences. *Journal of the American Society of Psychical Research*, 46.

[A.6]
STUDIES OF PARENT–CHILD TELEPATHY

Anderson, L. M. 1957. A further investigation of teacher–pupil attitudes and clairvoyance test results. *Journal of Parapsychology*, 21:81–97.

Anderson, M. L. 1957. Clairvoyance and teacher–pupil attitudes in fifth and sixth grades. *Journal of Parapsychology*, 21:1–12.

Anderson, M. L. 1958. A survey of work on ESP and teacher–pupil attitudes. *Journal of Parapsychology*, 22:246–268.

Anderson, M. L. 1958. ESP score level in relation to students' attitudes toward teacher-agents acting simultaneously. *Journal of Parapsychology*, 22:20–28.

Anderson, M., & E. Gregory. 1959. A two-year program and tests for clairvoyance and precognition with a class of public school pupils. *Journal of Parapsychology*, 23:149–77.

Anderson, M., & R. White. 1956. Teacher–pupil attitudes and clairvoyance test results. *Journal of Parapsychology*, 20:141–57.

Ehrenwald, J. 1956. Telepathy and the child–parent relationship. *Journal of the American Society of Psychical Research*, 48:43–55.

Eisenbud, J., et al. 1960. A further study of teacher–pupil attitudes and results on clairvoyance tests in the fifth and sixth grades. *Journal of the American Society for Psychical Research*, 54:72–80.

Louwerens, N. G. 1960. ESP experiments with nursery school children in the Netherlands. *Journal of Parapsychology*, 24:75–93.

Schwarz, B. E. 1961. Telepathic events in a child between 1 and 3 years of age. *International Journal of Parapsychology*, 4:5–52.

Schwarz, B. E. 1971. *Parent–Child Telepathy*. Garrett.

Van Busschbach, J. 1953. An investigation of extrasensory perception in school children. *Journal of Parapsychology*, 17:210–14.

Van Busschbach, J. 1955. A further report of an investigation of ESP in school children. *Journal of Parapsychology*, 20:71–80.

Van Busschbach, J. 1956. An investigation of ESP between teachers and pupils in American schools. *Journal of Parapsychology*, 20:71–80.

Van Busschbach, J. 1959. An investigation of ESP in the first and second grades of Dutch schools. *Journal of Parapsychology*, 23, 227–37.

Van Busschbach, J. 1961. An investigation of ESP in first and second grades in American schools. *Journal of Parapsychology*, 25:161–74.

[A.7]
EXPERIMENTS WITH SUGGESTION AT A DISTANCE

Between October 3, 1885 and May 6, 1886, the influential French psychiatrist Pierre Janet, with Frederic Myers, Dr. Gibert of LeHavre, and other medical people, conducted 25 experiments during which either Janet of Gibert tried to hypnotize a subject, Madame B. or Leonie, at a distance without her knowing they were doing so. The experimenters deemed 19 of these 25 experiments to be successful in that Madame B. performed the particular tasks the experimenters telepathically suggested. Janet made 35 more experiments of this kind in the autumn of 1886, and deemed 16 of them to be successful. A summary of the 25 experiments conducted by Janet and Gibert can be found in Myers's *Human Personality*, appendices to Chapter V, paragraph 568A. The following record of the Janet-Gibert study is taken from Myers's account.[†]

    L. L. Vasiliev, who was chairman of the physiology department at the University of Leningrad, conducted experiments for many years with hypnotic suggestion at a distance. These experiments took place during the 1920s and 1930s until Communist Party policy prevented their continuation, and were resumed during the 1960s. A record of them can be found in a well-written, carefully documented English-language edition of a Vasiliev book, *Experiments in Distant Influence*, published in 1976 by Dutton. This book contains a review of Vasiliev's work by the book's editor,

---

[†] Myers, F. 1954, p. 528.

No. of Experiments	Date	Operator	Hour when Given	Remarks	Success or Failure
1	1885 Oct. 3	Gibert	11.30 A.M.	She washes hands and wards off trance	½
2	" 9	do.	11.40 A.M.	Found entranced 11.45	1
3	" 14	do.	4.15 P.M.	Found entranced 4.30: had been asleep about 15 minutes	1
4	1886 Feb. 22	Janet		She washes hands and wards off trance	[½]
5	" 25	do.	5 P.M.	Asleep at once	1
6	" 26	do.		Mere discomfort observed	0
7	March 1	do.		do. do.	0
8	" 2	do.	3 P.M.	Found asleep at 4: has slept about an hour	1
9	" 4	do.		Will interrupted: trance coincident, but incomplete ...	1
10	" 5	do.	5–5.10 P.M.	Found asleep a few minutes afterwards	1
11	" 6	Gibert	8 P.M.	Found asleep 8.3	1
12	" 10	do.		Success—no details	1
13	" 14	Janet	3 P.M.	Success—no details	1
14	" 16	Gibert	9 P.M.	Brings her to his house: she leaves her house a few minutes after 9	1
15	April 18	Janet		Found asleep in 10 minutes	1
16	" 19	Gibert	4 P.M.	Found asleep 4.15	1
17	" 20	do.	8 P.M.	Made to come to his house	1
18	" 21	do.	5.50 P.M.	My case I.: trace too tardy	0
19	" 21	do.	11.35 P.M.	Attempt at trance during sleep: see my case I	0
20	" 22	do.	11 A.M.	Asleep 11.25: trance too tardy: my case II.: count as failure ...	0
21	" 22	do.	9 P.M.	Comes to his house: leaves her house 9.15: my case III	1
22	" 23	Janet	4.30 P.M.	Found asleep 5.5, says she has slept since 4.30: my case V	1
23	" 24	do.	3 P.M.	Found asleep 3.30, says she has slept since 3.5: my case VI	1
24	May 5	do.		Success—no details	1
25	" 6	do.		Success—no details	1
					19

Anita Gregory, as well as the following references to experimental studies of suggestion at a distance.

Bechterev, V. M. 1920. Experiments on the effects of "mental" influence on the behaviour of dogs. *Problems in the Study and Training of Personality*. Petrograd, pp. 230–265.

Bozzano, E. 1933. Considérations et hypothèses au sujet des phénomènens télépathiques. *Revue Métapsychique*, 3:145.

Caslian. Cited by Warcollier, R. 1926. La Télépathie experimentale. *Les Conferences de l'institute Metapsychique.* Paris.

Janet, P., & M. Gibert. 1886. Sur quelques phénomènes de somnambulisme. *Revue Philosophique,* I et II, 1886.

Janet, P. 1889. *L'automatisme Psychologique.* Paris. Alcan, p. 103.

Konstantinides, K. 1930. Telepathische Experimente zwigschen Athen, Paris, Warschau and Wien. *Transactions of the Fourth International Congress of Psychical Research.* Athens.

Ochorovicz, J. 1887. *De la suggestion mentale.* Paris.

Osty, E. 1925. La télépathie experimentale. *Revue Métapsychique,* 1:5.

Vasiliev, L. L. 1959. *Mysterious Phenomena of the Human Psyche.* Moscow State Political Publishing House.

Warcollier, R. 1927. La télépathie à très grande distance. *Compute rendu du III.* Congrès International des recherches psychiques. Paris.

[A.8]
PARANORMAL PHOTOGRAPHY

Since the discovery of daguerreotypy in the mid-nineteenth century, many people have claimed that images can be produced upon light-sensitized surfaces by paranormal influences. Discussing such claims, psychiatrist Jule Eisenbud wrote:

> The first well-defined allegedly paranormal image is said to have turned up in 1861 when William M. Mumler, the principal engraver of a leading Boston jewelry concern, was trying to get a picture of himself with a friend's camera. Mumler at first assumed the image to be the result of a previously exposed plate but later concluded that it was a "spirit-form," as which it was heralded by the Spiritualist press of New York and Boston as the first permanently visible proof of an afterlife. Mumler shortly thereafter set up a studio where, according to his own and other accounts, he was widely investigated by professional photographers, journalists, and others. When one or two of his purported spirit forms were identified as living persons, however, the inevitable allegations of trickery made it necessary for him to leave Boston. He settled in New York where soon afterward charges of his having swindled credulous persons by means of his spirit photographs were ordered to be brought against him by no less a personage than the mayor.
>
> In a well-publicized trial Mumler was acquitted, not only because of the lack of any concrete evidence against him but because of the number of persons who testified tellingly in his favor. It was brought out not only that in many instances he had nothing more to do with the spirit forms that turned up alongside the sitters than to lend his presence in the room and that in some cases no daugerreotypes or photographs of relatives corresponding to their inexplicable appearance on the exposed plates were known to exist.
>
> From Mumler to the present, more than two dozen persons in a half-dozen countries have claimed to do spirit photography, including *scotography*

(images caught directly on nonexposed film, without the mediation of a camera) and *psychography* (film messages allegedly in the handwriting of deceased persons).

No less than 35 references to spirit photography can be found in the *British Journal of Photography*, mostly in the 1870s. Numerous articles and notes—more than a score on Mrs. Emma Deane alone—appear in the journals and proceedings of the British SPR and the ASPR, while references to the subject in Spiritualist books and journals run into the hundreds.

Although so-called spirit photography has by no means disappeared from the scene, interest shifted in the early part of this century to a type of paranormal photography that did not necessarily involve discarnate entities and in which a wider range of images were obtained than the usual spirit-form extra. The term *thoughtography*, which has been applied to this branch of psychic photography, originated in 1910 when a sensitive was being tested by Tomokichi Fukurai in Japan for clairvoyance of a latent photographic image of a calligraphic character that had been accidentally imprinted by what Fukurai assumed to be the psychic concentration of the clairvoyant. This led to further trials to determine whether the sensitive could deliberately imprint film plates with characters chosen for her. Fukurai zeroed right in on the radiation hypothesis. He requested the sensitive to try to imprint only the middle plate in a triple-decker film sandwich placed inside two boxes. When the sensitive did exactly as requested, Fukurai went on to many types of experiments with this and other sensitives. The results of this work between 1910 and 1913 were published in his *Clairvoyance and Thoughtography*, which appeared in English in 1931.

Experiments somewhat similar to Fukurai's were carried out in France, England, and America. For some reason, however—perhaps the publication in 1933 in the *Proceedings of the Society for Psychical Research* of a damaging exposé of spirit photography—little more was heard of the subject [until a 1962 article that] claimed Ted Serios, a 42-year-old Chicago bellhop, had been putting images onto Polaroid film under conditions that precluded normal means of his doing so. Serios simply stared into a camera lens through a cardboard cylinder about 7/8-inch high and 3/4-inch wide and tripped the shutter. The cylinder was adopted at the suggestion of one investigator several years after the start of his work, to keep Serios's fingers away from the lens. Although the usual result was the expected blurred close-up of his face, this would occasionally be replaced by images of people, objects, or scenes that could not be accounted for under the conditions in force [Eisenbud, 1983, pp. 111–114].

Eisenbud has described his own studies of Ted Serios in:

Eisenbud, J. 1967. *The World of Ted Serios.* Morrow.
Eisenbud, J. 1967. The Cruel Cruel World of Ted Serios. *Popular Photography* (November).
Eisenbud, J. 1970. Light and the Serios Images. *Journal of the Society for Psychical Research,* 45:424–27.
Eisenbud, J. 1972. The Serios "Blackies" and Related Phenomena. *Journal of the American Society for Psychical Research,* 66:180–92.

Eisenbud, J. et al. 1967. Some Unusual Data from a Session with Ted Serios. *Journal of the American Society for Psychical Research*, 62:309–20.

Eisenbud, J. 1970. An Archeological Tour de Force with Ted Serios. *Journal of the American Society for Psychical Research*, 64:40–52.

Eisenbud, J., et al. 1970. Two Camera and Television Experiments with Ted Serios. *Journal of the American Society for Psychical Research*, 64:261–76.

For other accounts of paranormal photography, see:

Carrington, H. 1925. Experiences in Psychic Photography. *Journal of the American Society for Psychical Research*, 19:258–67.

Eisenbud, J. 1972. Some Notes on the Psychology of the Paranormal. *Journal of the American Society for Psychical Research*, 66:27–41.

Eisenbud, J. 1972. Gedanken zur Psychophotographie. *Zeitschrift fur Parapsychologie und Grenzgebiete der Psychologie*, 14:1–11.

Eisenbud, J. 1983. Psychic Photography and Thoughtography. *Parapsychology and the Unconscious*. North Atlantic Books.

Hyslop, J. Some Unusual Phenomena in Photography. *Proceedings of the American Society for Psychical Research*, 3:395–465.

Hyslop, J. 1915. Photographing the Invisible. *Journal of the American Society for Psychical Research*, 9:148–75.

Joire, P. 1916. *Psychical and Supernormal Phenomena*. Marlowe.

Oehler, P. 1962. The Psychic Photography of Ted Serios. *Fate*, 16:68–82.

Rindge, J., et al. An Investigation of Psychic Photography with the Veilleux Family. *New Horizons*, 1:28–32.

Stevenson, I., & J. Pratt, 1969. Exploratory Investigations of the Psychic Photography of Ted Serios. *Journal of the American Society for Psychical Research*, 63:352–54.

Taylor, J. 1893. Spirit Photography with Remarks on Fluorescence. *British Journal of Photography*, 40:167–69.

Warrick, F. 1939. *Experiments in Psychics*. Dutton.

APPENDIX B
SCIENTIFIC AND COMMONSENSE ASSUMPTIONS WHICH INHIBIT BELIEF THAT PARANORMAL EVENTS OCCUR

Philosopher C. D. Broad outlined these "basic limiting principles" that are commonly accepted either as self-evident or as established "by overwhelming and uniformly favourable empirical evidence." The assumptions underlying these principles inhibit belief that paranormal events occur. (Source: Broad, C. D. 1953. *Religion, Philosophy and Psychical Research.* Harcourt, Brace, pp. 9–12.)

(1) *General Principles of Causation.* (1.1) It is self-evidently impossible that an event should begin to have any effects before it has happened.

(1.2) It is impossible that an event which ends at a certain date should contribute to cause an event which begins at a later date unless the period between the two dates is occupied in one or other of the following ways: (i) The earlier event initiates a process of change, which continues throughout the period and at the end of it contributes to initiate the later event. Or (ii) the earlier event initiates some kind of structural modification which persists throughout the period. This begins to cooperate at the end of the period with some change which is then taking place, and together they cause the later event.

(1.3) It is impossible that an event, happening at a certain date and place, should produce an effect at a remote place unless a finite period elapses between the two events, and unless that period is occupied by a causal chain of events occurring successively at a series of points forming a continuous path between the two places.

(2) *Limitations on the Action of Mind on Matter.* It is impossible for an event in a person's mind to produce *directly* any change in the material world except certain changes in his own brain.

(3) *Dependence of Mind on Brain.* A necessary, even if not a sufficient, immediate condition of any mental event is an event in the brain of a living body. Each different mental event is immediately conditioned by a different brain-event.

(4) *Limitations on Ways of Acquiring Knowledge.* (4.1) It is impossible for a person to perceive a physical event or a material thing except by means of sensations which that event or thing produces in his mind. The object perceived is not the immediate cause of the sensations by which a person perceives it. The immediate cause of these is always a certain event in the percipient's brain; and the perceived object is (or is the seat of) a rather remote causal ancestor of this brain-event. The intermediate links in the causal chain are, first, a series of events in the space between the perceived object and the percipient's body; then an event in a receptor organ, such as his eye or ear; and then a series of events in the nerve connecting this receptor organ to his brain. When this causal chain is completed, and a sensory experience arises in the per-

cipient's mind, that experience is not a state of acquaintance with the perceived external object, either as it was at the moment when it initiated this sequence of events or as it now is. The qualitative and relational character of the sensation is wholly determined by the event in the brain which is its immediate condition; and the character of the latter is in part dependent on the nature and state of the afferent nerve, of the receptor organ, and of the medium between the receptor and the perceived object.

(4.2) It is impossible for A to know what experiences B is having or has had except in one or other of the following ways: (i) By hearing and understanding sentences, descriptive of that experience, uttered by B, or by reading and understanding such sentences, written by B, or reproductions or translations of them. (I include under these heading messages in Morse or any other artificial language which is understood by A.) (ii) By hearing and interpreting cries which B makes, or seeing and interpreting his gestures, facial expressions, etc. (iii) By seeing, and making conscious or unconscious inferences from, persistent material records, such as tools, pottery, pictures, etc., which B has made or used in the past. (I include under this head seeing copies or transcriptions, etc., of such objects.)

Similar remarks apply, mutatis mutandis, to the conditions under which A can acquire from B knowledge of facts which B knows or acquaintance with propositions which B contemplates. Suppose that B knows a certain fact or is contemplating a certain proposition. Then the only way in which A can acquire from B knowledge of that fact or acquaintance with that proposition is by B stating it in sentences or other symbolic expressions which A can understand, and by A perceiving those expressions themselves, or reproductions or translations of them, and interpreting them.

(4.3) It is impossible for a person to forecast, except by chance, that an event of such and such a kind will happen at such and such a place and time except under one or other of the following conditions: (i) By making an inference from data supplied to him by his present sensations, introspections, or memories, together with his knowledge of certain rules of sequence which have hitherto prevailed in nature. (ii) By accepting from others, whom he trusts, either such data or such rules or both, and then making his own inferences; or by accepting from others the inferences which they have made from data which they claim to have had and regularities which they claim to have verified. (iii) By non-inferential expectations, based on associations which have been formed by certain repeated sequences in his past experience and which are now stimulated by some present experience.

(4.4) It is impossible for a person to know or have reason to believe that an event of such and such a kind happened at such and such a place and time in the past except under one or another of the following conditions: (i) That the event was an experience which he himself had during the lifetime of his present body; that this left a trace in him which has lasted until now; and that this trace can be stimulated so as to give rise in him to a memory of the past experience. (ii) That the event was one which he witnessed during the lifetime of his present body; that the experience of witnessing it left a trace in him which has lasted till now; and that he now remembers the event witnessed, even though he may not be able to remember the experience of witnessing it. (iii) That the event was experienced or witnessed by someone else, who now remembers it and tells this person about it. (iv) That the event was experienced or witnessed by someone (whether this person himself or another), who made a record of it; and that it is now perceptible by and intelligible to this person. (These four methods may be summarized under the heads of present memory, or testimony based on present memory or on records of past perceptions or memories.) (v) Explicit or implicit inference, either made by the person himself or made by others and accepted by him on their authority, from data supplied by present sense-perception, introspection, or memory, together with knowledge of certain laws of nature.

I do not assert that these instances of basic limiting principles are exhaustive, or that they are all logically independent of each other. But I think that they will suffice as examples of important restrictive principles of very wide range, which are commonly accepted today by educated plain men and by scientists in Europe and America.

For a discussion of Broad's Basic Limiting Principle's, see: Braud, S. E. 1979. *ESP and Psychokinesis.* Temple University Press, 247–55.

Appendix C
Scientific Meditation Studies

References to the studies cited here are listed in the main bibliography.

LOWERED HEART RATE

Several studies have shown that Transcendental Meditation (TM), Zen Buddhist sitting, and other calming forms of meditation produce a lowering of resting heart rate. See: Bono 1984, Bagga & Gandhi 1983, Cummings, V. T. 1984, Pollard & Ashton 1982, Throll 1982, Cuthbert, Kristeller et al. 1981, Lang, Dehob, Meurer et al. 1979, Bauhofer 1978, Correy 1977, Routt 1977, Glueck & Stroebel 1975, Wallace & Benson 1972, Wallace, Benson, Wilson et al. 1971, and Wallace, R. K. 1970.

REDUCED BLOOD PRESSURE

There is good evidence that meditation can help lower blood pressure in people who are normal or moderately hypertensive. This finding has been replicated by several studies, some of which have produced systolic reductions of 25 mmHg or more. In some experiments, combining meditation with biofeedback or other relaxation techniques proved to be more effective than meditation alone. Several studies, however, have shown that relief from high blood pressure diminishes or disappears entirely if meditation is discontinued, and few people with acute hypertension have experienced marked lowering of blood pressure in experiments of this kind. Meditation can help to relax large muscle groups pressing on the circulatory system, and it might also relax small muscles that control the blood vessels themselves, so that the vessels' resulting elasticity helps reduce the pressure inside them. See: Delmonte 1984b, Wallace, Silver, Mills et al. 1983, Bagga & Gandhi 1983, Hafner 1982, Seer & Raeburn 1980, Surwit, Shapiro, Good et al. 1978, Pollack, Weber, Case et al. 1977, Simon, Oparil, & Kimball 1977, Blackwell, Bloomfield, Gartside et al. 1976, Stone & DeLeo 1976, Patel & North 1975, Patel 1975a, b, & c, Benson, Rosner, Marzetta et al. 1974b, Patel 1973, Deabler, Fidel, Dillenkoffer et al. 1973, Benson & Wallace 1972d, Datey, Deshmukh, Dalvi et al. 1969.

CHANGES IN CORTICAL ACTIVITY

Alterations of EEG alpha activity. More than 35 studies have shown that meditation produces marked increases of alpha activity (8 to 12 cycles per second). In several experiments, subjects' alpha waves increased in amplitude as they slowed in frequency. See: Gaylord et al. 1989, Taneli & Krahne 1987, Echenhofer & Coombs 1987, Delmonte 1984b, Daniels & Fernall 1984, Stigsby, Rodenburg, Moth et al. 1981, Lehrer, Schoicket, Carrington et al. 1980, Wachsmuth, Dolce, & Offenloch 1980, West, M.A., 1980a, Dostalek, Faber, Krasa et al. 1979, Corby, Ross, Zargone et al. 1978, Pelletier & Peper 1977, Elson, Hauri, & Cunis 1977, Kras 1977, Fenwick, Donaldson, Gillis et al. 1977, Glueck & Stroebel 1975, Tebecis 1975, Williams & West 1975, Woodfolk 1975, Banquet 1972 & 1973, Vassiliadis 1973, Benson, Malvea, Graham et al. 1973, Wallace, Benson, & Wilson 1971b, Akishige 1970, Wallace R. K., 1970, Kasamatsu & Hirai 1963, 1966, 1969a, & 1969b, Kamiya 1968 & 1969, Anand et al. 1961a, Hirai et al. 1959, Hirai 1960, Kasamatsu et al. 1957, Bagchi & Wenger 1957a & b, and Das & Gastaut 1955.

Alterations of EEG theta activity. More than 20 studies have shown that meditation produces strong bursts of frontally dominant theta waves (4 to 7 cycles per second), especially among advanced meditators.[†] See: Delmonte 1984b, West, M. A., 1979, Hebert & Lehmann 1977, Elson, Hauri, & Cunis 1977, Pelletier & Peper 1977, Fenwick, Donaldson, Gillis et al. 1977, Banquet & Sailhan 1974, Ghista, Mukherji, Nandagopal et al. 1976, Levine, P. H., 1976, Tebecis 1975, Glueck & Stroebel 1984, Krahne & Tenoli 1975, Hirai 1974, Banquet 1972 & 1973, Wallace & Benson 1972, Wallace, Benson, & Wilson 1971, Wallace, R. K., 1970, Kasamatsu & Hirai 1963 & 1966, Anand et al. 1961a, and Bagchi & Wenger 1958.

Bursts of EEG beta activity. During deep meditation, experienced subjects sometimes exhibit bursts of high-frequency beta waves (13 to 40 cycles per second). This sudden activity is often associated by the meditator with intense pleasure, ecstasy, or deep concentration, and it is usually accompanied by an acceleration of heart rate. The following studies have reported bursts of beta activity during meditation: Peper & Ancoli 1979, West, M.A., 1979 & 1980a, Surwillo & Hobson 1978, Corby, Roth, Zarcone et al. 1978, Fenwick, Donaldson, Gillis et al. 1977, Banquet 1973, Kasamatsu & Hirai 1963 & 1966, Anand et al. 1961a, and Das & Gastaut 1955.

Hemispheric synchronization in EEG records. More than 25 studies have shown that meditation can produce synchronization of brain-wave activity between the right and left hemispheres and the anterior and posterior parts of the brain. See: Jevning & O'Halloran 1984, Badawi et al. 1984, Orme-Johnson & Haynes 1981, Dillbeck & Bronson 1981, Dillbeck, Orme-Johnson, & Wallace 1981, Glueck & Stroebel 1975 & 1978, Corby, Roth, Zarcone et al. 1978, Bennett & Trinder 1977, Orme-Johnson 1977, Morse, Martin, Furst et al. 1977, Hebert & Lehmann 1977, Westcott 1977, Haynes, Mosley, McGowan 1975, Banquet & Sailhan 1974, Banquet 1972 & 1973, Wallace, Benson, & Wilson 1971, Wallace, R. K., 1970, Anand et al. 1961a, and Das & Gastaut 1955.

Dehabituation indicated by EEG records. The Sanskrit *anuraga,* constant freshness of perception, is said to be a primary result of yoga; and Zen Buddhist teachers describe the freedom from "perpetual averaging" that zazen helps to produce. Some religious

---

[†] Biofeedback researchers Elmer and Alyce Green have reported strong correlations between high incidence of theta activity and creative imagery (see chapter 16.4).

ecstatics, however, become so absorbed in trance that they inhibit or entirely suppress their responses to external stimuli. Bagchi and Wenger compared yogis and Zen masters in this regard and found significant differences in their EEG responses (see chapter 23.1). The yogis that Bagchi and Wenger studied typically habituated to repeated stimuli more rapidly and completely than Zen masters did in other studies, leading the two psychologists to speculate that the two types of discipline produced either inner absorption or heightened awareness of the outer world. For studies of habituation to stimuli during meditation, see: Delmonte 1984b, McEvoy, Frumkin, Harkins et al. 1980, Davidson, J. M., 1976, Williams & West 1975, Hirai 1974, Wada & Hamm 1974, Banquet 1973, Orme-Johnson 1973, Wallace, Benson, & Wilson 1971, Wallace, R. K., 1970, Akishige 1970, Kasamatsu & Hirai 1963 & 1966, Anand et al. 1961a, and Bagchi & Wenger 1957a & b.

METABOLIC AND RESPIRATORY CHANGES

More than 40 studies have shown that oxygen consumption, carbon dioxide elimination, and respiration rate are reduced during meditation. Some studies, moreover, have shown that meditation lowers oxygen consumption in some subjects working at a fixed intensity, and that meditators sometimes suspend breathing longer than do control subjects, without apparent ill effects. These studies strongly suggest that meditation lowers the body's need for energy and the oxygen to metabolize it. See: Kesterson 1986, Wolkove, Kreisman, Darragh et al. 1984, Morse, Cohen, Furst, & Martin 1984, Singh 1984, Cadarette, Hoffman, Caudill et al. 1982, Jevning, Wilson et al. 1978, Fenwick, Donaldson, Gillis et al. 1977, Peters, Benson, Porter et al. 1977, Benson, Dryer, Hartley et al. 1977, Dhanaraj & Singh 1977, Elson, Hauri, & Cunis et al. 1977, McDonagh & Egenes 1977, Corey 1977, Routt 1977, Davidson, J. M., 1976, Benson, Steinert, Greenwood et al. 1975, Glueck & Stroebel 1975, Woolfolk 1975, Beary & Benson 1974, Hirai 1974, Parulkar, Prabhavalker, & Bhall 1974, Benson, Klemchuk, & Graham 1974, Kanellakos & Lukas 1974, Banquet 1973, Treichel, Clinch, & Cran 1973, Wallace & Benson 1972, Russell, P. J., 1972, Watanabe, Shapiro, & Ochwartz 1972, Goyeche, Chilhara, & Shimizu 1972, Wallace, Benson, & Wilson 1971, Wallace, R. K., 1970, Allison 1970, Sugi & Akutsu 1968, Karambelkar, Vinekar, & Bhole 1968, Kasamatsu & Hirai 1963 & 1966, Anand et al. 1961a, Wenger & Bagchi 1961, Anand & Chhina 1961, and Bagchi & Wenger 1957a & b.

MUSCLE TENSION AND LACTATE REDUCTION

For evidence that meditation helps reduce muscle tension, see: Credidio 1982, Zaichkowsky & Kamen 1978, Morse, Martin, Furst et al. 1977, Delmonte 1979 & 1984b, Brandon 1983, Cangelosi 1981, Pelletier & Peper 1977, Haynes, Mosley, & McGowan 1975, Ikegami 1974, Gellhorn & Kiely 1972, and Das & Gastaut 1955.

High blood lactate concentrations have been associated with anxiety and high blood pressure, while the infusion of lactate in the blood has been found to produce symptoms of anxiety. Several studies have produced declines in blood lactate during meditation, among them: Bagga, Gandhi, & Bagga 1981, Jevning, Wilson et al. 1978, Jevning & Wilson 1977, Benson 1975, Benson, Malvea, Graham et al. 1973, Orme-

Johnson 1973, Wallace & Benson 1972, Wallace, Benson, Wilson et al. 1971, and Wallace, R. K., 1970.

SKIN-RESISTANCE CHANGES

Low skin resistance as measured by the galvanic skin response test, because it is caused in large part by anxiety-induced perspiration, is generally thought to be a reliable indicator of stress. Increased skin resistance, or lower frequency of spontaneous galvanic skin responses (another stress indicator) have been reported by: Bono 1984, Bagga & Gandhi 1983, Schwartz, Davidson, & Goleman 1978, Sinha, Prasad, Sharma et al. 1978, Pelletier & Peper 1977, Orme-Johnson & Farrow 1977, Farrow 1977, Laurie 1977, West, M. A., 1977, Smith, T. R., 1977, Glueck & Stroebel 1975, Walrath & Hamilton 1975, Woolfolk 1975, Orme-Johnson 1973, Benson, Malvea, Graham et al. 1973, Wallace & Benson 1972, Wallace, Benson, & Wilson 1971, Wallace, R. K., 1970, Akishige 1970, Karambelkar, Vinekar, & Bhole 1968, and Bagchi & Wenger 1957a & b.

SALIVARY CHANGES

D. R. Morse and associates (1982) found that 12 subjects often exhibited significant anxiety reduction by the end of experimental meditation sessions, as well as increased salivary translucency, decreased salivary protein, and lowered saliva bacteria levels. (The lowering of bacteria levels during relaxation indicates that stress contributes to dental caries and that relaxation can have an anticaries effect.) These results support previous findings by Morse about salivary changes produced by meditation. See: Morse, Martin, Furst et al. 1977, Morse, Schacterle, Furst et al. 1981, Morse 1976a & b, 1977a, b, & c, and Morse & Hilderbrand 1976.

IMPROVEMENTS IN RESPONSE TO STRESS

Goleman and Schwartz (1976) exposed 30 experienced meditators to a stressful film while measuring their responses by skin conductance and heart-rate tests, self-report, and personality tests. The heart rates of both experienced and inexperienced meditators recovered from stressful impacts more quickly than those of control subjects. In a study by Glueck and Stroebel (1975), meditators exhibited fewer chronic or inappropriate activations of the fight-or-flight response.

ALLEVIATION OF PAIN

Kabat-Zinn and associates (1987) studied 225 chronic-pain patients after training them in mindfulness meditation, and found significant physical and psychological improvements among them as recorded with the Pain Rating Index, measures of negative body image, and number of medical symptoms. Earlier, Kabat-Zinn and associates (1985) had trained 90 chronic-pain patients in mindfulness meditation and had found statistically significant reductions in their present-moment pain, negative body image, and activity caused by pain, anxiety, or depression. These improvements appeared to be independent of gender, source of referral, or type of pain. A comparison group of patients did not show significant improvement on these measures after traditional treatments. For other evidence that meditation can produce significant pain relief, see: Hustad & Carnes 1988, Kabat-Zinn, Lipworth, Sellers et al. 1984, Kabat-Zinn 1982, Kabat-Zinn & Burney 1981, Mills & Farrow 1981, Pelletier & Peper 1977, Lovell-Smith 1985, Buckler 1976, Katcher, Segal, & Beck 1984, Morse, Schacterle, Furst et al. 1981, Morse & Wilcko 1979, Morse 1976a & b, 1977a, b, & c, Goleman 1976, Anderson, D. A., 1984, Benson, Malvea, Graham et al. 1973, Fentress, Masek, Mehegan et al. 1986, Benson et al. 1984, and Kutz, Caudill, & Benson 1983.

POSITIVE BEHAVIORAL EFFECTS

Heightened perception. Contemporary studies have produced some evidence that the following capacities are improved by meditation:

- □ Visual sensitivity, as measured by tachistoscopic identification of light flashes or image sequences (Dillbeck 1977b & 1982), by the Ames Trapezoid Illusion Test (Martinetti 1976), and by other devices (Nolly 1975 and Heil 1983).
- □ Auditory acuity as measured by brainstem and cortical auditory evoked potentials (Wandhofer & Plattig 1973, Clements & Milstein 1977, Pirot 1977, and McEvoy, Frumkin, Harkins et al. 1980); and by other methods (Keithler 1981).
- □ The discrimination of musical tones (Pagano & Frumkin 1977).

For other evidence that meditation can improve perception, see: Forte, Brown, & Dysart 1984–85, Shapiro, D. H., & Giber 1978, Shaw & Kolb 1977, Davidson, Goleman, & Schwartz et al. 1976, Davidson, Schwartz, & Rothman 1976; and Udupa 1973. In summarizing one meditation study, research psychologist Daniel Brown wrote that

> practitioners of the mindfulness form of Buddhist meditation were tested for visual sensitivity before and immediately after a three-month retreat during which they practiced mindfulness meditation for sixteen hours each day. A control group composed of the staff at the retreat center was similarly tested. Visual sensitivity was defined in two ways: by a detection threshold based on the duration of simple light flashes and a discrimination threshold based on the interval between [them]. All light flashes were presented tachistoscopically and were of fixed luminance. After the retreat, practitioners could detect shorter single-light flashes and required a shorter interval to differentiate between successive flashes correctly. The control group did not change on either measure. Phenomenological reports indicate that mindfulness practice enables practitioners to become aware of some of the usually preattentive processes involved in visual detection. The results support the statements found in Buddhist texts on meditation concerning the changes in perception encountered during the practice of mindfulness (Brown, D., et al. 1984a & b).

Improvements in reaction time and responsive motor skill. See: Jedrczak, Toomey, Clements et al. 1986, Robertson 1983, Warshal 1980, Holt, Caruso, Riley et al. 1978, Sinha, Prasad, Sharma et al. 1978, Blackwell, Bloomfield, Gartside et al. 1976, Appelle & Oswald 1974, Shaw & Kolb 1977, Orme-Johnson 1973, and Blasdell 1977.

Deautomatization. Psychiatrist Arthur Deikman (1963, 1966a & b) found that meditating subjects perceived a blue vase to be more fluid and luminescent than they had when the experiment started. On the basis of this finding he suggested that meditation deautomatizes perception and emotion, thus facilitating the intense realness, unitive sense, and fluidity of mystical experience.

Increased field independence. The term *field independence* designates the capacity to make visual and kinesthetic discriminations in spite of misleading stimuli from the environment. Such capacity is correlated with independent judgment and a strong sense of body and self, and some studies have shown that it is increased by meditation. See: Jedrczak, Beresford, Clements et al. 1984, Shapiro & Giber 1978, Orme-Johnson & Granieri 1977, Abrams 1977b, Goleman 1976, Smith, J. C., 1975b, Pelletier 1974 & 1976, and Linden 1973.

Concentration/attention. See: Sabel 1980, Spanos, Rivers, & Gottlieb 1978, Moretti-Altuna 1987, Tomassetti 1985, Williams, R. D., 1985, Sinha, Prasad, Sharma et al. 1978, Kelton 1978, Goleman 1976, Davidson, Goleman, Schwartz et al. 1976, Davidson, Schwartz, & Rothman 1976, Walrath & Hamilton 1975, Orme-Johnson & Granieri 1977, Pelletier 1974, Linden 1973, Van Nuys 1973, Deikman 1971, Tart 1971, and Maupin 1965.

Empathy. See: Lesh 1970b, Reiman 1985, Shapiro, D. H., 1980a, Kornfield 1979, Walsh, R. N., 1978, Kohr 1977, Shapiro, D. H., & Giber 1978, Pelletier 1976a & 1978, Davidson, Goleman, Schwartz et al. 1976, Griggs 1976, Kubose 1976, Van den Berg & Mulder 1976, Leung 1973, Udupa 1973, Osis, Bokert, Carlson et al. 1973, Banquet 1973, Van Nuys 1973, Deikman 1966b, & Maupin 1965.

Regression in the service of the ego. The legend that Gautama Buddha witnessed his past lives before he attained enlightenment can be interpreted to mean that meditation facilitates liberation from conditionings of early experience. Some contemporary studies support this interpretation. Shafii (1973), for example, found that meditation returns some people to early fixation points, helping them reexperience developmental traumas on a silent, nonverbal level. This process frees psychic energy, Shafii proposed, increases freedom from earlier patterns of behavior, and promotes more openness to various kinds of learning. Maupin (1965) suggested that meditation brings about a sequence of regressive states; and others have reported that meditation increases adaptive regression. See: Kornfield 1979, Pelletier 1976a & 1978, Moles 1977, and Lesh 1970b.

Anxiety reduction. Participants in several meditation studies have experienced significant reductions of acute or chronic anxiety. Among the reasons for such relief, researchers have proposed: meditation's promotion of calm body-states and emotions; its facilitation of cathartic relief from past traumas; its strengthening of detachment from anxiety-inducing imagery or reflexive responses; and its reduction of scattered thoughts and feelings.[†] See: Muskatel, Woolfold, Carrington et al. 1984, Beiman et al. 1984, Heide & Borkovec 1983, Lehrer, Schoicket, Carrington et al. 1980, Lehrer, Woolfolk, Rooney et al. 1983, Deberry 1982, Woolfolk et al. 1982, Fling, Thomas, & Gallaher 1981, Throll 1981, Carrington et al. 1980, Raskin, Bali, & Peeke 1980, Kirsch & Henry 1979, Benson et al. 1978, Thomas & Abbas 1978, Davies 1977, Stern, M., 1977, Lazar, Farwell, & Farrow 1977, Ross 1977, Shapiro, D. H., 1976, Nidich, Seeman, & Dreskin 1973, Puryear, Cayce, & Thurston 1976, Davidson, Goleman, Schwartz et al. 1976, Goleman & Schwartz 1976, Smith, J. C., 1975, Girodo 1974, Hjelle 1974, and Vahia, Doongaji, Jest et al. 1973.

Relief from addiction. In more than 50 studies of meditation, subjects have reported significant relief from addiction to alcohol, cocaine, marijuana, and other drugs. Reasons suggested for this result include: meditation's promotion of relaxed well-being, which helps relieve the cravings that lead to addiction; its promotion of cathartic recall; and its strengthening of self-control in general. See: Clements et al. 1988, Klajner, Hartman, & Sobel 1984, Marlatt, Pagano, Rose et al. 1984, Wong, Brochin, & Gedron 1981, Parker & Gilbert 1978, Anderson, D. J., 1977, Winquist 1977, Brautigam 1977, Katz, D., 1977, Shafii, Lavely, & Jaffe 1974 & 1975, Marcus 1974, and Benson & Wallace 1972b.

Improvements in memory and intelligence. Several TM studies indicate that meditation helps improve intelligence, school grades, learning ability, and short- and long-term recall. See: Dillbeck, Assimakis, & Raimondi 1986, Jedrczak, Toomey, Clements et al. 1986, Verma, Jayashan, & Palani 1982, Lewis, J., 1978, Orme-Johnson & Granieri 1977, Abrams 1977a & b, Heaton & Orme-Johnson 1977, Collier 1977, Levin 1977, Glueck & Stroebel 1975, and Tjoa 1975.

Some researchers, however, have not found that this is the case: see Yuille & Sereda 1980.

CHANGES REVEALED PRIMARILY
BY SUBJECTIVE REPORT

Besides the observable physiological and behavioral effects noted above, meditation has produced results revealed by interviews, questionnaires, diary entries, and other devices to elicit subjective reports. Among such results, the following have been reported in several competent studies.

Improved perception of internal and external events. See: Forte, Brown, & Dysart 1984–85, Brown, D., et al. 1982–83, Kornfield 1979 & 1983, Goleman 1978–79, and Walsh, R. N., 1977.

[†] Research psychologists R. J. Davidson and Gary Schwartz (1984) suggested that the cognitive and somatic contributions to anxiety are affected in different ways by various meditation styles. Thus, forms of meditation that require breath-counting or repetition of mantrams in synchrony with breathing are particularly effective because they simultaneously attenuate both cognitive and somatic anxiety; while meditation primarily involving cognition (such as TM) probably has a greater effect upon the cognitive rather than the somatic components of anxiety.

Equanimity. See: Pickersgill & White 1984, Goleman 1976 & 1978–79, Kornfield 1979, Pelletier 1976b & 1978, Walsh, R. N., 1977, Davidson & Goleman 1977, Davidson, J. M., 1976, Woolfolk 1975, Hirai 1974, Boudreau 1972, Kasamatsu & Hirai 1966, and Anand et al. 1961a.

Detachment. See: Brown, D., et al. 1982–83, Goldstein 1982, Pelletier 1976a & 1978, Goleman 1977, Walsh, R. N., 1977, Davidson, J. M., 1976, and Mills & Campbell 1974.

Pleasure, bliss, and ecstasy. See: West, M. A., 1980b, Kornfield 1979, Goleman 1978–79, and Farrow 1977.

Altered body image and ego boundaries. See: Kornfield 1979 & 1983, Goleman 1978–79, Krippner & Maliszewski 1978, Piggins & Morgan 1977–78, Woolfolk, Carr-Kaffashan, McNulty et al. 1976, and Deikman 1966a.

Increased energy and excitement. See: Kornfield 1979, Krippner & Maliszewski 1978, Piggins & Morgan 1977–78, Davison, J. M., 1976, Shimano & Douglas 1975, and Maupin 1965.

Hallucinations and illusions. See: Kornfield 1979 & 1983, Goleman 1978–79, Walsh, R. N., 1978, Shimano & Douglas 1975, Kohr 1977, Osis, Bokert, Carlson et al. 1973, and Deikman 1966a.

Unusual dreams and increased dream recall. Reed (1978) analyzed the effect of meditation on the completeness and vividness of intentional dream recall, using approximately 400 subjects who recorded dreams for 28 consecutive days and voluntarily recorded the results. He found that when subjects meditated the day before dreaming, they had significantly greater completeness of dream recall on the following morning. See also: Kornfield 1979 & 1983, and Faber, Suayman, & Touyz 1978.

NEGATIVE EFFECTS

Psychologist Albert Ellis (1984) claimed that meditation can intensify obsessiveness and a tendency to dwell on trivial matters. He noted that some of his clients had gone into "dissociative semi-trance states by meditating," and that meditation often diverts people from doing things that overcome their disturbances. Some people might feel better during meditation, he wrote, while sabotaging their chances for lasting personal growth. Psychiatrist Roger Walsh (1979) reported several disturbing experiences during his own meditation, among them anxiety, tension, and anger; and Walsh & Rauche (1979) suggested that meditation can precipitate psychotic episodes in individuals with a history of schizophrenia. Carrington (1977) observed that prolonged meditation can induce symptoms that range in severity from insomnia to psychotic behavior. Lazarus (1976) reported that severe depression and schizophrenic breakdown can be precipitated by TM. French, Smid, & Ingalls (1975) reported that anxiety, tension, anger, and other disturbing experiences sometimes occur during TM. Carrington & Ephron (1975) described TM practitioners who felt themselves overwhelmed by negative and unpleasant thoughts during meditation. Otis (1974) reported that five patients suffered a recurrence of serious psychosomatic symptoms after commencing meditation. And Maupin (1969) noted that meditation can produce withdrawn, serene people who are not accessible to what is actually going on in their lives. These and other negative outcomes are noted in the comtemplative literature. The path to enlightenment is "sharp like a razor's edge," says the Katha Upanishad. Though the rewards of contemplative practice can be great, they do not come easily or automatically.

APPENDIX D

[D.1]
SPIRITUAL HEALING STUDIES AND PSYCHOKINESIS
STUDIES INVOLVING LIVING MATERIALS

This bibliography is based to a large extent upon an inventory compiled by Jerry Solf-vin, 1984. Mental healing. In *Advances in Parapsychological Research,* vol. 4. Mc-Farland.

PSYCHOKINESIS (PK) STUDIES

Biological Materials

Barry, J. 1968a. General and comparative study of the psychokinetic effect on fungus culture. *Journal of Parapsychology,* 32:237–43.

Barry, J. 1968b. PK on fungus growth. *Journal of Parapsychology,* 32:55 (Abstract).

Braud, W. G. 1974. Psychokinetic influence upon E. Coli bacteria. Unpublished manuscript, University of Houston.

Braud, W. G., et al. 1979. Experiments with Matthew Manning. *Journal of the Society for Psychical Research,* 50:199–223.

Brier, R. M. 1969. PK on a bio-electrical system. *Journal of Parapsychology,* 33:187–205.

Deamer, D. W. 1979. Observations using biological materials. In J. Mishlove (ed.), *A Month with Matthew Manning: Experiences and Experiments in Northern California During May–June, 1977.* San Francisco: Washington Research Center.

Haraldsson, E., & T. Thorsteinsson. 1973. Psychokinetic effects on yeast: An exploratory experiment. In W. G. Roll et al. (eds.), *Research in Parapsychology 1972.* Scarecrow Press.

Kief, H. 1972. A method for measuring PK ability with enzymes. In W. G. Roll et al. (eds.), *Proceedings of the Parapsychological Association No. 8,* pp. 19–20.

Lorenz, F. W. 1979. Experiments with Matthew Manning. In J. Mishlove (ed.), *A Month with Matthew Manning: Experiences and Experiments in Northern California During May–June, 1977.* Washington Research Center.

Metta, L. 1972. Psychokinesis of lepidopterous larvae. *Journal of Parapsychology,* 36:213–21.

Nash, C. B. 1982. Psychokinetic control of bacterial growth. *Journal of the Society for Psychical Research,* 51:217–21.

Nash, C. B., & C. S. Nash. 1967. The effect of paranormally conditioned solution on yeast fermentation. *Journal of Parapsychology,* 31:314.

Tedder, W. H., & M. L. Monty. 1981. Exploration of long-distance PK: A conceptual replication of the influence on a biological system. In W. Roll & J. Beloff (eds.), *Research in Parapsychology 1980.* Scarecrow Press, pp. 90–93.

Whole Plant Life Processes

Grad, B. 1967. The "laying on of hands": Implications for psychotherapy, gentling and placebo effect. *Journal of the American Society for Psychical Research,* 61: 286–305.

Lenington, S. 1979. Effects of holy water on the growth of radish plants. *Psychological Reports,* 45:381–82.

Nicholas, C. 1977. The effects of loving attention on plants. *New England Journal of Parapsychology,* 1:19–24.

Pauli, E. N. 1973. PK on living targets as related to sex, distance and time. In W. G. Roll et al. (eds.), *Research in Parapsychology 1972.* Scarecrow Press, pp. 68–70.

Vasse, P. 1950. Experiences de germination de planter: Methode du Professeur J. B. Rhine de Duke. *Revue Métaphysique (Nouvelle Serie),* 12:223–25.

Vasse, P., & C. Vasse. 1948. Influence de la pensee sur la croissance des plantes. *Revue Métaphysique (Nouvelle Serie),* 2:87.

Wallack, J. M. 1984. Testing for a psychokinetic effect on plants: Effect of "laying on" of hands on germinating corn seeds. *Psychological Reports,* 55:15–18.

Animal Life Processes

Bechterev, W., 1948. Direct influence of a person upon the behavior of animals. *Journal of Parapsychology,* 13:166–76.

Extra, J. 1971. Telepathie bij de rat. *Tijdschrift voor Parapsychologie,* 39:18–31.

Gruber, E. R. 1979. A study of conformance behavior involving rats and mice. Paper presented at the meeting of the Society for Psychical Research, Edinburgh, Scotland.

Nash, C., & C. Nash. 1981. Psi-influenced movement of chicks and mice onto a visual cliff. In W. Roll & J. Beloff (eds.), *Research in Parapsychology 1980.* Scarecrow Press, pp. 109–10.

Osis, K. 1952. A test of the occurrence of a psi effect between man and cat. *Journal of Parapsychology,* 16:233–56.

Randall, J. 1970. An attempt to detect psi effects with protozoa. *Journal of the Society for Psychical Research,* 45:294–96.

Richmond, K. 1952. Two series of PK tests on paramecia. *Journal of the Society for Psychical Research,* 36:577–88.

Human Life Processes

Braud, W. G. 1978. Allobiofeedback: Immediate feedback for a psychokinetic influence upon another person's physiology. In W. G. Roll (ed.), *Research in Parapsychology 1977.* Scarecrow Press, pp. 123–34.

Braud, W. G. 1986. PSI and PNI: Exploring the interface between parapsychology and psychoneuroimmunology. *Parapsychology Review,* 17(4): 1–5.

Braud, W. G. 1990. Distant mental influence of rate of hemolysis of red blood cells. *Journal of the American Society for Psychical Research,* 84:1–23.

Braud, W. G., G. Davis, & R. Wood. 1979. Experiments with Matthew Manning. *Journal of the Society for Psychical Research,* 50:199–223.

Braud, W. G., & M. Schlitz. 1983. Psychokinetic influence on electrodermal activity. *Journal of Parapsychology,* 47:95–119.

Braud, W. G., & M. Schlitz. 1989. A methodology for the objective study of transpersonal imagery. *Journal of Scientific Exploration.*

Gruber, E. R. 1979. Conformance behavior involving animal and human subjects. *European Journal of Parapsychology,* 3:36–50.

Vasiliev, L. L. 1976. *Experiments in Distant Influence.* Dutton.

MENTAL HEALING STUDIES

Inorganic Materials

Dean, D., & E. Brame. 1975. Physical changes in water by laying-on-of-hands. *Proceedings of the Second International Congress of Psychotronics Research.* Paris: Institut Metaphysique International, pp. 200–02.

Schwartz, S., et al. 1986. *Infrared Spectra Alteration in Water Proximate to the Palms of Therapeutic Practitioners.* Mobius Society.

Biological Materials

Edge, H. 1980. The effect of laying on of hands on an enzyme: An attempted replication. In W. Roll (ed.), *Research in Parapsychology 1979.* Scarecrow Press, pp. 137–39.

Grad, B. 1965. A telekinetic effect on yeast activity. *Journal of Parapsychology,* 29:285–86.

Kmetz, J. 1981. Effects of healing on cancer cells (Appendix). In D. Kraft, *Portrait of a Psychic Healer.* Putnam.

Knowles, F. W. 1954. Some investigations into psychic healing. *Journal of the American Society for Psychical Research,* 48:21–26.

Rubik, B., & E. Rauscher. 1980. Effects on motility behavior and growth rate of salmonella typhimurium in the presence of Olga Worrall. In W. G. Roll (ed.), *Research in Parapsychology 1979.* Scarecrow Press, pp. 140–42.

Smith, M. J. 1968. Paranormal effects on enzyme activity. *Journal of Parapsychology,* 32:281 (Abstract).

Smith, M. J. 1972. Paranormal effects on enzyme activity. *Human Dimensions,* 1:15–19.

Smith, M. J. 1977. The influence of "laying-on" of hands. In N. M. Regush (ed.), *Frontiers of Healing.* Avon.

Snel, F. 1980. PK influence on malignant cell growth. *Research Letter,* Parapsychology Laboratory, University of Utrecht, 10:19–27.

Whole Plant Life Processes

Grad, B. 1963. A telekinetic effect on plant growth. *International Journal of Parapsychology,* 5:117–33.

Grad, B. 1964. A telekinetic effect on plant growth, II. Experiments involving treatment of saline in stoppered bottles. *International Journal of Parapsychology,* 6:473–98.

Harary, S. B. 1975. A pilot study of the effects of psychically treated saline solution on the growth of seedlings. Unpublished manuscript, Psychical Research Foundation.

Knowles, F. W. 1954. Some investigations into psychic healing. *Journal of the American Society for Psychical Research,* 48:21–26.

Loehr, F. 1959. *The Power of Prayer on Plants.* Signet.

Miller, R. N. 1972. The positive effect of prayer on plants. *Psychic,* 3:24–25.

Solfvin, J. 1982. The effects of an induced "expectancy structure" on the growth of corn seedlings. *European Journal of Parapsychology,* 4:345–80.

Animal Life Processes

Grad, B. 1971–72. Some biological effects of the laying on of hands: A review of experiments with animals. *Journal of Pastoral Counseling,* 6:38–41.

Grad, B. 1977. Laboratory evidence of "laying on of hands." In N. Regush, (ed.), *Frontiers of Healing.* Avon.

Grad, B. 1979. Healing by the laying on of hands: A review of experiments. In D. Sobel, (ed.), *Ways of Health.* Harcourt, Brace, Jovanovich.

Grad, B., et al. 1961. An unorthodox method of treatment on wound healing in mice. *International Journal of Parapsychology,* 3:5–24.

Heaton, E. 1974. Mouse healing experiments. Unpublished manuscript, Foundation for Research on the Nature of Man, Durham, N.C.

Schlitz, M. J. 1982. PK on living systems: Further studies with anesthetized mice. *Journal of Parapsychology,* 46:51–52 (Abstract).

Solfvin, J. 1982. Psi expectancy effects in psychic healing studies with malarial mice. *European Journal of Parapsychology,* 4:48–64.

Watkins, G., & A. Watkins. 1971. Possible PK influence on the resuscitation of anesthetized mice. *Journal of Parapsychology,* 35:257–72.

Watkins, G. K., et al. 1973. Further studies on the resuscitation of anesthetized mice. In W. Roll, R. Morris, & J. Morris (eds.), *Research in Parapsychology 1972.* Scarecrow Press, pp. 157–59.

Wells, R., & J. Klein. 1972. A replication of a "psychic healing" paradigm. *Journal of Parapsychology,* 36:144–49.

Wells, R., & G. Watkins. 1975. Linger effects in several PK experiments. In J. Morris, W. Roll, & R. Morris (eds.), *Research in Parapsychology 1974.* Scarecrow Press, pp. 143–47.

Human Life Processes

Benor, D. J. 1984. Psychic healing. In J. Salmon (ed.), *Alternative Medicine: Popular and Policy Perspectives.* Tanstock.

Boguslawski, M. 1979. The use of therapeutic touch in nursing. *Journal of Continuing Education in Nursing,* 10:9–15.

Boguslawski, M. 1980. Therapeutic Touch: A facilitator of pain relief. *Topics in Clinical Nursing,* 2:27–37.

Collipp, P. J. 1969. The efficacy of prayer: A triple-blind study. *Medical Times,* 97:201–4.

Goodrich, J. 1974. Psychic healing—A pilot study. Doctoral dissertation, Union Graduate Shool, Cincinnati.

Heidt, P. 1981. Effective Therapeutic Touch on the anxiety level of hospitalized patients. *Nursing Research,* 35:101–5.

Hubacher, J., et al. 1975. A laboratory study of unorthodox healing. *Proceedings of the Second International Conference of Psychotronic Research.* Paris: Institut Metaphysique International.

Joyce, C., & R. Welldon. 1965. The objective efficacy of prayer: A double-blind clinical trial. *Journal of Chronic Diseases,* 18:367–77.

Keller, E. 1986. Effects of Therapeutic Touch on tension headache pain. *Nursing Research,* 35:101–5.

Knowles, F. W. 1954. Some investigations into psychic healing. *Journal of the American Society for Psychical Research,* 48:21–26.

Knowles, F. W. 1956. Psychic healing in organic disease. *Journal of the American Society for Psychical Research,* 50:110–17.

Krieger, D. 1974. Healing by the "laying-on" of hands as a facilitator of bioenergetic change: The response of in-vivo human hemoglobin. *Psychoenergetic Systems,* 1:121–29.

Krieger, D. 1975. Therapeutic Touch: The imprimatur of nursing. *American Journal of Nursing,* 75:784–87.

Krieger, D. 1979a. *Therapeutic Touch.* Prentice-Hall.

Krieger, D. 1979b. Therapeutic Touch: Searching for evidence of physiological change. *American Journal of Nursing* (Apr.), 79:660–62.

Krieger, D., et al. 1979. Physiologic indices of Therapeutic Touch. *American Journal of Nursing,* 4:660–62.

Kunz, D., & E. Peper. 1982–1983. Fields and their clinical implications. Parts I–III. *The American Theosophist,* 70:395–401; 71:19–21, 200–3.

Macrae, J. 1981. Therapeutic Touch: A way of life. In Borelli & Heidt (eds.), *Therapeutic Touch: A Book of Readings.* Springer.

Miller, L. 1979. An explanation of therapeutic touch: Using the science of unitary man. *Nursing Forum,* 18:278–87.

Quinn, J. 1982. An investigation of the effect of therapeutic touch done without physical contact or state anxiety of hospitalized cardiovascular patients. Dissertation Abstracts International (University Microfilms No. 82-26-788).

Quinn, J. 1984. Therapeutic Touch as energy exchange: Testing the theory. *Advances in Nursing Science* (Jan.), 6:42–49.

Quinn, J. 1988. Building a body of knowledge: Research on therapeutic touch 1974–1986. *Journal of Holistic Nursing,* 6:37–45.

Quinn, J. 1989. Future directions for therapeutic touch research. *Journal of Holistic Nursing,* 7:19–25.

Rehder, H. 1955. Wunderheilungen, ein experiment. *Hippokrates,* 26:577–80.

Richmond, K. 1946. Experiments in the relief of pain. *Journal of the Society for Psychical Research*, 33:194–200.

Winston, S. 1975. Research in psychic healing: A multivariate experiment. Doctoral dissertation, Union Graduate School, Cincinnati.

Wirth, D. 1990. The effect of non-contact Therapeutic Touch on the healing rate of full thickness dermal wounds. *Subtle Energies,* 1:1–20.

[D.2]
CROSS-CULTURAL STUDIES OF
SPIRITUAL HEALING

Edwards, F. 1983. Healing and transculturation in Xhosa Zionist practice. *Culture, Medicine and Psychiatry,* 7:177–98.

Harner, M. 1983. *The Jivaro.* University of California Press.

Katz, R. 1981. Education as transformation: Becoming a healer among the Kung and Fijians. *Harvard Educational Review,* 5:57–78.

Kreiger, D. 1972. The response in in-vivo human hemoglobin to an active healing therapy by direct laying-on of hands. *Human Dimensions,* Vol. 1 (Autumn).

Luckert, K. 1979. *Coyoteway: A Navajo Healing Ceremonial.* Trans. J. C. Cooke. University of Arizona Press and Museum of North Arizona Press.

Mischel, F. 1959. Faith healing and medical practice in the Southern Caribbean. *Southwestern Journal of Anthropology,* 15:407–17.

Opler, M. E. 1936. Some points of comparison and contrast between the treatment of functional disorders by Apache shamans and modern psychiatric practice. *American Journal of Psychiatry,* 92:1371–87.

Romano, O. I. 1965. Charismatic medicine, folk-healing, and folk sainthood. *American Anthropologist,* 67:1151–73.

Smith, J. 1972. The influence of enzyme growth by "laying-on-of-hands." In *The Dimensions of Healing: A Symposium.* Academy of Parapsychology and Medicine.

APPENDIX E
STUDIES OF NEAR-DEATH, OUT-OF-BODY, REINCARNATION-TYPE, AND OTHERWORLD EXPERIENCES

Near-Death Experience

Alvarado, C. 1989. Trends in the study of out-of-body experiences: An overview of developments since the nineteenth century. *Journal of Scientific Exploration*, 3:1:27–42.

Basford, T. 1990. *Near-Death Experiences: An Annotated Bibliography*. Garland.

Basterfield, K. 1986. Near-death experiences: An Australian survey. *Australian Institute of Psychic Research Bulletin*, 9:1–4.

Blackmore, S. 1988. Visions from the dying brain. *New Scientist* (May 5), 118:43–46.

Bluebond-Langner, M. 1978. *The Private Worlds of Dying Children*. Princeton University Press.

Cook, S. 1974. *Children and Dying: An Exploration and Selective Bibliographies*. Health Sciences Publishing Corporation.

Doore, G. 1990. *What Survives? Contemporary Explorations of Life after Death*. Jeremy P. Tarcher.

Flynn, C. P. 1986. *After the Beyond: Human Transformation and the Near-Death Experience*. Prentice-Hall.

Gabbard, G. O., et al. 1981. Do "near-death" experiences occur only near death? *Journal of Nervous and Mental Diseases*, 169:374–77.

Grey, M. 1985. *Return from Death. An Exploration of the Near-Death Experience*. Arkana.

Greyson, B. 1982. Near-death studies, 1981–1982. *Anabiosis: The Journal for Near-Death Studies*, 2:150–58.

Greyson, B. 1983. The near-death experience scale: Construction, reliability and validity. *Journal of Nervous and Mental Disease*, 171:6:369–75.

Grosso, M. 1982. Toward an explanation of near-death phenomena. In C. Lundahl (ed.), *A Collection of Near-Death Research Readings*. Nelson-Hall.

Krishnan, V. 1985. Near-death experiences: Evidence for survival. *Anabiosis: The Journal for Near-Death Studies* (Spring), 5:1:21–38.

Lundahl, C. R., ed. 1982. *A Collection of Near-Death Research Readings*. Nelson-Hall.

Moody, R. 1975. *Life after Life*.

Moody, R. A., Jr. 1976. *Life After Life? The Investigation of a Phenomenon: Survival of Bodily Death*. Stackpole Books.

Moody, R. A., Jr. 1977. *Reflections on Life After Life*. Bantam/Mockingbird.

Moody, R., & P. Perry. 1988. *The Light Beyond.* Bantam.

Morse, M. 1990. *Closer to the Light: Learning from the Near-Death Experiences of Children.* Ivy Books (Ballantine).

Morse, M., Castillo, P., Venecia, D., Milstein, J., and D. C. Tyler, 1986. Childhood near-death experiences. *American Journal of Diseases of Children* (Nov.), 140: 1110–14.

Morse, M., Conner, D., and D. Tyler, 1985. Near-death experiences in a pediatric population. *American Journal of Diseases of Children* (June), 139:595–600.

Morse, M. L., Venecia, D., Jr., and J. Milstein. 1989. Near-death experiences: A neurophysiologic explanatory model. *Journal of Near-Death Studies,* 8:45–53.

Noyes, R., Jr. 1972. The experience of dying. *Psychiatry,* 35:174–84.

Noyes, R., Jr. 1974. Near-death experiences: Their interpretation and significance. In R. Kastenbaum (ed.), *Between Life and Death.* Springer.

Osis, K. 1961. *Deathbed Observations by Physicians and Nurses.* Parapsychological Monographs, No. 3. Parapsychological Foundation, Inc.

Osis, K., & E. Haraldsson. 1986. *At the Hour of Death.* Hastings House.

Provonsha, J. 1981. *Is Death for Real: An Examination of Reported Near-Death Experiences in the Light of the Resurrection.* Pacific Press.

Ring, K. 1980. *Life at Death: A Scientific Investigation of the Near-Death Experience.* Coward, McCann & Geoghegan.

Ring, K. 1984. *Heading Toward Omega: In Search of the Meaning of the Near-Death Experience.* William Morrow.

Ring, K. 1989. Near-death experiences. In R. Kastenbaum, & B. Kastenbaum, *Encyclopedia on Death.* Oryx Press.

Sabom, M. 1982. *Recollections of Death.* Harper & Row.

Vicchio, S. 1981. Near-death experiences: A critical review of the literature and some questions for further help. *Anabiosis: The Journal for Near-Death Studies,* 1:66–87.

Zaleski, C. 1987. *Otherworld Journeys.* Oxford University Press.

Out-of-Body Experience

Alvarado, C. 1986. Research on spontaneous out-of-body experiences: A review of modern developments. 1960–1984. In B. Shapin & L. Coly (eds.), *Current Trends in Psi Research: Proceedings of an International Conference Held in New Orleans, Louisiana, August 13–14, 1984,* Parapsychology Foundation.

Bendit, L., & P. Payne. *The Psychic Sense.* Quest Books.

Blackmore, S. 1983. Are out-of-body experiences evidence for survival? *Anabiosis: The Journal for Near-Death Studies* (Dec.), 3:137–55.

Crookall, R. 1966. *Study and Practice of Astral Projection.* University Books.

Crookall, R. 1970. *Out-of-the-Body Experiences.* University Books.

Crookall, R. 1972. *Casebook of Astral Projection.* University Books.

Currie, I. 1978. *You Cannot Die: The Incredible Findings of a Century of Research on Death.* Methuen.

Fox, O. 1962. *Astral Projection.* University Books.

Gabbard, G., & S. Tremlow. 1984. *With the Eyes of the Mind: An Empiric Analysis of Out-of-Body States.* Praeger.

Green, C. 1968. *Out-of-the-Body Experiences.* Ballantine.

Irwin, H. J. 1985. *Flight of Mind. Psychological Study of the Out-of-Body Experience.* Scarecrow Press.

Josephs, A. 1983. Hemingway's out of body experience. *Hemingway Review*, 2(2): 11–17.

Mitchell, J. 1974. Out-of-body experiences and autoscopy. *The Osteopathic Physician*, 41:44–49.

Mitchell, J. 1981. *Out-of-Body Experiences: A Handbook.* McFarland.

Monroe, R. A. 1971. *Journeys Out of the Body.* Doubleday.

Muldoon, S., & H. Carrington. 1936. *The Case for Astral Projection.* Aries Press.

Muldoon, S., & H. Carrington. *The Phenomena of Astral Projection.* Samuel Weiser.

Muldoon, S., & H. Carrington. 1961. *The Projection of the Astral Body.* Samuel Weiser.

Myers, F. W. H. 1903, 1954. *Human Personality and Its Survival of Bodily Death.* Vol. 1. Longmans, Green, pp. 682–85.

Noyes, R., & R. Kletti. 1976. Depersonalization in the face of life-threatening danger: A description. *Psychiatry*, 39:19–27.

Osis, K. 1961. *Deathbed Observations by Physicians and Nurses.* Parapsychological Foundation.

Osis, K. 1978. Out-of-the-body experiences: A preliminary survey. Paper presented at the Parapsychology Association Annual Convention, St. Louis, Missouri (Aug.).

Oxenham, J. 1941. *Out of the Body.* Longmans, Greene & Co.

Rhine, J. B. 1960. Incorporeal personal agency: The prospect of a scientific solution. *Journal of Parapsychology* (Dec.), 24:4:279–309.

Shirley, R. 1965. *Mystery of the Human Double.* University Books.

Tart, C. T. 1969. *Altered States of Consciousness: A Book of Readings.* John Wiley & Sons.

Tart, C. T. 1974. Out-of-the-body experiences. In E. Mitchell (ed.), *Psychic Exploration.* Putnam.

Thouless, R. H. 1960. The empirical evidence for survival. *Journal of the American Society for Psychical Research* (Jan.), 54:23–32.

Twemlow, S., et al. 1982. The out-of-body experience: A phenomenological typology based on questionnaire responses. *American Journal of Psychiatry*, 139: 450–55.

Vieira, W. 1986. *Projeciologia: Survey of OBE Experiences.* Rio de Janeiro, Brazil: Editora Brasil-America S.A. Bibl: 699–771.

Reincarnation-Type Experience

Chari, C. T. 1962. "Buried memories" in survivalist research. *International Journal of Parapsychology* (Summer), 4:40–65.

Chari, C. T. 1978. Reincarnation Research: Method and interpretation. In M. Ebon (ed.), *Signet Handbook of Parapsychology.* New American Library.

Cook, E. W. 1986. Research on reincarnation-type cases: Present status and suggestion for future research. In K. R. Rao (ed.), *Case Studies in Parapsychology.* McFarland.

Matlock, G. 1989. Age and stimulus in past life memory cases: A study of published cases. *Journal of the American Society for Psychical Research* (Oct.), 83:4: 303–16.

Mills, A. 1988. A preliminary investigation of cases of reincarnation among the Beaver and Gitksan Indians. *Anthropologica* 30.

Mills, A. 1989. A replication study: Three cases of children in Northern India who are said to remember a previous life. *Journal of Scientific Exploration,* 3(2): 133–84.

Stevenson, I. 1966. *Twenty Cases Suggestive of Reincarnation.* University Press of Virginia.

Stevenson, I. 1970. Characteristics of cases of the reincarnation type in Turkey and their comparison with cases in two other cultures. *International Journal of Comparative Sociology* (Mar.), 11:11–17.

Stevenson, I. 1975–1983. *Cases of the Reincarnation Type.* Vols. 1–4. University Press of Virginia.

Stevenson, I. 1977. The explanatory value of the idea of reincarnation. *Journal of Nervous and Mental Diseases,* 164:305–26.

Stevenson, I. 1987. *Children Who Remember Previous Lives: A Question of Reincarnation.* University Press of Virginia.

APPENDIX F
EXPERIMENTAL STUDIES AND HISTORICAL ACCOUNTS REGARDING THE SUBTLE BODY

DeLubicz, I. S. 1981. *The Opening of the Way*. Inner Traditions.

Dodds, E. R. 1923. *Select Passages Illustrating Neoplatonism*. Ares, pp. 74–92.

Evans-Wentz, W. Y. 1958. *Tibetan Yoga and Secret Doctrines*. Oxford University Press.

Fraser-Harris, D. F. 1932. A psycho-physiological explanation of the so-called human 'aura.' *British Journal of Medical Psychology*, vol. 12, pt. 2 (Sept.), pp. 174–84.

Gurwitch, A. 1977. *Teoriya Biologicheskogo Polya* (Theory of the Biological Field). Selected Works. Moscow: Meditsina.

Inyushin, V. 1970. *Gorizonty Bioniki* (Horizons of Bionics). Kazakhstan.

Inyushin, V. 1978. *Elementy Teorii Biologicheskogo Polya* (Elements of the Theory of the Biological Field). Kazakh State University.

Kilner, W. J. 1965. *The Human Aura*. University Books.

Kolotilov, M., & M. Litvinov. 1986. Physics of the Biofield. *Znannya ta Pratsya*, Kiev, 1:8–9 (Jan.).

Mead, G. R. S. 1919, 1967. *The Doctrine of the Subtle Body in Western Tradition*. Theosophical Publishing House, Quest Book.

Nietzke, A. 1979. Portrait of an aura reader. *Human Behavior* (Feb.), 8:28–35.

Perevozchikov, A. 1986. Rainbow of human physical field (Interview with academician Yuri V. Gulyaev and Dr. Eduard E. Godik). *Tekhnika-Molodezhi* (Technology for Youth), 12:12–17. (Dec.).

Pierrakos, J. 1987. *Core Energetics*. Chap. 7.

Poortman, J. J. 1978. *Vehicles of Consciousness: The Concept of Hylic Pluralism*. Vols. 1–4. Theosophical Publishing House.

Powell, A. E. 1925. *The Etheric Double & Allied Phenomena*. Theosophical Publishing House.

Powell, A. E. 1927a. *The Astral Body*. Theosophical Publishing House.

Powell, A. E. 1927b. *The Mental Body*. Theosophical Publishing House.

Powell, A. E. 1928. *The Causal Body & the Ego*. Theosophical Publishing House.

Russell, E. 1975. Parapsychic luminosities. *Quadrant* (Winter), 8:2, pp. 58–61.

Shuman, E. 1979. Our glowing auras: New evidence for the body's energy forces. *Human Behavior* (Jan.), 8:39–40.

Taittiriya Upanishad. 1969. In S. Radhakrishnan, *The Principal Upanishads*. Humanities Press.

Taittiriya Upanishad. 1971. In R. E. Hume, *The Thirteen Principal Upanishads*. Oxford University Press.

Vilenskaya, L., ed. 1983. *PSI Research* (Dec.), 2:28–66.

APPENDIX G
CREATIVITY STUDIES

Barron, F. 1953a. Complexity-simplicity as a personality dimension. *Journal of Abnormal Social Psychology*, 48:163–72.

Barron, F. 1953b. Some personality correlates of independence of judgment. *Journal of Personality*, 21:478–85.

Barron, F. 1955. The disposition toward originality. *Journal of Abnormal Social Psychology*, 51:478–85.

Barron, F. 1957. Originality in relation to personality and intellect. *Journal of Personality*, 25:730–42.

Barron, F. 1958. The psychology of imagination. *Scientific American*, 199:150–56.

Barron, F. 1963a. *Creativity and Psychological Health*. Van Nostrand.

Barron, F. 1963b. The needs for order and for disorder as motives in creative activity. In C. Taylor & F. Barron, *Scientific Creativity*. Wiley.

Barron, F. 1969. *Creative Person and Creative Process*. Holt, Rinehart & Winston, pp. 75–78.

Barron, F. 1972. *Artists in the Making*. Seminar Press.

Barron, F. 1979. *The Shaping of Personality*. Harper & Row.

Barron, F. 1987. Putting creativity to work. In R. Sternberg (ed.), *The Nature of Creativity*. Cambridge University Press.

Barron, F., & D. Harrington. 1981. Creativity, intelligence, and personality. In M. Rosenzweig & L. Porter (eds.), *Annual Review of Psychology*. Annual Reviews.

Barron, F., & G. Welsh. 1952. Artistic perception as a factor in personality style: Its measurement by a figure-preference test. *Journal of Psychology*, 33:199–203.

Crawford, H. J. 1982. Hypnotizability, daydreaming style, imagery, vividness, and absorption: A multidimensional study. *Journal of Personality & Social Psychology*, 42:915–26.

Ghiselin, B. 1980. *The Creative Process*. New American Library.

Koestler, A. 1964. *The Act of Creation*. Macmillan.

Lindauer, M. S. 1983. Imagery and the Arts. In A. Sheikh (ed.), *Imagery—Current Theory, Research, and Application*. Wiley, pp. 268–506.

MacKinnon, D. W. 1960. Genus architectus creator varietas Americanus. *American Institutes of Architects Journal* (Sept.), pp. 31–35.

Shaw, G., & S. Belmore. 1982–83. The relationship between imagery and creativity. *Imagination, Cognition & Personality*, 2:115–23.

Torrance, E. P., et al. 1959. *Explorations in Creative Thinking in the Early School Years: I–XII*. Bureau of Educational Research, University of Minnesota.

Appendix H
Clinical and Experimental Studies of Progressive Relaxation

All of the works listed here are by Edmund Jacobson.

1911. Experiments on the inhibition of sensations. *Psychological Review,* 18:24–53.
1920. Reduction of nervous irritability and excitement by Progressive Relaxation. *Journal of Nervous & Mental Disease,* 53:282.
1920. Use of relaxation in hypertensive states. *New York Medical Journal,* 111:419.
1921. The use of experimental psychology in the practice of medicine. *Journal of the American Medical Association,* 77:342–47.
1924. The physiology of globus hystericus. *Journal of the American Medical Association,* 83:911–13.
1924. The technic of progressive relaxation. *Journal of Nervous & Mental Disease,* 60:568–78.
1925. Voluntary relaxation of the esophagus. *American Journal of Physiology,* 72:387–94.
1926. Response to a sudden unexpected stimulus. *Journal of Experimental Psychology,* 9:19–25.
1927. Action currents from muscular contractions during conscious processes. *Science,* 66:403.
1928. Differential relaxation during reading, writing, and other activities as tested by the knee-jerk. *American Journal of Physiology,* 86:675–93.
1930. Electrical measurements of neuromuscular states during mental activities. I. Imagination of movement involving skeletal muscle. *American Journal of Physiology,* 91:567–608.
1930. Electrical measurements of neuromuscular states during mental activities. II. Imagination and recollection of various muscular acts. *American Journal of Physiology,* 94:22–34.
1930. Electrical measurements of neuromuscular states during mental activities. III. Visual imagination and recollection. *American Journal of Physiology,* 95:694–702.
1930. Electrical measurements of neuromuscular states during mental activities. IV. Evidence of contraction of specific muscles during imagination. *American Journal of Physiology,* 95:703–712.
1931. Electrical measurements of neuromuscular states during mental activities. V. Variation of specific muscles contracting during imagination. *American Journal of Physiology,* 96:115–21.

1931. Electrical measurements of neuromuscular states during mental activities. VI. A note on mental activities concerning an amputated limb. *American Journal of Physiology*, 96:122–25.

1931. Electrical measurements of neuromuscular states during mental activities. VII. Imagination, recollection and abstract thinking involving the speech musculature. *American Journal of Physiology*, 97:200–09.

1932. Electrophysiology of mental activities. *American Journal of Psychology*. 44: 677–94.

1933. Measurement of the action-potentials in the peripheral nerves of man without anesthetic. *Proceedings of the Society for Experimental Biology & Medicine*, 30:713–15.

1934. Electrical measurements concerning muscular contraction (tonus) and the cultivation of relaxation in man—relaxation-times of individuals. *American Journal of Physiology*, 108:573–80.

1936. The course of relaxation in muscles of athletes. *American Journal of Psychology*, 48:98–108.

1939. Variations in blood pressure with skeletal muscle tension and relaxation. *Annals of Internal Medicine*, 12:1194–1212.

1939. Variations in blood pressure with skeletal muscle tension (action-potentials) in man: The influence of brief voluntary contractions. *American Journal of Physiology*, 126:546–47.

1940. Cultivated relaxation in "essential" hypertension. *Archives of Physical Therapy*, 21:645–54.

1940. The direct measurement of nervous and muscular states with the integrating neurovoltmeter (action-potential integrator). *American Journal of Psychiatry*, 97:513–23.

1940. Variation of blood pressure with skeletal muscle tension and relaxation: The heart beat. *Annals of Internal Medicine*, 13:1619–25.

1940. Variation of blood pressure with brief voluntary muscular contractions. *Journal of Laboratory & Clinical Medicine*, 25:1029–37.

1943. Cultivated relaxation for the elimination of "nervous breakdowns." *Archives of Physical Therapy*, 24:133–43.

1943. Muscular tension and the smoking of cigarettes. *American Journal of Psychology*, 56:559–74.

1952. Specialized electromyography in supplement to clinical observations during hyperkinetic states in man ("functional nervous conditions"). *Fed. Proc.*, 11:1.

1955. Neuromuscular control in man: Methods of self-direction in health and in disease. *American Journal of Psychology*, 68:549–61.

1974. *Progressive Relaxation.* University of Chicago Press (Midway reprint).

APPENDIX I
READINGS ON THE CULTURAL SHAPING
OF INDIVIDUAL DEVELOPMENT

Benthall, J., & T. Polhemus, eds. 1975. *The Body as a Medium of Expression.* Allen, Lane & Dutton.

Bourdieu, P. 1984. *Distinction—A Social Critique of the Judgement of Taste.* Harvard University Press.

Brown, N. O. 1968. *Life Against Death: The Psychoanalytical Meaning of History.* Vintage/Random House.

Chernin, K. 1981. *The Obsession: Reflections on the Tyranny of Slenderness.* Harper & Row.

Douglas, M. 1970. *Purity & Danger: An Analysis of Concepts of Pollution and Taboo.* Penguin.

Foucault, M. 1970. *The Order of Things: An Archeology of the Human Sciences.* Vintage/Random House.

Freud, S. *Collected Papers. Character & Culture.* Vol. 9. International Psychoanalytical Press.

Freud, S. *Totem & Taboo,* and *Civilization and Its Discontents.* Hogarth.

Griffin, S. 1978. *Woman and Nature.* Harper & Row.

Johnson, D. 1980. Somatic Platonism. *Somatics* (Autumn), 3:4–7.

Kunzle, D. 1981. *Fashions and Fetishism: A Social History of the Corset, Tight-Lacing, and Other Forms of Body-Sculpture in the West.* Rowman & Littlefield.

Merleau-Ponty, M. 1962. *The Phenomenology of Perception.* Trans. C. Smith. Routledge & Kegan Paul.

Polhemus, T., ed. 1978. *The Body Reader: Social Aspects of the Human Body.* Pantheon.

Reich, W. 1970. *The Mass Psychology of Fascism.* Trans. V. Carfagno. Farrar, Straus & Giroux.

Schilder, P. 1950. *The Image and Appearance of the Human Body: Studies in the Constructive Energies of the Psyche.* International Universities Press.

Tinbergen, N. 1974. Ethology and stress disease. *Science* (July), 185:20–27.

Appendix J

[J.1]
Processes Critical to Living Systems

SUBSYSTEMS WHICH PROCESS BOTH
MATTER-ENERGY AND INFORMATION

Source: James Grier Miller, 1978. *Living Systems*. McGraw-Hill, p. 3.

1. *Reproducer*, the subsystem which is capable of giving rise to other systems similar to the one it is in.

2. *Boundary*, the subsystem at the perimeter of a system that holds together the components which make up the system, protects them from environmental stresses, and excludes or permits entry to various sorts of matter-energy and information.

SUBSYSTEMS WHICH PROCESS MATTER-ENERGY

3. *Ingestor*, the subsystem which brings matter-energy across the system boundary from the environment.

4. *Distributor*, the subsystem which carries inputs from outside the system or outputs from its subsystems around the system to each component.

5. *Converter*, the subsystem which changes certain inputs to the system into forms more useful for the special processes of that particular system.

6. *Producer*, the subsystem which forms stable associations that endure for significant periods among matter-energy inputs to the system or outputs from its converter, the materials synthesized being for growth, damage repair, or replacement of components of the system, or for providing energy for moving or constituting the system's outputs of products or information markers to its suprasystem.

7. *Matter-energy storage*, the subsystem which retains in the system, for different periods of time, deposits of various sorts of matter-energy.

8. *Extruder*, the subsystem which transmits matter-energy out of the system in the forms of products or wastes.

9. *Motor*, the subsystem which moves the system or parts of it in relation to part or all of its environment or moves components of its environment in relation to each other.

10. *Supporter*, the subsystem which maintains the proper spatial relationships among components of the system, so that they can interact without weighting each other down or crowding each other.

SUBSYSTEMS WHICH PROCESS INFORMATION

11. *Input transducer*, the sensory subsystem which brings markers bearing information into the system, changing them to other matter-energy forms suitable for transmission within it.

12. *Internal transducer*, the sensory subsystem which receives, from subsystems or components within the system, markers bearing information about significant alterations in those subsystems or components, changing them to other matter-energy forms of a sort which can be transmitted within it.

13. *Channel and net*, the subsystem composed of a single route in physical space, or multiple interconnected routes, by which markers bearing information are transmitted to all parts of the system.

14. *Decoder*, the subsystem which alters the code of information input to it through the input transducer into a "private" code that can be used internally by the system.

15. *Associator*, the subsystem which carries out the first stage of the learning process, forming enduring associations among items of information in the system.

16. *Memory*, the subsystem which carries out the second stage of the learning process, storing various sorts of information in the system for different periods of time.

17. *Decider*, the executive subsystem which receives information inputs from all other subsystems and transmits to them information outputs that control the entire system.

18. *Encoder*, the subsystem which alters the code of information input to it from other information processing subsystems, from a "private" code used internally by the system into a "public" code which can be interpreted by other systems in its environment.

19. *Output transducer*, the subsystem which puts out markers bearing information from the system, changing markers within the system into other matter-energy forms which can be transmitted over channels in the system's environment.

[J.2]
SOME ANALOGIES BETWEEN ACTIVITIES IN DIFFERENT EVOLUTIONARY DOMAINS

□ As they arose, members of the inorganic, organic, and human domains came into relationship with: (1) other members (or products) of that domain—that is, other particles, organisms, or human beings; (2) the general constraints and opportunities provided by the evolutionary domain that preceded it—namely, universal conditions when the first atomic particles were formed, the earth's environment in the case of organisms, and the physical-biological milieus encountered by humans; and (3) the new (or emergent) principles, conditions, or fields activated (or created) by them, such as space-time by particles, living interactions by organisms, and highly articulated consciousness by humans. In a third evolutionary transcendence, the new members would similarly engage: (1) other metanormally developing individuals; (2) the physical-biological-human worlds in which they arose; and (3) the supernature they would activate.

□ New kinds of structure in each evolutionary domain often appear with great suddenness in relation to the subsequent duration of such structures. The first hydrogen atoms appeared, for example, in a period infinitesimally brief in comparison to the many billion years that have since elapsed; and presumably the first iron atoms were created in a particular supernova. Similarly, the first living cells on Earth appeared during a time-span that was comparatively short in relation to the 4-billion-year history of organisms on this planet; and animal species advanced to homo sapiens by a series of developmental jumps characterized by Simpson and other evolutionary theorists as "quantum evolution" or by Eldridge and Gould in terms of "punctuated equilibrium." Metanormal capacities, too, often appear with great suddenness, though their integrated realization takes more time to achieve.

□ Though inorganic, biological, and human forms develop by gradual steps from one type or stage of activity to the next, their development is also marked at times by sharp discontinuities. New elements are created suddenly, for example, in exploding stars. New species have not appeared by smoothly continuous stages, and thus have left many gaps in their fossil records. And metanormal capacities often appear to be discontinuous from their analogues in ordinary human functioning: mystical cognition, for example, is markedly different from ordinary mental activity; and unitive awareness of other people is fundamentally distinct (to those who experience it) from mere empathy for one's fellows.

□ As they arise, new structures, whether particles, atoms, molecules, cells, individual humans, or cultures, are integrated into the dynamic interactions of their respective domains—or they vanish. At each evolutionary level, those that fit survive. To last, metanormal capacities must be effectively integrated into the worlds ongoing life processes.

This list could be extended indefinitely. For discussions of other analogies between structures and processes in different evolutionary arenas, see James Grier Miller's *Living Systems* or the works of emergent evolutionists noted in chapter 7.2.

APPENDIX K

INCORRUPTIBLE CADAVERS OF ROMAN CATHOLIC SAINTS

This list is taken from Herbert Thurston's *The Physical Phenomena of Mysticism,*
chapter 22. The date after each name represents the year of death. *AA.SS.* stands for
the Bollandist *Acta Sanctorum,* and Process, *Summario* for the summary of evidence
found in official submissions to the Congregation of Rites for the saint's beatification
and canonization procedures. For a description of this inventory, and the meaning of
the A, B, and C classifications, see chapter 22.8.

Jan. 29 B	S. FRANCIS OF SALES (1622). Embalmed body found entire in 1632; only fragrant dust in 1656. Hamon, *Vie,* II, pp. 481–2. Heart preserved apart; "oil" distils from it. Bougaud, *Vie de S. Chantel* II. p. 566.
March 8 A	S. JOHN OF GOD (1550). In 1570 body, except for the tip of the nose, entire, and also fragrant. *AA.SS.* March. Vol. I. pp. 831 & 853.
March 9 A	S. FRANCES OF ROME (1440). Body exhumed four months after death; fresh and very fragrant. *AA.SS.* March, Vol. II, pp. 101 & 209.
March 28 B	S. JOHN CAPISTRAN (1456). Evidence of exhumation not satisfactory. *AA.SS.* Oct., Vol. X, pp. 432–536 and 915; but said to have been reliably identified with a still incorrupt body in 1765. See Leon, *Lives O.S.F.,* III, p. 419.
April 2 A	S. FRANCIS OF PAULA (1507). Supple and fragrant a week after death; body still entire in 1562 when burnt by the Huguenots. Dabert, *Vie,* pp. 443 & 463.
April 5 C	S. VINCENT FERRER (1419). Fragrant and supple after death, but in 1456 only bones and dust. Fages, *Vie,* II, p. 274; Notes, p. 416.
April 24 C	S. FIDELIS A SIGMARINGA (1661). No evidence of incorruption at exhumation eighteen months after death. F. della Scala, *Der H. Fidelis,* p. 179.
April 28 C	S. PAUL OF THE CROSS (1775). Only skeleton left when first exhumed in 1852, but fragrant and flexible twenty-four hours after death. Devine, *Life,* pp. 377–378.
May 5 C	S. PIUS V (1572). Viscera removed, but body remained supple, high-colored and like that of a living man for four days. At translation in 1588 nothing left but skeleton. *AA.SS.* May, Vol. I. pp. 695–697.

May 10 A	S. ANTONINUS OF FLORENCE (1459). Unburied for eight days, it remained flexible and fragrant; found in 1589 still incorrupt. *AA.SS.* May, Vol. I, pp. 328 and 360. Cf. T. Buonsegni, *Descrizzione*, etc. published at the time of the translation, p. 17.
May 15 C	S. JOHN BAP. DE LA SALLE (1719). No phenomena. Translation in 1734; only skeleton found. *Vie* (1876), II, p. 321.
May 17 A	S. PASCHAL BAYLON (1592). Covered with quicklime, but found entire and incorrupt nine months later. In 1611 surgeons declared preservation miraculous; fragrant. Staniforth, *Life*, pp. 183–189.
May 20 A	S. BERNARDINE OF SIENA (1444). Kept above ground twenty-six days after death; fragrant; copious discharge of blood from nostrils after twenty-four days. Incorrupt in 1472. Amadio, *Life*, pp. 287–289, 325. Still incorrupt in seventeenth century. *AA.SS.* May, Vol. V, p. 148.
May 26 A	S. PHILIP NERI (1595). Viscera removed but body apparently not embalmed. Found perfectly free from corruption eight months after death. Still sound and entire in 1599, 1602, and 1639. Bacci, *Life*, II, pp. 124–125, 130. Capecelatro, *Life*, II, pp. 465–466 and 487.
May 31 A	S. ANGELA MERICI (1540). Fragrant and flexible for thirty days; intact, incorrupt and sweet smelling in 1672. Still entire in 1867. B. O'Reilly, *Life*, pp. 247 and 253.
June 4 B	S. FRANCIS CARACCIOLO (1608). Flexible and fragrant, blood flowed when incision made. Embalmed, partly preserved in 1628. Cencelli, *Compendio*, p. 203.
June 12 C	S. JOHN A. S. FACUNDO (1479). No evidence of incorruption but extraordinary fragrance at translation in 1533. Valauri, *Vita*, p. 143.
June 21 C	S. ALOYSIUS GONZAGA (1591). Nothing left but skeleton in 1598; no phenomena. Cepari-Goldie *Life*, p. 244.
July 5 A	S. ANTHONY M. ZACCARIA (1539). Body remained entire, though kept above ground until 1566; then buried in damp earth, only skeleton in 1664. Teppa, *Vita*, p. 177.
July 18 A	S. CAMILLUS DE LELLIS (1614). Body soft and flexible until interment. At official recognition in 1625 still found fresh and supple like a living body. A copious exudation of fragrant liquid. Cecatelli, *Life*, I, p. 216.
July 19 A	S. VINCENT OF PAUL (1660). When exhumed in 1712 body incorrupt and entire, though eyes and nose had suffered. In 1737 flesh reduced to fragrant dust. Maynard, *Vie*, IV, pp. 370–371.
July 20 C	S. JEROME EMILIANI (1537). Alleged fragrance in 1566, but no other phenomena. *AA.SS.* Feb., Vol. II, p. 218.
July 31 C	S. IGNATIUS LOYOLA (1556). Viscera removed and body roughly embalmed. Translation 1568; no phenomena. Bartoli-Michel, *Vie*, II, p. 210. *AA.SS.* July, Vol. VII, p. 610.

Aug. 2 C	S. ALPHONSUS LIGUORI (1787). Apparently no phenomena except "ruddy countenance" before burial. Exhumed c. 1817. Berthe-Castle, *Life*, II, pp. 615, 683.
Aug. 7 C	S. CAJETAN (1547). Body seemingly thrown into a common pit with others. No phenomena known. *AA.SS.* Aug., Vol. II, p. 324; Maulde de la Claviere, *Vie*, p. 154.
Aug. 21 B	S. JANE FRANCES DE CHANTAL (1641). Embalmed; body found entire in 1722. Heart preserved separately; strange phenomena. Bougaud, *Vie*, pp. 538, 566, 584.
Aug. 27 B	S. JOSEPH CALASANCTIUS (1648). Viscera removed after death. Heart and tongue still remain fresh and supple as in life. Losada, *Vida*, p. 215.
Aug. 30 A	S. ROSE OF LIMA (1617). Body found entire, fresh-colored and fragrant, eighteen months after death. Still fragrant but wasted and desiccated in 1630. Feuillet, *Life*, pp. 156–157; *AA.SS.* Aug., Vol. V, pp. 987–989.
Sept. 5 A	S. LAWRENCE JUSTINIAN (1455). Body remained above ground and exposed to the air for sixty-seven days. There had been no embalming process, but it continued entire, fragrant and ruddy. *AA.SS.* Jan., Vol. I, p. 563.
Sept. 18 C	S. JOSEPH OF COPERTINO (1663). Embalmed in deference to suggestion of the Pope. No phenomena. *AA.SS.* Sept., Vol. V, p. 1043; Laing, *Life*, p. 118.
Sept. 22 A	S. THOMAS OF VILLANOVA (1555). Quite incorrupt in 1582; resolved into dust, but very fragrant at later translation. *AA.SS.* Sept., Vol. V, pp. 958 & 976.
Oct. 10 C	S. FRANCIS BORGIA (1572). No phenomena. Body not disturbed until 1617. Newly enshrined 1625. Suau, *Vie*, pp. 541–542.
Oct. 15 A	S. TERESA (1582). Minute description of its incorrupt state and marvelous fragrance by Ribera in 1588, confirmed by Gracian. Phenomena of heart. Ribera, *Life*, Bk. V, chs. i, ii, iii; Mir, *Vida*, II, pp. 815–817.
Oct 19 A	S. PETER OF ALCANTARA (1562). Incorrupt and fragrant in 1566, still intensely fragrant in 1616, but flesh consumed. *AA.SS.* Oct., Vol. VIII, pp. 641, 699, 783.
Oct 20 B	S. JOHN CANTIUS (1473). Said to have been found incorrupt in 1539, evidence not satisfactory. Fragrant dust in 1603. *AA.SS.* Oct., Vol. VIII, p. 1059.
Nov. 4 B	S. CHARLES BORROMEO (1584). Body (embalmed) to a large extent entire in 1606, despite damp and leaky coffin. Doctors consider preservation supernatural. Giussano, *Life*, II, p. 555. Body in 1880 still in same condition. Sylvain, *Vie*, III, pp. 387–388, 395.
Nov. 10 A	S. ANDREW AVELLINO (1608). Body found incorrupt a year after death. Curious phenomenon of blood remaining liquid and uncongealed. Fernandez Moreno, *Vida*, pp. 112–114.

Nov. 13 A	S. DIDACUS (1463). The body, dug up four days after death, remained above ground for six months supple and fragrant. It was still entire in 1562. Rottogmo. *Vita*, pp. 87–90.
Nov. 14 A	S. JOSAPHAT (1623). Martyred and thrown into the river on the Sunday, body fished up beautiful and fresh-colored on the Friday. In 1637 remained still almost completely incorrupt. Official verification in 1637, and again in 1674. Guepin, *Vie*, II, pp. 105, 355, 402.
Nov. 24 A	S. JOHN OF THE CROSS (1591). Body found incorrupt and fragrant nine months after death; bled when fingers cut off. Lime added; still incorrupt in 1859. Munoz y Garnica, *Vida*, pp. 229–300. D. Lewis, *Life*, p. 293.
Dec. 3 A	S. FRANCIS XAVIER (1552). Buried in the earth at Sancian, lime heaped on top. In Feb. 1553 body disinterred and found quite fresh as if just dead. Brought to Malacca, reburied, then in Dec. transferred to Goa. Formal medical attestation (Nov. 18, 1556) that it had not been embalmed but remained fresh, supple, and with natural color. Some parts of the body still supple in 1615, the greater portion desiccated. Brou, *Vie*, II, pp. 370, 385, 404.

[K.2]
LEVITATED ECSTATICS OF THE ROMAN CATHOLIC CHURCH WITH REFERENCES TO ACCOUNTS OF THEIR LIVES

Bd. Nicholas Factor, 1583 (Moreno, *Vida*, pp. 128–129).
Bd. Andrew Ibernon, 1602 (Process, *Summario*, pp. 319, 324–25, 331).
Bd. Gaspar de Bono, 1604 (P. A. Miloni, *Vita*, pp. 76–77).
Bd. Juan di Ribera, 1612 (V. Castrillo, *Vita*, p. 92).
St. Alphonsus Rodriguez, 1617 (*AA.SS.* Oct., Vol. xiii, p. 622).
Bd. Lorenzo da Brindisi, 1619 (B. da Coccaglia, *Ristretto*, pp. 136, 196).
Veronica Laparelli, 1620 (Process, *Summario*, pp. 138, 141).
St. Michael de Santis, 1625 (N. della Vergine, *Vita*, pp. 45–49, 56).
Bd. Bernard da Corleone, 1667 (B. Sanbenedetti, *Vita*, pp. 63, 72).
Maria Minima Strozzi, 1672 (*Vita*, p. 19).
Juana de la Encarnacion, 1705 (L. J. Zevallos, *Passion de Christo*, pp. 23–24).
Bonaventura Potentini, 1711 (*AA.SS.* Oct., Vol. xii, p. 154).
Bd. Francisco de Posadas, 1713 (V. Sopena, *Vita*, pp. 43–44).
Angiolo Paoli, 1720 (T. Cacciari, *Vita*, p. 147).
St. Pacificus di San Severino, 1721 (Melchiorri, *Vita*, p. 73).
Bd. Angelo di Acri, 1739 (*AA.SS.* Oct., Vol. xiii, pp. 661, 673).
Clara Isabella de Furnariis, 1744 (Process, *Summario*, p. 103).
Gertrude Salandri, 1748 (*Vita*, an anonymous but admirable biography, pp. 220–224).
St. Maria Francesca delle Cinque Piaghe, 1791 (B. Laviosa, *Vita*, p. 52).
Andrew Hubert Fournet, 1821 (Process, *Summario*, pp. 376, 395, 396, etc.).
J. B. Cottolengo, 1842 (Process, *Summario*, pp. 411, 412, 416).

NOTES

CHAPTER 2

[1] See, for example: Guenon 1942 & 1953; Schuon 1972, 1975, 1976, & 1984; Nasr 1989; Scholem 1961; Eliade 1972; Blofeld 1973 & 1979; Govinda 1966; and Smith, H., 1976.

[2] Wilber 1983, pp. 59–61.

[3] FitzGerald 1986; and Huxley 1952.

[4] Smart 1964, p. 220; and Deutsch, E., 1969, pp. 86–90.

[5] Broad 1953, p. 8.

[6] Braude 1986, p. 5.

[7] Braude 1986, p. 11.

[8] Shor 1979, p. 40.

[9] Eisenbud 1983, pp. 153, 155–56.

[10] Experimental conditioning of the immune system was reported in France during the 1920s and confirmed in the Soviet Union. It was accomplished for the first time in the United States by Ader and Cohen in 1975. See: Ader & Cohen 1975; and Metalnikov & Chorne 1926.

[11] National Academy of Sciences 1989.

[12] McMahon 1973 & 1974.

[13] Wybran et al. 1979.

[14] Morley et al. 1987.

[15] Healy et al. 1983.

[16] Morley et al. 1987; and Pert et al. 1985.

[17] Solomon & Moos 1964.

[18] For overviews of psychoneuroimmunology research, see: Ader et al. 1990; Solomon, G. F., 1987; and Locke et al. 1985.

[19] Cited in McMahon 1976, p. 180.

[20] Descartes 1637.

[21] McMahon & Sheikh (1984); McMahon 1973 & 1976.

CHAPTER 3

[1] Zoologist Edward Wilson, for example, has argued that "the mind will be more precisely explained as an epiphenomenon of the neuronal machinery of the brain" (Wilson, E. O., 1978a, p. 195). The sociobiology Wilson helped to establish would "serve as the antidiscipline of the social sciences [so that] the scientific materialism embodied in biology will, through a reexamination of the mind and the foundations of social behavior, serve as a kind of antidiscipline to the human-

ities" (Wilson, E. O., 1978a, p. 204). Having cannibalized psychology, "the new neurobiology will yield an enduring set of first principles for sociobiology" (Wilson, E. O., 1978b, p. 575). See also: Skinner 1974, p. 189.

[2] Dobzhansky & Ayala 1977, p. 9. In chapter 19.2, I discuss two sets of ideas related to the Dobzhansky-Ayala concept of evolutionary transcendence: first, doctrines of *emergence* that emphasize the novel processes, structures, and relational patterns that appear in evolution; and second, views labeled *panpsychic* because they see subjectivity, experience, or soul in the physical elements and in all living things. In the article cited here, Dobzhansky and Ayala affirmed the novelty emphasized by emergent evolutionists but did not thereby sanction panpsychism. During numerous seminars and conversations, Dobzhansky explicitly opposed the notion that inorganic matter had subjectivity.

[3] Simpson, G. G., 1960, p. 261.

[4] Simpson, G. G., 1960; and Ayala 1974.

[5] Ayala & Valentine 1979, p. 1.

[6] For a statement of creationist argument by one of its leading advocates, see Gish 1978.

[7] Gould 1983, pp. 254–55.

[8] Gould 1983, pp. 11–14.

CHAPTER 4

[1] Gleick 1988, p. 314.

[2] Corbin, H., 1978, pp. 104–105.

[3] Rogo 1970, pp. 14–15.

[4] Draeger & Smith 1969, pp. 123–25.

[5] Govinda 1966, pp. 77–78.

[6] Eckhart 1941, cited in Huxley 1970.

[7] Miller, J. G., 1978, pp. 1–4.

[8] Greeley 1987.

[9] Gallup Poll 1985.

[10] Greeley & McCready 1975.

[11] Greeley & McCready 1975.

[12] Koestler 1978, p. 5.

[13] MacLean 1962.

[14] MacLean 1958.

[15] Koestler 1978, p. 13.

CHAPTER 5

[1] Marais 1973.

[2] Murphy & Brodie 1973, pp. 19–22.

[3] Nieporte & Sauers 1968, p. 23.

[4] Zimmerman 1970, p. 161.

[5] Jones, R. F., 1970, pp. 23–25.

[6] Marais 1973.

[7] Milne & Milne 1972, p. 38.

[8] Buddhaghosa 1964, chap. 13.1–7, p. 447.

[9] Buddhaghosa 1964, p. 448.

[10] Marais 1973.

[11] Milne & Milne 1972, pp. 18–19.

[12] Redgrove 1987, chap. 2.

[13] Redgrove 1987, pp. 59–64.

[14] Ackerman 1990, pp. 44–45.

[15] Droscher 1969, p. 100.

[16] Ackerman 1990, pp. 49, 51.

[17] Ackerman 1990, pp. 157–61.

[18] Ackerman 1990, pp. 290–91.

[19] Ackerman 1990, pp. 291–92.

[20] Kapleau 1965, pp. 237–38.

[21] George 1932, p. 91.

[22] Eliade 1972.

[23] Gitananda 1973.

[24] Russell, N., 1980, pp. 52–56; and Attwater & Thurston 1981. For an analysis of extrasensory perception among Christian saints, see: White, R., 1981 & 1982.

[25] Rahner 1979, bk. 1, chap. 1, para. 9, p. 83.

[26] Rahner 1979, chaps. 6 & 7.

[27] Poulain 1950, pp. 88–90.

[28] Jung, *Collected Works*, vol. 12, p. 301, n. 26; vol. 14, para. 42; vol. 24, para. 41.

[29] Russell, E., 1975, pp. 58–61.

[30] Pierrakos 1987, p. 73.

[31] Schimberg 1947, pp. 38–39.

[32] Corbin, H., 1978, p. 104.

[33] Corbin touches upon the luminous phenomena of mystic states in other books as well, among them *Spiritual Body and Celestial Earth* (1977) and *Creative Imagination in the Sufism of Ibn Arabi* (1969).

[34] Thurston 1952, pp. 162–70.

[35] Williamson 1976.

[36] Bonington 1976, pp. 176–77.

[37] Williamson 1976.

[38] Williamson 1976, p. 318.

[39] Williamson 1976, p. 319.

[40] Bonington 1976, p. 177.

[41] Slocum 1956, pp. 39–42.

[42] Lindbergh 1953, pp. 389–90.

[43] Shackleton 1947, p. 211.

[44] Worsley 1977, p. 197.

[45] Gurney 1886, 1970, vol. 2, p. 102.

[46] Gurney 1886, 1970, vol. 2, p. 103.

[47] Cited in Rogo 1972, pp. 64–66.

[48] Thurston 1952, p. 224; and Rogo 1970, pp. 87–88.

[49] Rogo, pp. 90–91.

[50] Narayanaswami 1914.

[51] Gurney 1886, 1970, vol. 1, pp. 225–26.

[52] Arintero 1951, p. 278.

[53] Jones, F. P., 1979, p. 2.

[54] Murphy & White 1978.

[55] Narayanaswami 1914.

[56] Murphy & White 1978.

[57] Grof 1975; Masters & Houston 1966.

[58] Simonton, Simonton-Matthews, & Creighton 1978.

[59] Leonard 1982.

[60] Targ & Puthoff 1977, pp. 122f.

[61] Phillips 1980, p. 4.

[62] Phillips 1980, p. 239. As an appendix, Phillips published this list of articles and books describing micro-psi investigations by Besant, Leadbeater, and other experimenters: *Lucifer.* 1895. London (November); *The Theosophist,* 1907–1908, Vol. 29, Pts. I & II, 1908–1909, Vol. 30, Pts. I & II, 1924–1926, Vol. 45–47, and 1932–1933, Vol. 54; and Besant, B., & C. Leadbeater. 1908. *Occult Chemistry.* Madras, India: Theosophical Publishing House (2d ed., 1919, 3d ed., 1951).

[63] Schwarz 1971, pp. 28–39.

[64] Koufax 1966, pp. 213–15.

[65] Russell, N., 1980.

[66] Sri Aurobindo 1972, *Letters on Yoga.* Vol. 26 in *The Collected Works,* pp. 184–85.

[67] Nikhilananda 1942, p. 58.

[68] Freemantle 1964, p. 26.

[69] Johnson, R., 1959, pp. 83–84.

[70] See, for example: Vasiliev 1976.

[71] Simpson, G. G., 1960, pp. 256–57.

[72] Gopi Krishna 1971.

[73] Nikhilananda 1942, pp. 829–30.

[74] Katz, R., 1982a, p. 42.

[75] Eliade 1972, pp. 474–75.

[76] Gitananda 1973.

[77] Benson et al. 1982, p. 235. See also: Benson 1982 & 1984.

[78] Thurston 1952, pp. 210–12.

[79] Thurston 1952, p. 215.

[80] Katz, R., 1982b, p. 348.

[81] Sannella 1987, pp. 71–72.

[82] Sannella 1987.

[83] Murphy & White 1978.

[84] Brodie & Houston 1974, p. 182.

[85] Govinda 1966, p. 82.

[86] Govinda 1966, p. 91.

[87] David-Neel 1956, pp. 200–204.

[88] Eliade 1972, pp. 61, 243, 477–82. See also: Czaplicka 1914, p. 175; Kroeber 1899, pp. 265–67; Perry 1926, p. 396; Malinowski 1932, p. 239; and Layard 1930, pp. 501–24.

[89] Benson 1984, pp. 157–58.

[90] *De Servorum Dei beatificatione et Beatorum canonizatione,* III, pp. xlix, 9. In Thurston 1952, p. 17.

[91] Saint Teresa of Avila, *Life,* chap. 20:4.

[92] M. Mir, *Vida de Santa Teresa,* 1912, vol. 1, p. 286. In Thurston 1952, p. 12.

[93] Michener 1976, p. 446.

[94] *Time,* 1975.

[95] Buckle 1909, p. 203; Fodor 1964, pp. 24–29; and Haskins 1975, p. 84.

[96] Myers, F. W. H., 1954, vol. 1, pp. 682–85.

[97] Cited in Myers, F. W. H., 1954, pp. 292–95.

[98] Green, C., 1968, p. 25.

[99] Crookall 1972, pp. 11–12.

[100] Myers, F. W. H., 1954, vol. 2, pp. 551–52.

[101] Swedenborg 1960a, 1960b, & 1962; see also: Trowbridge 1970; and Van Dusen 1972, appendix.

[102] Many near-death experiences, too, include contact with other worlds. See: Ring 1980 & 1984; Moody 1975; Eliade 1972; Corbin, H., 1977; and Grof 1975.

[103] *New Catholic Encyclopedia* 1967, p. 203.

[104] Rush 1964.

[105] *Life* magazine, May 1959, p. 43.

[106] Beard & Schaap 1970, p. 85.

[107] Herrigel 1971, pp. 77–79.

[108] Shainberg 1989, p. 34.

[109] Murphy & White 1978, chap. 4.

[110] Thurston 1952, p. 143.

[111] Anthropologist Waldemar Bogoras, for example, observed a Chuckchee shaman who appeared to make his tent shake and pieces of wood fly through the air; and anthropologist Sergei Shirokogoroff reported similar phenomena he had witnessed or heard about among the Tungus. See: Bogoras 1904; and Shirokogoroff 1935.

[112] Murphy & Brodie 1973.

[113] Herrigel 1971, p. 92.

[114] Myers, F. W. H., 1954, vol. 1, p. xix.

[115] Myers, F. W. H., 1954, vol. 2, pp. 196–97.

[116] Sri Aurobindo 1970, *The Life Divine*, chap. 12.

[117] Elliott 1961, p. 38.

[118] Tanzer 1972.

[119] Dick-Read 1959, pp. 162–63.

[120] Descriptions of these and other cognitive processes can be found in *Frames of Mind: The Theory of Multiple Intelligences* by Howard Gardner (1985). Basic Books.

[121] James, W., 1902, lectures 16 & 17: Mysticism.

[122] Robinson 1978, pp. 113–15. See also: Robinson 1977a & 1977b.

[123] James, W., 1902, lectures 16 & 17: Mysticism; and lecture 20: Conclusions.

[124] James, W., 1902, lectures 16 & 17: Mysticism.

[125] Nietzsche, *Ecce Homo*, sec. 3.

[126] Myers, F. W. H., 1954, vol. 1, p. 71.

[127] Myers, F. W. H., 1954, vol. 1, pp. 110–11.

[128] Cited in Ghiselin 1980, pp. 44–45. Taken from Holmes, *The Life of Mozart*.

[129] Cited in Ghiselin 1980, pp. 36–37.

[130] Myers, F. W. H., 1954, vol. 1, pp. 95–96.

[131] Nietzsche, *Ecce Homo*, sec. 4.

[132] Myers, F. W. H., 1954, vol. 1, p. 73.

[133] See: James, W., 1902, lectures on The Sick Soul, The Divided Self, and Conversion; Maslow 1968, chapters The Need to Know and the Fear of Knowing, Cognition of Being in the Peak-Experiences, and Some Dangers of Being-Cognition; Maslow 1971, chapter Emotional Blocks to Creativity; Wilber's essays The Spectrum of Development, and The Spectrum of Psychopathology, in Wilber, Engler, & Brown 1986; Barron 1963b, chapters Unconscious and Preconscious Influences in the Making of Fiction, Unusual Realization and Changes in Consciousness, and Violence and Vitality; and Kubie 1958.

[134] Robinson, E., 1978, pp. 104–5.

[135] Assagioli 1973. See also: Leonard 1991.

[136] Cited in Huxley 1970, p. 170.

[137] Jackson 1975, p. 34.

[138] Novak 1976, p. 164.

[139] Csikszentmihalyi 1990, p. 4.

[140] Csikszentmihalyi 1988, p. 365.

[141] Jung 1961. The book's glossary contains definitions of synchronicity.

[142] Jung 1961, chap. 6.

[143] See: Jung, *The Structure and Dynamics of the Psyche.* In *The Collected Works,* vol. 8.

[144] Simpson, G. G., 1949, pp. 260–61.

[145] Leonard 1986, pp. 48–50.

[146] James, W., 1902, lectures 4 & 5.

[147] James, W., 1902, lectures 4 & 5.

[148] Sri Aurobindo, *The Collected Works,* vol. 22, pp. 285, 367–68.

[149] Nagao 1991, p. 9.

[150] Suzuki-roshi 1970, pp. 31, 133.

[151] *Enneads,* 6.4.14.

[152] *Enneads,* 4.8.1.

[153] These quotations are taken from Huxley 1970, pp. 11–12.

[154] von Urban 1951, pp. 74–90.

[155] Redgrove 1987, p. 128.

[156] Redgrove 1987, pp. 128–29.

[157] Leonard 1981, pp. 21–23.

[158] Sokol 1989. In Feuerstein 1989, pp. 115–20. There is a long-standing witness in many cultures to erotic experience such as this woman described. See, for example: Feuerstein 1989; Chia & Chia 1986; Whitmont 1982; Ricci 1979; Kinsley 1975; von Urban 1952; and Redgrove 1987.

[159] James, W., 1902, lectures 11, 12, & 13.

[160] James, W., 1902, lectures 11, 12, & 13.

[161] Nikhilananda 1942, p. 115.

[162] Nikhilananda 1942.

[163] Rolland 1910, bk. 2, pp. 312–13.

[164] Cited in Robinson, 1967, p. 95.

[165] James, W., 1902, lectures 14 & 15: The Value of Saintliness.

[166] According to Swami Saraswati, these seven chakras below the muladhara are named: atala, vitala, sutala, talahata, rasatala, mahatala, and patala. See: Saraswati 1973 & 1984. See also: Motoyama 1981; and Woodroffe 1972.

[167] See: Poortman 1978; and Mead, G. R. S., 1967.

[168] Corbin, H., 1977.

[169] Tulku Thondup Rinpoche 1989, p. 200.

[170] Corbin, H., 1977, pp. 176–268.

[171] In writing my novel *Jacob Atabet* (Jeremy P. Tarcher 1988), I drew upon such reports collected from seminars on transformative practice and from interviews with meditation students, martial artists, and athletes.

CHAPTER 6

[1] Mauss 1935 (1973 trans.), pp. 70–88.

[2] Below 1935; and Bailey 1942.

[3] Astrov 1950.

[4] Labarre 1947.

[5] Efron 1941. The two groups, Efron found, exhibited changes in gesture proportional to their degree of general assimilation.

[6] Hewes 1955 & 1957.

[7] Firth 1970; and Sapir 1949.

[8] Hall 1959; and Birdwhistell 1960, 1971, 1972, & 1975.

[9] Douglas 1971.

[10] Douglas 1970, p. 65.

[11] Bourdieu 1984.

[12] Fisher & Cleveland 1968.

[13] See, for example: Benthall & Polhemus 1975; Brown, N. O., 1968; Buytendijk 1957; Chernin 1981; Douglas 1970; Foucault 1970; Freud 1924–1950 & 1953–1974; Griffin, S., 1978; Johnson, D., 1980; Kunzle 1981; Mauss 1982; Merleau-Ponty 1962; Polhemus 1978; Reich 1970; Schilder 1950; and Tinbergen 1974.

[14] Huxley 1970; Stace 1987; Schuon 1984; and James, W., 1902.

[15] Katz, S., 1978.

[16] Katz, S., 1978, p. 40. For other discussions of culture's formative role in religious experience, see: Katz, S., 1983.

[17] Gardner, H., 1985.

[18] Tulving 1985; and Gazzaniga 1985.

CHAPTER 7

[1] Though some thinkers of Greek antiquity entertained ideas of general progress, they did not influence most Platonists, Neo-Platonists, or early Christian mystics. See: Nisbet 1980.

[2] Sri Aurobindo, *Sapta-Chatusthaya*, vol. 27, p. 366, in *The Collected Works*. For other discussions of the siddhis, see vol. 27, pp. 366–74, and references to *siddhi* in the index, vol. 30 of *The Collected Works*.

[3] Duchesne 1933; and Lecky 1975.

[4] Novak 1984, pp. 64–65.

[5] Novak 1984, p. 71.

[6] Pallis 1974, pp. 279–80.

7 Pallis 1974, pp. 275–79. For other arguments that Christian doctrines of grace and Buddhist doctrines of nonattainment have much in common, see: Novak 1981; and Smith, H., 1983a & 1983b.

8 Alexander, S., 1979; Morgan, C. L., 1923; Broad 1980; Needham 1937; and Polanyi 1964.

9 Goudge 1967a, pp. 474–75.

10 Goudge 1967a, p. 475.

11 Solomon, R., 1983; Miller, A. V., 1977; and Baillie 1967.

12 Sri Aurobindo, *The Life Divine*, bk. 1, chap. 18. Vol. 18 in *The Collected Works*.

13 Bergson 1911; Whitehead, A. N., 1978; and Hartshorne 1970.

14 Griffin, D. R., 1989, p. 39.

15 Whitehead, A. N., 1978, p. 23.

16 Whitehead, A. N., 1978, p. 24.

17 For a review of panpsychism in modern philosophy, see Hartshorne 1950.

18 Griffin, D. R., 1989, pp. 92–93.

19 Solomon, R., 1983; Miller, A. V., 1977; and Baillie 1967.

20 For a summary of Bergson's views of intellect and intuition, see: Goudge 1967b, pp. 287–95.

21 Goudge 1967b, pp. 287–95.

22 James, H., 1853, p. 479.

23 James, H., 1863, pp. 396–97, 425–526.

24 Young 1951, pp. 155–56.

25 Sri Aurobindo, *The Life Divine*, bk. 1, chap. 1. Vol. 18 in *The Collected Works*.

26 Young 1951, pp. 167–69.

27 Lovejoy 1964, p. 244.

28 Lovejoy's *The Great Chain of Being* includes a review of emanationist doctrine as it developed in Western culture from Plato's dialogues to the works of Novalis, Schelling, and other writers of the eighteenth and early nineteenth centuries.

29 Plotinus, *Enneads*. 6.9.7.

30 Cited in Huxley 1970, pp. 11–12.

31 For a review of metaphors regarding our human "Return to the Source," see: Metzner 1986, chap. 8.

32 Whitehead, A. N., 1978, pp. 362–63 (see index entry *eternal objects*).

33 Cross 1957.

34 Robinson, J., 1967, pp. 82–96. For a discussion of Whiteheadian and Hartshornian panentheism, see: Griffin, D. R., 1989, pp. 30, 84, 90, 143.

35 Laski 1990; Johnson, R., 1959; and Robinson, E., 1977 & 1978.

36 For a discussion of ways in which metaphors shape our common activities, see: Lakoff & Johnson 1980; and Lakoff 1987.

37 Kaufman 1974, chap. 11.

38 Kaufman cited some of these misquotations and compared them to their original version. See: Kaufman 1974, chap. 10: The Master Race.

39 Kaufman 1974, pp. 303–304.

CHAPTER 8

[1] *New Catholic Encyclopedia* 1967.

[2] Bynum 1991.

[3] Thomas Aquinas 1975, pp. 325–27.

[4] Bynum 1991, p. 55.

[5] Butterworth 1973, pp. xxiii–xxviii.

[6] Butterworth 1973, pp. 81–82, 139, 251–53.

[7] Guardini 1954, pp. 61–75. See also: Durrwell 1960, pp. 14–17, 173–80, 287–91, 357–59.

[8] Blofeld 1979, p. 73.

[9] Blofeld 1979, chap. 8.

[10] Eliade 1972; and Campbell 1959.

[11] Eliade 1972, pp. 44–46.

[12] Eliade 1972, pp. 47–48. Eliade quoted Spencer & Gillen, *The Native Tribes of Central Australia.*

[13] Eliade 1972, p. 47, n. 36.

[14] Eliade 1972, p. 56.

[15] Eliade 1972, p. 62.

CHAPTER 9

[1] Lewis, C. S., 1970, pp. 197–99. This is the second book in a space trilogy that begins with *Out of the Silent Planet* and concludes with *That Hideous Strength.*

[2] Thurston & Attwater 1981. See also: White, R., 1982.

[3] Blofeld 1979 & 1973.

[4] Corbin, H., 1977 & 1978.

[5] Broad 1923; Smythies 1967; and Smythies & Koestler 1969.

[6] Sirag 1985.

[7] Corbin, H., 1977, p. 213.

CHAPTER 10

[1] Gauld 1982.

[2] See, for example: Stevenson 1970 & 1975–1983; Thouless 1960; Chari 1962a & 1962b; and Lorimer 1984.

[3] Mead, G. R. S., 1967.

[4] Poortman 1978.

[5] Evans-Wentz 1949.

[6] Hick 1975.

[7] For other proposals about the afterlife by twentieth-century Christian thinkers, see: Moltmann 1967; Pannenberg 1970; and Rahner 1965.

CHAPTER 11

[1] Lifschutz 1957, pp. 529–30.

[2] Needles 1943.

[3] Deutsch 1951, f. 70.

[4] Hadley 1930, pp. 1101–11.

[5] Moody, R. L., 1946.

[6] Moody, R. L., 1946, p. 935.

[7] Moody, R. L., 1948.

[8] Graff & Wallerstein 1954.

[9] Gardner & Diamond 1955.

[10] Ratnoff & Agle 1968.

[11] Ratnoff 1969, p. 161.

[12] Ratnoff, 1969, pp. 162–63.

[13] Brown & Barglow 1971.

[14] Aldrich 1972.

[15] Kimball 1970.

[16] Good 1823.

[17] Steinberg 1949.

[18] Biven & Klinger 1937.

[19] Fried et al. 1951.

[20] Murray & Abraham 1978.

[21] Devane 1985.

[22] Brown & Barglow 1971.

[23] See: Lerner et al. 1967; Murray & Abraham 1978; Fried et al. 1951; and Kroger 1962.

[24] Pawlowski & Pawlowski 1958.

[25] More specifically, Brown and Barglow proposed that depression suppressed the production of biogenic amines via the cortical and limbic systems, thus impairing the release of luteinizing hormone releasing factor, follicle-stimulating hormone releasing factor, and prolactin inhibitory factor at the median eminence of the hypothalamus. As a consequence of these interactions, the two researchers suggested, circulating levels of luteinizing hormone and follicle-stimulating hormone from the anterior pituitary would be decreased so that ovulation and menstrual bleeding were inhibited. At the same time, increased prolactin levels caused by the reduction of prolactin inhibitory factor would stimulate both lactation and maintenance of a persistent corpus luteum (a ductless gland developed in the ovary by the reorganization of a Graafian follicle following ovulation), which would promote amenorrhea and the production of estrogen and progesterone to

keep the uterus in a quasi-pregnant condition (Brown & Barglow 1971). Supporting Brown and Barglow's theory, animal studies have shown that rats and rabbits have false pregnancies after cervical stimulation with glass rods, electrical stimulation of the cervical canal, electrical stimulation of the head, or mating with vasectomized males; while other studies have shown that a corpus luteum persists in rats and rabbits during false pregnancies (Abram et al. 1965; Lloyd 1968; Moulton 1943; and Bogdanove 1966).

Brown and Barglow's theory is open to question, though, because luteinizing hormone levels have varied significantly among women with pseudocyesis; because prolactin has not been shown to be luteotropic in humans; because A. Zarate and associates, 1974, found by laparoscopy that two women with pseudocyesis did not have a corpus luteum (though Moulton, 1943, had reported three women with false pregnancy who did); because some cases have displayed contrasting patterns of gonadotropin secretion; and because other studies have suggested that pseudocyesis is mediated by various combinations of hormonal changes. (See: Zarate et al. 1974; Starkman et al. 1985; Murray & Abraham 1978; and Moulton 1943.)

[26] Starkman et al. 1985, p. 53.

[27] Knight 1960.

[28] Evans & Seely 1984.

[29] Evans, W., 1951; and Aronson 1952.

[30] O'Regan & Hurley 1985, pp. 3–6.

[31] Braun 1983, p. 127.

[32] Braun 1983, p. 131.

[33] Goleman 1988.

[34] Coons et al. 1982.

[35] Putnam 1984.

[36] O'Regan & Hurley 1985, p. 20.

[37] Goleman 1988.

[38] Bliss 1984.

[39] Bliss 1984.

CHAPTER 12

[1] Shapiro & Morris 1978.

[2] Shapiro, A. K., 1959.

[3] Cousins 1977, p. 16.

[4] Beecher 1955, p. 1603.

[5] Cherry 1981, p. 60.

[6] Shapiro, A. K., 1964; Frank 1973; and Benson & Epstein 1975.

[7] Lasagna 1955; Jacobs & Nordan 1979; and Buckalew & Coffield 1982.

[8] Benson & McCallie 1979, p. 1428.

[9] Cures such as those evident in placebo-induced healing are apparent, too, in spontaneous remissions from disease. Though such remission has not been documented with nearly the care devoted to disease itself, it has been reported from time to time in leading medical journals. See: Gaylord & Clowes 1903; Handley 1909; Rohdenburg 1918; Boyd 1921 & 1957; Dunphy 1950; Willis 1952; Everson & Cole 1966; Sakiyama et al. 1984; and McClain et al. 1985.

From a survey of physician reports published in various languages between 1900 and 1965, Tilden Everson and W. H. Cole identified 176 cancer cures with enough documentation to suggest they had spontaneously regressed (Everson & Cole 1966). More recently, Brendan O'Regan and Caryle Hirshberg of California's Institute of Noetic Sciences have assembled some 3,500 articles from more than 860 medical journals in over 20 languages that deal with apparent remission cases (see: O'Regan & Hirshberg 1990). Of the regressions included in the O'Regan-Hirshberg database, some occurred with no medical intervention at all. "These," O'Regan wrote, "give us the strongest evidence that there is an extraordinary self-repair system lying dormant within us" (O'Regan 1987).

[10] Beecher 1961.

[11] Dimond et al. 1958; and Cobb et al. 1959.

[12] Benson & McCallie 1979, pp. 1420–27.

[13] Allington 1934.

[14] Bloch, 1927.

[15] Luparello et al. 1968.

[16] McFadden et al. 1969.

[17] Evans, F., 1984. For other studies of placebo-induced pain reduction, see: McGlashan et al. 1969; Lanz 1970; Sarles et al. 1977; Sturdevant et al. 1977; Gudjonsson & Spiro 1978; and Levine, J., et al. 1981.

In one pain reduction study, the experimenters compared four groups of patients who received placebos from four different physicians. Though all patients treated with a placebo experienced a significantly shorter period of pain than untreated patients, there were significant differences between the groups, which suggested that the character of the physician played an important role in the placebos' effectiveness.

[18] Cousins 1977, p. 12.

[19] Beecher 1955, p. 1604.

[20] McGlashan et al. 1969.

[21] Evans, F., 1984; and Plotkin & Rice 1981.

[22] Malitz & Kanzler 1971.

[23] Lasagna et al. 1955; and Trouton 1957.

[24] Brodeur 1965; Lasagna et al. 1955; Smith, G. M., & Beecher 1960; Nicassio et al. 1982; Steinmark & Borkovec 1974; Smith, H. W., 1978; Goodman 1969; Meyhoff et al. 1983; and Baker & Thorpe 1957.

[25] Beecher 1955.

[26] Wolf & Pinsky 1954.

[27] Aznar-Ramos et al. 1969.

[28] See, for example: Cahill & Belfer 1978. After finding that placebos did not enhance word association speed, Cahill and Belfer suggested that the value of placebos is peculiar to therapeutic situations in which the patient is suffering and needs relief. "To apply the term 'placebo effect' to changes experienced by persons not in the reduced psychological state brought on by physical or severe emotional stress," they wrote, "may be to distort the placebo concept beyond its usefulness for scientific inquiry." Other researchers and physicians have also questioned the ill-defined meanings now attached to the placebo concept. See, for example: Berg 1983; Evans, F., 1984; and Grunbaum 1985.

[29] However, in a study reported by Howard Brody of Michigan State University, 13 of 15 patients reported improvement in their neuroses after receiving pills that they knew to be placebos. See: Brody 1984.

[30] Knowles & Lucas 1960; Shapiro, A., 1970; Hurley 1985; Benson 1980; Benson & Epstein 1975; and Hahn 1985.

[31] Sternbach 1964.

[32] Hurley 1985.

[33] Bootzin 1985.

[34] Plotkin 1985.

[35] Levine, J., et al. 1978.

[36] Gracely et al. 1984; and Grevert & Goldstein 1985, pp. 332–48.

[37] White, L., et al. 1985, pp. 431–37.

CHAPTER 13

[1] For a discussion of differences and similarities among some of the activities these terms refer to, see: Haynes 1977.

[2] For multidimensional models of spiritual healing, see: Leshan 1975 and Gerber 1988.

[3] James, W., 1902.

[4] Smith, S., 1953.

[5] Smith, S., 1953. Hippocrates' family was associated with the temple of Asklepios on the island of Cos.

[6] This list was taken from a doctoral thesis on spiritual healing; see Geddes 1981.

[7] See, for example: Gardner, R., 1983.

[8] Finucane 1973, p. 343.

[9] Finucane 1973, p. 344.

[10] Finucane 1973, p. 344.

[11] Finucane 1973, p. 345.

[12] Laver 1978.

[13] Laver 1978.

[14] Laver 1978, pp. 43–44.

[15] Laver 1978, p. 45.

[16] Paulsen 1926.

[17] Lichtenstein 1925.

[18] Pattison et al. 1973; See also: Festinger et al. 1956.

[19] For information regarding the Bureau's archives, one may write: Bureau Médicale de Notre-Dame de Lourdes, Lourdes, France 65100. See also: Cranston 1988. For a skeptical view of alleged cures at the shrine, see: West, D. J., 1957.

[20] Carrel 1950.

[21] Carrel 1935.

[22] Aradi 1956, pp. 236–37.

[23] Dowling 1984.

[24] Salmon 1972.

[25] Mouren n.d.

[26] Taken from "Cures of Lourdes Recognized as Mircaculous by the Church," published by the Bureau Medicale, Lourdes, France.

[27] Katz, R., 1982a, p. 348.

[28] Katz, R., 1982a, p. 349.

[29] Katz, R., 1982a, p. 352.

[30] Katz, R., 1982a, p. 352.

[31] Richmond 1952.

[32] Vasse & Vasse 1948.

[33] Solfvin 1984.

[34] Grad 1963 & 1964.

[35] Grad 1965a.

[36] Grad 1965a.

[37] Grad 1965a.

[38] Watkins & Watkins 1971.

[39] Watkins et al. 1973; Wells & Klein 1972; and Wells & Watkins 1975.

[40] Krieger 1979a. See also: Krieger 1976; and Krieger et al. 1979.

[41] Quinn 1984. See also: Brody 1985b.

[42] Wirth 1990.

[43] Goodrich 1974.

[44] Winston 1975.

[45] Knowles, F. W., 1954 & 1956.

[46] Solfvin 1984.

[47] Byrd 1988.

[48] Braud & Schlitz 1989.

[49] Vasiliev 1976.

[50] Watkins et al. 1973.

[51] Deane & Brame 1975; Miller, R., 1977; Dean 1982; and Schwartz, S., et al. 1990.

CHAPTER 14

[1] Redgrove 1987.

[2] Wright, D., 1969, p. 22.

[3] Sacks 1990, pp. 6f.

[4] Keller 1954, pp. 100–110.

[5] Keller 1954, pp. 294–97.

[6] Lusseyran 1987, pp. 26–33, 94–95.

[7] Lusseyran 1987, pp. 169, 174–77.

[8] Lusseyran 1987, p. 312.

[9] Sacks 1985a & b.

[10] Briggs 1988.

CHAPTER 15

[1] See, for example: Hilgard, E., 1965a & 1975b; and Kihlstrom 1985.

[2] Hartmann 1896, p. 193.

[3] Pattie 1956. Pattie argued that Mesmer plagiarized Mead's treatise for his medical dissertation.

[4] Ellenberger 1970, pp. 57–69.

[5] Podmore 1963, pp. 1–6. Podmore drew upon Mesmer's autobiographical piece *Precis historique des Faits relatifs au Magnestisme animal* (1781) for his description of the healer's methods.

[6] Podmore 1963, p. 7.

[7] Franklin 1881. Cited in McConkey et al. 1985.

[8] Shor 1979, p. 21.

[9] Ellenberger 1970, pp. 70–74.

[10] Shor 1979, p. 23.

[11] Shor 1979, p. 26.

[12] Though the French Academy of Sciences had rejected Mesmer's ideas about magnetic fluid, a commission appointed by the Prussian government sanctioned theories of animal magnetism in 1816 (Ellenberger 1970, pp. 77–78).

[13] Ellenberger 1970, pp. 77–78. Ellenberger cited Kluge's 1811 work.

[14] Brentano 1837 & 1852. In Ellenberger 1970.

[15] Elliotson 1843.

[16] Esdaile 1846, pp. xxiii–xxiv.

[17] Esdaile 1852, pp. 148–49. The account was by F. W. Sims and was originally published in a British periodical, *The Englishman*, in 1846.

[18] Ellenberger 1970, p. 82.

[19] Shor 1979, p. 30.

[20] Ellenberger 1970, p. 85.

[21] Ellenberger 1970, pp. 90–91.

[22] Ronald Shor called Charcot's commitment to precise method "a point of departure for a continuing scientific evolution of hypnosis research."

[23] Ellenberger 1970.

[24] See, for example: Erickson 1983.

[25] For a review of studies that support these two interpretations, see: Kihlstrom 1985.

[26] Hilgard, E., 1965a and 1975b; and Kihlstrom 1985.

[27] Hilgard, E., 196;a & 1975b; and Kihlstrom 1985.

[28] Cunningham & Blum 1982; and Spanos & Saad 1984.

[29] Spiegel et al. 1985.

[30] Gurney 1888. In Myers, F. W. H., 1954, vol. 1, appendix to chap. 5.

[31] Myers, F. W. H., 1954, vol. 1, pp. 503–504.

[32] Myers, F. W. H., 1954, vol. 1, pp. 506–510. Myers excerpted two articles by Bramwell: Bramwell 1896 & 1900.

[33] Myers, F. W. H., 1954, vol. 1, p. 509.

[34] Myers, F. W. H., 1954, vol. 1, pp. 509–510.

[35] James, W., 1902.

[36] Wells, W., 1940; Edwards 1963; Hilgard, E., 1965a; and Dixon 1981.

[37] Ludwig & Lyle 1964.

[38] Gibbons 1974 & 1976.

[39] Banyai & Hilgard 1976.

[40] Malott & Goldstein 1980; and Malott 1984.

[41] Cardena 1988.

[42] Sarbin & Andersen 1967; Sarbin & Lim 1963.

[43] Spanos 1986. This article is accompanied by peer reviews from defenders of state and social-psychological hypnosis theories. For a defense of state theory, see: Hilgard, E., 1986.

[44] Barber 1965 & 1984.

[45] Hilgard, E., 1971.

[46] Commenting upon hypnosis simulation experiments, Orne and his colleagues David Dinges and Emily Orne wrote:

> To the extent that a simulating subject can duplicate the behavior of someone hypnotized, it is simply not possible to distinguish—based upon behavior—between a hypnosis effect due to changes in subjective experience versus one resulting from overt behavioral compliance. Simulators are in fact able to successfully mimic a remarkably broad range of hypnotic phenomena and to fool experienced clinicians into believing that they are hypnotized.
>
> [This fact] led Orne to conclude that the essence of hypnosis must be sought in the subjective experience of the individual. One useful approach seemed to be the identification of situations where the hypnotic experience

elicited counterintuitive behaviors; where the simulator did not anticipate what the hypnotized individual would do in response to a subjectively compelling experience. Although exceedingly difficult to find, there were a few hypnotic behaviors shown by some [truly hypnotized subjects] but by no simulators. Orne termed certain of these behaviors "trance logic" because they appeared to reflect a lack of critical judgment and greater tolerance for incongruity on the part of hypnotized persons (Orne et al. 1986).

[47] The consistent failure to find any phenomenon that cannot be observed elsewhere while occurring consistently in all hypnotic subjects causes some researchers to doubt that hypnosis is a special kind of consciousness. According to John Kihlstrom, though, this is a problem "only if hypnosis is considered to be a proper set defined by singly necessary and jointly sufficient attributes. [But] recent work in philosophy and psychology suggests that natural categories are best regarded as fuzzy sets whose instances are related by family resemblances, and that they are represented by a prototype whose features are only probabilistically associated with category membership." From this point of view, amnesia, trance logic, or any other phenomenon may be considered to be more or less characteristic of hypnosis, depending upon its frequency and strength among hypnotized subjects. See: Kihlstrom 1984 & 1985.

[48] See, for example: Kihlstrom 1985.

[49] Tart 1979, pp. 591–96.

[50] Tart 1979.

[51] Sherman 1971; Feldman 1976; and Cardeña 1988.

[52] Morgan, A. H., et al. 1974.

[53] Hilgard, J., 1974 & 1979; Wilson & Barber 1983; Bowers, P., 1979; and Tellegen & Atkinson 1974.

[54] Bowers, P., 1979. See also: Sheehan, P. W., 1979 & 1982.

[55] Barron 1953, 1955, 1957, 1958a, & 1958b; and MacKinnon 1960.

[56] Bowers, P., 1979.

[57] Bowers & Bowers 1979; Crawford 1981 & 1982a.

[58] Gur & Gur 1974.

[59] Graham & Pernicano 1979.

[60] Sackeim et al. 1979.

[61] These findings were also supported by a study that showed apparent shifts in cortical activation as measured by EEG alpha density from the left to the right hemisphere among hypnotizable subjects entering hypnosis. See: MacLeod-Morgan & Lack 1982.

[62] Tellegen & Atkinson, 1974.

[63] Graham 1975.

[64] Van Nuys 1973.

[65] Graham & Evans 1977.

[66] Graham & Evans 1977. The authors cite: Hilgard, E., 1965a.

[67] Karlin 1979.

[68] For evidence of perceptual flexibility and Gestalt-making abilities among Mexican shamans, see: Scweder 1972. For descriptions of inspired spontaneity among Zen masters, see: Suzuki 1959; and for an account of Morehei Uyeshiba's astonishing improvisations in aikido, see: Stevens, J., 1987.

[69] Wilson & Barber 1983, pp. 345–47. See also: Wilson & Barber 1981.

[70] Wilson & Barber 1983, p. 348.

[71] Wilson & Barber 1983, p. 349.

[72] Wilson & Barber 1983, pp. 352–53.

[73] Wilson & Barber 1983, pp. 354–56.

[74] Wilson & Barber 1983, pp. 356–57.

[75] Wilson & Barber 1983.

[76] Wilson & Barber 1983, pp. 359–64.

[77] Wilson & Barber 1983, p. 380.

[78] Galton 1883. In Wilson & Barber 1983.

[79] Jaensch 1930. In Wilson & Barber 1983.

[80] Vogt & Sultan 1977. In Wilson & Barber 1983.

[81] Luria 1968, p. 96. In Wilson & Barber 1983.

[82] Wilson & Barber 1983, p. 367.

[83] Lynn & Rhue 1988.

[84] Diamond, M., 1974.

[85] Wicramasekera 1976. Wicramasekera proposed a number of ways to increase hypnotizability, among them training for imagination; instructions that the subject be more adventuresome, both cognitively and interpersonally; alterations of the subject's perceptions through psychedelic drugs; sensory deprivation; and biofeedback training.

[86] See: Kihlstrom 1985.

[87] Braid 1846 & 1855.

[88] Braid 1855.

[89] Johnson, L., 1981; Johnson & Weight 1976; and Fromm et al. 1981.

[90] Hilgard, E., 1975a.

[91] Butler, B., 1954a, & 1954b, 1955a, & 1955b. See also: Lea et al. 1960.

[92] Cangello 1961 & 1962.

[93] Hilgard & Hilgard 1983. See also: Kroger 1963; Sacerdote 1970; and Erickson 1959.

[94] Hilgard & Hilgard 1983. In a study by L. S. Wolfe and J. Millet, 1,500 patients were given suggestions for postoperative pain relief while under chemical anesthesia, and half reported no discomfort after their operation (Wolfe & Millet 1960). In a study by R. E. Pearson, 43 patients received tape-recorded suggestions through earphones while they were deeply anesthetized to assist their postoperative responses, while 38 other patients heard placebo tapes that played music or were blank. Neither the patients, surgeons, nurses, nor other hospital personnel knew which tapes had been played in each instance, so that the entire experiment was conducted double-blind, and the results were determined by the length of

time each patient spent in the hospital. Since the patients who received the posthypnotic suggestions were discharged an average 2.42 days sooner than their controls, Pearson concluded that patients can hear and respond to instructions given during anesthesia (Pearson 1961).

[95] Hilgard, J., 1979.

[96] Hilgard & Hilgard 1983, p. 134.

[97] August 1961.

[98] Katchan & Belozerski 1940.

[99] Davidson, J. A., 1962.

[100] Rock et al. 1969.

[101] Jacoby 1960; Gottfredson 1973; and Morse & Wilcko 1979.

[102] Crasilneck 1979; Toomey & Sanders 1983; Sachs, L. B., et al. 1977; Zane 1966; Siegel 1979; John & Parrino 1983; Crasilneck et al. 1955; Dahinterova 1967; and Anderson, J., et al. 1975.

[103] Evans & Paul 1970; and Hilgard, E., 1967.

[104] Hilgard, E., 1975a.

[105] Shor 1967. See also: Lanzetta et al. 1976.

[106] Hilgard, E., 1975a.

[107] A study by Greene and Reyher, 1972, supports this observation in that it found no significant correlation between increased pain tolerance in hypnotic analgesia and measures of state and trait anxiety taken before and after exposure to shock. See: Greene & Reyher 1972.

[108] Hilgard, E., 1975a, pp. 220–21.

[109] See, for example: Spanos et al. 1974; and Spanos 1986.

[110] Hilgard, E., 1975a.

[111] Hilgard, E., 1973 & 1977.

[112] Spiegel et al. 1989.

[113] Hilgard, E., 1977, p. 50.

[114] Goldstein & Hilgard 1975; and Spiegel & Leonard 1983.

[115] Mason 1952 & 1955.

[116] Wink 1961.

[117] Kidd 1966; and Schneck 1966.

[118] Mullins et al. 1955.

[119] Tuke 1872.

[120] Bloch 1927.

[121] Rulison 1942; Ullman 1959; Ullman & Dudek 1960. See also: Asher 1956.

[122] Thomas, L., 1979.

[123] Clawson & Swade 1975. See also: Henry 1985.

[124] Lucas 1965; and Chaves 1980. See also: Newman 1971.

[125] Bishay et al. 1984; and LaBaw 1970.

[126] Bramwell 1903.

[127] Hadfield 1917.

[128] Chapman et al. 1959, p. 97.

[129] Ewin 1978 & 1979.

[130] Moore & Kaplan 1983. See also: Hammond et al. 1983.

[131] Gould & Tissler 1984; and Frankel & Misch 1973.

[132] Ikemi & Nakagawa 1962.

[133] LeHew 1970.

[134] *British Medical Journal* 1968.

[135] Diamond, H., 1959; Moorefield 1971; and Ben-Zvi et al. 1982. See also: Maher-Loughnan 1970; and Aronoff et al. 1975.

[136] Gross 1984.

[137] Mott & Roberts 1979.

[138] Crasilneck 1982.

[139] O'Brien et al. 1981; Bakal 1981; Scrignar 1981; Van Der Hart 1981; Cohen 1981; and Frutiger 1981.

[140] Honiotes 1977; LeCron 1969; Staib & Logan 1977; Willard 1977; Williams, J. E., 1974; and Erickson 1960.

[141] Johnson & Barber 1976.

[142] Bellis 1966.

[143] Hadfield 1917.

[144] Ullman 1947.

[145] Platonov 1959.

[146] Pattie 1941; Weitzenhoffer 1953; Barber 1961; Paul 1963; and Chertok 1981. See also: Doswald & Kreibich 1906; and Borelli 1953.

[147] Barber 1984.

[148] Agle et al. 1967. The authors suggested that a vasodilator polypeptide that promotes vascular permeability mediated their subjects' hypnotically induced bruising.

[149] Grabowska 1971.

[150] Dikel & Olness 1980.

[151] Hadfield 1920.

[152] Barber 1984, pp. 99–107.

[153] Klein & Spiegel 1989.

[154] Barber 1964; Hilgard, E., 1965a & 1975b; and Kihlstrom 1985.

[155] Kline 1952–53; Weitzenhoffer 1951; Erickson 1943; Davison & Singleton 1967; Graham & Leibowitz 1972; and Sheehan, E. P., et al. 1982.

[156] Crawford et al. 1979.

[157] Kramer & Tucker 1967.

[158] Schneck 1966.

[159] Wallace & Garrett 1973 & 1975; Wallace & Hoyenga 1980; Nash et al. 1984; Brenman et al. 1947; and Hilgard, J., 1974b.

[160] Barber 1965; Blum 1968 & 1979; Blum & Nash 1982; Blum et al. 1967, 1968a, 1968c, & 1971; Blum & Porter 1972; Blum & Barbour 1979; Blum & Green

1978; Blum & Wohl 1971; Crawford & Allen 1983. Crawford also reviewed other studies which suggested that hypnotically responsive subjects exhibit shifts during hypnosis away from verbal, detail-oriented cognitive functioning to more imaginal, nonanalytic, holistic strategies. See: Crawford 1982a & 1982b; Bower 1981; Bower et al. 1978 & 1981; Blum et al. 1968b & 1971; and Blum 1967.

161
Hadfield 1923; Wells, W. R., 1947; Ikai & Steinhaus 1961; Nicholson 1920. For a review of such studies conducted before 1966, see: Barber 1966.

162
Callen 1983. Eric Krenz described the use of Modified Autogenic Training, which utilizes hypnotic techniques, to improve performance with four athletes. Lee Pulos described the hypnotic techniques he used to improve Canadian athletes' mental attitudes and physical skills. And Jencks and Krenz described their use of posthypnotic suggestions for anxiety reduction and mental rehearsal in sport (Krenz 1984; Pulos 1969; and Jencks & Krenz 1984).

163
MacHovec 1986.

164
"Most published accounts of personal difficulties which arise following hypnosis," wrote psychiatrist Josephine Hilgard, "are in the context of therapy." Hilgard, J., 1974b.

165
Faw et al. 1968.

166
Hilgard, J., 1974b.

167
Coe & Ryken 1979.

168
Hilgard, J., 1974b.

169
People v. Hughes, 59 NY 2d 523, 466 NYS 2d 255, 543 NE 2d, 484 (1983). *State ex rel Collins v. Superior Court,* 132 Ariz. 180, 644 P. 2d 1266 (1982), supplemental opinion filed May 4, 1982. *People v. Hurd,* Sup. Ct., NJ, Somerset Co., April 2, 1980. *People v. Shirley,* 31 Cal. 3d 18, 641 P2d 775 (1982), modified 918a (1982). *People v. Guerra,* C-41916 Sup. Ct., Orange Co., CA, 1984. (Cited in: Spiegel 1989.)

170
Hilgard, E., 1965a & 1975b; and Kihlstrom 1985.

171
Van DeCastle 1969; and Schechter 1984.

172
Van DeCastle 1969.

173
Marquis de Puysegur 1811.

174
Esdaile 1849 & 1852.

175
Gurney & Sidgwick 1883. In Van De Castle 1969.

176
Janet 1886. See also: Myers, F. W. H., 1954, vol. 1, pp. 524–29.

177
Richet 1923.

178
Barrett 1911.

179
From a review by James 1903, in *Proceedings of the Society for Psychical Research,* vol. 18, pt. 46 (June), reprinted in James 1986. James accepted the presidency of the British Society for Psychical Research in 1894 and 1895, helped start its American counterpart, and wrote extensively on hypnosis and paranormal phenomena.

180
Vasiliev 1976.

181
Rhine 1934.

182
Van De Castle 1969.

[183] Schechter 1984.

[184] Aaronson 1966, 1967, 1969, & 1971.

[185] Sacerdote 1977.

[186] Such experience, Tart wrote:

> might represent a transition from the gestalt configuration we call hypnosis to a new configuration, a new state of consciousness. States of consciousness of this type have not been dealt with in Western scientific literature on hypnosis to any great extent, but sound similar to Eastern descriptions of consciousness in which time, space and ego are supposedly transcended, leaving pure awareness of the primal nothingness from which all manifested creation comes (Tart 1979. See also: Tart 1970).

[187] Thomas 1979.

CHAPTER 16

[1] Green & Green 1977, pp. 42–43.

[2] Bruce 1973, p. 111.

[3] Martony 1968; Platt 1947; and Pickett 1968.

[4] West & Savage 1918; Carter & Wedd 1918; and Ogden & Shock 1939.

[5] For a review of attitudes among experimental psychologists regarding the voluntary control of autonomic systems, see: Kimmel 1974.

[6] DiCara & Miller 1968 & 1969; Miller & DiCara 1967 & 1968; and Miller & Banuazizi 1968.

[7] Miller & Dworkin 1974.

[8] Basmajian 1963; Stoyva 1976a & 1976b; Budzynski et al. 1973; Kamiya 1969; and Kimmel 1974.

[9] Support for biofeedback training was also provided by a review of Russian work with visceral conditioning by G. Razran published in *Psychological Review* (Razran 1961). An earlier account of Russian work by Bykov (1957) and a later one by Adam (1967) reinforced the influence of Razran's review.

[10] Butler, F., 1978; and Basmajian 1989, p. v. For reviews of the biofeedback field, see: Green & Green 1977; Yates 1980; White & Tursky 1982; Hatch et al. 1987; and Basmajian 1989.

[11] Gilson & Mills 1941; Harrison & Mortensen 1962; and Basmajian 1963 & 1966.

[12] Basmajian 1963; Clamann 1970; Basmajian & Simard 1967; Harrison & Mortensen 1962; Carlsoo & Edfeldt 1963; Gray 1971; Scully & Basmajian 1969; Maton 1976; Johnson, C. P., 1976; and Simard & Ladd 1969.

[13] Simard & Ladd 1969. In Yates 1980, pp. 73–78; see also: Simard & Basmajian 1967; and Hefferline 1958.

[14] Lloyd & Shurley 1976; see also: Yates 1980, pp. 79–80. Lloyd and Shurley compared the effects of sensory isolation and nonisolation on SMU control in the

belly of the tibialis anterior on the right leg. After a two-week interval, half of the group first trained in a sensory isolation condition was switched to a nonisolation condition, and vice versa, while the remainder were retrained as before. In the initial training, the isolation group required only 27 trials on average to reach the criterion set, whereas the nonisolation group required an average of 129 trials. On retraining, the isolation subgroup that continued in isolation required fewer trials (13.4) to reach criterion, whereas the isolation subgroup switched to non-isolation training now required an average of 51 trials. On the other hand, the nonisolation subgroup switched to isolation needed only 42 trials to achieve criterion (Lloyd & Shurley 1976).

[15] For a review of experiments in which the noise reduction model has been used to account for differences in ESP scoring, see: Braud 1978b.

[16] Budzynski & Stoyva 1969; Green, E., et al. 1969; and Jacobs & Felton 1969.

[17] Carlsson & Gale 1976; Carlsson et al. 1975; Dohrmann & Laskin 1978; and Farrar 1976.

[18] Budzynski et al. 1970 & 1973; and Wicramasekera 1972 & 1973.

[19] Cleeland 1973; Brudny et al. 1974; Korein et al. 1976; and Korein & Brudny 1976.

[20] Basmajian 1989, pp. 91–167; Hatch 1987, pp. 1–41; Hardyck et al. 1966; Hardyck & Petrinovich 1969; Jacobson 1925; Daly & Johnson 1974; Roll 1973; Moller et al. 1973; Stevens, K. N., et al. 1974; Guralnick & Mott 1976; Yates 1963; Ballard et al. 1972; Peck 1977; Norton 1976; and Levee et al. 1976.

[21] Caton 1875; and Lubar 1989.

[22] Kamiya 1968 & 1969; Nowlis & Kamiya 1970. During experiments he reported in 1962 and 1963, T. B. Mulholland had shown that some people could learn to turn lights on and off by changing the electrical activity of their brains without knowing they were doing so, but his work did not have the celebrity or impact that Kamiya's did (Yates 1980, pp. 271–73).

[23] Peper & Mulholland 1970.

[24] Peper 1971a and 1971b.

[25] Peper 1970; Mulholland & Peper 1971; and Eason & Sadler 1977.

[26] Plotkin 1979.

[27] Plotkin 1979, pp. 1145–46.

[28] Walsh, D. H., 1974.

[29] Lubar 1989.

[30] Green, E., et al. 1970; Brown, B., 1971; Lubar 1989; and Green & Green 1977. The Greens have used yogic exercises and Autogenic Training in these studies to augment biofeedback training.

J. Beatty and his associates trained some subjects to decrease and some to increase theta waves, then tested their vigilance while detecting visual targets on a simulated radar system. As might be expected (given the fact that theta waves accompany drowsiness), subjects were more vigilant while suppressing theta activity than they were while performing the task in an unregulated manner, whereas subjects taught to augment it were less vigilant while doing so (Beatty et al. 1974).

[31] Sheer 1970, 1974, 1975, & 1976; Das & Gastaut 1955; and Bird et al. 1978a & 1975b.

[32] Peper 1971a & 1972.

[33] Eberlin & Mulholland 1976; O'Malley & Conners 1972; Davidson, R. J., et al. 1976; and Nowlis & Wortz 1973.

[34] Shearn 1962; Hnatiow & Lang 1965; Frazier 1966; Engel & Hansen 1966; Brener & Hothersall 1967; Engel & Chism 1967; and Levene et al. 1968.

[35] Lang, P. J., et al. 1967.

[36] Bergman & Johnson 1971; Blanchard et al. 1974 & 1975; Davidson & Schwartz 1976; Ray 1974; Ray & Lamb 1974; and Stephens et al. 1975.

[37] Lang & Twentyman 1974 & 1976; Lang, P. J., et al. 1975; Whitehead, W. E., et al. 1977. See also: Gatchel 1974; and Bell & Schwartz 1975.

[38] Engel & Bleecker 1974; Scott et al. 1973; Vaitl 1975; Bleecker & Engel 1973a & 1973b; Weiss & Engel 1971; Pickering & Miller 1977; and Pickering & Gorham 1975.

[39] Yates 1980, p. 192; Engel & Baile 1989; Engel et al. 1983; Glasgow et al. 1982; and Stoyva 1989.

[40] Gavalas 1967 & 1968; Rice 1966; Van Twyer & Kimmel 1966; Edelman 1970.

[41] Stern & Lewis 1968.

[42] See: Yates 1980; Stern & Kaplan 1967; Crider et al. 1966; Klinge 1972; Ikeda & Hirai 1976; and Wagner et al. 1974.

[43] Green & Green 1989; and Yates 1980.

[44] Hubel 1974; Whitehead, W. E., et al. 1975; Schuster 1968, 1974, & 1989; Schuster et al. 1974; Alva et al. 1976; and Engel et al. 1974.

[45] Green & Green 1986; and Patel 1973 & 1975b.

[46] Steiner & Dince 1981; see also: Surwit & Keefe 1983; and Thompson et al. 1983.

[47] Fahrion 1983, pp. 28–29.

[48] In Fahrion 1983, pp. 28–29.

[49] Green et al. 1969.

[50] Green & Green 1986.

[51] Lang, P. J., 1974; see also: Brener 1977.

[52] Shapiro & Surwit 1976.

[53] Shapiro & Surwit 1976.

[54] Blanchard et al. 1982; and Neff & Blanchard 1982. Blanchard et al. 1982 showed that high scorers on the Tellegen Absorption Scale who suffered from vascular headache improved more through relaxation training than with instrumented feedback, while low scorers on the scale experienced greater relief from the same affliction *with* biofeedback training. Blanchard suggested that high scorers on the Tellegen scale do better without the distraction of feedback signals because they can then focus their considerable powers of attention on the task at hand.

In another study to explore individual differences in relation to feedback training, Thomas McCanne and Kate Hathaway found that males who scored relatively well on a motor skills task increased their heart rates significantly with feedback and maintained their self-control without it, while males who scored poorly on the same skills task could not increase their heart rates *with or without* biofeedback. "These results," McCanne and Hathaway concluded, "suggest that ability to control striate musculature during motor skills tasks may be fundamentally similar to ability to increase heart rate." And some studies using the "locus-of-control" construct measured by the Rotter IE scale showed that subjects who were more internally oriented—that is, believed they controlled their own behavior rather than had it controlled by forces outside them—learned self-regulation skills faster than those who were externally oriented (McCanne & Hathaway 1984; Yates 1980; Denkowski et al. 1984; see also: Davidson et al. 1976).

[55] Brener & Hothersall 1967; Sroufe 1969; and Vandercar et al. 1977.

[56] Eason & Sadler 1977; Schwartz, G. E., 1975; and Nowlis & Kamiya 1970.

[57] Brener 1974; and Obrist et al. 1970.

[58] Hess 1925; Gellhorn 1967 & 1970; and Gellhorn & Kiely 1972.

[59] Yates 1980, p. 425.

CHAPTER 17

[1] See, for example: Herink 1980.

[2] Blanck & Blanck 1974 & 1979; Masterson 1981; Kernberg 1976; and Stone, M. H., 1980.

[3] Berne 1972; Nichols 1984; and Branden 1971.

[4] Beck et al. 1979; Kelley 1955; and Ellis 1973.

[5] Binswanger 1956; Boss 1963; May et al. 1958; and Bugental 1965.

[6] *Journal of Transpersonal Psychology;* Vaughan 1986; and Walsh & Vaughan 1980.

[7] See, for example: Kernberg 1975; Masterson 1981; and Gedo 1981.

[8] *American Psychiatric Association* 1987.

[9] Wilber, Engler, & Brown 1986, chaps. 3–5.

[10] Smith, M. L., et al. 1980.

[11] Schultz, D., 1978; Sheikh 1976 & 1984; Bornstein & Siprelle 1973; Singer & Switzer 1980; Jaffe & Bresler 1980; Meichenbaum 1977; Singer 1974; Ahsen 1968; Sheikh et al. 1979; Achterberg 1985; Achterberg & Lawlis 1978; and Simonton, Simonton-Matthews, & Creighton 1978.

[12] Sheikh & Jordan 1983.

[13] Janet 1898.

[14] Jung 1976. *The Symbolic Life.* In *The Collected Works.*

[15] Happich 1932.

[16] Singer 1974.

[17] Wolpe 1958 & 1969; Sheikh & Panagiotou 1975; Singer 1974; and Singer & Pope 1978.

[18] Stampfl & Levis 1967.

[19] Kazdin 1977 & 1978; Singer & Pope 1978; Little & Curran 1978; Wolpe & Lazarus 1966; and Goldfriend et al. 1974.

[20] Gendlin & Olsen 1970.

[21] Reyher 1977.

[22] Shorr 1972 & 1978.

[23] Sheikh & Jordan 1983.

[24] Ahsen 1968, 1972, & 1977; Dolan & Sheikh 1976; Panagiotou & Sheikh 1974; Sheikh 1978; and Sheikh & Jordan 1981 & 1983.

[25] Perls 1970; and Perls et al. 1951.

[26] Progoff 1963; Gerard 1964; and Masters & Houston 1972.

[27] Stekel 1921.

[28] Jones, E., 1957, vol. 3, p. 394.

[29] Jones, E., 1957, vol. 3, p. 394.

[30] Jones, E., 1957, vol. 3, p. 395.

[31] Jones, E., 1957, vol. 3, p. 395–96.

[32] Jones, E., 1957, vol. 3, p. 380.

[33] Jones, E., 1957, vol. 3, p. 381.

[34] For accounts of extranormal experience in Jung's life and therapeutic work, see: Jung 1961; Fodor 1971; and van der Post 1975.

[35] Jung 1975, pp. 538–42.

[36] Devereux 1953; Ehrenwald 1940, 1942, 1944, 1948a, 1948b, 1949, 1950a, 1950b, 1954, 1956, & 1975; Eisenbud 1954, 1955, 1967, 1970, & 1983; Meerloo 1948–1951, 1949, 1950, 1953; Pederson-Krag 1947 & 1948–1951; and Ullman 1948 & 1952.

[37] Eisenbud 1983, pp. 19–21.

[38] Ehrenwald 1975, pp. 235–36, 241–42.

[39] Eisenbud 1970, p. 234.

[40] Eisenbud 1970, p. 287.

[41] The *Journal of Transpersonal Psychology* may be ordered from P.O. Box 4437, Stanford, CA 94309.

[42] Sutich 1980.

[43] Wilber, Engler, & Brown 1986, chaps. 1, 3, & 4.

[44] Assagioli 1965 & 1973.

[45] Assagioli 1965 & 1973.

CHAPTER 18

[1] Hanna founded the American journal *Somatics* in 1976, and French psychiatrist Richard Meyer the French journal *Somatotherapie* in 1989 to review theoretical and practical work in the field. Hanna provided a description of Somatics in four articles entitled "What Is Somatics?" in *Somatics* (1986 & 1987).

[2] Personal communication. Johnson's views on somatic education are developed in Johnson, D., 1977 & 1983.

[3] For Alexander's description of this discovery, see: Alexander, F. M., 1969, pp. 139–69. See also: Barlow 1980; Jones, F. P., 1976; and Peterson 1946.

[4] Alexander, F. M., 1985.

[5] Jones, F. P., 1979.

[6] Jones, F. P., 1979, p. 151.

[7] Schultz & Luthe 1959.

[8] Schultz & Luthe 1959, p. 1.

[9] Schultz & Luthe 1959, pp. 8–95.

[10] Schultz & Luthe 1959, pp. 95–120.

[11] Schultz & Luthe 1959, pp. 120–24.

[12] Schultz & Luthe 1959, pp. 125–227.

[13] Schultz & Luthe 1959, pp. 248–73.

[14] Feldenkrais 1972.

[15] Feldenkrais 1972. See also: Feldenkrais 1985, pp. 1–53.

[16] Feldenkrais 1970, 1972, 1977, & 1985.

[17] Feitis 1978, pp. 34–35. See also: Rolf 1973.

[18] Johnson, D., 1977, pp. 129–30, 134–35.

[19] Silverman et al. 1973.

[20] Silverman and his colleagues used a "nearest neighbor ISO-data statistical analysis." See: Silverman et al. 1973; and Ball & Hall 1967.

[21] Silverman et al. 1973, pp. 11–12.

[22] Silverman, 1986. See also: Silverman et al. 1973.

[23] Hunt & Massey 1977, p. 209.

[24] Silverman et al. 1973; Silverman 1986; and Hunt & Massey 1977. In summarizing her study, Hunt concluded that "one major conclusion is warranted. All evidence points to improved organization and greater balance in the neuromuscular system with extensive positive implications for motor efficiency."

[25] Johnson, D., 1980, pp. 5–6.

[26] Jacobson 1974, p. 101.

[27] Jacobson 1974, p. 105.

[28] Jacobson 1974, pp. 107–109.

[29] Jacobson 1974, p. 110.

[30] Jacobson 1955, pp. 558–59.

[31] Jacobson 1925.

[32] Jacobson 1955, p. 560.

[33] Jacobson 1974, pp. 47–57.

[34] Jacobson 1974, pp. 82–83.

[35] Jacobson 1974, p. 97.

[36] Jacobson 1974, pp. 430–31.

[37] Hengstenberg 1985.

[38] Hanna 1983, pp. 158–59.

[39] Brooks 1974, p. 231.

[40] Gindler 1986–87.

[41] Gindler 1986–87, pp. 42–45.

[42] *Encyclopedia of Philosophy* 1967, vol. 7, p. 104.

[43] Reich 1972.

[44] Reich 1942, p. 102.

[45] Reich 1942, p. 146.

[46] Reich 1942, pp. 121–22.

[47] Reich 1972, pp. 194–96.

[48] Boadella 1985, pp. 116–20.

[49] Boadella 1985, p. 120. Ola Raknes, a Norwegian therapist who worked closely with Reich, listed these features of vegetative liveliness:

> i. The entire organism has a good tonus; the body stature is elastically erect; no cramps or jerks.
>
> ii. The skin is warm with a plentiful blood supply, the colour reddish or slightly tanned; the sweat may be warm.
>
> iii. The muscles can change between tension and relaxation, being, however, neither chronically contracted nor flaccid; peristalsis is easy; no constipation or haemorrhoids.
>
> iv. The facial features are lively and mobile, never set or mask-like. The eyes are clear with lively pupil reactions, and the eyeballs are neither protruding nor sunken.
>
> v. There is complete, deep expiration with a pause before new inspiration; free and easy movements of the chest.
>
> vi. The pulse is usually regular, calm and strong; normal blood pressure, neither too high nor too low (Raknes 1970).

[50] Reich was influenced by Friedrick Kraus, a Berlin internist who had studied the bioelectric potentials in human tissues and introduced the term *vegetative current* to designate the fluid convection process that produce changes in complexion and muscle tonus; the physiologist L. R. Muller, whose book on nervous physiology *Die Lebensnerven* describes the differing functions of the sympathetic and parasympathetic nervous systems; physiologists such as Stern who had shown the connections between fluid movements and electrical discharge in plants; and experimenters who had described plasmatic movement in the amoeba (See: Kraus 1926; Muller 1931). Reich made no claims, wrote David Boadella, that he had

> discovered any new facts during [his] intensive study of these biological processes. What he thought he had succeeded in doing was to reduce generally known reactions from a number of unrelated fields to a valid and fundamental biological formulation of the concept of "psychosomatic identity and antithesis." The expansion and contraction process in the amoeba

was functionally parallel to the processes performed in higher animals by the vastly more complicated network of vegetative nerves. The vagal system was the function essentially of libidinal expansion, of reaching out towards the world; the sympathetic was essentially the system of libidinal retreat, of drawing back from the world into oneself (Boadella 1985, p. 109).

CHAPTER 19

[1] Nabokov 1981; Water 1970; Scully 1972; and Underhill 1954.

[2] For these definitions I have drawn upon Casperson et al. 1985.

[3] Hippocrates, *Regimen* 3.73, 76, 78.

[4] Edelstein & Edelstein 1945.

[5] Hansen 1971; Harris 1964; Lang, M., 1977; and Caton 1900.

[6] Morris et al. 1953 & 1966. See also: Morris et al. 1973 & 1980.

[7] Kahn, H. A., 1963; Zukel et al. 1959; Cassel et al. 1971; Taylor, H. L., et al. 1962; and Menotti & Paddu 1976.

[8] Brunner et al. 1974; and Paffenbarger et al. 1970.

[9] Paffenbarger & Hyde 1984.

[10] *Public Health Reports* 1985a & 1985b.

[11] Stephens, T., et al. 1985.

[12] Stephens, T., et al. 1985. The eight studies were: the 1974 survey by the President's Council on Physical Fitness and Sports; the 1975 National Health Interview Survey supplement; the 1976 Fitness and Amateur Sport Canada survey; the 1978 Perrier study; the 1979 National Survey of Personal Health Practices and Consequences; the 1981 Canada Fitness Survey; the 1982 Behavioral Risk Factor survey conducted jointly by the Centers for Disease Control, 28 states, and the District of Columbia; and the Miller Lite survey (Stephens, T., et al. 1985, p. 147).

Summarizing the trends revealed by these surveys, Stephens et al. concluded that "some activities clearly gained participants during 1972–1983, namely, jogging-runnning, calisthenics, and swimming. This trend is corroborated by Paffenbarger, whose study of Harvard Alumni shows substantial increases in sports participation within cohorts and age groups for the period 1962–1977" (Stephens, T., et al. 1985, p. 153).

[13] Gallup Poll 1984.

[14] Powell & Paffenbarger 1985.

[15] Paffenbarger & Hyde 1984. See also: Paffenbarger et al. 1970.

[16] Paffenbarger et al. 1984.

[17] Paffenbarger et al. 1986.

[18] Castelli et al. 1986.

[19] Ekelund 1988.

[20] Blair et al. 1989. The findings produced by the epidemiological and longitudinal studies cited above have been supported by other studies. See, for example: Wilson, P., et al. 1986; and Slattery & Jacobs 1988.

[21] Siscovick et al. 1985. Ralph Paffenbarger, a principal researcher in both the Harvard and longshoremen studies, and his colleague Robert Hyde reflected the prevailing opinion among experts on fitness and exercise when they wrote:

> The body of clinical, epidemiological and experimental data accumulated over the past 30 years provides strong evidence that the inverse relationship between contemporary physical activity and risk of (coronary heart disease) stems from "protection" (i.e. a *cause* of reduced CHD incidence) rather than "selection" (an *effect* of CHD symptomatology or other elective practices). Acceptance of this concept is supported by investigations of the potential mechanisms whereby exercise induces numerous salutary results: cardiopulmonary fitness, improved blood lipid patterns, augmented fibrinolytic activity, altered platelet stickiness, increased insulin sensitivity.
> Pertinent findings to support a cause and effect relationship between regular physical activity and freedom from coronary heart disease include the following:

> □ A reduced risk of developing CHD relates to both occupational and leisure-time endurance practices of lifting, carrying, shoving, climbing, walking, playing sports, etc. A reduced CHD risk associated with any adequate level of current energy expenditure is lowered further if the exercise pattern includes sufficient vigorous activity or strenuous bursts of energy output.

> □ The relationship is dose-dependent over a considerable range, i.e., incidence, case-fatality ratio, mortality, and recurrence.

> □ The findings are consistent by age of subjects, by sex, and by clinical manifestation of disease (angina pectoris, myocardial infarction, sudden unexpected death, and total CHD death).

> □ The findings are persistent in successive increments of time, and in diverse populations.

> □ The findings for exercise influence are at least partially independent of other host and environmental characteristics associated with CHD risk (cigarette habit, blood pressure level, weight-for-height status, blood lipid pattern, prior existence of diabetes mellitus, familial history of CHD, etc.).

> □ Men physically active in prior times are, if anything, at increased risk of CHD unless they maintain their vigorous high activity pattern throughout adult life. In reverse order, men physically inactive in bygone years (as in adolescence or early childhood) are at decreased risk of CHD if their adult life acquires an adequate, sustained physically active lifestyle.

> □ Direct experimental evidence from exercise and diet studies as related to risk of CHD in monkeys supports the concept that long-term moderate exercise inhibits the development of CHD. Even on an atherogenic diet, animals that were exercised had less coronary atherosclerosis and its complications than those not exercised (Paffenbarger & Hyde 1984).

[22] Quoted in Brody 1985a.

[23] Siscovick et al. 1985.

[24] Brody 1985a.

[25] Richter & Kellner 1965; Zeldis et al. 1978; Snoeckx et al. 1982; and Morganroth et al. 1975.

[26] Hickson et al. 1985.

[27] Dowell 1983; McArdle et al. 1978; and Pechar et al. 1974.

[28] Saltin 1969.

[29] Scheuer & Tipton 1977.

[30] Klemola 1951.

[31] Saltin 1969.

[32] Kjellberg et al. 1949; Covertino 1982; and Falch & Stromme 1979.

[33] Shepherd 1967; Clausen & Trap-Jensen 1974; and Mahler & Loke 1985.

[34] McArdle 1986; Ekblom 1969; and Rowell 1974.

[35] Saltin et al. 1977.

[36] Pollack, M., et al. 1974.

[37] McArdle 1986.

[38] Ingjer 1978.

[39] Currens et al. 1961.

[40] Boyer & Kasch 1970; Hagberg et al. 1981; and Tipton 1984.

[41] Clausen et al. 1969; and Redwood et al. 1972.

[42] McArdle 1986.

[43] Powell et al. 1987.

[44] Aloai et al. 1978; Brewer et al. 1983; and Huddleston et al. 1980.

[45] Kriska et al. 1988.

[46] McArdle 1986.

[47] Goldberg et al. 1975; and Tipton et al. 1975.

[48] Goldberg et al. 1975.

[49] Kroll & Clarkson 1978.

[50] Pattengale & Holloszy 1967.

[51] McArdle 1986.

[52] Mole et al. 1971; Gollnick & Saltin 1973; and McArdle 1986.

[53] Wood et al. 1988.

[54] Wood et al. 1983 & 1988; Wilson, P., et al. 1986; Castelli et al. 1986; Nye et al. 1981; Huttunen et al. 1979; Wood & Haskell 1979; and Lehtonen et al. 1979.

[55] Hermansen 1969; McArdle 1986; and Mickelson & Hagerman 1982.

[56] McArdle 1986.

[57] Fintarone, et al. 1988 & 1989; Solomon 1991. See also: Pedersen 1991.

[58] Northoff and Berg 1991.

[59] La Perriere et al. 1991.

[60] Zanker and Kroczek 1991. See also: Uhlenbruck and Order 1991.

[61] Fitzgerald 1991.

[62] Paffenbarger et al. 1984.

[63] Frisch et al. 1985.

[64] Gerhardsson et al. 1986.

[65] Sonstroem 1984; Dishman 1985 & 1988; Taylor et al. 1985; Hughes 1984; Sime 1984; Morgan, W., 1982; Folkins & Sime 1981; and Nagle & Montoye 1981.

[66] Morgan & Goldston 1987. Various theories account for the positive mental-emotional results of exercise, among them the monoamine hypothesis that exercise stimulates the production of one or more brain monoamines that mediate improvements of mood; the thermogenic hypothesis that episodic elevations of core body temperature enhance psychophysical functioning; and the hypothesis that endorphins produced by exercise stimulate pleasurable feelings. For a discussion of these three hypotheses, see: Sachs & Buffone 1984; and Dishman 1988, chap. 4.

[67] Taylor, C., et al. 1985; Little 1979; Yates et al. 1983; and Thaxton 1982.

[68] Leonard 1990, p. 194.

[69] Murchie 1954, pp. 290–91.

[70] Leonard 1990, p. 196.

[71] Lindbergh, 1953. Lindbergh did not describe this experience in his first book, We, Pilot and Plane, published in 1927. The Spirit of St. Louis was published in 1953, 26 years later. Lindbergh did not acknowledge or appreciate the metaphysical and extrasensory aspects of his adventure, it seems, until many years had passed. His posthumously published book Autobiography of Values contains some of his reflections about these matters.

[72] Schweickart 1977, pp. 2–13.

[73] Cuddon 1979, p. 588.

[74] Suinn 1983 & 1985; and Nideffer 1976.

[75] Bannister 1956, p. 193.

[76] Murphy & White 1978.

[77] Doust 1973.

[78] Murphy & Brodie 1973.

[79] Nicklaus 1974, p. 79–80.

[80] Lazarus 1977.

[81] Suinn 1983.

[82] Taken as a whole, experimental studies do not provide evidence for the effectiveness of imagery rehearsal in sports as convincing as testimonies such as Nicklaus's, Zane's, or Schwarzenegger's. Though such studies have grown in number since the 1960s and now constitute a significant addition to the developing field of sports psychology, they have not provided evidence for imagery's effectiveness in promoting performance skills as robust as many firsthand accounts. This is the case in part because research on athletic mental practice suffers from the general limitation upon experimental studies of transformative activity discussed in chapters 2.2, 16.6, and 23.2—namely, that the subjects of such studies rarely practice more than a few weeks or months, hardly long enough to achieve dra-

matic improvement. Nevertheless, when psychologist Alan Richardson reviewed 25 studies of mental practice for motor skills in 1967, he found that 11 had produced statistically significant results, while 7 more had positive though statistically insignificant outcomes (Richardson 1967).

Assessing similar reviews by Oxendine 1968, C. Corbin 1972, and Nideffer 1976, Richard Suinn concluded that experimental research through the early 1980s indicated that: persons with experience in a task may profit more than novices from mental practice; that beginners seem to profit more from physical practice; that simple motor tasks may be more readily enhanced through mental practice than complex tasks; and if the subject is experienced or trained in a complex task, mental practice can contribute further improvements in performance (Suinn 1983).

[83] Huizinga 1950, p. 10.

CHAPTER 20

[1] Draeger & Smith 1980, p. 35.

[2] Draeger & Smith 1980, p. 91.

[3] Suzuki 1959, p. 200.

[4] Suzuki 1959, pp. 155–57.

[5] Uyeshiba, M. 1983, pp. 3–12. For vivid description of a contemporary aikidoist's spiritual transformation, see: Leonard 1986, "This isn't Richard."

[6] Stevens, J., 1987. See also: Kisshomaru 1977, 1983, 1984, & 1986; Kanemoto 1969; and *Aiki News*.

[7] Suzuki 1959, p. 149.

[8] See, for example: Barclay 1973; Gluck 1962; Oyama 1965; Ratti & Westbrook 1973; and Westbrook & Ratti 1970.

[9] Leonard 1990, pp. 64–68.

[10] Musashi 1974, pp. 89–90.

[11] Leonard 1990, p. 68.

[12] Random 1978, p. 22.

[13] Leonard 1990, pp. 62–63.

[14] *Subtle Energies* is the journal of the International Society for the Study of Subtle Energies and Energy Medicine, 356 Goldco Circle, Golden, CO 80401. The Society, which has more than 1,800 members from several nations, promotes "the study of informational systems and energies that interact with the human psyche and physiology."

[15] Suzuki 1959, p. 165.

[16] Suzuki 1959. Such capacity was also dramatized cinematically in *The Seven Samurai* when several warriors were tested as they entered a room to see which of them could anticipate surprise attack.

[17] Aikido student John Stevens told this and other stories about Uyeshiba's exceptional abilities; see Stevens 1985. See also: Stevens 1987, pp. 67–74.

[18] Draeger & Smith 1980, pp. 87, 120–31.

[19] Draeger & Smith 1980, pp. 32–33. See also: Waley 1956.

[20] Draeger & Smith 1980, p. 34.

[21] Draeger & Smith 1980, p. 90.

[22] Quoted in Levine, D., 1983.

[23] The repertoires of some martial arts have been complemented by recent invention, moreover; witness the explosive emergence of aikido techniques in America since 1970. See, for example: Leonard 1990; and Leonard & Kirsch 1983.

[24] Feld et al. 1979, p. 150.

[25] Hirata 1971; Theiss 1971; and Yura 1975.

[26] Rasch & O'Connell 1963.

[27] Rasch & Pierson 1963.

[28] Spaulding 1977.

[29] Bender 1972.

[30] Leonard 1986.

[31] Oh & Falkner 1984.

[32] Oh & Falkner 1984, pp. 137–38.

[33] Oh & Falkner 1984, pp. 138–39.

[34] Oh & Falkner 1984, pp. 149–51.

[35] Oh & Falkner 1984, p. 162.

[36] Oh & Falkner 1984, p. 163.

[37] Oh & Falkner 1984, p. 175.

[38] Heckler 1985. See chapter by J. Haller, pp. 155–65.

CHAPTER 21

[1] Duchesne 1933, p. 391.

[2] Lecky 1926, p. 110. See also: Noldeke 1892, p. 212.

[3] Durant 1950, p. 60.

[4] Russell, N., 1980, pp. 52–56, 63–79, 100.

[5] Russell, N., 1980. See also: Brown, P., 1971a & 1971b; Clarke 1912; Hardy 1931a & 1931b; Jones, A. H. M., 1964; MacKean 1920; Petrie 1924; Rousseau 1978; and Waddell 1957.

[6] Karjalainen 1921–27; and Sieroszewski 1902. Both cited in Eliade 1964.

[7] Shirokogoroff 1935; Rig-Veda VIII, 59,6; X, 136,2; Eliade 1964, pp. 412–13, 438, 474–76. See also: Abbott 1932; Dumezil 1930 & 1942; Eliade 1962; Propp 1949; and Webster 1948.

[8] Park 1934.

[9] Katz, R., 1982b.

[10] Eliade 1964, p. 54.

[11] Neihardt 1972, p. 226.

[12] Bogoras 1904.

[13] Nabokov 1981, p. 70.

[14] Nabokov 1981, pp. 92–93.

[15] Nimuendaju 1946.

[16] Nabokov 1981, p. 123.

[17] Nabokov 1981, p. 123. Nabokov drew upon Underhill's research. See: Underhill, R., et al. 1979; and Underhill, R., 1975.

[18] Nabokov 1981, p. 164–70. Nabokov published a number of photographs taken between the 1930s and 1970s that recorded Tarahumara runners in action.

[19] Blofeld 1966, p. 126.

[20] Blofeld 1973, pp. 92–96.

[21] Blofeld 1973, pp. 24–25, 36, 47.

[22] Hoffman, E., 1981, p. 122.

[23] Hoffman, E., pp. 179–180.

[24] Braid 1850.

[25] Braid 1850, pp. 11–16. Cited in Garbe 1900.

[26] Honigberger 1851.

[27] Braid 1850, p. 16. Cited in Garbe 1900.

[28] *Life of Sri Ramakrishna* 1924; Isherwood 1959; and Nikhilananda 1969.

[29] Nikhilananda 1969, p. 16.

[30] Isherwood 1959, p. 62.

[31] Isherwood 1959, pp. 76–77.

[32] Nikhilananda 1969, p. 69.

CHAPTER 22

[1] Attwater & Thurston 1981; Thurston 1952; Bynum 1987; Imbert-Gourbeyre 1894; *Acta Sanctorum* 1863; and *Analecta Bollandiana* 1882. See also: *Bibliotheca Sanctorum* 1961. (This includes accounts from canonization proceedings and other sources of 261 women officially recognized by the Catholic church as saints, blesseds, venerables, or servants of God who lived between 1200 and the present in Italy.)

[2] *New Catholic Encyclopedia* 1967.

[3] *The Book of Saints* 1942 (preface). Other descriptions of the canonization process are given in the *New Catholic Encyclopedia* 1967, vol. 3, pp. 56–61, and by Thurston in the appendix to *Butler's Lives of the Saints* entitled Beati and Sancti. See: Attwater & Thurston 1981, vol. 4, pp. 667–71.

[4] *Benedictae Papae XIV doctrina de Servorum Dei beatificatione et Beatorum canonizatione in synopsin* 1751.

[5] Attwater & Thurston 1981, vol. 1, p. vii.

[6] Crehan 1952.

[7] Knowles, D., 1963, p. 6.

[8] *New Catholic Encyclopedia* 1967, vol. 10, pp. 173–74.

[9] Imbert-Gourbeyre 1894. Imbert-Gourbeyre notes 321 cases of stigmatization that include both visible and invisible wounds.

[10] Thurston describes possible cases of stigmatization before Saint Francis. See: Thurston 1952, pp. 32–40.

[11] Thurston 1952, p. 44.

[12] Thurston 1952, p. 45.

[13] Thurston 1952, p. 45.

[14] Thurston 1952, pp. 45–46.

[15] Thurston 1952, p. 49. Thurston cited a description of Domenica Lazzari's stigmata by the Medical Superintendent of the Ospitale Civico-Militare in the city of Trent, which appeared in a medical journal of Milan, *Annali, Universali di Medicina*, 1837, vol. 84, pp. 255 seq.

[16] Schmoger 1976, vol. 1, p. 5.

[17] Schmoger 1976, vol. 1, p. 239.

[18] Schmoger 1976, vol. 1, p. 237–43.

[19] Germano di S. Stanislao. *Life of Gemma Galgani*. Cited in Thurston 1952, p. 53.

[20] Thurston 1952, p. 123.

[21] Molloy 1873. Cited in Thurston 1952, p. 55.

[22] Biot 1962, p. 33; and Lefebvre 1873.

[23] This description is taken in large part from a pamphlet entitled *Louise Lateau ou la Stigmatisee belge,* which was published in 1875 by *Progres Medical* in Paris, a journal that published Charcot's neuropsychiatric studies.

[24] Biot 1962, pp. 107–108.

[25] Biot 1962, p. 131. This is taken from Lefebvre 1873.

[26] Imbert-Gourbeyre 1894, vol. 2, p. 27. Cited in Thurston 1952, p. 64.

[27] Imbert-Gourbeyre 1894, vol. 2, p. 21. Cited in Thurston 1952, p. 64.

[28] Thurston 1952, p. 65.

[29] See, for example: St. Albans 1983; Gigliozzi 1965; and Hocht 1974.

[30] Thurston 1952, pp. 99–101.

[31] Festa 1949.

[32] Thurston 1952, p. 96.

[33] Thurston 1952, p. 96.

[34] Biot 1962, pp. 37–40; Lhermitte 1953; Steiner, J., 1963, 1973, & 1977; Gerlich 1929; Klauder 1938; and de Poray Madeyeski 1940.

[35] Schweikert 1981.

[36] Cummings 1956, pp. 938–40.

[37] Fisher & Kollar 1980.

[38] Early & Lifschutz 1974.

[39] St. Albans 1983.

[40] Klauder 1938.

[41] Gopi Krishna 1971; and Sannella 1987.

[42] Sannella 1987, pp. 7–13.

[43] Imbert-Gourbeyre 1894, vol. 2., p. 21. Cited in Thurston 1952, pp. 63–64.

[44] Thurston 1952, pp. 59–60.

[45] Thurston 1952, pp. 40–41, 73–78, 110–29.

[46] Thurston 1952, p. 341.

[47] Bell, R., 1985, chap. 2.

[48] Bynum 1987, p. 169. Bynum cites Raymund of Capua, *Life of Catherine of Siena*, pt. 2, chap. 5, p. 904.

[49] Bynum 1987, pp. 169, 194–207; and Bell, R., 1985, pp. 75–76, 100, 175.

[50] Thurston 1952, p. 350. Thurston cited the debates recorded in the *Bulletin de l'Académie royale de Médicine*, 1876, vol. 9.

[51] Thurston 1952, pp. 363–66. For more details of the commission's procedures and findings, see: Gerlich 1929, vol. 1, pp. 129–31.

[52] Johnston 1982, pp. 83–84.

[53] Thurston 1952, p. 169.

[54] Thurston 1952, pp. 162–70.

[55] Schimberg 1947, pp. 38–39.

[56] Thurston 1952, p. 163. See also: Treece 1989, chap. 4.

[57] Thurston 1952, p. 164.

[58] See, for example, Corbin, H., 1969 & 1977; as well as Thurston 1952, chap. 5, The Luminous Phenomena of Mysticism.

[59] Thurston 1952, pp. 212–18.

[60] Sguillante & Pagani 1748. *Vita della Ven. Serafina di Dio*. Cited in Thurston 1952, pp. 218–19.

[61] Thurston 1952, p. 220.

[62] Thurston 1955, pp. 112–15.

[63] Thurston 1952, pp. 222–24.

[64] Saint Teresa of Avila. *Book of Foundations*, chap. 28.

[65] Thurston 1952, p. 228.

[66] Salvatori, F. M. *Life of St. Veronica Guiliani*. Cited in Thurston 1952, pp. 228–29.

[67] Weber, B. *La V. Jeanne Marie de la Croix*. Cited in Thurston 1952, p. 229.

[68] Thurston 1952, pp. 230–32.

[69] St. Albans 1983, pp. 198–200.

[70] Thurston 1952, p. 224.

[71] See, for example: St. Albans 1983, pp. 198–200.

[72] Bynum 1987, p. 126.

[73] Bynum 1987, p. 145–46.

[74] Thurston 1952, p. 233.

[75] Cruz 1977; Thurston 1952, pp. 233–94.

[76] Thurston 1952, pp. 245–46.

[77] Excerpt from a letter dated May 16, 1952, by Harry Rowe, mortuary director of Forest Lawn Memorial Park, Los Angeles. *Paramahansa Yogananda-In Memoriam* 1983. See: Yogananda 1985.

[78] Meyrick. *Life of St. Walburga*, p. 36. Cited in Thurston 1952, p. 268.

[79] Vassall-Phillips. *Life of St. Gerard Majella*, pp. 185–86. Cited in Thurston 1952, p. 268.

[80] *The Life of St. Hugh of Lincoln*. Cited in Thurston 1952, p. 269.

[81] Grasset. *Life of St. Catherine of Bologna*. Cited in Thurston 1952, p. 285; *Acta Sanctorum*, Mar., vol. 3, pp. 864c, 866b; and Thurston 1952, pp. 289–93.

[82] Strub 1967, p. 85.

[83] For an account of remarkable physical contortions, see: Thurston 1952, pp. 194–98.

[84] Thurston 1952, p. 198. Thurston drew upon a printed *Summarium* (1747) for the Congregation of Rites.

[85] Thurston 1952, pp. 199–200.

[86] Canon Buti, 1745, *Vita della Madre Costante Maria Castreca*; B. M. Borghigiani, *Vita della Venerable Sposa di Gesu, Suor Domenica dal Paradiso*; and *Compendio della Vita B. Stefana Quinzana*, 1784. All references cited in Thurston 1952, pp. 200–201.

[87] Imbert-Gourbeyre 1894, vol. 2, pp. 131–36. See also: Thurston 1952, pp. 192–208.

[88] Saint Teresa of Avila. *Autobiography*, chap. 20.

[89] Cited in Thurston 1952, pp. 10–11, from a translation of Saint Teresa's autobiography by David Lewis, which Thurston modified in accordance with a facsimile edition of Saint Teresa's *Autobiography* in her own handwriting.

[90] M. Mir, *Vida de Santa Teresa*, 1912, vol. 1, p. 286. Cited in Thurston 1952, p. 12.

[91] *Acta Sanctorum*, Oct., vol. 7, p. 399.

[92] Marchese, 1717, *Vita della v. Serva di Dio Suor Maria Villani*. Cited in Thurston 1952, p. 13.

[93] Cited in Thurston 1952, p. 17.

[94] Thurston 1952, pp. 17–18.

[95] Benson 1984, pp. 156–58.

[96] Monnin, *Life of the Cure d'Ars*, vol. 2, p. 394. Cited in Thurston 1952, p. 143.

[97] Thurston 1952, pp. 146–48.

[98] Gerlich 1929, p. 167.

[99] Carty 1974, p. 55.

[100] Fahsel 1932. Cited in Thurston 1952, p. 114; and Teodorowitz 1940.

[101] White, R., 1982.

[102] White, R., 1981.

[103] White, R., 1982.

[104] Poulain 1957, pp. 166–67.

[105] Poulain 1957.

[106] Thurston 1955, p. 112–15.

[107] Bynum 1987, pp. 294–95.

CHAPTER 23

[1] Behanan 1937, pp. 225–49. Behanan's study was reported again by Walter Miles with some details added 27 years later. See: Miles 1964.

[2] Behanan 1937, pp. 186–212.

[3] Brosse 1946.

[4] Bagchi & Wenger 1957a & 1957b; Wenger & Bagchi 1961; Wenger et al. 1961; and Bagchi 1969.

[5] Wenger & Bagchi 1961.

[6] Wenger & Bagchi 1961.

[7] Wenger & Bagchi 1961.

[8] Das & Gastaut 1955.

[9] Gellhorn & Kiely 1972; and Benson 1975.

[10] Anand et al. 1961a.

[11] Fischer 1971; and Davidson, J. M., 1976.

[12] Green & Green 1977.

[13] McClure 1959.

[14] Vakil 1950.

[15] Hoenig 1968.

[16] Kothari, Bordia, & Gupta 1973.

[17] Kasamatsu & Hirai 1966.

[18] For descriptions of this and similar research, see: Kasamatsu et al. 1957; Hirai 1960; and Kasamatsu & Hirai 1963.

[19] Green & Green 1977, p. 240.

[20] Green & Green 1977, p. 242.

[21] Wallace, R., 1970; Wallace, Benson, & Wilson 1971; and Wallace & Benson 1972.

[22] Shapiro, D. H., 1982.

[23] Davidson, J. M., 1976; Fischer 1971 & 1976; and Gellhorn & Kiely 1972.

CHAPTER 24

[1] Jacobson 1955.

[2] Benson 1976.

[3] Leonard, G. From a report to Esalen Institute's Study of Exceptional Functioning.

[4] Myers, F. W. H., 1954, vol. 2, pp. 311–13.

[5] James, W., 1902, postscript.

[6] Demille 1976.

[7] Nimuendaju 1946.

[8] *Patanjali's Yoga Sutras* 1972.

[9] Suzuki, 1965a & 1965b.

[10] Assagioli 1965 & 1973.

[11] Sri Aurobindo, vols. 20–24 in *The Collected Works.*

CHAPTER 25

[1] See, for example: FitzGerald 1986; Appel 1983; and Deikman 1990.

[2] Myers, F. W. H., 1954, chaps. 1, 2, & 9; James, W., 1902, lecture 1: Religion & Neurology, lectures 6 & 7: The Sick Soul, and lecture 8: The Divided-Self and the Process of Its Unification; Jung 1961, chap. 6: Confrontation with the Unconscious, chap. 10: Visions. See also: Jung 1970a & 1970b, vols. 13 & 14; Thurston 1952, chap. 2: Stigmata; Assagioli 1965, chap. 2: Self-Realization and Psychological Disturbances; Maslow 1968, chap. 4: Defense and Growth, chap. 5: The Need to Know and the Fear of Knowing, and chap. 8: Some Dangers of Being-Cognition; and Hillman 1975, chap. 2: Pathologizing or Falling Apart, chap. 4: Dehumanizing or Soul-Making.

[3] Wilber, Engler, & Brown 1986, chap. 3: The Spectrum of Development, chap. 4: The Spectrum of Psychopathology.

[4] Saint John of the Cross, *The Dark Night of the Soul,* chaps. 2–4; and Sri Aurobindo, *Letters on Yoga,* pt. 1, sec. 5 & 6 in vol. 22 of *The Collected Works,* and pt. 4 in vol. 24 of *The Collected Works.* See also: Thurston 1952, chap. 2.

[5] Dillon 1977, pp. 76, 301.

[6] Jung 1961. See also: Jung 1970a & 1970b; Ogilvy 1979; Campbell 1973, chap. 3: Transformation of the Hero (in this chapter Campbell describes symbolic representations of the hero as warrior, lover, emperor, tyrant, world redeemer, saint, and hero); and Hillman 1975.

[7] Hillman 1972, p. 264.

[8] Sri Aurobindo, *The Collected Works,* vol. 22, sec. 2: Integral Yoga & Other Paths, sec. 5: Planes and Parts of the Being, vol. 23, sec. 2: Synthetic Method & Integral Yoga; Nikhilananda 1942, chap. 39: The Master's Reminiscences, chap. 44: The Master on Himself and His Experience.

CHAPTER 26

[1] Leonard & Kirsch 1983.

[2] Barron 1953, 1955, 1957, 1958b, 1963a, 1963c, 1972a, 1972b, 1979, & 1987; Barron & Welsh 1952; Ghiselin 1980; MacKinnon 1960.

[3] Barber & Wilson 1979; Barber 1984; Hilgard, E., 1965b; Hilgard, J., 1974a; Bowers & Bowers 1979; and Crawford 1981.

[4] Koestler 1964; Lindauer 1983; Ghiselin 1980; Crawford 1982a & 1982b; Shaw & Belmore 1982–83.

[5] Shafii 1973; Maupin 1965; Kornfield 1979; Cowger & Torrance 1982; and Kubose & Umemoto 1980.

[6] Leonard 1991.

[7] See: Walsh & Vaughan 1980; Benson 1983; Bower 1981; Brown & Engler 1986, pp. 191–218; Brown, D., et al. 1984; Deikman 1963; Erickson & Rossi 1974–1980; Iaquierdo 1984; Langer 1988; Maslow 1968; Rossi 1972 & 1985; Tart 1972 & 1983; Zornetzer 1978; Overton 1978; and Rossi & Ryan 1986.

[8] Green described Swami Rama's methods of self-control during seminars at the Menninger Foundation and Esalen Institute. See: Green & Green 1977, pp. 158–61, 197–218.

[9] Eliade 1972, pp. 33–144; Saint John of the Cross, *Dark Night of the Soul*; and Suzuki-roshi 1970, pp. 31–40, 55–56, 80–85.

[10] Experimental research has also shown that suggestions of oneness can help improve performance and psychological integration. Research psychologist Lloyd Silverman and his colleagues found, for example, that the phrase "Mommie and I are one," repeatedly given to subjects with a tachistoscope so that they were not conscious of it, helped them lose weight, stop smoking, raise school grades, or improve their general mental health. Silverman interpreted these findings in psychoanalytic terms, suggesting that they were produced by the satisfaction of "compensatory needs arising from failures of individuation."

[11] Sri Aurobindo, *The Life Divine*, vol. 18 in *The Collected Works*. See also: Vol. 30 in *The Collected Works*, index references to *sleep*.

BIBLIOGRAPHY

Aaronson, B. S. 1966. Behavior and the place names of time. *American Journal of Clinical and Experimental Hypnosis,* 9:1–17.

Aaronson, B. S. 1967. Mystic and schizophreniform states and the experience of depth. *Journal for the Scientific Study of Religion,* 6: 246–52.

Aaronson, B. S. 1969. The Hypnotic Induction of the Void. Presented at the American Society of Clinical Hypnosis, San Francisco.

Aaronson, B. S. 1971. Time, time stance, and existence. *Studium Generale,* Springer-Verlag, 24:369–87.

Abbott, J. 1932. *The Keys of Power: A Study of Indian Ritual and Belief.* London.

Abram, C., et al. 1965. Paper read before the 47th Meeting of the American Endocrine Society, New York.

Abrams, A. I. 1977a. The effects of meditation on elementary school students. *Dissertation Abstracts International,* 37 (9-A), 5:689.

Abrams, A. I. 1977b. Paired-associate learning and recall: A pilot study of the Transcendental Meditation program. In D. W. Orme-Johnson and J. T. Farrow (eds.), *Scientific Research on the Transcendental Meditation Program: Collected Papers.* Vol. 1. New York: MERU Press.

Achterberg, J. 1985. *Imagery in Healing: Shamanism and Modern Science.* Shambhala.

Achterberg, J., & G. Lawlis. 1978. *Imagery of Cancer: A Diagnostic Tool for the Process of Disease.* Champaign, IL: Institute for Personality and Ability Testing.

Ackerman, D. 1990. *A Natural History of the Senses.* Random House.

Acta Sanctorum. 1863. J. Carnandet, ed. Paris.

Adam, G. 1967. *Interoception and Behavior.* Budapest.

Ader, R., & N. Cohen. 1975. Behaviorally conditioned immune-suppression. *Psychosomatic Medicine,* 37:333–40.

Ader, R., et al. 1990. *Psychoneuroimmunology II.* Academic Press.

Agle, D. P., et al. 1967. Studies in autoerythrocyte sensitization: The induction of purpuric lesions by hypnotic suggestion. *Psychosomatic Medicine,* 29:491–503.

Ahsen, A. 1968. *Basic Concepts in Eidetic Psychotherapy.* Brandon House.

Ahsen, A. 1972. *Eidetic Parents Test and Analysis.* Brandon House.

Ahsen, A. 1977. *Psycheye:Self-Analytic Consciousness.* Brandon House.

Akishige, Y., ed. 1970. *Psychological Studies on Zen.* Tokyo: Zen Institute of the Komazawa University.

Aldrich, C. K. 1972. A case of recurrent pseudocyesis. *Perspectives in Biological Medicine,* 16:11–21.

Alexander, F. M. 1969. *The Resurrection of the Body.* University Books.

Alexander, F. M. 1985. Reprint. *Constructive Conscious Control of the Individual.* Centerline Press, pp. 5–6. (Originally published in 1923, D. P. Dutton.)

Alexander, S. 1979. Reprint. *Space, Time and Deity.* Peter Smith.

Allington, H. V. 1934. Sulpharsphenamine in the treatment of warts. *Archives of Dermatology and Syphilology,* 29:687–90.

Allison, J. 1970. Respiratory changes during Transcendental Meditation. *Lancet,* 1: 7651.

Aloai, J. F., et al. 1978. Skeletal mass and body composition in marathon runners. *Metabolism,* 27:1973.

Alva, J., et al. 1976. Reflex and electromyographic abnormalities associated with fecal incontinence. *Gastroenterology,* 51:101.

Alvarado, C. 1983. Paranormal faces: The Belmez case. *Theta,* 11:2 (Summer).

American Psychiatric Association. 1987 (3d ed.). *Diagnostic and Statistical Manual.*

Analecta Bollandiana. 1882. Brussels: Société des Bollandistes.

Anand, B., & G. Chhina. 1961. Investigations on yogis claiming to stop their heart beats. *Indian Journal of Medical Research,* 49:90–94.

Anand, B. K., et al. 1961a. Some aspects of electroencephalographic studies in yogis. *Electroencephalography & Clinical Neurophysiology,* 13:452–56.

Anand, B. K., et al. 1961b. Studies on Sri Ramanand Yogi during his stay in an air-tight box. *Indian Journal of Medical Research,* 49:82–89.

Anderson, D. A. 1984. Meditation as a treatment for primary dysmenorrhea among women with high and low absorption scores. *Dissertation Abstracts International,* 45:(1-B), 341.

Anderson, D. J. 1977. Transcendental Meditation as an alternative to heroin abuse in servicemen. *American Journal of Psychiatry,* 134:1308.

Anderson, J., et al. 1975. Migraine hypnotherapy. *International Journal of Clinical & Experimental Hypnosis,* 23:48–58.

Appel, W. 1983. *Cults in America: Programmed for Paradise.* Holt, Rinehart & Winston.

Appelle, S., & L. Oswald. 1974. Simple reaction time as a function of alertness and prior mental activity. *Perceptual & Motor Skills,* 38:1263–68.

Aquínas, T. 1975. *Summa Contra Gentiles.* Notre Dame Press.

Aradi, Z. 1956. *The Book of Miracles.* Farrar, Straus & Cudahy.

Arintero, J. G., Rev. 1951. *The Mystical Evolution in the Development and Vitality of the Church.* Herder.

Aronoff, G. M., et al. 1975. Hypnotherapy in the treatment of bronchial asthma. *Annals of Allergy,* 34:356–62.

Aronson, G. J. 1952. Delusion of pregnancy in a male homosexual with abdominal cancer. *Bulletin of the Menninger Clinic,* 16:159–66.

Arrian. 1933 [1976]. *Anabasis Alexandri.* (Two volumes.) Loeb Classical Library, Harvard University Press.

Asher, R. 1956. Respectable hypnosis. *British Medical Journal,* 1:309–312.

Assagioli, R. 1965. *Psychosynthesis: A Collection of Basic Writings.* Viking.

Assagioli, R. 1973. *The Act of Will.* Penguin.

Astrov, M. 1950. The concept of motion as the psychological leitmotif of Navaho life and literature. *Journal of American Folklore,* 63:45–56.

Attwater, D., & H. Thurston, eds. 1981. *Butler's Lives of the Saints.* Christian Classics.

August, R. V. 1961. *Hypnosis in Obstetrics.* McGraw-Hill.

Ayala, F. J. 1974. The concept of biological progress. In F. Ayala & T. Dobzhansky (eds.), *Studies in the Philosophy of Biology.* Macmillan.

Ayala, F., & J. Valentine. 1979. *Evolving: The Theory and Processes of Organic Evolution.* Benjamin-Cummings.

Aznar-Ramos, R., et al. 1969. Incidence of side effects with contraceptive placebo. *American Journal of Obstetrics and Gynecology,* 105:1144–49.

Badawi, K., et al. 1984. Electrophysiologic characteristics of respiratory suspension periods occurring during the practice of the Transcendental Meditation program. *Psychosomatic Medicine,* 46:267–76.

Bagchi, B. K. 1969. Mysticism and mist in India. *Journal of the American Society of Psychosomatic Dentistry and Medicine,* 16:73–87.

Bagchi, B., & M. Wenger. 1957a. Electrophysiological correlates of some yogi exercises. *Electroencephalography & Clinical Neurophysiology,* 7:132–49.

Bagchi, B., & M. Wenger. 1957b. Electro-physiological correlates of some yogi exercises. First International Congress of Neurological Sciences, Brussels; and 1959, Vol. 3, *EEG Clinical Neurophysiology and Epilepsy.* Pergamon, pp. 132–49.

Bagchi, B., & M. Wenger. 1958. Simultaneous EEG and other recordings during some yogic practices. *Electroencephalography & Clinical Neurophysiology,* 10:193.

Bagga, O., & A. Gandhi. 1983. A comparative study of the effect of Transcendental Meditation and Shavasana practice on the cardiovascular system. *Indian Heart Journal,* 35:39–45.

Bagga, O., A. Gandhi, & S. Bagga. 1981. A study of the effect of Transcendental Meditation and yoga on blood glucose, lactic acid, cholesterol and total lipids. *Journal of Clinical Chemistry & Clinical Biochemistry,* 19:607–608.

Bailey, F. L. 1942. Navaho motor habits. *American Anthropologist,* 44:210–34.

Baillie, J. B., trans. 1910. 1967 reprint. *Phenomenology of Mind.* Harper.

Bair, J. H. 1901. Development of voluntary control. *Psychological Review,* 8:474–510.

Bakal, P. A. 1981. Hypnotherapy for flight phobia. *American Journal of Clinical Hypnosis,* 23:248.

Baker, A., & J. Thorpe. 1957. Placebo response. *Archives of Neurology and Psychiatry,* 78:57–60.

Ball, G., & D. Hall. 1967. A clustering technique for summarizing multivariate data. *Behavioral Science,* 12:2.

Ballard, P., et al. 1972. Arrest of a disabling eye disorder using biofeedback. *Psychophysiology,* 9:271 (Abstract).

Bannister, R. 1956. *First Four Minutes.* Sportsmans Book Club.

Banquet, J. P. 1972. EEG and meditation. *Electroencephalography & Clinical Neurophysiology,* 33:454.

Banquet, J. P. 1973. Spectral analysis of the EEG in meditation. *Electroencephalography & Clinical Neurophysiology,* 35:143–51.

Banquet, J., & M. Sailhan. 1974. EEG analysis of spontaneous and induced states of consciousness. *Revue d'Electroencephalographie et de Neurophysiologie Clinique,* 4:445–53.

Banyai, E., & E. Hilgard. 1976. A comparison of active-alert hypnotic induction with traditional relaxation induction. *Journal of Abnormal Psychology,* 85:218–24.

Barber, T. X. 1961. Physiological effects of "hypnosis." *Psychological Bulletin,* 58:390–419.

Barber, T. X. 1964. Hypnotic "colorblindness," "blindness," and "deafness." *Diseases of the Nervous System,* 25:529–37.

Barber, T. X. 1965. Physiological effects of "hypnotic suggestions": A critical review of recent research (1960–64). *Psychological Bulletin,* 63:201–22.

Barber, T. X. 1966. The effects of "hypnosis" and motivational suggestions on strength and endurance: A critical review of research studies. *British Journal of Social and Clinical Psychology* 5:42–50.

Barber, T. X. 1978. Hypnosis, suggestions, and psychosomatic phenomena: A new look from the standpoint of recent experimental studies. *American Journal of Clinical Hypnosis,* 21:13–27.

Barber, T. X. 1984. Changing "unchangeable" bodily processes by (hypnotic) suggestions: A new look at hypnosis, cognitions, imagining, and the mind-body problem. In A. Sheikh (ed.), *Imagination and Healing.* Baywood.

Barber, T. X., & S. Wilson. 1979. Guided imagining and hypnosis: Theoretical and empirical overlap and convergence in a new creative imagination scale. In A. Sheikh & J. Shaffter (eds.), *The Potential of Fantasy and Imagination.* Brandon House.

Barber, T. X., et al. 1964. The effect of hypnotic and non-hypnotic suggestions on parotid gland response to gustatory stimuli. *Psychosomatic Medicine,* 26:374–80.

Barclay, G. 1973. *Mind over Matter.* Bobbs-Merrill.

Barlow, W. 1980. *The Alexander Technique.* Warner.

Barr, B., & H. Benson. 1984. The relaxation response and cardiovascular disorders. *Behavioral Medicine Update,* 6:28–30.

Barrett, W. 1911. *Psychical Research.* Henry Holt.

Barron, F. 1953. Complexity-simplicity as a personality dimension. *Journal of Abnormal Social Psychology,* 48:163–72.

Barron, F. 1955. The disposition toward originality. *Journal of Abnormal Social Psychology,* 51:478–85.

Barron, F. 1957. Originality in relation to personality and intellect. *Journal of Personality,* 25:730–42.

Barron, F. 1958a. The needs for order and for disorder as motives in creative activity. In C. Taylor et al., *The Second (1957) Research Conference on the Identification of Creative Scientific Talent.* University of Utah Press.

Barron, F. 1958b. The psychology of imagination. *Scientific American*, 199:150–56.

Barron, F. 1963a. Creativity (psychology of). *Encyclopaedia Britannica.* University of Chicago Press.

Barron, F. 1963b. *Creativity and Psychological Health.* Van Nostrand.

Barron, F. 1963c. The needs for order and for disorder as motives in creative activity. In C. Taylor & F. Barron, *Scientific Creativity.* Wiley.

Barron, F. 1972. *Artists in the Making.* Seminar Press.

Barron, F. 1979. *The Shaping of Personality.* Harper & Row.

Barron, F. 1987. Putting creativity to work. In R. Sternberg (ed.), *The Nature of Creativity.* Cambridge University Press.

Barron, F., & G. Welsh. 1952. Artistic perception as a factor in personality style: Its measurement by a figure-preference test. *Journal of Psychology,* 33:199–203.

Barry, J. 1968a. General and comparative study of the psychokinetic effect on fungus culture. *Journal of Parapsychology,* 32:237–43.

Barry, J. 1968b. PK on fungus growth. *Journal of Parapsychology,* 32:55 (Abstract).

Basmajian, J. V. 1963. Control and training of individual motor units. *Science,* 141:440–41.

Basmajian, J. V. 1966. Conscious control of single nerve cells. *New Scientist* (Dec. 12), pp. 662–64.

Basmajian, J. V., ed. 1989. *Biofeedback: Principles and Practice for Clinicians.* Williams & Wilkins.

Basmajian, J., & T. Simard. 1967. Effects of distracting movements on the control of trained motor units. *American Journal of Physical Medicine,* 46:1427–49.

Bateson, G. 1975. *Steps Toward an Ecology of Mind.* Ballantine.

Bateson, G., & M. Mead. 1942. *Balinese Character: A Photographic Analysis.* Vol. 2. Special Publications of the New York Academy of Sciences.

Bauhofer, V. 1978. Physiological cardiovascular effects of the Transcendental Meditation technique. Doctoral dissertation, Julius-Maximilian University, Wurzburg, West Germany.

Beard, F., & D. Schaap. 1970. *Pro: Frank Beard on the Golf Tour.* World Press.

Beary, J., & H. Benson. 1974. A simple physiologic technique which elicits the hypometabolic changes of the relaxation response. *Psychosomatic Medicine,* 36:115–20.

Beatty, J., et al. 1974. Operant control of occipital theta rhythm affects performance in a radar monitoring task. *Science,* 183:871–73.

Bechterev, W. 1948. Direct influence of a person upon the behavior of animals. *Journal of Parapsychology,* 13:166–76.

Beck, A., et al. 1979. *Cognitive Therapy of Depression.* Guilford.

Beecher, H. K. 1955. The powerful placebo. *Journal of the American Medical Association,* 159:1603–4.

Beecher, H. K. 1961. Surgery as placebo. *Journal of the American Medical Association,* 176:1102–7.

Behanan, K. 1937. *Yoga: A Scientific Evaluation.* Dover.

Beiman, et al. 1984. The relationship of client characteristics to outcome for transcendental meditation, behavior therapy, and self-relaxation. In D. H. Shapiro & R. Walsh (eds.), *Meditation: Classic & Contemporary Perspectives.* Aldine.

Bell, I., & G. Schwartz. 1975. Voluntary control and reactivity of human heart rate. *Psychophysiology,* 12:339–48.

Bell, R. 1985. *Holy Anorexia.* University of Chicago Press.

Bellah, R., et al. 1986. *Habits of the Heart.* Harper & Row.

Bellah, R., et al. 1991. *The Good Society.* Alfred A. Knopf.

Bellis, J. M. 1966. Hypnotic pseudo-sunburn. *American Journal of Clinical Hypnosis,* 8:310–12.

Below, J. 1935. The Balinese temper. *Character and Personality,* 4:120–46.

Ben-Zvi, Z., et al. 1982. Hypnosis for exercise-induced asthma. *American Review of Respiratory Diseases,* 125:392–95.

Bender, B. 1972. Turn on your mind to turn off pain. *Probe* (Dec.), pp. 8–15.

Benedictae Papae XIV doctrina de Servorum Dei beatificatione et Beatorum canonizatione in synopsin. 1751. Emmanual de Azevedo, S. J. Sacrorum Rituum Consultore. In Benedictine Monks of St. Augustine's Abbey, Ramsgate (eds.), *The Books of Saints* (3d ed. 1942), Preface. Macmillan.

Bennett, J., & J. Trinder. 1977. Hemispheric laterality and cognitive style associated with Transcendental Meditation. *Psychophysiology,* 14:293–94.

Benson, H. 1975. *The Relaxation Response.* William Morrow.

Benson, H. 1976. The relaxation response and cardiovascular disease. *Chest, Heart, Stroke Journal,* 1:28–31.

Benson, H. 1980. The placebo effect. *Harvard Medical School Health Letter* (Aug.).

Benson, H. 1982. Body temperature changes during the practice of gTum-mo yoga. (Letter) *Nature,* 298:402.

Benson, H. 1983. The relaxation response: Its subjective and objective historical precedents and physiology. *Trends in Neuroscience* (July), pp. 281–84.

Benson, H. 1984. *Beyond the Relaxation Response.* Times Books.

Benson, H., T. Dryer, L. Hartley et al. 1977. Decreased oxygen consumption at a fixed work intensity with simultaneous elicitation of the relaxation response. *Clinical Research,* 25:453.

Benson, H., & M. Epstein. 1975. The placebo effect: A neglected asset in the care of patients. *Journal of the American Medical Association,* 232:1225–27.

Benson, H., & I. Goodale. 1981. The relaxation response: Your inborn capacity to counteract the harmful effects of stress. *Journal of Florida Medical Association,* 68:265–67.

Benson, H., H. Klemchuk, & J. Graham. 1974. The usefulness of the relaxation response in the therapy of headache. *Headache,* 14:49–52.

Benson, H., B. Malvea, J. Graham et al. 1973. Physiologic correlates of meditation and their clinical effects in headache: An ongoing investigation. *Headache,* 13: 1:23–24.

Benson, H., & D. McCallie. 1979. Angina pectoris and the placebo effect. *New England Journal of Medicine,* 300:1424–29.

Benson, H., B. Rosner, B. Marzetta et al. 1974a. Decreased blood pressure in borderline hypertensive subjects who practiced meditation. *Journal of Chronic Disease,* 27:163–69.

Benson, H., B. Rosner, & B. Marzetta et al. 1974b. Decreased blood pressure in untreated borderline hypertensive subjects who regularly elicited the relaxation response. *Clinical Research,* 22:262.

Benson, H., R. Steinert, M. Greenwood et al. 1975. Continuous measurement of O_2 consumption and CO_2 elimination during a wakeful hypometabolic state. *Journal of Human Stress,* 1:37–44.

Benson, H., & R. Wallace. 1972a. Decreased blood pressure in hypertensive subjects who practiced meditation. *Circulation,* 46:1:130.

Benson, H., & R. Wallace. 1972b. Decreased drug abuse with Transcendental Meditation: A study of 1,862 subjects. In Zarafonetis, J. D. (eds.), *Drug Abuse. Proceedings of the International Conference, 1972.* Philadelphia: Lea & Febiger.

Benson, H., et al. 1971. Decreased systolic blood pressure through operant conditioning techniques in patients with essential hypertension. *Science,* 173:740–42.

Benson, H., et al. 1978. Treatment of anxiety: A comparison of the usefulness of self-hypnosis and a meditational relaxation technique. *Psychotherapy & Psychosomatics,* 30:229–42.

Benson, H., et al. 1982. Body temperature changes during the practice of gTum-mo yoga. *Nature,* 295.

Benson, H., et al. 1984. Pain and relaxation response. In P. Wall & R. Melzack (eds.), *Textbook of Pain.* Churchill Livingston.

Benthall, J., & T. Polhemus, eds. 1975. *The Body as a Medium of Expression.* Allen, Lane & Dutton.

Berenson, B. 1950. *Aesthetics and History.* London.

Berg, A. O. 1983. The placebo effect reconsidered. *Journal of Family Practice,* 17: 647–50.

Bergman, J., & H. Johnson. 1971. The effects of instructional set and autonomic perception on cardiac control. *Psychophysiology,* 8:180–90.

Bergson, H. 1911. *Creative Evolution.* Trans. A. Mitchell. Henry Holt.

Berne, E. 1972. *What Do You Say After You Say Hello?* Bantam.

Besant, A.,& C. W. Leadbeater. 1908. *Occult Chemistry.* Theosophical Publishing House.

Bhole, M. V., et al. 1967. Underground burial or Bhoogarbha samadhi. *Yoga-Mimamsa,* 10:12–16.

Bibliotheca Sanctorum. 1961. 12 vols. Rome.

Binswanger, L. 1956. Existential analysis and pychotherapy. In F. Fromm-Reichmann & J. Moreno (eds.), *Progress in Psychotherapy.* Grune & Stratton.

Biot, R. 1962. *The Enigma of the Stigmata.* Hawthorn.

Bird, B. L., et al. 1978a. Behavioral and electroencephalographic correlates of 40-Hz EEG biofeedback training in humans. *Biofeedback and Self-Regulation,* 3: 13–28.

Bird, B. L., et al. 1978b. Biofeedback training of 40-Hz EEG in humans. *Biofeedback and Self-Regulation,* 3:1–11.

Birdwhistell, R. 1960. Kinesics and communication. In E. Carpenter & M. McLuhan (eds.), *Explorations in Communication.* Beacon.

Birdwhistell, R. 1971. *Kinesics and Context: Essays on Body-Motion Communication.* Penguin.

Birdwhistell, R. 1972. Kinesics. In D. Sills (ed.), *International Encyclopedia of the Social Sciences,* vol. 8. Macmillan.

Birdwhistell, R. 1975. In J. Benthall & T. Polhemus (eds.), *The Body as a Medium of Expression.* Allen Lane, Dutton.

Birk, L., et al. 1966. Operant electrodermal conditioning under partial curarization. *Journal of Comparative & Physiological Psychology,* 62:165–66.

Bishay, E. G., et al. 1984. Hypnotic control of upper gastrointestinal hemorrhage: A case report. *American Journal of Clinical Hypnosis,* 27:22–25.

Biven, G., & M. Klinger. 1937. *Pseudocyesis.* Principia Press.

Blackwell, B., S. Bloomfield, P. Gartside et al. 1976. Transcendental Meditation in hypertension: Individual response patterns. *Lancet,* 1:223–36.

Blair, S. N., et al. 1989. Physical fitness and all-cause mortality. *Journal of the American Medical Association* (Nov. 3), 262:2395–2401.

Blanchard, E., et al. 1982. The prediction by psychological tests of headache patients' response to treatment with relaxation and biofeedback. In *Self-Regulation Strategies: Efficacy and Mechanisms.* Biofeedback Society of America 13th Annual Meeting. Chicago, Mar. 5–8.

Blanchard, E. B., et al. 1974. Differential effects of feedback and reinforcement in voluntary acceleration of human heart rate. *Perceptual & Motor Skills,* 38: 683–91.

Blanchard, E. B., et al. 1975. Long-term instructional control of heart rate without exteroceptive feedback. *Journal of General Psychology,* 92:291–92.

Blanck, G., & R. Blanck. 1974. *Ego Psychology: Theory and Practice.* Columbia University Press.

Blanck, G., & R. Blanck. 1979. *Ego Psychology II: Psychoanalytic Developmental Psychology.* Columbia University Press.

Blasdell, K. A. 1977. The effects of the Transcendental Meditation technique upon a complex perceptual-motor task. In D. W. Orme-Johnson and J. T. Farrow (eds.), *Scientific Research on the Transcendental Meditation Program: Collected Papers.* Vol. 1. New York: MERU Press.

Bleecker, E., & B. Engel. 1973. Learned control of cardiac rate and cardiac conduction in the Wolff-Parkinson-White syndrome. *New England Journal of Medicine,* 288:560–62.

Bleecker, E., & B. Engel. 1973. Learned control of ventricular rate in patients with atrial fibrillation. *Psychosomatic Medicine,* 35:161–75.

Bliss, E. 1984. Spontaneous self-hypnosis in multiple personality disorder. *Psychiatric Clinics of North America,* 7:137.

Bloch, B. 1927. Ueber die Heilung der Warzen durch Suggestion. *Klinl. Wchnschr,* 48: 2271, 2320.

Blofeld, J. 1966. *People of the Sun.* Hutchinson.

Blofeld, J. 1973. *The Secret and Sublime.* Dutton.

Blofeld, J. 1979. *Taoism: The Quest for Immortality.* Unwin.

Blum, G. S. 1967. Experimental observations on the contextual nature of hypnosis. *International Journal of Clinical & Experimental Hypnosis,* 15:160–71.

Blum, G. S. 1968. Effects of hypnotically controlled strength of registraton vs. rehearsal. *Psychonomic Science,* 10:351–52.

Blum, G. S. 1979. Hypnotic programming techniques in psychological experiments. In E. Fromm & R. Shor (eds.), *Hypnosis: Developments in Research and New Perspectives.* Aldine.

Blum, G. S., & J. S. Barbour. 1979. Selective inattention to anxiety-linked stimuli. *Journal of Experimental Psychology: General,* 108:182–224.

Blum, G. S., & M. Green. 1978. The effects of mood upon imaginal thought. *Journal of Personality Assessment,* 42:227–32.

Blum, G. S., & J. K. Nash. 1982. EEG correlates of posthypnotically controlled degrees of cognitive arousal. *Memory & Cognition,* 10:475–78.

Blum, G. S., & M. L. Porter. 1972. The capacity for rapid shifts in level of mental concentration. *Quarterly Journal of Experimental Psychology,* 24:431–38.

Blum, G. S., & M. L. Porter. 1973. The capacity for selective concentration on color versus form of consonants. *Cognitive Psychology,* 5:47–70.

Blum, G. S., & B. M. Wohl. 1971. An experimental analysis of the nature and operation of anxiety. *Journal of Abnormal Psychology,* 78:1–8.

Blum, G. S., et al. 1967. Cognitive arousal: The evolution of a model. *Journal of Personality and Social Psychology,* 5:138–51.

Blum, G. S., et al. 1968a. Effects of interference and cognitive arousal upon the processing of organized thought. *Journal of Abnormal Psychology,* 73:610–14.

Blum, G. S., et al. 1968b. Overcoming interference in short-term memory through distinctive mental contexts. *Psychonomic Science,* 11:73–74.

Blum, G. S., et al. 1968c. Studies of cognitive reverberation: Replications and extensions. *Behavioral Science,* 13:171–77.

Blum, G. S., et al. 1971. Distinctive mental contexts in long-term memory. *International Journal of Clinical & Experimental Hypnosis,* 19:117–33.

Boadella, D. 1985. *Wilhelm Reich.* Arkana.

Bogdanove, E. 1966. Preservation of functional corpora lutea in the rat by estrogen treatment. *Endocrinology,* 79:1011–15.

Bogoras, W. 1904. *The Chuckchee.* Cited in M. Eliade, 1964, *Shamanism.* Princeton University Press, Bollingen Series.

Bonington, C. 1976. *Everest the Hard Way.* Random House.

Bono, J. 1984. Psychological assessment of Transcendental Meditation. In D. H. Shapiro & R. Walsh (eds.), *Meditation: Classic & Contemporary Perspectives.* Aldine.

Book of Saints. 1942 (3d ed.). Benedictine Monks of Saint Augustine's Abbey, Ramsgate (eds.). Macmillan.

Bootzin, R. R. 1985. The role of expectancy in behavior change. In L. White et al. (eds.), *Placebo: Theory, Research, & Mechanisms.* Guilford Press.

Borelli, S. 1953. Psychische Einflusse and Reactive Hauterscheinungen. *Muenchener Medizinische Wochenschrift,* 95:1078–82. In A. Sheikh, 1984, *Imagination and Healing.* Baywood.

Bornstein, P., & C. Sipprelle. 1973. Clinical Application of Induced Anxiety in the Treatment of Obesity. Paper presented at the Southeastern Psychological Association Meeting, Apr. 6.

Boss, M. 1963. *Psychoanalysis and Daseinsanalysis.* Basic Books.

Boudreau, L. 1972. Transcendental Meditation and yoga as reciprocal inhibitors. *Journal of Behavior Therapy and Experimental Psychology,* 3:97–98.

Bourdieu, P. 1984. *Distinction—A Social Critique of the Judgement of Taste.* Harvard University Press.

Bower, G. H. 1981. Mood and memory. *American Psychologist,* 36:129–48.

Bower, G. H., et al. 1978. Emotional mood as a context for learning and recall. *Journal of Verbal Learning & Verbal Behavior,* 17:573–85.

Bower, G. H., et al. 1981. Selectivity of learning caused by affective states. *Journal of Experimental Psychology: General,* 110:451–73.

Bowers, K. S., & P. Bowers. 1979. Hypnosis and creativity: A theoretical and empirical reapproachment. In E. Fromm & R. Shorr (eds.), *Hypnosis: Developments in Research and New Perspectives.* Aldine.

Bowers, P. 1979. Hypnosis and creativity: The search for the missing link. *Journal of Abnormal Psychology,* 88:564–72.

Boyd, W. 1921. Tissue resistance in malignant disease. *Surgery, Gynecology and Obstetrics,* 32:306.

Boyd, W. 1957. The spontaneous regression of cancer. *Journal of the Canadian Association of Radiology,* 8:45.

Boyer, J., & F. Kasch. 1970. Exercise therapy in hypertensive men. *Journal of the American Medical Association,* 211:1668.

Bozzano, E. *Phenomenes Psychiques au Moment de la Mort.* Alcan.

Braid, J. 1846. *The Power of Mind over the Body.* John Churchill.

Braid, J. 1850. *Observations on Trance or Human Hibernation.* London and Edinburgh.

Braid, J. 1855. *The Physiology of Fascination and the Critics Criticized.* Grant & Co.

Bramwell, J. M. 1903. *Hypnotism: Its History, Practice and Theory.* Grant Richards.

Bramwell, M. 1896. Personally observed hypnotic phenomena. *Proceedings of the Society for Psychical Research,* 12:176–203.

Bramwell, M. 1900. Hypnotic and post-hypnotic appreciation of time: Secondary and multiple personalities. *Brain* (Summer).

Branden, N. 1971. *The Psychology of Self-Esteem.* Bantam.

Brandon, J. E. 1983. A comparative evaluation of three relaxation training procedures. *Dissertation Abstracts International,* 43 (7-A), 2:279.

Braud, W. 1974. Psychokinetic influence upon E. Coli bacteria. Unpublished manuscript, University of Houston.

Braud, W. 1978a. Allobiofeedback: Immediate feedback for a psychokinetic influence upon another person's physiology. In W. G. Roll (ed.), *Research in Parapsychology 1977.* Scarecrow Press.

Braud, W. 1978b. Psi conducive conditioning: Explorations and interpretations. In B. Shapin & L. Coly (eds.), *Psi & States of Awareness.*

Braud, W. 1979. Conformance behavior involving living systems. In *Research in Parapsychology 1978.* Scarecrow Press.

Braud, W. 1988. Distant mental influence of rate of hemolysis of human red blood cells. *Proceedings of Presented Papers,* 31st Annual Parapsychological Association Convention, Montreal (Aug).

Braud, W., & P. Dennis. 1989. Geophysical variables and behavior: LVIII. Autonomic activity, hemolysis, and biological psychokinesis: Possible relationships with geomagnetic field activity. *Perceptual & Motor Skills,* 68:1243–54.

Braud, W., & M. Schlitz. 1989. A methodology for the objective study of transpersonal imagery. *Journal of Scientific Exploration,* 3:51.

Braud, W., et al. 1979. Experiments with Matthew Manning. *Journal of the Society for Psychical Research,* 50:199–223.

Braude, S. E. 1979. *ESP and Psychokinesis.* Temple University Press.

Braude, S. E. 1986. *The Limits of Influence.* Routledge & Kegan Paul.

Braun, B. 1983. Psychophysiologic phenomena in multiple personality and hypnosis. *American Journal of Clinical Hypnosis,* 26:124–35.

Brautigam, E. 1977. Effects of the Transcendental Meditation program on drug abusers: A prospective study. In D. W. Orme-Johnson & J. T. Farrow (eds.), *Scientific Research on the Transcendental Meditation Program: Collected Papers.* Vol. 1. New York: MERU Press.

Brende, J. O. 1984. The psychophysiologic manifestations of dissociation. *Psychiatric Clinics of North America,* 7:41–49.

Brener, J. 1974. A general model of voluntary control applied to the phenomena of learned cardiovascular change. In P. A. Obrist et al. (eds.), *Cardiovascular Psychophysiology.* Aldine.

Brener, J. 1977. Sensory and perceptual determinants of voluntary visceral control. In G. E. Schwartz & J. Beatty (eds.), *Biofeedback: Theory and Research.* Academic Press.

Brener, J., & D. Hothersall. 1966. Heart rate control under conditions of augmented sensory feedback. *Psychophysiology,* 3:23–28.

Brener, J., & D. Hothersall. 1967. Paced respiration and heart rate control. *Psychophysiology,* 4:1–6.

Brenman, M., et al. 1947. Alterations in the state of the ego in hypnosis. *Bulletin of the Menninger Clinic,* 11:60–66. In M. Nash et al. 1984. The direct hypnotic suggestion of altered mind/body perception. *American Journal of Clinical Hypnosis,* 27:96.

Brentano, C. 1837. *Das bittere Leiden unseres Herrn Jesu Christi. Nach den Betrachtungen der gottseligen Anna Katherina Emmerich.* Sulzbach: Seidel.

Brentano, C. 1852. Leben der heiligen Jungfrau Maria. *Nach den Betrachtungen der gottseligen Anna Katherina Emmerich.* Literarisch-artistische Anstalt. In H. Ellenberger, 1970, *The Discovery of the Unconscious.* Basic Books.

Brewer, V., et al. 1983. Role of exercise in prevention of involutional bone loss. *Medicine & Science in Sports & Exercise,* 15:455.

Brier, R. M. 1969. PK on a bio-electrical system. *Journal of Parapsychology,* 33:187–205.

Briggs, J. 1988. Madness and the mirror-maker's nightmare. In *Fire in the Crucible.* St. Martin's.

British Medical Journal. 1968. Hypnosis for asthma—A controlled trial. A report to the Research Committee of the British Tuberculosis Association, 4:71–76.

Broad, C. D. 1923. *Scientific Thought.* Routledge & Kegan Paul.

Broad, C. D. 1925, 1980. *The Mind and Its Place in Nature.* Routledge & Kegan Paul.

Broad, C. D. 1953. *Religion, Philosophy and Psychical Research.* Harcourt, Brace.

Brodeur, D. W. 1965. The effects of stimulant and tranquilizer placebos on healthy subjects in a real-life situation. *Psychopharmacologia,* 7:445–52.

Brodie, J., & J. Houston. 1974. *Open Field.* Houghton Mifflin.

Brody, H. 1984. Placebos work without trickery. *Medical Tribune,* 22:3. (Michigan State University, College of Human Medicine, East Lansing, MI.)

Brody, J. 1985a. Benefits and dangers of exercise. *New York Times* (July 16).

Brody, J. 1985b. Laying-on-of-hands gains new respect. *New York Times* (March 26).

Brooks, C. V. 1974. *Sensory Awareness.* Viking.

Brosse, T. 1946. A psycho-physiological study. *Main Currents in Modern Thought,* 4: 77–84.

Brown, B. 1970. Recognition of aspects of consciousness through association with EEG alpha activity represented by a light signal. *Psychophysiology,* 6:442–52.

Brown, B. 1971. Awareness of EEG-subjective activity relationships detected within a closed feedback system. *Psychophysiology,* 7:451–64.

Brown, D., & J. Engler. 1986. The stages of mindfulness meditation: A validation study. Part II. Discussion. In K. Wilber, J. Engler, & D. Brown (eds.), *Transformations of Consciousness.* Shambhala, New Science Library.

Brown, D., et al. 1982–83. Phenomenological differences among self-hypnosis, mindfulness meditation, and imaging. *Imagination, Cognition & Personality,* 2: 291–309.

Brown, D., et al. 1984a. Differences in visual sensitivity among mindfulness meditators and nonmeditators. *Perceptual & Motor Skills,* 58:727–33.

Brown, D., et al. 1984b. Visual sensitivity and mindfulness meditation. *Perceptual & Motor Skills,* 58:775–84.

Brown, E., & P. Barglow. 1971. Pseudocyesis: A paradigm for psychophysiological interactions. *Archives of General Psychiatry,* 24:221.

Brown, N. O. 1968. *Life Against Death: The Psychoanalytical Meaning of History.* Vintage/Random House.

Brown, N. O. 1991. *Apocalypse and/or Metamorphosis.* University of California Press.

Brown, P. 1971a. The rise and function of the holy man in late antiquity. *Journal of Roman Studies,* 61:80–101.

Brown, P. 1971b. *The World of Late Antiquity.* London.

Bruce, R. V. 1973. *Alexander Graham Bell and the Conquest of Solitude.* Little, Brown.

Brudny, J., et al. 1974. Spasmodic torticollis: Treatment by feedback display of the EMG. *Archives of Physical Medicine and Rehabilitation,* 55:403–8.

Brunner, D., et al. 1974. Physical activity at work and the incidence of myocardial infarction, angina pectoris and death due to ischemic heart disease: An epidemiological study in Israeli collective settlements (Kibbutzim). *Journal of Chronic Diseases,* 27:217–33.

Brunt, P. A. 1976. Introduction to his translation of Books I–IV of Arrian's *Anabasis Alexandri.* Loeb Classical Library, Harvard University Press.

Buckalew, L., & K. Coffield. 1982. An investigation of drug expectancy as a function of capsule color and size and preparation form. *Journal of Clinical Psychopharmacology,* 2:245–48.

Bucke, R. M. 1961. *Cosmic Consciousness.* University Books.

Buckle, R. 1909. *Nijinsky.* Simon & Schuster.

Buckler, W. 1976. Transcendental Meditation (Letter). *Canadian Medical Association Journal,* 115:607.

Buddhaghosa. 1964. *Visuddhimagga. (The Path to Purification).* Trans. B. Nyanamoli. Gunasena & Co.

Budge, E. A. W., trans. 1901. *The Book of the Dead.* Chicago & London.

Budzynski, T., & J. Stoyva. 1969. An instrument for producing deep muscle relaxation by means of analog information feedback. *Journal of Applied Behavior Analysis,* 2:231–37.

Budzynski, T. H., et al. 1970. Feedback-induced muscle relaxation: Application to tension headache. *Journal of Behavior Therapy and Experimental Psychiatry,* 1: 205–11.

Budzynski, T. H., et al. 1973. EMG biofeedback and tension headache: A controlled outcome study. *Psychosomatic Medicine,* 35:484–96.

Bugental, J. 1965. *The Search for Authenticity.* Holt, Rinehart & Winston.

Butler, B. 1954. The use of hypnosis in the care of the cancer patient. *Cancer,* 7:1–14.

Butler, B. 1954/1955. The use of hypnosis in the care of the cancer patient. *British Journal of Medical Hypnotism.*

Butler, F. 1978. *Biofeedback: A Survey of the Literature.* Plenum.

Butterworth, C. W. 1973. *Origen—On First Principles.* Peter Smith.

Buytendijk, 1957. *Attitude et mouvements: Etude fonctionelle du mouvement humaine.* Desclee de Brouwer.

Bykov, K. M. 1957. *The Cerebral Cortex and the Internal Organs.* Chemical Publishing Company.

Bynum, C. 1987. *Holy Feast and Holy Fast.* University of California Press.

Bynum, C. 1991. Material continuity, personal survival, and the resurrection of the body: A scholastic discussion in its medieval and modern contexts. *History of Religions,* 30:52–54.

Byrd, R. 1988. Positive therapeutic effects of intercessory prayer in a coronary care unit population. *Southern Medical Journal,* 81:826–29.

Cadarette, B., J. Hoffman, M. Caudill et al. 1982. Effect of the relaxation response on selected cardiorespiratory response during physical exercise. *Medicine & Science in Sports & Exercise,* 14:117.

Cahill, M., and P. Belfer. 1978. Word association times, felt effects, and personality characteristics of science students given a placebo energizer. *Psychological Reports,* 42:231–38.

Callen, K. E. 1983. Auto-hypnosis in long distance runners. *American Journal of Clinical Hypnosis,* 26:30.

Campbell, J. 1959. *The Masks of God: Primitive Mythology.* Viking.

Campbell, J. 1973. *Hero with a Thousand Faces.* Princeton University Press, Bollingen Series.

Cangello, V. W. 1961. The use of the hypnotic suggestion for relief in malignant disease. *International Journal of Clinical & Experimental Hypnosis,* 9:17–22.

Cangello, V. W. 1962. Hypnosis for the patient with cancer. *American Journal of Clinical Hypnosis,* 4:215–26.

Cangelosi, A. 1981. The differential effects of three relaxation techniques: A physiological comparison. *Dissertation Abstracts International,* 42 (1-B), 418.

Cardeña, E. 1988. The Phenomenology of Quiescent and Physically Active Deep Hypnosis. Paper presented in the symposium Phenomenological Experiences of Hypnosis, 39th Annual Meeting of the Society for Clinical and Experimental Hypnosis, Asheville, NC (Nov).

Carlsoo, S., & A. Edfeldt. 1963. Attempts at muscle control with visual and auditory impulses as auxiliary stimuli. *Scandinavian Journal of Psychology,* 4:231–35.

Carlsson, S., & E. Gale. 1976. Biofeedback treatment for muscle pain associated with the temporomandibular joint. *Journal of Behavior Therapy and Experimental Psychiatry,* 7:383–85.

Carlsson, S., & E. Gale. 1977. Biofeedback in the treatment of long-term temporomandibular joint pain. *Biofeedback and Self-Regulation,* 2:161–71.

Carlsson, S., et al. 1975. Treatment of temporomandibular joint syndrome with biofeedback training. *Journal of the American Dental Association,* 91:602–5.

Carnandet, J. Editio Novissima 1863–19_ . *Acta Sanctorum quotquot toto orbe coluntur, vel a catholicis scriptoribus cele brantur quae ex latinis et graecis, aliarumque gentium antiquis monumentis collegit, digessit, notis illustravit Joannes Bollandus . . . servata primigenia scriptorum phrasi. Operam et studium contulit Godefridus Henschenius. . . .* Paris: V. Palme.

Carrel, A. 1935. *Man the Unknown.* Harper.

Carrel, A. 1950. *Voyage to Lourdes.* Harper.

Carrington, H. 1909. *Eusapia Palladino and Her Phenomena.* Dodge.

Carrington, P. 1977. *Freedom in Meditation.* Anchor Press/Doubleday.

Carrington, P., & H. Ephron. 1975. Meditation and psychoanalysis. *Journal of American Academy of Psychoanalysis,* 3:43–57.

Carrington, P., et al. 1980. The use of meditation-relaxation techniques for the management of stress in a working population. *Journal of Occupational Medicine,* 22:221–31.

Carter, E., & A. Wedd. 1918. Report of a case of paroxysmal tachycardia characterized by unusual control of the fast rhythm. *Archives of Internal Medicine,* 22:571–80.

Carty, C. M. 1974. *Who Is Theresa Neumann?* Tan Books.

Casperson, C. J., et al. 1985. Physical activity, exercise, and physical fitness: Definitions and distinctions for health-related research. *Public Health Reports,* 100: 126–31.

Cassel, J., et al. 1971. Occupation and physical activity and coronary heart disease. *Archives of Internal Medicine,* 128:920–28.

Castelli, W. P., et al. 1986. Incidence of coronary heart disease and lipoprotein cholesterol levels (The Framingham Study). *Journal of the American Medical Association* (Nov. 28), 256:2835–38.

Caton, R. 1875. The electrical currents of the brain. *British Journal of Medicine,* 2:278.

Caton, R. 1900. *The Temples and Rituals of Asklepios at Epidauros and Athens.* C. J. Clay.

Channon, L. 1984. Extrasensory communication in hypnosis: Some uncomfortable speculations. *Australian Journal of Clinical & Experimental Hypnosis,* 12: 23–29.

Chapman, L., et al. 1959. Increased inflammatory reaction induced by central nervous system activity. *Transactions of the Association of American Physicians,* 72:84–109.

Chari, C. T. 1962. "Buried memories" in survivalist research. *International Journal of Parapsychology* (Summer), 4:40–65.

Chaves, J. F. 1980. Hypnotic Control of Surgical Bleeding. Paper presented at Annual Meeting of the American Psychological Association, Montreal (Sept).

Chernin, K. 1981. *The Obsession: Reflections on the Tyranny of Slenderness.* Harper & Row.

Cherry, L. 1981. The power of the empty pill. *Science Digest* (Sept.), 89:60.

Chertok, L. 1959. *Psychosomatic Methods in Painless Childbirth: History, Theory & Practice.* Pergamon.

Chertok, L. 1981. *Sense and Nonsense in Psychotherapy: The Challenge of Hypnosis.* Trans. R. H. Ahrenfeldt. Pergamon.

Chia, M., & M. Chia. 1986. *Cultivating Female Sexual Energy.* Healing Tao Books.

Chow, D., & R. Spangler. 1977. *Kung Fu: Philosophy and Technique.* Doubleday.

Clamann, H. P. 1970. Activity of single motor units during isometric tension. *Neurology*, 20:254–60.

Clarke, S. 1912. *Christian Antiquities of the Nile Valley*. Oxford University Press.

Clausen, J., & J. Trap-Jensen. 1974. Arteriohepatic venous oxygen difference and heart rate during initial phases of exercise. *Journal of Applied Physiology*, 37: 716–19.

Clausen, J., et al. 1969. Physical training in the management of coronary artery disease. *Circulation*, 40:143.

Clawson, T., & R. Swade. 1975. The hypnotic control of blood flow and pain: The cure of warts and the potential for the use of hypnosis in the treatment of cancer. *American Journal of Clinical Hypnosis*, 17:160–69.

Cleeland, C. S. 1973. Behavioral technics in the modification of spasmodic torticollis. *Neurology*, 23:1241–47.

Clements, G., & S. Milstein. 1977. Auditory thresholds in advanced participants in the Transcendental Meditation program. In D. W. Orme-Johnson and J. T. Farrow (eds.), *Scientific Research on the Transcendental Meditation Program: Collected Papers*. Vol. 1. MERU Press.

Clements, G., et al. 1988. The use of the Transcendental Meditation Program in the prevention of drug-abuse in the treatment of drug-addicted persons. *Bulletin of Narcotics*, 40:51–56.

Clews, E. 1939. *Pueblo Indian Religion*. University of Chicago Press.

Cobb, L. A., et al. 1959. An evaluation of internal-mammary-artery ligation by a double-blind technique. *New England Journal of Medicine*, 260:1115–18.

Coe, W., & K. Ryken. 1979. Hypnosis and risks to human subjects. *American Psychologist*, 34: 673–781.

Cohen, S. B. 1981. Phobia of bovine sounds. *American Journal of Clinical Hypnosis*, 23:266.

Collier, R. W. 1977. The effect of the Transcendental Meditation program upon university academic attainment. In D. W. Orme-Johnson and J. T. Farrow (eds.), *Scientific Research on the Transcendental Meditation Program: Collected Papers*. Vol. 1. MERU Press.

Coons, P. M., et al. 1982. EEG studies of two multiple personalities and a control. *Archives of General Psychiatry* (July), 39:823.

Cooper, M., & M. Aygen. 1979. A relaxation technique in the management of hypercholesterolemia. *Journal of Human Stress*, 5:24–27.

Corbin, C. 1972. Mental practice. In W. Morgan (ed.), *Ergogenic Aids and Muscular Performance*. Academic Press.

Corbin, H. 1969. *Creative Imagination in the Sufism of Ibn Arabi*. Princeton University Press.

Corbin, H. 1977. *Spiritual Body and Celestial Earth*. Princeton University Press.

Corbin, H. 1978. *The Man of Light in Iranian Sufism*. Shambhala.

Corby, J., W. Roth, V. Zarcone et al. 1978. Psychophysiological correlates of the practice of tantric yoga meditation. *Archives of General Psychiatry*, 35:571–77.

Corcoran, J. 1983. *Martial Arts: Traditions, History & People*. Gallery Books.

Corey, P. W. 1977. Airway conductance and oxygen consumption changes associated with practice of the Transcendental Meditation technique. In D. W. Orme-Johnson & J. T. Farrow (eds.), *Scientific Research on the Transcendental Meditation Program: Collected Papers.* Vol. 1. MERU Press.

Cousins, N. 1977. The mysterious placebo. *Saturday Review* (Oct. 1), p. 12.

Covertino, V. 1982. Heart rate and sweat rate responses associated with exercise-induced hypervolemia. *Medicine & Science in Sports & Exercise,* 15:77.

Cowger, E., & E. Torrance. 1982. Further examination of the quality of changes in creative functioning resulting from meditation (Zazen) training. *Creative Child & Adult Quarterly,* 7:211–17.

Craig, K. D. 1969. Physiological arousal as a function of imagined, vicarious, and direct stress experiences. *Journal of Abnormal Psychology,* 73:513–20.

Cranston, R. 1988. *The Miracle of Lourdes.* Doubleday.

Crasilneck, H. B. 1979. Hypnosis in the control of chronic low-back pain. *American Journal of Clinical Hypnosis,* 2:71–78.

Crasilneck, H. B. 1982. A follow-up study in the use of hypnotherapy in the treatment of psychogenic impotency. *American Journal of Clinical Hypnosis,* 25:52.

Crasilneck, H. B., et al. 1955. Use of hypnosis in the management of patients with burns. *Journal of the American Medical Association,* 158:103–6.

Crawford, H. J. 1981. Hypnotic susceptibility as related to gestalt closure tasks. *Journal of Personality and Social Psychology,* 40:376–83.

Crawford, H. J. 1982a. Hypnotizability, daydreaming style, imagery, vividness, and absorption: A multidimensional study. *Journal of Personality and Social Psychology,* 42:915–26.

Crawford, H. J. 1982b. Cognitive processing during hypnosis: Much unfinished business. *Res. Comm. Psychology, Psychiatry & Behavior,* 7:169–79.

Crawford, H. J., & S. N. Allen. 1983. Enhanced visual memory during hypnosis as mediated by hypnotic responsiveness and cognitive strategies. *Journal of Experimental Psychology: General,* 112:662–85.

Crawford, H. J., et al. 1979. Hypnotic deafness: A psychophysical study of responses to tone intensity as modified by hypnosis. *American Journal of Psychology,* 92:193–214.

Crawfurd, R. 1911. *The King's Evil.* Oxford.

Credidio, S. G. 1982. Comparative effectiveness of patterned biofeedback vs. meditation training on EMG and skin temperature changes. *Behavior Research & Therapy,* 20:233–41.

Crehan, J., S.J. 1952. *Father Thurston.* Sheed and Ward.

Crider, A., et al. 1966. Studies on the reinforcement of spontaneous electrodermal activity. *Journal of Comparative & Physiological Psychology,* 61:20–27.

Crompton, P. 1975. *Kung Fu: Theory and Practice.* Pagurian Press.

Crookall, R. 1972. *Casebook of Astral Projection.* University Books.

Cross, F. L., ed. 1957. *The Oxford Dictionary of the Christian Church.* Oxford University Press.

Cruz, J. C. 1977. *The Incorruptibles.* Tan Books.

Csikszentmihalyi, M. 1988. *Optimal Experience.* Cambridge University Press.

Csikszentmihalyi, M. 1990. *Flow: The Psychology of Optimal Experience.* Harper & Row.

Cuddon, J. A. 1979. *The International Dictionary of Sports and Games.* Schocken.

Cummings, J. 1956. *The Priest* (Nov.), pp. 938–40.

Cummings, V. T. 1984. The effects of endurance training and progressive relaxation-meditation on the physiological response to stress. *Dissertation Abstracts International,* 45:451.

Cunningham, P., & G. Blum. 1982. Further evidence that hypnotically induced color blindness does not mimic congenital defects. *Journal of Abnormal Psychology,* 91:139–43.

Currens, J., et al. 1961. Half a century of running. Clinical, physiological and autopsy findings in the case of Clarence DeMar ("Mr. Marathon"). *New England Journal of Medicine,* 265:988–93.

Cuthbert, B., J. Kristeller et al. 1981. Strategies of arousal control: Biofeedback, meditation and motivation. *Journal of Experimental Psychology: General,* 110:518–46.

Czaplicka, M. A. 1914. *Aboriginal Siberia: A Study in Social Anthropology.* London.

Dahinterova, J. 1967. Some experiences with the use of hypnosis in the treatment of burns. *International Journal of Clinical & Experimental Hypnosis,* 15:49–53.

Daly, D., & H. Johnson. 1974. Instrumental modification of hypernasal voice quality in retarded children: Case reports. *Journal of Speech and Hearing Disorders,* 39:500–507.

Daniels, E., & B. Fernall. 1984. Continuous EEG measurement to determine the onset of a relaxation response during a prolonged run. *Medicine & Science in Sports & Exercise,* 16:182.

Darwin, C. *The Descent of Man.* Random House.

Darwin, C. 1872 (1965 reprint). *The Expression of the Emotions in Man and Animals.* University of Chicago Press.

Darwin, C. *The Origin of Species.* Random House.

Das, N., & H. Gastaut. 1955. Variations in the electrical activity of the brain, heart, skeletal muscles during yogic meditation and trance. *Electroencephalography & Clinical Neurophysiology,* 6:211–19.

Datey, K., S. Deshmukh, S. Dalvi et al. 1969. Shavasan: A yogic exercise in the management of hypertension. *Angiology,* 20:325–33.

David-Neel, A. 1956. *Magic and Mystery in Tibet.* University Books.

Davidson, J. A. 1962. An assessment of the value of hypnosis in pregnancy and labour. *British Medical Journal,* 4:951–53.

Davidson, J. M. 1976. The physiology of meditation and mystical states of consciousness. *Perspectives in Biology & Medicine,* 19:345–80.

Davidson, R., & D. Goleman. 1977. The role of attention in meditation and hypnosis: A psychobiological perspective on transformations of consciousness. *International Journal of Clinical & Experimental Hypnosis,* 25:291–308.

Davidson, R., D. Schwartz, & L. Rothman. 1976. Attentional style under self-regulation of mode specific attention: An electroencephalographic study. *Journal of Abnormal Psychology*, 85:611–21.

Davidson, R., & G. Schwartz. 1976. Patterns of cerebral lateralization during cardiac biofeedback versus the self-regulation of emotion: Sex differences. *Psychophysiology*, 13:62–68.

Davidson, R., & G. Schwartz. 1984. Matching relaxation therapies to types of anxiety: A patterning approach. In D. H. Shapiro & R. Walsh (eds.), *Meditation: Classic and Contemporary Perspectives*. Aldine.

Davidson, R., D. Goleman, G. Schwartz et al. 1976. Attentional and affective concomitants of meditation: A cross-sectional study. *Journal of Abnormal Psychology*, 85:235–38.

Davidson, R. J., et al. 1976. Sex differences in patterns of EEG asymmetry. *Biological Psychology*, 4:119–38.

Davies, J. 1977. The Transcendental Meditation program and progressive relaxation: Comparative effects on trait anxiety and self-actualization. In D. W. Orme-Johnson & J. T. Farrow (eds.), *Scientific Research on the Transcendental Meditation Program: Collected Papers*. Vol. 1. MERU Press.

Davison, G. C., & L. Singleton. 1967. A preliminary report of improved vision under hypnosis. *International Journal of Clinical & Experimental Hypnosis*, 15: 57–62.

de Poray Madeyeski, B. 1940. *Le cas de la visionnaire stigmatisee Therese Neumann de Konnersreuth, etude analytique et critique du probleme*. Lethielleux.

de Puysegur, M. 1811. *Recherchers, experience et observations physiologiques sur l'homme dans l'etat de somnambulism nature*. J. G. Dentu.

Deabler, H., E. Fidel, R. Dillenkoffer et al. 1973. The use of relaxation and hypnosis in lowering high blood pressure. *American Journal of Clinical Hypnosis*, 16: 75–83.

Dean, D. 1982. An examination of infra-red and ultra-violet techniques to test for changes in water following the laying-on of hands. Doctoral dissertation, Saybrook Institute, San Francisco. Dean, D., & E. Brame. 1975. Physical changes in water by laying-on-of-hands. *Proceedings of the Second International Conference of Psychotronic Research*. Paris: Institut Metaphysique International, pp. 220–301.

Deberry, S. 1982. The effects of meditation-relaxation on anxiety and depression in geriatric population. *Psychotherapy Theory, Research and Practice*, 19:512–21.

Deikman, A. J. 1963. Experimental meditation. *Journal of Nervous & Mental Disease*, 136:329–43.

Deikman, A. J. 1966a. De-automatization and the mystic experience. *Psychiatry*, 29: 324–38.

Deikman, A. J. 1966b. Implications of experimentally produced contemplative meditation. *Journal of Nervous & Mental Disease*, 142:101–16.

Deikman, A. J. 1971. Bimodal consciousness. *Archives of General Psychiatry*, 25: 481–89.

Deikman, A. J. 1990. *The Wrong Way Home: Uncovering the Patterns of Cult Behavior in American Society.* Beacon Press.

Delmonte, M. M. 1979. Pilot study on conditional relaxation during simulation meditation. *Psychological Reports,* 45:44–49.

Delmonte, M. M. 1981. Expectation and meditation. *Psychological Reports,* 49: 699–709.

Delmonte, M. M. 1984. Electrocortical activity and related phenomena associated with meditation practice: A literature review. *International Journal of Neuroscience,* 24:217–31.

Delmonte, M. M. 1984. Factors influencing the regularity of meditation practice in a clinical population. *British Journal of Medical Psychology,* 57:275–78.

DeMille, R. 1976. *Castaneda's Journey: The Power & The Allegory.* Capra Press.

Denkowski, K. M. et al. 1984. Predictors of success in the EMG biofeedback training of hyperactive male children. *Biofeedback & Self-Regulation,* 9:253–64.

Department of Health and Human Services. 1980. *Promoting Health/Preventing Disease: Objectives for the Nation.* (Fall). Washington, DC: U.S. Government Printing Office.

Descartes, R. 1637. Discours de la méthode. In N. K. Smith (trans.), 1958, *Descartes' Philosophical Writings.* Random House.

Desoille, R. 1961. *Theorie et pratique du reve eveille dirige.* Mont Blanc.

Deutsch, E. 1969. *Advaita Vedanta, a Philosophical Reconstruction.* University of Hawaii Press.

Deutsch, H. 1951. *Psycho-Analysis of the Neuroses.* Hogarth.

Devane, G., et al. 1985. Opioid peptides in pseudocyesis. *Obstetrics & Gynecology,* 65:187.

Devereux, G. 1953. *Psychoanalysis and the Occult,* chaps. 3–8. International Universities Press.

Dhanaraj, V., & M. Singh. 1977. Reduction in metabolic rate during the practice of the Transcendental Meditation technique. In D. W. Orme-Johnson & J. T. Farrow (eds.), *Scientific Research on the Transcendental Meditation Program: Collected Papers.* Vol. 1. MERU Press.

Diamond, H. 1959. Hypnosis in children: The complete cure of forty cases of asthma. *American Journal of Clinical Hypnosis,* 3:123–29.

Diamond, M. 1974. Modification of hypnotizability: A review. *Psychological Bulletin,* 81:180–98.

Dick-Read, G. 1959. *Childbirth Without Fear.* Harper & Row.

DiCara, L., & N. Miller. 1968. Changes in heart rate instrumentally learned by curarized rats as avoidance responses. *Journal of Comparative & Physiological Psychology,* 65:8–12.

DiCara, L., & N. Miller. 1969. Transfer of instrumentally learned heart-rate changes from curarized to noncurarized state: Implication for a mediational hypothesis. *Journal of Comparative & Physiological Psychology,* 68:159–62.

Dikel, W., & K. Olness. 1980. Self-hypnosis, biofeedback, and voluntary peripheral temperature control in children. *Pediatrics,* 66: 335–40.

Dillbeck, M. C. 1977a. The effect of the Transcendental Meditation technique on anxiety level. *Journal of Clinical Psychology,* 33:1076–78.

Dillbeck, M. C. 1977b. The effects of the TM technique on visual perception and verbal problem solving. *Dissertation Abstracts International,* 37(10-B), 319–20.

Dillbeck, M. C. 1982. Meditation and flexibility of visual perception and verbal problem solving. *Memory and Cognition,* 10:207–15.

Dillbeck, M., P. Assimakis, & D. Raimondi. 1986. Longitudinal effects of the Transcendental Meditation and TM-sidhi program on cognitive ability and cognitive style. *Perceptual & Motor Skills,* 62/3:731–38.

Dillbeck, M., & E. Bronson. 1981. Short-term longitudinal effects of the Transcendental Meditation technique on EEG power and coherence. *International Journal of Neuroscience,* 14:147–51.

Dillbeck, M., D. Orme-Johnson, & R. K. Wallace. 1981. Frontal EEG coherence, H-reflect recovery, concept learning and the TM-sidhi program. *International Journal of Neuroscience,* 15:151–57.

Dillon, J. 1977. *The Middle Platonists.* Cornell University Press.

Dimond, E. G., et al. 1958. Evaluation of internal mammary artery ligation and sham procedure in angina pectoris. *Circulation,* 18:712–13.

Dingwall, E. J. 1962. *Some Human Oddities.* University Books.

Dishman, R. K. 1985. Medical psychology in exercise and sports. *Medical Clinics of North America,* 69:123–43.

Dishman, R. 1988. *Exercise Adherence.* Human Kinetics.

Dixon, N. 1981. *Preconscious Processing.* Wiley.

Dobereiner, P. 1976. The day Joe Ezar called his shots for a remarkable 64. *Golf Digest* (Nov.), 27:67.

Dobzhansky, T., & F. Ayala. 1977. *Humankind: A Product of Evolutionary Transcendence.* Witwaterstrand University Press.

Dohrmann, R., & D. Laskin. 1978. An evaluation of electromyographic biofeedback in the treatment of myofascial pain-dysfunction syndrome. *Journal of the American Dental Association,* 96:656–62.

Dolan, A., & A. Sheikh. 1976. Eidetics: A visual approach to psychotherapy. *Psychologia,* 19:210–19.

Dostalek, C., J. Faber, E. Krasa et al. 1979. Yoga meditation effect on the EEG and EMG activity. *Activitas Nervos Superior* (Praha), 21:41.

Doswald, D., & K. Kreibich. 1906. Zur Frage der posthypnotischen Hautphanomene. *Monatshefte Praktik Dermatologie,* 43:634–40. In A. Sheikh (ed.), 1984, *Imagination and Healing.* Baywood.

Douglas, M. 1970. *Natural Symbols.* Pantheon.

Douglas, M. 1971. Do dogs laugh? A cross-cultural approach to body symbolism. *Journal of Psychosomatic Research,* 15:387–90.

Doust, D. 1973. Opening the mystical door of perception in sport. *Sunday London Times* (Nov. 4).

Dowell, R. 1983. Cardiac adaptations to exercise. In R. Terjung (ed.), *Exercise and Sport Sciences Reviews*. Vol. 11. American College of Sports Medicine. Franklin Institute Press.

Dowling, St. J. 1984. Lourdes cures and their medical assessment. *Journal of the Royal Society of Medicine*, 77:635–36.

Draeger, D. 1973a. *Classical Budo*. Weatherhill.

Draeger, D. 1973b. *Classical Bujutsu*. Weatherhill.

Draeger, D. 1974. *Modern Bujutsu and Budo*. Weatherhill.

Draeger, D., & R. Smith. 1980. *Comprehensive Asian Fighting Arts*. Kodansha International Ltd.

Dreher, E. R. 1974. The effects of hatha yoga and judo on personality and self-concept profiles on college men and women. *Dissertation Abstracts International*, 34 (8-A), 4833–34.

Droscher, V. 1969. *The Magic of the Senses*. E. P. Dutton.

Duchesne, L. 1933. *Early History of the Christian Church*. John Murray.

Dudek, S. Z. 1967. Suggestion and play therapy in the cure of warts in children: A pilot study. *Journal of Nervous & Mental Disease*, 145:37–42.

Dumezil, G. 1930. *Legendes sur les Nartes. Suives de Cinq Notes Mythologiques*. Paris.

Dumezil, G. 1942. *Horace et les Curiaces (Les Mythes Romains)*. Paris.

Dunphy, J. 1950. Some observations on the natural behavior of cancer in man. *New England Journal of Medicine*, 242:167.

Durant, W. 1950. The Age of Faith. In *The Story of Civilization*, Vol. 4. Simon & Schuster.

Durrwell, R. X. 1960. *The Resurrection*. Sheed & Ward.

Duthie, R. B., L. Hope, & D. G. Barker. 1978. Selected personality traits of martial artists as measured by the adjective checklist. *Perceptual & Motor Skills*, 47: 71–76.

Duval, P. 1971. Exploratory experiments with ants. *Journal of Parapsychology*, pp. 35–58 (Abstract).

Early, L., & J. Lifschutz. 1974. A case of stigmata. *Archives of General Psychiatry* (Feb.), 30:197–200.

Eason, R., & R. Sadler. 1977. Relationship between voluntary control of alpha activity level through auditory feedback and degree of eye convergence. *Bulletin of the Psychonomic Society*, 9:21–24.

Eberlin, P., & T. Mulholland. 1976. Bilateral differences in parietal-occipital EEG induced by contingent visual feedback. *Psychophysiology*, 13:212–18.

Echenhofer, F., & M. Coombs. 1987. A brief review of research and controversies in EEG biofeedback and meditation. *Journal of Transpersonal Psychology*, 19: 161–71.

Eddy, M. B. 1971. *Science and Health with Key to the Scriptures*. First Church of Christ Scientist.

Edelman, R. I. 1970. Effects of differential afferent feedback on instrumental GSR conditioning. *Journal of Psychology*, 74:3–14.

Edelstein, E., & L. Edelstein. 1945. *Asclepius: A Collection and Interpretation of the Testimonies.* Vol. 1. Johns Hopkins Press.

Edge, H. 1980. The effect of laying on of hands on an enzyme: An attempted replication. In W. Roll et al. (eds.), *Research in Parapsychology 1979.* Scarecrow Press.

Edwards, G. 1963. Duration of post-hypnotic effect. *British Journal of Psychiatry,* 109:259–66.

Efron, D. 1941. *Gesture and Environment.* King's Crown Press.

Ehrenwald, J. 1940. Psychopathological aspects of telepathy. *Proceedings of the Society for Psychical Research,* 46:224–44.

Ehrenwald, J. 1942. Telepathy in dreams. *British Journal of Medical Psychology,* 19: 313–23.

Ehrenwald, J. 1944. Telepathy in the psychoanalytic situation. *British Journal of Medical Psychology,* 20:51–62.

Ehrenwald, J. 1948a. Neurobiological aspects of telepathy. *Journal of the American Society of Psychical Research,* 42:132–41.

Ehrenwald, J. 1948b. *Telepathy and Medical Psychology.* W. W. Norton.

Ehrenwald, J. 1949. Quest for psychics and psychical phenomena in psychiatric studies of personality. *Psychiatric Quarterly,* 23:236–47.

Ehrenwald, J. 1950a. Presumptively telepathic incidents during analysis. *Psychiatric Quarterly,* 24:726–43.

Ehrenwald, J. 1950b. Psychotherapy and the telepathy hypothesis. *American Journal of Psychotherapy,* 4:51–79.

Ehrenwald, J. 1956. Telepathy and the child-parent relationship. *Journal of the American Society of Psychical Research,* 48:43–55.

Ehrenwald, J. 1956. Telepathy: Concepts, criteria, and consequences. *Psychiatric Quarterly,* 30:425–49.

Ehrenwald, J. 1975. *New Dimensions of Deep Analysis.* Arno.

Eisenbud, J. 1954. Behavioral correspondences to normally unpredictable future events. *Psychoanalytic Quarterly,* 23:205–33, 355–87.

Eisenbud, J. 1955. On the use of the PSI hypothesis in psychoanalysis. *International Journal of Psychoanalysis,* 36: 1–5.

Eisenbud, J. 1967. *The World of Ted Serios.* Morrow.

Eisenbud, J. 1970a. Light and the Serios images. *Journal of the Society for Psychical Research,* 45:424–27.

Eisenbud, J. 1970b. *PSI and Psychoanalysis.* Grune & Stratton.

Eisenbud, J. 1972a. Gedanken zur Psychophotographie. *Zeitschrift fur Parapsychologie und Grenzgebiete der Psychologie,* 14:1–11.

Eisenbud, 1972b. The Serios "blackies" and related phenomena. *Journal of the American Society for Psychical Research,* 66:180–92.

Eisenbud, J. 1977. Paranormal photography. In B. Wolman (ed.), *Handbook of Parapsychology.* Van Nostrand Reinhold.

Eisenbud, J. 1982. *Paranormal Foreknowledge: Problems and Perplexities.* Human Sciences Press.

Eisenbud, J. 1983. *Parapsychology and the Unconscious.* North Atlantic Books.

Eisenbud, J., et al. 1967. Some unusual data from a session with Ted Serios. *Journal of the American Society for Psychical Research,* 61:241–53.

Eisenbud, J., et al. 1968. Two experiments with Ted Serios. *Journal of the American Society for Psychical Research,* 62:309–20.

Eisenbud, J., et al. 1970a. An archeological tour de force with Ted Serios. *Journal of the American Society for Psychical Research,* 64:40–52.

Eisenbud, J., et al. 1970b. Two camera and television experiments with Ted Serios. *Journal of the American Society for Psychical Research,* 64:261–76.

Ekblom, B. 1969. Effect of physical training on oxygen transport system in man. *Acta Physiol. Scand.,* 77 (Suppl. 328):1–45.

Ekelund, L. G. 1988. Physical fitness as a predictor of cardiovascular mortality in asymptomatic North American men. *New England Journal of Medicine,* 319: 1379.

Eldridge, N., & S. Gould. 1972. Punctuated equilibria: An alternative to phyletic gradualism. In T. J. Schopf (ed.), *Models of Paleobiology.* Freeman & Coopers.

Eliade, M. 1962. *The Forge and the Crucible.* University of Chicago Press.

Eliade, M. 1964. *Shamanism.* Princeton University Press, Bollingen Series.

Ellenberger, H. 1970. *The Discovery of the Unconscious.* Basic Books.

Elliott, H. 1961. *The Herb Elliott Story.* Thomas Nelson.

Elliotson, J. 1843. *Numerous Cases of Surgical Operations Without Pain in the Mesmeric State.* Lea & Blanchard.

Ellis, A. 1973. *Humanistic Psychotherapy: The Rational-Emotive Approach.* McGraw-Hill.

Ellis, A. 1984. The Place of Meditation in Cognitive Behavior Therapy and Rational Emotive Therapy. In D. Shapiro & R. Walsh (eds.). *Meditation: Classic Contemporary Perspectives.* Aldine.

Elson, B., P. Hauri, & D. Cunis. 1977. Physiological changes in yoga meditation. *Psychophysiology,* 14:55–57.

Encyclopedia of Philosophy. 1967. Macmillan.

Engel, B., & W. Baile. 1989. Behavioral applications in the treatment of patients with cardiovascular disorders. In J. V. Basmajian (ed.), *Biofeedback—Principles and Practice for Clinicians.* Williams & Wilkins.

Engel, B., & E. Bleecker. 1974. Application of operant conditioning techniques to the control of the cardiac arrhythmias. In P. Obrist et al. (eds.), *Cardiovascular Psychophysiology: Current Issues in Response Mechanisms, Biofeedback and Methodology.* Aldine.

Engel, B., & R. Chism. 1967. Operant conditioning of heart rate speeding. *Psychophysiology,* 3:418–26.

Engel, B., & S. Hansen. 1966. Operant conditioning of heart rate slowing. *Psychophysiology,* 3:176–87.

Engel, B., et al. 1974. Operant conditioning of rectosphincteric responses in the treatment of fecal incontinence. *New England Journal of Medicine,* 290:646–49.

Engel, B., et al. 1983. Behavioral treatment of high blood pressure. III. Follow-up results and treatment recommendations. *Psychosomatic Medicine*, 45:23–29.

Engler, J. 1986. Therapeutic aims in psychotherapy and meditation. In K. Wilber, J. Engler, & D. Brown (eds.), *Transformations of Consciousness*. Shambhala, New Science Library.

Erickson, M. H. 1943. Hypnotic investigation of psychosomatic phenomena: Psychosomatic interrelationships studied by experimental hypnosis. *Psychosomatic Medicine*, 5:51–58.

Erickson, M. H. 1959. Hypnosis in painful terminal illnesses. *American Journal of Clinical Hypnosis*, 1:117–22.

Erickson, M. H. 1960. Breast development possibly influenced by hypnosis: Two instances and the psychotherapeutic results. *American Journal of Clinical Hypnosis*, 2:157–59.

Erickson, M. H. 1983. *Seminars, Workshops, and Lectures*, ed. E. Rossi, M. Regan, & F. Sharp. Irvington.

Erickson, M., & E. Rossi. 1974–1980. Varieties of hypnotic amnesia. In E. Rossi (ed.), *The Collected Papers of Milton H. Erickson on Hypnosis III. Hypnotic Investigation of Psychodynamic Processes*. Irvington.

Esdaile, J. 1846. *Mesmerism in India and its Practical Application in Surgery and Medicine*. Longman, Brown, Green, and Longmans. Reprinted in 1975, Arno Press.

Esdaile, J. 1849. Testimony to the reality of clairvoyance. In a letter to Dr. Elliotson in *The Zoist*, 7:213–23.

Esdaile, J. 1852. *Natural and Mesmeric Clairvoyance*. Hippolyte, Bailliere.

Evans, A. 1921. On a Minoan bronze group of a galloping bull and acrobatic figure from Crete. *Journal of Hellenic Studies*, 41:247–59.

Evans, D., & T. Seely. 1984. Pseudocyesis in the male. *Journal of Nervous & Mental Disease*, 172:38.

Evans, F. 1963. The structure of hypnosis: A factor-analytic investigation. Doctoral thesis, University of Sydney, Australia. In E. R. Hilgard. 1965. Hypnosis. *Annual Review of Psychology*, 16:157–80.

Evans, F. 1984. Unravelling placebo effects. *Advances*, 1:16.

Evans, M., & G. Paul. 1970. Effects of hypnotically suggested analgesia on physiological and subjective responses to cold stress. *Journal of Consulting & Clinical Psychology*, 35:362–71.

Evans, W. 1951. Simulated pregnancy in a male. *Psychoanalytic Quarterly*, 20:165–78.

Evans-Wentz, W. 1949. *Tibetan Book of the Dead*. Oxford University Press.

Everson, T., & W. Cole. 1966. *Spontaneous Regression of Cancer*. Saunders.

Ewin, D. 1978. Clinical use of hypnosis for attenuation of burn depth. Cited in A. Sheikh. 1984. *Imagination and Healing*. Baywood.

Ewin, D. 1979. Hypnosis in burn therapy. In G. Burrows et al. (eds.), *Hypnosis 1979*. Elsevier.

Extra, J. 1971. Telepathie bij de rat. *Tijdschrift voor Parapsychologie*, 39:18–31.

Eysenck, H. 1952. The effects of psychotherapy: An evaluation. *Journal of Consulting Psychology*, 16:319–24.

Faber, P., G. Suayman, & S. Touyz. 1978. Meditation and archetypal content on nocturnal dreams. *Journal of Analytical Psychology*, 23:1–22.

Fahrion, S. 1983. Presidential address at the meeting of the Biofeedback Society of America, Denver.

Fahsel, K. 1932. *Konnersreuth; Tatsachen und Gedanken*. Cited in Thurston, 1952.

Falch, D., & S. Stromme. 1979. Pulmonary blood volume and interventricular circulation time in physically trained and untrained subjects. *European Journal of Applied Physiology*, 40:211–18.

Farrar, W. 1976. Using electromyographic biofeedback in treating orofacial dyskinesia. *Journal of Prosthetic Dentistry*, 35:384–87.

Farrow, J. 1976. Physiological changes associated with transcendental consciousness, the state of least excitation of consciousness. In D. W. Orme-Johnson & J. T. Farrow (eds.), *Scientific Research on the Transcendental Meditation Program: Collected Papers*. Vol. 1. MERU Press.

Faw, V., et al. 1968. Psychopathological effects of hypnosis. *International Journal of Clinical & Experimental Hypnosis*, 16: 26–37.

Feitis, R. 1978. *Ida Rolf Talks about Rolfing and Physical Reality*. Harper & Row.

Feld, M., et al. 1979. The physics of karate. *Scientific American* (Apr.).

Feldenkrais, M. 1970. *Body and Mature Behavior*. International Universities Press.

Feldenkrais, M. 1972. *Awareness Through Movement*. Harper & Row.

Feldenkrais, M. 1977. *The Case of Nora*. Harper & Row.

Feldenkrais, M. 1985. *The Potent Self*. Harper & Row.

Feldman, B. 1976. A phenomenological and clinical inquiry into deep hypnosis. Doctoral dissertation, University of California, Berkeley.

Feltz, D., & D. Landers. 1983. The effects of mental practice on motor skill learning and performance: A meta-analysis. *Journal of Sport Psychology*, 5:25–26.

Fenichel, O. 1945. *Psychoanalytic Theory of Neurosis*. Routledge & Kegan Paul.

Fentress, D., B. Masek, J. Mehegan et al. 1986. Biofeedback and relaxation-response training in the treatment of pediatric migraine. *Developmental Medicine and Child Neurology*, 28:139–46.

Fenwick, P., S. Donaldson, L. Gillis et al. 1977. Metabolic and EEG changes during Transcendental Meditation: An explanation. *Biological Psychology*, 5:101–18.

Ferenczi, S. 1926. An attempted explanation of hysterical stigmatization. In *Further Contributions of the Theory and Technique of Psychoanalysis*. Woolf.

Festa, G. 1949. *Misteri di Scienza e Luci di Fedi: le Stigmate del Padre Pio da Pietrelcina*. V. Ferri.

Festinger, L., et al. 1956. *When Prochecy Fails*. University of Minnesota Press.

Feuerstein, G. 1989. *Enlightened Sexuality*. Crossing Press.

Fiatarone, M. A., J. Morley, E. Bloom, D. Benton, T. Makinodan, & G. Solomon. 1988. Endogenous opioids and the exercise-induced augmentation of natural killer cell activity. *Journal of Laboratory and Clinical Medicine*, 112:544–52.

Fiatarone, M. A., J. Morley, E. Bloom, D. Benton, G. Solomon, & T. Makinodan. 1989. The effect of exercise on natural killer cell activity in young and old subjects. *Journal of Gerontology*, 44:M37–45.

Finucane, R. 1973. Faith healing in medieval England: Miracles at saints' shrines. *Psychiatry*, 36:341–46.

Firth, R. 1970. Postures and gestures of respect. In J. Pouillon & P. Maranda, *Échanges et Communications; mélanges offerts a Claude Levi-Strauss a l'occasion de son 6oeme anniversaire*. The Hague.

Fischer, R. A. 1971. A cartography of the ecstatic and meditative states. *Science*, 174: 897–904.

Fischer, R. A. 1976. Transformations of consciousness: A cartography II: The perception-meditation continuum. *Confina Psychiatrica*, 19:1–23.

Fisher, J., & E. Kollar. 1980. Investigation of a stigmatic. *Southern Medical Journal* (Nov.), 73:1461–66.

Fisher, S., & S. Cleveland. 1968. Cultural differences in boundary characteristics. In *Body Image and Personality*. Van Nostrand.

FitzGerald, F. 1986. *Cities on a Hill*. Simon & Schuster.

Fitzgerald, L. 1991. Overtraining increases the susceptibility to infection. *International Journal of Sports Medicine*, 12:S5–S8.

Fling, S., A. Thomas, & M. Gallaher. 1981. Participant characteristics and the effects of two types of meditation vs. quiet sitting. *Journal of Clinical Psychology*, 37: 784–90.

Fodor, N. 1964. The riddle of Nijinsky. In *Between Two Worlds*. Parker.

Fodor, N. 1971. *Freud, Jung and Occultism*. University Books.

Folkins, C., & W. Sime. 1981. Physical fitness training and mental health. *American Psychologist*, 36:373–89.

Forte, M., D. Brown, & M. Dysart. 1984–85. Through the looking glass: Phenomenological reports of advanced meditators at visual threshold. *Imagination, Cognition & Personality*, 4:323–28.

Foucault, M. 1970. *The Order of Things: An Archeology of the Human Sciences*. Vintage/Random House.

Fowler, R., & H. Kimmel. 1962. Operant conditioning of the GSR. *Journal of Experimental Psychology*, 63:563–67.

Frank, J. 1973. *Persuasion and Healing: A Comparative Study of Psychotherapy*. Johns Hopkins University Press.

Frankel, F., & R. Misch. 1973. Hypnosis in a case of long-standing psoriasis in a person with character problems. *International Journal of Clinical & Experimental Hypnosis*, 21:121–30.

Franklin, B. 1881 (2d ed.). *The Life of Benjamin Franklin, Written by Himself*. Ed. John Bigelow. Vol. 3. Lippincott.

Frazier, T. W. 1966. Avoidance conditioning of heart rate in humans. *Psychophysiology*, 3:188–202.

Freedman, D., & P. Van Nieuwenhuizen. 1985. The hidden dimensions of spacetime. *Scientific American* (March).

Freemantle, Ann. 1964. *The Protestant Mystics.* Little Brown.

French, A., A. Smid, & E. Ingalls. 1975. Transcendental Meditation, altered reality testing and behavioral change: A case report. *Journal of Nervous & Mental Disease,* 161:55–58.

Fretigny, R., & A. Virel. 1968. *L'imagerie mentale.* Mont-Blanc.

Freud, S. 1924–1950. *Collected Papers. Character & Culture.* Vol. 9. International Psychoanalytical Press.

Freud, S. 1953–1974. *The Standard Edition of the Complete Psychological Works of Sigmund Freud.* Hogarth.

Freud, S. Dreams and Telepathy. Volume 18 of *The Complete Psychological Works.* Hogarth.

Freud, S. Dreams and Occultism. Volume 22 of *The Complete Psychological Works.* Hogarth.

Fried, P., et al. 1951. Pseudocyesis. A psychosomatic study in gynecology. *Journal of the American Medical Association,* 145:1329–35.

Frisch, R., et al. 1985. Lower prevalence of breast cancer and cancers of the reproductive system among former college athletes compared to non-athletes. *British Journal of Cancer,* 52:885–91.

Fromm, E., et al. 1981. The phenomena and characteristics of self-hypnosis. *International Journal of Clinical & Experimental Hypnosis,* 29:189–246.

Frutiger, A. D. 1981. Treatment of penetration phobia through the combined use of systematic desensitization and hypnosis: A case study. *American Journal of Clinical Hypnosis,* 23:269.

Gallup Poll. 1984. Six of ten adults exercise daily. Los Angeles Times Syndicate (May).

Gallup Poll. 1985. Press release, May 14.

Galton, F. 1883. *Inquiries into Human Faculty and Its Development.* Dent.

Garbe, R. 1900. On the voluntary trance of Indian fakirs. *The Monist,* 10:490–500.

Gardner, F., & L. Diamond. 1955. Autoerythrocyte sensitization: A form of purpura producing painful bruising following autosensitization to red blood cells in certain women. *Blood,* 10:675.

Gardner, H. 1985. *Frames of Mind: The Theory of Multiple Intelligences.* Basic Books.

Gardner, R. 1983. Miracles of healing in Anglo-Celtic Northumbria as recorded by the Venerable Bede and his contemporaries: A reappraisal in the light of twentieth century experience. *British Medical Journal,* 287:1927–33.

Garfield, S., & R. Kurtz. 1974. A survey of clinical psychologists: Characteristics, activities and orientation. *Clinical Psychologist,* 28:7–10.

Garfield, S., & R. Kurtz. 1976. Clinical psychologists in the 1970s. *American Psychologist,* 31:1–9.

Gatchel, R. J. 1974. Frequency of feedback and learned heart rate control. *Journal of Experimental Psychology,* 103:274–83.

Gauld, A. 1982. *Mediumship and Survival.* Heinemann.

Gavalas, R. J. 1967. Operant reinforcement of an autonomic response: Two studies. *Journal of the Experimental Analysis of Behavior,* 10:119–30.

Gavalas, R. 1968. Operant reinforcement of a skeletally mediated autonomic response: Uncoupling of the two responses. *Psychonomic Science,* 11:195–196.

Gaylord, C., et al. 1989. The effects of the Transcendental Meditation technique and progressive muscle relaxation on EEG coherence, stress reactivity and mental health in black adults. *International Journal of Neuroscience,* 46:77–86.

Gaylord, H., & G. Clowes. 1903. On spontaneous cure of cancer. *Surgery, Gynecology and Obstetrics,* 2:633.

Gazzaniga, M. 1985. *The Social Brain: Discovering the Networks of the Mind.* Basic Books.

Gebser, J. 1985. *The Ever-Present Origin.* Ohio University Press.

Geddes, F. 1981. Healing training in the church. Doctoral thesis, San Francisco Theological Seminary.

Gedo, J. 1981. *Advances in Clinical Psychoanalysis.* International Universities Press.

Geley, G. 1927. *Clairvoyance and Materialization.* Fisher & Unwin.

Gellhorn, E. 1967. *Principles of Autonomic-Somatic Interactions: Physiological Basis and Psychological and Clinical Implications.* University of Minnesota Press.

Gellhorn, E. 1970. The Emotions and the Ergotropic and Trophotropic Systems. *Psychol. Forsch.,* 34:48–94.

Gellhorn, E., & W. Kiely. 1972. Mystical states of consciousness: Neurophysiological and clinical aspects. *Journal of Nervous & Mental Disease,* 154:399–405.

Gendlin, E. T. 1978. *Focusing.* Everest House.

Gendlin, E., & L. Olsen. 1970. The use of imagery in experiential focusing. *Psychotherapy: Theory, Research and Practice,* 7:221–23.

George, A., ed. 1932. *The Complete Poetical Works of Wordsworth.* Houghton Mifflin.

Gerard, R. 1964. *Psychosynthesis: A Psychotherapy for the Whole Man.* New York: Psychosynthesis Research Foundation.

Gerber, R. 1988. From magnetic passes to spiritual healing: A multidimensional model of healing energies. In *Vibrational Medicine,* Bear & Co.

Gerhardsson, M., et al. 1986. Sedentary jobs and colon cancer. *American Journal of Epidemiology,* 123:775.

Gerlich, F. 1929. *Die Stigmatisierte Therese Neumann.*

Germano di S. Stanislao. *Life of Gemma Galgani.* Trans. A. M. O'Sullivan. Sand & Co.

Ghiselin, B. 1980. *The Creative Process.* New American Library.

Ghista, D., A. Mukherji, D. Nandagopal et al. 1976. Physiological characteristics of the "meditative state" during intuitional practice (the Ananda Marga system of meditation) and its therapeutic value. *Medical & Biological Engineering,* 14:209–13.

Gibbons, D. 1974. Hyperempiria, a new "altered state of consciousness" induced by suggestion. *Perceptual & Motor Skills,* 39:47–53.

Gibbons, D. 1976. Hypnotic vs. hyperempiric induction procedures: An experimental comparison. *Perceptual & Motor Skills,* 42:834.

Gigliozzi, G. 1965. *Padre Pio*. Phaedri Publishing.

Gilbey, J. 1963. *Secret Fighting Arts of the World*. Tuttle.

Gilson, A., & W. Mills. 1941. Activities of single motor units in man during slight voluntary efforts. *American Journal of Physiology*, 133:658–69.

Gindler, E. 1986–87. Gymnastik for everyone. *Somatics*, vol. 6, no. 1 (Autumn/Winter).

Girodo, M. 1974. Yoga meditation and flooding in the treatment of anxiety neurosis. *Journal of Behavior Therapy & Experimental Psychiatry*, 5:157–60.

Gish, D. 1978. *Evolution? The Fossils Say No!* Creation-Lite Publishers.

Gitananda, Swami. 1973. An Evaluation of Siddhis and Riddhis. Paper prepared for the Second International Yoga Festival, London (Aug.).

Glasgow, M., et al. 1982. Behavioral treatment of high bood pressure. II. Acute and sustained effects of relaxation and systolic blood pressure biofeedback. *Psychosomatic Medicine*, 44:155–70.

Gleick, J. 1988. *Chaos*. Viking.

Gluck, J. 1962. *Zen Combat*. Ballantine Books.

Glueck, B., & C. Stroebel. 1975. Biofeedback as meditation in the treatment of psychiatric illnesses. *Comprehensive Psychiatry*, 16:303–21.

Glueck B., & C. Stroebel. 1978. Meditation in the treatment of psychiatric illness. In A. Sugurman & R. Tarter (eds.), *Expanding Dimensions of Consciousness*. Springer.

Glueck, B., & C. Stroebel. 1984. Psychophysiological correlates of meditation: EEG changes during meditation. In D. H. Shapiro and R. Walsh (eds.), *Meditation: Classic & Contemporary Perspectives*. Aldine.

Goldberg, A., et al. 1975. Mechanism of work-induced hypertrophy of skeletal muscle. *Medicine & Science in Sports & Exercise*, 7:185.

Goldfriend, M., et al. 1974. Systematic rational restructuring as a self-control technique. *Behavior Therapy*, 5:247–54.

Goldstein, A., & E. Hilgard. 1975. Lack of influence of the morphine antagonist naloxone on hypnotic analgesia. *Proceedings of the National Academy of Sciences*, 72:2041–43.

Goldstein, J. 1982. *The Experience of Insight*. Shambhala.

Goleman, D. 1976a. Meditation and Consciousness: An Asian approach to mental health. *American Journal of Psychotherapy*, 30:41–54.

Goleman, D. 1976b. Meditation helps break the stress spiral. *Psychology Today* (February).

Goleman, D. 1977. *The Varieties of the Meditative Experience*. E. P. Dutton.

Goleman, D. 1978–79. A taxonomy of meditation-specific altered states. *Journal of Altered States of Consciousness*, 4:203–13.

Goleman, D. 1988. Probing the enigma of multiple personality. *New York Times* (June 28), pp. C1, C13.

Goleman, D., & G. Schwartz. 1976. Meditation as an intervention in stress reactivity. *Journal of Consulting & Clinical Psychology*, 44:456–66.

Gollnick, P., & B. Saltin. 1973. Biochemical adaptation to exercise: Anaerobic metabolism. In J. Wilmore (ed.), *Exercise and Sport Sciences Reviews.* Vol. 1. Academic.

Good, J. M. 1823. *Physiological System of Nosology with Corrected and Simplified Nomenclature.* Wells & Lilly.

Goodman, N. 1969. Triiodothyronine and placebo in the treatment of obesity. *Medical Annals of the District of Columbia,* 38:658–76.

Goodrich, J. 1974. Psychic healing—A pilot study. Doctoral dissertation, Union Graduate School, Cincinnati.

Gopi Krishna. 1971. *Kundalini—The Evolutionary Energy in Man.* Shambhala.

Gottfredson, D. 1973. Hypnosis as an anesthetic in dentistry. *Dissertation Abstracts International,* 33:(7-B),3303.

Goudge, T. A. 1967a. Emergent evolutionism. In *The Encyclopedia of Philosophy.* Macmillan.

Goudge, T. A. 1967b. Henri Bergson. In *The Encyclopedia of Philosophy.* Macmillan.

Gould, S., & D. Tissler. 1984. The use of hypnosis in the treatment of herpes simplex II. *American Journal of Clinical Hypnosis,* 26:171–74.

Gould, S. J. 1977. *Ontogeny and Phylogeny.* Harvard University Press.

Gould, S. J. 1983. Evolution as fact and theory. In *Hen's Teeth and Horse's Toes.* Norton.

Govinda, Lama. 1966. *The Way of the White Clouds.* Hutchinson.

Goyeche, J., T. Chilhara, & H. Shimizu. 1972. Two concentration methods: A preliminary comparison. *Psychologia,* 15:110–11.

Grabowska, M. 1971. The effect of hypnosis and hypnotic suggestion on the blood flow in the extremities. *Polish Medical Journal,* 10:1044–51.

Gracely, R., et al. 1984. Placebo and naloxone can alter post-surgical pain by separate mechanisms. *Nature,* 306:264–65.

Grad, B. 1963. A telekinetic effect on plant growth. *International Journal of Parapsychology,* 5:117–33.

Grad, B. 1964. A telekinetic effect on plant growth. II. Experiments involving treatment of saline in stoppered bottles. *International Journal of Parapsychology,* 6: 473–98.

Grad, B. 1965a. Some biological effects of the "laying-on of hands": A review of experiments with animals and plants. *Journal of the American Society for Psychical Research,* 59:95–127.

Grad, B. 1965b. A telekinetic effect on yeast activity. *Journal of Parapsychology,* 29: 285–86.

Grad, B. 1977. Laboratory evidence of "laying on of hands." In N. Regush (ed.), *Frontiers of Healing.* Avon.

Grad, B., et al. 1961. The influence of an unorthodox method of treatment on wound healing in mice. *International Journal of Parapsychology,* 3:5–24.

Graff, N., & R. Wallerstein. 1954. Unusual wheal reaction in a tattoo. *Psychosomatic Medicine,* 16:513–14.

Graham, C. 1975. Hypnosis and attention. In L. E. Unestahl (ed.), *Hypnosis in the Seventies*. Veje Forlag.

Graham, C., & F. Evans. 1977. Hypnotizability and the deployment of waking attention. *Journal of Abnormal Psychology*, 86:633.

Graham, C., & H. Leibowitz. 1972. The effect of suggestion on visual acuity. *International Journal of Clinical & Experimental Hypnosis*, 20:169–86.

Graham, K., & K. Pernicano. 1979. Laterality, hypnosis, and the autokinetic effect. *American Journal of Clinical Hypnosis*, 22:79–84.

Gray, E. 1971. Conscious control of motor units in a tonic muscle: The effect of motor unit training. *American Journal of Physical Medicine*, 50:34–40.

Greatrakes, V. 1666. *A Brief Account of Mr. Valentine Greatrakes and Divers of the Strange Cures by Him Lately Performed*. London.

Greeley, A. 1987. Mysticism goes mainstream. *American Health* (Jan./Feb.).

Greeley, A., & W. McCready. 1975. Are we a nation of mystics? *New York Times Magazine* (Jan. 26).

Green, C. 1968. *Out-of-the Body Experiences*. Ballantine.

Green, E., & A. Green. 1977. *Beyond Biofeedback*. Dell.

Green, E., & A. Green. 1986. Biofeedback and states of consciousness. In B. Wolman & M. Ullman (eds.), *Handbook of States of Consciousness*. Van Nostrand Reinhold.

Green, E., & A. Green. 1989. General and specific applications of thermal biofeedback. In J. V. Basmajian (ed.), *Biofeedback—Principles and Practice for Clinicians*. Williams & Wilkins.

Green, E., et al. 1969. Feedback technique for deep relaxation. *Psychophysiology*, 6:371–77.

Green, E., et al. 1970. Voluntary control of internal states: Psychological and physiological. *Journal of Transpersonal Psychology*, 2:1–26.

Greene, R., & J. Reyher. 1972. Pain tolerance in hypnotic analgesic and imagination states. *Journal of Abnormal Psychology*, 79:29–38.

Grevert, P., & A. Goldstein. 1985. Placebo analgesia, naloxone, and the role of endogenous opioids. In L. White et al. (eds.), *Placebo: Theory, Research & Mechanisms*. Guilford Press.

Griffin, D. R. 1989. *God & Religion in the Postmodern World*. State University of New York Press.

Griffin, D. R. & H. Smith. 1989. *Primordial Truth and Post-Modern Theology*. State University of New York Press.

Griffin, S. 1978. *Woman and Nature*. Harper & Row.

Griggs, S. T. 1976. A preliminary study into the effect of TM on empathy. Doctoral dissertation, United States International University.

Grof, S. 1975. *Realms of the Human Unconscious*. Viking.

Gross, M. 1984. Hypnosis in the therapy of anorexia nervosa. *American Journal of Clinical Hypnosis*, 26:175.

Gruber, E. R. 1979a. Conformance behavior involving animal and human subjects. *European Journal of Parapsychology*, 3:36–50.

Gruber, E. R. 1979b. A Study of Conformance Behavior Involving Rats and Mice. Paper presented at the meeting of the Society for Psychical Research, Edinburgh, Scotland.

Grunbaum, A. 1985. Explication and implications of the placebo concept. In L. White et al. (eds.), *Placebo: Theory, Research & Mechanisms.* Guilford Press.

Guardini, R. 1954. "The Spiritual Body." *The Last Things.* London: Burns & Oates.

Gudjonsson & Spiro. 1978. Response to placebos in ulcer disease. *American Journal of Medicine,* 65:399–402.

Guenon, R. 1942. *The Crisis of the Modern World.* Luzac.

Guenon, R. 1953. *The Reign of Quantity.* Luzac.

Gur, R., & R. E. Gur. 1974. Handedness, sex and eyedness as moderating variables in the relation between hypnotic susceptibility and functional brain asymmetry. *Journal of Abnormal Psychology,* 83:635–43.

Guralnick, M. J., & D. Mott. 1976. Biofeedback training with a learning disabled child. *Perceptual & Motor Skills,* 42:27–30.

Gurney, E. 1887. Peculiarities of certain post-hypnotic states. *Proceedings of the Society for Psychical Research,* 4:268–323.

Gurney, E. 1888. Recent experiments in hypnotism. *Proceedings of the Society for Psychical Research,* 5:3–17.

Gurney, E., F. Myers, & F. Podmore. 1886 (1970 reprint). *Phantasms of the Living.* Vols. 1 & 2. Scholars' Facsimiles & Reprints.

Gurney, E., & H. Sidgwick. 1883. Third report of the Committee on Mesmerism. *Proceedings of the Society for Psychical Research,* 2:12–19. In R. L. Van De Castle, 1969, The facilitation of ESP through hypnosis. *American Journal of Clinical Hypnosis,* 12:37–56.

Hadfield, J. A. 1917. The influence of hypnotic suggestion on inflammatory conditions. *Lancet,* 2:678–79. Cited in A. Sheikh, 1984, *Imagination and Healing.* Baywood.

Hadfield, J. A. 1920. The influence of suggestion on body temperature. *Lancet,* 2: 68–69.

Hadfield, J. A. 1923. *The Psychology of Power.* Macmillan.

Hadley, E. 1930. Axillary 'menstruation' in a male. *American Journal of Psychiatry,* 9:1101–11.

Hafner, R. 1982. Psychological treatment of essential hypertension: A controlled comparison of meditation and meditation plus biofeedback. *Biofeedback and Self-Regulation,* 7:305–16.

Hagberg, J., et al. 1981. Effect of exercise training on the blood pressure and hemodynamics of adolescent hypertensives. *American Journal of Cardiology,* 52: 763.

Hahn, R. 1985. A sociocultural model of illness & healing. In L. White et al. (eds.), *Placebo: Theory, Research & Mechanisms.* Guilford Press.

Hall, E. 1959. *The Silent Language.* Fawcett Publication.

Hammond, D. C., et al. 1983. Hypnotic analgesia with burns: An initial study. *American Journal of Clinical Hypnosis*, 26:56.

Handley, W. 1909. The natural cure of cancer. *British Medical Journal*, 1:582.

Hanna, T. 1983. *The Body of Life*. Knopf.

Hansen, E. 1971. *The Attalids of Pergamum*. Cornell University Press.

Happich, C. 1932. Das Bildbewusstsein als Ansatzstelle psychischer Behandling. *Zbl. Psychotherap.*, 5:663–67.

Haraldsson, E., & T. Thorsteinsson. 1973. Psychokinetic effects on yeast: An exploratory experiment. In W. G. Roll et al. (eds.), *Research in Parapsychology 1972*. Scarecrow Press.

Hardy, E. R. 1931a. *Christian Egypt: Church and People*. New York.

Hardy, E. R. 1931b. *Large Estates of Byzantine Egypt*. Oxford University Press.

Hardyck, C., & L. Petrinovich. 1969. Treatment of subvocal speech during reading. *Journal of Reading*, 12:361–68, 419–22.

Hardyck, C., et al. 1966. Feedback of speech muscle activity during silent reading: Rapid extinction. *Science*, 154:1467–68.

Harner, Michael J. 1972. *The Jívaro: People of the Sacred Waterfalls*. Doubleday/Natural History Press.

Harner, Michael J. 1973. *Hallucinogens and Shamanism*. Oxford University Press.

Harner, Michael J. 1980. *The Way of the Shaman*. Harper & Row.

Harris, H. 1964. *Greek Athletes and Athletics*. Hutchinson.

Harrison, T. 1981. *The Marks of the Cross*. Darton, Longman & Todd.

Harrison, V., & O. Mortensen. 1962. Identification and voluntary control of single motor unit activity in the tibialis anterior muscle. *Anatomical Record*, 144:109–16.

Hartmann, F. 1896. *The Life of Phlippus Theophrastus Bombast of Hohenheim: Known by the Name of Paracelsus and the Substance of His Teachings*. Kegan Paul.

Hartshorne, C. 1950. Panpsychism. In Vergilius Ferm (ed.), *A History of Philosophical Systems*. The Philosophical Library.

Hartshorne, C. 1970. *Creative Synthesis and Philosophic Method*. Open Court.

Haskins, J. 1975. *Doctor J*. Doubleday.

Hatch, J., et al., eds. 1987. *Biofeedback Studies in Clinical Efficacy*. Plenum Press.

Haxthausen, H. 1936. The pathogenesis of hysterical skin-affections. *British Journal of Dermatology and Syphilology*, 48:563–67.

Haynes, R. 1977. Faith healing and psychic healing: Are they the same? *Parapsychology Review* (July-Aug.).

Haynes, S., D. Mosley, & W. McGowan. 1975. Relaxation training and biofeedback in the reduction of frontalis muscle tension. *Psychophysiology*, 12:547–52.

Healy, D. L., et al. 1983. *Science*, 222:1353–55.

Heaton, D., & D. Orme-Johnson. 1977. Transcendental Meditation program and academic achievement. In D. W. Orme-Johnson & J. T. Farrow (eds.), *Scientific Research on the Transcendental Meditation Program: Collected Papers*. Vol. 1. MERU Press.

Hebert, J., & D. Lehmann. 1977. Theta bursts: An EEG pattern in normal subjects practicing the Transcendental Meditation technique. *Electroencephalography & Clinical Neurophysiology*, 42:397–405.

Heckler, R. S., ed. 1985. *Aikido and the New Warrior*. North Atlantic Books.

Hefferline, R. F. 1958. The role of proprioception in the control of behavior. *Transactions of the New York Academy of Sciences*, 20:739–64.

Heide, F., & T. Borkovec. 1983. Relaxation-induced anxiety: Paradoxical anxiety enhancement due to relaxation training. *Journal of Consulting & Clinical Psychology*, 51:171–82.

Heil, J. D. 1983. Visual imagery change during meditation training. *Dissertation Abstracts International*, 43:7:2338.

Hengstenberg, E. 1985. *The Charlotte Selver Foundation Bulletin*. Caldwell, NJ. (Summer), 12:12.

Henry, L. 1985. Preoperative suggestion reduces blood loss. *Human Aspects of Anesthesia* (Jan./Feb).

Herink, R. 1980. *The Psychotherapy Handbook*. New American Library.

Hermansen, L. 1969. Anaerobic energy release. *Medicine & Science in Sports & Exercise*, 1:32.

Herrigel, E. 1971. *Zen in the Art of Archery*. Vintage Books (Random House).

Hertz, R. 1909. The pre-eminence of the right hand: A study in religious polarity. In R. & C. Needham, trans., 1960, *Death and the Right Hand*. Cohen & West.

Hess, W. R. 1925. *On the Relations Between Psychic and Vegetative Functions*. Schwabe.

Hess, W. R. 1926. Funktionsgesetze des vegetativen Nervensystems. *Klin. Wschr.*, pp. 1353–54.

Hess, W. R. 1953. *Diencephalon. Autonomic and Extrapyramidal Functions*. Grune & Stratton.

Hess, W. R. 1954. The diencephalic sleep centre. In *Brain Mechanisms and Consciousness*. Charles C. Thomas.

Hewes, G. 1955. World distribution of certain postural habits. *American Anthropologist*, 57:231–44.

Hewes, G. 1957. The anthropology of posture. *Scientific American*, 196:123–32.

Heywood, R. 1961. *Beyond the Reach of Sense*. Dutton.

Hick, J. 1975. *Death and Eternal Life*. Harper.

Hickson, R., et al. 1985. Reduced training intensities and loss of aerobic power, endurance and cardiac growth. *Journal of Applied Physiology*, 58:492.

Hilgard, E. 1965a. Hypnosis. *Annual Review of Psychology*, 16:157–80.

Hilgard, E. 1965b. *Hypnotic Susceptibility*. Harcourt, Brace & World.

Hilgard, E. 1967. A quantitative study of pain and its reduction through hypnotic suggestion. *Proceedings of the National Academy of Sciences*, 57:1581–86.

Hilgard, E. 1971. Hypnotic phenomena: The struggle for scientific acceptance. *American Scientist*, 59:574–75.

Hilgard, E. 1973. A neodissociation interpretation of pain reduction in hypnosis. *Psychological Review,* 80:396–411.

Hilgard, E. 1975a. The alleviation of pain by hypnosis. *Pain,* 1:213–31.

Hilgard, E. 1975b. Hypnosis. *Annual Review of Psychology,* 26:19–44.

Hilgard, E. 1977. The problem of divided consciousness: A neodissociation interpretation. *New York Academy of Sciences,* 296:48–59.

Hilgard, E. 1986. *Divided Consciousness.* Wiley.

Hilgard, E., & J. Hilgard. 1983. *Hypnosis in the Relief of Pain.* William Kaufmann.

Hilgard, J. 1974a. Imaginative involvement: Some characteristics of the highly hypnotizable and the non-hypnotizable. *International Journal of Clinical & Experimental Hypnosis,* 22:138–56.

Hilgard, J. 1974b. Sequelae to hypnosis. *International Journal of Clinical & Experimental Hypnosis,* 22:281–98.

Hilgard, J. 1979. *Personality and Hypnosis: A Study of Imaginative Involvement.* University of Chicago Press.

Hillman, J. 1972. *The Myth of Analysis.* Harper & Row.

Hillman, J. 1975. *Revisioning Psychology.* Harper Colophon.

Hinde, R. A., ed. 1972. *Non Verbal Communication.* Cambridge University Press.

Hirai, T. 1960. Electroencephalographic study on the Zen meditation (zazen): EEG changes during the concentrated relaxation. *Psychiatrica et Neurologia Japanica,* 62:76–105.

Hirai, T. 1974. *The Psychophysiology of Zen.* Igaku Shoin.

Hirai, T., et al. 1959. EEG and Zen Buddhism: EEG changes in the course of meditation. *Electroencephalography and Clinical Neurophysiology,* 18:52–53.

Hirata, K. 1971. Karate. In L. Larson (ed.), *Encyclopedia of Sports, Sciences and Medicine.* Macmillan.

Hjelle, L. A. 1974. TM and psychological health. *Perceptual & Motor Skills,* 39:623–28.

Hnatiow, M. J. 1971. Learned control of heart rate and blood pressure. *Perceptual & Motor Skills,* 33:219–26.

Hnatiow, M., & P. Lang. 1965. Learned stabilization of cardiac rate. *Psychophysiology,* 1:330–36.

Hocht, J. M. 1974. *Von Franziskus zu Pater Pio und Therese Neumann.* Christiana-Verlag.

Hoenig, J. 1968. Medical research on yoga. *Confinia Psychiatrica,* 11:69–89.

Hoffman, E. 1981. *The Way of Splendor.* Shambhala.

Hoffman, J., P. Arns, & G. Stainbrook et al. 1981. Effect of the relaxation response on oxygen consumption during exercise. *Clinical Research,* 29:207.

Holt, W., J. Caruso, J. Riley et al. 1978. Transcendental Meditation versus pseudo meditation on visual choice reaction time. *Perceptual & Motor Skills,* 46:726.

Honigberger, J. M. 1851. *Fruchte aus dem Morgenlande.* Wien. Republished as *Thirty-Five Years in the East* in London, 1852. Cited in R. Garbe. 1900. On the voluntary trance of Indian fakirs. *The Monist,* 10:492–500.

Honiotes, G. J. 1977. Hypnosis and breast enlargement—A pilot study. *Journal of the International Society for Professional Hypnosis*, 6:8–12. In A. Sheikh (ed.), 1984. *Imagination & Healing*. Baywood.

Horning, J. C. 1932. Nervous pregnancy in the dog. *Veterinarian Medicine*, 27:24–31.

Horowitz, M. 1970. *Image Formation and Cognition*. Appleton.

Horowitz, M. 1974. *Image Techniques in Psychotherapy*. Behavioral Science Tape Library.

Howard, F., & F. Patry. 1935. *Mental Health*. Harper.

Hubel, K. A. 1974. Voluntary control of gastrointestinal function: Operant conditioning and biofeedback. *Gastroenterology*, 66:1085–90.

Huddleston, A. L., et al. 1980. Bone mass in lifetime tennis athletes. *Journal of the American Medical Association*, 244:1107.

Hughes, J. 1984. Psychological effects of habitual aerobic exercise: A critical review. *Preventive Medicine*, 13:66–78.

Huizinga, J. 1950. *Homo Ludens*. Beacon.

Hunt, H. 1985. Cognitions and states of consciousness: The necessity for empirical study of ordinary and nonordinary consciousness for contemporary cognitive psychology. *Perceptual & Motor Skills*, 60:239–82.

Hunt, V., & W. Massey. 1977. Electromyographic evaluation of Structural Integration techniques. *Psychoenergetic Systems*, 2:209.

Hunt, V., et al. 1977. *Project Report: A Study of Structural Integration from Neuromuscular, Energy Field, and Emotional Approaches*. Rolf Institute.

Hurley, T. J. 1985. Keys to placebo power. *Investigations*, 2:16. Sausalito, CA: Institute of Noetic Sciences.

Hustad, P., & J. Carnes. 1988. The effectiveness of walking meditation on EMG readings in chronic pain patients. *Biofeedback & Self-Regulation*, 13:69.

Huttunen, J., et al. 1979. Effect of moderate physical exercise on serum lipoproteins. *Circulation*, 60:1220.

Huxley, A. 1952. *Devils of Loudoun*. Harper & Row.

Huxley, A. 1970. *The Perennial Philosophy*. Harper Colophon Books.

Iaquierdo, I. 1984. Endogenous state-dependency: Memory depends on the relation between the neurohumoral and hormonal states present after training at the time of testing. In G. Lynch et al. (eds.), *Neurobiology of Learning and Memory*. Guilford Press.

Ikai, M., & A. H. Steinhaus. 1961. Some factors modifying the expression of human strength. *Journal of Applied Physiology*, 16:157–63.

Ikeda, Y., & H. Hirai. 1976. Voluntary control of electrodermal activity in relation to imagery and internal perception scores. *Psychophysiology*, 13:330–33.

Ikegami, R. 1974. Psychological study of Zen posture. *Bulletin of the Faculty of Literature of Kyushu University*, 5:105–35.

Ikemi, Y., & S. Nakagawa. 1962. A psychosomatic study of contagious dermatitis. *Kyushu Journal of Medical Science*, 13:335–50. In A. Sheikh (ed.), 1984, *Imagination & Healing*. Baywood.

Imbert-Gourbeyre, A. 1894. *La Stigmatisation.* Vols. 1 & 2. Clermont-Ferrand. Cited in Thurston 1952.

Ingalls, A. 1939. Fire-walking. *Scientific American* (March).

Ingjer, F. 1978. Maximal aerobic power related to the capillary supply of the quadriceps femoris muscle in man. *Acta Physiol. Scand.*, 104:238–40.

Isherwood, C. 1959. *Ramakrishna and His Disciples.* Simon & Schuster.

Jackson, I. 1975. *Yoga and the Athlete.* World Publication.

Jacobs, A., & G. Felton. 1969. Visual feedback of myoelectric output to facilitate muscle relaxation in normal persons and patients with neck injuries. *Archives of Physical Medicine and Rehabilitation*, 50:34–39.

Jacobs, K., & F. Nordan. 1979. Classification of placebo drugs: Effect of color. *Perceptual & Motor Skills*, 49:367–72.

Jacobson, E. 1925. Voluntary relaxation of the esophagus. *American Journal of Physiology*, 72:387–94.

Jacobson, E. 1939. Variations in blood pressure with skeletal muscle tension and relaxation. *Annals of Internal Medicine*, 12:1194–1212.

Jacobson, E. 1955. Neuromuscular control in man: Methods of self-direction in health and in disease. *American Journal of Psychology*, 68:549–61.

Jacobson, E. 1974. *Progressive Relaxation.* University of Chicago Press (Midway reprint).

Jacobson, E., & F. Kraft. 1942. Contraction potentials (right quadriceps femoris) in man during reading. *American Journal of Physiology*, 137:1–5.

Jacoby, J. D. 1960. Statistical report on general practice; hypnodontics; tape recorder conditioning. *International Journal of Clinical & Experimental Hypnosis*, 3: 115–19.

Jaensch, E. R. 1930. *Eidetic Imagery and Typological Methods of Investigation.* Kegan Paul, Trench, Trubner.

Jaffe, D., & D. Bresler. 1980. Guided imagery: Healing through the mind's eye. In J. E. Shorr et al. (eds.), *Imagery: Its Many Dimensions and Applications.* Plenum.

James, H. 1853. The works of Sir William Hamilton. *Putnam's Magazine*, vol. 2 (Nov.).

James, H. 1863. *Substance and Shadow.* Ticknor and Fields.

James, W. 1902. *The Varieties of Religious Experience.* Random House.

James, W. 1903. Review of *Human Personality and Its Survival of Bodily Death* by F. W. H. Myers. In *Proceedings of the Society for Psychical Research*, London. Vol. 18, pt. 46 (June).

James, W. 1976. *Essays in Radical Empiricism.* In *The Works of William James.* Harvard University Press.

James, W. 1986. *Essays in Psychical Research.* In *The Works of William James.* Harvard University Press.

Janet, P. 1886. Deuxieme note sur le sommeil provoque a distance. *Revue Philosophique*, 22:212–23.

Janet, P. 1898. *Nervoses et Idees Fixes.* Alcan.

Jaynes, J. 1976. *The Origins of Consciousness in the Breakdown of the Bicameral Mind.* Houghton Mifflin.

Jedrczak, A., M. Beresford, G. Clements et al. 1984. The TM-sidhi programme and field independence. *Perceptual & Motor Skills,* 59:999–1000.

Jedrczak, A., M. Toomey, G. Clements et al. 1986. The TM-Sidhi programme, age, and brief tests of perceptual-motor speed and nonverbal intelligence. *Journal of Clinical Psychology,* 42:161–64.

Jellinek, A. 1949. Spontaneous imagery: A new psychotherapeutic approach. *American Journal of Psychotherapy,* 3:372–91.

Jencks, B., & E. Krenz. 1984. Hypnosis in sports. In A. H. Smith & W. D. Wester (eds.), *Comprehensive Clinical Hypnosis.* Lippincott.

Jevning, R., & J. O'Halloran. 1984. Metabolic effects of Transcendental Meditation: Toward a new paradigm of curobiology. In D. H. Shapiro & R. Walsh (eds.), *Meditation: Classic and Conemporary Perspectives.* Aldine.

Jevning, R., & A. Wilson. 1977. Altered red cell metabolism in TM. *Psychophysiology,* 14:94.

Jevning, R., A. F. Wilson et al. 1978. Redistribution of blood flow in acute hypometabolic behavior. *American Journal of Physiology,* 235:89–92.

John, M. E., & J. P. Parrino. 1983. Ericksonian hypnotherapy of intractable shoulder pain. *American Journal of Clinical Hypnosis,* 26:26–29.

Johnson, C. P. 1976. Analysis of five tests commonly used in determining the ability to control single motor units. *American Journal of Physical Medicine,* 55: 113–21.

Johnson, D. 1977. *The Protean Body.* Harper Colophon.

Johnson, D. 1980. Somatic Platonism. *Somatics* (Autumn), 3:4–7.

Johnson, D. 1983. *Body.* Beacon.

Johnson, D. 1986/87. Principles versus techniques: Towards the unity of the Somatics field. *Somatics* (Autumn/Winter) 6: 4–8.

Johnson, L. S. 1981. Current research in self-hypnotic phenomenology: The Chicago Paradigm. *International Journal of Clinical and Experimental Hypnosis,* 24: 247–58.

Johnson, L., & D. Weight. 1976. Self-hypnosis versus heterohypnosis: Experimental and behavioral comparison. *Journal of Abnormal Psychology,* 85:523–26.

Johnson, R. 1959. *Watcher on the Hills: A Study of Some Mystical Experiences of Ordinary People.* Harper.

Johnson, R. F. Q., & T. X. Barber. 1976. Hypnotic suggestions for blister formation: Subjective and physiological effects. *American Journal of Clinical Hypnosis,* 18:172–81.

Johnston, F. 1982. *Alexandrina, the Agony and the Glory.* Tan.

Jones, A. H. M. 1964. *The Later Roman Empire.* Oxford.

Jones, E. 1957. *The Life and Work of Sigmund Freud.* Vols. 1–3. Basic Books.

Jones, F. P. 1976. *Body Awareness in Action.* Schocken.

Jones, R. F. 1970. "You learn the art of invisibility." *Sports Illustrated* (Nov. 16).

Jones, R. T. 1966. *Bobby Jones on Golf.* Doubleday.

Jordan, C., & K. Lenington. 1979. Physiological correlates of eidetic imagery and induced anxiety. *Journal of Mental Imagery,* 3:31–42.

Journal of Transpersonal Psychology. Published by Transpersonal Institute, P.O. Box 4437, Stanford, CA 94309.

Jung, C. G. 1961. *Memories, Dreams, and Reflections.* Ed. A. Jaffe, trans. R. & C. Winston. Pantheon.

Jung, C. G. *The Collected Works.* Princeton University Press, Bollingen Series.

Jung, C. G. *Psychology and Alchemy.* Vol. 12 in *The Collected Works.*

Jung, C. G. *The Structure and Dynamics of the Psyche.* Princeton University Press, Bollingen Series.

Jung, C. G. 1970a. *Alchemical Studies.* Vol. 13 in *The Collected Works.*

Jung, C. G. 1970b. *Mysterium Coniunctionis.* Vol. 14 in *The Collected Works.*

Jung, C. G. 1975. *C. G. Jung Letters.* Ed. G. Adler. Princeton University Press, Bollingen Series.

Jung, C. G. 1976. *The Symbolic Life.* Vol. 18 in *The Collected Works.*

Kabat-Zinn, J. 1982. An outpatient program in behavioral medicine for chronic pain patients based on the practice of mindfulness meditation: Theoretical considerations and preliminary results. *General Hospital Psychiatry,* 4:33–47.

Kabat-Zinn, J., & R. Burney. 1981. The clinical use of awareness meditation in the self-regulation of chronic pain. (Abstract.) *Pain* 1(Suppl):S273.

Kabat-Zinn, J., L. Lipworth, & R. Burney. 1985. The clinical use of mindfulness meditation for the self-regulation of chronic pain. *Journal of Behavioral Medicine,* 8:163–90.

Kabat-Zinn, J., L. Lipworth, R. Burney et al. 1987. Four-year follow-up of a meditation-based program for the self-regulation of chronic pain: Treatment outcomes and compliance. *Clinical Journal of Pain,* 2:159–73.

Kabat-Zinn, J., L. Lipworth, W. Sellers et al. 1984. Reproducibility and four-year follow-up of a training program in mindfulness meditation for the self-regulation of chronic pain. *Pain Supplement,* 2:303.

Kahn, H. A. 1963. The relationship of reported coronary heart disease mortality to physical activity work. *American Journal of Public Health,* 53:1058–67.

Kamiya, J. 1968. Conscious control of brain waves. *Psychology Today,* 1:57–60.

Kamiya, J. 1969. Operant control of the EEG alpha rhythm and some of its reported effects on consciousness. In C. Tart (ed.), *Altered States of Consciousness.* Julian.

Kanellakos, D., & J. Lukas. 1974. *The Psychobiology of Transcendental Meditation: A Literature Review.* W. A. Benjamin.

Kanemoto, S. 1969. *Bu no shinjin* (True Man of Martial Valor). Tama Publishing Co.

Kapleau, P. 1965. *The Three Pillars of Zen.* Beacon.

Karambelkar, P. V., S. L. Vinekar, & M. V. Bhole. 1968. Studies on human subjects staying in an air-tight pit. *Indian Journal of Medical Research,* 56:1282–88.

Karjalainen, K. F. 1921–27. *Die Religion der Jugra-Volker.* Vol. 3. Cited in M. Eliade 1964. *Shamanism.* Princeton University Press, Bollingen Series.

Karlin, R. 1979. Hypnotizability and attention. *Journal of Abnormal Psychology,* 88: 92–95.

Kasamatsu, A., & T. Hirai. 1963. Science of Zazen. *Psychologia,* 6:86–91.

Kasamatsu, A., & T. Hirai. 1966. An EEG study of Zen meditation. *Folia Psychiatrica et Neurologia Japonica,* 20:315–36.

Kasamatsu, A., & T. Hirai. 1969a. An EEG study on the Zen meditation (zazen). *Psychologia,* 12:205–25.

Kasamatsu, A., & T. Hirai. 1969b. An electroencephalographic study of Zen meditation. In C. Tart (ed.), *Altered States of Consciousness.* Wiley.

Kasamatsu, A., et al. 1957. The EEG of "Zen" and "Yoga" practitioners. *Electroencephalography & Clinical Neurophysiology,* Supp. 9, 51–52.

Katchan, F. A., & G. Belozerski. 1940. Obstetrical analgesia by hypnosis and suggestion. Cited in L. Chertok, 1959, *Psychosomatic Methods in Painless Childbirth: History, Theory & Practice.* Pergamon.

Katcher, A., H. Segal, & A. Beck. 1984. Comparison of contemplation and hypnosis for the reduction of anxiety and discomfort during dental surgery. *American Journal of Clinical Hypnosis,* 27:14–21.

Katz, D. 1977. Decreased drug use and prevention of drug use through the Transcendental Meditation Program. In D. W. Orme-Johnson & J. T. Farrow (eds.), *Scientific Research on the Transcendental Meditation Program: Collected Papers.* Vol. 1. MERU Press.

Katz, R. 1982a. Accepting "boiling energy." *Ethos,* 10:344–67.

Katz, R. 1982b. *Boiling Energy.* Harvard University Press.

Katz, S., ed. 1978. *Mysticism and Philosophical Analysis.* Oxford University Press.

Katz, S. ed. 1983. *Mysticism and Religious Traditions.* Oxford University Press.

Kaufman, W. 1974. *Nietzsche.* Princeton University Press.

Kavanaugh, K., and O. Rodriguez, trans. 1976. In *St. Teresa of Avila. Book of Her Life.* Institute of Carmelite Studies.

Kazdin, A. E. 1977. Research issues in covert conditioning. *Cognitive Therapy & Research,* 1:45–49.

Kazdin, A. E. 1978. Covert conditioning: The therapeutic application of imagined rehearsal. In J. Singer & K. Pope (eds.), *The Power of Human Imagination.* Plenum.

Keens, S. 1986. *Faces of the Enemy.* Harper & Row.

Keithler, M. A. 1981. The influence of the Transcendental Meditation program and personality variables on auditory thresholds and cardiorespiratory responding. *Dissertation Abstracts International,* 42:1662–63.

Keller, H. 1954. *The Story of My Life.* Doubleday.

Kelley, G. 1955. *The Psychology of Personal Constructs.* Vols. 1 & 2. Norton.

Kelton, J. J. 1978. Perceptual and cognitive processes in meditation. *Dissertation Abstracts International,* 38:3:931.

Kepecs, J. G. 1954. Observations on screens and barriers in the mind. *Psychoanalytic Quarterly,* 23:62–77.

Kernberg, O. 1975. *Borderline Conditions and Pathological Narcissism.* Jason Aronson.

Kernberg, O. 1976. *Object Relations Theory and Clinical Psychoanalysis.* Jason Aronson.

Kesterson, J. B. 1986. Changes in respiratory patterns and control during the practice of the Transcendental Meditation technique. *Dissertation Abstracts International,* 47:4:337.

Kief, H. 1972. A method for measuring PK ability with enzymes. In W. G. Roll et al. (eds.), *Proceedings of the Parapychological Association No. 8.*

Kidd, C. 1966. Congenital ichthyosiform erythrodermia treated by hypnosis. *British Journal of Dermatology,* 78:101–5.

Kierkegaard, S. *Philosophical Fragments: Johannes Climacus.* Ed. & trans. H. Hong & E. Hong. Concluded unscientific postscript to the *Philosophical Fragments.* Princeton University Press.

Kihlstrom, J. F. 1984. Conscious, subconscious, unconscious: A cognitive view. In K. Bowers & D. Meichenbaum (eds.), *The Unconscious: Reconsidered.* Wiley.

Kihlstrom, J. R. 1985. Hypnosis. *Annual Review of Psychology,* 36:385–418.

Kimball, C. P. 1970. A case of pseudocyesis caused by roots. *American Journal of Obstetrics and Gynecology,* 107:801–3.

Kimmel, E., & H. Kimmel. 1963. A replication of operant conditioning of the GSR. *Journal of Experimental Psychology,* 65:212–13.

Kimmel, H. D. 1974. Instrumental conditioning of autonomically mediated responses in human beings. *American Psychologist,* 29:325–35.

Kimmel, H., & F. Hill. 1960. Operant conditioning of the GSR. *Psychological Reports,* 7:555–62.

Kindlon, D. J. 1983. Comparison of use of meditation and rest in treatment of test anxiety. *Psychological Reports,* 53:931–38.

King, J. T. 1920. An instance of voluntary acceleration of the pulse. *Johns Hopkins Bulletin,* 2:303–5.

Kinsley, D. 1975. *The Sword and the Flute.* University of California Press.

Kirsch, I., & D. Henry. 1979. Self-desensitization and meditation in the reduction of public speaking anxiety. *Journal of Consulting & Clinical Psychology,* 47: 536–41.

Kisshomaru, U. 1977. *Aikido kaiso Uehiba Morihei den* (Founder of Aikido: The Biography of Morihei Uyeshiba). Kodansha.

Kisshomaru, U. 1983. *Aikido kaiso* (Founder of Aikido). Kodansha.

Kisshomaru, U. 1984. *Aikido no kokoro* (The Heart of Aikido). An English translation of this work was published by Kodansha International in 1984 as *The Spirit of Aikido.*

Kisshomaru, U. 1986. *Aikido shintai* (The Truth of Aikido). Kodansha.

Kjellberg, S., et al. 1949. Increase in the amount of hemoglobin and blood volume in connection with physical training. *Acta Physiol. Scand.*, 19:146.

Klajner, F., L. Hartman, & M. Sobell. 1984. Treatment of substance abuse by relaxation training: A review of its rationale, efficacy and mechanisms. *Addictive Behaviors*, 9:41–55.

Klauder, J. V. 1938. Stigmatization. *Archives of Dermatology and Syphilology* (Apr.), 37:658.

Klein, K., & D. Spiegel. 1989. Modulation of gastric acid secretion by hypnosis. *Gastroenterology*, 96:1383–7.

Klemola, E. 1951. Cardiographic observations of 650 Finnish athletes. *An. Med. Fin.*, 40:121–32.

Kline, M. V. 1952–53. The transcendence of waking visual discrimination capacity with hypnosis: A preliminary case report. *British Journal of Medical Hypnotism*, 4:32–33.

Klinge, V. 1972. Effects of exteroceptive feedback and instructions on control of spontaneous galvanic skin response. *Psychophysiology*, 9:305–17.

Klinger, E. 1980. Therapy and the flow of thought. In J. E. Shorr et al. (eds.), *Imagery: Its Many Dimensions and Applications*. Plenum.

Kmetz, J. 1981. Effects of healing on cancer cells (Appendix). In D. Kraft, *Portrait of a Psychic Healer*. Putnam.

Knight, J. A. 1960. False pregnancy in a male. *Psychosomatic Medicine*, 22:260–66.

Knowles, D. 1963. *Great Historical Enterprises*. Nelson.

Knowles, F. W. 1954. Some investigations into psychic healing. *Journal of the American Society for Psychical Research*, 48:21–26.

Knowles, F. W. 1956. Psychic healing in organic disease. *Journal of the American Society for Psychical Research*, 50:110–17.

Knowles, J., & C. Lucas. 1960. Experimental studies of the placebo response. *Journal of Mental Science*, 106:231–40.

Koestler, A. 1960. *The Lotus and the Robot*. Hutchinson.

Koestler, A. 1964. *The Act of Creation*. Macmillan.

Koestler, A. 1978. *Janus: A Summing Up*. Random House.

Koestler, A., & J. R. Smythies (eds.), 1969. *Beyond Reductionism: The Alpbach Symposium*. Macmillan.

Kohr, R. L. 1977. Dimensionality in meditative experience: A replication. *Journal of Transpersonal Psychology*, 9:193–203.

Konzak, B., & F. Boudreau. 1984. Martial arts training and mental health: An exercise in self-help. *Canada's Mental Health*, 32:2–8.

Korein, J., & J. Brudny. 1976. Integrated EMG feedback in the management of spasmodic torticollis and focal dystonia: A prospective study of 80 patients. In M. D. Yahr (ed.), *The Basal Ganglia*. Raven.

Korein, J., et al. 1976. Sensory feedback therapy of spasmodic torticollis and dystonia: Results in treatment of 55 patients. In R. Eldridge & S. Fahn (eds.), *Advances in Neurology*. Raven.

Kornfield, J. 1979. Intensive insight meditation: A phenomenological study. *Journal of Transpersonal Psychology,* 11:41–58.

Kornfield, J. 1983. The psychology of mindfulness meditation. *Dissertation Abstracts International,* 44:610.

Kothari, L. K., A. Bordia, & O. P. Gupta. 1973. Studies on a yogi during an eight-day confinement in a sealed underground pit. *Indian Journal of Medical Research,* 61:1645–50.

Koufax, S., & E. Linn. 1966. *Koufax.* Viking.

Krahne, W., & G. Tenoli. 1975. EEG and Transcendental Meditation. *Pfleuger's Archives,* 359:R93.

Kramer, E., & G. R. Tucker. 1967. Hypnotically suggested deafness and delayed auditory feedback. *International Journal of Clinical & Experimental Hypnosis,* 15: 37–43.

Kras, D. J. 1977. The Transcendental Meditation technique and EEG alpha activity. In D. W. Orme-Johnson and J. T. Farrow (eds.), *Scientific Research on the Transcendental Meditation Program: Collected Papers.* Vol. 1. MERU Press.

Kraus, F. 1926. *Kliniche Syzyologie: Allgemeine und Spezielle Pathologie der Person.* Thieme.

Krenz, E. W. 1984. Improving competitive performance with hypnotic suggestions and modified autogenic training: Case reports. *American Journal of Clinical Hypnosis,* 27:58–63.

Kretschmer, E. 1922. *Kretschmer's Textbook of Medical Psychology.* Oxford University Press.

Krieger, D. 1975. Therapeutic Touch: The imprimatur of nursing. *American Journal of Nursing,* 75:784–87.

Krieger, D. 1976. Healing by the laying-on-of-hands as a facilitator of bioenergetic change: The response of in-vivo human hemoglobin. *Psychoenergetic Systems,* 1:121–29.

Krieger, D. 1979a. *Therapeutic Touch.* Prentice-Hall.

Krieger, D. 1979b. Therapeutic Touch: Searching for evidence of physiological change. *American Journal of Nursing* (Apr.), 79:660–62.

Krieger, D., et al. 1979. Physiologic indices of Therapeutic Touch. *American Journal of Nursing,* 4:660–62.

Krippner, S., & M. Maliszewski. 1978. Meditation and the creative process. *Journal of Indian Psychology,* 1:40–58.

Kriska, A., et al. 1988. The assessment of historical physical activity and its relation to adult bone parameters. *American Journal of Epidemiology,* 127:1053.

Kroger, W. 1962. *Psychosomatic Obstetrics, Gynecology and Endocrinology.* Charles C. Thomas.

Kroger, W. S. 1963. *Clinical & Experimental Hypnosis.* Lippincott.

Kroll, W., & B. Carlson. 1970. Discriminant function and hierarchical grouping analysis of karate participants' personality profiles. In Morgan (ed.), *Contemporary Readings in Sport Psychology.* Charles C. Thomas.

Kubie, L. 1943. The use of induced hypnogogic reveries in the recovery of repressed amnesic data. *Bulletin of the Menninger Clinic,* 7:172–83.

Kubie, L. 1958. *Neurotic Distortion of the Creative Process.* University of Kansas.

Kubose, S. 1976. An experimental investigation of psychological aspects of meditation. *Psychologia,* 19:1–10.

Kubose, S., & T. Umemoto. 1980. Creativity and the Zen koan. *Psychologia,* 23:1–9.

Kunzle, D. 1981. *Fashion and Fetishism: A Social History of the Corset, Tight-Lacing, and Other Forms of Body-Sculpture in the West.* Rowman & Littlefield.

Kutz, I., M. Caudill, & H. Benson. 1983. The role of relaxation in behavioral therapies of chronic pain. In J. Stein & C. Winfield (eds.), *Pain Management.* Little, Brown.

Labarre, W. 1947. The cultural basis of emotions and gestures. *Journal of Personality,* 16:49–68.

LaBaw, W. L. 1970. Regular use of suggestibility by pediatric bleedings. *Haematologia,* 4:419–25.

Lakoff, G. 1987. *Women, Fire and Other Dangerous Things.* University of Chicago Press.

Lakoff, G., & M. Johnson. 1980. *Metaphors We Live By.* University of Chicago Press.

Lamaze, F. 1958. *Painless Childbirth: Psychoprophylactic Method.* Burke.

Lang, M. 1977. *Cure and Cult in Ancient Corinth.* American School of Classical Studies at Athens.

Lang, P. J. 1974. Learned control of human heart rate in a computer directed environment. In P. A. Obrist et al. (eds.), *Cardiovascular Psychophysiology.* Aldine.

Lang, P., & C. Twentyman. 1974. Learning to control heart rate: Binary vs. analogue feedback. *Psychophysiology,* 11:616–29.

Lang, P., & C. Twentyman. 1976. Learning to control heart rate: Effects of varying incentive and criterion of success on task performance. *Psychophysiology,* 13:378–85.

Lang, P. J., et al. 1967. Effects of feedback and instructional set on the control of cardiac rate variability. *Journal of Experimental Psychology,* 75:425–31.

Lang, P. J., et al. 1975. Differential effects of heart rate modification training on college students, older males, and patients with ischemic heart disease. *Psychosomatic Medicine,* 37:429–46.

Lang, R., K. Dehob, K. Meurer et al. 1979. Sympathetic activity and TM. *Journal of Neural Transmission,* 44:117–35.

Langer, E. 1988. Minding matters: The consequences of mindlessness/mindfulness. In L. Berkowitz (ed.), *Advances in Experimental Social Psychology.* Academic.

Langley, S. P. 1901. The fire-walk ceremony in Tahiti. *Nature,* (May).

Lanz, B. 1970. Effects of placebo administration on the postoperative course of oral surgery. *Swedish Dental Journal,* 63:621–26.

Lanzetta, J., et al. 1976. Effects of non-verbal dissimulation on emotional experience and autonomic arousal. *Journal of Personality and Social Psychology,* 33:354–70.

La Perriere, A., M. Fletcher, M. Antoni, et al. 1991. Aerobic exercise training in an AIDS risk group. *International Journal of Sports Medicine,* 12:S53–S57.

Lasagna, L. 1955. Placebos. *Scientific American,* 193:68–71.

Lasagna, L., et al. 1955. Drug-induced mood changes in man. I. Observations on healthy subjects, chronically ill patients, and postaddicts. *Journal of the American Medical Association,* 157:1006–20.

Laski, M. 1990 (reprint). *Ecstasy.* Tarcher.

Laurie, G. 1977. An investigation into the changes in skin resistance during the Transcendental Meditation technique. In D. W. Orme-Johnson and J. T. Farrow (eds.), *Scientific Research on the Transcendental Meditation Program: Collected Papers.* Vol. 1. MERU Press.

Laver, A. B. 1978. Miracles no wonder! The Mesmeric phenomena and organic cures of Valentine Greatrakes. *Journal of the History of Medicine and Allied Sciences* (Jan.), 33:35–46.

Layard, J. W. 1930. Malekula: Flying tricksters, ghosts, gods and epileptics. *Journal of the Royal Anthropological Institute,* London.

Lazar, Z., L. Farwell, & J. Farrow. 1977. The effects of the Transcendental Meditation program on anxiety, drug abuse, cigarette smoking and alcohol consumption. In D. W. Orme-Johnson and J. T. Farrow (eds.), *Scientific Research on the Transcendental Meditation Program: Collected Papers.* Vol. 1. MERU Press.

Lazarus, A. 1976. Psychiatric problems precipitated by Transcendental Meditation. *Psychological Reports,* 39:601–2.

Lazarus, A. 1977. *In the Mind's Eye.* Lawson.

Lea, P., et al. 1960. The hypnotic control of intractable pain. *American Journal of Clinical Hypnosis,* 3:3–8.

Lecky, W. E. 1926. *History of European Morals.* Vols. 1 & 2. Reprinted by Arno.

LeCron, L. M. 1969. Breast development through hypnotic suggestion. *Journal of the American Society of Psychosomatic Dentistry & Medicine,* 16:58–61.

LeCron, L. M. 1969. Breast development through hypnotic suggestion. *Journal of the American Society of Psychosomatic Dentistry & Medicine,* 16:58–61.

Lefebvre, F. 1873 (2d ed.). *Louise Lateau de Bois-D'Haine. Sa Vie. Ses Extases. Ses Stigmates, Étude Médicale.* Ch. Peeters, Editeur.

LeHew, J. L. 1970. The use of hypnosis in the treatment of musculo-skeletal disorders. *American Journal of Clinical Hypnosis,* 13:131–34.

Lehrer, P., A. Schoicket, P. Carrington et al. 1980. Psychophysiological and cognitive responses to stressful stimuli in subjects practicing progressive relaxation and clinically standardized meditation. *Behavior Research & Therapy,* 18:293–303.

Lehrer, P., R. Woolfolk, A. Rooney et al. 1983. Progressive relaxation and meditation. *Behavior Research Therapy,* 21:651–62.

Lehrer, P., et al. 1983. Progressive Relaxation and meditation. *Behavior Research Therapy,* 21:651–62.

Lehtonen, A., et al. 1979. The effects of exercise on high density (HDL) lipoprotein apoproteins. *Acta Physiol. Scand.,* 106:487.

Leonard, G. 1982. *Report to the Esalen Institute Study of Exceptional Functioning* (Dec.).

Leonard, G. 1986. *The Transformation.* Tarcher.

Leonard, G. 1990. *The Ultimate Athlete.* North Atlantic.

Leonard, G. 1991. *Mastery.* Dutton.

Leonard, G. 1991. *The Silent Pulse.* Arkana.

Leonard, G., & J. Kirsch. 1983. *Energy Training: A Manual for Trainers.* Mill Valley, CA: Leonard Energy Training Institute, Aikido of Tamalpais.

Lerner, B., et al. 1967. On the need to be pregnant. *International Journal of Psychoanalysis,* 48:288–97.

Lesh, T. V. 1970a. Zen and psychotherapy: A partially annotated bibliography. *Journal of Humanistic Psychology,* 10:75–83.

Lesh, T. V. 1970b. Zen meditation and the development of empathy in counselors. *Journal of Humanistic Psychology,* 10:39–74.

Leshan, L. 1975. *The Medium, the Mystic, and the Physicist.* Pt. 2. Ballantine Books.

Leuner, H. 1977. Guided affective imagery: An account of its development. *Journal of Mental Imagery,* 1:73–92.

Leuner, H. 1978. Basic principles and therapeutic efficacy of guided affective imagery. In J. Singer & K. Pope (eds.), *The Power of Human Imagination.* Plenum.

Leung, P. 1973. Comparative effects of training in external and internal concentration on two counseling behaviors. *Journal of Counseling Psychology,* 20:227–34.

Levee, J. R., et al. 1976. Electromyographic biofeedback for relief of tension in the facial and throat muscles of a wood-wind musician. *Biofeedback and Self-Regulation,* 1:113–20.

Levene, H., et al. 1968. Differential operant conditioning of heart rate. *Psychosomatic Medicine,* 30:837–45.

Levin, S. 1977. The Transcendental Meditation technique in secondary education. *Dissertation Abstracts International,* 38:706–7.

Levine, D. 1983. *The Liberal Arts and the Martial Arts.* Gallery Books.

Levine, J., et al. 1978. The mechanism of placebo analgesia. *Lancet,* 2:654–57.

Levine, J. D., et al. 1981. Analgesic responses to morphine and placebo in individuals with postoperative pain. *Pain,* 10:379–89.

Levine, P. H. 1976. Analysis of the EEG by COSPAR: Application to TM. In J. I. Martin (ed.), *Proceedings of the San Diego Biomedical Symposium,* 15:237–47.

Lewis, C. S. 1970. *Perelandra.* Macmillan.

Lewis, J. 1978. The effects of a group meditation technique upon degree of test anxiety and level of digit-letter retention in high school students. *Dissertation Abstracts International,* 38:6015–16.

Lhermitte, J. 1953. *Mystiques et Faux Mystiques.* Bloud et Gay.

Libby, B. 1974. *Foyt.* Hawthorn Books.

Lichtenstein, M. 1925. *Jewish Science and Health.* Jewish Science Publishing.

Life of Sri Ramakrishna. 1924. Advaita Ashrama.

Lifschutz, J. 1957. Hysterical stigmatization. *American Journal of Psychiatry*, 114: 529–30.

Lindauer, M. S. 1983. Imagery and the arts. In A. Sheikh (ed.), *Imagery—Current Theory, Research, and Application*. Wiley.

Lindbergh, C. 1953. *The Spirit of St. Louis*. Charles Scribner's.

Linden, W. 1973. Practicing of meditation of school children and their levels of field dependence-independence, test anxiety, and reading achievement. *Journal of Consulting and Clinical Psychology*, 41:139–43.

Lindsley, D. B. 1935. Electrical activity of human motor units during voluntary contraction. *American Journal of Physiology*, 114:90–99.

Little, J. C. 1979. Neurotic illness in fitness fanatics. *Psychiatric Annals*, 9:49–56.

Little, L., & P. Curran. 1978. Covert sensitization: A clinical procedure in need of some explanations. *Psychological Bulletin*, 85:513–31.

Lloyd, A., & J. Shurley. 1976. The effects of sensory perceptual isolation on single motor unit conditioning. *Psychophysiology*, 13:340–44.

Lloyd, C. W. 1968. The ovaries. In R. Williams (ed.), *Textbook of Endocrinology*. W. B. Saunders.

Locke, S., et al. 1985. *Foundations of Psychoneuroimmunology*. Aldine.

Loehr, F. 1959. *The Power of Prayer on Plants*. Signet.

Lord, R. A. 1957. A note on stigmata. *American Imago*, 14:299–301.

Lorenz, F. W. 1979. Experiments with Matthew Manning. In J. Mishlove (ed.), *A Month with Matthew Manning: Experiences and Experiments in Northern California During May–June, 1977*. Washington Research Center.

Lorimer, D. 1984. *Survival? Body, Mind & Death in the Light of Psychic Experience*. Routledge & Kegan Paul.

Lovejoy, A. 1964. *The Great Chain of Being*. Harvard University Press.

Lovell-Smith, H. D. 1985. Transcendental Meditation and three cases of migraine. *New Zealand Medical Journal*, 98:443–45.

Lubar, J. F. 1989. Electroencephalographic biofeedback and neurological applications. In J. V. Basmajian (ed.), *Biofeedback—Principles and Practice for Clinicians*. Williams & Wilkins.

Lucas, O. N. 1965. Dental extractions in the hemophiliac: Control of the emotional factors by hypnosis. *American Journal of Clinical Hypnosis*, 7:301–7.

Ludwig, A., & W. Lyle. 1964. Tension induction and the hyperalert trance. *Journal of Abnormal and Social Psychology*, 69:70–76.

Luparello, T., et al. 1968. Influences of suggestion on airway reactivity in asthmatic subjects. *Psychosomatic Medicine*, 30:819–25.

Luria, A. R. 1968. *The Mind of a Mnemonist*. Basic Books.

Lusseyran, J. 1987. *And There Was Light*. Parabola.

Lynn, S., & J. Rhue. 1988. Fantasy proneness. *American Psychologist*, 45:1–43.

MacDonald, R. G., et al. 1976. *Preliminary Physical Measurements of Psychophysical Effects Associated with Three Alleged Psychic Healers*. Washington Research Center.

MacHovec, F. 1986. *Hypnosis Complications: Prevention and Risk Management.* Thomas.

MacKean, W. H. 1920. *Christian Monasticism in Egypt to the Close of the Fourth Century.* London.

MacKinnon, D. W. 1960. Genus architectus creator varietas Americanus. *American Institute of Architects Journal* (Sept.), pp. 31–35.

Maclean, P. 1958. *American Journal of Medicine,* 25:611–26.

MacLean, P. 1962. New findings relevant to the evolution of psychosexual functions of the brain. *Journal of Nervous & Mental Disease,* 135:4.

MacLeod-Morgan, C., & L. Lack. 1982. Hemispheric specificity: A physiological concomitant of hypnotizability. *Psychophysiology,* 19:687–90.

Maher-Loughnan, G. P. 1970. Hypnosis and autohypnosis for the treatment of asthma. *International Journal of Clinical & Experimental Hypnosis,* 18:1–14.

Mahler, D., & J. Loke. 1985. The physiology of marathon running. *The Physician and Sports Medicine,* 13:85–97.

Malinowski, B. 1932. *The Argonauts of the Pacific.* London.

Malitz, S., & M. Kanzler. 1971. Are antidepressants better than placebo? *American Journal of Psychiatry,* 127:1605–11.

Malott, J. 1984. Active-alert hypnosis: Replication and extension of previous research. *Journal of Abnormal Psychology,* 93:246–49.

Malott, J. K., & M. Goldstein. 1980. Active-alert hypnotic induction: Effect of motor activity upon responsiveness to suggestions. Unpublished manuscript.

Marais, E. 1973. *The Soul of the Ape.* Penguin.

Marcus, J. B. 1974. Transcendental Meditation: A new method of reducing drug abuse. *Drug Forum,* 3:113–36.

Marlatt, G., R. Pagano, R. Rose et al. 1984. Effects of meditation and relaxation training upon alcohol use on male social drinkers. In D. H. Shapiro and R. Walsh (eds.), *Meditation: Classic and Contemporary Perspectives.* Aldine.

Marquis de Puysegur. 1811. *Recherchers, experience et observations physiologiques sur l'homme dans l'etat de somnambulism naturel.* J. G. Dentu.

Martinetti, P. 1976. Influence of Transcendental Meditation on Perceptual Illusion. *Perceptual & Motor Skills,* 43:922.

Martony, J. 1968. On the correction of the voice pitch level for severely hard of hearing subjects. *American Annals of the Deaf,* 114:195–202.

Maslow, A. 1964. *Religions, Values and Peak-Experiences.* Ohio State University Press.

Maslow, A. 1966. *The Psychology of Science.* Harper & Row.

Maslow, A. 1968 (2d ed.). *Toward a Psychology of Being.* Van Nostrand Reinhold.

Maslow, A. 1971. *The Farther Reaches of Human Nature.* Viking.

Mason, A. A. 1952. A case of congenital ichthyosiform erythrodermia of Brocq treated by hypnosis. *British Medical Journal,* 2:422–23.

Mason, A. A. 1955. Ichthyosis and hypnosis. *British Medical Journal,* 2:57–58.

Mason, A. A. 1960. Hypnosis and suggestion in the treatment of allergic phenomena. *Acta Allergologica*, 15:332–38.

Mason, A., & S. Black. 1958. Allergic skin responses abolished under treatment of asthma and hay fever by hypnosis. *Lancet*, 1:877–80.

Masters, R. E. L., & J. Houston. 1966. *The Varieties of Psychedelic Experience*. Holt, Rinehart & Winston.

Masters, R., & J. Houston. 1972. *Mind Games*. Viking.

Masterson, J. 1981. *The Narcissistic and Borderline Disorders*. Brunner/Mazel.

Maton, B. 1976. Motor unit differentiation and integrated surface EMG in voluntary isometric contraction. *European Journal of Applied Physiology*, 35:149–57.

Maupin, E. W. 1965. Individual differences in response to a Zen meditation exercise. *Journal of Consulting Psychology*, 29:139–45.

Maupin, E. W. 1969. On meditation. In C. T. Tart (ed.), *Altered States of Consciousness: A Book of Readings*. Wiley.

Mauss, M. 1935. Les techniques du corps. *Journal de Psychologie Normale et Pathologique*. Vol. 32. English trans. B. Brewster, in *Economy and Society*, Feb. 1973, pp. 70–88.

Mauss, M. 1982. Effect physique chez l'individu de l'idee de mort suggeree par la collectivite. *Journal de Psychologie Normale et Pathologique*, 23:653–69.

May, J., & H. Johnson. 1973. Physiological activity to internally-elicited arousal and inhibitory thoughts. *Journal of Abnormal Psychology*, 82:239–45.

May, R., et al., eds. 1958. *Existence*. Basic Books.

McArdle, W. D. 1986. *Exercise Physiology*. Lea & Febiger.

McArdle, W., et al. 1978. Specificity of run training on VO$_2$Max and heart rate changes during running and swimming. *Medicine & Science in Sports & Exercise*, 10:16.

McCanne, T., & K. Hathaway. 1984. Individual differences in motor skills ability affect the self-regulation of heart rate. *Biofeedback & Self-Regulation*, 9:241–52.

McClain, K., et al. 1985. Spontaneous remission of Burkett's lymphoma associated with herpes zoster infection. *American Journal of Pediatric Hematology & Oncology*, 7:11–19.

McClure, C. M. 1959. Cardiac arrest through volition. *California Medicine*, 90:440–41.

McConkey, K., et al. 1985. Benjamin Franklin and mesmerism. *International Journal of Clinical & Experimental Hypnosis*, 33:122–30.

McDonagh, J., & T. Egenes. 1977. The Transcendental Meditation technique and temperature homeostasis. In D. W. Orme-Johnson & J. T. Farrow (eds.), *Scientific Research on the Transcendental Meditation Program: Collected Papers*. Vol. 1. MERU Press.

McEvoy, T. M., L. R. Frumkin, S. W. Harkins et al. 1980. Effects of meditation on brainstem auditory evoked potentials. *International Journal of Neuroscience*, 10:165–70.

McFadden, E. R., et al. 1969. The mechanism of action of suggestion in the induction of acute asthma attacks. *Psychosomatic Medicine*, 31:134–43.

McGlashan, T. H., et al. 1969. The nature of hypnotic analgesia and placebo response to experimental pain. *Psychosomatic Medicine,* 31:227–46.

McGuigan, F. J. 1971. Covert linguistic behavior in deaf subjects during thinking. *Journal of Comparative & Physiological Psychology,* 75:417–20.

McIntyre, A. 1984. *After Virtue.* University of Notre Dame Press.

McKinney, S. 1975. How I broke the world's speed record. *Ski Magazine* (Spring), 39:7.

McMahon, C. E. 1973. Images as motives and motivators: A historical perspective. *American Journal of Psychology,* 86:465–90.

McMahon, C. E. 1974. Voluntary control of "involuntary functions": The approach of the Stoics. *Psychophysiology,* 11:710–14.

McMahon, C. E. 1976. The role of imagination in the disease process: Pre-Cartesian history. *Psychological Medicine,* 6:180–83.

McMahon, C., & A. Sheikh. 1984. Imagination in disease and healing processes: A historical perspective. In A. Sheikh (ed.), *Imagination and Healing.* Baywood.

Mead, G. R. S. 1919, 1967. *The Doctrine of the Subtle Body in Western Tradition.* Theosophical Publishing House, Quest Book.

Mead, M. 1956. Personal character and the cultural milieu. In D. Harring (ed.), *On the Implications for Anthropology of the Geselling Approach to Maturation.* Syracuse University Press.

Mead, M., & F. MacGregor. 1951. *Growth and Culture: A Photographic Study of Balinese Childhood.* Putnam.

Meerloo, J. A. 1948–51. Papers read to the Medical Section, American Society of Psychical Research.

Meerloo, J. A. 1949. Telepathy as a form of archaic communication. *Psychiatric Quarterly,* 23:691–704.

Meerloo, J. A. 1950. *Patterns of Panic.* International Universities Press.

Meerloo, J. A. 1953. *Communication and Conversation.* International Universities Press.

Meichenbaum, D. 1977. *Cognitive-Behavior Modification: An Integrative Approach.* Plenum.

Menotti, A., & V. Paddu. 1976. Death rates among the Italian railroad employees, with special reference to coronary heart disease and physical activity at work. *Environmental Research,* 11:331–42.

Merleau-Ponty, M. 1962. *The Phenomenology of Perception.* Trans. C. Smith. Routledge & Kegan Paul.

Metalnikov, S., & V. Chorne. 1926. The role of conditioned reflex in immunity. *Annals Pasteur Institute,* 11:1–8.

Metta, L. 1972. Psychokinesis of lepidopterous larvae. *Journal of Parapsychology,* 36:213–21.

Metzner, R. 1986. *Opening to Inner Light.* Tarcher.

Meyhoff, H. H., et al. 1983. Placebo—the drug of choice in female motor urge incontinence? *British Journal of Urology,* 55:34–37.

Michael, D. N. 1973. *On Learning to Plan—and Planning to Learn.* Jossey-Bass, Inc.

Michael, D. N. 1983. Competence and compassion in an age of uncertainty. *World Future Society Bulletin*, XVII:1–6 (Jan/Feb).

Michael, D. N. 1991. Leadership's shadow: The dilemma of denial. *Futures* (Jan./Feb.) pp. 69–79.

Michener, J. 1976. *Sports in America*. Random House.

Mickelson, T., & F. Hagerman. 1982. Anaerobic threshold measurements of elite oarsmen. *Medicine & Science in Sports & Exercise*, 14:44.

Miles, W. 1964. Oxygen consumption during Yoga-type breathing patterns. *Journal of Applied Physiology*, 19:75–82.

Miller, A. V., trans. 1977. *Phenomenology of Spirit by G. W. F. Hegel*. With an analysis of the text and foreword by J. N. Findlay. Oxford University Press.

Miller, G. A., et al. 1960. *Plans and the Structure of Behavior*. Holt.

Miller, J., & D. Shanklin. 1976. *Pure Golf*. Doubleday.

Miller, J. G. 1978. *Living Systems*. McGraw-Hill.

Miller Lite Report on American Attitudes Toward Sports. 1983. Research & Forecasts, Inc.

Miller, N., & A. Banuazizi. 1968. Instrumental learning by curarized rats of a specific visceral response, intestinal or cardiac. *Journal of Comparative & Physiological Psychology*, 65:1–7.

Miller, N., & L. DiCara. 1967. Instrumental learning of heart-rate changes in curarized rats: Shaping and specificity to discriminative stimulus. *Journal of Comparative & Physiological Psychology*, 63:12–19.

Miller, N., & L. DiCara. 1968. Instrumental learning of urine formation by rats: Changes in renal blood flow. *American Journal of Physiology*, 215:677–83.

Miller, N., & B. Dworkin. 1974. Visceral learning: Recent difficulties with curarized rats and significant problems for human research. In P. A. Obrist (ed.), *Cardiovascular Psychophysiology*. Aldine.

Miller, R. 1977. Methods of detecting and measuring healing energies. In J. White & S. Krippner (eds.), *Future Science*. Anchor/Doubleday.

Miller, R. N. 1972. The positive effect of prayer on plants. *Psychic*, 3:24–25.

Mills, G., & K. Campbell. 1974. A critique of Gellhorn and Kiely's mystical states of consciousness. *Journal of Nervous & Mental Disease*, 159:191.

Mills, W., & J. Farrow. 1981. The TM technique and acute experimental pain. *Psychosomatic Medicine*, 43:157–64.

Milne, L., & M. Milne. 1972. *The Senses of Animals and Men*. Atheneum.

Minick, M. 1974. *The Wisdom of Kung Fu*. Morrow.

Minsky, M. 1986. *Society of Mind*. Simon & Schuster.

Mishima, Y. 1970. *Sun and Steel*. Kodansha International.

Mole, P., et al. 1971. Adaptation of muscle to exercise. *Journal of Clinical Investigation*, 50:323.

Moles, E. A. 1977. Zen meditation: A study of regression in service of the ego. *Dissertation Abstracts International*, 38:871–72.

Molloy, G. 1873. *A Visit to Louise Lateau.* London.

Moltmann, J. 1967. *The Theology of Hope.* Harper.

Monroe, R. A. 1971. *Journeys out of the Body.* Doubleday.

Moody, R. 1975. *Life After Life.* Mockingbird Books.

Moody, R. L. 1946. Bodily changes during abreaction. *Lancet* (Dec. 28), pp. 934–35.

Moody, R. L. 1948. Bodily changes during abreaction. *Lancet* (letter of June 19), p. 964.

Moore, L., & J. Kaplan, 1983. Hypnotically accelerated burn wound healing. *American Journal of Clinical Hypnosis,* 26:16.

Moorefield, C. W. 1971. The use of hypnosis and behavior therapy in asthma. *American Journal of Clinical Hypnosis,* 13:162–68.

Moretti-Altuna, G. E. 1987. The effects of meditation versus medication in the treatment of attention deficit disorder with hyperactivity. *Dissertation Abstracts International,* 47:658.

Morgan, A. H., et al. 1974. The stability of hypnotic susceptibility: A longitudinal study. *International Journal of Clinical & Experimental Hypnosis,* 22:249–57.

Morgan, C. L. 1923. *Emergent Evolution.* Henry Holt.

Morgan, W. 1982. Psychological effects of exercise. *Behavioral Medical Update,* 4: 25–30.

Morgan, W., & S. Goldston, eds. 1987. *Exercise and Mental Health.* Hemisphere.

Morganroth, J., et al. 1975. Comparative left ventricular dimensions in trained athletes. *Annals of Internal Medicine,* 82:521–24.

Morley, J. E., et al. 1987. Neuropeptides: Conductors of the immune orchestra. *Life Sciences,* 41:527–44.

Morphis, O. L. 1961. Hypnosis and its use in controlling patients with malignancy. *American Journal of Roentgenology,* 85:897–900.

Morris, J. N. 1975. *Uses of Epidemiology.* Churchill Livingstone.

Morris, J. N., et al. 1953. Coronary heart disease and physical activity of work. *Lancet,* 2:1053–57.

Morris, J. N., et al. 1966. Incidence and prediction of ischaemic heart-disease in London busmen. *Lancet,* 2:552–59.

Morris, J. N., et al. 1973. Vigorous exercise in leisure-time and the incidence of coronary heart-disease. *Lancet,* 1:333–39.

Morris, J. N., et al. 1980. Vigorous exercise in leisure-time: Protection against coronary heart disease. *Lancet,* 2:1207–10.

Morse, D. R. 1976a. Meditation in dentistry. *General Dentistry,* 24:57–59.

Morse, D. R. 1976b. Use of a meditative state for hypnotic induction in the practice of endodontics. *Oral Surgery, Oral Medicine, Oral Pathology,* 41:664–72.

Morse, D. R. 1977a. An exploratory study of the use of meditation alone and in combination with hypnosis in clinical dentistry. *Journal of the American Society of Psychosomatic Dentistry & Medicine,* 24:113–20.

Morse, D. R. 1977b. Overcoming practice stress via meditation and hypnosis. *Dental Survey*, 53:32–36.

Morse, D. R. 1977c. Variety, exercise, and meditation can relieve practice stress. *Dental Studies*, 56:26–29.

Morse, D. R., L. Cohen, M. Furst, & J. Martin. 1984. A physiological evaluation of the yoga concept of respiratory control of autonomic nervous system activity. *International Journal of Psychosomatics*, 31:3–19.

Morse, D. R., & C. Hilderbrand. 1976. Case report: Use of TM in periodontal therapy. *Dental Survey*, 52:36–39.

Morse, D. R., J. S. Martin, M. L. Furst et al. 1977. Physiological and subjective evaluation of meditation, hypnosis relaxation. *Psychosomatic Medicine*, 39:304–24.

Morse, D. R., G. R. Schacterle, M. L. Furst et al. 1981. Stress, relaxation, and saliva: A pilot study involving endodontic patients. *Oral Surgery* (Sept.), pp. 308–13.

Morse, D. R., G. R. Schacterle, M. L. Furst et al. 1982. The effect of stress and meditation on salivary protein and bacteria: A review and pilot study. *Journal of Human Stress*, 8:31–39.

Morse, D. R., R. Schoor, & B. Cohen. 1984. Surgical and non-surgical dental treatments for a multi-allergic patient with meditation-hypnosis as the sole anesthetic: Case report. *International Journal of Psychosomatics*, 31:27–33.

Morse, D. R., & J. M. Wilcko. 1979. Nonsurgical endodontic therapy for a vital tooth with meditation-hypnosis as the sole anesthetic: A case report. *American Journal of Clinical Hypnosis*, 21:258–62.

Morse, M., & P. Jerry. 1990. *Closer to the Light: Learning from the Near-Death Experiences of Children.* Ivy Books (Ballantine).

Motoyama, H. 1981. *Theories of the Chakras: Bridge to Higher Consciousness.* Theosophical Publishing House.

Mott, T., & J. Roberts. 1979. Obesity and hypnosis: A review of the literature. *American Journal of Clinical Hypnosis*, 22:3.

Moulton, R. 1943. Psychosomatic implication of pseudocyesis. *Psychosomatic Medicine*, 4:376–89.

Mouren, P. n.d. *The Cure of M. Serge Perrin.* Report to the International Medical Committee of Lourdes. Bureau Médical. Lourdes, France.

Mowrer, O. H. 1977. Mental imagery: An indispensable psychological concept. *Journal of Mental Imagery*, 1:303–26.

Mowrer, O., & W. Mowrer. 1938. Enuresis: A method for its study and treatment. *American Journal of Orthopsychiatry*, 8:436–59.

Muldoon, S., and H. Carrington. 1961. *The Projection of the Astral Body.* Rider.

Mulholland, T., & E. Peper. 1971. Occipital alpha and accommodative vergence, pursuit tracking, and fast eye movements. *Psychophysiology*, 8:556–75.

Muller, L. 1931. *Die Lebensnerven.* Springer.

Mullins, J. F., et al. 1955. Pachyonychia congenita: A review and new approach to treatment. *Archives of Dermatology*, 71:265–68.

Murchie, G. 1954. *Song of the Sky.* Riverside Press.

Murphy, M. J. 1973. Explorations in the use of group meditation with persons in psychotherapy. *Dissertation Abstracts International,* 33(12-B): 6089.

Murphy, M. 1977. *Jacob Atabet.* Tarcher.

Murphy, M., & J. Brodie. 1973. "I experience a kind of clarity." *Intellectual Digest* (Jan.), 3:5:19–22.

Murphy, M., & S. Donovan. 1990. *The Physical and Psychological Effects of Meditation.* Study of Exceptional Functioning, Esalen Institute.

Murphy, M., & R. White. 1978. *The Psychic Side of Sports.* Addison-Wesley.

Murray, J., & G. Abraham. 1978. Pseudocyesis: A review. *Obstetrics and Gynecology,* 51:627–31.

Musashi, M. 1974. *A Book of Five Rings.* Trans. V. Harris. Overlook Press.

Muskatel, N., R. Woolfold, P. Carrington et al. 1984. Effect of meditation training on aspects of coronary-prone behavior. *Perceptual & Motor Skills,* 58:515–18.

Myers, F. W. H. 1903, 1954. *Human Personality and Its Survival of Bodily Death.* Vols. 1 & 2. Longmans, Green.

Myers, T., & E. Eisner. 1974. *An Experimental Evaluation of the Effects of Karate and Meditation.* American Institutes for Research (Oct.).

Nabokov, P. 1981. *Indian Running.* Capra.

Nagao, G. 1991. *Madhyamika and Yogacard.* Trans. by L. Kawamura. State University of New York Press.

Nagel, E. 1979. *The Structure of Science.* Hackett.

Nagle, F., & H. Montoye. 1981. *Exercise in Health & Disease.* Charles C. Thomas.

Narayanaswami Aiyar, ed. 1914 trans. *Thirty Minor Upanishads.* Madras, India.

Nash, C. B. 1982. Psychokinetic control of bacterial growth. *Journal of the Society for Psychical Research,* 51:217–21.

Nash, C. B., & C. S. Nash. 1967. The effect of paranormally conditioned solution on yeast fermentation. *Journal of Parapsychology,* 31:314.

Nash, M., et al. 1984. The direct hypnotic suggestion of altered mind/body perception. *American Journal of Clinical Hypnosis,* 27: 95–102.

Nasr, S. H. 1989. *Knowledge of the Sacred.* State University of New York Press.

National Academy of Sciences Institute of Medicine. 1989. *Behavioral Influences on the Endocrine and Immune Systems. A Research Briefing.*

Needham, J. 1937. *Integrative Levels: A Revaluation of the Idea of Progress.* Oxford University Press.

Needles, W. 1943. Stigmata occurring in the course of psychoanalysis. *Psychoanalytic Quarterly,* 12:23–39.

Neff, D., & E. Blanchard. 1982. The relationship between capacity for absorption and headache patients' response to relaxation and biofeedback treatment. In *Self-Regulation Strategies: Efficacy and Mechanisms.* Biofeedback Society of America 13th Annual Meeting, Chicago, Mar. 5–8.

Neihardt, J. 1972. *Black Elk Speaks.* Pocket Books, Simon & Schuster.

New Catholic Encyclopedia. 1967. McGraw-Hill.

Newman, M. 1971. Hypnotic handling of the chronic bleeder in extraction: A case report. *American Journal of Clinical Hypnosis,* 14:126–27.

Nicassio, P. M., et al. 1982. Progressive relaxation, EMG biofeedback and biofeedback placebo in the treatment of sleep-onset insomnia. *British Journal of Medical Psychology*, 55:159–66.

Nichols, M. 1984. *Family Therapy*. Gardner.

Nicholson, N. C. 1920. Notes on muscular work during hypnosis. *Johns Hopkins Hospital Bulletin*, 31:89–91.

Nicklaus, J. 1974. *Golf My Way*. Simon & Schuster.

Nideffer, R. 1976. *The Inner Athlete*. Thomas Crowell.

Nidich, S., W. Seeman, & T. Dreskin. 1973. Influence of Transcendental Meditation on a measure of self-actualization: A replication. *Journal of Counseling Psychology*, 20:565–66.

Nieporte, T., & D. Sauers. 1968. *Mind over Golf*. Doubleday.

Nikhilananda, S. 1942. Introduction. *The Gospel of Sri Ramakrishna*. Ramakrishna-Vivekananda Center.

Nimuendaju, C. 1946. *Eastern Tinbira*. University Publications in American Archeology and Ethnology.

Nisbet, R. 1980. *The History of the Idea of Progress*. Basic Books.

Noldeke, T. 1892. *Sketches from Eastern History*. London.

Nolly, G. A. 1975. The immediate after-effects of meditation on perceptual awareness. *Dissertation Abstracts International*, 36(20B): 919.

Northoff, H., & A. Berg. 1991. Immunologic mediators as parameters of the reaction to strenuous exercise. *International Journal of Sports Medicine*, 12:S9–S15.

Norton, G. R. 1976. Biofeedback treatment of long-standing eye closure reactions. *Journal of Behavior Therapy and Experimental Psychiatry*, 7:279–80.

Novak, M. 1976. *The Joy of Sports*. Basic Books.

Novak, P. 1981. Empty willing: Contemplative being-in-the-world in St. John of the Cross and Dogen. Doctoral Dissertation, University Microfilms, Ann Arbor, MI.

Novak, P. 1984. The dynamics of the will in Buddhist and Christian practice. In *Buddhist-Christian Studies: 4*. East-West Religions Project, University of Hawaii.

Nowlis, D., & J. Kamiya. 1970. The control of electroencephalographic alpha rhythms through auditory feedback and the associated mental activity. *Psychophysiology*, 6:476–84.

Nowlis, D., & E. Wortz. 1973. Control of the ratio of midline parietal to midline frontal EEG alpha rhythms through auditory feedback. *Perceptual & Motor Skills*, 37:815–24.

O'Brien, R. M., et al. 1981. Augmentation of systematic desensitization of snake phobia through posthypnotic dream suggestion. *American Journal of Clinical Hypnosis*, 23:231.

Obrist, P. A., et al. 1970. The cardiac-somatic relationship: Some reformulations. *Psychophysiology*, 6:569–87.

Office of the Assistant Secretary for Health and Surgeon General. 1979. *Healthy People: The Surgeon General's Report on Health Promotion and Disease Preven-*

tion. DHEW (PHS) Publication No. 79-55071. U.S. Government Printing Office, Washington, DC.

Ogden, E., & N. Shock. 1939. Voluntary hypercirculation. *American Journal of the Medical Sciences,* 198:329–42.

Ogilvy, J. 1979. *Many Dimensional Man.* Harper Colophon.

Oh, S., & D. Falkner. 1984. *Sadaharu Oh: A Zen Way of Baseball.* Times Books.

O'Malley, J., & K. Conners. 1972. The effect of unilateral alpha training on visual evoked response in a dyslexic adolescent. *Psychophysiology,* 9:467–70.

Onetto, B., & G. Elquin. 1964. Psychokinesis in experimental tumerogenesis. *Journal of Parapsychology,* 30:220.

O'Regan, B. 1987. *Healing, Remission and Miracle Cures.* Institute of Noetic Sciences Special Report.

O'Regan, B., & C. Hirshberg. 1990. *Spontaneous Remission: An Annotated Bibliography of Selected Articles and Case Reports from the World Medical Literature.* Institute of Noetic Sciences.

O'Regan, B., & T. Hurley. 1985. Multiple personality—mirrors of a new model of mind? *Investigations.* Institute of Noetic Sciences.

Orme-Johnson, D. W. 1973. Autonomic stability and Transcendental Meditation. *Psychosomatic Medicine,* 35:341–49.

Orme-Johnson, D. W. 1977. Coherence during transcendental consciousness. *Electroencepahlography & Clinical Neurophysiology,* 43:581.

Orme-Johnson, D. W., & J. T. Farrow, eds. 1976. *Scientific Research on the Transcendental Meditation Program: Collected Papers.* Vol. 1. MERU Press.

Orme-Johnson, D., & B. Granieri. 1976. The effects of the age of enlightenment governor training courses on field independence, creativity, intelligence, and behavioral flexibility. In D. W. Orme-Johnson & J. T. Farrow (eds.), *Scientific Research on the Transcendental Meditation Program: Collected Papers.* Vol. 1. MERU Press.

Orme-Johnson, D., & C. Haynes. 1981. EEG phase coherence, pure consciousness, creativity, and TM-sidhi experiences. *International Journal of Neuroscience,* 13:211–17.

Orne, M., et al. 1986. Hypnotic experience: A cognitive social-psychological reality. Open peer commentary in N. Spanos, Hypnotic behavior: A social-psychological interpretation of amnesia, analgesia, and "trance logic." *Behavioral and Brain Sciences,* 9:477–78.

Ornish, D., et al. 1990. Can lifestyle changes reverse coronary heart disease? *Lancet,* 336:129–35.

Osis, K. 1952. A test of the occurrence of a psi effect between man and cat. *Journal of Parapsychology,* 16:233–56.

Osis, K., E. Bokert, M. Carlson et al. 1973. Dimensions of the meditative experience. *Journal of Transpersonal Psychology,* 5:109–35.

Osis, K., & E. Haraldsson. 1986. *At the Hour of Death.* Hastings House.

Otis, L. S. 1984. Adverse effects of transcendental meditation. In D. H. Shapiro and R. W. Walsh (eds.), *Meditation: Contemporary and Classical Perspectives.* Aldine.

Otto, R. 1960. *Mysticism East & West.* Macmillan.

Overton, D. 1978. Major theories of state-dependent learning. In B. Ho et al. (eds.), *Drug Discrimination and State-Dependent Learning.* Academic.

Oxendine, J. 1968. *Psychology of Motor Learning.* Meredith.

Oyama, M. 1965. *This Is Karate.* Japan Publications.

Paffenbarger, R., & R. Hyde. 1984. Exercise in the prevention of coronary heart disease. *Preventive Medicine, 13:5.*

Paffenbarger, R., et al. 1970. Work activity of longshoremen as related to death from coronary heart disease and stroke. *New England Journal of Medicine, 282:* 1109–14.

Paffenbarger, R., et al. 1984. A natural history of athleticism and cardiovascular health. *Journal of the American Medical Association, 252:491.*

Paffenbarger, R. S, et al. 1986. Physical activity, all-cause mortality, and longevity of college alumni. *New England Journal of Medicine, 314:607–9.*

Pagano, R., & L. Frumkin. 1977. The effect of Transcendental Meditation on right hemispheric functioning. *Biofeedback and Self-Regulation, 2:407–15.*

Paivio, A. 1973. Psychophysiological correlates of imagery. In F. McGuigan & R. Schoonover (eds.), *The Psychophysiology of Thinking.* Academic.

Pallis, M. 1974. Is there room for grace in Buddhism? In *Sword of Gnosis.* Penguin.

Panagiotou, N., & A. Sheikh. 1974. Eidetic psychotherapy: Introduction and evaluation. *International Journal of Social Psychiatry, 20:231–41.*

Panagiotou, N., & A. Sheikh. 1977. The image and the unconscious. *International Journal of Social Psychiatry, 23:169–86.*

Pannenberg, W. 1970. *What Is Man?* Fortress Press.

Papebroch, D. A memoir, de vita, operibus et virtutibus Joannis Bollandi. *Acta Sanctorum,* t. I Martii. Cited in D. Knowles, 1963, *Great Historical Enterprises.* Nelson.

Paramahansa Yogananda—In Memoriam. 1983. Self-Realization Fellowship.

Park, W. 1934. Paviotso shamanism. *American Anthropologist* (Jan.–Mar.), 36: 98–113.

Parker, J., & G. Gilbert. 1978. Anxiety management in alcoholics: A study of generalized effects of relaxation techniques. *Addictive Behavior, 3:123–27.*

Parr, J. 1976. *The Superwives.* Coward-McCann & Geoghegan.

Parulkar, V., S. Prabhavalker, & J. Bhall. 1974. Observations on some physiological effects of Transcendental Meditation. *Indian Journal of Medical Science, 28:* 156–58.

Patanjali's Yoga Sutras. 1972. Translated into English by I. K. Taimni as *The Science of Yoga.* Theosophical Publishing House.

Patel, C. H. 1973. Yoga and biofeedback in the management of hypertension. *Lancet,* 2:1053–55.

Patel, C. H. 1975a. Yoga and biofeedback in the management of hypertension. *Journal of Psychosomatic Research*, 19:355–60.

Patel, C. H. 1975b. Yoga and biofeedback in the management of "stress" in hypertensive patients. *Clinical Science & Molecular Medicine*, 48:171–74.

Patel, C. H. 1975c. Twelve-month follow-up of yoga and biofeedback in the management of hypertension. *Lancet*, 2:62–64.

Patel, C., & W. North. 1975. Randomized controlled trial of yoga and biofeedback in management of hypertension. *Lancet*, 2:93–95.

Pattengale, P., & J. Holloszy. 1967. Augmentation of skeletal muscle myoglobin by programs of treadmill running. *American Journal of Physiology*, 213:783.

Pattie, F. A. 1941. The production of blisters by hypnotic suggestion: A review. *Journal of Abnormal and Social Psychology*, 36:62–72.

Pattie, F. A. 1956. Mesmer's medical dissertation and its debt to Mead's De Imperio Solis et Lunae. *Journal of Medicine and Allied Sciences*, 11:275–87.

Pattison, E., et al. 1973. Faith healing. *Journal of Nervous & Mental Disease*, 157: 397–409.

Paul, G. L. 1963. The production of blisters by hypnotic suggestion: Another look. *Psychosomatic Medicine*, 25:233–44.

Paulsen, A. E. 1926. Religious healing. *Mental Hygiene*, 10:541–95.

Pawlowski, E., & M. Pawlowski. 1958. Unconscious and abortive aspects of pseudocyesis. *Wisconsin Medical Journal*, 57:437–40.

Pearson, R. E. 1961. Responses to suggestions given under general anesthesia. *American Journal of Clinical Hypnosis*, 4:106–14.

Pechar, G., et al. 1974. Specificity of cardio-respiratory adaptation to bicycle and treadmill training. *Journal of Applied Physiology*, 36:753.

Peck, D. J. 1977. The use of EMG feedback in the treatment of severe case of blepharospasm. *Biofeedback and Self-Regulation*, 2:273–77.

Pedersen, B. K. 1991. Influence of physical activity on the cellular immune system: Mechanisms of action. *International Journal of Sports Medicine*, 12:S23–S29.

Pedersen-Krag, G. 1947. Telepathy and repression. *Psychoanalytic Quarterly*, 6: 61–68.

Pedersen-Krag, G. 1948–1951. Papers read to the Medical Section, American Society of Psychical Research.

Pelletier, K. 1974. Influence of Transcendental Meditation upon autokinetic perception. *Perceptual & Motor Skills*, 39:1031–34.

Pelletier, K. 1976. The effects of the Transcendental Meditation program on perceptual style: Increased field independence. In D. W. Orme-Johnson & J. T. Farrow (eds.), *Scientific Research on the Transcendental Meditation Program: Collected Papers*. Vol. 1. MERU Press.

Pelletier, K. 1978. *Toward a Science of Consciousness*. Delacorte.

Pelletier, K., & C. Garfield. 1977. Meditative states of consciousness. In P. Zimbardo and C. Maslach (eds.), *Psychology of Our Times*. Scott, Foresman.

Pelletier, K., & E. Peper. 1977. Alpha EEG feedback as a means for pain control. *Journal of Clinical & Experimental Hypnosis*, 25:361–71.

Peper, E. 1970. Feedback regulation of the alpha electroencephalogram activity through control of internal and external parameters. *Kybernetik,* 7:107–12.

Peper, E. 1971a. Comment on feedback training of parietal-occipital alpha asymmetry in normal human subjects. *Kybernetik,* 9:156–58.

Peper, E. 1971b. Reduction of efferent motor commands during alpha feedback as a facilitator of EEG alpha and a precondition for changes in consciousness. *Kybernetik,* 9:226–31.

Peper, E. 1972. Localized EEG alpha feedback training: A possible technique for mapping subjective, conscious, and behavioral experiences. *Kybernetik,* 11: 166–69.

Peper, E., & S. Ancoli. 1979. The two end points of an EEG continuum of meditation: Alpha/theta and fast beta. In E. Peper, S. Ancoli, & M. Quinn, *Mind/Body Integrations: Essential Readings in Biofeedback.* Plenum.

Peper, E., & T. Mulholland. 1970. Methodological and theoretical problems in the voluntary control of electroencephalographic occipital alpha by the subject. *Kybernetik,* 7:10–13.

Perls, F. 1988. *Gestalt Therapy Verbatim.* Center for Gestalt Development.

Perls, F., et al. 1951. *Gestalt Therapy.* Dell.

Perry, W. J. 1926. *The Children of the Sun: A Study of the Early History of Civilization.* London.

Pert, C., et al. 1985. Neuropeptides and their receptors: A psychosomatic network. *Journal of Immunology,* 135:820–26.

Peters, R., H. Benson, D. Porter et al. 1977. Daily relaxation response breaks in a working population: 1. Health, Performance & Well-Being. *American Journal of Public Health,* 67:946–53.

Peterson, W. F. 1946. *Hippocratic Wisdom: For Him Who Wishes to Pursue Properly the Science of Medicine.* Charles C. Thomas.

Petrie, W. M. F. 1909. *Personal Religion in Egypt Before Christianity.* London.

Petrie, W. M. F. 1924. Egypt under Roman rule. In *A History of Egypt.* Vol. 5. J. G. Milne.

Phillips, S. H. 1986. *Aurobindo's Philosophy of Brahman.* E. J. Brill.

Phillips, S. H. 1989. "Mutable God": Hartshorne and Indian theism. In R. Kane & S. Phillips (eds.), *Hartshorne: Process Philosophy and Theology.* University of New York Press.

Phillips, S. M. 1980. *Extra-Sensory Perception of Quarks.* Theosophical Publishing House.

Pickering, T., & G. Gorham. 1975. Learned heart-rate control by a patient with a ventricular parasystolic rhythm. *Lancet* (Feb.), pp. 252–53.

Pickering, T., & N. Miller. 1977. Learned voluntary control of heart rate and rhythm in two subjects with premature ventricular contractions. *British Heart Journal,* 39:152–59.

Pickersgill, M., & W. White. 1984. Subjective states during Transcendental Meditation. *International Journal of Psychophysiology,* 23:229–30.

Pickett, J. M. 1968. Recent research on speech—analyzing aids for the deaf. I.E.E.E. *Transaction on Audio and Electroacoustics*, AU16:227–34.

Piddington, J. G. 1902. The fire-walk in Mauritius. *Journal of Society for Psychical Research*, Vol. 10 (190), (June), pp. 250–53.

Pierrakos, J. 1987. *Core Energetics*. Chap. 7. LifeRhythm.

Piggins D., & D. Morgan. 1977–78. Perceptual phenomena resulting from steady visual fixation and repeated auditory input under experimental conditions and in meditation. *Journal of Altered States of Consciousness*, 3:197–203.

Pinsent, J. 1983. "Bull-leaping" in Minoan society. *Proceedings of the Cambridge Colloquium*, 1981:263.

Pirot, M. 1977. The effects of the Transcendental Meditation technique upon auditory discrimination. In D. W. Orme-Johnson & J. T. Farrow (eds.), *Scientific Research on the Transcendental Meditation Program: Collected Papers*. Vol. 1. MERU Press.

Plato. *Timaeus*. Jowett translation.

Platonov, K. 1959. *The Word as a Psychological and Therapeutic Factor*. Foreign Language Publishing House. In A. Sheikh, 1984, *Imagination and Healing*. Baywood.

Platt, J. H. 1947. The Bell reduced visible symbol method for teaching speech to the deaf. *Journal of Speech Disorders*, 12:381–86.

Plotinus. *Enneads* XX. Stephen MacKenna translation.

Plotkin, W. 1979. The alpha experience revisited: Biofeedback in the transformation of psychological state. *Psychological Bulletin*, 86:1132–48.

Plotkin, W. 1985. A psychological approach to placebo: The role of faith in therapy and treatment. In L. White et al. (eds.), *Placebo: Theory, Research & Mechanisms*. Guilford Press.

Plotkin, W., & K. Rice. 1981. Biofeedback as a placebo: Anxiety reduction facilitated by training in either suppression or enhancement of alpha brainwaves. *Journal of Consulting and Clinical Psychiatry*, 49:590–96.

Podmore, F. 1963. *From Mesmer to Christian Science*. University Books.

Polanyi, M. 1964. *Personal Knowledge*. Harper Torchbook.

Polhemus, T., ed. 1978. *The Body Reader: Social Aspects of the Human Body*. Pantheon.

Pollack, A., M. Weber, D. Case et al. 1977. Limitations of Transcendental Meditation in the treatment of essential hypertension. *Lancet* 1:71–73.

Pollack, M., et al. 1974. Physiological characteristics of champion American track athletes 40 to 75 years of age. *Journal of Gerontology*, 29:645–49.

Pollard, G., & R. Ashton. 1982. Heart rate decrease: A comparison of feedback modalities and biofeedback with other procedures. *Biological Psychology*, 14:245–57.

Poortman, J. J. 1978. *Vehicles of Consciousness: The Concept of Hylic Pluralism*. Vols. 1–4. Theosophical Publishing House.

Pope, K., & J. Singer. 1978. *The Stream of Consciousness*. Plenum.

Poulain, A., S.J. 1950. *The Graces of Interior Prayer*. Routledge & Kegan Paul.

Powell, K., & R. Paffenbarger. 1985. Workshop on epidemiologic and public health aspects of physical activity and exercise: A summary. *Public Health Reports,* 100:118–26.

Powell, K. E., et al. 1987. Physical activity and the incidence of coronary heart disease. *Annual Review Public Health,* 8:253–87.

Price, G. R. 1955. Science and the supernatural. *Science,* 122:359–67.

Price, H. H. 1939. Haunting and the "psychic ether" hypotheses: With some preliminary reflections on the present condition and possible future of psychical research. *Proceedings of the Society for Psychical Research,* 45:307–43.

Progoff, I. 1963. *The Symbolic and the Real.* Julian Press.

Propp, V. I. 1949. *Le radici storiche dei racconti di fate.* Turin.

Public Health Reports. 1985a. 100:113–224.

Public Health Reports. 1985b. 113:202 (Mar.-Apr.).

Pulos, L. 1969. Hypnosis and Think Training with Athletes. Paper presented at the 12th Annual Scientific Meeting, American Society of Clinical Hypnosis, San Francisco.

Puryear, H., C. Cayce, & M. Thurston. 1976. Anxiety reduction associated with meditation: Home study. *Perceptual & Motor Skills,* 43:527–31.

Putnam, F. 1984. The psychophysiologic investigation of multiple personality disorder. *Psychiatric Clinics of North America,* 7:31–39.

Pyecha, J. 1970. Comparative effects of judo and selected physical education activities on male university freshman personality traits. *Research Quarterly,* 41: 425–31.

Quinn, J. 1982. An investigation of the effect of therapeutic touch without physical contact on state anxiety of hospitalized cardiovascular patients. *Dissertation Abstracts International,* University Microfilms, No. DA 82-26-788.

Quinn, J. 1984. Therapeutic Touch as energy exchange: Testing the theory. *Advances in Nursing Science* (Jan.), 6:42–49.

Quinn, J. 1988. Building a body of knowledge: Research on Therapeutic Touch. *Journal of Holistic Nursing,* 6:37–45.

Radhakrishnan, S., & C. Moore. 1957. *A Source Book in Indian Philosophy.* Princeton University Press.

Radin, D. 1989. Searching for "signatures" in anomalous human-machine interaction data: A neural network approach. *Journal of Scientific Exploration,* 3:185–200.

Rahner, K. 1965. *On the Theology of Death.* Herder & Herder.

Rahner, K. 1979. Experiencing the Spirit: Source of Theology. In *Theological Investigations.* Vol. 16. Seabury Press.

Raknes, O. 1970. The orgonomic concept of health and its social consequences. In *Wilhelm Reich and Orgonomy.* St. Martin's.

Ramazzini, B. 1964. *Diseases of Workers.* Trans. W. C. Wright. Hafner.

Randall, J. 1970. An attempt to detect psi effects with protozoa. *Journal of the Society for Psychical Research,* 45:294–96.

Random, M. 1978. *The Martial Arts.* Octopus.

Rao, K., & J. Palmer. 1987. The anomaly called psi: Recent research and criticism. *Behavioral and Brain Sciences,* 10(4):539–51.

Rasch, P., & E. O'Connell. 1963. TPS scores of experienced karate students. *Research Quarterly,* 34:108–10.

Rasch, P., & W. Pierson. 1963. Reaction and movement time of experienced karate-ka. *Research Quarterly,* 34:242–43.

Raskin, M., L. Bali, & H. Peeke. 1980. Muscle biofeedback and Transcendental Meditation: A controlled evaluation of efficacy in the treatment of chronic anxiety. *Archives of General Psychiatry,* 37:93–97.

Ratnoff, O. 1969. Stigmata: Where mind and body meet. *Medical Times* (June), 97:161.

Ratnoff, O., and D. Agle. 1968. Psychogenic purpura: A re-evaluation of the syndrome of autoerythrocyte sensitization. *Medicine,* 47:475–500.

Ratti, O., & A. Westbrook. 1973. *Secrets of the Samurai.* Tuttle.

Rauramaa, R. et al. 1984. Effects of mild physical exercise on serum lipoprotein and metabolites of arachidomic acid. *British Medical Journal,* 288:603–06.

Rauramaa, R. et. al. 1986. Inhibition of platelet aggregation by moderate-intensity physical exercise: A randomized clinical trial in overweight men. *Circulation,* 74:939–44.

Ray, W. J. 1974. The relationship of locus of control, self-report measures, and feedback to the voluntary control of heart rate. *Psychophysiology,* 11:527–34.

Ray, W., & S. Lamb. 1974. Locus of control and the voluntary control of heart rate. *Psychosomatic Medicine,* 36:180–82.

Razran, G. 1961. The observable unconscious and the inferable conscious in current Soviet psychophysiology. *Psychological Review,* 68:81–147.

Redgrove, P. 1987. *The Black Goddess and the Unseen Real.* Grove.

Redwood, D., et al. 1972. Circulatory and symptomatic effects of physical training in patients with coronary-artery disease and angina pectoris. *New England Journal of Medicine,* 286:959.

Reed, H. 1978. Improved dream recall association with meditation. *Journal of Clinical Psychology,* 34:150–56.

Rehyer, J., & W. Smeltzer. 1968. Uncovering properties of visual imagery and verbal association: A comparative study. *Journal of Abnormal Psychology,* 73:218–22.

Reich, W. 1942. *The Function of the Orgasm.* Orgone Institute.

Reich, W. 1970. *The Mass Psychology of Fascism.* Trans. V. Carfagno. Farrar, Straus & Giroux.

Reich, W. 1972. *Character Analysis.* Trans. V. Carfagno. Farrar, Straus & Giroux.

Reiman, J. W. 1985. The impact of meditative attentional training on measures of select attentional parameters and on measures of client perceived counselor empathy. *Dissertations Abstracts International,* 46:1569.

Research report: Setzer's sanctuary effect. 1980. *Spiritual Frontiers,* 12:20–23.

Reyher, J. 1963. Free imagery, an uncovering procedure. *Journal of Clinical Psychology,* 19:454–59.

Reyher, J. 1977. Spontaneous visual imagery: Implications for psychoanalysis, psychopathology, and psychotherapy. *Journal of Mental Imagery*, 2:253.

Rhine, J. B. 1934. *Extra-sensory Perception*. Bruce Humphries.

Rhine, L. 1961. *Hidden Channels of the Mind*. Chap. 8. William Sloane.

Rhine, L. 1967. *ESP in Life and Lab*. Collier.

Ricci, F. 1979. *Tantra: Rites of Love*. Rizzoli International.

Rice, D. G. 1966. Operant conditioning and associated electromyogram responses. *Journal of Experimental Psychology*, 71:908–12.

Richardson, A. 1967. Mental practice: A review and discussion. Part I. *The Research Quarterly*, 38:95–107.

Richardson, A. 1983. Imagery: Definition and types. In A. Sheikh (ed.), *Imagery— Current Theory, Research and Application*. Wiley.

Richet, C. 1923. *Thirty Years of Psychical Research*. Macmillan.

Richmond, N. 1952. Two series of PK tests on paramecia. *Journal of the Society for Psychical Research*, 36:577–88.

Richter, B., & A. Kellner. 1965. Hypertrophy of the human heart at the level of fine structure: Analysis of two postulates. *Journal of Cell Biology*, 18:195.

Ring, K. 1980. *Life at Death*. Coward, McCann & Geoghegan.

Ring, K. 1984. *Heading Toward Omega*. William Morrow.

Robertson, D. W. 1983. The short and long range effects of the Transcendental Meditation technique on fractionated reaction time. *Journal of Sports Medicine*, 23:113–20.

Robinson, E. 1977a. *The Original Vision: A Study of the Religious Experience of Childhood*. Religious Experience Research Unit, Oxford University.

Robinson, E., ed. 1977b. *This Time-Bound Ladder: Ten Dialogues on Religious Experience*. Religious Experience Research Unit, Oxford University.

Robinson, E. 1978. *Living the Question*. Religious Experience Research Unit, Oxford University.

Robinson, J. 1967. *Exploration into God*. Stanford University Press.

Rocard, Y. 1964. Actions of a very weak magnetic gradient: The reflex of the dowser. In M. Barnothy (ed.), *Biological Effects of Magnetic Fields*. Plenum.

Rock, N., et al. 1969. Hypnosis with untrained, nonvolunteer patients in labor. *International Journal of Clinical & Experimental Hypnosis*, 17:25–36.

Rogo, D. S. 1970. *Nad: A Study of Some Unusual "Other-World" Experiences*. University Books.

Rogo, D. S. 1972. A psychic study of "The Music of the Spheres." In *Nad*. Vol. 2. University Books.

Rohdenburg, G. 1918. Fluctuations in the growth energy of malignant tumors in man with especial reference to spontaneous recession. *Journal of Cancer Research*, 3:193.

Rolf, I. P. 1973. A contribution to the understanding of stress. *Confinia Psychiatrica*, 16:69–79.

Roll, D. L. 1973. Modification of nasal resonance in cleft-palate children by informative feedback. *Journal of Applied Behavior Analysis*, 6:397–403.

Rolland, R. 1910. *Jean Christophe*. Henry Holt.

Ross, J. 1977. The effects of the Transcendental Meditation program on anxiety, neuroticism, and psychoticism. In D. W. Orme-Johnson & J. T. Farrow (eds.), *Scientific Research on the Transcendental Meditation Program: Collected Papers*. Vol. 1. MERU Press.

Rossi, E. 1972, 1985. *Dreams and the Growth of Personality: Expanding Awareness in Psychotherapy*. Brunner/Mazel.

Rossi, E., & M. Ryan, eds. 1986. Mind-body communication in hypnosis. In *The Seminars, Lectures, and Workshops of Milton H. Erickson*. Vol. 3. Irvington.

Rothpearl, A. 1980. Personality traits in martial artists: A descriptive approach. *Perceptual & Motor Skills*, 50:395–401.

Rousseau, P. 1978. *Ascetics, Authority, and the Church in the Age of Jerome and Cassian*. Oxford University Press.

Routt, T. 1977. Low normal heart and respiration rates in individuals practicing the Transcendental Meditation Technique. In D. W. Orme-Johnson & J. T. Farrow (eds.), *Scientific Research on the Transcendental Meditation Program: Collected Papers*. Vol. 1. MERU Press.

Rowell, L. 1974. Human cardiovascular adjustments to exercise and thermal stress. *Physiology Review*, 54:75.

Rubik, B., & E. Rauscher. 1980. Effects on motility behavior and growth rate of salmonella typhimurium in the presence of Olga Worrall. In W. G. Roll (ed.), *Research in Parapsychology 1979*. Scarecrow Press.

Rulison, R. H. 1942. Warts: A statistical study of nine hundred and twenty-one cases. *Archives of Dermatology and Syphilology*, 46:66–81.

Rush, J. H. 1964. *New Directions in Parapsychological Research*. Parapsychology Foundation.

Russell, E. 1975. Parapsychic luminosities. *Quadrant* (Winter), 8:2, pp. 58–61.

Russell, N., trans. 1980. *The Lives of the Desert Fathers*. A translation of the *Historia Monachorum in Aegypto*. Mowbray.

Russell, P. J. 1972. Transcendental Meditation. *Lancet*, 1:1125.

Sabel, B. 1980. TM and concentration ability. *Perceptual & Motor Skills*, 50:799–802 (Pt. 1).

Sacerdote, P. 1970. Theory and practice of pain control in malignancy and other protracted or recurring painful illnesses. *International Journal of Clinical & Experimental Hypnosis*, 3:160–80.

Sacerdote, P. 1977. Applications of hypnotically elicited mystical states to the treatment of physical and emotional pain. *International Journal of Clinical & Experimental Hypnosis*, 25:309–24.

Sachs, L. B., et al. 1977. Hypnotic self-regulation of chronic pain. *American Journal of Clinical Hypnosis*, 20:106–13.

Sachs, M., & G. Buffone, eds. 1984. Running therapy and psychology: A selected bibliography. In *Running as Therapy: An Integrated Approach*. University of Nebraska Press.

Sackeim, H. A., et al. 1979. Classroom seating and hypnotic susceptibility. *Journal of Abnormal Psychology,* 88:81–84.

Sacks, O. 1985a. The autist artist. Chap. 24 in *The Man Who Mistook His Wife for a Hat.* Summit.

Sacks, O. 1985b. The twins. *The New York Review of Books* (Feb. 28), pp. 16–20.

Sacks, O. 1990. *Seeing Voices.* Harper Perennial.

Saint John of the Cross. 1952. *The Complete Works of John of the Cross.* Vols. 1–3. Trans. P. Silverio deSanta Teresa & E. A. Peers. Newman Press.

Saint Teresa of Avila. *Book of Foundations.* Trans. E. A. Peers. Sheed & Ward.

Saint Teresa of Avila. 1972. *The Complete Works of Saint Teresa of Jesus.* Vols. 1–3. Trans. E. A. Peers. Sheed & Ward.

Saint Teresa of Avila. 1976. *The Collected Works.* Trans. K. Kavanaugh. Institute of Carmelite Studies.

Saint Thomas Aquinas. 1975. *Summa Contra Gentiles,* Books 1–4. Trans. Charles O'Neil. University of Notre Dame.

Sakiyama, R., et al. 1984. Cyclic Cushing's syndrome. *American Journal of Medicine,* 77:944.

Salmon, M.-M. 1972. *The Extraordinary Cure of Vittorio Micheli.* Report to the International Medical Committee of Lourdes. Bureau Médical, Lourdes, France.

Salter, A. 1949. *Conditioned Reflex Therapy.* Farrar, Straus.

Saltin, B. 1969. Physiological effects of physical conditioning. *Medicine & Science in Sports & Exercise,* 1:50.

Saltin, B., et al. 1977. Fiber types and metabolic potentials of skeletal muscles in sedentary man and endurance runners. *Annals of New York Academy of Sciences,* 301:3.

Sannella, L. 1987. *The Kundalini Experience.* Integral Publishing.

Sapir, E. 1949. In D. G. Mandelbaum (ed.), *Selected Writings of Edward Sapir.* University of California Press.

Saraswati, S. 1973. *Kundalini Yoga.* Bihar School of Yoga.

Saraswati, S. 1984. *Kundalini Tantra.* Bihar School of Yoga.

Sarbin, T. R., & M. Andersen. 1967. Role-theoretical analysis of hypnotic behavior. In J. E. Gordon (ed.), *Handbook of Clinical and Experimental Hypnosis.* Macmillan.

Sarbin, T. R., & W. Coe. 1972. *Hypnosis: A Social Psychological Analysis of Influence Communication.* Holt.

Sarbin, T. R., & D. Lim. 1963. Some evidence in support of the role-taking hypothesis in hypnosis. *International Journal of Clinical & Experimental Hypnosis,* 11: 98–103.

Sarles, H., et al. 1977. A study of the variations in the response regarding duodenal ulcer when treated with placebo by different investigators. *Digestion,* 16: 289–92.

Satya Prakash Singh. 1972. *Sri Aurobindo and Whitehead on the Nature of God.* Vigyan Prakashan.

Satyanarayanamurthi, G. G., & B. P. Shastry. 1958. Preliminary scientific investigation into some of the unusually physiological manifestations acquired as a result of yogic practices in India. *Wien. Z. Nervenheilk,* 15:239–49.

Schachtel, E. G. 1959. *Metamorphosis: On the Development of Affect, Perception, Attention, and Memory.* Basic Books.

Schafer, E. A., et al. 1886. On the rhythm of muscular response to volitional impulses in man. *Journal of Physiology,* 7:111–19.

Schechter, E. I. 1984. Hypnotic induction vs. control conditions: Illustrating an approach to the evaluation of replicability in parapsychological data. *Journal of the American Society for Psychical Research,* 78:1–27.

Scheuer, J., & C. Tipton. 1977. Cardiovascular adaptations to physical training. *Annual Review Physiology,* 39:221–51.

Schilder, P. 1950. *The Image and Appearance of the Human Body: Studies in the Constructive Energies of the Psyche.* International Universities Press.

Schimberg, A. P. 1947. *The Story of Therese Neumann.* Bruce.

Schmidt, K. E. 1976. Transcendental Meditation. *British Medical Journal,* 1:459.

Schmoger, C. E. 1976. *The Life of Anne Catherine Emmerich.* Vols. 1 & 2. Tan Books.

Schneck, J. M. 1966. A study of alterations in body sensations during hypnoanalysis. *International Journal of Clinical & Experimental Hypnosis,* 14:216–31.

Scholem, G. 1961. *Jewish Mysticism.* Schocken Books.

Schultz, D. 1978. Imagery and the control of depression. In J. Singer & K. Pope (eds.), *The Power of Human Imagination.* Plenum.

Schultz, J., & W. Luthe. 1959. *Autogenic Training.* Grune & Stratton.

Schuon, F. 1972. *Understanding Islam.* Harcourt Brace.

Schuon, F. 1975. *Logic and Transcendence.* Harper & Row.

Schuon, F. 1976. *Islam and the Perennial Philosophy.* World of Islam Festival Publishing.

Schuon, F. 1984. *The Transcendent Unity of Religions.* Quest Book.

Schuster, M. M. 1968. Motor action of the rectum and anal sphincters in continence. In C. Code & C. Prosser (eds.), *Handbook of Physiology.* Williams & Wilkins.

Schuster, M. M. 1974. Operant conditioning in gastrointestinal dysfunction. *Hospital Practice,* 9:135.

Schuster, M. 1989. Biofeedback control of gastrointestinal motility. In J.V. Basmajian (ed.), *Biofeedback—Principles and Practice for Clinicians.* Williams & Wilkins.

Schuster, M. M., et al. 1974. Biofeedback control of lower esophageal sphincter contraction in man. In E. E. Daniel (ed.), *Fourth International Symposium on Gastrointestinal Motility.* Mitchell Press.

Schwartz, G. E. 1975. Biofeedback, self-regulation, and the patterning of physiological processes. *American Scientist,* 63:314–24.

Schwartz, G. E., et al. 1976. Heart rate regulation as skill learning: Strength-endurance versus cardiac reaction time. *Psychophysiology,* 13:472–78.

Schwartz, G., R. Davidson, & D. Goleman. 1978. Patterning of cognitive and somatic processes in self-regulation of anxiety. *Psychosomatic Medicine,* 40:321–28.

Schwartz, S., et al. 1990. Infrared spectra alteration in water proximate to the palms of therapeutic practitioners. *Subtle Energies,* 1:1:43–72.

Schwarz, B. E. 1971. *Parent–Child Telepathy.* Garrett.

Schweickart, R. 1977. No frames, no boundaries. In *Earth's Answer.* Lindisfarne Books.

Schweikert, D. T. 1981. Stigmata. *Catholic Digest* (Oct.).

Scott, R. W., et al. 1973. A shaping procedure for heart rate control in chronic tachycardia. *Perceptual & Motor Skills,* 37:327–38.

Scrignar, C. B. 1981. Rapid treatment of contamination phobia with hand-washing compulsion by flooding with hypnosis. *American Journal of Clinical Hypnosis,* 23:252.

Scully, H., & J. Basmajian. 1969. Motor-unit training and influence of manual skill. *Psychophysiology,* 5:625–32.

Scully, V. 1972. *Pueblo: Mountain, Village, Dance.* Viking.

Scweder, R. A. 1972. Aspects of cognition in Zinacatenco shamans: Experimental results. In W. Lessa & E. Vogt (eds.), *Reader in Comparative Religion.* Harper & Row.

Seabourne, T., R. Weinberg, & A. Jackson. 1984. Effect of individualized practice and training of visuo-motor behavior rehearsal in enhancing karate performance. *Journal of Sport Behavior,* 7:58–67.

Seer, P., & J. Raeburn. 1980. Meditation training and essential hypertension: A methodological study. *Journal of Behavioral Medicine,* 3:59–70.

Selzer, R. 1976. *Mortal Lessons.* Simon & Schuster.

Servadio, E. 1932. Proofs and counter-proofs concerning human "fluid." *Journal of the American Society for Psychical Research,* 26:168–73, 199–206.

Sguillante and Pagani. 1748. *Vita della Serafina di Dio.* Cited in Thurston 1952.

Shackleton, E. H. 1947. *South.* Macmillan.

Shafii, M. 1973. Silence in the service of ego: Psychoanalytic study of meditation. *International Journal of Psychoanalysis,* 54:431–43.

Shafii, M., R. Lavely, & R. Jaffe. 1974. Meditation and marijuana. *American Journal of Psychiatry,* 131:60–63.

Shafii, M., R. Lavely, & R. Jaffe. 1975. Meditation and the prevention of alcohol abuse. *American Journal of Psychiatry,* 132:924–45.

Shainberg, L. 1989. Finding the zone. *New York Times Magazine* (Apr. 9).

Shapiro, A. K. 1959. The placebo effect in the history of medical treatment: Implications for psychiatry. *American Journal of Psychiatry,* 116:298–304.

Shapiro, A. K. 1964. Factors contributing to the placebo effect: Their implications for psychotherapy. *American Journal of Psychotherapy* (Supplement), 18:73–88.

Shapiro, A. 1970. Placebo effects in psychotherapy and psychoanalysis. *Journal of Clinical Pharmacology* (Mar./Apr.), 73–78.

Shapiro, A. K., & L. A. Morris. 1978. The placebo effect in medical and psychological therapies. In *Handbook of Psychotherapy and Behavior Change.* Wiley.

Shapiro, D., J. Shapiro, R. Walsh et al. 1982. Effects of intensive meditation on sex role identification: Implications for a control model of psychological health. *Psychological Reports,* 51:44–46.

Shapiro, D., & R. Surwit. 1976. Learned control of physiological function and disease. In H. Leitenberg (ed.), *Handbook of Behavior Modification and Behavior Therapy.* Prentice-Hall.

Shapiro, D., et al. 1964. Differentiation of an autonomic response through operant reinforcement. *Psychonomic Science,* 1:147–48.

Shapiro, D., et al. 1969. Effects of feedback and reinforcement on the control of human systolic blood pressure. *Science,* 163:588–89.

Shapiro, D., et al. 1970. Control of blood pressure in man by operant conditioning. *Circulation Research* (Supplement 1), 27:27–32.

Shapiro, D. H. 1976. Zen meditation and behavioral self-management applied to a case of generalized anxiety. *Psychologia,* 19:134–38.

Shapiro, D. H. 1978. Meditation and the East: The Zen Master. In D. H. Shapiro, *Precision Nirvana.* Prentice-Hall.

Shapiro, D. H. 1980a. Meditation and holistic medicine. In A. Hastings, J. Fadiman, & J. Gordon (eds.), *Holistic Medicine.* National Institute of Mental Health.

Shapiro, D. H. 1980b. *Meditation: Self-Regulation Strategy and Altered States of Consciousness: A Scientific Personal Exploration.* Aldine.

Shapiro, D. H. 1982. Overview: Clinical and physiological comparison of meditation and other self-control strategies. *American Journal of Psychiatry,* 139:267–74.

Shapiro, D. H., & D. Giber. 1978. Meditation and psychotherapeutic effects: Self-regulation strategy and altered states of consciousness. *Archives of General Psychiatry,* 35:294–302.

Shapiro, D. L. 1970. The significance of the visual image in psychotherapy. *Psychotherapy: Theory, Research, and Practice,* 7:209–12.

Shaw, G., & S. Belmore. 1982–83. The relationship between imagery and creativity. *Imagination, Cognition & Personality,* 2:115–23.

Shaw, R., & D. Kolb. 1977. Reaction time following the Transcendental Meditation technique. In D. W. Orme-Johnson & J. T. Farrow (eds.), *Scientific Research on the Transcendental Meditation Program: Collected Papers.* Vol. 1. MERU Press.

Shearn, D. W. 1962. Operant conditioning of heart rate. *Science,* 137:530–31.

Sheehan, E. P., et al. 1982. A signal detection study of the effects of suggested improvement on the monocular visual acuity of myopes. *International Journal of Clinical & Experimental Hypnosis,* 30:138–46.

Sheehan, P. W. 1968. Some Comment on the Nature of Visual Imagery: A Problem of Affect. Paper presented at the meeting of the International Society for Mental Imagery Techniques, Geneva, Switzerland.

Sheehan, P. W. 1979. Hypnosis and the process of imagination. In E. Fromm & R. Shor (eds.), *Hypnosis: Developments in Research and New Perspectives.* Aldine.

Sheehan, P. W. 1982. Imagery and hypnosis—forging a link, at least in part. *Research Communications in Psychology, Psychiatry and Behavior,* 7:257–72.

Sheer, D. E. 1970. Electrophysiological correlates of memory consolidation. In G. Ungar (ed.), *Molecular Mechanisms in Memory and Learning*. Plenum.

Sheer, D. E. 1974. Electroencephalographic studies in learning disabilities. In H. Eichenwald & A. Talbot (eds.), *The Learning Disabled Child*. University of Texas Health Sciences Center.

Sheer, D. E. 1975. Biofeedback training of 40-Hz EEG and behavior. In N. Burch & H. Altschuler (eds.), *Behavior & Brain Electrical Activity*. Plenum.

Sheer, D. E. 1976. Focused arousal and 40 Hz EEG. In R. M. Knights & D. Bakker (eds.), *The Neuropsychology of Learning Disorders*. University Park.

Sheikh, A. 1976. Treatment of insomnia through eidetic imagery: A new technique. *Perceptual & Motor Skills*, 43:994.

Sheikh, A. 1978. Eidetic psychotherapy. In J. Singer & K. Pope (eds.), *The Power of Human Imagination*. Plenum.

Sheikh, A., ed. 1984. *Imagination and Healing*. Baywood.

Sheikh, A., & C. Jordan. 1981. Eidetic psychotherapy. In R. J. Corsini (ed.), *Handbook of Innovative Psychotherapies*. Wiley.

Sheikh, A., & C. Jordan. 1983. Clinical uses of mental imagery. In A. Sheikh (ed.), *Imagery—Current Theory, Research, and Application*. Wiley.

Sheikh, A., & N. Panagiotou. 1975. Use of mental imagery in psychotherapy: A critical review. *Perceptual & Motor Skills*, 41:555–85.

Sheikh, A., et al. 1979. Psychosomatics and mental imagery: A brief review. In A. Sheikh & J. Shaffter (eds.), *The Potential of Fantasy and Imagination*. Brandon House.

Shepard, R. N. 1978. The mental image. *American Psychologist*, 33:125–37.

Shepherd, J. 1967. Behavior of resistance and capacity vessels in human limbs during exercise. *Circulation Research*, 20 (Supplement I):70.

Sherman, S. 1971. Very deep hypnosis: An experiential and electroencephalographic investigation. Doctoral dissertation, Stanford University.

Shimano, E., & D. Douglas. 1975. On research in Zen. *American Journal of Psychiatry*, 132:1300–2.

Shirokogoroff, S. 1935. *Psychomental Complex of the Tungus*. London. Cited in M. Eliade, 1964, *Shamanism*. Princeton University Press, Bollingen Series.

Shor, R. E. 1959. Hypnosis and the concept of the generalized reality-orientation. *American Journal of Psychotherapy*, 13:582–602.

Shor, R. E. 1962. Three dimensions of hypnotic depth. *International Journal of Clinical & Experimental Hypnosis*, 10:23–38.

Shor, R. E. 1967. Physiological effects of painful stimulation during hypnotic analgesia. In J. E. Gordon (ed.), *Handbook of Clinical and Experimental Hypnosis*. Macmillan.

Shor, R. 1979. The fundamental problem in hypnosis research as viewed from historic perspectives. In E. Fromm & R. Shor (eds.), *Hypnosis: Developments in Research & New Perspectives*. Aldine.

Shorr, J. E. 1972. *Psycho-Imagination Therapy: The Integration of Phenomenology and Imagination*. Intercontinental Medical Book Corp.

745

Shorr, J. E. 1978. Clinical use of categories of therapeutic imagery. In J. Singer & K. Pope (eds.), *The Power of Human Imagination*. Plenum.

Shostrom, E. L. 1966. *Personal Orientation Inventory*. Educational and Industrial Testing Service.

Siegel, E. F. 1979. Control of phantom limb pain by hypnosis. *American Journal of Clinical Hypnosis*, 21:285–86.

Sieroszewski, Q. 1902. Du Chamanisme d'après les croyances des Yakoutes. *RHR*, 46:317. Cited in M. Eliade, 1964, *Shamanism*. Princeton University Press, Bollingen Series.

Silverman, J. 1986. *Agnews Project Report to the Rolf Institute*. Boulder, CO.

Silverman, J. et al. 1973. Stress, stimulus intensity control, and the Structural Integration technique. *Confinia Psychiatrica*, 16:201–19.

Silverman, L. 1982. *The Search for Oneness*. International University Press.

Simard, T., & J. Basmajian. 1967. Methods in training the conscious control of motor units. *Archives of Physical Medicine & Rehabilitation*, 48:1:12–19.

Simard, T., & H. Ladd. 1969. Pre-orthotic training: An electromyographic study in normal adults. *American Journal of Physical Medicine*, 48:301–12.

Sime, W. 1984. Psychological benefits of exercise. *Advances*, 1:15–20.

Simon, D., S. Oparil, & C. Kimball. 1977. The Transcendental Meditation program and essential hypertension. In D. W. Orme-Johnson & J. T. Farrow (eds.), *Scientific Research on the Transcendental Meditation Program: Collected Papers*. Vol. 1. MERU Press.

Simonton, C., S. Simonton-Matthews, & J. Creighton. 1978. *Getting Well Again*. Tarcher.

Simpson, G. G. 1949. *The Meaning of Evolution*. Yale University Press.

Simpson, G. G. 1953. *The Major Features of Evolution*. Columbia University Press.

Simpson, G. G. 1960. *The Meaning of Evolution*. Yale University Press.

Simpson, H. M., & A. Paivio. 1966. Changes in pupil size during an imagery task without motor involvement. *Psychonomic Science*, 5:405–6.

Sinclair-Geiben, A., & D. Chalmers. 1959. Evaluation of treatment of warts by hypnosis. *Lancet*, 2:480–82.

Singer, J. L. 1974. *Imagery and Daydream Methods in Psychotherapy and Behavior Modification*. Academic.

Singer, J., & K. Pope. 1978. The use of imagery and fantasy techniques in psychotherapy. In J. Singer & K. Pope (eds.), *The Power of Human Imagination*. Plenum.

Singer, J., & E. Switzer. 1980. *Mind Play: The Creative Uses of Imagery*. Prentice-Hall.

Singh, B. S. 1984. Ventilatory response to CO_2. II. Studies in neurotic psychiatric patients and practitioners of transcendental meditation. *Psychosomatic Medicine*, 46:347–62.

Sinha, S., S. Prasad, K. Sharma et al. 1978. An experimental study of cognitive control and arousal processes during meditation. *Psychologia*, 21:227–30.

Sipprelle, C. N. 1967. Induced anxiety. *Psychotherapy: Theory, Research and Practice*, 4:36–40.

Sirag, S.-P. 1985. Unpublished paper presented to a colloquium of physicists and mathematicians at Georgetown University.

Siscovick, D., et al. 1985. The disease-specific benefits and risks of physical activity and exercise. *Public Health Reports,* 100:180–88.

Skinner, B. F. 1974. *About Behaviorism.* Alfred Knopf.

Slattery, M., & D. Jacobs. 1988. Physical fitness and cardiovascular disease mortality (The U.S. Railroad Study). *American Journal of Epidemiology,* 123:571–80.

Slocum, J. 1956. *Sailing Alone Around the World.* Dover.

Smart, N. 1964. *Doctrine and Argument in Indian Philosophy.* London.

Smith, G. M., & H. K. Beecher. 1960. Amphetamine, secobarbital, and athletic performance. II. Subjective evaluations of performance, mood, and physical states. *Journal of the American Medical Association,* 172:1502–14.

Smith, H. 1976. *Forgotten Truth.* Harper & Row.

Smith, H. 1983a. In defense of spiritual discipline. In J. Duerlinger (ed.), *Ultimate Reality and Spiritual Discipline.* Rose of Sharon Press.

Smith, H. 1983b. Spiritual Discipline in Zen and Comparative Perspective. Paper presented at the International Symposium for Religious Philosophy, Institute for Zen Studies, Hanazono College, Kyoto, Japan, Mar. 26–30.

Smith, H. 1987. Is there a perennial philosophy? *Journal of the American Academy of Religion,* pp. 553–66.

Smith, H. 1989. *Beyond the Post-Modern Mind.* Theosophical Publishing House.

Smith, H. W. 1978. Effects of set on subjects' interpretation of placebo marijuana effects. *Social Science and Medicine,* 12:107–9.

Smith, J. C. 1975a. Meditation as psychotherapy: A review of the literature. *Psychological Bulletin,* 32:553–64.

Smith, J. C. 1975b. Psychotherapeutic effects of Transcendental Meditation with controls for expectation of relief and daily sitting. *Journal of Consulting and Clinical Psychology,* 44:630–37.

Smith, M. J. 1968. Paranormal effects on enzyme activity. *Journal of Parapsychology,* 32:281 (Abstract).

Smith, M. J. 1972. Paranormal effect on enzyme activity. *Human Dimensions,* 1:15–19.

Smith, M. J. 1977. The influence of "laying-on" of hands. In N. M. Regush (ed.), *Frontiers of Healing.* Avon.

Smith, M. L., et al. 1980. *The Benefits of Psychotherapy.* Johns Hopkins University Press.

Smith, O. C. 1934. Action potentials from single motor units in voluntary contraction. *American Journal of Physiology,* 108:629–38.

Smith, R. 1974. *Hsing-I.* Kodansha International.

Smith, S. 1953. Magic, medicine and religion. *British Medical Journal* (Apr. 18), p. 848.

Smith, T. R. 1977. The Transcendental Meditation technique and skin resistance response to loud tones. In D. W. Orme-Johnson & J. T. Farrow (eds.), *Scientific Research on the Transcendental Meditation Program: Collected Papers.* Vol. 1. MERU Press.

Smythies, J. R., ed. 1967. *Science and ESP*. Routledge & Kegan Paul.

Snel, F. 1980. PK influence on malignant cell growth. *Research Letter,* Parapsychology Laboratory, University of Utrecht, 10:19–27.

Snel, F., & B. Millar. 1982. PK with the enzyme trypsin. Unpublished manuscript.

Snoeckx, L., et al. 1982. Echocardiographic dimensions in athletes in relation to their training programs. *MSSE,* 14:426–34.

Sokol, D. 1989. Spiritual breakthroughs in sex. In G. Feuerstein, 1989, *Enlightened Sexuality*. Crossing Press.

Solfvin, J. 1984. Mental healing. In S. Krippner (ed.), *Advances in Parapsychological Research*. Vol. 4. McFarland.

Solomon, G. F. 1987. Psychoneuroimmunology: Interactions between the central nervous system and immune system. *Journal of Neuroscience Research,* 18:1–9.

Solomon, G. F. 1991. Psychosocial factors, exercise, and immunity: Athletes, elderly persons, and AIDS patients. *International Journal of Sports Medicine,* 12: S50–S52.

Solomon, G. F., & R. Moos. 1964. Emotions, immunity, and disease: A speculative theoretical integration. *Archives of General Psychiatry,* 11:657–74.

Solomon, R. C. 1983. *In the Spirit of Hegel: A Study of G. W. F. Hegel's Phenomenology of Spirit*. Oxford University Press.

Sonstroem, R. J. 1984. Exercise and self-esteem. In R. Terjung (ed.), *Exercise and Sport Sciences Reviews*. Vol. 12, American College of Sports Medicine Series. Heath.

Spanos, N. 1986. Hypnotic behavior: A social-psychological interpretation of amnesia, analgesia and "trancelogic." *Behavioral and Brain Sciences,* 9:449–500.

Spanos, N., J. Gottlieb, & S. Rivers. 1980. The effects of short-term meditation practice on hypnotic responsivity. *Psychological Record,* 30:343–48.

Spanos, N., S. Rivers, & J. Gottlieb. 1978. Hypnotic responsivity, meditation and laterality of eye movements. *Journal of Abnormal Psychology,* 87:566–69.

Spanos, N., & C. Saad. 1984. Prism adaptation in hypnotically limb-anesthetized subjects: More deconfirming data. *Perceptual & Motor Skills,* 59:379–86.

Spanos, N., et al. 1974. Cognition and self-control: Cognitive control of painful input. In H. London & R. Nisbett (eds.), *Thought and Feeling: Cognitive Alteration of Feeling States*. Aldine.

Spaulding, S. 1977. Right from the start. *Karate Illustrated* (Mar.), pp. 26–27.

Spiegel, D. 1989. Uses and abuses of hypnosis. *Integr. Psychiatry,* 6:218.

Spiegel, D., & A. Leonard. 1983. Naloxone fails to reverse hypnotic alleviation of chronic pain. *Psychopharmacology,* 81:140–43.

Spiegel, D., et al. 1985. Hypnotic hallucination alters evoked potentials. *Journal of Abnormal Psychology,* 94:249–55.

Spiegel, D., et al. 1989. Hypnotic alteration of somatosensory perception. *American Journal of Psychiatry,* 146:752.

Sri Aurobindo. 1970. *The Life Divine*. Vols. 18 & 19 in *The Collected Works*. Pondicherry, India: Sri Aurobindo Ashram.

Sri Aurobindo. 1970–76. *The Collected Works.* Sri Aurobindo Ashram.

Sri Aurobindo. 1972. *Letters on Yoga.* Vol. 22 in *The Collected Works.* Pondicherry, India: Sri Aurobindo Ashram.

Sri Ramakrishna. 1969. *The Gospel of Sri Ramakrishna.* Trans. Swami Nikhilananda. Ramakrishna-Vivekananda Center.

Sroufe, L. A. 1969. Learned stabilization of cardiac rate with respiration experimentally controlled. *Journal of Experimental Psychology,* 81:391–93.

Sroufe, L. A. 1971. Effects of depth and rate of breathing on heart rate and heart rate variability. *Psychophysiology,* 8:648–55.

St. Albans, S. 1983. *Magic of a Mystic.* Clarkson Potter.

Stace, W. T. 1987. *Mysticism & Philosophy.* Tarcher.

Staib, A., & D. Logan. 1977. Hypnotic stimulation of breast growth. *American Journal of Clinical Hypnosis,* 19:201–8.

Stampfl, T., & D. Levis. 1967. Essentials of therapy: A learning theory-based psychodynamic behavioral therapy. *Journal of Abnormal Psychology,* 72:496–503.

Starkman, M., et al. 1985. Pseudocyesis: Psychologic and neuroendocrine interrelationships. *Psychosomatic Medicine,* 47:46–57.

Stebbins, G. L. 1969. *The Basis of Progressive Evolution.* University of North Carolina Press.

Steinberg, A., et al. 1949. Psychoendocrine relationship in pseudocyesis. *Psychosomatic Medicine,* 8:176–79.

Steiner, J. 1963. *Therese Neumann: La Stigmatisee de Konnersreuth.* Editions Meddens.

Steiner, J. 1973, 1977. *Visionen der Therese Neumann.* Vols. 1 & 2. Schnell & Steiner.

Steiner, S., & W. Dince. 1981. Biofeedback efficacy studies. *Biofeedback & Self-Regulation,* 6:275–88.

Steinmark, S., and T. Borkovec. 1974. Active and placebo treatment effects on moderate insomnia under counterdemand and positive demand instructions. *Journal of Abnormal Psychology,* 83:157–63.

Stekel, W. 1921. *Der Telepathische Traum.* Johannes Baum Verlag.

Stephens, J. H., et al. 1975. Psychological and physiological variables associated with large magnitude voluntary heart rate changes. *Psychophysiology,* 12:381–87.

Stephens, T., et al. 1985. A descriptive epidemiology of leisure-time physical activity. *Public Health Reports,* 100:149.

Stern, M. 1977. The effects of the Transcendental Meditation program on trait anxiety. In D. W. Orme-Johnson & J. T. Farrow (eds.), *Scientific Research on the Transcendental Meditation Program: Collected Papers.* Vol. 1. MERU Press.

Stern, R., & B. Kaplan. 1967. Galvanic skin response: Voluntary control and externalization. *Journal of Psychosomatic Research,* 10:349–53.

Stern, R., & N. Lewis. 1968. Ability of actors to control their GSR's and express emotions. *Psychophysiology,* 4:294–99.

Sternbach, R. 1964. The effects of instructional sets on autonomic responsivity. *Psychophysiology,* 1:67–72.

Stevens, J. 1985. The founder, Uyeshiba Morehei. In *Aikido and the New Warrior,* 1985. North Atlantic.

Stevens, J. 1987. *Abundant Peace: The Biography of Morehei Uyeshiba.* Shambhala.

Stevens, K. N., et al. 1974. Use of a visual display of nasalization to facilitate training of velar control for deaf speakers (Report No. 2899). Bolt Beranek & Newman.

Stevenson, I. 1970. Characteristics of cases of the reincarnation type in Turkey and their comparison with cases in two other cultures. *International Journal of Comparative Sociology* (March), 11:11–17.

Stevenson, I. 1975–1983. *Cases of the Reincarnation Type.* Vols. 1–4. University Press of Virginia.

Stigsby, B., J. C. Rodenburg, & H. Moth. 1981. Electroencephalographic findings during mantra meditation (Transcendental Meditation): A controlled, quantitative study of experienced meditators. *Electroencephalography & Clinical Neurophysiology,* 51:434–42.

Stone, M. H. 1980. *The Borderline Syndromes: Constitution, Personality and Adaptation.* McGraw-Hill.

Stone, R., & J. DeLeo. 1976. Psychotherapeutic control of hypertension. *New England Journal of Medicine,* 2:80–84.

Stoyva, J. 1976. A psychophysiological model of stress disorders as a rationale for biofeedback training. In F. J. McGuigan (ed.), *Tension Control: Proceedings of the Second Meeting of the American Association for the Advancement of Tension Control.* University Publications.

Stoyva, J. 1976. Self-regulation and the stress-related disorders: A perspective on biofeedback. In D. I. Mostofsky (ed.), *Behavior Control and Modification of Physiological Activity.* Prentice-Hall.

Stoyva, J. M. 1989. Autogenic Training and biofeedback combined: A reliable method for the induction of general relaxation. In J. V. Basmajian (ed.), *Biofeedback—Principles and Practice for Clinicians.* Williams & Wilkins.

Strub, C. 1967 (4th ed.). *The Principles and Practices of Embalming.* L. G. Darko Frederick.

Sturdevant, R., et al. 1977. Antacid and placebo produced similar pain relief in duodenal ulcer patients. *Gastroenterology,* 72:1–5.

Sugi, Y., & K. Akutsu. 1968. Studies on respiration and energy: Metabolism during sitting in zazan. *Research Journal of Physiology,* 12:190–206.

Suinn, R. 1976. Body thinking: Psychology for Olympic athletes. *Psychology Today* (July), 10:38–43.

Suinn, R. 1983. Imagery in sports. In A. Sheik (ed.), *Imagery: Current Theory, Research, and Application.* Wiley.

Suinn, R. M. 1985. Imagery rehearsal: Application to performance enhancement. *Behavior Therapist,* 8:155–59.

Sulzberger, M., & J. Wolf. 1934. The treatment of warts by suggestion. *Medical Record,* 140:552–57.

Surwillo, W., & D. Hobson. 1978. Brain electrical activity during prayer. *Psychological Reports,* 43:135–43.

Surwit, R., & F. Keefe. 1983. The blind leading the blind: Problems with the "double-blind" design in clinical biofeedback research. *Biofeedback & Self-Regulation,* 8:1–2.

Surwit, R., D. Shapiro, M. Good et al. 1978. Comparison of cardiovascular biofeedback, neuromuscular biofeedback, and meditation in the treatment of borderline essential hypertension. *Journal of Consulting & Clinical Psychology,* 46:252–63.

Sutich, A. 1980. Transpersonal psychotherapy. In S. Boorstein, *Transpersonal Psychotherapy.* Science & Behavior Books.

Suzuki, D. T. 1959. *Zen and Japanese Culture.* Princeton University Press, Bollingen Series.

Suzuki, D. T. 1965a. *The Training of the Zen Buddhist Monk.* University Books.

Suzuki, D. T. 1965b. *Zen and Japanese Culture.* Princeton University Press.

Suzuki-roshi, S. 1970. *Zen Mind, Beginner's Mind.* Weatherhill.

Swami Gitananda. 1973. An Evaluation of Siddhis and Riddhis. Paper presented to the Second International Yoga Festival, London.

Swedenborg, E. 1960a. *Arcana Coelestia.* Twelve vols. Swedenborg Society.

Swedenborg, E. 1960b. *Heaven & Hell.* Swedenborg Foundation.

Swedenborg, E. 1962. *The Spiritual Diary.* Vols. 1–5. Swedenborg Society.

Taimni, I. K. 1961. *The Science of Yoga.* Quest Books.

Taneli, B., & W. Krahne. 1987. EEG changes of Transcendental Meditation practitioners. *Advances in Biological Psychiatry,* 16:41–71.

Tanzer, D. 1972. *Why Natural Childbirth?* Doubleday.

Tarchanoff, J. R. 1885. Uber die willkurliche acceleration der herzschlage beim menschen (Voluntary acceleration of the heart beat in man). *Pfluger's Archive der Gesamten Physiologie,* 35:109–35. In A. Yates, 1980, *Biofeedback and the Modification of Behavior.* Plenum.

Targ, R., & K. Harary. 1984. *The Mind Race.* Villard.

Targ, R., & H. Puthoff. 1977. *Mind-Reach.* Delta.

Tart, C. T. 1966. Types of hypnotic dreams and their relation to hypnotic depth. *Journal of Abnormal Psychology,* 71:377–82.

Tart, C. T. 1967. The control of nocturnal dreaming by means of posthypnotic suggestion. *Parapsychology,* (Sept.), pp. 184–89.

Tart, C. T. 1967. A second psychophysiological study of out-of-the-body experiences in a gifted subject. *Parapsychology,* (Dec.), pp. 251–58.

Tart, C. T. 1970. Marijuana intoxication: common experiences. *Nature,* 226: 701–4.

Tart, C. T. 1970. Transpersonal potentialities of deep hypnosis. *Journal of Transpersonal Psychology,* 2:27–40.

Tart, C. T. 1971. A psychologist's experience with Transcendental Meditation. *Journal of Transpersonal Psychology,* 3:135–40.

Tart, C. T. 1972. States of consciousness and state specific sciences. *Science*, 176:1203–10.

Tart, C. T. 1975. *The Application of Learning Theory to ESP Performance*. Parapsychology Foundation.

Tart, C. T. 1975. *Transpersonal Psychologies*. Harper & Row.

Tart, C. T. 1976. *Learning to Use Extrasensory Perception*. University of Chicago Press.

Tart, C. T. 1977. Toward conscious control of psi through immediate feedback training: Some considerations of internal processes. *The Journal of the American Society for Psychical Research*, 71:375–407.

Tart, C. T. 1979. Measuring the depth of an altered state of consciousness, with particular reference to self-report scales of hypnotic depth. In E. Fromm & R. Shor (eds.), *Hypnosis: Developments in Research and New Perspectives*. Aldine.

Tart, C. T., H. Puthoff, & R. Targ. 1979. *Mind at Large*. Praeger.

Tart, C. T. 1983. *States of Consciousness*. Psychological Processes.

Tart, C. T. 1984. Moscow-San Francisco remote viewing experiment. *Psi Research*, No. 3/4 (Sept./Dec.).

Tart, C. T. 1987. The world simulation process in waking and dreaming: A systems analysis of structure. *Journal of Mental Imagery*, 11:145–58.

Tart, C. T. 1990. *Altered States of Consciousness*. (3rd ed.) Harper-Collins.

Tart, C. T. 1990. Multiple personality, altered states and virtual reality: The world simulation process approach. *Dissociation*, 3:222–33.

Tart, C. T., & A. Deikman. 1991. Mindfulness, spiritual seeking, and psychotherapy. *The Journal of Transpersonal Psychology*, 23:29–52.

Taylor, C., et al. 1985. The relation of physical activity and exercise to mental health. *Public Health Reports*, 100:195–202.

Taylor, H. L., et al. 1962. Death rates among physically active sedentary employees of the railroad industry. *American Journal of Public Health*, 52:1697–1707.

Tebecis, A. K. 1975. A controlled study of the EEG during Transcendental Meditation: Comparison with hypnosis. *Folia Psychiatrica et Neurologica Japonica*, 29:305–13.

Tedder, W. H., & M. L. Monty. 1981. Exploration of long-distance PK: A conceptual replication of the influence on a biological system. In W. Roll & J. Beloff (eds.), *Research in Parapsychology 1980*. Scarecrow Press.

Teilhard de Chardin. 1959. *The Phenomenon of Man*. Harper.

Tellegen, A., & G. Atkinson. 1974. Openness to absorbing and self-altering experiences ("absorption"), a trait related to hypnotic susceptibility. *Journal of Abnormal Psychology*, 83:268–77.

Teodorowitz, J. 1940. *Mystical Phenomena in the Life of Theresa Neumann*. Trans. Rev. R. Kraus. B. Herder Book Co.

Terray, L. 1964. *The Borders of the Impossible*. Doubleday.

Thaxton, L. 1982. Physiological and psychological effects of short-term exercise addiction on habitual runners. *Journal of Sport Psychology*, 4:73–80.

Theiss, F. B. 1971. Karate, a scientific view. *Samurai* (Summer), pp. 42–47.

Thomas, D., & K. Abbas. 1978. Comparison of TM and Progressive Relaxation in reducing anxiety. *British Medical Journal*, 2:1749.

Thomas, L. 1979. Warts. *Human Nature*, 2:58–59.

Thompson, J. K., et al. 1983. The control issue in biofeedback training. *Biofeedback & Self-Regulation*, 8:153–64.

Thompson, K. 1991. *Angels and Aliens*. Addison-Wesley.

Thouless, R. H. 1960. The empirical evidence for survival. *Journal of the American Society for Psychical Research* (Jan.), 54:23–32.

Throll, D. 1981. Transcendental Meditation and Progressive Relaxation: Their psychological effects. *Journal of Clinical Psychology*, 37:776–81.

Throll, D. 1982. Transcendental Meditation and Progressive Relaxation: Their physiological effects. *Journal of Clinical Psychology*, 38:522–30.

Thurston, H. 1935. *The Church and Spiritualism*. Bruce.

Thurston, H. 1952. *The Physical Phenomena of Mysticism*. Burns Oates.

Thurston, H. 1955. *Surprising Mystics*. Henry Regnery.

Thurston, H., and D. Attwater. 1981 ed. *Butler's Lives of the Saints*. Vols. 1–4. Christian Classics.

Time magazine. 1975. Baryshnikov: Gotta dance. (May 19), pp. 44–50.

Tinbergen, N. 1974. Ethology and stress disease. *Science* (July), 185:20–27.

Tipton, C. 1984. Exercise training and hypertension. In R. Terjung (ed.), *Exercise and Sport Sciences Reviews*. Vol. 12. American College of Sports Medicine. Heath.

Tipton, C. M., et al. 1975. The influence of physical activity on ligaments and tendons. *Medicine & Science in Sports & Exercise*, 7:165.

Tjoa, A. 1975. Meditation, neuroticism and intelligence: A follow-up. *Gedrag: Tijdschrift voor Psychologie*, 3:167–82.

Tomassetti, J. T. 1985. An investigation of the effects of EMG biofeedback training and relaxation training on dimensions of attention and learning of hyperactive children. *Dissertation Abstracts International*, 45:2081.

Toomey, T., & S. Sanders. 1983. Group hypnotherapy as an active control strategy in chronic pain. *American Journal of Clinical Hypnosis*, 26:20–25.

Tower, R., & J. Singer. 1981. The measurement of imagery: How can it be clinically useful? In P. Kendall & S. Holland (eds.), *Cognitive-Behavioral Interventions: Assessment Methods*. Academic.

Treece, P. 1989. *The Sanctified Body*. Chap. 4. Doubleday.

Treichel, M., N. Clinch, & M. Cran. 1973. The metabolic effects of Transcendental Meditation. *The Physiologist*, 16:471.

Trinick, J. 1967. *The Fire-Tried Stone*. Stuart & Watkins.

Trouton, D. S. 1957. Placebos and their psychological effects. *Journal of Mental Science*, 103:344–54.

Trowbridge, G. 1970. *Swedenborg: Life and Teaching*. Swedenborg Foundation.

Tuke, D. H. 1872. *Illustrations of the Influence of the Mind Upon the Body in Health and Disease Designed to Elucidate the Action of the Imagination*. J. & A. Churchill.

Tulku Thondup Rinpoche. 1989. *Buddha Mind. An Anthology of Longchen Rabjam's Writings on Dzogpa Chenpo*. Ed. H. Talbott. Snow Lion Publications.

Tulpule, T. 1971. Yogic exercises in the management of ischemic heart disease. *Indian Heart Journal*, 23:259–64.

Tulving, E. 1985. How many memory systems are there? *American Psychologist*, 40: 385–98.

Udupa, K. N. 1973. Certain studies in psychological and biochemical responses to the practice of hatha yoga in young normal volunteers. *Indian Journal of Medical Research*, 61:237–44.

Uhlenbruck, G., & U. Order. 1991. Can endurance sports stimulate immune mechanisms against cancer and metastasis? *International Journal of Sports Medicine*, 12:S636–68.

Ullman, M. 1947. Herpes simplex and second degree burn induced under hypnosis. *American Journal of Psychiatry*, 103:828–30.

Ullman, M. 1948. Communication. *Journal of the American Society of Psychical Research*, 42.

Ullman, M. 1952. On the nature of resistence to psi experiences. *Journal of the American Society of Psychical Research*, 46.

Ullman, M. 1959. On the psyche and warts: I. Suggestion and warts: A review and comments. *Psychosomatic Medicine*, 21:473–88.

Ullman, M., and S. Dudek. 1960. On the psyche and warts: II. Hypnotic suggestion and warts. *Psychosomatic Medicine*, 22:68–76.

Underhill, E. 1961. *Mysticism*. Dutton.

Underhill, R. 1954. *Workaday Life of the Pueblos*. Bureau of Indian Affairs.

Underhill, R. 1975. The salt pilgrimage. In *Teachings from the American Earth*. Liveright.

Underhill, R., et al. 1979. *Rainhouse and Ocean: Speeches for the Papago Year*. Museum of Northern Arizona.

Uyeshiba, M. 1983. Excerpts from the writings and transcribed lectures of the founder. *Aiki News* (Jan.), 52:3–12.

Vahia, H., D. Doongaji, D. Jest et al. 1973. Further experience with the therapy based upon concepts of Patanjali in the treatment of psychiatric disorders. *Indian Journal of Psychiatry*, 15:32–37.

Vaitl, D. 1975. Biofeedback-Einsatz in der Behandlung einer Patientin mit Sinustachykardie. In H. Legewie & L. Nusselt (eds.), *Biofeedback Therapie*. Urban and Schwarzenberg.

Vakil, R. J. 1950. Remarkable feat of endurance by a yogi priest. *Lancet* (Dec. 23), p. 871.

Van De Castle, R. L. 1969. The facilitation of ESP through hypnosis. *American Journal of Clinical Hypnosis*, 12:37–56.

Van den Berg, W., & B. Mulder. 1976. Psychological research on the effects of the TM technique on a number of personality variables. *Gedrag: Tijdschrift voor Psychologie*, 4:206–18.

Van Der Hart, O. 1981. Treatment of a phobia for dead birds: A case report. *American Journal of Clinical Hypnosis*, 23:263.

van der Post, L. 1975. *Jung and the Story of Our Time*. Pantheon.

Van Dusen, W. 1972. *The Natural Depth in Man*. Harper & Row.

Van Nuys, D. 1973. Meditation, attention and hypnotic susceptibility: A correlational study. *International Journal of Clinical & Experimental Hypnosis*, 21: 59–69.

Van Twyer, H., & H. Kimmel. 1966. Operant conditioning of the GSR with concomitant measurement of two somatic variables. *Journal of Experimental Psychology*, 72:841–46.

Vandercar, D. H., et al. 1977. Instrumental conditioning of human heart rate during free and controlled respiration. *Biological Psychology*, 5:221–31.

Vasiliev, L. L. 1976. *Experiments in Distant Influence*. Dutton.

Vasse, P., & C. Vasse. 1948. Influence de la pensee sur la croissance des plantes. *Nouvelle Serie*, 2:87.

Vassiliadis, A. 1973. Physiological effects of Transcendental Meditation: A longitudinal study. In D. Kanellakos & J. Lukas (eds.), *Psychobiology of Transcendental Meditation: A Literature Review*. Stanford Research Institute.

Vaughan, F. 1986. *The Inward Arc*. Shambhala, New Science Library.

Verma, I., B. Jayashan, & M. Palani. 1982. Effect of Transcendental Meditation on the performance of some cognitive psychological tests. *International Journal of Medical Research*, 7:136–43.

Vermes, G. 1978. *Jesus the Jew*. Fortress Press.

Vilenskaya, L. 1980. On PK and related subjects' research in the USSR. In W. Uphoff & M. Uphoff, *Mind over Matter*. New Frontiers Center.

Vogt, D., & G. Sultan. 1977. *Reality Revealed: The Theory of Multidimensional Reality*. Vector Associates.

Von Schrenck Notzing, A. 1975. *Phenomena of Materialization*. Arno.

von Urban, R. 1952. *Sex Perfection*. Rider.

Wachsmuth, D., T. Dolce, & K. Offenloch. 1980. Computerized analysis of the EEG during Transcendental Meditation and sleep. *Electroencephalography & Clinical Neurophysiology*, 48:39.

Wada, J., & A. Hamm. 1974. Electrographic glimpse of meditative state: Chronological observations of cerebral evoked response. *Electroencephalography & Clinical Neurophysiology*, 37:201.

Waddell, H. 1957. *The Desert Fathers*. University of Michigan Press.

Wagner, C., et al. 1974. Mutidimensional locus of control and voluntary control of GSR. *Perceptual & Motor Skills*, 39:1142.

Wainwright, W. 1981. *Mysticism*. University of Wisconsin.

Waley, A. 1956. *The Way and Its Power*. George Allen and Unwin Ltd.

Walker, J. 1980. The amateur scientist. *Scientific American* (July), pp. 150–61.

Wallace, B., & J. Garrett. 1973. Reduced felt arm sensation effects on visual adaptation. *Perception and Psychophysiology*, 14:597–600.

Wallace, B., & J. Garrett. 1975. Perceptual adaptation with selective reductions of felt sensations. *Perception*, 4:437–45.

Wallace, B., & K. Hoyenga. 1980. Production of proprioceptive errors with induced hypnotic anesthesia. *International Journal of Clinical & Experimental Hypnosis*, 28:140–47.

Wallace, R., H. Benson, A. Gattozzi et al. 1971. Physiological effects of a meditation technique and a suggestion for curbing drug abuse. *Mental Health Program Reports*, National Institute of Mental Health.

Wallace, R., H. Benson, & A. Wilson. 1971. A wakeful hypometabolic state. *American Journal of Physiology*, 221:795–99.

Wallace, R., H. Benson, A. Wilson et al. 1971. Decreased blood lactate during TM. *Federation Proceedings*, 30:376.

Wallace, R., J. Silver, P. Mills et al. 1983. Systolic blood-pressure and long-term practice of the Transcendental Meditation and TM-sidhi program: Effects of TM on systolic blood pressure. *Psychosomatic Medicine*, 45:41–46.

Wallace, R. K. 1970. Physiological effects of Transcendental Meditation. *Science*, 167:1751–54.

Wallace, R. K., & H. Benson. 1972. The physiology of meditation. *Scientific American*, 226:84–90.

Wallace, R. K., et al. 1971. A wakeful hypometabolic physiologic state. *American Journal of Physiology*, 221:795–99.

Walrath, L., & D. Hamilton. 1975. Autonomic correlates of meditation and hypnosis. *American Journal of Clinical Hypnosis*, 17:190–97.

Walsh, D. H. 1974. Interactive effects of alpha feedback and instructional set on subjective state. *Psychophysiology*, 11:428–35.

Walsh, R. N. 1977. Initial meditative experience: Part I. *Journal of Transpersonal Psychology*, 9:151–92.

Walsh, R. N. 1978. Initial meditative experience: Part II. *Journal of Transpersonal Psychology*, 10:1–28.

Walsh, R. N. 1979. Meditation research: An introduction and review. *Journal of Transpersonal Psychology*, 11:161–74.

Walsh, R., D. Goleman, et al. 1978. Meditation: Aspects of research and practice. *Journal of Humanistic Psychology*, 10:2.

Walsh, R., & L. Rauche. 1979. Precipitation of acute psychotic episodes by intensive meditation in individuals with a history of schizophrenia. *American Journal of Psychiatry*, 136:1085–86.

Walsh, R., & F. Vaughan, eds. 1980. *Beyond Ego: Transpersonal Dimensions in Psychology*. Tarcher.

Wandhoefer, A., & K. Plattig. 1973. Stimulus-linked DC-shift and auditory evoked potentials in Transcendental Meditation. *Pfleuger's Archives*, 343:R79.

Warshal, D. 1980. Effects of the TM technique on normal and Jendrassik reflex time. *Perceptual & Motor Skills*, 50:1103–6.

Watanabe, T., D. Shapiro, & G. Ochwartz. 1972. Meditation as an anoxic state: A critical review and theory. *Psychophysiologia*, 9:29.

Water, F. 1970. *Masked Gods.* Ballantine.

Watkins, G., & A. Watkins. 1971. Possible PK influence on the resuscitation of anesthetized mice. *Journal of Parapsychology,* 35:257–72.

Watkins, G. K., et al. 1973. Further studies on the resuscitation of anesthetized mice. In W. Roll, R. Morris, & J. Morris (eds.), *Research in Parapsychology 1972.* Scarecrow Press.

Webster, H. 1948. *Magic: A Sociological Study.* Stanford University Press.

Weinstein, D., & R. Bell. 1982. *Saints and Society: The Two Worlds of Western Christendom, 1000–1700.* University of Chicago.

Weiss, T., & B. Engel. 1971. Operant conditioning of heart rate in patients with premature ventricular contractions. *Psychosomatic Medicine,* 33:301–21.

Weitzenhoffer, A. M. 1951. The discriminatory recognition of visual patterns under hypnosis. *Journal of Abnormal Social Psychology,* 46:388–97.

Weitzenhoffer, A. M. 1953. *Hypnotism: An Objective Study in Suggestibility.* John Wiley.

Weitzenhoffer, A. M., & B. Sjoberg. 1961. Suggestibility with and without "induction of hypnosis." *Journal of Nervous & Mental Disease,* 132:204–20.

Wells, R., and J. Klein. 1972. A replication of a "psychic healing" paradigm. *Journal of Parapsychology,* 36:144–49.

Wells, R., and G. Watkins. 1975. Linger effects in several PK experiments. In J. Morris, W. Roll, & R. Morris (eds.), *Research in Parapsychology 1974.* Scarecrow Press.

Wells, W. 1940. The extent and duration of post-hypnotic amnesia. *Journal of Psychology,* 9:137–51.

Wells, W. R. 1947. Expectancy versus performance in hypnosis. *Journal of General Psychology,* 35:99–119.

Welwood, J. 1976. Exploring mind: Form, emptiness and beyond. *Journal of Transpersonal Psychology,* 8:89–99.

Wenger, M. A., & B. K. Bagchi. 1961. Studies of autonomic functions in practitioners of yoga in India. *Behavioral Science,* 6:312–23.

Wenger, M. A., et al. 1961. Experiments in Indian on "voluntary" control of the heart and pulse. *Circulation,* 24:6:1319–25.

Wepukhulu, H. S. 1973. Soccer soothsayer. *Africa Report* (Nov.-Dec.), 18:23.

West, D. J. 1957. *Eleven Lourdes Miracles.* Duckworth.

West, H., & W. Savage. 1918. Voluntary acceleration of the heart beat. *Archives of Internal Medicine,* 22:290–95.

West, M. A. 1977. Changes in skin resistance in subjects resting, reading, listening to music, or practicing the Transcendental Meditation technique. In D. W. Orme-Johnson & J. T. Farrow (eds.), *Scientific Research on the Transcendental Meditation Program: Collected Papers.* Vol. 1. MERU Press.

West, M. A. 1979. Physiological effects of meditation: A longitudinal study. *British Journal of Social & Clinical Psychology,* 18:219–26.

West, M. A. 1980a. Meditation and the EEG. *Psychological Medicine*, 10:369–75.

West, M. A. 1980b. The psychosomatics of meditation. *Journal of Psychosomatic Research*, 24:265–73.

Westbrook, A., & O. Ratti. 1970. *Aikido and the Dynamic Sphere.* Charles E. Tuttle.

Westcott, M. 1977. Hemispheric symmetry of the EEG during the Transcendental Meditation technique. In D. W. Orme-Johnson & J. T. Farrow (eds.), *Scientific Research on the Transcendental Meditation Program: Collected Papers.* Vol. 1. MERU Press.

White, L., & B. Tursky. 1982. *Clinical Biofeedback—Efficacy and Mechanisms.* Guilford Press.

White, L., et al., eds. 1985. *Placebo: Theory, Research and Mechanisms.* Guilford Press.

White, R. 1981. Saintly psi. *Journal of Religion and Psychical Research* (July & Oct.), pp. 157–67.

White, R. 1982. An analysis of ESP phenomena in the saints. *Parapsychology Review* (Jan.-Feb.), p. 15.

White, R. A. 1941. A preface to a theory of hypnotism. *Journal of Abnormal and Social Psychology*, 36:477–506.

Whitehead, A. N. 1978. Corrected edition ed. D. Griffin & D. Sherburne. *Process and Reality.* Free Press.

Whitehead, W. E., et al. 1975. Modification of human gastric acid secretion with operant-conditioning procedures. *Journal of Applied Behavior Analysis*, 8: 147–56.

Whitehead, W. E., et al. 1977. Relation of heart rate control to heartbeat perception. *Biofeedback & Self-Regulation*, 2:371–92.

Whitlock, F., & J. Hynes. 1978. Religious stigmatization: An historical and psychophysiological enquiry. *Psychological Medicine*, 8:185–202.

Whitmont, E. 1982. *Return of the Goddess.* Crossroad.

Wicramasekera, I. 1972. Electromyographic feedback training and tension headache: Preliminary observations. *American Journal of Clinical Hypnosis*, 15: 83–85.

Wicramasekera, I. 1973. The application of verbal instructions and EMG feedback training to the management of tension headache—preliminary observations. *Headache*, 13:74–76.

Wicramasekera, I. 1976. *Biofeedback, Behavior Therapy and Hypnosis.* Nelson Hall.

Wiklund, N. 1984. On the assessment of evidence for psi. *European Journal of Parapsychology*, 5:245–59.

Wilber, K. 1983. *Eye to Eye.* Doubleday Anchor Book.

Wilber, K., J. Engler, & D. Brown, eds. 1986. *Transformations of Consciousness.* Shambhala, New Science Library.

Wilber, T. K. 1988. Attitudes and cancer: What kind of help really helps? *Journal of Transpersonal Psychology*, 20:49–59.

Willard, R. D. 1977. Breast enlargement through visual imagery and hypnosis. *American Journal of Clinical Hypnosis*, 19:195–200.

Williams, J. E. 1974. Stimulation of breast growth by hypnosis. *Journal of Sex Research,* 10:316–26.

Williams, P., & M. West. 1975. EEG responses to photic stimulation in persons experienced at meditation. *Electroencephalography & Clinical Neurophysiology,* 39:519–22.

Williams, R., et al. 1980. Physical conditioning augments the fibrinolytac response to venous occlusion in healthy adults. *New England Journal of Medicine,* 302:987–91.

Williams, R. D. 1985. The effects of shamatha meditation on attentional and imaginal variables. *Dissertation Abstracts International,* 46:319–20.

Williamson, C. J. 1976. The Everest message. *Journal of the Society for Psychical Research* (Sept.), 48:318–20.

Willis, R. 1952. *The Spread of Tumors in the Human Body.* Butterworth.

Wilson, E. O. 1978a. *On Human Nature.* Harvard University Press.

Wilson, E. O. 1978b. *Sociobiology: The New Synthesis.* Harvard University Press.

Wilson, P., et al. 1986. Assessment methods for physical activity and physical fitness in population studies: Report of NHLBI Workshop. *American Heart Journal,* 111:1177–92.

Wilson, S. C., & T. X. Barber. 1981. Vivid fantasy and hallucinatory abilities in the life histories of excellent hypnotic subjects ("somnambules"): Preliminary report with female subjects. In E. Klinger (ed.), *Imagery: Concepts, Results & Applications.* Plenum.

Wilson, S. C., & T. X. Barber. 1983. The fantasy-prone personality: Implications for understanding imagery, hypnosis, and parapsychological phenomena. In A. Sheikh (ed.), *Imagery—Current Theory, Research, & Application.* Wiley.

Wink, C. A. S. 1961. Congenital ichthyosiform erythrodermia treated by hypnosis: Report of two cases. *British Medical Journal,* 2:741–43.

Winquist, W. 1977. The Transcendental Meditation program & drug abuse: A retrospective study. In D. W. Orme-Johnson & J. T. Farrow (eds.), *Scientific Research on the Transcendental Meditation Program: Collected Papers,* Vol. 1. M.E.R.U. Press.

Winston, S. 1975. Research in psychic healing: A multivariate experiment. Doctoral dissertation, Union Graduate School, Cincinnati.

Wirth, D. 1990. The effect of non-contact therapeutic touch on the healing of full thickness dermal wounds. *Subtle Energies,* 1:1–20.

Wolf, S., & R. Pinsky. 1954. Effects of placebo administration and occurrence of toxic reactions. *Journal of the American Medical Association,* 155:339–41.

Wolfe, L. S., & J. Millet. 1960. Control of post-operative pain by suggestion under general anesthesia. *American Journal of Clinical Hypnosis,* 3:109–12.

Wolkove, N., H. Kreisman, D. Darragh et al. 1984. Effect of Transcendental Meditation on breathing and respiratory control. *Journal of Applied Physiology,* 56:607–12.

Wolpe, J. 1958. *Psychotherapy by Reciprocal Inhibition.* Stanford University Press.

Wolpe, J. 1969. *The Practice of Behavior Therapy.* Pergamon.

Wolpe, J., & A. Lazarus. 1966. *Behavior Therapy Techniques.* Pergamon.

Wong, M., N. Brochin, & K. Gendron. 1981. Effects of meditation on anxiety and chemical dependency. *Journal of Drug Education,* 11:91–105.

Wood, P., & W. Haskell. 1979. The effect of exercise on plasma high density lipoproteins. *Lipids,* 14:417.

Wood, P. D., et al. 1983. Increased exercise level and plasma lipoprotein concentration. A one-year randomized controlled study in sedentary, middle-aged men. *Metabolism,* 32:31.

Wood, P. D., et al. 1988. Changes in plasma lipids and lipoproteins in overweight men during weight loss through dieting as compared with exercise. *New England Journal of Medicine,* 319:1173.

Woodroffe, J. 1972. *The Serpent Power.* Ganesh.

Woodworth, R. S. 1901. On the voluntary control of the force of movement. *Psychological Review,* 8:350–59.

Woolfolk, R. 1975. Psychophysiological correlates of meditation. *Archives of General Psychiatry,* 32:1326–33.

Woolfolk, R., L. Carr-Kaffashan, T. McNulty et al. 1976. Meditation training as a treatment of insomnia. *Behavior Therapy,* 7:359–66.

Woolfolk, R., et al. 1982. Effects of progressive relaxation and meditation on cognitive and somatic manifestations of daily stress. *Behavior Research Therapy,* 20: 461–67.

Worsely, F. 1977. *Shackleton's Boat Journey.* Norton.

Wright, D. 1969. *Deafness.* Stein & Day.

Wright, E. 1966. *The New Childbirth.* Hart.

Wybran, J., et al. 1979. *Journal of Immunology,* 123:1068–70.

Yates, A. 1980. *Biofeedback and the Modfication of Behavior.* Plenum.

Yates, A., et al. 1983. Running—an analogue of anorexia? *New England Journal of Medicine,* 308:251–55.

Yates, A. J. 1963. Recent empirical and theoretical approaches to the experimental manipulation of speech in normal subjects and in stammers. *Behavior Research and Therapy,* 1:95–119.

Yogananda, S. 1985. *Autobiography of a Yogi.* Self-Realization Fellowship.

Young, F. H. 1951. *The Philosophy of Henry James, Sr.* Bookman.

Younger, J. 1976. Bronze Age representations of Aegean bull-leaping. *American Journal of Archaeology,* 80:127–37.

Yuille, J. C., & L. Sereda. 1980. Positive effects of meditation: A limited generalization. *Journal of Applied Psychology,* 65:333–40.

Yura, H. T. 1975. A physicist looks at karate. *Samurai* (Jan.), pp. 24–27.

Zaichkowsky, L., & R. Kamen. 1978. Biofeedback and meditation: Effects on muscle tension and locus of control. *Perceptual & Motor Skills,* 45:955–58.

Zaleski, C. 1987. *Otherworld Journeys.* Oxford University Press.

Zamarra, G., I. Besseghini, & S. Wettenberg. 1977. The effects of the Transcendental Meditation program on the exercise performance of patients with angina pectoris. In D. W. Orme-Johnson & J. T. Farrow (eds.), *Scientific Research on the Transcendental Meditation Program: Collected Papers*. Vol. 1. MERU Press.

Zane, M. D. 1966. The hypnotic situation and changes in ulcer pain. *Inter. J. Clin. Exp. Hypnosis,* 14:292–304.

Zanker, K., & R. Kroczek. 1991. Looking along the track of the psychoneuroimmunologic axis for missing links in cancer progression. *International Journal of Sports Medicine,* 12:S58–S62.

Zarate, A., et al. 1974. Gonadotropin and prolactin secretion in human pseudocyesis. *Ann. Endocrinology (Paris),* 35:445–50.

Zeldis, S., et al. 1978. Cardiac hypertrophy in response to dynamic conditioning in female athletes. *Journal of Applied Physiology,* 44:849.

Zimmerman, P. 1970. *A Thinking Man's Guide to Pro Football.* Dutton.

Zornetzer, S. 1978. Neurotransmitter modulation and memory: A new neuropharmacological phrenology? In M. Lipton et al. (eds.), *Psychopharmacology: A Generation of Progress.* Raven.

Zukel, et al. 1959. A short-term community study of the epidemiology of coronary heart disease: A preliminary report of the North Dakota Study. *American Journal of Public Health,* 49:1630–39.

Permissions
and Copyrights

INDEX

About the Author

MICHAEL MURPHY is co-founder and Chairman of the Board of Esalen Institute, and author of three novels—*Golf in the Kingdom, Jacob Atabet,* and *An End to Ordinary History*—as well as a work of nonfiction—*The Psychic Side of Sports,* co-authored with Rhea White. During his thirty-year involvement in the human potential movement, he and his work have been profiled in the *New Yorker* and featured in many magazines and journals worldwide. He was born in Salinas, California, attended Stanford University, and lived for over a year at the Sri Aurobindo Ashram in Pondicherry, India. In 1980, he began Esalen's Soviet-American Exchange Program, which was the premiere diplomacy vehicle for citizen-to-citizen Russian-American relations. In 1990, Boris Yeltsin's first visit to America was initiated by the institute.